Instrumentation and Sensors for the Food Industry

Industrial Instrumentation Series
Series Editor: Dr B. E. Noltingk

Published with The Institute of Measurement and Control

The Industrial Instrumentation Series of books describes, authoritatively and in detail, the instrumentation available in the major industries, and the ways in which instruments are put to use. Classical instruments and the latest applications of new technology in the field are covered. Emphasis is placed on those aspects of an industry (such as measurements required or environmental constraints) that particularly influence the instrumentation used.

Each volume is written by experienced practitioners with specialist knowledge of instruments for particular applications. Each book contains extensive lists of references. Thus every title provides a comprehensive and profound treatment of the practical use of instrumentation. This makes the series essential for anyone involved in the specification, design, supply, installation, use, maintenance and repair of instruments.

The series will cover all major industries in which instrumentation is used and all important issues relating to the use of instruments in industry.

The Institute of Measurement and Control was founded in the UK in 1944 as the Society of Instrument Technology and took its present name in 1968. It was incorporated by Royal Charter in 1975 with the object 'to promote for the public benefit by all available means the general advancement of the science and practice of measurement and control technology and its application'. The Institute of Measurement and Control provides routes to Engineering Council status as Chartered and Incorporated Engineers and Engineering Technicians. The Institute is the UK member organization of the International Measurement Confederation (IMEKO).

Other Titles in the Industrial Instrumentation Series

Analytical Instrumentation for the Water Industry
T. R. Crompton
0 7506 1139 1 324pp 1991

Reliability in Instrumentation and Control
J. C. Cluley
0 7506 0737 8 200pp 1992

Power Station Instrumentation
Edited by M. W. Jervis
0 7506 1196 0 700pp 1993

Flow, Level and Pressure Movement in the Water Industry
G. Fowles
0 7506 1047 6 240pp 1993

Instrumentation and Sensors for the Food Industry

Edited by

Erika Kress-Rogers

Series Editor: B. E. NOLTINGK

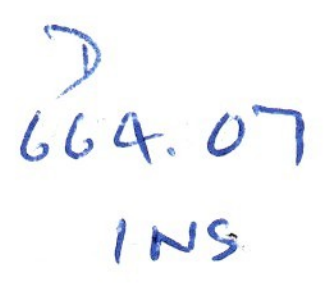

Published in association with The Institute of Measurement and Control

Butterworth-Heinemann Ltd
Linacre House, Jordan Hill, Oxford OX2 8DP

 PART OF REED INTERNATIONAL BOOKS

OXFORD LONDON BOSTON
MUNICH NEW DELHI SINGAPORE SYDNEY
TOKYO TORONTO WELLINGTON

First published 1993

British Library Cataloguing in Publication Data
A catalogue record for this book is available from the British Library

ISBN 0 7506 1153 7

Typeset by MS Filmsetting Limited, Frome, Somerset
Printed and bound in Great Britain

Contents

Preface xix

Acknowledgements xxv

List of contributors xxvii

1 Instrumentation for Food Quality Assurance
Erika Kress-Rogers 1
1 Introduction 2
1.1 The Role of Quality Assurance in the Food Industry 2
1.2 On-line, At-line and Off-line Instrumentation 3
1.3 Technology Transfer: Opportunities and Pitfalls 9
2 Challenging Conditions for Sensors 11
2.1 Complex and Variable Samples 11
2.2 Hostile Conditions and Stringent Hygiene Requirements 14
2.3 Non-Contact Techniques and Robotic Sampling and Conditioning 16
3 Interpreting the Readings 17
3.1 Measured Variables and Target Variables 17
3.2 Relationship Between On-line and QC-Laboratory Methods 18
3.3 Data Processing Approaches 20
3.4 The Marker Approach – Novel Sensors 20
3.5 Instrumentation as an Interdisciplinary Subject 21
4 Measurement Types 22
4.1 Target Variables 22
4.2 Instrumental Methods 23
4.3 Fringe Benefits 27
5 Further Reading 27
Appendix: Measurement Types 29
References 34

Part I In-line Measurement for the Control of Food Processing Operations

2 Principles of Colour Measurement for Food
Douglas B. MacDougall 39
1 Introduction 39
2 Colour Vision: Trichromatic Detection 40
3 Influence of Ambient Light and Food Structure 43

3.1 Adaptation 43
3.2 Appearance 44
3.3 Absorption and Scatter 45
4 Colour Description 45
4.1 The CIE System 45
4.2 Uniform Colour Space 48
4.3 Further Terminology 49
5 Instrumentation 50
6 Examples 53
6.1 Fresh Meat 53
6.2 Orange Juice 56
Acknowledgements 58
References 59

3 Colour Measurement of Foods by Colour Reflectance Instrumentation
C. J. B. Brimelow and C. A. Groesbeck 63
1 Introduction: Food Colour and Quality 64
2 Colour Measurement Principles: Brief Introduction 65
2.1 Tristimulus Colorimetry 65
2.2 Colour Scales and Colour Difference Formulae 66
3 Colour Measurement Methodology 68
3.1 Standardization of the Instrument 69
3.2 Selection, Preparation and Presentation of the samples 72
3.2.1 Opaque Powders, Granules and Flakes 75
3.2.2 Opaque Particulate or Lumpy Materials 76
3.2.3 Large Area Solid Foods 76
3.2.4 Pastes and Slurries 77
3.2.5 Liquids 77
3.3 Setting the Instrumental Variables 78
3.4 Determination of Colour Values 79
4 Colour Measurement of Typical Food Materials 80
4.1 Powders, Granules and Flakes 80
4.2 Particulate and Lumpy Solids 87
4.3 Large Area Solid Foods 88
4.4 Pastes and Slurries 90
4.5 Liquids 91
5 Conclusions 92
Acknowledgement 93
References 93

4 Sorting by Colour in the Food Industry
J. M. Low and W. S. Maughan 97
1 Introduction 98
2 What is a Sorting Machine? 98

3 Assessment of Food Particles for Colour Sorting 100
3.1 Spectrophotometry 100
3.2 Monochromatic Sorting 102
3.3 Bichromatic Sorting 102
3.4 Dual Monochromatic Sorting 103
3.5 Fluorescence Techniques 104
3.6 Near Infrared Techniques 104
4 The Optical Inspection System 104
4.1 Introduction 104
4.2 Illumination 105
4.3 Background and Aperture 107
4.4 Optical Filters 108
4.5 Detectors 110
5 Completing The Sorting System 111
5.1 Feed 111
5.2 Separation 111
5.3 Cleaning and Dust Extraction 112
5.4 Electronics Systems 113
5.4.1 User Interfaces 113
5.4.2 Mapping Techniques 114
6 Future Trends: Computer Vision Systems 114
7 Using a Colour Sorter 116
8 Further Reading 118

5 Compositional Analysis Using Near Infrared Absorption Spectroscopy
Ian B. Benson 121
1 Introduction 121
2 Theory of Near Infrared Absorption Spectroscopy 123
3 Instrumentation 128
3.1 On-line Near Infrared Instrumentation 128
3.2 Laboratory NIR Instrumentation 133
3.3 NIR Measurement Characteristics 135
3.4 Instrumentation and Installation in the Food Industry 140
4 Applications in the Food Industry 142
4.1 Moisture Measurements in Foodstuffs 142
4.2 Multi-component Analysis of Food Products 149
4.3 Quality Control for Food Packaging Materials 156
5 Practical Aspects of the Calibration of On-Line NIR Instruments 157
5.1 Calibration Methods 158
5.2 On-Line Installation and Sampling Procedures for Cross-Checking 161
6 Conclusion and What the Future Holds 164
References 165

6 Practical Aspects of Infrared Thermometry
G. Beynon 167
1 Introduction 167
2 Radiation Thermometers 168
2.1 Collection 168
2.2 Detectors 170
2.3 Processing Electronics 172
2.4 Housings and Configurations 173
2.5 Line Scans and Thermal Imagers 175
3 Measurement Principles 178
3.1 The Black Body 178
3.2 Non-black Target 179
3.3 Background Temperature 179
3.4 Emissivity 181
3.5 Examples 182
4 Miscellaneous Techniques 184
4.1 Enhancing Emissivity 185
4.2 Avoiding the Need to Know Emissivity 185
4.3 Inaccessible Targets 186
References 188

7 Microwave Measurements of Product Variables
M. Kent 189
1 Introduction 190
2 Advantages and Disadvantages of Microwave Techniques 191
2.1 Advantages 191
2.2 Disadvantages 192
3 Dielectric Properties and their Parameters 193
3.1 Polarization 193
3.2 Dielectric Dispersion 194
3.3 Dielectric Dispersion of Water in Foodstuffs 196
3.4 Temperature Effects 198
3.5 Bulk Density 200
3.6 Conductivity 201
4 Methods for Measurement of Dielectric Properties 202
4.1 Attenuation Measurements 202
4.2 Resonant Methods 203
4.3 Reflectance Measurements 204
5 Dielectric Properties and Measurements of Bulk Density and Composition 205
5.1 Density Measurement and Compensation in Moisture Measurement 205
5.2 Fat and Water 211
5.3 Sugar Solutions 215
5.4 Other Mixtures of Materials 217

5.5 Salt Concentration 217
6 Material Structure 217
6.1 Particle Shape and Distribution 217
6.2 Particle Size 218
7 Apparatus for Microwave Measurement 219
7.1 Attenuation 219
7.2 Phase 220
7.3 Simultaneous Phase and Attenuation 221
7.4 Resonant Systems 223
7.5 Reflectance 223
8 Sensors 224
8.1 Horn Antennae 224
8.2 Stripline Sensors 227
8.3 Stripline Antennae 228
8.4 Open Ended Transmission Lines 229
9 Areas for Development 230
Appendix 1: Some Manufacturers of Microwave Moisture Measurement Instruments 231
Appendix 2: List of Symbols 231
References 232

8 Ultrasound Propagation in Foods and Ambient Gases: Principles and Applications
Erika Kress-Rogers 237
1 Introduction 238
2 Overview of Ultrasound Applications 239
2.1 What is Measured? 239
2.2 Communication, Detection and Location 240
2.3 Level and Flow Rate Measurement 242
2.4 Non-destructive Testing 243
2.5 Concentration Measurement 244
2.6 Passive Ultrasound Equipment: Acoustic Emission Monitoring 245
2.7 High-power Ultrasound Equipment 246
3 Speed of Sound 247
3.1 Velocity of Propagation for Bulk Longitudinal Waves 247
3.2 Other Ultrasonic Wave Types 248
3.3 Speed of Sound in Gases 250
3.4 Speed of Sound in Liquids 250
3.4.1 Influence of Solutes, Solvents and Temperature 250
3.4.2 Influence of Dispersed Particles, Droplets and Bubbles 256
3.4.3 Comparison of Methods for Composition Monitoring 264
3.5 Speed of Sound in Solids 266

4 Acoustic impedance 267
4.1 Definition 267
4.2 Implications for Velocity Measurement 268
4.3 Measuring Impedance Instead of Velocity 269
5 Attenuation 270
5.1 Sources of Attenuation 270
5.2 Relaxation 271
5.3 Scattering 273
5.3.1 Hindrance or Help? 273
5.3.2 Concentration of Suspensions 277
5.3.3 Juice Stability and Produce Defects 277
5.3.4 Flow Metering 277
5.3.5 Emulsions – the Three Contributions to Scattering Losses 278
6 Conclusions 280
Appendix 1: Ultrasound Measurement Applications in and for the Food Industry 281
Appendix 2: List of Symbols 283
References 284

9 Ultrasonic Instrumentation in the Food Industry
Nicholas Denbow 289
1 Introduction 290
2 Low Frequency Techniques 291
2.1 Level Measurement by Echo Ranging 291
2.1.1 Calculating or measuring the speed of sound in the headspace 292
2.1.2 Pulse transmission, reflection and reception 295
2.1.3 Microprocessor-based electronics 296
2.1.4 Liquids applications 297
2.1.5 Solids applications 298
2.1.6 Accuracy 299
2.2 Beam-break Detectors as Counters 301
2.3 Future Developments in Low-frequency Instrumentation 301
3 High-frequency Techniques 302
3.1 Flow Measurement Systems 302
3.1.1 Time-of-flight Flow Measurement 303
3.1.2 Doppler Flow Measurement 304
3.2 Liquid Level Switches 306
3.2.1 Ultrasonic-transmission-based Level Detectors 307
3.2.2 Pulse-attenuation-based Level Detectors 311
3.3 Liquid Level Measurement Systems 312
3.4 Suspended solids and interface detection systems 312

3.4.1 Suspended solids measurement 313
3.4.2 Liquid Identification and Interface Detection 315
3.5 Concentration Measurement Systems 316
3.6 Future Developments in High-frequency Instrumentation 318
4 Contacts for Further Information 318
References 319

Part II Instrumental Techniques in the Quality Control Laboratory

10 Rheological Measurements
B. M. McKenna 323
1 Introduction 323
2 Relevance of Rheological Properties of Foods 324
2.1 The Consumer's Perception 325
2.2 The Requirements of the Processor 325
3 Basic Rheology 327
4 Measurement Systems 331
4.1 Capillary Viscometers 331
4.1.1 Theory 331
4.1.2 Instruments 334
4.2 Rotary Viscometers 336
4.2.1 Concentric Cylinder Viscometers 336
4.2.2 Cone and Plate Viscometers 338
4.2.3 Other Rotary Viscometers 339
5 On-line Measurement Systems 340
6 Instrument Selection 342
Appendix: List of Symbols 344
References 345

11 Modern Methods of Texture Measurement
D. Kilcast and A. Eves 349
1 Introduction 349
2 Current Methods of Texture Measurement 350
2.1 Sensory Texture Assessment 350
2.2 Instrumental Texture Measurement 352
2.2.1 Empirical Methods 352
2.2.2 Imitative Methods 353
2.2.3 Fundamental Methods 357
3 Physiological Aspects of Chewing 358
3.1 Monitoring Human Subjects 358
4 Texture Measurement by Electromyography 359

4.1 Instrumentation and Measurement Procedure 360
4.2 Reproducibility 363
4.2.1 Between Subjects 363
4.2.2 Between Occasions 363
4.3 Data Interpretation 364
4.3.1 Relationship to Biting Force 364
4.3.2 Relationship to Sensory Data 365
5 Future Developments in Texture Measurement 369
6 Conclusions 372
References 372

12 Water Activity and its Measurement in Food
Wolfgang Rödel 375
1 Definition 376
2 Significance of Water Activity 377
2.1 Effect of Water Activity on Food Quality 377
2.2 Effect of Water Activity on Food Stability 378
2.2.1 Water Activity Tolerance of Bacteria, Yeasts and Moulds 378
2.2.2 Water Activity Tolerance of Trichinae and Bovine Bladderworms 382
2.3 Legal Requirements 384
3 Water Activity Levels in Food and their Control 385
3.1 Water Activity Levels in Food of Animal Origin 385
3.2 Water Activity Levels in Food of Vegetable Origin 386
3.3 Control of the Water Activity Level 388
3.4 Example: Regulating Raw Sausage Ripening by Controlling the Water Activity Level 389
4 Measuring the Water Activity Level 390
4.1 Background 390
4.2 Water Activity as a Function of Temperature 390
4.3 Influence of Equilibration Periods and Sample Properties 392
4.4 Instrument Calibration 393
5 Measurement Technique 394
5.1 Manometric Method 394
5.2 Gravimetric Method 395
5.3 Psychrometric Method 395
5.4 Hygrometric Methods 396
5.4.1 Salt Method 396
5.4.2 Thread Hygrometers 396
5.4.3 Electric Hygrometers 397
5.5 Thermometric Technique 403
6 Conclusions 405
References 405

13 Instrumental Methods in the Chemical Quality Control Laboratory
P. T. Slack 417
1 Introduction 418
2 Versatile Instruments 419
2.1 Gas and Liquid Chromatography 420
2.2 Limitations of Gas and Liquid Chromatography 421
2.3 Supercritical Fluid Chromatography 422
2.4 Capillary Electrophoresis 423
3 The Trend towards Dedication 424
3.1 GLC System for the Analysis of Pesticides: Nordion Analysis System 424
3.2 HPLC System for the Assay of Ions and Sugars: Dionex 2000i Series 427
3.2.1 Configuration for the Detection of Anions 428
3.2.2 Configuration for the Detection of Sugars 429
3.3 Ion Chromatograph for Sulphite: Wescan Sulphite Analyser 432
3.4 Enzyme Electrode Analysers for Sugars and Alcohols: YSI Model 27 and Model 2000 Analysers 433
4 Dedicated Instruments 435
5 Dedicated Instruments for the Determination of Moisture, Ash and Fat 436
5.1 Moisture by Drying Methods 436
5.2 Moisture by Karl Fischer Titration 437
5.3 Ash by Combustion Methods 438
5.4 Fat by Solvent Extraction 438
5.5 System Based on the Karl Fischer Procedure: Baird and Tatlock Turbotitrator 439
5.6 System Based on Oven Drying: Computrac Max 50 Moisture Analyser 440
5.7 Instrument for Moisture Determination by Oven Drying and Ash by Combustion: Leco MAC-400 442
5.8 Instrument for Moisture by Microwave Drying and Fat by Solvent Extraction: CEM Meat Analysis System 443
5.9 Instrument for Fat Determination by Solvent Extraction: Foss-Let System 445
6 Dedicated Instruments for Nitrogen Determinations: for the Calculation of Protein and Meat Content 445
6.1 Instrument for Nitrogen Determination by Acid Digestion: 446
6.1.1 Kjel-Foss Automatic 446
6.1.2 Kjeltec Auto/Labtec System 448
6.1.3 Büchi Kjeldahl System 449

6.2 Instrument for Nitrogen Determination by Combustion: Leco FP-228 452
7 Further Development of Instrumentation for Food Analytical Laboratories 454
References 454

14 Impedance Techniques for Microbial Assay
D. M. Gibson and A. C. Jason 457
Nomenclature 458
1 Introduction 459
2 Rapid Microbiological Methods: an Overview 460
2.1 Impedance as an Indicator of Microbial Load 464
3 Principles of Electrical Conductance Methods 464
3.1 Impedance and its Component Variables 464
3.2 Cell Design and Geometry and Composition of Electrodes 467
3.3 Relationship between Test Cell Conductance and Bacterial Growth 469
3.4 Bacterial Number Resolution of Analysers 471
3.5 Temperature Control 472
4 Capacitance versus Conductance Measurement 474
4.1 Selection of Display Variable Early in the Growth Cycle 474
4.2 Mechanisms of Changes Observed Later in the Growth Cycle 474
5 Instrument Design 478
5.1 Method of Measurement 478
5.2 Multiplexed Cell Switching 479
5.3 Commercial Instrumentation 480
6 The Evaluation of Conductance Data 483
6.1 Bacterial Growth in Batch Culture 483
6.2 Determination of Inocula 484
6.3 Choice of Growth Media 486
6.4 Correlation with Conventional Microbiological Data 488
7 Future Possibilities 491
References 493

15 Impedance Microbiology in Food Quality Control
Mike L. Arnott 499
1 Introduction 499
2 Development of Protocols 501
3 Application to the Detection of Micro-organisms in Foods 503
3.1 Correlation Procedures 503
3.2 Total Counts 509
3.3 Selective Procedures 510

3.4 Shelf Life Prediction 514
4 Choice of Instrumentation 515
5 Future Developments 516
References 517

Part III New Sensors for Applications in the Food Industry 521

16 The Marker Concept: Frying Oil Monitor and Meat Freshness Sensor
Erika Kress-Rogers 523
1 Introduction to the Concept and Overview on Applications 523
2 A Novel *In situ* Monitor for Frying Oil 528
2.1 The Need for a New Probe 528
2.2 Development of the Probe Concept 531
2.3 Study to Assess Feasibility and Define Specifications 532
2.4 Characteristics of the New Probe 545
2.5 Protocol for the Trial with Prototype 1 549
2.6 Results of the Trial with the New Probe 552
2.7 Conclusions and Outlook 553
3 A Novel Knife-Type Probe for Meat Freshness 558
3.1 Development of the Probe Concept 558
3.2 Study to Assess Feasibility and Define Specifications 561
3.3 Characteristics of the New Probe 563
3.4 Trials with the New Probe 564
3.5 Conclusions and Outlook 571
Acknowledgements 575
References 577

17 Chemosensors, Biosensors and Immunosensors
Erika Kress-Rogers 581
1 Introduction 582
1.1 Chemical Sensor Types and their Building Blocks 582
1.2 Recent Developments 587
1.3 Sensor Specifications for Food Applications 588
2 Chemically Sensitive Semiconductor Devices: Solid-State Sensors for pH, Acidity, Ions, Gases and Volatiles 589
2.1 Introduction to Novel pH Sensors 589
2.2 Alternative Approaches to pH Measurement 591
2.2.1 Solid Contact to a Glass Membrane 591
2.2.2 Solid Contacts to Other Ion-Sensitive Membranes 594
2.2.3 Electrodes Based on Electronically Conducting Metal Oxides 595
2.2.4 Microelectronic Devices 595
2.2.5 Fibre-Optic Chemical Sensors 598

2.3 Microelectronic Chemical Sensors Based on the FET: Principles, Characteristics and Applications 599
2.3.1 Introduction to the ISFET and the CHEMFET 599
2.3.2 Nomenclature: MOS and MeOx, IC and Smart Sensor 600
2.3.3 How Do the ISFET and the CHEMFET Work? 601
2.3.4 Ion-Sensitive Membranes for the ISFET Family 603
2.3.5 The REFET Reference Device 605
2.3.6 Development for Industrial Applications 606
2.3.7 A Commercial ISFET/REFET System for pH Measurement in the Food Industry 608
2.3.8 Trials of pH ISFET Probes in Meat Processing 608
2.3.9 Acidity Sensors 608
2.4 Gas Sensing Semiconductor Devices 611
2.4.1 Introduction 611
2.4.2 Gas Sensor Characteristics 613
2.4.3 Gas Sensor Arrays for the Measurement of Odours 616
2.5 Conclusions 617
References for Sections 2.1 to 2.3 618
References for Section 2.4 621
3 Amperometric, Potentiometric and Thermometric Biosensors 622
3.1 Introduction: Biosensor Types 622
3.2 Biosensors Based on Amperometric Mediated Enzyme Electrode (AMEE) 626
3.3 Biosensors Based on Amperometric Indirect Dual-Membrane Enzyme Electrode (AIDMEE) 633
3.4 Biosensors Based on the FET: the ENFET 636
3.5 The Enzyme Thermistor and Related Devices (ENTHERM) 638
References for Sections 3.1 to 3.3 639
References for Section 3.4 642
References for Section 3.5 643
4 Chemically Sensitive Optical and Acoustic Devices 644
4.1 Introduction 644
4.2 The Surface Plasmon Resonance (SPR) Device 647
4.3 The Total Internal Reflectance (TIR) Device 647
4.4 Fibre-Optic Probe (FOP) Devices 649
4.5 Piezocrystal Balance Devices 652
4.6 Surface Acoustic Wave (SAW) Devices 654

References for Section 4 655
5 Applying Sensors 659
5.1 Introduction 659
5.2 Pattern Recognition by Multivariate Analysis or Neural Networks 660
5.3 Robotics and Flow Injection Analysis 662
5.4 Choice of Instrumentation Type 664
References for Section 5 665
General Reading on Chemical Sensors and Biosensors 668

Appendix A Glossary: Terms in Instrumentation and Sensors Technology 671
Appendix B Ancillary Tables 693

Addendum: Trial with Prototype 2 of the *In situ* Monitor for Frying Oil 714
Index and Dictionary of Abbreviations 715

Preface

The Optimization of Industrial Food Processing

The enjoyable, everyday food in an industrialized society relies not only on agriculture and fishing but also on efficient food processing. Few people mill their own cereals, preserve and store their fruit and vegetables from harvest to consumption, churn their butter, ferment milk into yoghurt and cereals into beer, turn meat carcasses into joints, sausages and pâtés, or pound mustard seeds to prepare meal accompaniments. Few wish to restrict themselves to the local products in season and most prefer instead the variety that modern food production, processing and distribution can offer. Many use ready-prepared meals so as to spend their evenings with the children or with friends rather than in the kitchen.

In the early stages of the industrialization of food processing, the competition between manufactured goods centred mainly on the price at which they were offered; now quality and safety are in the foreground. A wider range of attractive food products has become affordable for a large proportion of the population through advances in food science and technology together with the development of a diverse range of efficient large-scale processing plant.

Many traditional batch processes have now been replaced by automated production methods, helped by the introduction of advanced process control systems in the 1980s. The signal processing and actuating capacity of process control systems is now adequate. The full potential of these systems, however, can be realized only if they are supplied with full and up-to-date information on the process to allow feedback or feedforward control. The development and knowledgeable application of sensors and instruments have become the key elements in meeting the consumer's expectations in the food industry to provide affordable, enjoyable, safe and nutritious products.

This has prompted the development of a wider range of sensors and instruments suitable for on-line and at-line measurements in the food industry, and also of modern instruments for the quality control (QC) laboratory. Many of the new instruments rely on a complex interaction with the food in order to determine properties of the food itself (such as composition) during processing. They extend the range of data inputs for 'sensible' process control (equipped with senses) well beyond the measurement of pressure, temperature, level and flow rate.

Other new instruments widen the range of applications for the measurement of the established variables, now allowing the reliable measurement of flow rate or temperature in food processes where this was previously impossible. Progress has also been made in the development of instruments for the assessment of food freshness and food safety, so that results are now often available within a day and a higher proportion of food ingredients and products can be screened to ensure good manufacturing practice.

In the choice of instrumentation, an analysis of the processing operation as a whole, together with an overview of the characteristics of the sensors and instruments available for on-line, at-line and QC laboratory measurements, will be the basis of optimum process control design. On-line and off-line instrumentation interlink in guiding process control and are therefore both included here. Calibration samples need to be chosen and correctly prepared, and a representative sampling technique and suitable reference methods must be selected.

For the reliable installation, calibration and operation of the new instruments, and for the correct interpretation of their readings, it is essential to understand the principles underlying the functioning of the instruments, the properties of the food and its processing environment, and their interplay. This approach also helps in assessing the many novel sensors and instrumental techniques now emerging to provide better long-term planning of process control optimization.

Special Application Details for Instrumentation

Instrument engineers coming from the aerospace, defence, nuclear or petrochemical sectors sometimes underestimate the challenges of designing sensors and instruments for the food industry. They find adequate challenges for their skills when they encounter a wide range of temperatures, pressures and pH values; mixing paddles continuously scraping container walls where a sensor is to be mounted in contact with the product; the rejection of guards around fragile sensor components as germ traps; and a limitation of the choice of engineering materials to those compatible with food hygiene considerations. The occasional fracture of a sensor in the chemical industry may be an inconvenience; in the food industry it is a major incident when any sharp fragments, however small, are lost into the process stream, requiring the screening or safe disposal of many thousands of food product items.

A standard procedure for the maintenance of hygienic conditions in food processing is cleaning-in-place (CIP). This may sound harmless enough, but the periodic flushing of the food processing system with hot caustic soda (NaOH) solutions or pressurized steam places restrictions on the design of contacting sensors, particularly in the development of chemical sensors. Instruments based on non-contact methods are especially attractive to the food industry, being both intrinsically hygienic and easy to maintain. Such instruments are covered in the first part of this book.

In some applications, hostile conditions and restricted access to the contents of a process vessel are the main challenges, for example in a cooker extruder which allows the continuous production of intricately shaped and textured snack foods at a throughput rate of 400 kg/hour. High pressures, high temperatures, a feed/mixer screw scraping the interior surface of the heavy metal barrel and sometimes abrasive raw materials combine here to render the construction of reliable sensors difficult, even for pressure and food mix temperature.

More often, it is the variable and complex nature of the food itself that presents problems in the design and application of instruments. This is the case for non-contact volume and mass flow rate metering of many foods. The most interesting problems arise, however, in the measurement of food properties such as composition or rheology. An interdisciplinary approach is needed here to take account of the interaction between the instrumental method and the chemical and physical properties of the food and its environment (beyond the variables to be determined). This applies also to the assessment of food freshness or conversely to the determination and prediction of changes due to microbial activity or oxidative processes. A further aspect is the perception of the consumer which needs to be represented in instruments for the assessment of appearance and texture.

A rapid accurate measurement is often needed to maintain specifications within narrow margins. A pH value or water activity above specifications could lead to food spoilage during storage and distribution; a deviation to lower values could reduce the palatability. Too little preservative could endanger food safety; too much would be unacceptable to many consumers. Too high a water content could be infringing legal requirements or be associated with a water activity above specifications (with implications for food stability); too little water could result in an unattractive texture and an uncompetitive price for the food product.

Line speeds in automated continuous food processing and packaging are high, and this is both a motivation for the application of on-line instrumentation (or of rapid at-line methods) and a challenge in the design of instruments for this purpose. A further constraint in the design of instrumentation for the food industry is the fact that the price of the sensor or instrument will be important in the purchasing decision. Whereas the aircraft constructor may well buy the best instruments at any price, the food industry cannot afford to do so.

Instrument Types and Aspects

Instruments relying particularly on an interaction with the food or an environment typical for the food industry are described in this book. Practical applications already established are discussed and newly emerging applications are introduced. The considerations that will allow the best use of the interplay of the instrumental method, the food and the process are outlined as a basis for the successful development and implementation of instrument applications. Both on-line and QC laboratory instruments are included as they have to interlink in guiding process control.

Instrument users often wonder why the flood of novel sensors and measurement techniques described in scientific and technical journals or at conference results in a mere trickle of novel commercial instruments. This has been the case particularly in the field of biosensors and chemical sensors based on microelectronic devices where rapid developments have taken place in recent years. Part III of the book illustrates the complex and expensive process of developing a novel instrument from concept to commercial fruition with the help of two examples. The basis of recent commercial instrument developments based on novel chemical sensors and the feasibility of further food applications are also examined there.

For each instrument type, the underlying principles are described with emphasis on aspects relevant to food applications. The authors show the significance of the variables to be determined, and identify the variables actually measured (unless identical) and their relation to the desired information about the food product. Considerations in the design, choice, calibration and running of instruments within a given group are discussed and illustrated with examples.

Aspects covered include hygienic design (e.g. flush fitting sensor heads or choice of non-contact techniques) or the adaptation of techniques to the variable nature of food ingredients. (In three cases, two chapters deal with different aspects of the same technique.) Factors influencing the accuracy and reliability of the technique (for a particular group of food products if applicable) are spelled out and compared with alternative techniques where applicable.

Instrument systems requiring a high computing capacity (such as real-time image acquisition and processing), employing ionizing radiation (such as gamma-ray density gauges) or relying on principles beyond the realms of classical physics are omitted to allow a full description of the instruments covered.

The Authors' Background

To promote an interdisciplinary understanding, these aspects are discussed here by scientists and engineers from a wide range of backgrounds including electronics, physics, chemistry, microbiology, food science and food technology. Their professional experience spans an equally wide range of areas within the fields of the development and application of instrumental methods for the food industry. The authors have worked in the management and optimization of quality control and quality assurance in the food industry, in the development of new techniques for this area, in applications development or overall management at an instrument supplier's laboratory, or in a research institute or association in close contact with the food industry.

It would be difficult to find a single author with full and detailed knowledge and practical experience in all the aspects of physics, electronics, chemistry, microbiology, food science, food technology and process control that are relevant to instrumentation in the food industry. Nor would it necessarily be helpful to set up a committee of authors to compose a text together. Instead, each chapter reflects the particular expertise of the author(s) based on their scientific or engineering background and their professional experience acquired in the practical application or development of instruments.

Aims and Scope

For a wide range of established and emerging instrument types, this book treats the underlying principles and their implications for industrial applications. It sets out the complementary roles and characteristics of both the on-line and at-line instrumentation linked to the process control system and of the off-line instruments in the quality control laboratory. The significance of the measured variables for quality assurance and process management and the technical and commercial factors that determine the success or failure of an instrument are considered.

The book is intended to assist engineers and managers responsible for process optimization and quality assurance in the food industry in choosing, setting-up and maintaining instruments and in using their readings to best effect. It is also intended for use by engineers in the instrumentation sector who develop new instruments, adapt existing instruments for new applications or liaise with instrument users. In the choice and installation of an instrument in a process line, the effective cooperation between instrument supplier and user is essential and this book aims to promote this by facilitating the communication between engineers and managers, from different backgrounds.

The chosen approach is also designed to help advanced students of instrument engineering, food science, physics or biochemistry who seek an introduction to instrumentation in the processing industries. Further, the book will be of interest to scientists active in research and pre-commercial development in the fields of process engineering, industrial instrumentation and process control.

Erika Kress-Rogers

Acknowledgements

Firstly, I should like to thank the contributing authors for making their knowledge and practical experience available. For his friendly advice throughout the editing of this book, my thanks are due to Dr Ben Noltingk, the editor of the Butterworth-Heinemann series on industrial instrumentation. My warmest thanks are due to my family, Clive and Katya, for their patience and support during the preparation of this book.

I am indebted to the Leatherhead Food Research Association (LFRA) and its member companies, both food manufacturers and instrument suppliers, for giving me the opportunity to learn about the food industry, its instrumentation requirements and various intriguing problems and solutions in the development of sensors and instrumental techniques. In particular, I should like to thank the members of the LFRA Process Instrumentation Panel, who gave me the opportunity to study microwave moisture measurement techniques, and the members of the Oils and Fats Panel, who supported the development of a novel frying oil monitor with their continued interest. Further, I am grateful to all those representatives of LFRA member companies who have patiently filled in questionnaires and answered queries on their instrument requirements and specifications.

My thanks are also due to the Ministry of Agriculture, Fisheries and Food, which supported a study on technology transfer that allowed me to survey the new generation of sensors and to assess the potential applications for the food industry. The Ministry, together with the Research Coordination Committee (now the Science and Technology Policy Committee) of the LFRA, also supported the project resulting from this survey, namely the development of sensor prototypes for meat freshness and frying oil quality, feasibility studies on other potential sensor adaptations, and the identification of further development opportunities.

For an enjoyable and productive collaboration in these projects, I should like to thank the Torry Research Station, the GEC Marconi Research Centre and the Cranfield Biotechnology Centre. Instrumentation and technical advice from Thorn EMI Central Research Laboratories allowed the study on the potential applications of microelectronic chemical sensors. The collaboration among LFRA staff with specialist experience in the areas of physics, analytical chemistry, trace analysis (GC-MS, enzyme assay, immunoassay), oils and fats science and technology, applied microbiology, process technology and product technology has been essential in the above projects. Further acknowledgements and references are given in the relevant chapters.

Erika Kress-Rogers

Contributors

Mike L. Arnott is a Senior Scientist at Reading Scientific Services Ltd. Before joining RSSL recently, he has been responsible for quality control tasks at Cadbury Schweppes Ltd. His special interest is in the development of rapid microbiological methods, particularly ELISA and impedance techniques. He has over 20 years experience in applied food microbiology covering the baking, canning, dairy, meat, confectionery and soft drinks industries. During this period, he has been involved in quality assurance, factory hygiene auditing and in the implementation of good manufacturing procedures (GMP).

His background is in applied microbiology (Birmingham Polytechnic, UK).

Ian B. Benson is Manager of the Technical Support Group of Infrared Engineering Ltd (Maldon, Essex, UK), providing technical service support and user training in the theory and application of NIR analysis. In his earlier position as Applications Engineering Manager, his responsibilities have included the development of the company's on-line near-infrared (NIR) applications for the determination of fat, moisture, protein, sugars and nicotine content and of film thickness.

His background is in chemistry; spectroscopic studies, X-ray crystallography and synthesis of organometallic complexes (University of Bristol).

Geoff Beynon is Divisional Director (Technical) at Land Infrared (Sheffield, UK). In more than ten years with Land Infrared, Geoff Beynon has been involved in the design of a multiplicity of instruments for non-contact temperature measurement for applications ranging from military aircraft engines to microwave ovens.

His background is in physics; astronomy (University of Oxford, University of Sussex, UK).

Chris J. B. Brimelow is Vice President of Research and Development for Nestlé/Westreco (Westreco Corporate Headquarters, New Milford, Connecticut, USA). His main responsibility is in strategic planning of R&D but he also has special responsibilities for food quality and safety aspects. Before moving to the USA, he worked at Nestec (Vevey, Switzerland) where he coordinated R&D projects worldwide relating to refrigerated savoury products and to food safety. Earlier, as Deputy Director of R&D at Londreco (Hayes, Middlesex, UK), he was responsible particularly for projects concerning instrumentation relating to food quality and safety.

He has worked on the on-line and off-line measurement of compositional and physical properties of food materials, including water, protein and lipid contents, water activity, bulk density and colour. The latter work has led to improved raw material colour specifications for food materials and, as part of the COST90-bis programme, to the creation of a European Community standard colour tile for tomato colour together with methodology for its use.

His background is in chemistry; analytical chemistry (University of Nottingham and London University, UK).

Nicholas J. Denbow is Technical Marketing Manager at KDG Mobrey Ltd (Slough, Berkshire, UK). He has been responsible for the company's ultrasonic instrumentation developments for application in liquids. He has been concerned particularly with the development of ultrasonic instruments for switching, measurement and control related to liquid level and for the measurement of liquid flow rate. He has also developed ultrasonic monitoring systems for slurries and sludges.

His background is the study of physics and electrical sciences at Cambridge University (UK), with subsequent engineering qualifications.

Anita Eves has worked at the Leatherhead Food Research Association (Leatherhead, Surrey, UK) as a Senior Scientist. She has recently joined the Natural Resources Institute (Chatham, UK). At the LFRA, she has worked in the areas of food texture measurement, including electromyography methods, consumer choice studies and the influence of food irradiation on sensory properties.

Her background is in food science; physical methods of food texture assessment (University of Reading, UK).

Donald M. Gibson is Head of Microbiology at Torry Research Station, Ministry of Agriculture, Fisheries and Food (Aberdeen, Scotland, UK). He leads a team carrying out research and development on food safety and quality particularly in the area of rapid and novel microbial assay methods. Previously he was part of a team which devised and developed automated instruments for conductimetric rapid microbial assays. He has since carried out extensive work on the application of these instruments for the detection of various micro-organisms.

His background is in biochemistry; microbial biochemistry (University of Aberdeen).

Cheryl A. Groesbeck is a Research Scientist at Nestlé Westreco Inc. (New Milford, Connecticut, USA). Projects at Westreco have included studies and measurements on the colour of various food products as part of efforts to improve appearance during the shelf-life. The studies have included evaluation of the influence on colour of factors such as raw materials properties, product formulation, processing and packaging.

Collaborative projects have involved method development for establishing colour protocols, development of colour specifications, and the training of staff in the use of colour measurement instrumentation.

Her background is in food science and technology; characterization of spoilage microorganisms in food (University of Tennessee; Michigan State University).

Alfred C. Jason has held the position of Assistant Director, and earlier that of Head of Physics Section, at Torry Research Station, Ministry of Agriculture, Fisheries and Food (Aberdeen, Scotland, UK). Retired, he is now a consultant physicist to the Department of Microbiology of the Grampian Health Board, Aberdeen Royal Infirmary.

At Torry Research Station since 1975, he has developed conductance measurement techniques for the monitoring of microbial growth and has also been involved in the mathematical modelling of bacterial growth. His work for the Health Board is concerned particularly with applications of conductance methodology.

His background is in physics; cosmic ray physics, terminal ballistics, glaciology, road safety research and food research.

Michael Kent is Head of Physics Section at Torry Research Station, Ministry of Agriculture, Fisheries and Food, (Aberdeen, Scotland, UK). At Torry, he has carried out studies of the dielectric properties of foods and on the applications of dielectric properties to composition measurement. This work has included the development of a handheld probe for the fat content of whole fish based on a microwave stripline sensor. He has been a major contributor to COST90 and COST90bis, the European Projects on the Physical Properties of Foods. As part of the COST programme, he has compiled bibliographies and a data compilation on the Electrical and Dielectric Properties of Food Materials and on the Colour and Optical Properties of Foods.

His background is in physics; electron spin resonance (Aberdeen, UK).

David Kilcast is the Manager of the Sensory Analysis and Food Texture Section at the Leatherhead Food Research Association (Leatherhead, Surrey, UK). Responsibilities at the LFRA have included the evaluation of instrumental methods for the assessment of texture attributes. He has developed electromyography-based methods aimed at providing a texture assessment more relevant to consumer perception than conventional instrumental methods. Wider interests include the identification of relationships of instrumental data to sensory data and the implementation of computerized data acquisition systems for sensory analysis.

His background is in physical-organic chemistry (University of St Andrews, Scotland).

Erika Kress-Rogers* is a Principal Scientist at the Leatherhead Food Research Association (Leatherhead, Surrey, UK). Work at the LFRA has included a technology transfer study on the principles and characteristics of novel sensors developed or under development for other sectors, and on their potential applications in the food industry. She has coordinated a subsequent collaborative project on the development of novel sensors for the food industry. Prototype developments of this inter-laboratory project include an in-situ probe for frying oil quality based on a vibrating tube viscometer and a meat freshness probe based on a biosensor array. Earlier projects were concerned with the in-line determination of water content based on two-variable microwave measurement and with the potential applications of ultrasound measurements in the food industry.

Her background is in experimental solid-state physics; semiconductor physics (Universität Karlsruhe, Germany, University of Oxford, UK, as Rhodes Scholar).

John M. Low is the Technical Director of Sortex Ltd (London, UK). He is responsible for the company's research and development programme and for the design of automatic sorting machines. His previous experience is in the area of optical systems for defence applications and of sonar image processing techniques.

His background is in electronics; medical physics (Robert Gordon's Institute of Technology (Aberdeen, GB); University of Aberdeen).

William S. Maughan is a Research Consultant at Sortex Ltd (London, UK). He has over 20 years experience in industrial research and development for the design of automatic sorting machines. Earlier, as a Research Fellow at Imperial College (London, UK), he has led a research group in the field of sorting techniques.

His background is in physics; optics (University of Durham, UK).

Douglas B. MacDougall is Lecturer in Food Science at the University of Reading in the Department of Food Science and Technology (Reading, Berkshire, UK). Earlier, he has worked at the AFRC (Agriculture and Food Research Council) Institute of Food Research (IFR) at Reading, and as Head of Colour Group and Development Section Project Leader at the AFRC IFR (formerly the Meat Research Institute) in Langford (Bristol, Avon, UK).

Projects at the Institutes of Food Research have included studies on the effects of animal stress, slaughter practice and meat handling on the colour appearance, translucency and colour stability of fresh meat; development of instruments for on-line measurement of meat faults; mathematical modelling of the colour and texture of composite meat products. Studies on quality measurement, colour appearance and visual appeal of foods as affected by composition, storage conditions and display illumination fidelity continue at Reading University.

* Erika Kress-Rogers is now with ATI Sensor Applications Limited (Surrey, U.K.).

H[illegible] science; colour stability of food (Royal College of Sci[illegible]gy, now Strathclyde University, Glasgow, Scotland; Ru[illegible] State University of New Jersey, New Brunswick, USA).

Brian M. McKenna is Professor of Food Science at University College Dublin (Dublin, Ireland). Earlier, he held positions as a Lecturer in Food Engineering at University College Dublin and as a Research Officer at the National Dairy Research Centre (Ireland).

He was a Member of the COST90 international project on food rheology and Chairman of the COST90bis international working group on mechanical properties of foods. He has also compiled a bibliography on Solid and Liquid Properties of Foods as part of the COST programme.

With a background in chemical engineering, he carried research responsibilities for teaching and research in food engineering and food physics at University College Dublin before his remit was widened to food science in general.

Wolfgang Rödel is Director and Professor at the Federal Centre for Meat Research in the Institute for Microbiology, Toxicology and Histology (Kulmbach, Germany).

Projects at the Federal Centre for Meat Research have included studies on measurement methods for the determination of water activity in meat and meat products, the development of novel instrumental techniques for water activity measurement in meat product, studies on the importance of water activity for the microbial stability of meat products, development of novel instrumental techniques for controlling ripening processes of fermented sausages and hams, studies on air-ionization apparatus and air-ozonization in meat-cooling rooms, studies on the measurement and the importance of the redox potential in meat products.

His background is in veterinary medicine (Free University of Berlin, Germany). He is a veterinary specialist for food hygiene.

Philip T. Slack is a Principal Scientist at the Leatherhead Food Research Association (Leatherhead, Surrey, UK). Earlier he has worked at the Brewing Research Foundation (Nutfield, Surrey, UK).

Work at the LFRA has included the coordination of a collaborative research programme in the area of the identification of toxic contaminants, the compilation of the LFRA's Analytical Methods Manual and work on a variety of research projects in the area of analytical chemistry. As part of his earlier employment at the BRF, he has studied the role of lipids in brewing, the effect of malt enzymes on barley starch granules and of hydrophobic proteins on the head of beer.

His background is in biology; plant physiology and biochemistry (University of London).

1 [illegible]entation for Food Qua[illegible] Assurance

Erika Kress-Rogers

Contents

1 Introduction

1.1 The Role of Quality Assurance in the Food Industry

Quality control is essential in the food industry, and efficient quality assurance is becoming increasingly important. The consumer expects a wide range of competitively priced food products of consistently high quality. Each food item has to be safe, wholesome and attractive in appearance, taste and texture, and needs to be consistent with the product image. Variations within the same batch or between batches will have to be kept to a minimum as they are often interpreted by the consumer as indicating a fault, even when the differing product is of high quality.

The availability, quality and price of raw materials will place conditions on the food manufacturing operation, as will the prevailing structure of the retailing sector. More and more frequently, the product palette has to be adapted to changes in tastes and nutritional ideas, and to the appearance of competing products on the market. In the manufacture of each new product, there is the challenge of getting it right first time. Increasingly, food processing operations are technology-based rather than skill-based. Legislation on food composition and labelling will also play a role. Changes in legislation are driven by consumer demands and by international harmonization.

Food processing has a long history (Georgala 1989) and has always had two main purposes. The first is the conversion of agricultural products (or of fished, hunted and gathered foods) into palatable, attractive, digestible and safe foods. Cereals, for example, are virtually inedible without prior milling and cooking or fermenting; some fruits and pulses are toxic without prior cooking; and large proportions of the Asian and African populations can consume lactose only after conversion to lactate by fermentation. The second purpose is the preservation of foods for availability out of season, for years of lean harvests, and for transport to areas distant from agricultural producers.

The assessment of food still centres on its taste, aroma, appearance and nutritional value, and on its safety and stability.

Optimized process control plays an essential part in maintaining the commercial viability of a food manufacturing operation in the face of changes in the food market and in the structure of the food industry. Advances in microelectronics have provided fast data processing and have made efficient process control systems possible. In the 1980s, programmable logic controllers (PLCs) were widely installed in the food industry. Massive control centres were designed earlier for integral plant control; these centres are now being replaced by distributed control systems (McFarlane 1983; Vidal 1988).

Whichever control system is used, it still has to make do with a small number of continuously updated product variables, and often relies largely on inputs at long time intervals and with long delays depending on the assay time and the distance to the QC laboratory. The effective application of both established and novel sensors and instruments will play a key role in gaining the full benefit of the potential that modern control systems offer.

Table 1.1 shows the sizes and growth rates of food and drink market sectors in the UK.

1.2 On-line, At-line and Off-line Instrumentation

For optimum quality assurance the manufacturer requires cost-effective methods for the rapid assessment, and preferably the on-line measurement, of the chemical and physical properties and the microbial status of raw materials, process streams and end products. Monitoring during the processing operation helps prevent expensive rework or disposal of out-of-specification product.

Tight control is needed for variables that influence the stability of the end product towards microbial spoilage or oxidative rancidity. This

Table 1.1 *Food and drink market sectors in the UK* (Bailey et al. 1991)
(a) Market sectors in 1990 by value (£ million)

*	Alcoholic beverages (inc. duty)	21 864
1	Meat and meat products	9 882
2	Dairy products (inc. ice cream)	5 884
3	Soft drinks	5 286
4	Fresh fruit and vegetables	4 598
5	Bakery products	3 790
6	Confectionery	3 700
7	Frozen foods	2 250
8	Canned foods	1 686
9	Fish and fish products	1 606
10	Snack foods	1 287
11	Hot beverages	1 281
12	Cereal products	1 230
13	Oils and fats products	971
14	Ready meals	715
15	Meal accompaniments	678
16	Sweeteners, preserves	535

Table 1.1 *continued*
(b) Increase of market sector values in the UK from 1989 to 1990 (per cent)

*	Alcoholic beverages	8.2
1	Frozen foods	11.2
2	Soft drinks	10.9
3	Ready meals	10.5
4	Confectionery	8.8
5	Fresh fruit and vegetables	8.2
6	Snack foods	6.5
7	Cereal products	5.4
8	Meal accompaniments	4.3
9	Hot beverages	4.0
10	Dairy products	3.5
11	Meat and meat products	3.5
12	Fish and fish products	3.3
13	Bakery products	2.1
14	Oils and fats products	1.4
15	Canned foods	1.1
16	Sweeteners, preserves	0.6

* As much of the market value of the alcoholic beverages sector is the duty, the value has not been used to rank this sector.

Comments

1 For the cost/benefit assessment for a potential instrument development, the market values listed here need to be seen together with other factors such as the relative values of raw materials and final product or the growth rate and profit margins in a sector. For snack foods, for example, the added value would in general be higher than for canned foods or meat.
2 All market sectors increased in value, but not all growth rates exceeded the rate of inflation.
3 Categories are not mutually exclusive.

concerns particularly the monitoring of temperature profiles during heat processing and storage, the control of cleaning-in-place procedures, and the measurement of the pH, water activity, solute concentration and preservative levels of the product. Water activity, usually measured as equilibrium relative humidity (ERH), cannot be measured rapidly. From an on-line measurement of the moisture content, the ERH can be deduced if the isotherm is well defined.

The trend towards continuous automated production in place of batch processing necessitates tight feedback loops based on on-line monitoring methods or, failing that, on rapid at-line techniques. Even when a laboratory method provides a result within one hour of taking a sample from the line, over a tonne of product or over 10 000 jars, tins or packs of food may already have passed the production line. The cost of rework or disposal for such a quantity is considerable. Alternatively, excessive safety margins with respect to legal requirements or customer specifications on the minimum content of expensive ingredients will lead to an uncompetitively priced product.

Prolonged holding times to await the outcome of assays, as a regular part of the process, lessen the benefits of continuous processing. Nevertheless, holding times of around eight hours are currently observed prior to filling certain sterilized foods, for example, in order to await test results from impedance monitoring for microbial assessment (Chapters 14, 15). Refinements of this technique, based on more sensitive oscillometric detection of impedance changes with microbial growth, for example, are being investigated in order to shorten the assay time (Cossar et al. 1990).

The advances in plant for automated continuous production and in the signal processing capabilities of process control systems have stimulated progress in the development of many novel sensors and instruments for the food industry, often by technology transfer from other industrial sectors or from the clinical sector (Kress-Rogers 1985; 1986). These have since matured; sensor concepts have been developed into prototypes, and instrument types already available in the early 1980s have become more versatile and can now be reliably applied to a wider range of foods and processing situations or determine a wider range of target variables (Figure 1.1).

With the help of these advances in on-line and at-line instrumentation, quality assurance (QA) is increasingly employed in the management of manufacturing operations. The quality control (QC) laboratory supports QA by checking and updating the calibration of on-line and at-line instrumentation and by providing a wide range of analyses and assessments that are not feasible for QA implementation.

The variables measured on-line and those measured off-line in the QC laboratory do not necessarily coincide. The process stream at the on-line measurement point will often be quite different from the sample taken to the laboratory, either due to changes during sampling and transporting, or because the laboratory test measures properties of the end product, whereas the on-line instrument measures precursors of these, or other properties of the process stream or the process environment that will determine the relevant properties of the product.

When the time taken for a QC laboratory result exceeds a day, as would be the case for many microbiological tests or trace analysis assays for toxins, it is often impractical to hold the food in quarantine during this time, as a perishable food may be well into its shelf life by the time the result is available. Even when prolonged holding times can be observed, it is not usually possible to provide 100 per cent screening of the product with

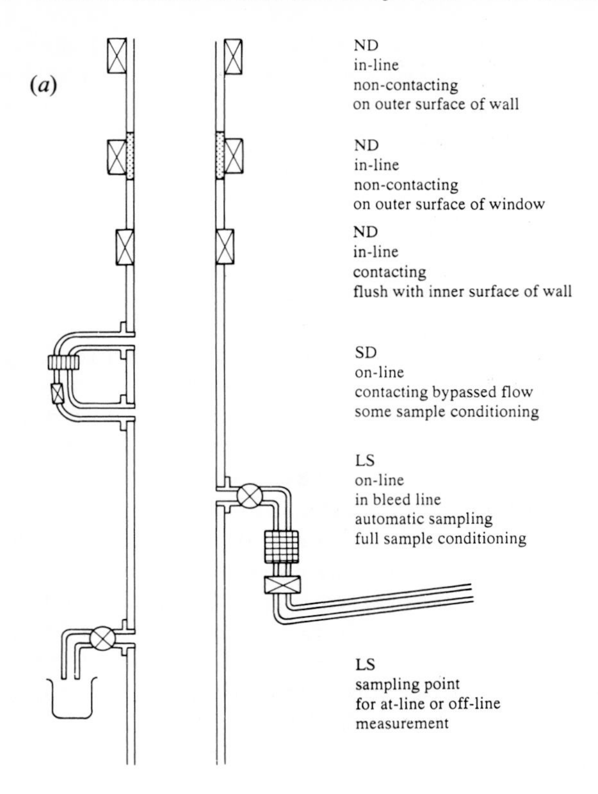

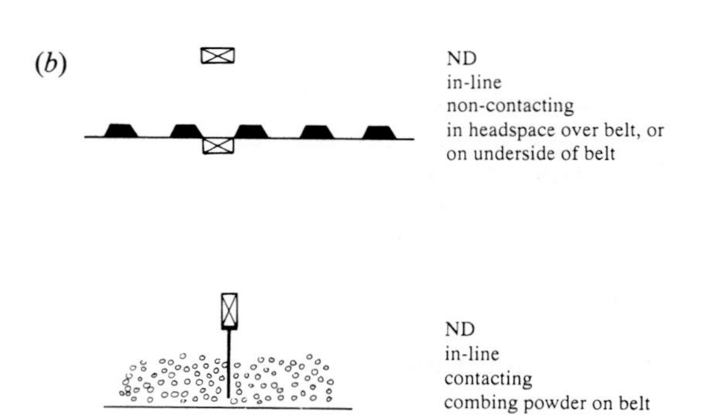

Figure 1.1 *Sensor configurations: (a) sensors on continuous processing lines (b) sensors on conveyor belts (c) sensors in batch processes (d) handheld sensors. The window material will depend on the instrument principle, for example Teflon for microwave transmission. Conditions for at-line measurements (not on-line, but in the production area) are more stringent than for off-*

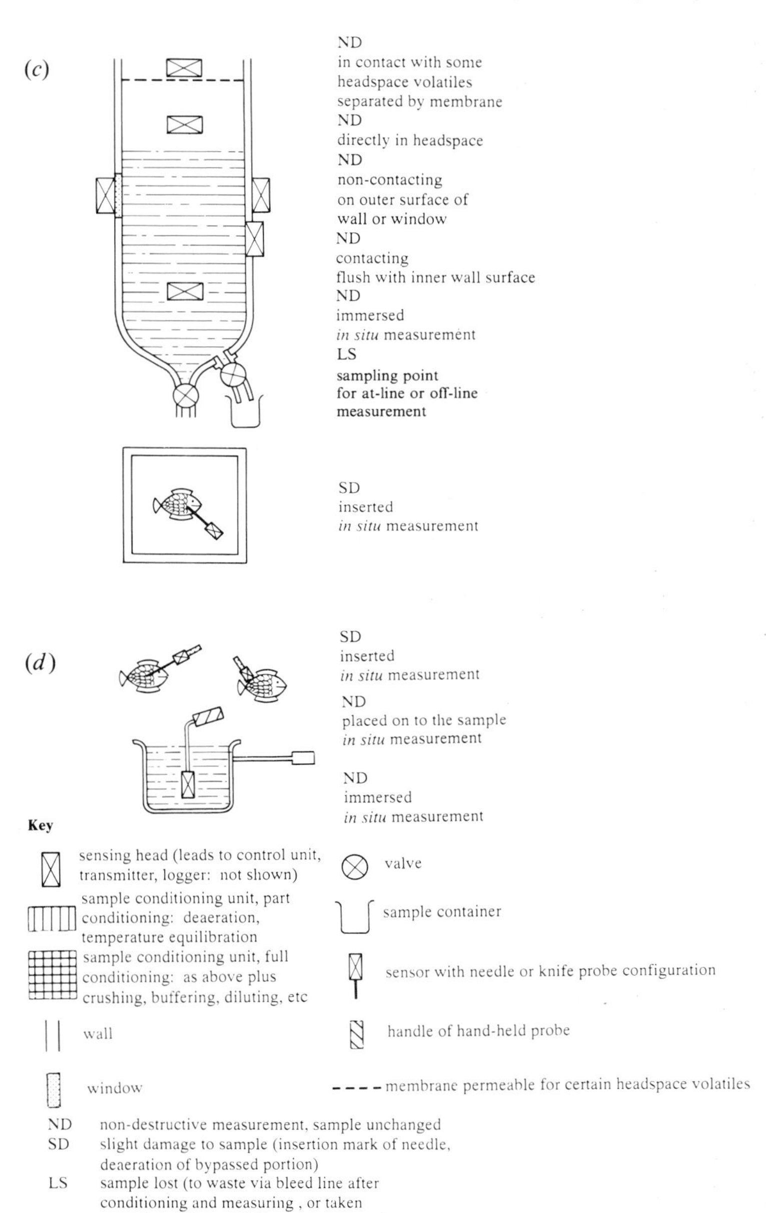

line measurements (in the QC laboratory). At-line instrumentation and accessories should be free of glass components (potential foreign body hazard) and of toxic reagents that are not fully contained at all times. Also, mechanical robustness, tolerance of the processing environment (for example, of steam) and simple and rapid operation are essential

QC methods, and so a negative result is no absolute guarantee that the whole production volume is 'clean'.

The test then becomes a means of checking that good manufacturing practice (GMP) is being observed, and the process has to be analysed to define the product and process variables that can be monitored and controlled in order to minimize the possibility of manufacturing a product having too high a microbial load, carrying pathogens or containing toxins. This approach is known as hazard analysis critical control point (HACCP) system.

Important for the stability of foods towards microbial spoilage are product properties such as the water activity and the pH as well as the microbial load and the concentration of preservatives and nutrients (Chapters 12–15, 17). The integrity of the food packaging is also vital, and in modern modified atmosphere packs (MAPs) the initial headspace gas composition and its retention during distribution and storage will be relevant. The adherence to appropriate storage temperatures (and ambient humidities) throughout the shelf life needs to be ensured.

An important process variable influencing the shelf life is the time–temperature profile of the process stream and, related to this, the excess pressure in the headspace. Also relevant are the concentration and temperature of cleaning liquids and their efficient application to process plant surfaces (Chapters 8, 9).

In conventional cooking and canning operations, heating the interior of a solid food item (or a highly viscous liquid) relies on thermal conduction, often resulting in overcooking of the outer layers in order to ensure adequate temperatures in the centre. This is not to say that high surface temperatures are not desirable in processes such as roasting, where the Maillard reaction provides a range of flavours and colours in the presence of reducing sugars and amino acids at elevated temperatures. The flavour changes caused by prolonged boiling are, however, usually considered undesirable.

Microwave or radiofrequency waves, on the other hand, can penetrate food and heat deeper layers directly. Direct ohmic heating is also possible by mounting electrodes in contact with a conductive food, and ohmic heaters allowing continuous automated heat processing are available.

With these methods, it is possible to retain more of the flavour and vitamins of the food, and yet to ensure a given minimum temperature to be reached throughout. In order to optimize such processes for the manufacture of products that combine adequate cooking, pasteurization or sterilization (as required for the product) with good flavour retention, analysis of the spatial distribution of the time–temperature profiles is necessary.

Several variants of time–temperature integration are used to assess the effect of heat processing on a food. The most common is the F_0 value, which expresses the degree of sterilization of a food. The F_0 value (expressed in minutes) is obtained by calculating the integral

$$F_0 = \int L \, dt$$

where $\lg L = (T - T_{ref})/Z$ defines the lethality L, and the temperature T has been measured in the coldest part of the food. For F_0 evaluation, $T_{ref} = 121$ °C and $Z = 10$.

For canned foods and ultra heat treated (UHT) products, F_0 values of 3 to 18 are used, depending on the types and numbers of spores present. This treatment results in commercial sterility, that is the remaining micro-organisms will not cause spoilage or disease or have a detrimental effect on the product quality during its stated shelf life (usually in excess of six months) (Lewis 1987).

Other values of T_{ref} and Z apply for the loss of nutrients by protein denaturation, vitamin destruction and certain other chemical reactions. A cook value can be defined, in analogy to the sterilization value F_0, to quantify the degree of cooking or overcooking and thus predict the loss of quality (flavour, nutrient levels) by heat processing.

Figure 1.2 summarizes the rôles of on-line, at-line and off-line instrumentation in process management, quality assurance and quality control. Measurements relevant for product safety, stability and quality and for process management are listed in Tables 1.2a and 1.2b. Instrument requirements and measurements for special concepts are given in Tables 1.2c and 1.2d.

1.3 Technology Transfer: Opportunities and Pitfalls

Instruments for measurements in quality control and in the control of processing operations in the food industry are often the result of technology transfer from other industries. The history of such new introductions has, in some cases, been characterized by initial successes, followed by a phase of disappointment with the instrument performance when the range of applications was widened. Subsequently, lost confidence had to be regained by defining the range of suitable application areas and by adapting the instrument or the setting-up and running procedures to particular applications. To avoid setbacks, it is necessary to understand both the instrument design and its underlying principles as well as the properties of the food and its processing environment.

Problems have, for instance, been experienced with some early applications of ultrasound flow meters in the food industry. These flow meters have the attraction of providing a non-contact measurement which facilitates maintenance and is intrinsically hygienic. However, for certain food process streams, unacceptable errors in the readings were observed until it was recognized that special designs or other types of flow meters were needed for samples with non-Newtonian flow profiles, or containing large particulates with flow rates differing from that of the carrier liquid, or

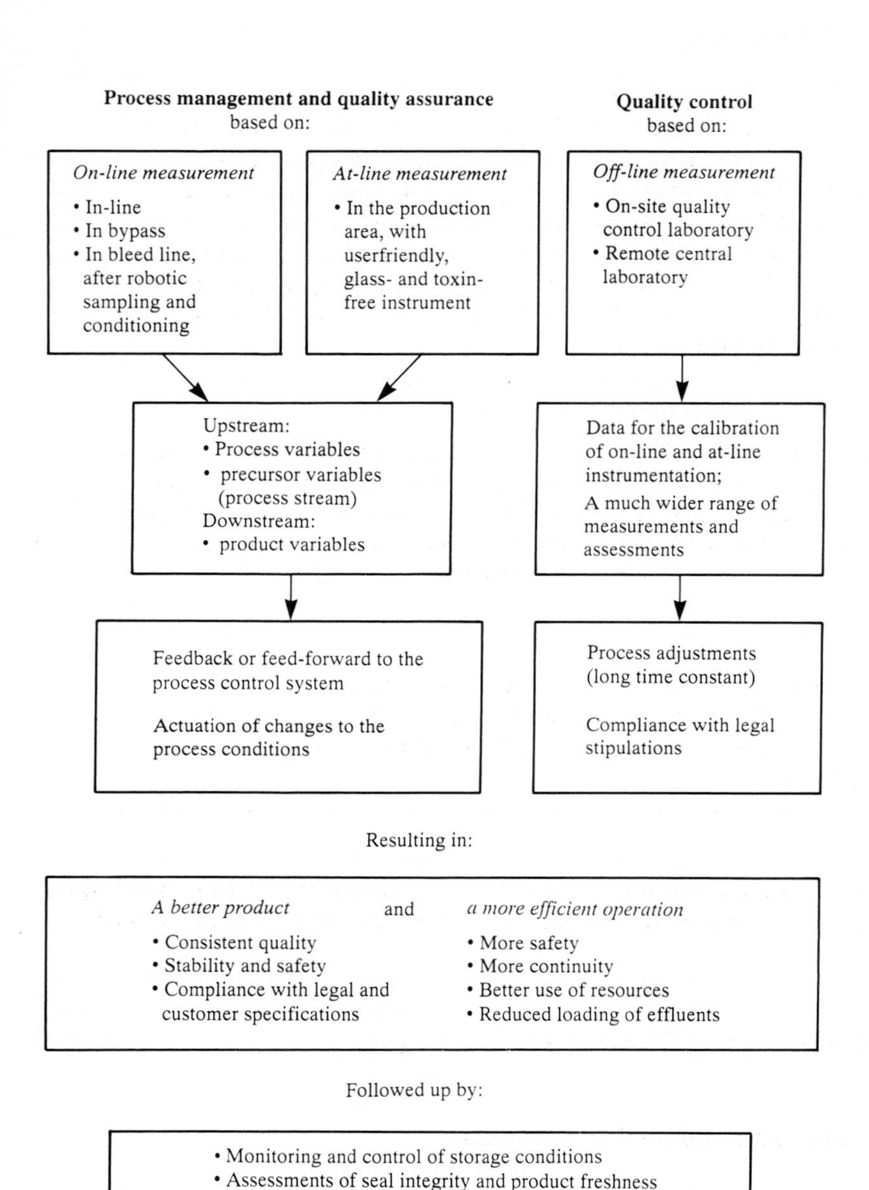

Figure 1.2 *Quality assurance (QA) and quality control (QC) in food processing operations*

Notes for Figure 1.2

Actuation of process changes

A programmable Logic Controller (PLC) may be used to operate actuators that cause changes in process conditions in dependence of a measured variable. Where complex relationships exist between the measured variables and the process, an expert system can provide an automatic evaluation of a set of measured values and a decision on corrections to the process conditions. Combined inspection/sorting systems are used to identify and remove, for example, products that contain foreign bodies or that are mis-shaped.

In-situ measurements in batch processes

The scheme needs to be adapted for industrial batch processing operations, for batch processes in catering establishments and for measurements in food distribution. In-situ measurements with dip- or stab-probes, or with instruments installed during the batch process or permanently installed in a processing vessel or storage container can be used here.

Calibration

The choice of the reference method can influence the calibration. Systematically different values can be obtained, for example, between drying and titration methods for moisture determination in the laboratory.

---

where high attenuation of the ultrasound signal by the food liquid restricted the sampled flow volume to the outer layer (Chapters 8–10).

2 Challenging Conditions for Sensors

2.1 Complex and Variable Samples

Many foods are highly complex in their chemical composition and in their physical structure. Gaseous, liquid and solid phases often coexist in the same product. Each phase may incorporate many different chemical compounds. One phase can be finely dispersed in another, or samples can be highly inhomogeneous or even largely separated. Within the liquid portion, fat and water may be combined in an emulsion, or even in a double emulsion. Water can be present as free water or bound in many different ways: as water of crystallization, bound to protein or starch molecules, entrapped in biopolymer networks or absorbed on solid surfaces of porous food powder particles. Active enzymes may be present, either in the tissues of fresh meat or produce, or within the cells of the microbial flora.

Samples in the food industry are, moreover, very diverse and highly

Table 1.2a *Measurements in Quality Assurance and Quality Control*

Measuring properties relevant for product quality
- appearance (colour, gloss, shape)
- texture, mouthfeel, pouring characteristics
- aroma, taste
- nutritional value
- functional properties
- composition according to specifications

Screening for product safety

chemical contamination
(agricultural residues, endogenic toxins, . . .)
microbial contamination
(total load, presence of pathogens and spoilage organisms, . . .)
foreign matter
(metal or glass fragments, insects, stones, . . .)
unwanted matter
(nutshells, fruit calices, . . .)

Assessing product stability towards
- chemical reactions
  (such as oxidative rancidity)
- microbial growth
  (due to inappropriate pH, water activity, preservative concentration, either in the product as a whole or in a small region within the product)
- microbial or chemical contamination
  (due to defective or inappropriate packaging)
  (including the migration of compounds in the packaging material into the food)
- migration of water or fat
  (between pastry shell and filling, between food and environment)
- loss of protective atmosphere
  (due to defective seal)
  (for products packed under a modified atmosphere designed to suppress microbial growth or oxidation)

Table 1.2b *Measurements in Process Management*

Objectives
- ensure safety and continuity of the processing operation
- maintain conditions for in-spec. products
- use resources efficiently (labour, raw materials, energy, machinery)
- reduce loading of effluents (e.g. of waste water with organic matter)

Measurements
- pressure
- temperature (also spatial distribution of temperature and time integral over temperature),
- pH
- mass and volume flow rates of liquids and particulate solids
- fill levels of liquids and particulate solids
- bulk density, weight
- apparent viscosity

Also wanted, but more difficult to achieve on-line:
- chemical composition (gross and fine)
- complex rheological properties (yield value, elasticity, ...)
- particle, droplet, bubble size, (average size and distribution)
- volatiles evolved in cooking, baking, roasting, drying operations

---

Table 1.2c *Instrumentation Requirements, on-line*

- hygienic sensing head
- contaminant-free (no reagents, no microbes)
- no foreign body hazard (no fragile glass components)
- robust
- CIP (cleaning-in-place) tolerant if permanently installed on-line (alternatives for specific chemical measurements: instruments with disposable sensing element which must be easily replaced and inexpensive, or, in certain applications, robust, easily cleanable dip-probes or stab-probes for in situ measurements)
- reproducibility in accordance with task
- reliable, low maintenance effort
- suitable for complex chemical and physical sample properties
- total cost (capital, maintenance, running) in good proportion to benefits

---

Table 1.2d *Measurements for Special Concepts*

HACCP – Hazard analysis critical control points
On-line measurements
- pressure, temperature (spatially resolved, time integral)
- relative humidity
- product pH
- product solute content
- strength and surface coverage of solutions used for periodic cleaning of machinery

Off-line measurements
- water activity (as equilibrium relative humidity)
- pH (spatially resolved)
- preservative concentration
- microbial contamination of ingredients including water
- microbial contamination on machinery and on other surfaces in the production area

Marker (Indicator) Approach
For the on-line, at-line or in situ assessment of
- microbial pre-spoilage status
- oxidative rancidity status
- level of heat-induced deterioration
- progress of ripening or conditioning
- browning potential
- end of heat processing operation

Measured are chemical or physical variables that have first been identified as indicative of the complex condition of interest. Usually, a given marker (or indicator) variable will be valid for a particular group of products only. (See Section 3.4 and Chapter 16.)

variable. The season, the region of origin, the harvesting and storage conditions as well as the processing steps (such as the fermentation of cocoa beans) will all influence the properties of the raw materials. New food processing technologies are being introduced to provide an ever wider range of food products that require frequent adaptation to changing consumer preferences and market structures.

2.2 Hostile Conditions and Stringent Hygiene Requirements

The pH extends over a wide range, with low values for vinegar or citrus fruit juices and high values for caustic cleaning solutions used regularly in-line. A wide range of pH values is also encountered in the monitoring of effluents, that is waste liquids formed in washing raw materials or in cleaning container surfaces, for example.

Temperatures vary from freeze-drying conditions (−50 °C or lower) to hot frying fat conditions (up to 250 °C) and roasting operations (320 °C or higher). Processing and packaging under vacuum is employed, and excess pressure is used in cooking and canning operations. A retort would typically operate with pressures of 60–600 kPa, that is 0.6–6 bar (McFarlane 1983, see Appendix B, Tables 3 and 4).

Particularly severe conditions can prevail in the cooker extruder, where both high pressures (over 10 MPa, that is over 100 bar) and high temperatures (around 200 °C) can be encountered. Moreover, the inner barrel surface is scraped by the extruder screw, and access to the food mix within the barrel or in the extruder head is certainly restricted. The food mix itself can be quite abrasive in the early part of its passage through the extrusion cooker. Maize grits, for example, may be present, expanding later on in the fashion of popcorn. These conditions present a challenge even for the design of pressure and temperature (p/T) probes. (For a description of extrusion cookers see McFarlane 1983; O'Connor 1987; Wiedmann and Strecker 1988.)

Nevertheless, sensors for the measurement of moisture and other variables are under development for this hostile environment. Radiofrequency open-ended coaxial probes have been designed to fit into the openings foreseen for the bolt-type p/T probes designed for extrusion cookers, and a microwave stripline has been constructed for mounting in an extrusion head (Chapter 7).

In general, the conditions in the food industry are more favourable than in the cooker extruder. A common challenge for *in situ* sensors is, however, the cleaning-in-place (CIP) procedures used in many processing systems in the food industry (Kessler and Weichner 1989). These usually involve flushing with hot caustic soda solutions (NaOH) which can corrode probe surfaces, and this is particularly unfavourable for many chemical sensors. High-pressure steam cleaning is another effective CIP procedure; this will challenge the mechanical and thermal stability of a sensing head.

The strict hygiene standards in the food industry also demand that in-line probes in contact with the sample must have crevice-free surfaces. This applies both to the sensing head and to the mounting flange area. For aseptic processes, any sensor surfaces in contact with the sample need to be tolerant to CIP procedures. In fermentation processes, the use of a disposable sterilized sensor can be an option.

Any danger of chemical contamination of the food by sensor reagents or components of slight solubility must be eliminated. The introduction of foreign bodies, particularly glass or metal fragments, in the case of damage to the sensor, must also be prevented. Food powders with a very low moisture content can accumulate high electrostatic charges, and sensors that may come into contact with such powders (typically starch-based products) must be designed to minimize the risk of a dust explosion.

The transducer and electronics may have to withstand exposure to water, steam or airborne dust. Occasionally, they may be enrobed in chocolate or coated with a thin film of condensed polymerized frying oil. Sensors in contact with food or food volatiles are often subject to fouling by proteins, fats or starch particles (Kessler and Weichner 1989).

Electromagnetic interference (EMI) will be encountered in industrial microwave ovens or in direct ohmic heating appliances. There will also be electromagnetic noise and mechanical vibrations from pumps, hoppers and other plant. Rotating mixer paddles scraping the walls of a vessel may be in the way of a radiated signal or restrict the positioning of a wall-mounted probe.

2.3 Non-Contact Techniques and Robotic Sampling and Conditioning

Given the often hostile conditions for invasive sensors, either during the processing of foods or during the periodic cleaning operations, and the always stringent hygiene and other food safety requirements, non-contact measurements are particularly attractive to the food industry. These can be based on the interaction of electromagnetic waves, including gamma-rays, light, infrared radiation, microwaves or radiofrequency waves, or of ultrasound signals with the sample.

Such methods do, however, require an awareness of the nature of this interaction of the applied signal with the food, its headspace and container. This understanding is needed at all stages of the instrument development, in the choice of suitable applications and installation points, during the setting-up and calibration procedures (including the preparation of training samples), in the running of the instrument and in the evaluation of the readings (Chapters 5–9).

To develop the wide range of sensors desirable for in-line measurement in the food industry would be prohibitively expensive. Not only different target variables, but also different analytical ranges and variable chemical and physical environments, would have to be catered for. The recognition of the cost that would be associated with the development of in-line sensors for a wide range of chemical, physical and microbial properties, each for a wide range of diverse applications, has led to an interest in techniques that make the best of the sensors available.

Robotic sampling and sample preparation systems allow rapid measurements at short intervals by enabling the use of sensors that would otherwise be confined to laboratory applications. This approach has been implemented particularly in Japan. An example is shown in Figure 1.3 (see also Chapter 17, Section 5).

3 Interpreting the Readings

3.1 Measured Variables and Target Variables

In non-contact measurements, the relationship between the measured variables and the target variables is often complex, so that a given calibration will apply only to a limited range of food products and processing conditions. For instance, a water content measurement based on near infrared reflection analysis will have to rely on a predictable relationship between surface moisture and average bulk moisture content; or, the monitoring of solute concentration by a measurement of ultrasound velocity depends on a constant composition of both the solute and the carrier liquid (and on compensation for temperature changes, as is the case with most measurement methods). Care in the setting up and calibration are essential, as is the choice of appropriate applications. Non-contact in-line techniques will then provide highly reliable continuous measurements that allow process adjustments before an out-of-specification situation arises (Chapters 5, 7–9).

In contact measurements also, the measured variable is not always the target variable. For example, pH is often measured as an indicator of acid concentration (provided that the acid composition is known). Ion activity is often measured in place of ion concentration. Frying oil samples are taken for an at-line measurement of colour or free fatty acid (FFA) content in order to infer the degree of frying-induced polymerization and oxidation. Yet, both colour and FFA content are highly dependent on other factors such as the oil type, the food fried and the frying conditions. (See Chapters 16 and 17 on oil quality and pH, respectively).

In the QC laboratory, the assay of chemical composition can involve

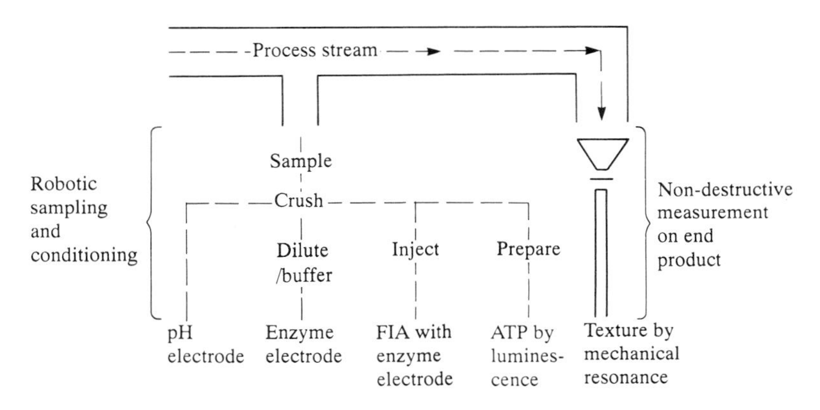

Figure 1.3 Robotics approach

deductions from the proportion (by weight) of sample becoming volatile or dissolved under certain conditions. Clearly, such assay types, and many others, need to take into account the nature of the sample. Indeed, the official methods prescribe sample-specific assay procedures (Chapter 13).

In addition to these relationships between measured and target variables, the significance of a target variable for the manufacturing operation needs to be considered. This can reside in assuring the safety and stability of the food, the enjoyment in handling and eating it, and the efficiency in producing it or in complying with legal regulations or customer specifications. These aspects are discussed in Section 4.1 of this chapter.

3.2 Relationship Between In-Line and QC Laboratory Methods

Differences between the readings of the in-line instrument and the off-line quality control (QC) laboratory results are, at times, unjustly blamed on the in-line method. The QC reference assay can be applied only to a small fraction of the sample volume passing the sampling point, and this alone can lead to a result differing from an in-line method that provides 100 per cent screening. Moreover, different laboratory methods will often give systematically differing results between them for a particular food sample type, even though they may give identical results for other sample types, (see, for example, Figure 1.4).

Often, the definition of the measured variable depends on the laboratory reference method used. For example, for an oven drying procedure, moisture is defined as that part of the sample that will be driven off at the applied temperature and pressure. For a titration procedure, on the other hand, the relevant part of the water is that which can be extracted from the food matrix (or dissolved and dispersed together with food solids) and brought into contact with the reagent. For this reason, official analysis protocols exist for different food types; for accurate results to be achieved, an instrument used in the QC laboratory needs to be calibrated for each food type against the official method (Chapter 13).

In some situations, the comparison between the on-line reading and the chemical or rheological QC result, for example, is not strictly valid because the sample changes on removal from the line and during the preparation and analysis steps in the QC laboratory. Oxidation, thinning, thickening, fermentation or other changes can occur during this time, and homogenization steps can cause the breaking up of tissue cells, leading to changes in the composition of the juices and also exposing the cell contents to oxidation or other reactions.

When changing from a process control system, updated from time to time by data from the QC laboratory, to a system relying primarily on continuous feedback from in-line instrumentation, the significance of the

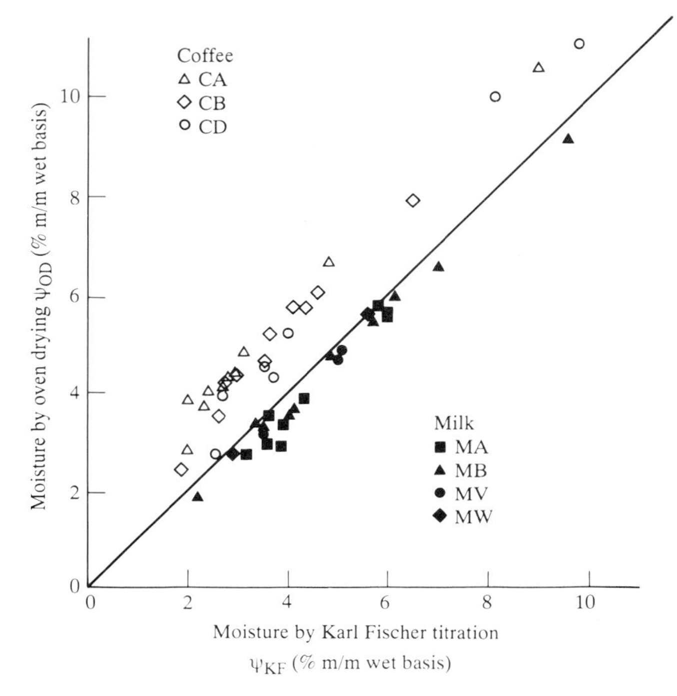

Figure 1.4 *Comparison of laboratory methods for moisture measurement. Oven drying at 102°C to constant weight leads to an increased apparent moisture content when non-water volatiles are present. This can be avoided by vacuum drying at a lower temperature. When lactose is present, incomplete dehydration can decrease the apparent moisture content by oven drying. This is avoidable by moistening prior to oven drying. Symbols mark different brands of instant coffee and milk powder (Kress-Rogers and Kent 1986).*

measured properties for the food processing operation needs to be well defined. It may well be favourable to replace an off-line measurement for a given set of variables with an on-line method determining a different set of variables, at a point upstream from the existing QC sampling point. Feedforward control based on the continuous on-line monitoring of the process stream before and during processing can often provide tighter control of the end product properties than delayed feedback control based on an off-line measurement of the end properties themselves.

Based on an analysis of the process as a whole, together with an awareness of the options for on-line and at-line instrumentation, an effective data input system for process control purposes can be designed. A complementary range of laboratory instruments will provide calibra- ··n updates and a wider range of measurements.

3.3 Data Processing Approaches

When several on-line measurements are carried out simultaneously, there is not necessarily a one-to-one relationship between each measured variable (or a combination of these) and a corresponding control parameter to be adjusted. The design of the response pattern of the control system can then be based on a variety of data processing approaches. Ideally, but rarely, there is a mathematical model based on sound physico-chemical principles, which links the measured variables to the relevant properties of the end product and to the process parameters that can be adjusted. For other cases, data interpretation based on neural networks (Jansson 1991) and control decisions based on fuzzy logic (Eerikainen et al. 1988) have advanced recently. (See also the entry on fuzzy logic in Appendix A of this book.)

Neural networks in particular have attracted much attention in recent years. Here, several on-line sensors that are not specific for an intended property are combined. Extensive output data from these sensors and associated quality control results are then collected over a period of weeks or months. These data are subjected to analysis with neural networks. This can result in a signal processing mode learnt by the neural network. The output of the educated network will then be indicative of a characteristic of the finished product such as taste or aroma, although none of the contributing sensors necessarily has a well-defined physico-chemical relationship to the characteristic of interest.

There is a temptation to employ such systems as a replacement for the development of sensors and instruments that give an output with a well-defined relationship to the process and product. However, the short-term saving in avoiding sensor development can be outweighed by the cost of frequent updating of the signal evaluation procedure for non-specific measurements. This updating procedure may involve the collection of process measurement data and associated product assessments by QC laboratory tests and taste panels over many weeks, and may be required whenever raw materials or recipes change. When the need for an update is not recognized in time, the system will be unreliable, and the quality of the product will be adversely affected. Although sophisticated data evaluation systems will play an important role in the future, they will be complementary to rather than a replacement for measurement techniques based on sound physico-chemical principles. (See also Chapter 17, Section 5.2)

3.4 The Marker Approach: Novel Sensors

In the case of lengthy tests for properties such as freshness, the progress of microbial or oxidative degradation processes, or taste and texture, efforts are increasingly made to identify and measure chemical or physical indicator variables linked to the condition. This can be described as the marker approach. Prototype probes for meat freshness and frying oil quality assessment have already been developed, based on a biosensor array and a vibrating rod viscometer, respectively (Chapters 17, 16).

Novel chemical sensors, including biosensors, have been developed in the last three decades, and the pace of this development has accelerated over recent years. Some biosensors have already been adapted for food applications, particularly those using the marker concept. Many sophisticated biosensor adaptations for the food industry are still at the prototype stage. This is consistent with the delay between the research and application stages of clinical biosensors, for example, which have received attention and funding earlier and more lavishly. In the absence of fully optimized biosensors for food applications, many current practical biosensor applications in the food industry employ the robotic approach (see above).

The microelectronic pH and ion sensors (primarily ion-selective FETs or ISFETs) had a hesitant commercial development. Now, a commercial instrument designed for the food industry is available and this could solve many of the problems experienced with traditional pH and ion probes (hazard of glass membrane, laborious maintenance etc.). Fibre-optic probes for chemical sensing have made rapid progress over recent years. They are immune to EMI in microwave ovens, and can be configured as robust and mechanically flexible remote probes. Optical sensors are also the basis of novel immunosensors that are now approaching the commercial stage (Chapter 17).

3.5 Instrumentation as an Interdisciplinary Subject

The development and implementation of successful new measurement applications are tasks requiring effective interdisciplinary communication and cooperation. This can be accomplished only when instrument suppliers and users have a common basis of understanding of the interplay between the physical, chemical and microbial properties of the food and its environment on the one hand, and the physics, chemistry and electronics governing the instrument characteristics on the other. An awareness of the nature of the reference method and of the significance of the target variable for the food processing operation is also needed.

In moisture measurement, for example, one has to establish whether the total water content is most relevant, or whether only water molecules with a high degree of mobility play a role. This will depend on whether the measurement is undertaken on account of legal requirements regarding the maximum water content of e.g. margarine, for the price definition of commodities such as wheat, or with a view to the influence of the water content on the product texture or its relationship with the water activity.

The last-mentioned factor determines the stability towards microbial spoilage and water migration from pie filling to pie shell, for example. Water activity is usually determined by a measurement of the equilibrium relative humidity of the sample, which by definition cannot be measured instantaneously. Instead, the water content can be measured in-line, and the water activity can then be deduced for a given product with a given processing history. Moisture and microbes are both ubiquitous, and improved instrumentation for their determination has been a prime concern for the food industry for many years (Chapters 12; 5, 7–9, 14–15).

From the instrument engineer's point of view, it may seem surprising, at first glance, to find methods based on a physical interaction of the sample with the instrument (using near infrared, microwave or ultrasound waves, or employing mechanical stress) and those based on chemical interactions presented together (compare Chapters 5–10 with Chapters 13, 17). The food scientist, on the other hand, may prefer to consider instruments for the determination of physical properties (rheology, colour, density, particle size) of the sample separately from those for the assay of chemical or microbiological properties. Process control engineers may wish to separate on-line and at-line instrumentation from off-line instruments.

However, in formulating a system of quality assurance and quality control procedures for a food manufacturing process, it is advantageous to be aware of the alternative techniques, each with its own characteristics, for the determination of a particular sample property (for example, for sugars, see Chapters 5, 8, 9, 13, 17; for water content see Chapters 5, 7, 8, 9, 13). A number of instrument types allow the measurement of both physical and chemical properties. For example, bulk density and water content can be determined simultaneously by a microwave technique (Chapter 7; Kress-Rogers and Kent 1987); film thickness and composition can be measured with near infrared instruments (Chapter 5).

Also, there is often a choice between process control based on measuring a certain set of variables upstream and in-line (for example, pressure, temperature, pH), and that based on measuring another set downstream and off-line (for example, glucose/sucrose ratio). As a further option, some off-line instruments can be used at-line with the help of robotic sampling and sample preparation, thus providing rapid feedback based on a wider range of measured variables. On-line and QC laboratory methods have to interlink in guiding process control, and are therefore presented together in this book, with the emphasis on the interplay between the instrumental method, the food and the manufacturing process.

4 Measurement Types

4.1 Target Variables

The on-line or rapid at-line measurement of variables relevant to the eating quality, the wholesomeness and the safety of the food product is an increasing concern of the food industry, and this is reflected in the topics covered in this book. Colour and other aspects of the appearance are dominant in determining the first impression and influence the choice of food products by the consumer. Aroma, taste, and texture or mouthfeel influence the enjoyment of the food and determine whether the consumer comes back for more of the product. The rheological properties also affect the handling characteristics of the product both in processing and in the hands of the consumer.

All these food properties can be assessed by panels of trained persons, or animals in the case of petfood. However, for frequent control purposes this is not practical, and instruments for the measurement of colour attributes, of variables related to the chemical composition (either proximate analysis or more specific and sensitive measurements) and of those related to the physical structure of the food (rheology, particle size, new methods in texture assessment) are used if equipment of the desired specifications is available (Chapters 2, 3, 4, 10, 11, 13, 17).

The assessment of wholesomeness and safety is not generally possible or advisable with the human senses, and instruments or laboratory tests will always be required for these food properties, including the content of nutrients, microbial load, chemical contaminants, pathogenic microbes and foreign bodies. Not only the current status but also the stability of the product during the intended shelf life needs to be assessed. (On the shelf life of foods, see for example Chapter 7 in Stewart and Amerine 1982.) To this end, a determination of the product's water activity, pH, packaging integrity, existing flora, content of preservatives, antioxidants content and other factors is undertaken.

The temperature of a food needs to be monitored in heat processing (such as sterilization), tempering and conditioning and also in storage. Spatial resolution is often needed, particularly information on the hottest and coldest points of the sample. A time–temperature profile including integral, maximum and minimum is also an important measurement target. Remote infrared thermometry has the advantage of providing a non-contact measurement and is also the basis of thermal imaging systems (Chapter 6). A recent development for depth profiling is the distributed fibre-optic temperature probe.

Table 1.3 shows the significance for the manufacturing operation of measurements of a range of variables.

4.2 Instrumental Methods

On-line instruments based on the interaction of electromagnetic waves or ultrasound with the sample are discussed in the first part of the book. In the second part, instruments in the quality control laboratory are treated. In the third part, the emerging range of chemical sensors, including biosen-

Table 1.3 *The significance of measurements for the manufacturing operation*

	Food safety and stability			*Food quality*					*Production efficiency*		*Compliance*		
	Micro-bial aspects	*Chemical aspects*	*Physical aspects*	*Nutri-tional aspects*	*Appear-ance*	*Texture Consist-ency*	*Aroma*	*Taste*	*Ingred-ients manage-ment*	*Process management*	*Legal con-ditions*	*Customer specifi-cations*	*Label-ling regula-tions*
Colour					×							×	
Sorting by colour	×	×	×	×	×	×	×	×			×	×	×
Temperature	×	×	×							×			
Temperature – time integral	×	×	×	×	×	×	×	×			×	×	1
Particle, droplet or bubble size			×		×	×			2	×		×	
Solid/liquid ratio and crystal size			×		×	×				×		×	
Bulk density			×		×	×			×	×		×	×
Rheology			×		×	×			3	×		×	
Texture						×				×		×	
Water activity	×	×	×		4	4	4	4		×	×	×	1
Water content	5	5	5	×	×	×		×	×	×	×	×	×
Proximates: fat, protein, carbohydrate, ash				×	×	×	i	×	×	×	×	×	×
pH	×	×					i	i		×	×	×	1
Acidity				×			×	×				×	
Sodium, potassium, calcium				×				×				×	×

Specific sugars	6		×	×	×	×		×				×	
Total reducing sugars					7		7	7		×			
Solute concentration	6		6	×	×	×		×		×		×	
Alcohols				×				×			×	×	×
Preservatives, antioxidants	×	×		8	8		8	8			×	×	×
Vitamins, trace minerals				×				×			×	×	×
Emulsifiers, gelling agents			×		×	×				×		×	×
Flavourings, colourings				×	×		×					×	×
Toxins, residues		×									×	×	×
Volatiles during cooking							×			×			
Volatiles during storage	×	×		×			×				×	×	
Microbial load, contamination	×	×			×		×	×			×	×	
Biomass, functional	×	×	×	×	×	×	×	×	×	×		×	×
Chilled meat freshness	×	×		×	×		×			×	×	×	1
Frying oil quality		×		×	×	×	×	×	×	×	×	×	1

Key for Table 1.3:
× direct link; i strong indirect link; 1 influencing storage conditions or use-by date; 2 influencing packing density; 3 influencing enrobing thickness; 4 governing microbial growth and metabolism as well as chemical reactions and water migration processes; thereby influencing composition and texture; 5 linked to water activity for a given food with a given process history; water content is more easily measured at-line and can be used as an indicator of water activity under controlled conditions; 6 solutes such as sugars or salts reduce water activity and have been traditionally used to preserve foods (for example jam, bacon); 7 at high temperatures, reducing sugars react with amino acids to form a wide range of compounds that flavour and colour roasted and baked products (Maillard reaction); 8 for example, antioxidants reduce the rancidity of oils and the browning of produce;

Authenticity assessment
the comparison of the sample's free fatty acid composition, ultraviolet absorption spectrum or other 'fingerprint'-type assays with the pattern expected for the claimed food type and origin is used in the authenticity assessment of foods.

sors, is presented, and the development of an *in situ* viscometer for frying oil is discussed as an example for the identification of an indicator variable that is more amenable to on-line implementation than current chemical reference methods for quality assessment.

Instruments that require an understanding of their interaction with the food and the manufacturing process as a condition for successful application are the subject of this book. In order to provide a useful description of principles and food-industry-specific application details for such instruments, it has been necessary to choose certain instrument types to the exclusion of others.

Only instrumental methods that can be fully understood on the basis of classical physics have been included in this book; thus instruments based on nuclear magnetic resonance (NMR) or electron spin resonance (ESR) are not represented. This is not to imply the lesser importance of these methods. In fact, the development of new applications of NMR and NMR imaging (NMRI) to the food industry is progressing rapidly, and many new instruments should emerge in this area in the next few years (see, for example, Levine and Slade 1991, pp. 405–626). For the recognition of irradiated foods, ESR applications are being developed.

Pulsed NMR techniques are already widely used in quality control in the food industry for the determination of water and lipids content and of the solid/liquid ratio of fat. Bench-top instruments for these applications are readily available and are being further optimized for a wider range of food applications (including the investigation of water 'binding' to biopolymers) and more convenient operation. Furthermore, research into the NMR characteristics of foods and the development of novel magnet designs for NMR instruments are now underway. This will lead to new application areas for NMR measurement in the food industry. One area is the on-line implementation of bulk NMR measurements as they are currently undertaken in QC laboratories. These are to become feasible for a much wider range of process lines than is possible at present. They will aid in the control of many operations in food processing, including baking, drying, concentrating, freezing, thawing and tempering. The other area is the use of NMR imaging in product development and quality control tasks. This latter measurement type will, in the medium term, be restricted to central laboratories owing to the cost and complexity of the equipment.

Also omitted in this book are those instruments that rely on ionizing radiation (such as x-ray foreign body detection systems) or on a substantial computing capacity, as, for example, image acquisition and analysis systems. Real-time processing and automatic evaluation of visible light, x-ray, infrared or other images is an area that is rapidly expanding. The underlying principles and application details relevant to the food industry for these systems could best be presented in a book dedicated to this area alone.

Appendix 1 to this chapter summarizes measurement types.

4.3 Fringe Benefits

The dielectric properties of food, discussed here in the context of measurements based on microwaves, are also relevant in the application of such radiation at much higher intensities to the processing of foods in industrial and domestic microwave ovens (Ohlsson 1988). Indeed, many dielectric data for foods have been acquired in order to predict microwave heating characteristics. For food scientists, a familiarity with the principles underlying dielectric measurement affords the added benefit of helping in the analysis of processing characteristics. To a lesser extent this applies to ultrasound, which also has applications both as a low-intensity signal used in measurement and as a high-intensity irradiation employed in sonoprocessing and sonochemistry. The overlap between the food properties relevant for measurement and those relevant for processing is smaller here. Ultrasound processing is therefore briefly discussed as a separate topic (Section 2.7 of Chapter 8).

5 Further Reading

The emphasis in this book is primarily on the measurement of food properties as this requires a particular understanding of the interplay between the instrumental method and the food and its environment. However, this understanding is also required in the measurement of certain process variables such as the volume and mass flow rates of food liquids with complex rheologies. Ultrasonic measurement of volume flow rate and of liquid and solid level is discussed in this book (Chapters 8, 9).

Volume flow rates do not always provide adequate information for materials with uneven aeration or temperature distribution, and are also difficult to measure for non-Newtonian liquids or inhomogeneous materials. For materials such as molten chocolate, mass flow rate can now be successfully measured with Coriolis force flow rate meters (developed by Exac and by Micromotion). These have the form of a U-shaped tube (or two parallel U-tubes), which can be flanged into the pipeline without restricting the flow. The Coriolis technique has recently been reviewed by McKenzie (1990), with details on one of the two embodiments.

A straight vibrating tube section can be flanged into pipes for measuring the mass density of the process stream. The tube section is vibrationally isolated from the remainder of the pipe by bellows. Another sensor based on mechanical resonance is the vibrating rod probe for the measurement of level, density or viscosity. Such probes are available in dipstick form or in flange mountings. This sensor family is represented in the book in the form of a new probe for the *in situ* determination of frying oil quality, based on a measurement of the viscosity of the hot oil by a vibrating tube viscometer. A review of the principles underlying the function of a wide range of mechanical resonance sensors has been given by Langdon (1985) (Chapter 16).

For the measurement of temperature, a non-contact method that also lends itself to combination with linescan and imaging techniques is presented in this book (Chapter 6). Fibre-optic thermometers and quartz temperature sensors (shear mode resonators or tuning fork devices), which have become available for industrial applications in recent years, are reviewed by Schaefer (1989). Updated overview tables on the specifications of conventional contact thermometers such as thermocouples are published regularly by transducer magazines. This is also the case for strain gauge pressure transducers. Information on special miniature designs for pressure and temperature transducers that can be mounted inside sample cans during heat processing can be obtained from the suppliers.

Humidity sensors have been discussed as part of instruments for the measurement of water activity. Similar sensors can also be used in monitoring ambient humidity in many storage or processing areas. In the latter applications, the required accuracy is less stringent; however, the emphasis is on robustness and long-term stability with minimal maintenance. High temperatures often need to be tolerated, for example in oven flues (McFarlane 1983). Inexpensive miniature humidity sensors based on solid-state devices are now part of a number of domestic appliances (such as some makes of tumble dryers or microwave ovens).

An overview of the automatic control of food manufacturing processes is given by McFarlane (1983). He describes processes in the areas of raw materials handling, recipe dispensing, pre-processing, cooking processes, biochemical processes, finishing and packaging. The range and tolerance for the controlled variables are given for each process. There are also books giving details on processing operations in one particular sector such as fish canning (Wheaton and Lawson 1985). Some recent developments in the control and optimization of food processes are presented in symposium proceedings edited by Renard and Bimbenet (1988) and by Spiess and Schubert (1990).

A comprehensive compilation of instruments, each described briefly without industry-specific aspects or applications details, is available in the form of a handbook on general instrumentation (Noltingk 1988). Often, instrument manufacturers will supply application details and related literature.

An introduction to food science and food technology has been given by Stewart and Amerine (1982). Information on the physical properties of foods can be found in the books by Lewis (1987), Jowitt et al. (1987), Singh and Medina (1988) and Okos (1986).

Appendix: Measurement Types

Measurement Types Discussed in This Book

Interaction of Electromagnetic Waves with Foods or Containers

Range description	*Order of magnitude of: frequencies (equivalent energy) wavelengths*	*Measurement applications*
Soft ultraviolet, visible, near infrared	$\sim 10^{15}$ to 10^{14} Hz ~ 100 nm to 1 μm	(Authenticity of foods) Sorting by 'colour'
Visible	$\sim 10^{15}$ Hz 400 to 700 nm	Colour (Optical imaging, for example to identify defective items or to measure dimensions)
Near infrared	$\sim 10^{14}$ Hz (equivalent temperature of radiating black body: ~ 1000 K) 700 to 2500 nm	Water content Fat, protein and carbohydrate content Caffeine contents of 1% m/m or higher Thickness of coated or laminated films on packaging materials
Mid infrared	$\sim 10^{13}$ Hz (equivalent temperature of radiating black body: ~ 100 K) 2.5 to 30 μm	Volatiles in headspace
Near and mid infrared	1 to 15 μm	Remote temperature measurement Thermal linescan (Thermal imaging)
Microwaves	$\sim 10^{11}$ to 10^{9} Hz ~ 1 to 10 cm	Water content of powdered or granular material Water content of low-

		and intermediate-moisture foods
		Water content of high-moisture foods
		Simultaneous bulk density measurement
		Fat content
		Ratio polar/non-polar liquids content
Radiowaves, upper range	$\sim 10^8$ to 10^7 Hz ~ 1 to 10 m	Water content
		Salt content
		Particle size, shape, distribution
		Bulk density

Note Measurements based on ionizing radiation are not included here (for these see later in this appendix). Some applications involving spectroscopic or imaging techniques (not covered in this book) are given in parentheses in the column 'Measurement applications' on pages 29–30.

Interaction of Ultrasound Waves with Food or Container

Common frequency and wavelength, order of magnitude:

Path through gas:	~ 100 kHz	~ 3 mm
Path through liquid:	~ 1 MHz	~ 1.5 mm
Path through solid:	~ 10 MHz	~ 0.6 mm

Applications:

Fill level of liquids or solids
Volume flow rate of liquids
Location of interfaces between: liquid and second liquid; or liquid and foam; or fat and lean meat
Solute content of a liquid
Distinction between two liquids
Counting of food packs
Suspended solids (in liquid) mass fraction
Dispersed droplets or bubbles (in liquid) volume fraction ⋆
Size distribution of dispersed droplets or bubbles (in liquid) ⋆
Creaming, sedimentation ⋆
Solid/liquid ratio ⋆
Temperature ⋆
Density of a liquid (as Z/c: see Chapter 8) ⋆

⋆ Pre-commercial, as far as food applications are concerned.

Acoustic emission monitoring

Range:
  Audiosound and ultrasound
Applications:
  Powder flow *
  Cleaning fluid impact *
  Cooker extruder performance *

* Pre-commercial, as far as food applications are concerned.

Interaction with Chemical Sensors, Including Biosensors and Immunosensors

Principle:
  Ion-sensitive membrane, inorganic catalyst, enzyme, antibody, specific adsorbent, etc. immobilized on base transducer, with electrochemical, optical, thermal or acoustic base transducers

Applications:
  pH (by ISFET)
  pH (by fibre-optic probe) *
  Inorganic ions (by ISFET) *
  Acidity (by microtitrator, ISFET-based) *
  Gases (by CHEMFET, MeOx or organic semiconductors)
  Specific sugars, alcohols, amines etc. (by biosensors)
  Vitamins, toxins and specific microbes (by immunosensors) *

* Pre-commercial, as far as food applications are concerned.

Properties under Mechanical Stress

Method	*Determined property*
Rheological analysis	Viscosity, yield point Changes of the rheological properties with shear rate and time
Texture assessment	Texture attributes (crisp, juicy, rubbery, mealy, tough, etc.)
Mechanical resonance dipstick probes	Level, density, viscosity, flow rate Determination of frying oil quality
Resonating tube section	Mass density of a processing stream
Coriolis force meter	Mass flow rate

Monitoring Headspace Humidity or Broth Impedance

Method	*Determined property*
Equilibrium relative humidity (ERH) measurement	Water activity
Electrical impedance or culture broth (at 2–10 kHz) monitoring (as method in rapid microbiology)	Microbial load

Observation of Drying, Extraction, Reaction and Chromatographic Separation

Method	*Analyte*
Drying or titration	Water content
Combustion	Ash content
Solvent extraction	Fat content
Acid digestion, distillation, and then titration	Nitrogen → protein content
Combustion, and then thermal conductivity detection	Nitrogen → protein content
HPLC (High Performance Liquid Chromatography)	Specific sugars Ions (nitrite, nitrate, bromide, chloride, sulphate)
IC (Ion Chromatography)	Sulphite
GLC (Gas–Liquid Chromatography)	Pesticides
GC–MS (Gas Chromatography – Mass Spectrometry)	Taints, flavours, spoilage indicators, toxins

Further Measurement Types

(not covered in this book)

Interaction of Electromagnetic Waves with Foods or Containers

Applications using ionizing radiation are listed here.

Range description	*Order of magnitude of: frequencies (equivalent energy) wavelengths*	*Measurement applications*

Gamma-rays (hard Roentgen rays)	$\sim 10^{20}$ to 10^{19} Hz (equivalent energy: $\sim$1 meV to 100 keV) $\sim$1 to 10 pm	Density
X-rays (soft Roentgen rays)	$\sim 10^{18}$ to 10^{17} Hz $\sim$100 pm to 1 nm (100 pm = 1 Å)	Foreign body detection (with linescan or imaging techniques)

Further Measurements Probing the Properties of Foods

Method	*Determined property*
Nuclear magnetic resonance (NMR)	Water or fat content Solid/liquid ratio
Electromagnetic induction	Metallic foreign body
Electromagnetic flow meter	Flow rate of conductive liquids
Electrical conductivity (simple electrode pair at 1–5 kHz)	Water quality
Electrical conductivity (multi-electrode system or inductive coupling to overcome deposits)	Solution strength

Note The measurement of water content and solution strength is discussed in Chapters 5, 7–9, 13; that of solid/liquid ratio in Chapter 8. The separation of foreign bodies by optical sorting is described in Chapter 4; impedance monitoring as a rapid microbial method in Chapters 14, 15.

Instruments Not Interacting with Complex Food Properties

Sensor	*Determined property*
Humidity sensors, based on a measurement of: polymer or oxide capacitance (not suitable in ovens where ammonia and volatile oils are present) dew point	Ambient humidity
Thermocouple probe	Temperature (whether the contact measurement is representative for the bulk of the food will depend on the thermal conductivity and heating pattern)

Fibre-optic probe (with fluorescent compound on tip)	Temperature
Strain gauge on diaphragm	Pressure in the headspace
Strain gauge configured for hydrostatic head measurement	Level of liquids
Capacitance probes	Level alarm
Mechanical flow rate meters (differential pressure type, positive displacement type)	Flow rate of liquids
Load cell arrangements	Container weight

Note Remote thermometry is covered in Chapter 6, ultrasonic flow rate and level metering in Chapters 8, 9, the measurement of equilibrium relative humidity in Chapter 12.

References

Bailey, L., Boyle, C. and Hilliam, M. (1991) The UK food and drinks report: market, industry and new product trends. *Leatherhead Food Research Association Special Report*, April 1991

Cossar, J. D., Blake-Coleman, B. C., Ramsay, C. and Atkinson, T. A. (1990) Oscillometric instrument for the non-invasive detection of low-level microbial activity. *Biosensors and Bioelectronics*, **5**, 273–289

Eerikainen, T., Linko, S. and Linko, P. (1988) The potential of fuzzy logic in optimization and control: fuzzy reasoning in extrusion cooker control. In *Automatic Control and Optimization of Food Processes*, eds M. Renard and J. J. Bimbenet, Elsevier Applied Science, London, 183–200

Georgala, D. L. (1989) Modern food processing. In *Food Processing*, Proceedings of the Ninth British Nutrition Foundation Branch Conference, ed. R. C. Cottrell, Parthenon Publishing Group, Carnforth and New Jersey

Jansson, P. A. (1991) Neural networks: an overview. *Analytical Chemistry*, **63**, 357A–362A

Jowitt, R., Escher, F., Kent, M., McKenna, B. and Roques, M. (eds) (1987) *Physical Properties of Foods vol. 2*, COST 90bis Final Seminar Proceedings. Elsevier Applied Science, London and New York (arranged by the Commission of the European Communities)

Kessler, H. G. and Weichner, K. (eds) (1989) *Fouling and Cleaning in Food Processing*, Proceedings of the Third International Conference on Fouling and Cleaning in Food Processing. Druckerei Walch, Augsburg. Distributed by: Heinz G. Kessler, Institut für Wissenschaft und Verfahrenstechnik in der Milchverarbeitung, Technische Universität München, D-8050 Freising-Weihenstephan, Germany; Daryl B. Lund, Department of Food Science, PO Box 231, Rutgers University, New Brunswick, NJ 08903, USA

Kress-Rogers, E. (1985) Technology transfer. II: The new generation of sensors. *Leatherhead Food Research Association Scientific and Technical Survey*, 150

Kress-Rogers, E. (1986) Instrumentation in the food industry. Part I: Chemical, biochemical and immunochemical determinands. Part II: Physical determinands in quality and process control. *Journal of Physics E: Scientific Instruments*, **19**, 13–21, 105–109

Kress-Rogers, E. and Kent, M. (1986) Two-parameter microwave technique for measurement of powder moisture and density. *Leatherhead Food Research Association Research Report*, 553

Kress-Rogers, E. and Kent, M. (1987) Microwave measurement of powder moisture and density. *Journal of Food Engineering*, **6**, 345–376

Langdon, R. M. (1985) Resonator sensors – a review. *Journal of Physics E: Scientific Instruments*, **18**, 103–115

Levine, H. and Slade, L. (1991) *Water Relationships in Food*, Plenum Press, New York and London

Lewis, M. J. (1987) *Physical Properties of Foods and Food Processing Systems*. Ellis Horwood, Chichester and VCH, Weinheim

McFarlane, I. (1983) *Automatic Control of Food Manufacturing Processes*. Applied Science Publishers, London and New York

McKenzie, G. (1990) Mass flow measurement. *Sensor Review*, July, 129–132

Noltingk, B. E. (ed) (1988) *Instrumentation Reference Book*. Butterworth-Heinemann, Oxford

O'Connor, C. (1987) *Extrusion Technology for the Food Industry*. Elsevier Applied Science, London and New York

Ohlsson, T. (1988) Dielectric properties and microwave processing. In *Food Properties and Computer-Aided Engineering of Food Processing Systems*, Proceedings NATO Workshop, Porto, eds R. P. Singh and A. G. Medina, Kluwer, Dordrecht, Boston and London, 73–92

Okos, M. R. (ed) (1986) *Physical and Chemical Properties of Food*. American Society of Agricultural Engineering, Michigan

Renard, M. and Bimbenet, J. J. (eds) (1988) *Automatic Control and Optimization of Food Processes*. Elsevier Applied Science, London and New York

Schaefer, W. (1989) Temperature sensors: new technologies on their way to industrial application. *Sensors and Actuators*, **17**, 27–37

Singh, R. P. and Medina, A. G. (1988) *Food Properties and Computer-Aided Engineering of Food Processing Systems*, Proceedings NATO Workshop, Porto, October 1988, Kluwer, Dordrecht, Boston and London

Spiess, W. E. L. and Schubert, H. (1990) *Engineering and Food*, vols 1–3, Elsevier Applied Science, London

Stewart, G. F. and Amerine, M. A. (1982) *Introduction to Food Science and Technology*. Academic Press, New York

Vidal, P. (1988) Automatization and optimization des procédès de l'industrie alimentaire. In *Automatic Control and Optimization of Food Processes*, eds M. Renard and J. J. Bimbenet, Elsevier Applied Science, London, 3–16

Wheaton, F. W. and Lawson, T. B. (1985) *Processing Aquatic Food Products*. Wiley, New York

Wiedmann, W. and Strecker, J. (1988) Process control of cooker-extruders. In *Automatic Control and Optimization of Food Processes*, eds M. Renard and J. J. Bimbenet, Elsevier Applied Science, London, 201–214

Part I

In-Line Measurement for the Control of Food Processing Operations

2 Principles of Colour Measurement for Food

Douglas B. MacDougall

Contents

1	Introduction	39
2	Colour Vision: Trichromatic Detection	40
3	Influence of Ambient Light and Food Structure	43
	3.1 Adaptation	43
	3.2 Appearance	44
	3.3 Absorption and Scatter	45
4	Colour Description	45
	4.1 The CIE System	45
	4.2 Uniform Colour Space	48
	4.3 Further Terminology	49
5	Instrumentation	50
6	Examples	53
	6.1 Fresh Meat	53
	6.2 Orange Juice	56
Acknowledgements		58
References		59

1 Introduction

Vision, the most studied of all the human senses, has intrigued investigators for as long as man's speculations or accomplishments have been recorded. Modern studies into the mechanism of vision and human colour perception began in the seventeenth century with the recognition that the eye's lens must somehow project an image on to the back of the eye. Newton's classic experiments on the refraction of light led him to conclude that the rainbow did not possess colour; rather it was the spectrum's rays that produced the sensation (Wright 1967). The rationality of arranging colours into orderly systems based on Newton's seven rainbow colours has resulted in many attempts to create colour atlases with equal spacing of adjacent colours; for example, both the widely used Munsell system and the newer Swedish natural colour space are designed for uniform colour order presentation. The former is based on five hues and the latter on the six unique perceptions of black, white, red, green, yellow and blue (Hard 1970).

From the nineteenth-century experiments of Maxwell, Young and Helmholtz in mixing coloured lights (MacAdam 1970), it became clear that people with normal colour vision must have at least three retinal pigments in their eyes. By the late 1920s the eye's sensitivity to light relative to wavelength was established, the so-called 'standard observer' defined (Wright 1980) and the Commission Internationale de l'Éclairage (CIE) system of colour measurement constructed. Since then, considerable improvements have been incorporated into the system to make it nearly visually uniform. Now, with the development of the computer, complex colour measurements and calculations are not only possible but are used routinely for such industrial processes as paint formulation (Best 1987) and the prediction of the appearance of dyed textiles (McLaren 1986; McDonald 1987). Instrumental colour measurement is increasingly being used for ingredient standardization and product quality control in the food industry.

The three interacting factors required to classify, estimate and measure the colour appearance of any object in a scene are an understanding of the human visual process, the effect of light on the environment and the nature of the material observed.

2 Colour Vision: Trichromatic Detection

The sensation of colour vision can be thought of as the sum of the responses recognized in the brain from the signals detected by the eye of the scene viewed. The sensation is perceived as though it were projected out into the world from which it originated. This can lead to the error of imputing to the scene the sensations it generated. Sensations exist in the mind and not in the external world which produces them.

Human eyes, with their near circular field of view, are composed of three membranes (Figure 2.1a). The outer membrane, the sclera, is continuous posteriorly with the sheath of the optic nerve and anteriorly with the cornea. The iris and the ciliary body, which suspends the lens, arise out of the middle layer, the choroid, which contains the capillary network. The inner membrane, the retina, lines the inside of the posterior of the eye. The first step in the visual process is the control of the amount of light entering the eye by the iris. The flux passes through the lens to be focused on to the fovea in the central region of the retina where it is detected. The signal is amplified (Normann and Werblin 1974) and then transmitted through the visual pathway (Rodieck 1979) for interpretation in the cortex of the brain

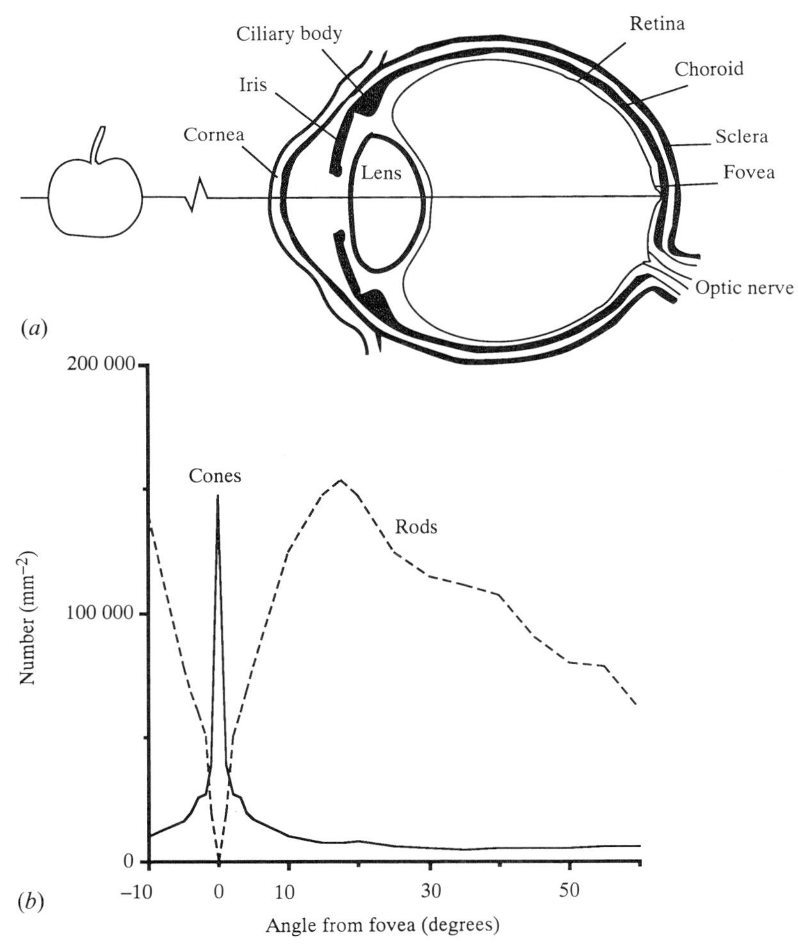

Figure 2.1 *Detection of light in the eye: (a) structure of the human eye (b) distribution of cones and rods in the temporal side of the retina (nasal side is similar except for the blind spot between 12° and 18° from the fovea)*

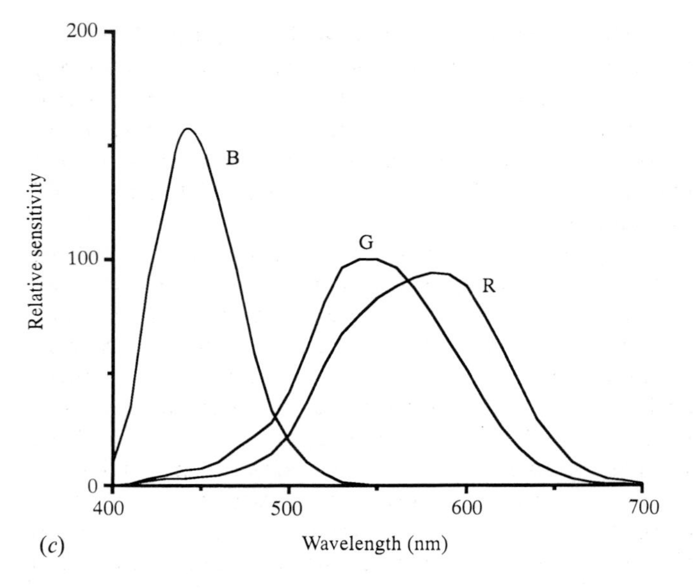

Figure 2.1 – continued. *(c) spectral sensitivities of blue (B), green (G) and red (R) cone pigments*

(Zeki 1980; Hubel 1988). The retina has two types of light detecting receptors, the cones and the rods, so named because of their shape. The 'photoptic' colour detecting cones, which are sensitive to three wavelength ranges, are densely packed in the centre of the fovea, which occupies <2° of the visual field and is the basis of the so-called CIE 1931 2° standard colorimetric observer. The 'scotoptic' colourless detecting rods, which are over 100 times more sensitive to light than the cones, increase in density to 20° from the fovea and then decrease towards the periphery of vision (Figure 2.1b). The 1964 10° supplementary standard observer was created to accommodate the improvement in colour discrimination as the visual angle increases beyond 2°; some rods are included in the detecting field.

Light energy, focused on the retina, is converted into electrical signals by conformational changes in the photopigments in the multifolded disk-shaped structures in the outer segments of the rods and cones (Wald 1968; Hurvich 1981; Jacobs 1981; Stryer 1988). Although only the rod pigment rhodopsin has been characterized, determination of the spectral absorption of the cone pigments has been possible using retinal densitometry with colour blind observers deficient in one pigment (Smith and Pokorny 1975). Although no unique set of human cone sensitivity curves has been established, the best set is probably that of Estévez (1982) (Figure 2.1c). Rhodopsin absorbs maximally at 505 nm and the so-called blue (B), green (G) and red (R) cone pigments at approximately 440, 540 and 570 nm respectively. The sensitivity range of B marginally overlaps G and R between 450 and 550 nm, whereas the G and R functions substantially overlap, displaced from each other by only 20 to 30 nm. Thus monochromatic light at 580 nm which is near maximum for R appears yellow and not red because of the combined contributions of G and R. Increasing the wavelength to >600 nm increases the contribution of R relative to G and the perceived colour becomes more red.

Cone vision is trichromatic; this means that any coloured light can be matched by a suitable mixture of red, green and blue primary lights, which is what would be expected from a three-receptor system. The actual colour-matching functions depend on the wavelengths used as primaries. Although the initial detection of the stimulus is trichromatic, subsequent post-retinal processing gives rise to achromatic lightness/darkness and coloured red/green and blue/yellow opponent mechanisms (Hurvich 1981). The lightness component consists of a weighted summation of all three cone pigment absorptions, whereas it is the degree of differences among the B, G, R absorptions that generates the opponent colour mechanism. However, the neural linkages among the pigment cone signals are not in simple one-to-one opposition. The simplest scheme that can be constructed is that the red/green opponent response is red activated by absorption of B plus R and green activated by G; yellow is activated by G plus R in opposition to blue activated by B.

3 Influence of Ambient Light and Food Structure

3.1 Adaptation

Adaptation conditions the visual system to the chromatic nature of the surroundings as affected by the quality and intensity of the illumination. It compensates for changes in the spectral power distribution of the light and serves to keep the eye in balance (Boynton 1979). The magnitude of adjustment that chromatic adaptation has on visual experience is not usually realized because of the limitations of human memory for individual colours and the phenomenon of colour constancy (Brill and West 1986). White objects are recognized as white over a vast range of light intensity, from bright unshaded sunlight to the relatively dim levels of light found in room interiors, while coloured objects tend to have similar, but not necessarily identical, colours under most types of white illumination. Studies into the phenomena of adaptation which elicit this near constancy of colour appearance have been concerned mostly with predicting the changes that occur to colour recognition when lamp type and output are altered (Bartleson 1979a). Lightness and contrast among greys are affected by luminance while colourfulness increases with the level of illumination and varies with the spectral emission of the lamp and its colour temperature (Hunt 1977). This is further discussed in Section 6.1 on fresh meat, where the degree of red enhancement of lamp spectra is shown to affect the perception of product attractiveness. Models of cone adaptation response have been used to predict the consequences of changing lamp spectra on object appearance (Bartleson 1979b; Nayatani et al. 1986; Hunt 1987a). The concept of colourfulness has been used to construct grids of constant hue from which other grids can be derived for other illuminants (Pointer 1980; 1982). Such models use logarithmic and hyperbolic functions to mimic the physiological mechanisms involved. Hunt's (1987a) model can be used to predict the changes that occur in object colours at any level of illumination for a wide range of backgrounds in the realistic situation where the eye's fixation wanders. These effects of light quality on colour perception illustrate the difficulties in separating the concept of vision from that of appearance. Vision is modulated by the light from the scene, while the characteristics of appearance are modified by the light incident upon the object. Hence systems for classifying appearance and procedures devised for measuring colour must take account of the nature, quality and quantity of the light as it affects the observer's perception.

3.2 Appearance

Colour is usually considered the most important attribute of any food's appearance (Francis and Clydesdale 1975) especially if it is associated with other aspects of food quality, for example the ripening of fruit or the visible deterioration which occurs when a food spoils. However, in addition to colour specification, the nature of internally scattered light and the distribution of surface reflectance are required for a more complete description of appearance. The food's structure and pigmentation interact to affect both translucency/opacity and colour; for example, small changes in scatter may produce larger changes in colour than are attributable to change in pigment concentration (MacDougall 1982).

The characterization of an object's appearance is accomplished in two stages. The first is physical and the second is psychological. The physical characteristics are the size, shape and uniformity of the object along with the type and degree of pigmentation and the nature of the object's structure that attenuates light. The physical information is converted to the psychological by translating the object's reflectance or transmittance spectrum into the tristimulus values and then to a defined colour space.

Foods have an infinite variety of appearance characteristics. Their surfaces may be diffuse, glossy, irregular or flat. They may be transparent, hazy, translucent or opaque and their colour may be uniform or patchy. Hence, colour measuring procedures for foods often have to be modified from those used in the measurement of flat opaque surfaces such as paint and paper for which most colour measuring instruments are designed. Different instrument optical geometries will lead to difficulties in sample presentation, producing different colour values for the same material. The inclusion or exclusion of surface specular reflection in the measurement depends not only on its importance as a characteristic of the food but also on the design of the detector-sensor unit in the instrument. Lateral transmittance of light through translucent materials affects their reflectance (Atkins and Billmeyer 1966; Hunter and Harold 1988; MacDougall 1988) and must be allowed for in the assessment of such products as tomato paste (Brimelow 1987) because the ratio of absorption to scatter varies with aperture area and the concentration of components in the product (Best 1987; MacDougall 1987).

3.3 Absorption and Scatter

The reflection of light from opaque and translucent objects depends on the ratio of absorption to scatter as affected by pigmentation, refractive index and the light scattering properties of the material. The Kubelka-Munk method of separating subsurface absorption and scatter (Kubelka 1948) is fully illustrated by Judd and Wyszecki (1975). It relates reflectivity R_∞, i.e. reflectance at infinite thickness, to the coefficients of absorption K and scatter S by

$$K/S = (1 - R_\infty)^2/2R_\infty$$

which can be calculated from the reflectance of thin layers on white and black backgrounds. If K and S are required for prediction purposes, the accuracy of their measurement can be improved by appropriate correction factors for surface reflection (Saunderson 1942). Colour calculated from R_∞ with separate estimation of the specular component as gloss is usually sufficient information to describe opaque objects, but for translucent materials K and S are also necessary.

4 Colour Description

4.1 The CIE System

The CIE system of colour measurement transforms the reflection or transmission spectrum of the object into three-dimensional space using the spectral power distribution of the illuminant and the colour-matching functions of the standard observer. The mathematical procedure is given in any standard text on colour, for example Wright (1980), Judd and Wyszecki (1975) and Hunt (1987b). The system is based on the trichromatic principle but, instead of using 'real' red, green and blue primaries with their necessity for negative matching, it uses 'imaginary' positive primaries X, Y, and Z. Primary Y, known as luminous reflectance or transmittance, contains the entire lightness stimulus. Every colour can be located uniquely in the 1931 CIE colour space by Y and its chromaticity coordinates $x = X/(X+Y+Z)$ and $y = Y/(X+Y+Z)$, provided the illuminant and the observer are defined.

Until recently the illuminant representative of daylight was defined by the CIE as source C, but it has now been superseded by D65, i.e. an illuminant which includes an ultraviolet component and has a colour temperature of 6500 K. The colour temperatures of lamps and daylight range from approximately 3000 K for tungsten filament lamps and 4000 K for warm white fluorescent to 5500 K for sunlight and 6500 K for average daylight. Because the original 2° colour-matching functions apply strictly only to small objects, i.e. equivalent to a 15 mm diameter circle viewed at a distance of 45 cm, the CIE has added a 10° observer (Figure 2.2) where the object diameter is increased to 75 mm. The current trend in colour measurement is to use D65 and the 10° observer. The 1986 CIE recommended procedures for colorimetry are included in the ASTM Standards (1987) and also in Hunt (1987b) along with the weighting factors for several practical illuminants (Rigg 1987). These include representative fluorescent

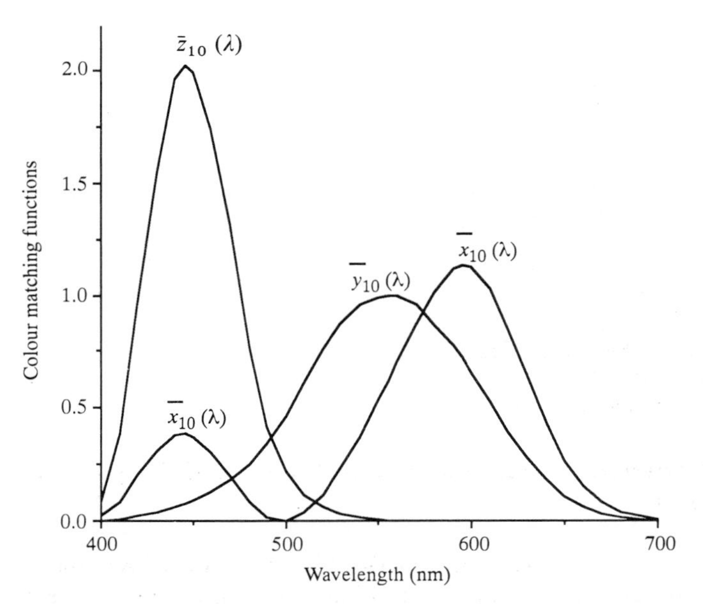

Figure 2.2 *Colour matching functions of the CIE 10° standard observer*

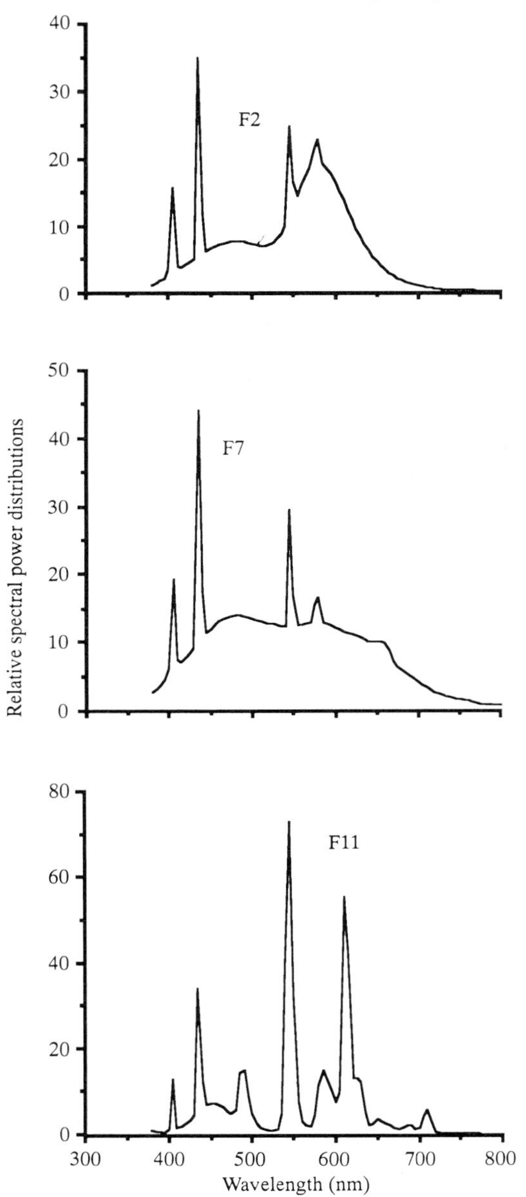

***Figure 2.3** Relative spectral power distributions of preferred CIE representative fluoresent lamps*

lamps, of which F2 is a typical lamp at 4230 K but with a low colour rendering index of 64 (Figure 2.3). The colour rendering index R_a is a measure of the efficiency of a lamp at a given colour temperature to render

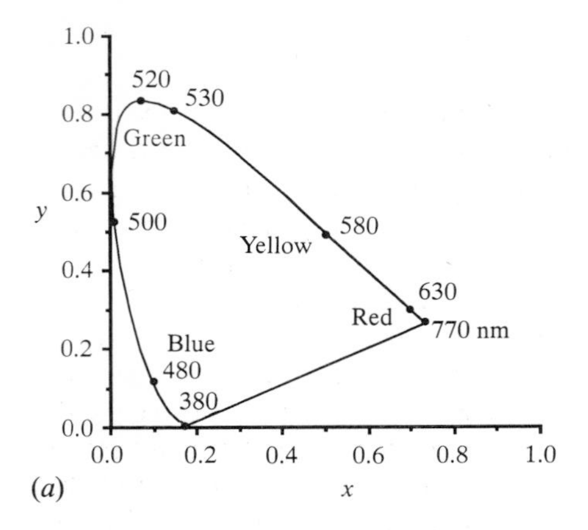

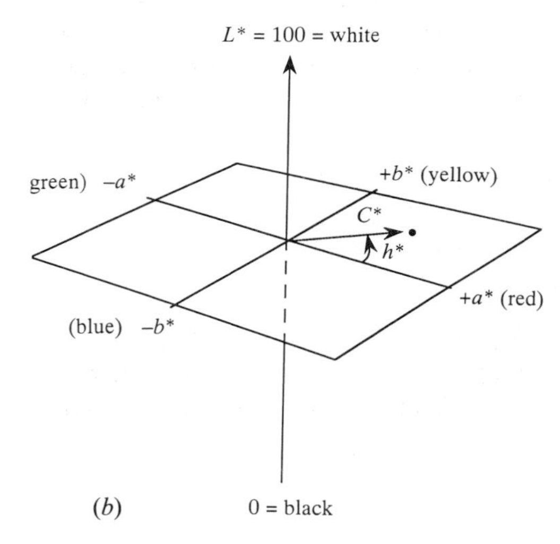

***Figure 2.4** Colour diagrams: (a) CIE 1931 chromaticity diagram showing non-uniformity of spacing of red, yellow, green and blue unique hues (b) CIELAB uniform diagram showing relationship of red/green (a^*+/−) and yellow/blue (b^*+/−) opponent coordinates to lightness L^*, chroma C^* and hue angle h^**

the true appearance of Munsell colours. The broadband lamp F7 has the same colour temperature (6500 K) and chromaticity coordinates as D65 and, because of its flatter spectrum, it has a high R_a of 90. The triband lamp F11 (4000 K) also has a moderately high R_a of 80, but its main advantage is its much improved efficiency in energy utilization.

4.2 Uniform Colour Space

The original 1931 CIE Y, x, y system of colour measurement is not visually uniform (Figure 2.4a). Constant hue and chroma are distorted and equal visual distances increase several-fold from purple-red to green. Improved spacing has been accomplished by both linear and non-linear transformations of Y, x, y (Billmeyer and Saltzman 1981). Near uniform colour spaces of practical importance are the Hunter and the CIELUV and CIELAB spaces. In the Hunter (1958) L, a, b colour space the lightness coordinate L is the square root of the tristimulus value Y, and a, and b are the red/green and yellow/blue opponent coordinates. In the 1976 CIELUV and CIELAB spaces (Robertson 1977), which were an attempt to reduce the many scales then in use to two, the lightness coordinate L^* is the same for both but the spaces use different concepts in their construction. The CIE L^*, a^*, b^* space (Figure 2.4b), known as CIELAB, has generally replaced the Hunter space for industrial applications. The improvement is due to the non-linear cube root transformation of the 1931 tristimulus values which approximates the visual spacing of the coloured samples in the Munsell system. The formulae are

$$L^* = \begin{cases} 116(Y/Y_n)^{1/3} - 16 & \text{for } Y/Y_n > 0.008\,856 \\ 903.3(Y/Y_n) & \text{for } Y/Y_n < 0.008\,856 \end{cases}$$
$$a^* = 500\,[(X/X_n)^{1/3} - (Y/Y_n)^{1/3}]$$
$$b^* = 200[(Y/Y_n)^{1/3} - (Z/Z_n)^{1/3}]$$

where X_n, Y_n, Z_n refer to the nominally white object colour stimulus.

Newer scales for determining small differences (BSI 1988), as used in the textile industry, are now included in the software of some automatic colour measuring spectrophotometers.

4.3 Further Terminology

Colour terms can be divided into the subjective and the objective (Hunt 1978). The subjective, the psychosensorial, are brightness, lightness, hue, saturation, chroma and colourfulness. Colourfulness, a recently introduced term, is that aspect of visual sensation according to which an area appears to exhibit more or less chromatic colour. Although hue is easily understood as that attribute described by colour names – red, green, purple etc. – the difference between saturation and chroma is less easily comprehended. Saturation is colourfulness judged in proportion to its brightness, whereas chroma is colourfulness relative to the brightness of its surroundings. A similar difference exists between lightness and brightness. Lightness is relative brightness. Lightness is unaffected by illumination level because it is the proportion of the light reflected, whereas the sensation of brightness increases with increase in illumination.

The objective terms, the psychophysical, refer to the stimulus and are evaluated from spectral power distributions, the reflectance or transmittance of the object and observer response. They provide the basis for the psychometric qualities which correspond more nearly to those perceived. For CIELAB space the more important terms are lightness L^*, hue $h^* = \tan^{-1}(b^*/a^*)$ and chroma $C^* = (a^{*2} + b^{*2})^{1/2}$. Total colour differences ΔE^* can be expressed either as the coordinates of colour space or as the correlates of lightness, chroma and hue. Hence

$$\Delta E^* = [(\Delta L^*)^2 + (\Delta a^*)^2 + (\Delta b^*)^2]^{1/2}$$

or

$$\Delta E^* = [(\Delta L^*)^2 + (\Delta C^*)^2 + (\Delta H^*)^2]^{1/2}$$

where ΔH^* is used rather than Δh^* because the latter is angular. For small colour differences away from the L^* axis, if h^* is expressed in degrees, then

$$\Delta H^* = C^* \Delta h^* (\pi/180)$$

A comparison of the major colour scales together with the associated terminology is given in Table 2.1.

5 Instrumentation

Since colour is a psychological phenomenon, its measurement is based on human colour perception. Hence, photoelectric instruments must be corrected for both lighting and human visual response, while visual techniques must use observers with ‘normal’ colour vision under defined lighting. Examples of direct visual assessment are colour atlases for broad definition of the location of colours in colour space, collections or sets of printed or painted coloured papers specific to products or processes, and visual matching instruments which use coloured filters. Typical of the former are the Munsell and Swedish NCS atlases which are structured on uniform colour space, and the Pantone collections of printer’s colours with defined ink mixtures printed from 10 to 100 per cent tinting strength. Probably the best known of the visual matching instruments is the Lovibond tintometer in which the object, under specified illumination, is viewed and matched against a series of coloured filters interposed over a white background by the observer.

Photoelectric colour measuring instruments can be divided into two classes, trichromatic colorimeters and spectrophotometers. The most successful of the early trichromatic colorimeters was developed in the 1940s by Hunter (1958). It comprised a light source and three wideband red, green and blue filters to approximate CIE standard illuminant C and the 2° observer. The tristimulus values obtained were transformed into Hunter L, a, b colour space. Until the advent of the computer and the photodiode

Table 2.1 *Overview on colour description systems and notation*

CIE system (1931)

This is based on the imaginary positive primaries X, Y, Z (transformed from real red, green and blue functions of trichromatic stimulus detection which may contain negative values).
In CIE space, colour is located (assuming defined illuminant and observer) by (Y, x, y), where

Y	luminous reflectance or transmittance (containing the entire lightness stimulus)
x, y	chromaticity coordinates $x = X/(X+Y+Z)$ $y = Y/(X+Y+Z)$

CIE space is not visually uniform.

Hunter Lab System (1958)

In Lab space, colour is defined by (L, a, b), where

L	correlate of lightness
a, b	red/green and yellow/blue opponent coordinate correlates $L = 10\ Y^{1/2}$ $a = [17.5\ (1.02\ X - Y)]/Y^{1/2}$ $b = [7.0\ (Y - 0.847\ Z)]/Y^{1/2}$

CIELAB system (1976)

In CIELAB space, colour is defined by ($L^\star$, $a^\star$, $b^\star$), where

$L^\star$	visually uniform lightness
$a^\star, b^\star$	visually uniform chromaticness coordinates

$L^\star = 116\ (Y/Y_n)^{1/3} - 16$ for $Y/Y_n > 0.008856$
$L^\star = 903.3\ (Y/Y_n)$ for $Y/Y_n < 0.008856$
$a^\star = 500\ [(X/X_n)^{1/3} - (Y/Y_n)^{1/3}]$
$b^\star = 200\ [(Y/Y_n)^{1/3} - (Z/Z_n)^{1/3}]$

where X_n, Y_n, Z_n are the values of X, Y, Z for the reference white.

Further terms used are

$h^\star = \tan^{-1}(b^\star/a^\star)$	hue
$C^\star = (a^{\star 2} + b^{\star 2})^{1/2}$	chroma

Other terms, such as adaptation, appearance, vision, colourfulness, saturation and brightness, are discussed in Sections 3.1, 3.2 and 4.3.

such instruments were much less expensive than spectrophotometers and, although absolute accuracy may have been poor, they were very good at measuring the small colour differences demanded for industrial process control (Patterson 1987). The more modern tristimulus instruments are linked to computers with automatic calibration and the provision of a number of colour spaces. Such instruments may be supplied with a selection of sensor heads of different illuminating geometries to allow measurement of a wide range of product types depending on the nature and dimensions of their surfaces. A range of hand-held lightweight colorimeters with optical geometries comparable in function with the larger bench instruments is now manufactured by Minolta. Their compactness is a direct result of the use of high energy pulsed xenon arc lamps combined with filtered silicon detectors and microchip circuitry. The comparative inexpensiveness of such instruments, with their built-in memories and choice of colour scales, has resulted in increased in-line colour measurement in all branches of industry where colour control is necessary or desirable.

The most accurate colour measuring instrument is the spectrophotometer. Reflectance instruments are usually fitted with an integrating sphere with the choice of including or excluding the specular component of reflectance. Care must be exercised in deciding which geometry is appropriate for particular applications. The diffuse component of reflectance from subsurface absorption and scatter is wavelength dependent, whereas the specular component is not. For materials with glossy surfaces the inclusion of the specular will increase measured reflectance which, when translated into colour space, can lead to large discrepancies in the interpretation of visual lightness, as usually viewed, and to a lesser extent of the chromaticness of the colour. For example, highly glossy black tiles used for instrument calibration have tristimulus Y values of approximately 0.3 when the specular is excluded but 4.5 when included. The consequence of this difference in Y of 4 per cent produces a specular excluded uniform lightness L^* of 3 and an included L^* of >25. For medium grey and white tiles the excluded to included Y values are approximately 25 to 29 and 78 to 82 repectively, which give L^* values of approximately 57 to 61 and 91 to 92 respectively. Hence the near constant effect of 4 per cent on Y from the specular reflectance produces a decreasing effect from black to white from >20 to about 1 per cent in L^*. The CIE recommends that colorimetric specifications of opaque materials should be obtained with one of the following conditions of illumination viewing geometries and should be specified in any report:

- 45°/0° or 0°/45°, specular excluded
- diffuse/0° or 0°/diffuse, specular included or excluded.

However, the spectrophotometers most commonly used for measuring colour do not have identical geometries. Three typical instruments have been compared by Patterson (1987), who points out that probably the biggest source of differences among the instruments can be traced to the specular component. Hunt (1987b) suggests that if measurements are to be compared it is better to include the specular because of the considerable variation in the area of gloss traps used in different spheres, but the more nearly correct measurements in relation to practical visual observation is with the specular excluded (Best 1987). For computer match prediction of pigmented materials, e.g. paint formulation, the total reflection (i.e. specular included) is preferred. This restriction does not usually apply to tristimulus colorimeters which normally exclude the specular component of reflectance.

Another important source of variation among tristimulus colorimeters and spectrophotometers is the area of the viewing aperture and the relative area of the illuminating light spot, which affects both the direction and the amount of light returned from translucent materials. MacDougall (1987) has demonstrated that translucent suspensions of milk exhibit a tenfold decrease in K/S for an increase in aperture area from 5 to 20 mm. Best (1987) states that accurate determination of K and S by measuring thin layers on black and white backgrounds requires that the ratio of the aperture area to the thickness of the sample must be considerably greater than 10, a criterion unlikely to be met for most foods. One further source of potential error, in addition to those associated with instrument geometry and sample structure, is the wavelength interval used to calculate the tristimulus values. Although the CIE (1986) specifies the standard observer at 5 nm intervals from 380 to 780 nm, such accuracy is not required for most practical purposes. For 10 nm accuracy the intermediate 10 nm values from the 5 nm tables should be used. However, the CIE has not yet officially recommended the use of 20 nm intervals, although many modern colour spectrophotometers detect at 20 nm intervals. Tables of weighting functions at 20 nm intervals for both the CIE illuminants and a variety of fluorescent lights have been calculated and are published in the up-to-date colour textbooks cited in this chapter. Errors attributable to wavelength interval are likely to be less important than those from instrument geometry, except when estimating the effects of narrowband emission lamps on materials with several absorption bands. Here the 20 nm interval may prove to be insufficient.

6 Examples

The progress of pigment oxidation in fresh meat and the effects of illumination on orange juice are given as examples of the interaction of absorption and scatter on measured colour and visual appearance.

6.1 Fresh Meat

On exposure to air the purple ferrous haem pigment myoglobin on the surface of freshly cut meat oxygenates to the bright red covalent complex oxymyoglobin. During refrigerated display, oxymyoglobin oxidizes to brownish green metmyoglobin (MacDougall 1982). Twenty per cent dilution of surface oxymyoglobin with metmyoglobin causes the product to be rejected at retail because of its faded colour (Hood and Riordan 1973). The changes in the mean reflectance spectra of over 100 packages of beef overwrapped with oxygen permeable film and held in the light at $<5°C$ over a period of 1 week are shown in Figure 2.5. As the pigment oxidizes there is an increase in reflectance in the green region of the spectrum as the alpha and beta absorption bands decrease and a distinct loss in reflectance in the red region with development of the metmyoglobin absorption band at 630 nm. The changes in colour, calculated for CIELAB and D65, are shown in Figure 2.6. As meat fades there is a small loss in lightness L^*, accompanied by much greater changes in the chromaticness coordinates a^* and b^*. The loss in a^* and gain in b^* can be interpreted as an increase in the

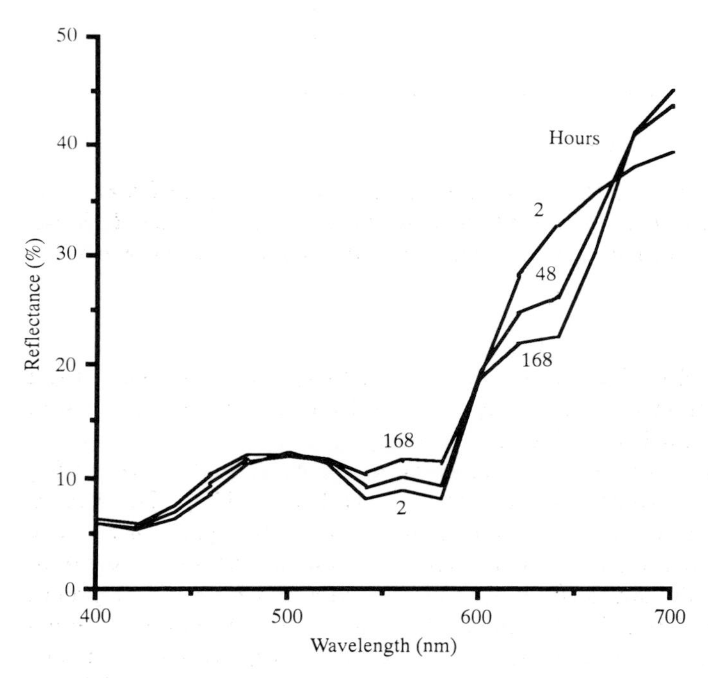

Figure 2.5 *Reflectance spectra of fresh beef during oxidation of oxymyoglobin to metmyoglobin obtained on a diode array spectrophotometer at 20 nm intervals: means of over 100 samples wrapped in oxygen permeable film and stored at $<5°C$ under 1000 lux fluorescent illumination for 1 week*

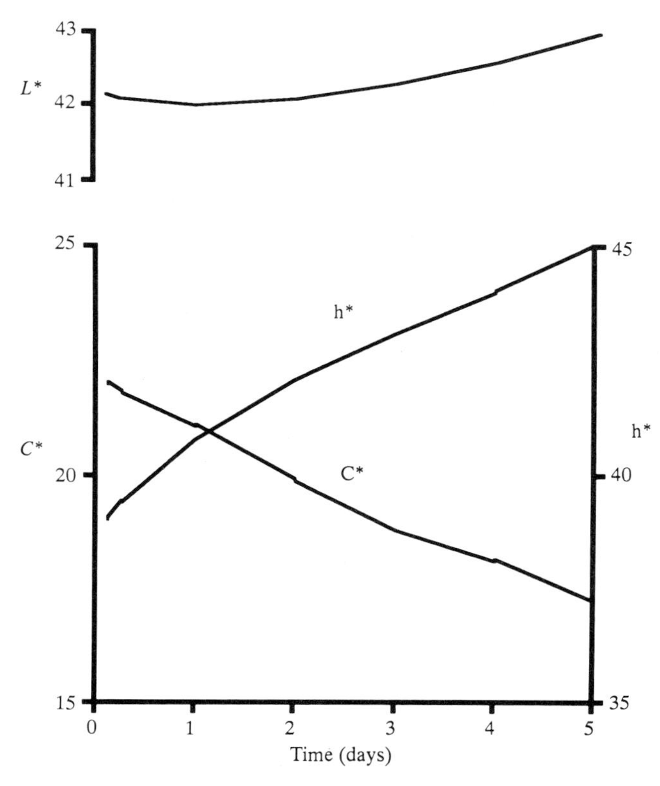

Figure 2.6 *Progressive changes in lightness L^*, hue angle h^* and chroma C^* calculated from spectra of wrapped fresh beef stored at $<5°C$ under 1000 lux fluorescent illumination during oxidation of surface oxymyoglobin to metmyoglobin*

hue angle h^* in the direction of yellow; however, since it occurs with a concomitant loss in chroma C^*, it is recognized as a more grey or dull colour. Dull yellow is perceived as brown.

The appearance of meat is greatly affected by the colour rendering properties of the lamps used for display (Halstead 1978). Fluorescent lamps recommended by the lamp industry for displaying meat have enhanced red emission which maintains the preferred colour of oxymyoglobin and visually shifts the early stages of metmyoglobin development from brown towards red. This effect of red enhancement on meat colours has been shown to elicit a greater visual colour change in making brown appear red than in making red appear more red (MacDougall and Moncrieff 1988). For some people the flattering of red enhanced lamps may make meat appear too red.

The ICS Micro Match spectrophotometer used to measure these samples is equipped with the option of using alternative illuminants to calculate CIELAB. The estimated changes in meat colour attributable to different illuminants after 1 and 4 days' exposure (Table 2.2.) illustrate the effect that light quality has on lightness, hue and chroma. The changes in colour produced by the differences in colour rendering among some of the lamps are equivalent to that which occurs after 4 days' fading, that is $\Delta L^* \leq 1, \Delta O^* \leq -4$ and $\Delta h^* \leq 7$. There was little change in L^* on changing illuminant, but the large changes in C^* and h^* illustrate the effects of decreasing lamp colour temperature, altering flattery and improving fidelity. A decrease in colour temperature from D65, as red emission increases, generally increases C^*, that is the colour is perceived as brighter or more intense with the observer adapted to white, but h^* may become more brown (more positive) or more red (more negative) as influenced by both colour temperature and the lamp's spectral bandwidth which affects fidelity.

6.2 Orange Juice

Translucent suspensions are difficult to measure, and direct unobserved interpretation of instrumental data can lead to confusion because of the way the incident light is dispersed in the sample. Most consistent results are obtained if the instrument aperture is large relative to the incident beam (Kent 1987; MacDougall 1987). The effects of optical geometry on colour and the Kubelka-Munk absorption K (mm^{-1}) and scatter S (mm^{-1}) coefficients for orange juice were studied by MacDougall (1983), who

Table 2.2 *Calculated changes in L^*, C^* and h^* from D65 to other lamps for the fresh beef spectra shown in Figure 2.5*

Storage time		*Difference in colour from D65*					
		ΔA	ΔWWF	ΔNFL	ΔCWF	$\Delta 83$	$\Delta 84$
2 hours	ΔL^*	2.9	2.0	0.6	2.0	2.2	1.2
	ΔC^*	7.3	7.1	2.2	7.1	7.7	4.9
	Δh^*	−2.3	−1.2	3.9	6.5	3.3	−2.3
4 days	ΔL^*	2.3	1.6	0.6	−0.2	1.7	0.9
	ΔC^*	5.7	5.5	2.1	−0.7	6.3	3.8
	Δh^*	−5.9	−1.9	4.5	9.1	2.8	−3.5

Tungsten lamp: A. Fluorescent lamps: WWF, warm white; NFL, natural; CWF, cool white; 83, triband at 3000 K; 84, triband at 4000 K.

found Y_{∞}, the luminous reflectance at calculated infinite thickness, increased by 50 per cent if the aperature diameter was increased from 2 cm to 5 cm while the incident beam was maintained at 1 cm.

The effect of dilution of fourfold orange juice concentrate on the reflectance spectra obtained on 4 cm thick samples in thin walled polystyrene bottles is shown in Figure 2.7. As can be seen, 4 cm is practically equivalent to infinite thickness. Kubelka-Munk absorption and scatter coefficients were calculated from 2 mm thick samples on black and white backgrounds. On dilution, S decreased for X, Y and Z as the suspension became more translucent. K for Z decreased to approach the much lower near constant values of K for X and Y (Figure 2.8). This decrease in K for Z is as anticipated for a blue absorbing pigment. The effect of loss of scattering power on dilution was to reduce Y_{∞} and hence lower L^*. The most dilute juice, therefore, is instrumentally the darkest, and the most concentrated is the lightest (Figure 2.9). However, this is not what is perceived. Glasses of orange juice viewed with overhead illumination range from pale yellow for concentrations less than 1 to deep orange at a concentration of 4, which is opposite to that determined instrumentally. For strongly scattering coloured materials in dilute suspension, measured colour, even supplemented by information on scatter, is inadequate to fully describe appearance. The instrument does not measure what the observer sees because light is reflected from a limited solid angle, whereas the

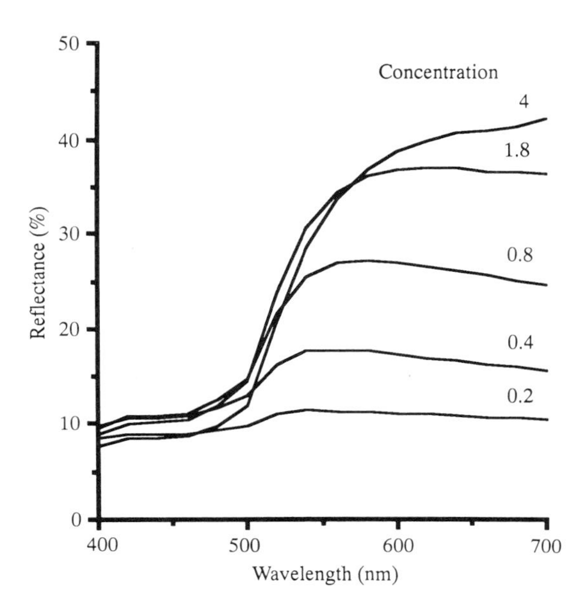

Figure 2.7 *Reflectance spectra of concentrated and diluted orange juice at a path length of 4 cm, equal to infinite thickness: reconstituted juice at normal concentration = 1.0*

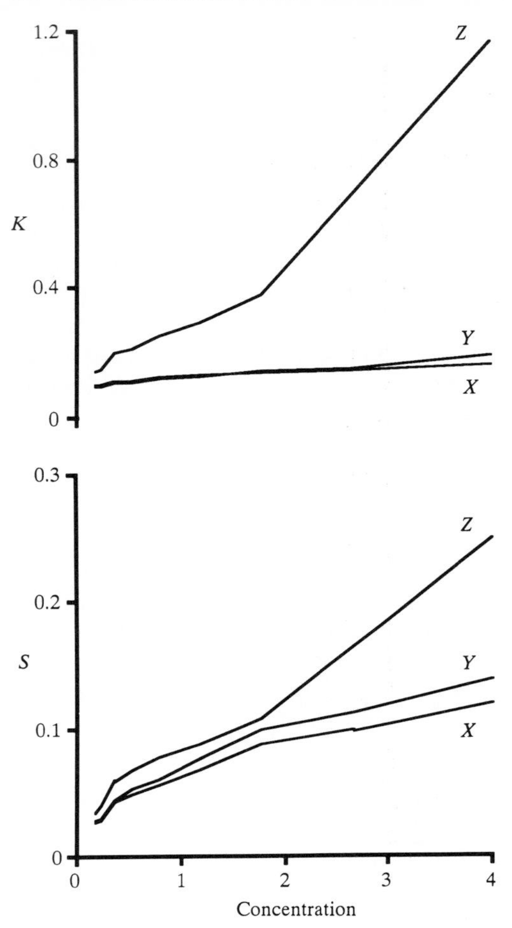

***Figure 2.8** Kubelka-Munk absorption K and scatter S coefficients for tristimulus values X, Y and Z for concentrated and diluted orange juice: values (mm^{-1}) calculated from reflectance spectra obtained from 2 mm path length cells with black and white backgrounds*

observer's perception is influenced by the multidirectionality of illumination which makes coloured translucent materials appear to glow.

Acknowledgements

The author wishes to thank Miss E. Morgan for her assistance in the measurement of the orange juice and Miss J. I. Aaron, Miss E. Morgan and Miss J. Wilcocks in the preparation of the figures.

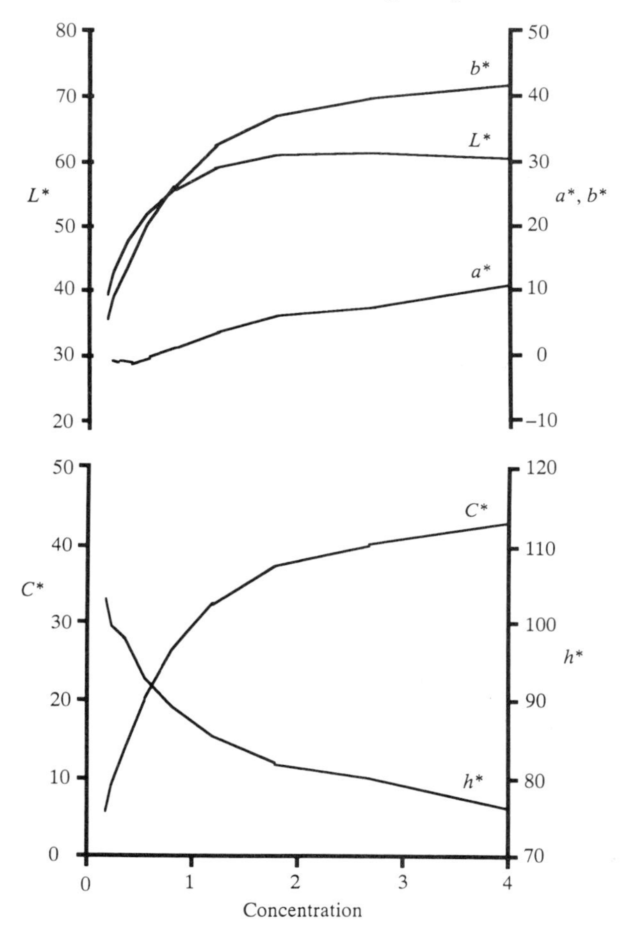

Figure 2.9 *Changes in lightness L^*, opponent coordinates a^* and b^*, hue angle h^* and chroma C^* calculated from reflectance spectra of concentrated and diluted orange juice at 4 cm path length*

References

ASTM (1987) Standards on color and appearance measurement. American Society for Testing and Materials, Philadelphia

Atkins, J.T. and Billmeyer, F. W. (1966) Edge-loss errors in reflectance and transmittance measurement of translucent materials. *Material Research Standards*, **6**, 564–569

Bartleson, C. J. (1979a) Changes in color appearance with variations in chromatic adaptation. *Color Research and Application*, **4**, 119–138

Bartleson, C. J. (1979b) Predicting corresponding colors with changes in adaptation. *Color Research and Application*, **4**, 143–155

Best, R. P. (1987) Computer match prediction – pigments. In *Colour Physics for Industry*, ed. R. McDonald, 186–210, Society of Dyers and Colourists, Bradford

Billmeyer, F. W. and Saltzman, M. (1981) *Principles of Color Technology*, 2nd edn. Wiley, New York

Boynton, R. M. (1979) *Human Color Vision*. Holt, Rinehart and Winston, New York

Brill, M. H. and West, G. (1986) Chromatic adaptation and colour constancy; a possible dichotomy. *Color Research and Application*, **11**, 196–204

Brimelow, C. J. B. (1987) Measurement of tomato paste color: investigation of some method variables. In *Physical properties of foods*, vol. 2, eds. R. Jowitt, F. Escher, M. Kent, B. McKenna and M. Roques, 295–317, Elsevier, London

BSI (1988) British Standard method for calculation of small colour differences (BS6923). British Standards Institution, London

CIE (1986) *Colorimetry*, 2nd edn. CIE publication 15.2, Commission Internationale de l'Éclairage, Vienna

Estévez, O. (1982) A better colorimetric standard observer for color-vision studies. The Stiles and Burch 2° color-matching functions. *Color Research and Application*, **7**, 131–134

Francis, F. J. and Clydesdale, F. M. (1975) *Food Colorimetry: Theory and Applications*. AVI, Westport, CT

Halstead, M. B. (1978) Colour rendering: past, present and future. In *AIC 77*, 97–127, Adam Hilger, London

Hard, A. (1970) Qualitative attributes of colour perception. In *AIC Color 69*, vol. 1, 351, Musterschmidt, Gottingen

Hood, D. E. and Riordan, E. B. (1973) Discoloration in pre-packaged beef: measurement by reflectance spectrophotometry and shopper discrimination. *Journal of Food Technology*, **8**, 333–343

Hubel, D. H. (1988) *Eye, Brain and Vision*. Freeman, New York

Hunt, R. W. G. (1977) Specification of color appearance. Effects of changes in viewing conditions. *Color Research and Application*, **2**, 109–120

Hunt, R. W. G. (1978) Color terminology. *Color Research and Application*, **3**, 79–87

Hunt, R. W. G. (1987a) A model of colour vision for predicting colour appearance in various viewing conditions. *Color Research and Application*, **6**, 297–314

Hunt, R. W. G. (1987b) *Measuring Colour*. Ellis Horwood, Chichester

Hunter, R. S. (1958) Photoelectric color difference meter. *Journal of the Optical Society of America*, **48**, 985–995

Hunter, R. S. and Harold, R. W. (1988) *The Measurement of Appearance*, 2nd edn. Wiley, New York

Hurvich, L. M. (1981) *Color Vision*. Sinaver, Sunderland, MA
Jacobs, G. H. (1981) *Comparative Color Vision*. Academic, New York
Judd, D. B. and Wyszecki, G. (1975) *Color in Business, Science and Industry*, 3rd edn. Wiley, New York
Kent, M. (1987) Collaborative measurements on the colour of light-scattering foods. In *Physical Properties of Foods*, vol. 2, eds. R. Jowitt, F. Escher, M. Kent, B. McKenna and M. Roques, 277–294, Elsevier, London
Kubelka, P. (1948) New contributions to the optics of intensely light scattering materials. *Journal of the Optical Society of America*, **38**, 448–457
MacAdam, D. L. (1970) *Sources of Color Science*. MIT, Cambridge, MA
MacDougall, D. B. (1982) Changes in colour and opacity of meat. *Food Chemistry*, **9**, 75–88
MacDougall, D. B. (1983) Instrumental assessment of the appearance of foods. In *Sensory Quality in Foods and Beverages: its Definition, Measurement and Control*, eds. A. A. Williams and R. K. Atkin, 121–139, Ellis Horwood, Chichester
MacDougall, D. B. (1987) Optical measurements and visual assessment of translucent foods. In *Physical Properties of Foods*, vol. 2, eds. R. Jowitt, F. Escher, M. Kent, B. McKenna and M. Roques, 319–330, Elsevier, London
MacDougall, D. B. (1988) Colour vision and appearance measurement. In *Sensory Analysis of Foods*, ed. J. R. Piggott, 103–130, Elsevier, London
MacDougall, D. B. and Moncrieff, C. B. (1988) Influence of flattering and tri-band illumination on preferred redness-pinkness of bacon. In *Food Acceptability*, ed. D. M. H. Thomson, 443–458, Elsevier, London
McDonald, R. (1987) Computer match prediction – dyes. In *Colour Physics for Industry*, ed. R McDonald, 116–185, Society of Dyers and Colourists, Bradford
McLaren, K. (1986) *The Colour Science of Dyes and Pigments*, 2nd edn. Adam Hilger, Bristol
Nayatani, Y., Takahama, K. and Sobagaki, H. (1986) Prediction of color appearance under various adapting conditions. *Color Research and Application*, **11**, 62–71
Normann, R. A. and Werblin, F. S. (1974) Control of retinal sensitivity. 1: Light and dark adaptation of vertebrate rods and cones. *Journal of General Physiology*, **63**, 37–61
Patterson, D. (1987) Instruments for the measurement of the colour of transparent and opaque objects. In *Colour Physics for Industry*, ed. R. McDonald, 35–62, Society of Dyers and Colourists, Bradford
Pointer, M. R. (1980) The concept of colourfulness and its use for deriving grids for assessing colour appearance. *Color Research and Application*, **2**, 99–107

Pointer, M. R. (1982) Analysis of colour-appearance grids and chromatic-adaptation transforms. *Color Research and Application*, **7**, 113–118

Rigg, B. (1987) Colorimetry and the CIE system. In *Colour Physics for Industry*, ed. R. McDonald, 63–96, Society of Dyers and Colourists, Bradford

Robertson, A. R. (1977) The CIE 1987 color-difference formulae. *Color Research and Application*, **2**, 7–11

Rodieck, R. W. (1979) Visual pathways. *Annual Reviews in Neuroscience*, **2**, 193–225

Saunderson, J. L. (1942) Calculation of the color of pigmented plastics. *Journal of the Optical Society of America*, **32**, 727–736

Smith, C. V. and Pokorny, J. (1975) Spectral sensitivity of the foveal cone photopigments between 400 and 500 nm. *Vision Research*, **15**, 161–171

Stryer, L. (1988) Molecular basis of visual excitation. *Cold Spring Harbour Symposia on Quantitative Biology*, **53**, 283–294

Wald, G. (1968) The molecular basis of visual excitation. *Nature*, **219**, 800–807

Wright, W. D. (1967) *The Rays Are Not Coloured*. Adam Hilger, London

Wright, W. D. (1980) *The Measurement of Colour*, 5th edn. Adam Hilger, London

Zeki, S. (1980) The representation of colours in the cerebral cortex. *Nature*, **284**, 412–418

3 Colour Measurement of Foods by Colour Reflectance Instrumentation

C. J. B. Brimelow and C. A. Groesbeck

Contents

1 Introduction: Food Colour and Quality

The first judgement of a food's quality is more often than not dependent on its various appearance characteristics, such as its colour, surface structure and shape. Colour in particular is an important sensory attribute. It is sometimes defined as the sensation experienced when energy in the form of radiation within the visible spectrum falls upon the retina of the eye (Francis and Clydesdale 1975). But this definition does not give the full picture, since it is known that the brain further processes and conditions the signals received from the rods and the three different types of cones on the retina, in order to rationalize the colour of an object despite changes in the level and colour of the surrounding lighting. This process of adaptation allows objects to be recognized as having virtually the same colour under very different conditions. Colour constancy is only an approximate quality, however, and this is one of the reasons for resorting to instrumental methods for measuring colour under standardized conditions.

The food industry uses colour measurement for a number of reasons, all aimed at reducing the variability caused by subjective analysis. Firstly, there is a need to ensure a good colour quality in the product going to the consumer. Pangborn (1967) has stated that 'In foods, colours are identified with previously experienced quality and serve as instant indicators of good or bad, according to the product and its intended use.' Consumers associate an acceptable colour with acceptable flavour, safety, nutrition and level of satisfaction (Christensen 1983). A perceived inferior colour quality or a colour variation from unit to unit of a branded product are factors not well appreciated by the consumer. The manufacturer, therefore, has the difficult task of providing a product of the expected colour quality at perhaps three different times of judgement: at purchase point if the product is packed in a transparent container; at the unpacking point in the kitchen; and at the consumer's plate after preparation and serving.

The first judgement, in the retail outlet, is particularly important because there are often many reference colours available in the form of competitor products, labelling decor, advertising graphics etc. In the case of products unwrapped in the home, colour expectation and colour memory become significant factors in the quality judgement. The colour of the food material on the consumer's plate is influenced by the manner of in-home preparation, in particular by the cooking procedure.

A second use for colour measurement is in the development of new foods or improved versions of existing foods, in cases where the colour is one of the major quality attributes. Here the reference would be the original product, and the aim would be to improve colour either by changing the manufacturing process, or by incorporating better colour-imparting ingredients in the recipe, or by reducing colour deterioration during storage.

A third use for food colour measurement is in the estimation of pigment concentration or tinctorial strength. Clydesdale (1977) has pointed out that in certain situations it is not possible to use optical density readings at absorbence maxima to do this job and that colorimetry often gives a better result.

A fourth use of colour measurement of foods is to provide a buying criterion in the purchase of certain raw materials or semifabricates, in cases where colour is an important indicator of quality. Examples of this application are in the purchase of tomato paste or citrus juices, where the colour gives an indication both of the original fruit quality and of the efficiency of the manufacturing process.

There are a bewildering variety of methods and instruments available to the food technologist in the field of colour measurement. When one is approaching the subject for the first time or when attempting to devise a method for a material outside the normal experience, the wealth of possibilities available sometimes makes the choice difficult. It is the purpose of this chapter to attempt to identify a systematic approach in order to ease the task. The approach is concerned primarily with the use of tristimulus colorimetry.

2 Colour Measurement Principles: Brief Introduction

For a detailed treatment of the principles of colour measurement, the reader is referred to Chapter 2 by D. B. MacDougall, in this book. Here we give a brief, practically oriented introduction to the subject.

2.1 Tristimulus colorimetry

The scientific basis for the measurement of colour is the existence of three different types of response signals in the human eye. Though four different types of receptor – ρ, γ and β cones, and rods – have been identified, the messages from these are encoded (in a way still not fully understood) to give the three types of signal.

In historic work carried out in the late 1920s and early 1930s by Wright (1969) and Guild, the performance response characteristics of a standard human eye (the standard observer) to different spectrum colour light sources were established. This work formed the basis of the Commission Internationale d'Éclairage system (CIE 1931). In order to be able to relate the results obtained from different combinations of primary colour stimuli, linear transformation equations were used to convert Wright's and Guild's curves to more usable functions. This resulted in the definition of three integral primaries, X, Y and Z (see Figure 3.1). These primaries do not exist as real lights, but they encompass all colours and are, therefore, mathematically very useful.

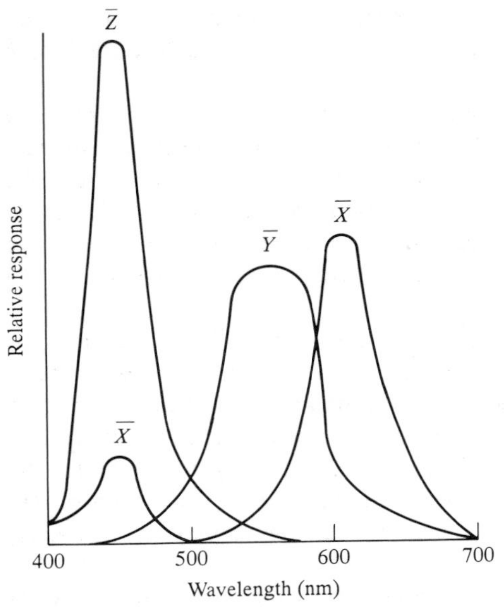

Figure 3.1 *Colour matching response functions of the human eye for the original CIE 2° observer; see Chapter 2, Table 2.1 on notation. (Francis and Clydesdale 1975)*

In a tristimulus colorimeter, three or four filters duplicate the response of the standard observer. The filters correspond to the three primary colours in the spectrum (red, green and blue) and they can be combined to match most colours. The more sophisticated instruments carry a fourth filter to simulate the blue part of the CIE $\bar{X}$ function shown in Figure 3.1. The other essential parts of a tristimulus colorimeter are a white light source, an array of photometers, and an integral calculator (see Figure 3.2). Nowadays, the integral calculator is often a computer which can also carry out data transformations between CIE and other colour scale systems or between different standard white light sources or white diffusers. A good summary of the transformation formulae employed is given by Hunter and Harold (1987).

2.2 Colour Scales and Colour Difference Formulae

The aim of the more popular of the colour scales is to represent colours in a three-dimensional colour space, such that similar visual colour differences are reflected by approximately similar distances in the colour space. Such colour scales are known as uniform scales. The original CIE scales, represented diagrammatically in Figure 3.3, were not intended for identifying the colours of objects and are certainly not uniform in spacing colours according to their visual differences. Probably the most popular of the uniform colour scales in use in the food industry is the Judd-Hunter L, a, b solid, depicted in Figure 3.4.

Hunter (1958) elaborated this scale with three purposes in mind: to be able to sense the scale directly with a tristimulus colorimeter; to provide terms which represented the red/green, blue/yellow and dark/light responses of the human brain; and to give a colour solid with visual uniformity across the solid. The latter aim is impossible to achieve, but the L, a, b scale comes close, as has been demonstrated by Hunter and Harold (1987).

In the Judd-Hunter solid, a colour can be represented by L (lightness or darkness, 0 = perfect black to 100 = perfect white), a ($-a$ = greenness, $+a$ = redness) and b ($-b$ = blueness, $+b$ = yellowness). The a and b axes are both scaled from -100 to $+100$, though higher saturation values are not always achievable in reality. Sometimes, in order to simplify the representation of a colour to two-dimensional space, the a and b values are transformed to θ, which is the angle between the line joining the point (a_1, b_1) in Judd-Hunter space to the origin and the green/red axis. If saturation is neglected, the colour can then be represented by the two independent quantities L and θ, lightness and hue angle (see Figure 3.5).

Judd (1939) proposed that in order to calculate differences between colours in a colour solid, Euclidean geometry could be used such that the

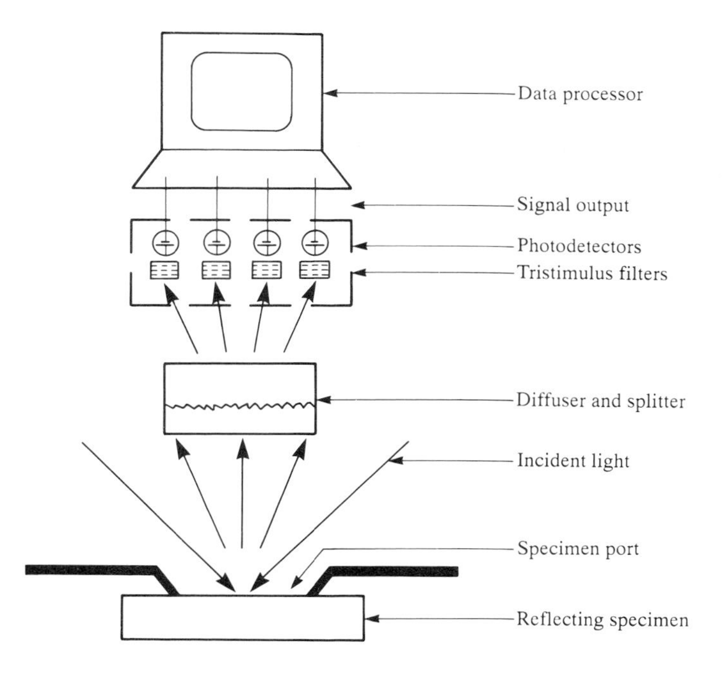

Figure 3.2 *Essential components of a tristimulus colorimeter*

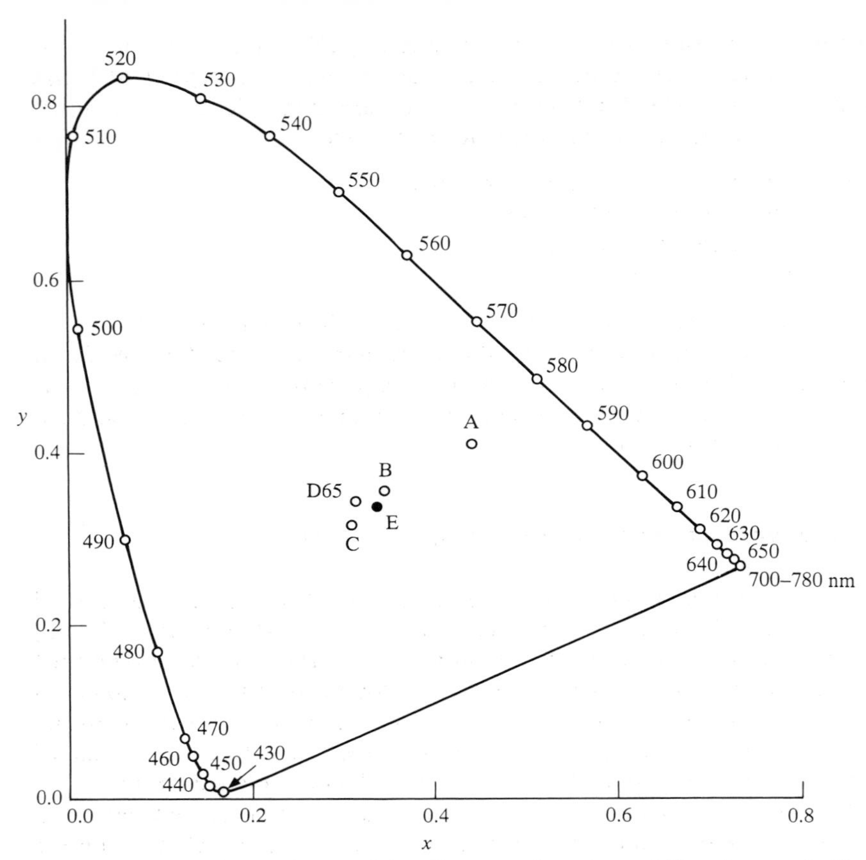

Figure 3.3 *CIE x, y chromaticity diagram: locations of illuminants A, B, C, D65 and E (equal energy) are also shown*

'Euclidean distances between points are proportional to the size or perceptibility, ΔE, of the colour difference between the colours represented by the points.' In the L, a, b colour solid, therefore,

$$\Delta E = [(\Delta L)^2 + (\Delta a)^2 + (\Delta b)^2]^{1/2}$$

and across most of the solid a value for ΔE of around unity indicates the maximum acceptable colour difference in the industry.

3 Colour Measurement Methodology

All methods for the instrumental measurement of food colour involve four stages: (1) standardization of the instrument, (2) selection, preparation and presentation of the sample to the instrument, (3) setting the instrument variables, and (4) determination of the colour values. Bilmeyer (1981) has suggested that deviations from absolute accuracy in colour measurement are caused by, in decreasing order of importance,

- sample preparation and operator variables
- different reference white standards
- variation in colour reference standards
- instrument conditions.

As can be understood, there are many variable factors in each of the four stages of a colour measurement method, which can affect the final result. In order to be able to compare results for a particular food material, these variables should be recognized and efforts made to eliminate, or at least reduce, their effects.

3.1 Standardization of the Instrument

Colour standards provide the references against which the colours of materials can be instrumentally compared. They fall into two classes, primary standards and secondary standards. Primary standards are pressed powder tablets of fresh MgO, $BaSO_4$ or halon G-80 (pressed tetrafluoroethylene resin manufactured by the Allied Chemical Corporation) maintained by governmental standards agencies such as the National Bureau of Standards in Washington DC, USA or the National Physical Laboratory in Middlesex, England. These white standards are measured against a theoretical perfect white diffuser by means of an auxiliary sphere to derive an absolute reflectance value. Unfortunately, at the moment there is

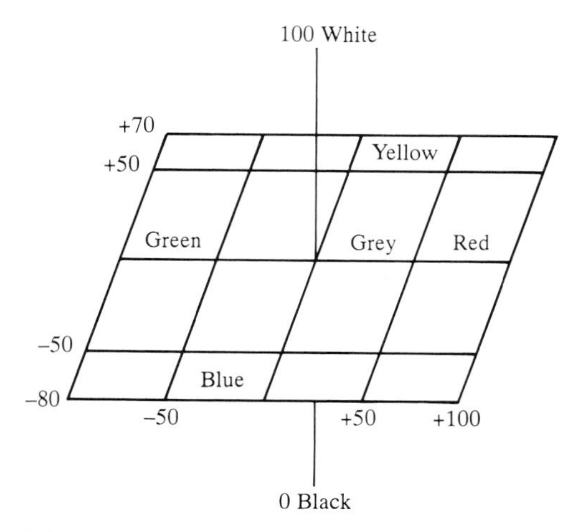

Figure 3.4 *Judd-Hunter colour solid*

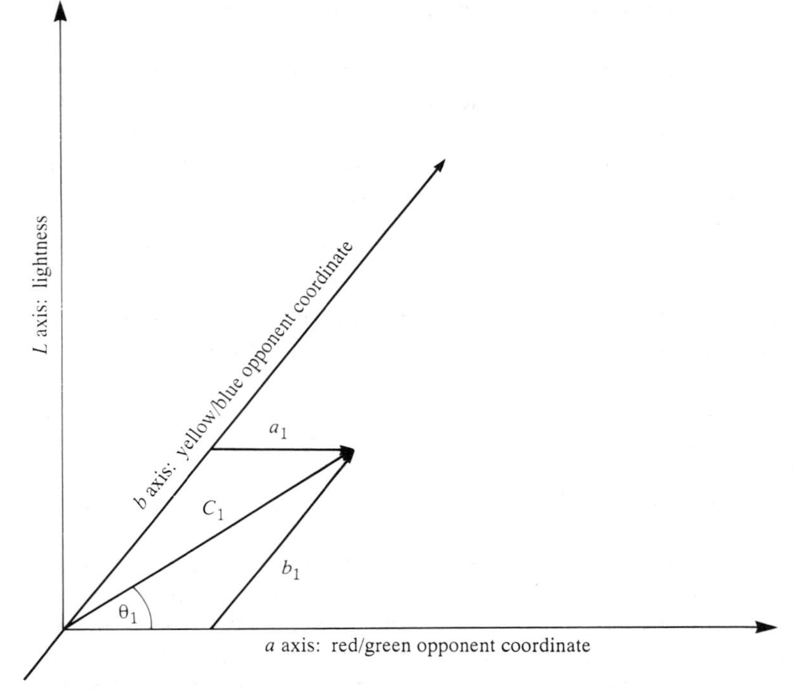

Figure 3.5 *Representation of angle θ_1 formed in the a–b plane by the a axis and the line from the origin to point (a_1, b_1): the hue angle θ is sometimes used together with L to give a simplified representation of colour, neglecting saturation C*

not a complete consensus between the various standards institutes on the perfect white diffuser values.

Secondary standards can be subdivided into a number of groups. For example, Hunter and Harold (1987) have described seven types of secondary standards, in the following terms:

1 Master standards calibrated by reference to primary standards and being in the form of porcelain or opaque glass panels, ceramic tiles, acrylic painted chips etc. These are maintained with extreme care, usually at the standards institutes, and are used only to calibrate other secondary standards.
2 Working standards used as a reference for a group of laboratory instrument standards. These are usually ceramic or porcelain tiles.
3 Instrument standards used to calibrate a particular instrument.
4 'Hitching-post' standards which are used in place of a white standard and which are close in colour to that of the samples to be measured. This

technique is used to minimize errors caused by scale-interval defects in the instrument.

5 Specific calibration standards which are permanent specimens or standards used for measurements carried out on impermanent products. To derive these standards the impermanent product is first measured on a reference instrument. The product can then be used to calibrate a second instrument on which the permanent standard is also read. This reading then becomes the assigned values of the permanent standard.
6 Diagnostic standards used for instrument checking or fault finding in inaccurate instruments.
7 Material comparison standards, which are usually samples of the materials under examination and provide limiting or target values.

It can be seen that, in all except the last two cases, the colour values of the standards are traceable back via the working and master standards to the primary standard.

Secondary standards should be treated with care. Colour values can be affected by the presence of dirt, grease, moisture, fingerprints or surface scratches on the viewing area. Ceramic standards should be cleaned carefully and regularly by washing in mild anionic detergent, rinsing in hot distilled water and blotting dry with a lint-free paper towel. When not in use, they should be maintained at a constant standard temperature away from light in a sealed container. Despite even these precautions, it is known that some surfaces can be attacked by atmospheric pollutants or may undergo changes during storage. Painted or printed colour chips should never be cleaned. Freezer storage is often practised to extend the life of colour chips (Huey 1965).

Before a standard is used to calibrate an instrument, the instrument should be thoroughly warmed up and stabilized. Colour standards should be placed under the instrument port in the same manner each time a calibration is carried out, to avoid the effects of any surface directionality in the colour layer of the standard on the calibration values. Because some of the colour standards exhibit thermochromicity, i.e. their colour values change as the temperature changes, it is advised that calibration is carried out with rapidity in order to avoid the instrument lamp raising the temperature of the standard. The subject of thermochromicity of the colour standard has been described by Brimelow (1987) with reference to measurement of tomato colour.

Kent and Smith (1987) reported the results of an inter-laboratory study on measurement of colour standards. They came to the obvious and important conclusion that in order to transfer and compare colour data from one laboratory to another, which is certainly necessary when colour is being used as a buying criterion, the measuring system and the colour standard have to be carefully defined. Variables inherent in the measuring system will be described in the next two sections.

3.2 Selection, Preparation and Presentation of the Samples

The sample itself is the key to good colour measurement. In optical terms, samples can be divided into four main classifications: opaque objects, metal-like objects, transparent objects and translucent objects.

With opaque objects, a class into which most food materials fall, light is either directly reflected from the surface as a white highlight (gloss) or absorbed, scattered and diffusely reflected from the surface (perceived colour). Opaque materials do not transmit light. The relative amounts of gloss and colour perceived depend, of course, on the surface properties, but also on the viewing angle. In order to compare colour differences independently of the gloss characteristics, the optimal positions of the light source and the detector (e.g. the eye) relative to the specimen surface are 45° and 0° or 0° and 45°.

In the case of metal-like objects, virtually all the light is reflected as gloss and very little is diffused. This effect is known as specular reflectance. Very often the reflectance is at a dominant wavelength, which provides colour gloss, rather than at all wavelengths which would provide white highlights. Foods, though they may sometimes have a high gloss component, do not fall into this class of metal-like materials.

Transparent objects exhibit colour by regular transmission, that is they absorb light and then transmit it preferentially at different wavelengths along the path of the light beam. Some liquid food materials and a few solid food materials are transparent, such as vegetable oils, fruit serums and wines, certain sugar confectionery, fruit jellies etc.

The class of materials between transparent and opaque, to which many food materials belong, is called translucent. Some of the light falling on a translucent object can pass through it in the direction of the light, but the remainder is diffused within it and emerges in different directions (see Figure 3.6). The light scattering properties of hazy or turbid food materials are a complex subject which has been covered in some depth by MacDougall (1982; 1983; 1987). It is clear that with this class of material, instrument/object geometry plays a larger part in the colour values obtained than is the case with the other classes. MacDougall (1987) has illustrated this point using concentrated and diluted orange juices as test samples (see Figure 3.7).

When reflectance measurements are carried out on opaque or translucent food materials, a number of problems exist in the selection and preparation of the samples. The first is that the sample areas as presented to the instrument may not be representative of the surface of the bulk of the material under test simply by reason of its presentation method. All the tristimulus instruments in common use are designed to accommodate samples which are flat and, therefore, reflect light from one plane only. Most food materials are not flat and some degree of compromise has to be made in order to create viewing areas which are. It is often the case that it is advantageous deliberately to change the surface characteristics of the sample in order to enhance colour differences between specimens. André and Pauli (1978), for example, pulverized dehydrated pasta into a powder and then tableted the powder for colour measurement in order to minimize variation caused by differences in the shapes and surface properties of commercially available products.

A second important problem is that during the process of creation of the flat viewing area, other factors come into play which may affect surface reflection. These factors include the ambient conditions of temperature and humidity applying during this process; the variables within the actual method used (e.g. the blade sharpness during cutting, the technique of polishing, the pressure applied during pressing); chemical reactions (e.g. browning effects) or physical reactions (e.g. settling effects) occurring at the surface up to and during the measurement event; and the settlement of vapour, dust or film-forming agents on the surface after its formation.

A third problem is the general non-uniformity of the surface colours of

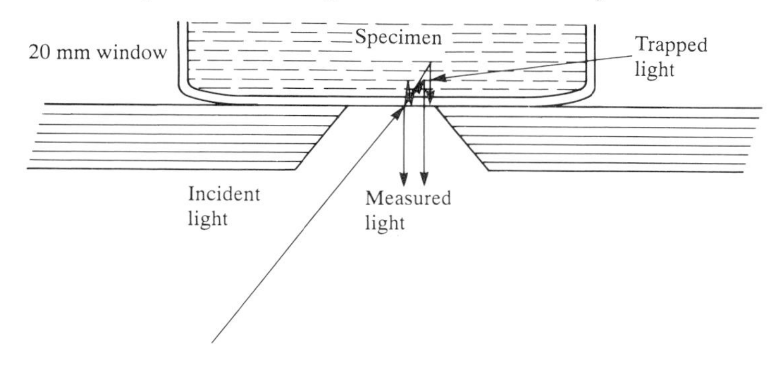

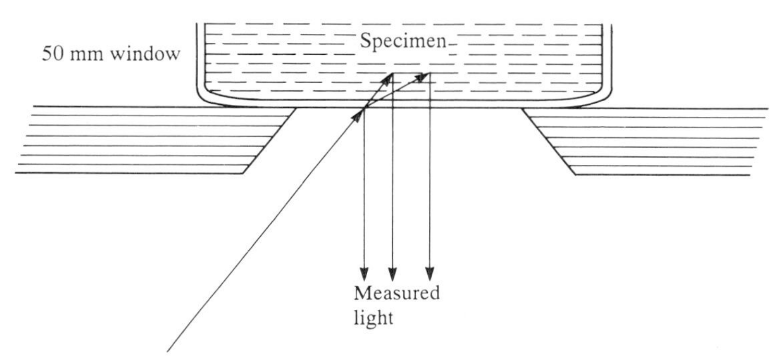

Figure 3.6 *Translucent foods diffuse the incident light, resulting in lost measurement of trapped light unless a large area viewing aperture is utilized to measure most of the diffused reflected light*

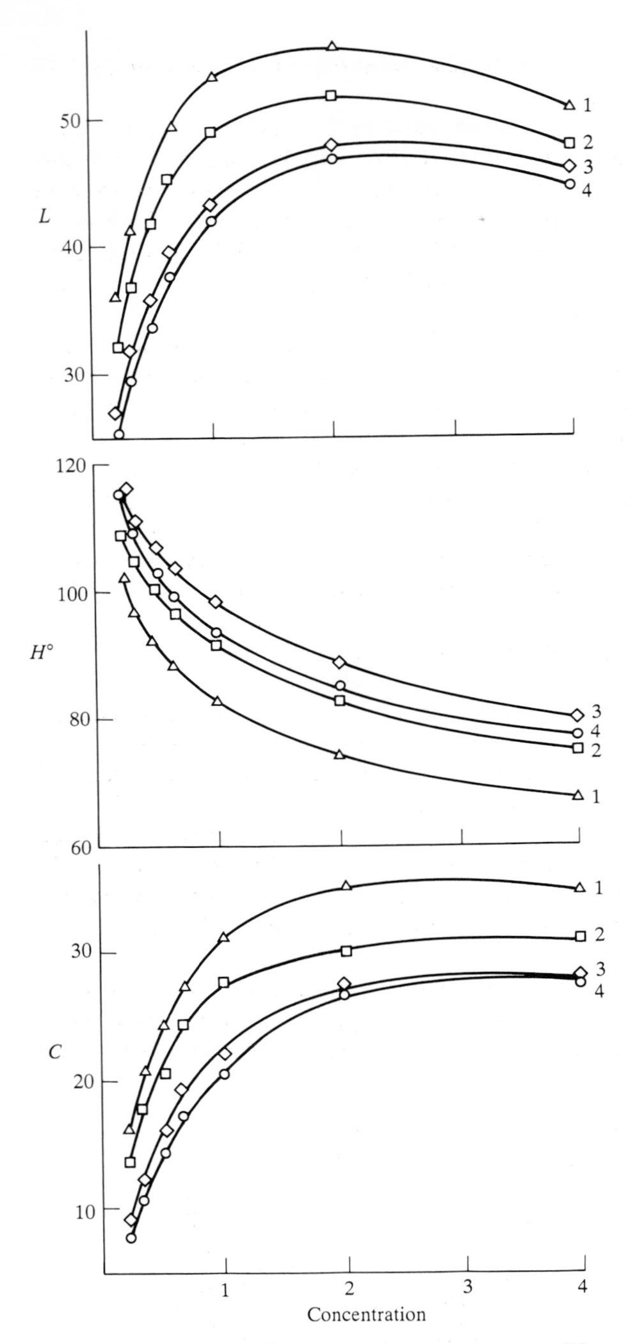

Figure 3.7 *Effect of presentation geometry on Hunter lightness L, hue angle H° and chroma C for concentrated and diluted orange juice (MacDougall 1983): reconstituted juice at normal concentration = 1.0. 1 △ small spot, 45°/5 cm diameter, 0°; 2 □ 4 cm diameter, 45°/4 cm diameter, 0°; 3 ◇ 1.5 cm², near 0°/5 cm² port, sphere; 4 ○ small spot, 45°/2 cm diameter, 0°*

foodstuffs. This means that it may be necessary to make repeated measurements in different places on a surface or make measurements of different samples from a bulk in order to achieve statistically significant average values. One ingenious solution to the problem of making repeated measurements on a surface has been suggested by Francis (1952) for apples. By rotating a single fruit rapidly in a small viewing aperture, an average surface colour was obtained.

One further problem is the physical size of the food sample. Some specimens may simply be too small to measure successfully as single items, though recently one or two instruments have appeared with very small aperture viewing heads which permit measurements on areas of around 3 mm diameter. It is, however, more usual to bulk such samples and then employ large area sample cups and large diameter instrument apertures when making the measurements.

Some general rules for the preparation of different types of food materials for reflectance colour measurement are as follows.

3.2.1 Opaque Powders, Granules and Flakes

It is usual to make multiple measurements on such materials and average the results. Bulk amounts are normally sieved prior to subsampling in order to reduce errors due to size variation in the particles. Hunter and Harold (1987) have clearly illustrated that the larger the particle size, the less the light is scattered and therefore the more coloured the sample appears. Smaller particles scatter and reflect more light and the sample therefore appears lighter (see Figure 3.8). Many of the errors encountered in comparing the colours of powders and granules relate to the particle size effect.

One of three methods of sample preparation of powders can be utilized. Samples can be poured into cups with optically flat and transparent bases and measurements made through the base. Alternatively, samples can be

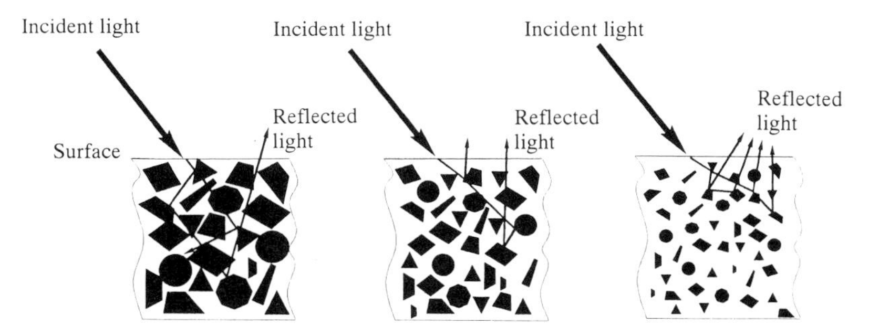

Figure 3.8 *Decreasing particle size increases light scatter, causing the sample to appear lighter in colour*

poured into a container with sides of equal height and the top can be scraped flat with a straight edge. Measurements are made from above. Thirdly, a powder compactor can be used. This is a device which can create a disk specimen from a sample of known weight and bulk density by applying a standardized force. Measurements can then be made from above or below.

3.2.2 Opaque Particulate or Lumpy Materials

Materials from 5 mm to 20 mm in size can be very difficult to measure with good repeatability, particularly if they are non-regular in shape. As has been mentioned, instruments have recently appeared on the market with small diameter viewing heads, such as the Minolta CR 121. This instrument has been used in the authors' laboratory to measure the colour of particulate pasta products such as ravioli. Normally, however, it is the practice to place particulate or lumpy materials into a large container with an optically flat and clear base, trying to ensure that the packing is as solid as possible. Measurements are then usually made from below.

An alternative method of presentation has been evolved which is useful in situations when repeated measurements have to be made on solid materials which have a high oil or fat content. These materials may smear or coat the surfaces of a glass cell. The method utilizes clear PVC film (e.g. Saran Wrap). The film is stretched across the viewing port of the colorimeter and the sample is placed on the film. Measurement is made from below. After each measurement, the specimen can be quickly wrapped up in the film and discarded, thus obviating the need for frequent washing of valuable glass sample cups (Mabon 1989).

3.2.3 Large Area Solid Foods

When colour measurements are made on solid opaque foods, such as whole fruits or thick slices of red meat, the specimen field should be flat and uniform. If a natural flat surface is not available, then it is sometimes acceptable to form such a surface by cutting or pressing. The latter can be achieved by sandwiching the sample between two plates of glass, one of which should be optically flat and transparent. Viewing is carried out through the transparent plate. If the food does not lend itself to cutting or pressing, then one other possibility is to rotate the food object in front of the viewing port of the instrument (Francis 1952).

Solid foods often show directionality. When the directionality is obvious, the specimen should be measured in the same orientation for each test. When the directionality is haphazard or undefined, one useful technique is to turn the sample through 90° between the two readings and average the results.

Many solid foods are translucent, for example, cheeses, butters, fatty meats and high water content fruits and vegetables. The effects of translucency on colour readings are large, particularly when such materials are being measured in thin slice form (for example, bacon, other sliced meats and sliced fruits). It is essential to standardize the preparation procedures for such samples. Normally a standard white backing plate would be employed and the sample thickness would also be standardized. Specimens are often viewed through an optically clean and flat plate placed on the surface of the solid. Some of the effects of the preparative variables on the colour values of sliced meats have been discussed by MacDougall (1982).

A few food solids are transparent and they can be measured by transmittance procedures. Specimens of known thickness are held against the viewing port of sphere-type colorimeters for measurement.

3.2.4 Pastes and Slurries

Food pastes and slurries are normally translucent materials. It is usual with such materials to introduce them into cells with optically clear and flat bases, to such a depth that all the light is trapped and diffused within the sample and none escapes through the sample. Measurements are made from below. However, in some instances, particularly when the aim is to compare lightness values in dark samples or when a standard background is desirable, a special cell equipped with a ring insert of known height is employed. The sample slurry is poured into the cell to fill it to above the ring and a disk with a white diffusing surface is pushed down to nest on the ring. This fixes the light path length of the sample when it is viewed from below.

Often it is advisable to cover the specimen cell with a black lined box in order to prevent extraneous room light from passing into the specimen. Care should be taken when introducing pastes and slurries into test cells that no air is entrapped in the sample. When pastes and slurries are prepared by mincing followed by diluting with water, the mincing conditions must be standardized and also the dilution factor. Samples should be well stirred and then measured quickly to reduce the effects of sedimentation and oxidation. Finally, the thermochromic properties of certain food materials should not be forgotten, and sample temperatures should often be standardized.

The effects of some of these preparative factors – dilution, test temperature and time, presence or absence of a black cover – on the colour values of tomato pastes have been discussed by Brimelow (1987). As already noted, MacDougall (1983) has illustrated the effects of the dilution factor on colour values in the case of orange juices (see Figure 3.7).

3.2.5 Liquids

Liquids can be translucent (milk, syrups, fruit juices, egg yolk) or transparent (wines, vinegars). Translucent liquids are measured in the same way as translucent slurries and pastes.

Transparent liquids are measured by pouring into optically clear fixed path length cells with parallel sides and then taking transmission readings in sphere colorimeters. Deeply coloured liquids such as dark treacles are measured in short path length cells of the order of 2 mm, whilst lightly coloured liquids such as white wines are measured in long path length cells up to 50 mm.

3.3 Setting the Instrumental Variables

The instrumental variables include the choice of light source, the measuring geometry and the relative sizes of the specimen area and the illuminated area.

When measurements are made of the colour of a food, it is necessary to specify the illuminant used as part of the viewing conditions. In 1931 the CIE established various standard illuminants which have characteristics close to natural light sources, but which also can be reproduced easily for use in the laboratory or in instrumentation. Illuminant A represents light close to that given by a tungsten filament lamp, which might for example be operating during subjective assessments of colour in the home. Illuminant B represents direct sunlight, whereas illuminant C represents average daylight from the total sky, which again might be present during assessments in the home. More recently, the CIE (1979) has proposed a D series of illuminants which more closely represent daylight than do B or C and also incorporate an ultraviolet range. The D series correlates with various daylight colour temperatures; the most commonly used is the D65 illuminant, approximating to a colour temperature of 6500 K.

Despite the evolvement of the D series illuminants, most colour measurements of foods are still made and reported using illuminant C. It should be borne in mind, however, that if a subjective assessment of the colour of a particular food is normally carried out under, say, a tungsten filament light or under direct sunlight, then it may be preferable to carry out the instrumental measurement under similar lighting conditions. It should be noted also that a colour which matches another under one illuminant may not match it under a different illuminant. This phenomenon is known as metamerism.

The choice of the remaining instrumental factors, such as measuring geometry and the relative sizes of the specimen viewing area and the illuminated areas, are very much influenced by the optical classification of the specimen. Therefore, in order to determine the colour values of a product it is necessary to decide whether the prepared sample is opaque, translucent or transparent. Transparency is relatively simple to assess: when a clear image of a light source can be seen through the sample it can essentially be considered as transparent. Translucency is more difficult to determine as the dividing line between opacity and some translucency is imprecise. A simple test is to measure the specimen with maximum area of illumination and maximum viewing aperture and then to repeat the measurement using the same viewing aperture but a reduced area of illumination. If there is a large increase in the lightness reading (the L value in the L, a, b scale), then the product is translucent.

The viewing geometry must be specified as part of the instrumental conditions. It is generally agreed that with food samples most applications can be covered without much compromise using 0°/45° or 45°/0° viewing geometry (source/detector). This is the preferred geometry when examining the reflected colour of translucent samples. It is also satisfactory for opaque samples, provided the samples being compared are of equal texture or gloss. Even samples with different texture or gloss characteristics can be compared using this geometry, provided that appearance differences rather than simply colour differences are being assessed. 0°/45° or 45°/0° geometry cannot be used for transparent samples or for measuring transmittance through translucent samples.

One source of error arising from the viewing geometry occurs through failing to define whether specular reflectance is included or not. With 0°/45° or 45°/0° geometries, very little specular reflectance is included in the total reflected light detected, even with glossy but non-smooth samples. However, with spherical (d/8°) geometry, specular reflectance will be present in the reflected light from glossy samples and the decision has to be made to include it or exclude it.

The choice of illuminated area size and aperture size is very dependent on the nature and the optical class of the specimen. Small area illumination is used on small samples, on samples having only small areas of flatness and on highly translucent and uniform samples. In the latter case, Hunter and Christie (1978) have recommended that the small area illumination is used together with large aperture viewing in order to capture that proportion of the light scattered by the sample and then returned to the detector from outside the illuminated area (see Figure 3.6). As has been mentioned already, particulate solids are best measured using large area illumination and large viewing apertures.

3.4 Determination of Colour Values

The last stage in obtaining a colour measurement is the conversion of the raw signal to colour values in one or other of the colour scales. In modern instrumentation, software packages are capable of carrying out these conversions with speed and accuracy, and can even make calculations to predict comparative colour results for different light sources.

Since calculations are based on reference white values, it is a point of concern that there are still differences between national standards for the perfect white diffuser.

4 Colour Measurement of Typical Food Materials

In this section, some typical applications of food colour measurement will be described by reference to recent work in this area.

The work will be treated under the headings utilized earlier in the chapter: powders, granules and flakes; particulate and lumpy materials; large area solid foods; pastes and slurries; and liquids. For further details on the preparation procedures, the sample temperature, the sample presentation, the instrument illumination and aperture diameters, the colour standard and the colour scale employed in each case, the reader may refer to Tables 3.1 to 3.5, respectively. The instruments utilized in each case are briefly summarized in Table 3.6.

4.1 Powders, Granules and Flakes (Table 3.1)

Johnston et al. (1980) reported that several researchers had compared colorimetric values with visual judgements for durum semolina, but that no one had compared colorimetric data with extracted pigment content. They found that semolina reflectance *b* values were highly correlated with pigment content as determined by pigment extraction and estimation using the standard AACC water-saturated butanol method, particularly when the objective was to segregate highly pigmented semolinas from low pigmented ones. The authors pointed out that reflectance colour measurement in these types of applications could be a genuine time-saving alternative to the lengthy pigment extraction technique.

Seakins (1971) developed a method to measure the colour of casein by immersing it in a liquid of similar refractive index in order to overcome the effects of particle size. Reflectance measurement of dry sample preparations resulted in poor separation of colour grades because reflectance increased as particle size decreased. However, when transmittance values were measured on the liquid suspension, the colour grades were clearly differentiated.

In order to minimize grind size effects in ground coffees of different roasts, Little and MacKinney (1956) standardized on a very fine grind with maximum lightness value.

Berset and Caniaux (1983), working with parsley, found that measurements on the ground leaves gave better reproducibility than measurements on the whole leaves. The authors surmised that grinding the leaves released

Table 3.1 *Colour measurement of powders, granules and flakes*

Reference	*Sample*	*Preparation*	*Temperature (°C)*	*Presentation (viewed from above/below)*	*Illumination/aperture diameters (mm)*	*Standard*	*Instrument*[a]	*Measurement*[b]
Johnston et al. (1980)	Durum semolina	Equilibrate to 15% H_2O, mill to semolina flour	NM	20 g samples	NM	NM	A(1)	b
Seakins (1971)	Casein	Sieve to 6 different particle ranges. Suspend samples in benzene	NM	0.5 cm light path	T[c]	Transmittance standard at 455 nm	B	% T at 455 nm
Little and MacKinney (1956)	Coffee beans at different roasts	Beans ground in commercial coffee grinder, sieved to 5 size ranges	NM	Ground coffee in clear glass cell	NM	White tile, neutral grey tile	C	Y
Berset and Caniaux (1983)	Dried parsley leaves	Leaves ground 1 min in ball grinder, sieved to 80–250 μm particles	NM	NM	NM	NM	D	L^*, a^*, b^* X, Y, Z, x, y
Habib and Brown (1956)	Potato crisps	Crisps size reduced by milling 5 min in Oster mill	NM	Viewing cell 89 mm high by 51 mm diameter. Sample depth of 76 mm (below)	NM/51	Ivory tile	E	L, a, b
Grieder (1989)	Herbs	Mix 5 g herbs with 5 g sand using automated mortar mill	20	Black metal cell with glass base, 55 mm diameter by 60 mm deep (below)	51/55	White tile	A	$L, a, b, \Delta E, \Delta C$

[a]For instrument codes see Table 3.6; [b]For colour scales see Chapter 2, Table 2.1.; [c]Transmission cell; NM: not mentioned in the reference.

Table 3.2 *Colour measurement of particulate and lumpy solids*

Reference	*Sample*	*Preparation*	*Temperature (°C)*	*Presentation (viewed from above/below)*	*Illumination/aperture diameters (mm)*	*Standard*	*Instrument*[a]	*Measurement*[b]
Bergthaller et al. (1983)	Potato cubes, dried	Commercially available cubes packed into cell to depth of 60 mm	NM	Dull black Al cell 60 mm internal diameter, height 60 mm, with optical glass base 1.2 mm thick (below)	50/NM	White enamel tile	G	Y, L, a, b
Baardseth et al. (1988)	Dry sausage, salami type	Slices, 3 mm thick	NM	No cover used, measured against black background	50/50 25/32	White tile	H(1) A(4)	L^*, a^*, b^*
Lee at al. (1988)	Frozen green beans	Boiled 5 min from frozen. Drained 1 min	NM	10 cm square acrylic cell	Large/NM	White tile	A	L, a, b
André and Pauli (1978)	Dried vermicelli (of various shapes)	Pasta ground, sieved through 0.28 mesh, 10 g of powder formed to a tablet 40 mm diameter by 10 mm thick	NM	Faces of tablets measured directly on sensor port	NM/10	White enamel tile	G	X, Y, Z x, y
Miller and Burns (1971)	Peanut hulls (internal colour)	Peanuts harvested, shelled, and sorted subjectively into mature or immature categories	NM	Peanut hull halves directly on narrow beam sensor port	Narrow/NM	Porcelain coloured tile	I	Rd[c], a, b
Authors' laboratory (1988)	Fresh pasta, angel hair and tortellini	Pasta portions packed into 89 mm diameter cell to cover optical glass base, flat against glass when possible	20/22	Reflectance through glass base. Black cover on cell (below)	50/89	White tile and black glass	I(3)	L, a, b

[a]For instrument codes see Table 3.6; [b]For colour scales see Chapter 2, Table 2.1; [c]Luminosity value for some older instruments. Rd = Y; NM: not mentioned in the reference.

Table 3.3 *Colour measurement of large solids*

Reference	*Sample*	*Preparation*	*Temperature (°C)*	*Presentation (viewed from above/below)*	*Illumination/aperture diameters (mm)*	*Standard*	*Instrument*[a]	*Measurement*[b]
Konstance et al. (1988)	Beef chuck	Sliced 2.54 cm thick	25	Glass cup (below)	NM/19	White plate	I(1)	ΔE, ΔC, YI[c]
Eagerman et al. (1978)	Beef semimembranosus muscle	Sliced 12 × 8 × 1 cm	0	NM	NM	NM	K	R (632, 614 nm)
Freedman and Francis (1984)	Fruit jellies	Juice-sucrose-pectin jellies	NM	4 × 5 cm transmission cell, 2 cm light path	T	Distilled water	A(5)	L, θ
Nagle et al. (1979); Reeves (1987)	Whole red peppers	Blanched whole fruit	NM	Measured around the shoulder of the fruit	NM	Colour plate	I(4)	L, θ
Shewfelt et al. (1989)	Fresh tomatoes	Whole, measured 'as is'	NM	Measured at 8 equidistant locations on surface of fruit	NM	Pink tile	I(6)	L, a, b θ, chroma
Lin et al. (1989)	Apples, fresh	Whole, measured 'as is'	NM	Measured at 8 equidistant locations on surface of fruit	NM	Pink tile	I(6)	L, a, b, θ
Sapers and Douglas (1987)	Raw apples and pears	Plug (22 mm diameter) removed from halved fruit. Transverse cut across plug just prior to measurement	20	Fresh transverse surface of plug centred over instrument aperture	Large/19	White tile	I(1)	L, a, b, X, Y, Z
Baardseth et al. (1988)	Potatoes, raw	Cut lengthwise, covered with polyethylene film	NM	Readings taken immediately through poly film	8/8 50/50 25/32	White tile	H(2) H(1) A(4)	L^*
Baardseth et al. (1988)	Raw cauliflower	Cauliflower was measured 'as is' directly at port	NM	No cover used	50/50 25/32	White tile White tile	H(1) A(4)	L^*, a^*, b^* L^*, a^*, b^*

[a]For instrument codes see Table 3.6; [b]For colour scales see Chapter 2, Table 2.1; [c]Yellowness index YI = 100/Y (1.277 X − 1.06Z) NM: not mentioned in the reference. T: transmission cell.

Table 3.4 *Colour measurement of pastes, purées and slurries*

Reference	Sample	Preparation	Temperature (°C)	Presentation (viewed from above/below)	Illumination/aperture diameters (mm)	Standard	Instrument[a]	Measurement[b]
Baardseth et al. (1988)	Mashed potatoes and yams	Commercially available samples	NM	Plexiglass cell, 10 cm cube with 3 mm quartz glass front	50/50 25/32	White tile	H(1) A(4)	L^*, a^*, b^*
Nagle et al. (1979); Reeves (1987)	Red pepper purée	Milled whole fruit	NM	Sample cell with white top plate (below)	NM	Colour plate	I(4)	L, a, b
Francis (1985)	Blueberry toppings	Puréed blueberries incorporated in topping mix with 15% or 30% fruit content. Formulations pasteurized	NM	3 cm depth in 6 cm diameter cell (below)	NM	White tile	I(1)	L, θ
Clydesdale and Francis (1969)	Spinach purée	Milled spinach processed to F_0 4.9	NM	NM	NM	Grey tile, pressed Ba_2SO_4	A C L	X, Y, Z L, a, b G, R, B
Huang et al. (1970)	Squash purée	Commercially available samples	NM	6.4 cm diameter cells, sample thickness between 2 and 8 mm, backed by black or white papers	14/47 47/47	Pressed Ba_2SO_4, pink tile	L A	X, Y, Z
Silva et al. (1989)	Sweet potato mash	Various potato sizes were canned or steamed, then frozen or canned	NM	NM	NM	Orange tile	A	L, a, b $\theta, \Delta C$
Bosset et al. (1986)	Yoghurt	Commercially available samples	NM	500 ml measuring cell, yoghurts stirred before measurement	NM	NM	F	L, a, b
Brimelow (1987)	Tomato paste	Dilute to 12% TSS[c]	20	50 mm depth in 62.5 mm diameter cell (below). Black cover	51/51	Red tile	A(2)	a/b

[a]For instrument codes see Table 3.6; [b]For colour scales see Chapter 2, Table 2.1; [c]TSS: total soluble solids (%) measured on the sugar scale; NM: not mentioned in

Table 3.5 *Colour measurement of liquids*

Reference	*Sample*	*Preparation*	*Temperature (°C)*	*Presentation (viewed from above/below)*	*Illumination/aperture diameters (mm)*	*Standard*	*Instrument*[a]	*Measurement*[b]
Francis (1985)	Blueberry beverages	Puréed blueberries incorporated in beverage mix with 5% and 12% fruit content. Formulation pasteurized	NM	2 cm transmission cell	T	Distilled water	I(2)	L, θ
Francis and Clydesdale (1972)	Egg Yolk	Homogenize	NM	20 mm depth glass (below)	10/NM	NM	A	a, b
Calvi and Francis (1978)	Carbonated grape beverage	Grape anthocyanins incorporated in carbonated beverage formulation	NM	1 cm transmission cell	T	Distilled water	I(5)	a/L
Skrede et al. (1985)	Blackcurrant syrups	Dilution of the syrups 1 to 4 with H_2O	NM	1 cm transmission cell	T	Water	J	L, a, b
Pilando et al. (1985)	Strawberry wine	Thawed berries, sugar and yeast fermented to wine, then racked and bottled with SO_2. wine stored in the dark	NM	Transmission in a 1.0 × 5.0 × 5.0 cm lucite cell, with gloss excluded	T	White tile	A(5)	L, a, b
Desarzens et al. (1983)	Liquid milk products	Cell filled with ca 500 ml of liquid	NM	Reflectance in 500 ml sample cell	NM	White tile	F	L, a, b
MacDougall (1987)	Evaporated milk, Orange juice, Tomato paste	Dilution with distilled water to obtain various specified solids contents, giving opaque to translucent solutions	NM	Measuring cell containing over 5 cm depth. Black cover	50 down to 10/50 down to 5	White tile	A(2)	L, a, b, y

[a]For instrument codes see Table 3.6; [b]For colour scales see Chapter 2, Table 2.1; NM: not mentioned in the reference; T: transmission cell.

Table 3.6 *Colour measurement instruments*

Code for Tables 3.1–3.5	*Instrument*	*Geometry*	*Illuminant source*
A	HunterLab Color Difference Meter D25	45°/0°	C
(1)	D25A	45°/0°	C
(2)	D25M	Circumferential 45°/0°	C
(3)	DL25	Circumferential 45°/0°	NM
(4)	D25P	Integrating sphere 0°/*d*	C
(5)	D25 with transmission	Transmission	NM
B	Hilger Biochem	Transmission	NM
C	Colormaster Differential Colorimeter	Tristimulus reflectance	C
D	DU COLOR Color Difference Meter (Neotec)	Circumferential illumination	NM
E	Hunter Color and Color Difference Meter	Tristimulus reflectance	NM
F	Macbeth Spectrophotometer MS-2000	Integrating sphere 0°/*d*	C
G	MOMCOLOR D (MOM Hungarian Optical Works)	Tristimulus colorimeter	C
H	Minolta Chroma Meter		
(1)	CR-110	*d*/0°	C
(2)	CR-100	*d*/0°	C
I	Gardner Automatic Color and Color Difference Meter	Tristimulus colorimeter	NM
(1)	XL-23	Tristimulus colorimeter	C
(2)	XL-23 with transmission	Transmission	NM
(3)	Colorguard 2000/45 sensor	Circumferential 45°/0°	C
(4)	XL-10A	NM	C
(5)	XL-10 with transmission	Transmission	NM
(6)	XL-845	Tristimulus colorimeter	NM
J	Shimadzu 300-UV	Transmission	C
K	Bausch & Lamb Spectronic 20 with reflectance attachment	NM	NM
L	GE Recording Spectrophotometer	NM	NM

NM: not mentioned in the reference.

from interior leaf cells chlorophyll pigments which were of a different type and more highly pigmented than those in the surface cells. It was shown that the hue function correlated highly with extracted chlorophyll a and b content. These workers also suggested that their colorimetric procedures could be utilized in the quality control of parsley drying and storage processes.

In order to overcome the problem of colour variation across potato crisps, Habib and Brown (1956) ground them down to a uniform colour and then introduced the size reduced samples into large volume sample cells using a defined tap-down procedure. Colour differences among four different potato varieties and different storage/conditioning environments were compared over storage time. It was shown by carrying out multiple correlations against Hunter *L* values that reducing sugars and free amino nitrogen were important in colour formation in the final chips.

A novel approach to the preparation of flaky materials for colour measurement has been suggested by Grieder (1989). Working with herbs, he utilized a pulverization procedure with sand. This aided the dispersion and trapping of oils, pigments and moisture expressed from the interior cells of the test materials during the grinding process.

4.2 Particulate and Lumpy Solids (Table 3.2)

The subject of colour measurement of dehydrated potato products, such as sliced or diced potato, has been well covered by Bergthaller and his team. In the case of diced potato, for example, Bergthaller et al. (1983) packed the cubes into a dull black aluminium cell with an optically clear glass base. The *l*, *a* and *b* values of the potato cubes were compared with sensory data and a high correlation between the two sets of data was obtained.

A number of different food materials were examined by Baardseth et al. (1988) as part of a comparative exercise involving two colour measuring instruments, the Minolta and the HunterLab D25P. One of the subject materials was salami-type dry sausage, which was prepared and sliced thinly for colour measurement. The sausage slices had a varied appearance, with visible fat particles of 1 to 3 mm diameter present on the surface. Correlation coefficients between instruments of over 0.9 were obtained, with much of the residual non-predictability being related to the inconsistencies in the slice surfaces.

Lee et al. (1988) investigated frozen cut green beans given various blanch treatments. The beans were boiled, drained and placed in a 10 cm square acrylic measuring cell. Five replicates per treatment were measured. Results revealed that significant differences were found in the Hunter *a* values, with beans blanched at temperatures above 82°C showing a greener colour than beans blanched at temperatures below 82°C.

In order to eliminate problems associated with the differences in surface properties and shapes of commercially available dehydrated vermicelli, André and Pauli (1978) ground the samples down to powders, then formed tablets from the powders using a hydraulic press. The 40 mm diameter by 10 mm thick tablets could be tested directly on the head of the colorimeter. Good correlations were found between colour coordinate values and the β-carotene content derived from the egg component of the pastas. Pastas made from different flours could also be differentiated in terms of colour values.

A non-sacrificial index of peanut kernel maturity has been derived by Miller and Burns (1971) using a colorimeter procedure. Individual cotyledons were placed on the port of a narrow light beam instrument. The half hulls thus acted as miniature hemispherical light scattering chambers. Ageing of peanuts is accompanied by darkening of the vein cells near the internal hull surface, and this change was measured by a change in the lightness scale readings obtained.

Fresh pasta colour has been measured at the authors' laboratory (1988), the subject materials being spinach and egg varieties of ravioli, fettuccine, angel hair and tortellini. Samples were provided packed under gas atmosphere. It was found that food substances of this type required immediate measurement after removal from the package since surface colour changes started to occur as soon as the specimens were exposed to air. The effect was most noticeable with thinly cut angel hair. Samples were placed quickly into large area glass cells and measured at a large aperture instrument port within 30 seconds of removal from the package.

4.3 Large Area Solid Foods (Table 3.3)

Sapers and Douglas (1987) measured enzymatic browning effects at cut surfaces of pears and apples, by cutting 22 mm diameter plugs from the samples using a cork borer and then placing the plugs on a glass cell over the instrument port. Browning could then be followed with time by measuring the L and a values.

A similar preparation technique has been utilized by Konstance et al. (1988) on beef chuck. In this case, samples were taken from 2.54 cm thick slices of chuck using a cork borer. The cylinders were then inserted in a previously bored rubber stopper placed on a glass cup. Components of the chuck meat such as lean, fat, bone and ligament could be separated in terms of ΔE and colour values.

Though this chapter is essentially on the subject of tristimulus colorimetry, it is important to recognize that occasionally reflectance spectrocolorimetry can be a useful technique for the measurement of colour changes in certain foods. One example is the colour change from red to brown occurring in meat due to oxidative/reductive reactions involving myoglobin. This colour change has a direct effect on the marketability of cut meat pieces. Eagerman et al. (1978) have followed the colour change with time in beef *semimembranosus* muscle. Approximately 1 cm thick slices were placed against the port of a spectrocolorimeter and the reflectance spectra measured between 600 and 650 nm. The formation of a depression in the region of the spectral curve around 632 nm is an indicator of brown metmyoglobin pigment production.

The use of transmittance colorimetry on solid materials is illustrated by the work of Freedman and Francis (1984) who have measured the colour of jellies. The jellies were heated sufficiently to pour into 2 cm light path cells. Lightness and hue angle were used to follow the effects of storage time, with or without the presence of ascorbic acid in the formulations.

Measurements on whole fruits have been carried out by Nagle et al. (1979) and Reeves (1987) for capsicums, Shewfelt et al. (1989) for fresh tomatoes and Lin et al. (1989) for apples. In all cases the fruits were measured directly on the instruments, taking a number of readings around the circumference of each specimen. High correlations were obtained between colour values of the whole fruit and extracted xanthophyll pigment in red peppers. Shewfelt et al. (1989) showed that for whole tomatoes there was a significant difference between colour values (colour difference ΔE, and difference in chroma $[a^2+b^2]^{1/2}$) due to cultivar, storage treatment and evaluation period. Simulated warehouse storage conditions were found to delay ripening colour changes.

Significant red colour loss and increase in browning were detected in modified atmosphere packaged apples by Lin et al. (1989). The unpackaged apples had more yellow colour loss than shrink-wrapped apples. The authors extracted three pigments from apple skin and showed that the highest correlation was obtained between cyanidin-3-galactoside and Hunter *a* values.

A useful technique for raw fresh cut potatoes has been employed by Baardseth et al. (1988). The pieces were wrapped in polyethylene film to retard browning and surface drying before placing on the instrument port; slices were then measured immediately. The L^* values were compared between two instruments. Significant correlations were obtained between results from the Minolta large measuring area instrument (CR-110) and the HunterLab large measuring area instrument, as well as between the small area Minolta (CR-100) and the HunterLab instrument.

Raw cauliflower was also measured by Baardseth et al. (1988). Measurements were made directly without extensive sample preparation using both large area measuring apertures on the Minolta (CR-110) and HunterLab units for comparison. The colour values obtained for cauliflower were found to overlap with mashed potato measurements, and the authors suggest that data from these two food commodities could be combined to develop a regression equation that would allow a researcher to predict lightness and yellowness from the onset of treatment of these two materials.

4.4 Pastes and Slurries (Table 3.4)

This group of materials offers the most problems in colour measurement terms, because of the influence of translucency and light trapping effects on the colour results.

Baardseth et al. (1988) reported that the prediction index for yam colour measurements was the least precise in the red/green parameter than for all the other foods evaluated in the paper, such as mashed potato, raw potato and raw cauliflower. The authors suggested that homogeneity was a key factor in obtaining the most precise prediction index, and that in the case of jams, therefore, variations in the fruit pieces may have been a factor in the loss of precision.

Nagle et al. (1979) homogenized pepper pods, the seeds and stems having been removed, prior to introduction of the sample into a cup provided with a white backing. Care was taken to allow the purées to settle; they were then vibrated to remove air bubbles. Good correlations were obtained between colour values and both xanthophyll and total pigment concentrations.

The influence of blueberry content on topping mix colour has been investigated by Francis (1985). Blueberries were added to topping formulations as a purée. The formulations were then pasteurized and cooled before conventional measurement in glass cells allowing 'infinite' sample thickness.

Clydesdale and Francis (1969) measured the colour of spinach purées formed by blanching fresh spinach, comminuting, deaerating and finally processing to F_0 4.9 in glass jars (F_0 being the time-temperature sterilization value, related to minutes at 121°C). Measurements were made on a tristimulus colorimeter 'hitched' to a grey colour tile standard.

Squash purées have been measured by Huang et al. (1970) using reflectance techniques in conjunction with the Kubelka-Munk concept. In this concept, reflectance measurements of thin layers of material are made with both white and black backgrounds. A reflectance value at infinite thickness can be calculated as well as ratio values for light scattering and absorption. In the work on squash purées, a tristimulus colorimeter was employed previously 'hitched' to a pink standard. Good correlations were achieved between visual ratings and colour values, and between visual ratings and Kubelka-Munk absorption/scattering ratios.

Silva et al. (1989) examined sweet potatoes prepared by various methods. All samples were mashed or puréed for chemical and physical analyses. The authors utilized an orange colour tile for 'hitching' standardization of a Hunter D25 instrument. Colour differentiation of the process variables in the *L*, *a*, hue angle and saturation or chroma measurements could be used, the authors suggested, in predicting the quality of these types of products. The lower *L*, *a* and saturation values of canned samples versus their frozen equivalents were indicative of the more extensive heating which the canned products received.

In an extensive study to evaluate the effects of packaging materials, light and storage on solid, whole milk yoghurt, Bosset et al. (1986) measured colour changes in the yoghurt in a specially designed 500 ml capacity cell for the Macbeth Spectrophotocolorimeter. Their results indicated that changes in the L, a and b colour values could be related to exposure to light through the various packages. In particular, the authors reported that Hunter a, b values were sensitive indicators of product changes due to photodegradation.

Finally, an exhaustive study of some of the variables affecting the measurement of tomato paste colour has been carried out by Brimelow (1987). The effects of such factors as paste dilution, test duration, instrument aperture size and illumination area, sample temperature, standardization hitching-post tile temperature, and the presence or absence of extraneous light were all examined. A standard colour measurement procedure was advocated for this important food commodity.

4.5 Liquids (Table 3.5)

Many clear or lightly translucent liquids are measured by transmission techniques, as already discussed. Thus Francis (1985) measured the colour of blueberry beverages in 2 cm cells, Calvi and Francis (1978) measured grape beverages in 1 cm cells, Skrede et al. (1985) measured blackcurrant syrups in 1 cm cells after diluting to 12° Brix, and Pilando et al. (1985) measured strawberry wine in 1 cm cells.

The study of Skrede et al. (1985) illustrates how colour measurement can be used to predict the shelf-life of a particular food material. Commercial blackcurrant syrups were packed in various packaging materials with different barrier properties. Storage lives varied between 2 months in high density polyethylene, 12 months in polyester and 18 months in glass. Combinations of Hunter L, a and b values were found to give acceptable predictive abilities for visual colour.

Francis (1985) found in the blueberry beverages that an increase of blueberry content from 5 to 12 per cent did not result in a corresponding increase in visual colour. Since the berries are highly pigmented, he surmised that a much greater increase in pigment would be required to achieve an observable change in the beverage colour. Calvi and Francis (1978) were able to correlate the anthocyanins content of the extracted concord grape filter trim with the a/L index of grape beverages at various storage temperatures. Pilando et al. (1985) found that both anthocyanin content and Hunter L values were highly correlated with sensory panel judgements when assessing the colour quality of strawberry wine.

Translucent liquids have been measured by Francis and Clydesdale (1972) (egg yolks), Desarzens et al. (1983) (liquid milk products), MacDougall (1987) (evaporated milk, orange juice) and others. These materials are measured by conventional reflectance techniques, ensuring usually that samples are at 'infinite' thickness.

Francis and Clydesdale (1972) urged that egg yolks should be homogenized before colour measurement. This is because yolks are formed in a layered manner and poultry feed variations can, therefore, cause colour variations within each yolk. Desarzens et al. (1983) had a large capacity sample cell especially made to use with the Macbeth MS-2000 for liquid milk products. In a study of several types of liquid milk products, the authors were able to establish a relationship between measured colour changes in the products and a decrease of riboflavin content during storage exposure to light. MacDougall (1987) has shown that instrumental variables, such as aperture size and illumination area, have a large effect on the colour values obtained for a number of translucent liquids.

5 Conclusions

It is clear that to be able to compare colour measurements on different samples of the same food materials, either within a laboratory or between different laboratories, the experimental procedure must be carefully defined and standardized.

Despite the obviousness of this statement, it is surprising how little information is given in much of the available literature on colour measurement, concerning the following essential areas:

- the choice of the viewing geometry and illumination type (which is associated with the selection of the instrument)
- the calibration of the instrument, including defining the calibration standards
- the selection of the instrument variables of aperture and illumination areas
- the selection, preparation and presentation of the samples
- the defining of other relevant details of the test procedure, such as the temperature.

Hunter (1987) has given a very useful checklist, in standardized format, which can be used as a reminder when documenting the variables in an experimental procedure for measuring the colour of a food material.

When at a loss as to how to proceed with a new problem in food colour measurement, it may be as well to keep in mind a number of pieces of good advice given by Francis and Clydesdale (1975):

1 To measure the visual impact of a food sample, the golden rule is to perform the measurement on the sample as the consumer sees it, with as little sample preparation as possible.
2 To predict the content of a chemical colour component in a food material

by colorimetry, it is best to ensure when setting up the predictive curve that the only variable is the colour component.

3 Having obtained a good correlation between a particular colour function and the visual judgement in one application it is not safe to assume, without further visual testing, that the colour function will give an equally good correlation in the case of another food material.

Acknowledgement

One of the authors, C. J. B. B. would like to acknowledge the contribution to colour measurement techniques of the Analytical Laboratory Section at Nestlé/Londreco Ltd., in particular M. Wilsch. Much of the practical advice in this chapter is based on the experiences of this excellent team.

References

André, L. and Pauli, P. (1978) The objective testing of dehydrated vermicellis with reference to colour and β-carotene content. *Hungarian Scientific Instruments*, **42,** 7–12

Baardseth, P., Skrede, G., Naes, T., Thomassen, M. S., Iversen, A. and Kaaber, L. (1988) A comparison of CIE (1976) $L^*a^*b^*$ values obtained from two different instruments on several food commodities. *Journal of Food Science*, **53** (6), 1737–1742

Bergthaller, W., Kempf, W. and Schaller, A. (1983) Instrumental measurement and sensory evaluation of the colour of dried potato cubes. *Internationale Zeitschrift fuer Lebensmittel-Technologie und Verfahrenstecknik*, **34** (3), 174, 177–178, 180, 182–183.

Berset, C. and Caniaux, P. (1983) Relationship between color evaluation and chlorophyllian pigment content in dried parsley leaves. *Journal of Food Science*, **48,** 1854–1857, 1877

Bilmeyer, Jr., F. W. (1981) In: *Golden Jubilee of Colour in the CIE* (Commission Internationale de l'Eclairage), Proceedings of a Symposium held by The Colour Group (GB), Imperial College, London

Bosset, J. O., Daget, N., Desarzens, C., Dieffenbacher, A., Fluckiger, E., Lavanchy, P., Nick, B., Pauchard, J. P. and Tagliaferri, E. (1986) The influence of light transmittance and gas permeability of various packing materials on the quality of whole natural yoghurt during storage. In *Food Packaging and Preservation: Theory and Practice*, ed. M. Mathlouthi, 235–270, Elsevier, London

Brimelow, C. J. B. (1987) Measurement of tomato paste colour: investigation of some method variables. In *Physical Properties of Foods*, vol. 2, eds R. Jowitt, F. Escher, M. Kent, B. McKenna and M. Roques, 295–317, Elsevier, London

Calvi, J. P. and Francis, F. J. (1978) Stability of concord grape (*V. labrusca*) anthocyanins in model systems. *Journal of Food Science*, **43** (5), 1448–1456

Christensen, C. M. (1983) Effects of color on aroma, flavor and texture judgements of foods. *Journal of Food Science*, **48,** 787–790

CIE (1931) *Proceedings of the Eighth Session*, Cambridge, England, Bureau Central de la CIE, Paris

CIE (1979) *Colorimetry*. CIE Publication 15 (E–1.3.1), Bureau Central de la CIE, Paris

Clydesdale, F. M. (1977) Color measurement. In *Current Aspects of Food Colorants*, ed. T. E. Furia, 1–17, CRC, West Palm Beach, FL

Clydesdale, F. M. and Francis, F. J. (1969) Colorimetry of foods. 1: Correlation of raw transformed and reduced data with visual rankings for spinach purée. *Journal of Food Science*, **34,** 349–352

Desarzens, C., Bosset, J. O. and Blanc, B. (1983) La photo dégradation de lait et de quelques produits laitiers. Part I: Altérations de la couleur, de goût et de la teneur en quelques vitamines. *Lebensmittel-Wissenschaft und Technologie*, **17,** 241–247

Eagerman, B. A., Clydesdale, F. M. and Francis, F. J. (1978) A rapid reflectance procedure for following myoglobin oxidative or reductive changes in intact beef muscle. *Journal of Food Science*, **43,** 468–469

Francis, F. J. (1952) A method for measuring the skin colour of apples. *Proceedings of the American Society of Horticultural Science*, **60**, 213–220

Francis, F. J. (1985) Blueberries as a colorant ingredient in food products. *Journal of Food Science*, **50,** 754–756

Francis, F. J. and Clydesdale, F. M. (1972) Colour measurement of foods. XXXV: Miscellaneous. Part V: egg yolks. *Food Products Development*, Aug./Sept., 47, 48, 55

Francis, F. J. and Clydesdale, F. M. (1975) *Food Colorimetry: Theory and Applications*, Chapter 1. AVI, Westport, CT

Freedman, L. and Francis, F. J. (1984) Effect of ascorbic acid on color of jellies. *Journal of Food Science*, **49,** 1212–1213

Grieder, H. (1989) Personal communication

Habib, A. T. and Brown, H. D. (1956) Factors influencing the color of potato chips. *Food Technology*, **7,** 332–336

Huang, I. L., Francis, F. J. and Clydesdale, F. M. (1970) Colorimetry of foods. 2: Color measurement of squash purée using the Kubelka-Munk concept. *Journal of Food Science*, **35,** 315–317

Huey, S. J. (1965) Low temperature storage of color standards panels. *Color Engineering*, **3,** Sept./Oct., 24–27

Hunter, R. S. (1958) Photoelectric color difference meter. *Journal of the Optical Society of America*, **48,** 985–995

Hunter, R. S. (1987) Objective methods for food appearance assessment. In *Objective Methods in Food Quality Assessment*, ed. J. G. Kapsalis, 137–153, CRC, Boca Raton, FL

Hunter, R. S. and Christie, J. S. (1978) Geometrical factors in colour evaluation of purées, pastes and granular food specimens. *Journal of Food Protection*, **41** (9), 726–729

Hunter, R. S. and Harold, R. W. (1987) *The Measurement of Appearance*. Wiley, New York

Johnston, R. A., Quick, J. S. and Donnelly, B. J. (1980) Note on comparison of pigment extraction and reflectance colorimeter methods for evaluating semolina colour. *Cereal Chemistry*, **57** (6), 447–448

Judd, D. B. (1939) Specification of uniform color tolerances for textiles. *Textile Research*, **IX** (7), 171 (8).

Kent, M. and Smith, G. C. (1987) Collaborative experiments in colour measurement. In *Physical Properties of Foods*, vol. 2, eds R. Jowitt, F. Escher, M. Kent, B. McKenna and M. Roques, 251–276, Elsevier, London

Konstance, R. P., Heiland, W. K. and Craig, Jr., J. C. (1988) Component recognition in beef chuck using colorimetric determination. *Journal of Food Science*, **53** (3), 971–972

Lee, C. Y., Smith, N. L. and Hawbecker, D. E. (1988) Enzyme activity and quality of frozen green beans as affected by blanching and storage. *Journal of Food Quality*, **11** (4), 279–287

Lin, T. Y., Kohler, P. E. and Shewfelt, R. L. (1989) Stability of anthocyanins in the skin of starkrimson apples stored unpackaged, under heat shrinkable wrap and in-package modified atmosphere. *Journal of Food Science*, **54** (2), 405–407

Little, A. C. and MacKinney, G. (1956) On the color of coffee. *Food Technology*, **10,** 503–506

Mabon, T. (1989) BYK-Gardner. Personal communication

MacDougall, D. B. (1982) Changes in the colour and opacity of meat. *Food Chemistry*, **9,** 75–88

MacDougall, D. B. (1983) Instrumental assessment of the appearance of foods. In *Sensory Quality in Foods and Beverages: its Definition, Measurement and Control*, eds A. A. Williams and R. K. Atkin, 121–139, Ellis Horwood, Chichester

MacDougall, D. B. (1987) Optical measurements and visual assessment of translucent foods. In *Physical Properties of Foods*, vol. 2, eds R. Jowitt, F. Escher, M. Kent, B. McKenna and M. Roques, 319–330, Elsevier, London

Miller, O. H. and Burns, E. E. (1971) Internal color of Spanish peanut hulls as an index of kernel maturity. *Journal of Food Science*, **36,** 669–670

Nagle, B. J., Villanon, B. and Burns, E. E. (1979) Color evaluation of selected capsicums. *Journal of Food Science*, **44** (2), 416–418

Pangborn, R. H. (1967) Some aspects of chemo-reception in human nutrition. In *The Chemical Senses and Nutrition*, eds M. R. Kane and O. Maller, Chapter 4, John Hopkins, Baltimore

Pilando, L. S., Wrolstad, R. E. and Heatherbell, D. A. (1985) Influence of fruit composition, maturity and mold contamination on the color and appearance of strawberry wine. *Journal of Food Science*, **50,** 1121–1125

Reeves, M. J. (1987) Re-evaluation of capsicum color data. Journal of Food Science, **52** (4), 1047–1049

Sapers, G. M. and Douglas, Jr., F. W. (1987) Measurement of enzymatic browning at cut surfaces and in juice of raw apples and pear fruits. *Journal of Food Science*, **52** (5), 1258–1262, 1285

Seakins, J. M. (1971) Colour evaluation of transparent dairy products in powder form. *New Zealand Journal of Dairy Science and Technology*, **6** (1), 24–25

Shewfelt, R. L., Brecht, J. K., Beverly, R. B. and Garner, J. C. (1989) Modifications of conditions at wholesale warehouse to improve quality of fresh-market tomatoes. *Journal of Food Quality*, **11** (5), 397–409

Silva, J. L., Yazid, M. D., Ali, M. D. and Ammerman, G. R. (1989) Effect of processing method on products made from sweet potato mash. *Journal of Food Quality*, **11** (5), 387–396

Skrede, G. (1985) Color quality of blackcurrant syrups during storage evaluated by Hunter *L, a, b* values. *Journal of Food Science*, **50,** 514–517, 525

Wright, W. D. (1969) *The Measurement of Colour*, 4th edn. Van Nostrand Reinhold, London

4 Sorting by Colour in the Food Industry

J. M. Low and W. S. Maughan

Contents

1 Introduction

Food has colour, and good food is often a different colour from bad food. This in itself is not a very remarkable statement but its implications for the food industry are very significant. The human perception of colour has proved very effective in the determination of quality in many foods, and the sorting of food particles using the human eye and hand has been, and still is, widely practised.

Machines designed to replace the hand sorter obviously need to duplicate some of the functions of the human eye, brain and hand. Automatic colour sorting is often referred to as 'electronic sorting', as it was the advent of electronics that made possible the duplication of the eye and brain functions.

Experiments with automatic sorting equipment started immediately after World War II and colour sorting machines gained early acceptance in the food industry. Demand for automatic sorting machines continues to increase owing to the escalating costs of hand sorting coupled with the higher quality requirements being imposed on food producers. There is also an increasing realization of the importance of sorting in the reduction of health hazards arising from contaminated food, whether due to the food itself or the presence of foreign bodies such as stones or wood, and the avoidance of associated product liability claims which are escalating dramatically in both number and value.

2 What is a Sorting Machine?

Colour sorters generally consist of four principal systems, as shown in Figure 4.1:

1 A *feed system*, which presents food particles to the subsequent systems in a controlled manner. The illustration shows a feed system which involves a conveyor belt to align the particles and ensure that they pass through the machine at a constant velocity. However, feed systems can also be based upon inclined chutes, contra-rotating rollers and different combinations of these features.
2 An *inspection system*, which measures the reflectivity of each particle. The inspection components are housed in an optical chamber, through which or past which the particles travel. Particles do not come into contact with any part of the optical chamber and are separated from it by glass screens or windows. The optical chamber contains one or more lens and detector units, depending on the number of directions from which the particles are viewed. Lamp units, designed to provide even

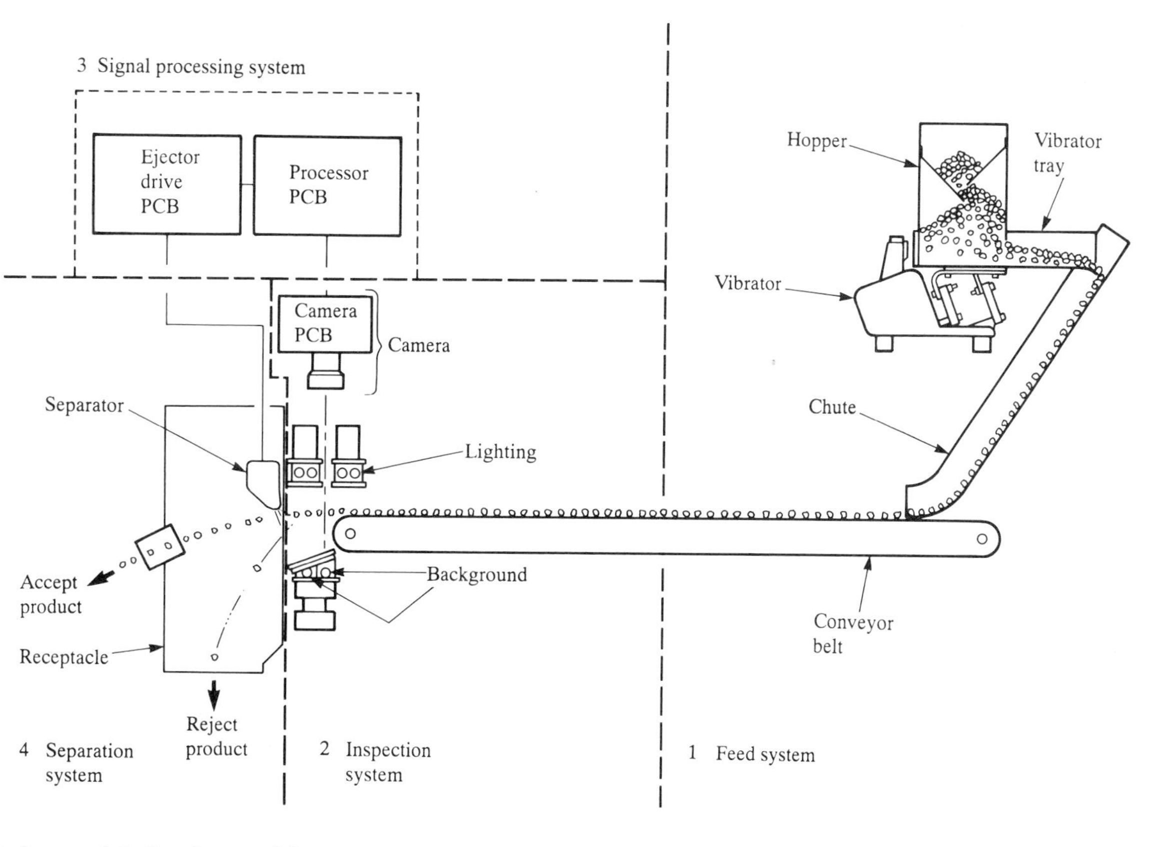

Figure 4.1 Sorting machine systems

and consistent illumination of particles, are also usually contained within the optical chamber.

3 A *signal processing system*, in which electrical signals from the detector(s) are amplified and fed to decision making circuitry which classifies particles as either 'accept' or 'reject'.

4 A *separation system*, capable of physically separating particles classified as reject from those classified as accept. Usually, separation takes place while the particles are in free fall; accept particles are allowed to continue along their normal trajectory, and rejects are deflected from it into a receptacle. Deflection is usually achieved by emitting short bursts of compressed air through nozzles aimed directly at the rejects, although large or heavy particles may require some sort of piston-operated device to deflect the rejects.

The size, cost and complexity of sorting machines varies depending on the size range of particles to be handled, the throughput requirement and the complexity of optical measurement.

Machines are employed in sorting particles as small as mustard seeds; however, rice grains are among the smallest particles to be sorted on a large commercial scale. At the other end of the size range are vegetables and fruit such as potatoes and peaches.

Seeds are usually sorted on a single- or double-chute machine at a throughput of typically 12 kg/hour. A higher throughput can be achieved on a multi-chute or a conveyor belt machine; rice at 4 tonnes/hour and peas at 10 tonnes/hour are typical examples.

The simplest type of optical measurement is taken from one direction and is of a single wavelength band. More complicated optical measurements involve two or more wavebands and may be made from several directions around the particle.

The products that can be handled by today's automatic sorting machines include seeds, coffee, rice, breakfast cereals, nuts and pulses; fresh, frozen and dehydrated vegetables; cherries and tomatoes; and biscuits and confectionery. Foreign material such as stones, sticks and organic matter can be removed, as well as particles with defects such as discoloration and damaged skin.

Figures 4.2 and 4.3 show two typical sorting machines.

3 Assessment of Food Particles for Colour Sorting

3.1 Spectrophotometry

To determine whether a particulate food product is suitable for colour sorting, and which type of sorting machine and optical configuration is

Figure 4.2 *A typical chute machine*

Figure 4.3 *A typical broad belt machine*

most suitable, samples of both acceptable and unacceptable produce must be measured and assessed in the laboratory.

Actually the term 'colour sorting' is a misnomer, possibly arising from the effect on the product of sorting. The criterion used when assessing food particles for sorting, and what the sorting machine measures when it inspects the product, is spectral reflectivity at particular wavelengths, rather than the colour as a whole. Figures 4.4 and 4.5 illustrate typical spectral curves obtained from a product. Intensity varies from black (0) to white (100), and wavelength covers the visible spectrum (400 to 700 nm) and extends into the near infrared. Ideally there will be a part of the spectrum where intensity values for all acceptable produce are either higher or lower than values for all unacceptable material; if so, this part of the spectrum can be used as a basis for sorting. In practice the spectrum is normally restricted to a narrow waveband in the sorting machine by the use of optical filters.

Conventional spectrophotometry in the food industry involves the measurement of carefully prepared surfaces under standard optical conditions and illumination. However, practical industrial sorting machines must deal with naturally occurring surfaces viewed under non-ideal light conditions. To obtain the best possible result for a given product, special computerized reflection spectrophotometers have been developed which enable the appropriate characteristics of the naturally occurring surfaces to be measured. Diffuse spherical broadband lighting is used to uniformly

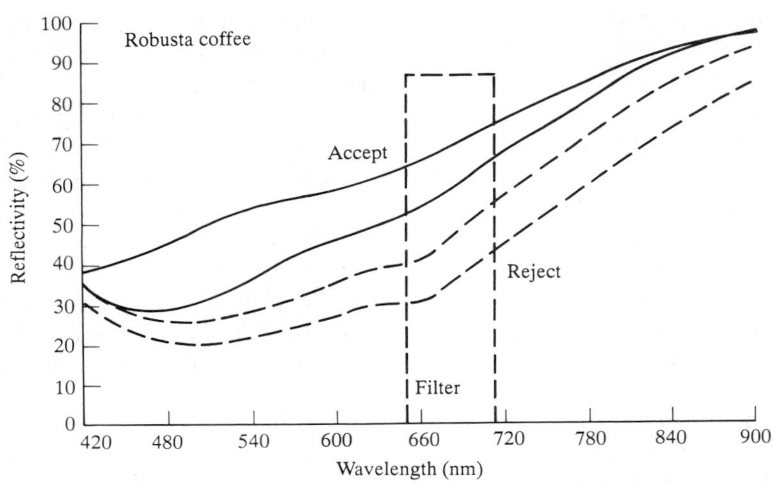

Figure 4.4 *Spectral curves – monochromatic sorting; response curves for —— 'accept' samples, --- 'reject' samples*

illuminate the particle under test. The reflected light is then passed through a computer controlled scanning monochromator which splits the light into its constituent wavelengths.

The output is measured using a suitable detector and the data are fed into the computer. When the equipment is appropriately calibrated, the results can be plotted showing the variation of reflectivity with wavelength for both acceptable and defective product.

3.2 Monochromatic Sorting

Sorting that uses the form of optical measurement described in the preceding section is known as monochromatic sorting. It is so called because it is based on the measurement of reflectivity in a single isolated band of wavelengths. For it to be effective, there must be a distinct difference in reflectivity within the selected waveband between all the acceptable particles and all the reject particles. See Figure 4.4.

Removing dark, rotten peanuts and removing black peck from rice are typical applications of monochromatic sorting.

3.3 Bichromatic Sorting

Unfortunately, it is not always possible to find a single section of the spectrum where the intensity levels of accept and reject material are clearly separated, and in these cases it is necessary to resort to a more complicated procedure. This is known as bichromatic sorting, and involves measurement in two different wavelength bands. Figure 4.5 shows two sets of spectral response curves obtained from green arabica coffee. One set of curves (solid curves) represents the lightest and darkest of acceptable beans, the other (dashed curves) represents the lightest and darkest of discoloured beans; clearly, there is no wavelength where there is a complete separation of the two sets of curves. However, it can be seen that between 510 nm and 660 nm the difference in the gradients of the two sets of curves is at its greatest. If measurement A is taken at 540 nm and measurement B is taken at 660 nm, and the ratio A:B is calculated, then this will give an unambiguous separation of the two particle types. (In principal, measurement A could be taken at 510 nm, but this would yield a lower signal intensity.)

Bichromatic sorting involves duplication of optical and detection components, the addition of light splitting devices and more complex signal processing. Therefore it is used only when a simple monochromatic measurement is unable to give a satisfactory result.

3.4 Dual Monochromatic Sorting

A third type of measurement, similar to bichromatic in that two wavebands are observed but without employing the ratio of the two measurements, is known as dual monochromatic. As its name implies, this system sorts monochromatically in each of two separate wavebands. This type of sorting is used when it is necessary to reject two types of defect or, for instance,

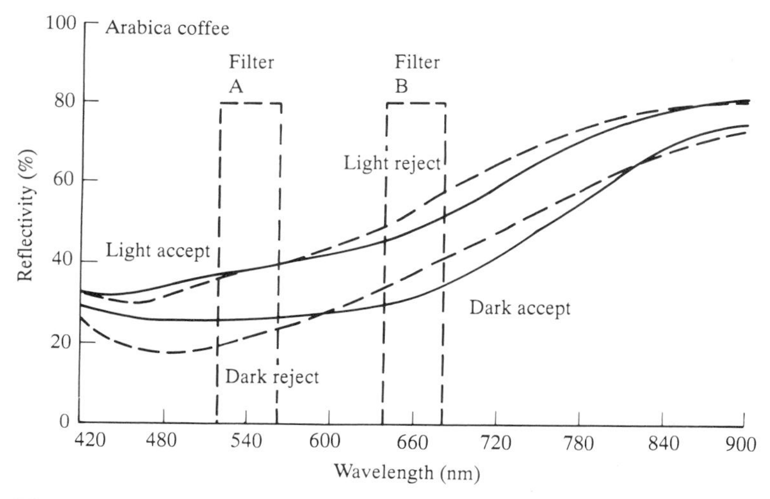

Figure 4.5 *Spectral curves – bichromatic sorting; response curves for —— 'accept' samples, --- 'reject' samples*

defects and foreign material, each of which exhibit different spectral characteristics.

An example of dual monochromatic sorting is found with white beans: maize is rejected by detecting blue reflectance, and white stones are rejected using near infrared.

Some practical separations require both monochromatic and bichromatic decisions for the successful removal of all defects. Therefore automatic sorting machines are available which are capable of making both types of measurement simultaneously. The technique is clearly extendable to trichromatic measurement, but few practical machines have been introduced with this feature. However, colour television cameras with fixed red, green and blue filters (RGB) have been used for special applications.

3.5 Fluorescence Techniques

Of course, not all bad food is a different colour from good food. It has been found that certain non-visible defects (bacteria for example) fluoresce when irradiated with long-wave ultraviolet light (350 nm), and this property may be used as a basis for sorting.

This technique was originally developed for removing 'stinkers' from green arabica coffee beans, but has found applications in sorting peanuts, almonds and cranberries.

3.6 Near Infrared Techniques

Recently the wavelength range used by sorting machines has been extended further into the infrared region where water absorption and other chemical effects play an important part in determining the reflectivity characteristics of food particles.

Bichromatic infrared machines are proving particularly effective in removing shell fragments from a variety of tree nuts.

4 The Optical Inspection System

4.1 Introduction

At any given wavelength, the characteristics of the electrical signal from the detector will depend on the particle reflectivity and size, the light source, the properties of any filters fitted and the nature of the detector itself. The range of wavelengths measured by the machine is defined by the choice of these last three elements.

Traditionally the light source and detector in a given machine were fixed, but the filters could be changed to accommodate different produce. In modern equipment it is often possible to fit lighting with different spectral characteristics and even to change the type of detector, as well as easily exchange filters.

Once the spectral feature or features which are to be used as a basis for separation have been identified, the relevant wavelength bands must be isolated by selection of appropriate filters and illumination. A primary objective of selecting filters and lighting is to obtain the maximum possible signal to noise ratio from the detector at the required wavelengths, and the minimum possible signal at all other wavelengths.

4.2 Illumination

When dealing with irregularly shaped particles, uniform diffuse illumination is necessary to minimize the occurrence of highlights and shadows, as these would obviously detract from the measurement of true surface reflectivity. Ideally, at the point of measurement, the particle should be surrounded by a spherical surface of uniform brightness; in practice, however, a number of factors preclude the attainment of this ideal. Firstly, if the inspection chamber is designed so that the product particles pass through it, there must be provision for entry and exit ports. Secondly, the position of the optical components will result in areas of different brightness compared with the main chamber wall. Thirdly, the use of light sources of finite size leads to non-uniform illumination.

A further problem that can arise, even with a perfect diffuse illumination sphere, is that of specular reflection. If a particle with a diffuse reflective surface is placed in such a sphere then its true colour will be observed. However, if the particle surface is not diffuse, specular reflection will occur, giving highlights which do not exhibit the true colour of the surface. Clearly the highlights can adversely affect the optical system and hence result in the incorrect classification of a particle.

In practice, illumination is provided by either fluorescent tubes or incandescent filament bulbs, or a combination of both. A number of lamps are arranged to give as uniform a distribution of light as possible; with incandescent lamps, screens are often positioned to diffuse the high-intensity point of light emitted by the filament.

Fluorescent tubes can be produced with different spectral characteristics, extending from the ultraviolet to the far red, as shown in Figure 4.6. The advantages of the fluorescent tube are its cool operation, long life and diffuse light. Its disadvantages are low red emission and the need for a special power supply to prevent flicker.

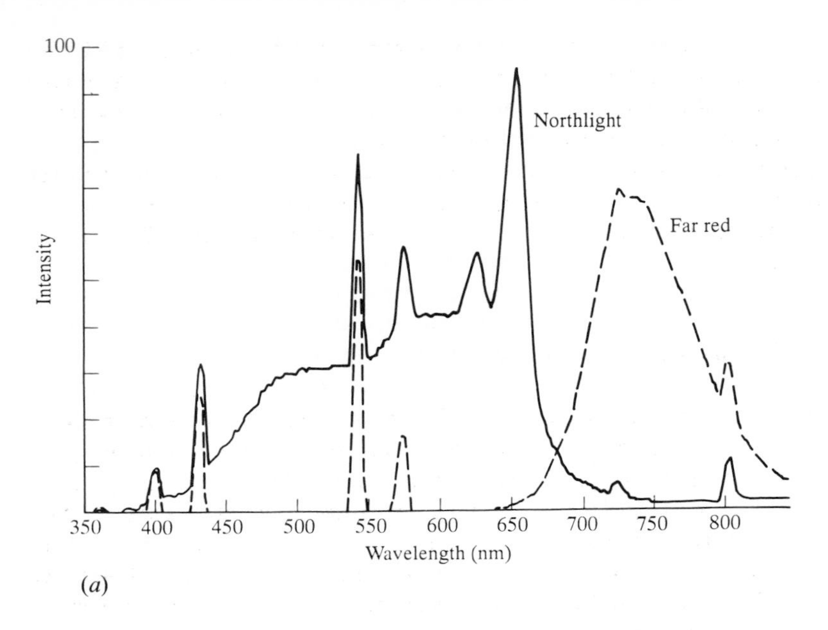

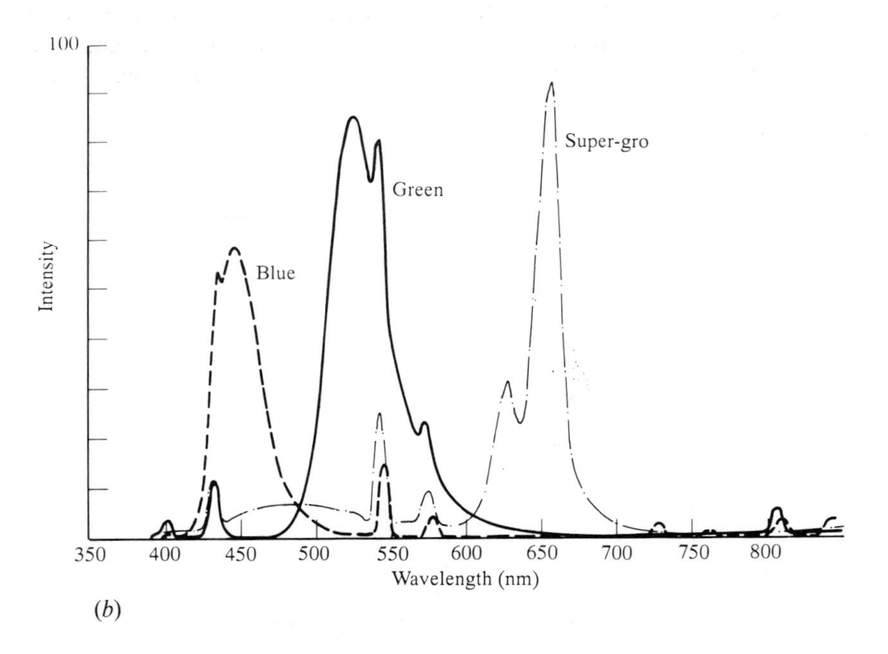

Figure 4.6 *Spectral characteristics of fluorescent tubes*

Incandescent lamps inherently have a wide spectral range from blue to the near infrared, as shown in Figure 4.7. These lamps have the advantage of wide spectral emission and DC operation, but suffer from being point sources which dissipate large amounts of heat.

In general, the use of fluorescent tubes is favoured except in cases where a deep red or near infrared measurement is required. The wider spectral range required for bichromatic machines necessitates the use of incandescent lamps.

4.3 Background and Aperture

Having arranged for the particle to be illuminated in the optimum manner, it is necessary to consider the best optical arrangements for obtaining the necessary data. The simplest form of inspection system views the particles against an illuminated background and through a small aperture. The brightness of the background is adjusted so that the optical system measures the same average value with or without product. This is known as a 'matched' background because it matches the average brightness of product, including any defects.

The advantage of matching the background is that measurement of reflectivity is independent of particle size. For example, consider the case of a stream of particles containing rejects which are darker than accepts. If a matched background is used then whenever a defect passes across the

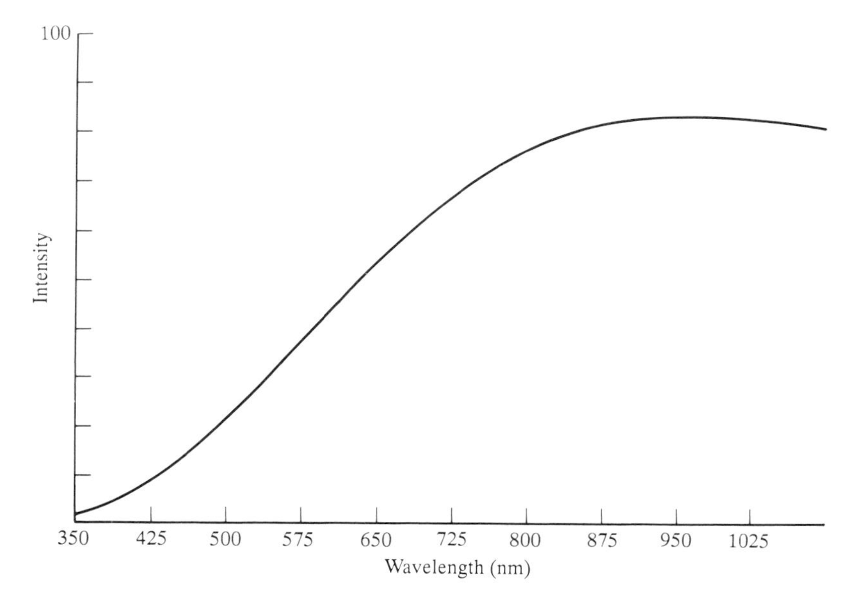

Figure 4.7 *Spectral characteristics of an incandescent lamp*

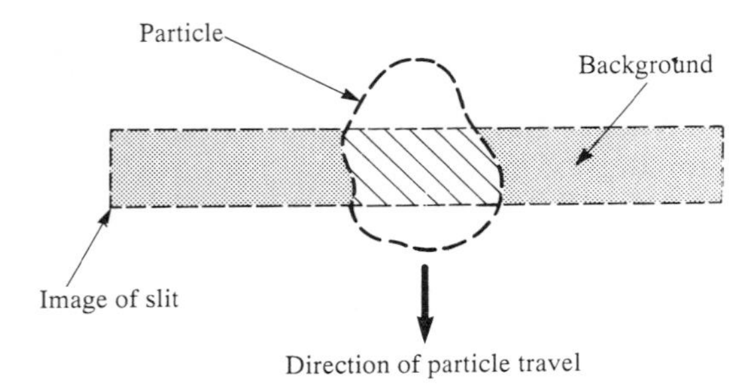

Figure 4.8 *Viewing aperture*

aperture a decrease in signal amplitude will result, but a light particle will cause an increase in amplitude; thus an unambiguous decision can be made by the electronics. If, on the other hand, a background lighter than all particles were to be used, then all particles would give a decrease in signal; in particular, small dark defects could give signals identical to those of large light particles and hence the two could not be separated. Note that the brightness of the light reflected from a particle through the aperture is the product of the size of the particle and its reflectivity, including any area of discoloration.

The background may be either an opaque surface reflecting light from suitably located lamps, or a translucent surface transmitting light from rear-mounted lamps or LEDs.

The aperture takes the form of a rectangular slit, as shown in Figure 4.8. The width of the slit must be sufficient to allow for scatter in the trajectories of the particles and for the range of particle sizes. The height of the slit is kept to a minimum, consistent with sufficient signal, in order to give maximum resolution and for accurate timing of the delay between detection of a defect and rejection of a particle.

4.4 Optical Filters

A wide range of coloured glass and interference optical filters is available, making it a simple matter to select suitable filters. Three basic classes of filter are used:

- low pass, transmitting only below a certain wavelength
- high pass, transmitting only above a certain wavelength
- bandpass, transmitting only within a band of wavelengths

Figure 4.9 illustrates the response of typical filters used in sorting machines.

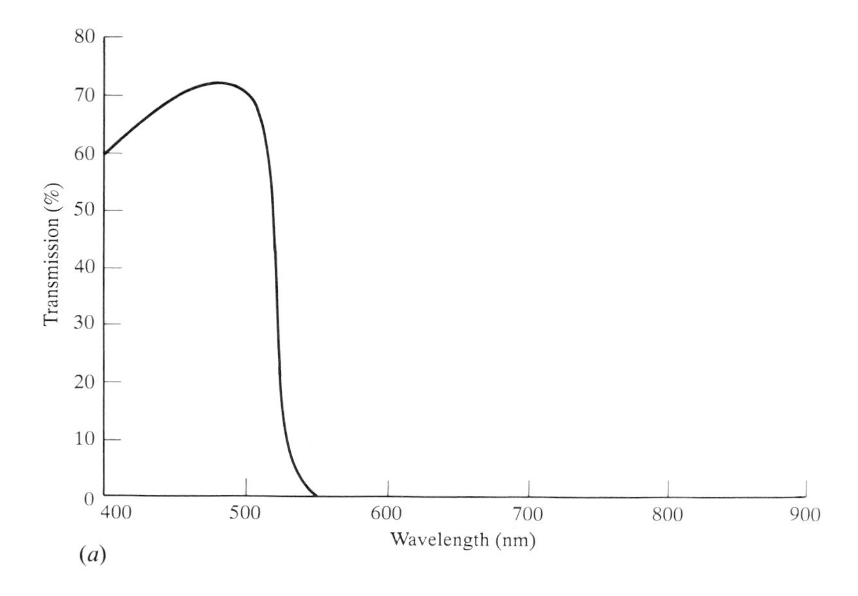

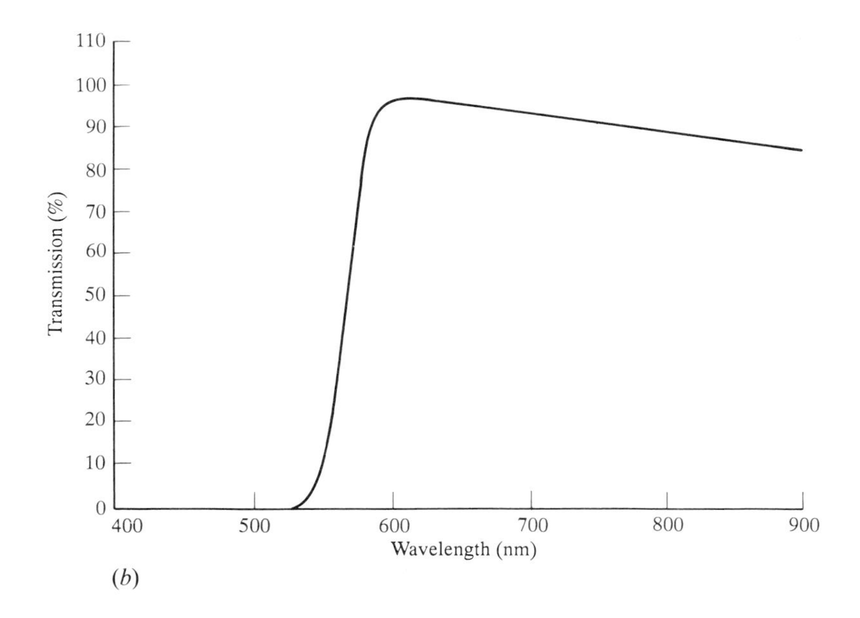

Figure 4.9 *(a) Low-pass filter (b) high-pass filter*

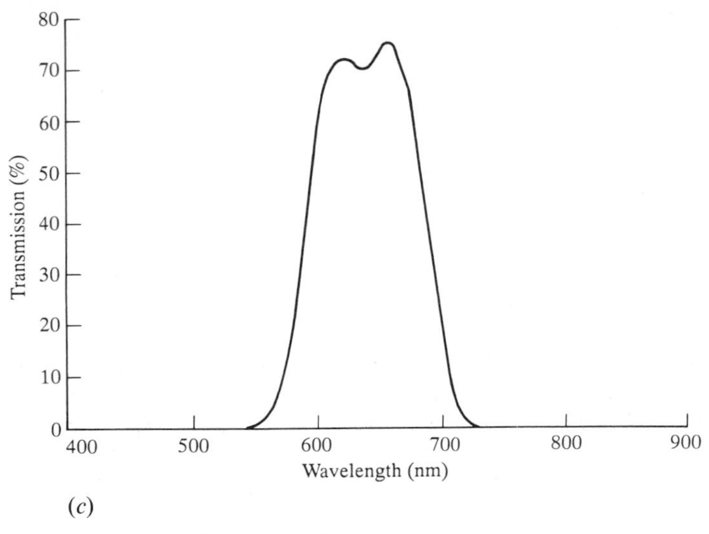

Figure 4.9 *(c) bandpass filter*

4.5 Detectors

For a number of years the photomultiplier tube was the best detector of visible radiation. It has a good signal to noise ratio which enables detection of low light levels, and a satisfactory blue response. However, it has the disadvantages of fragile mechanical construction, limited life, high operating voltage and poor deep red and near infrared response.

Most modern colour sorting machines use a solid-state photodiode as the detector. The photodiode has the advantages of comparative cheapness, mechanical robustness and an indefinite life. However, it has a poor blue response when compared with the photomultiplier, which is still used for critical blue or ultraviolet measurement.

The use of a single detector collecting the light from a slit view (see Section 4.3) results in an optical resolution which is inferior along the length of the slit compared with that across the slit. In order to overcome this shortcoming, the slit view can be split into a number of independent views called pixels. This can be achieved by using a detector consisting of a linear array of photodiodes each of which gives an output signal which can be processed to give a sorting decision.

When the number of pixels required becomes significant, for example when viewing a wide conveyor belt, a linear CCD array can be used. This device incorporates internal multiplexing circuits which result in a single output signal which is composed of the signals from all the pixels in a sequential manner. This type of output is particularly suitable for microprocessor-based machines.

5 Completing the Sorting System

5.1 Feed

A product feeding system in a sorting machine has to provide three functions:

- Metering, to ensure that the optimum number of particles per unit time is presented to the inspection system.
- Acceleration to a constant presentation speed. The time taken for particles to travel from the inspection point to the separation point must be constant in order for the operation of the separation device to be accurately timed. Presentation speeds can be as high as 4 m/s and the delay between detection and separation lies between 0.5 and 100 ms.
- Alignment, to ensure a controlled trajectory through the inspection and separation points.

In practice, the metering requirement is met by a vibrating tray feeder mounted below the output of a holding hopper. A variety of systems have been developed for providing the acceleration and alignment requirements, and the following are commonly used:

- inclined chute
- narrow grooved belt
- contra-rotating rollers
- flat belt single layer.

The first three methods above channel and separate the particles, allowing an all-round view of their surface. The flat belt system presents the particles in a single layer, restricting the view but enabling a much higher throughput of product to be achieved. Throughput is increased in channel feed systems by providing two or more channels.

With conveyor belt feed systems the inspection components may be mounted above the belt, with the belt serving as the background, or after the end of the belt so that the product is viewed in free fall as in channel systems. The latter arrangement also enables the product to be viewed from below as well as above.

5.2 Separation

The usual form of separating device is the high-speed solenoid valve which releases short bursts of compressed air through a nozzle. These devices exhibit the essential features of rapid action, reliability and mechanical strength; the fastest operates at a frequency of 1 kHz.

Because the action of the air blast on a reject could cause dust particles and skin fragments to be blown around, it is usual to arrange for the separation point to be outside the inspection area so as to minimize the effect on the optics. The necessary time delay between inspection and separation is generated by the electronic circuits.

For certain large or heavy objects a solenoid valve may be used to control a pneumatically operated flap or plunger to deflect the reject. Specialized rejectors for product pulp remove rejects by suction and are mounted above a flat belt, downstream from the inspection unit.

5.3 Cleaning and Dust Extraction

The application of optical techniques in a dusty, dirty or wet industrial environment poses major difficulties for designers. Considerable effort has gone into designing sorting machines suitable for use in such conditions.

To protect the optical components from dirt or moisture, they are separated and sealed from the inspection point through which product passes by a glass window. The position of this window in the optical path should be such that any small particles which may settle on the surface are out of focus and therefore cause the minimum of disturbance to the optical signal.

Nevertheless it is essential that this window is kept as clear as possible and a number of facilities to achieve this may be provided on the machine.

Firstly, the product being fed to the machine should be as dry and dust free as possible, however the action of storing it in a hopper and feeding it on a vibrating tray may well create some dust and therefore a dust extraction nozzle may be fitted at the end of the vibrating tray. In the case of a chute feed, the top of the chute may be perforated so that air can be passed through the product stream to remove dust particles.

Secondly, the actual window can be cleaned by means of compressed air jets which (a) provide a continuous curtain of air to prevent particles settling on the surface and/or (b) a periodic high pressure blast which removes any particles that have managed to reach the window.

Thirdly, the window may be periodically wiped by a pneumatically actuated blade or brush. In more advanced machines this may be combined with an air blowdown facility.

Finally any dirt created by the action of the ejector blast on the particles, may be drawn away from the window area by the provision of a dust extraction nozzle below the optical box.

Similar precautions may be taken where a wet product is to be sorted, with water jets and wiper blades being fitted to the optical window.

5.4 Electronics Systems

The electronic systems in sorting machines have progressed from the simple analogue circuits of the early machines to the sophisticated microprocessor based circuits found in the present generation of machines.

The microprocessor machine exhibits two important advantages over its predecessors in that it is very much easier to set up and that it will maintain a consistent level of performance without operator intervention.

Most of the setting up of the sorting parameters can be done by the machine itself including in some cases the ability of the machine to 'learn' the differences between good and bad product. However, the operator is always given the opportunity to fine-tune the final result.

If required the machine will track the average colour of the product so that, even though the product colour may change with time, the machine will continue to remove only the predefined abnormal particles. Alternatively the machine may be provided with a calibration plate which is periodically placed in the optical view so that the machine is able to correct for any measurement drift that has occurred. These facilities result in a much more stable performance when compared with that of machines which require periodic human attention.

Once a machine has been set up for a particular product then all the settings can be stored in memory. This can be repeated for a number of different products and then, at a later time, the machine can be made ready to sort any of these products simply by recalling the appropriate settings from the memory.

As the machine has a memory capability this can be utilized to provide useful information, such as the number of rejects that have occurred in a certain time, to the operator and where appropriate to the plant monitoring computer. Information can also be provided to assist with preventative maintenance.

Because of the greatly increased signal processing available, the sorting decision can be based on other parameters as well as reflectivity. For example the minimum size of the discoloration necessary for a particle to be classed as defect can be defined.

As well as providing signals to external monitoring devices the machines can be controlled via a computer link from a remote position.

5.4.1 User Interfaces

Microprocessor control and visual display devices have greatly simplified the procedure for setting up and operating sorting machines, as well as enabling more comprehensive fault information to be displayed to the user.

A typical machine will have a keypad and a display unit (Figure 4.10) which will enable the operator to set up and control the machine by means of a series of menus. In addition the display unit will provide the operator

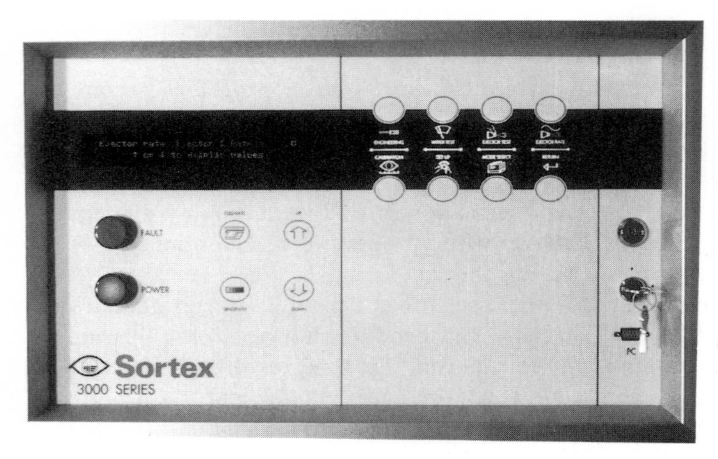

Figure 4.10 Keypad and display panel

with information regarding the settings of the machine while it is sorting, together with details of any faults that may occur.

5.4.2 Mapping Techniques

The microprocessor-based machine is able to utilize a more sophisticated method of defining the separation of reject product from accept.

The conventional bichromatic sorting machine using two wavelength bands, say green and red, made a decision based on the ratio of the two signals as well as the intensity of the individual signals. The situation can be represented by a two dimensional map as shown in Figure 4.11(a). Clearly, the defined accept region is triangular in shape whilst the actual region is not and therefore it is impossible to set the machine without some accept or reject product being on the wrong side of the lines.

However, with the microprocessor machine it is possible to define the boundary between accept and reject to match the true shape of the actual product colour distribution as shown in Figure 4.11(b). This method enables the microprocessor machine to remove a far greater range of defects, more accurately, without the penalty of removing a large amount of accept product.

6 Future Trends: Computer Vision Systems

Computer vision systems are being used increasingly in general manufacturing, but the demands of the food industry are generally much greater. At

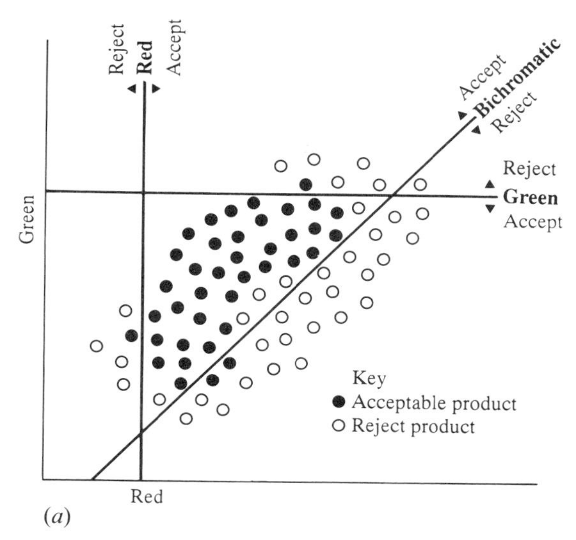

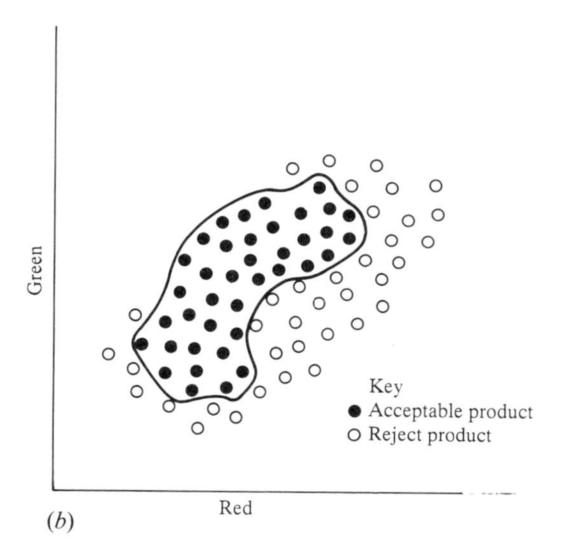

Figure 4.11 *(a) Conventional map (b) microprocessor map*

present, there is only a limited range of computer vision equipment available for use in the food industry. However, in the future this is likely to change.

Two factors are limiting the rate at which computer vision systems are being introduced to the food industry:

1 The data processing rates required in a high-speed food sorting machine are very much higher than those in a similar inspection machine for manufactured items.
2 The development of improved materials handling and separation systems is not keeping pace with the dramatic advances being made in computer technology.

The computerized sorting system of the future may comprise a TV camera and a computer with a large capacity for information storage. Some form of interface would connect the computer to a series of robot graders and sorters. Items to be inspected and sorted would pass along a belt where they would be viewed by the camera. The computer would identify the individual items and could assess them according to several criteria, including colour, shape and size. Devices would then be actuated to separate the items into size and quality grades.

A computer vision system potentially offers many benefits over a conventional colour sorter. The ability to sort food on the basis of several different criteria would be a primary advantage. Flexibility is another distinct advantage of such a system. A machine which is busy sorting potatoes could easily be reprogrammed to examine oranges, or indeed anything else which could pass along the conveyor. Different products could be accommodated by simple software changes instead of major adaptations to hardware. However, for the immediate future, the most likely application of advances in computer technology is the improvement in performance of the present generation of sorting machines.

7 Using a Colour Sorter

There is a common misconception that a colour sorter can remove all defects from a given batch of product. In practice this is impossible. A colour sorter will reduce the concentration of defective product, but it can never be 100 per cent effective.

All colour sorters are bound to remove some acceptable product and fail to remove some of the defective product. There are a number of possible causes for this limitation. In some cases, the physical size or the colour difference of the individual defect may be too small for the machine to be able to see it. The defective item will then be allowed to pass through the machine as if it were unblemished. In other cases, the machine may detect a defect and remove the item, only for it to bounce back into the accept stream after it has been ejected.

Machines can be adjusted to optimize their performance. However, changing a given parameter may have some disadvantages in addition to the

Table 4.1 *Typical machine performance figures*

Product	*Machine*	*Purpose*	*Throughput (per machine)*	*Typical efficiency (removal of defects)*
Cherries in liquid containing SO_2	2-channel bichromatic grooved belts	Remove spotted and off-colour cherries	800 kg/h	85% spotted 95% off colour
Green coffee	6-channel bichromatic chutes	Remove defective beans and foreign material	900 kg/h	90%
Parboiled rice	40-channel monochromatic chutes	Remove spotted, discoloured rice and foreign material	3 500 kg/h	87%
White Michigan beans	12-channel monochromatic chutes	Remove discoloured beans and foreign material	1 500 kg/h	97%
Frozen peas	Dual monochromatic 900 mm broad belt	Remove foreign material	10 000 kg/h	99%
Roast hazelnuts	2-channel infrared visible bichromatic grooved belts	Remove shell and rotten nuts	300 kg/h	99.5%

intended benefits. There is normally a compromise between achieving a high sorting efficiency and the amount of good material that is rejected.

Sensitivity is one of the principal parameters that the operator can change. Increasing the sensitivity will result in the machine rejecting more defective material but a great proportion of good product will also be rejected as the sensitivity threshold approaches the average product colour.

There are limits to throughput within which a colour sorter should work. If the product flow is increased above the upper limit, or decreased below the lower limit, optimum performance cannot be obtained. Within these limits a general rule can be applied. Increasing the flow of product through the machine will result in more good product being lost.

The colour sorter will normally achieve its best performance if the product to be sorted has been properly cleaned and size-graded prior to sorting. Size-grading product prior to sorting will allow a more precise setting of the ejection system and will allow better accept quality without increasing the amount of good product rejected. Proper cleaning of the product prior to sorting will reduce the build up of dust on optical surfaces and should improve performance between periodic cleaning of the machine.

Table 4.1 illustrates some typical performance figures for a variety of products sorted on different machines.

8 Further Reading

Anon. (1981) Sorting frozen commodities electronically, *Quick Frozen Foods International*, **167**

Anon. (1982) It grades Japanese white rice at 800 million grains an hour, *Food Processing Industry*, **18**

Anon. (1987) Electronic sorting reduces labour costs, *Food Technology in New Zealand*, **47**

Gaffney, J. J. (1976) *Quality Detection in Foods*, American Society of Agricultural Engineers, St Joseph, Michigan

Janson-Smith, D. (1988) Minder machines for flawless food, *New Scientist*, **120,** 19

Maughan, W. S. (1974) Automatic Optical Sorting Machines, *Electro-Optical Equipment in Mechanical Handling and Sorting*, **27,** Institute of Mechanical Engineers, London

McClure, W. F. (1975) Design of a high speed fibre optic blueberry sorter, *Transactions of the ASAE*, **487,** USA

Mohsenin, N. N. (1984) *Electromagnetic Radiation Properties of Foods and Agricultural Products*, **201,** Gordon and Breach Science Publishers, New York

Stanley, *et al.* (1987) Fiber optic colour grading of whole peeled tomatoes, *Can. Inst. Food Sci. Technology*, **20,** No 1, 38
Swientik, R. J. (1987) Machine Vision Systems, *Food Processing*, **68**
Telford, A. P. Control of aflotoxins in peanuts by segregation and sorting, *Food Technology in Australia*, **34,** 214

5 Compositional Analysis Using Near Infrared Absorption Spectroscopy

Ian B. Benson

Contents

1 Introduction 121
2 Theory of Near Infrared Absorption Spectroscopy 123
3 Instrumentation 128
3.1 On-line Near Infrared Instrumentation 128
3.2 Laboratory NIR Instrumentation 133
3.3 NIR Measurement Characteristics 135
3.4 Instrumentation and Installation in the Food Industry 140
4 Applications in the Food Industry 142
4.1 Moisture Measurements in Foodstuffs 142
4.2 Multi-component Analysis of Food Products 149
4.3 Quality Control for Food Packaging Materials 156
5 Practical Aspects of the Calibration of On-line NIR Instruments 157
5.1 Calibration Methods 158
5.2 On-line Installation and Sampling Procedures for Cross-checking 161
6 Conclusion and What the Future Holds 164
References 165

1 Introduction

The moisture content of coffee beans, the sugar and fat content of chocolate, the thickness of polypropylene film food wrapping, and even the coating weight of the protective lacquers a few micrometres thick that are applied to metal foil packaging, are clearly a diverse range of products and measurement requirements – but remarkably they are all amenable to analysis by near infrared (NIR) absorption spectroscopy! Furthermore, these applications are not just restricted to laboratory NIR analysis but can be readily performed on-line directly, monitoring the industrial process as the products are being manufactured.

Near infrared spectroscopy is by no means a new analytical technique, and many examples of its application can be found in the literature going back as far as the 1950s and before. Specifically, early laboratory applications of NIR concentrated upon quantitative and qualitative studies of liquids and solvent mixtures, and such applications are cited in a review by Kaye in 1954. From the same era McCallum (1961) and Miller (1949) provide extensive general coverage of NIR spectroscopy, while Willis (1979) reviews industrial analytical applications of the technique. The most significant and active area of development over the last twenty years has been the application of NIR for the compositional analysis of solids, particularly cereals and foodstuffs where the product is analysed by a reflectance type measurement. The chemical complexities of most foodstuffs or natural products, however, have necessitated the accompanying development of a variety of statistical and mathematical signal processing methods. These techniques have enabled useful measurements to be obtained from spectral reflectance data which characteristically show many overlapping, broad absorption features of seemingly low information content. Uses of NIR within the food and cereals industries are far too numerous to be comprehensively listed, and therefore a selection of references to pertinent reviews and papers is given (Star et al. 1981; Norris and Hart 1965; Osborne and Fearn 1986; Norris and Williams 1987; Hunt et al. 1977; Biggs 1979). These generally refer to laboratory based analysis, although on-line analysis is discussed in other reports (Edgar and Hindle 1971; Bruton 1970; Benson 1989).

Strangely, NIR is a subject which generally receives poor coverage in university science syllabi, with the exception perhaps of food science courses; yet as suggested above there seems little justification for its omission, given its potential widespread application. The development of the technique has therefore largely depended upon the industrial sector and a number of industrially oriented research organizations.

NIR measurement has broadly evolved on two fronts, these being laboratory and on-line application of the technique. This division has resulted from the very different demands that the two approaches place upon the instrument design and specification. Laboratory measurement has the benefit of offering very controlled measurement conditions. The product can be appropriately prepared; for example it can be ground to a specific particle size and consistently presented to the instrument, usually in some form of windowed cell. Also an acceptable time for each measurement may be 30 seconds or more, which is obviously faster than the laboratory wet chemical equivalent, but slow in terms of a continuous on-line analysis.

The on-line sensor must operate very much more quickly, making perhaps several measurements per second, which places great emphasis upon optimizing instrument signal to noise ratio. For an on-line measurement, product presentation can rarely be controlled. The distance of the product from the sensor will almost certainly fluctuate, and the quantity of product passing the point of measurement may vary. Equally, the form of the product is rarely uniform; there may be variations in the particle size of powders and fibrous materials, and even subtle colour changes. The on-line sensor must therefore be designed to ignore such variations and to provide an output that is only affected by the parameter being measured. Finally, instrumentation must be industrially robust and built of optical components able to tolerate a range of adverse operating environments.

The application requirements for off-line and on-line measurement have also widely differed over the years. The most active areas of on-line analysis have been concerned with measurements where, if the component concentration changes, a corrective action can be taken in the process. The most obvious example is moisture, where dryers or conditioners can be readily controlled. Off-line analysis, on the other hand, has tended to concentrate upon providing a rapid replacement for some of the time consuming wet chemical analyses such as those for protein, fat or carbohydrates.

The author's expertise and interest has been strongly biased to on-line measurement and the following text will therefore concentrate on this area. However, off-line measurement possibilities will be discussed and a novel approach to the concept of NIR laboratory analysis, using many of the key features derived from an on-line philosophy, will be described.

While moisture measurement remains a dominant on-line requirement, new highly accurate digital instrumentation has revolutionized the possibilities for on-line analysis, as will be shown. Today, many of the conventional laboratory based NIR analyses can be performed on-line at least as well as in the laboratory.

Theory and instrumentation are discussed, often illustrated with details for moisture measurement, but equally applicable to multi-component analysis. Emphasis has been placed upon describing the characteristics and limitations of the NIR technique, and providing the reader with many actual examples of its application in the food industry for both moisture and non-moisture applications.

2 Theory of Near Infrared Absorption Spectroscopy

Infrared light is part of the broad spectrum of energy known as electromagnetic radiation. Figure 5.1 shows the relative wavelengths and energies in the electromagnetic spectrum that are used in spectroscopy. While X-rays are of extremely high energy, capable of promoting inner electron transitions in high atomic number elements, the infrared region is of relatively low energy and consequently, on interaction with molecules, will only cause inter-atomic vibrations. Near infrared spectroscopy is concerned with a specific region of the infrared, namely the 1–3 micrometres (μm) (10 000 to 3333 cm^{-1}) range, adjacent to the red end of the visible spectrum.

Electromagnetic radiation can be represented as two fields, one electric and the other magnetic, oscillating together but at right angles to each other (Figure 5.2). The frequency of this wave oscillation in the infrared region is about 10^{13}–10^{14} Hz, which is of the same order as the natural mechanical vibrational frequencies of many chemical groups. (The wavelength λ (m) of

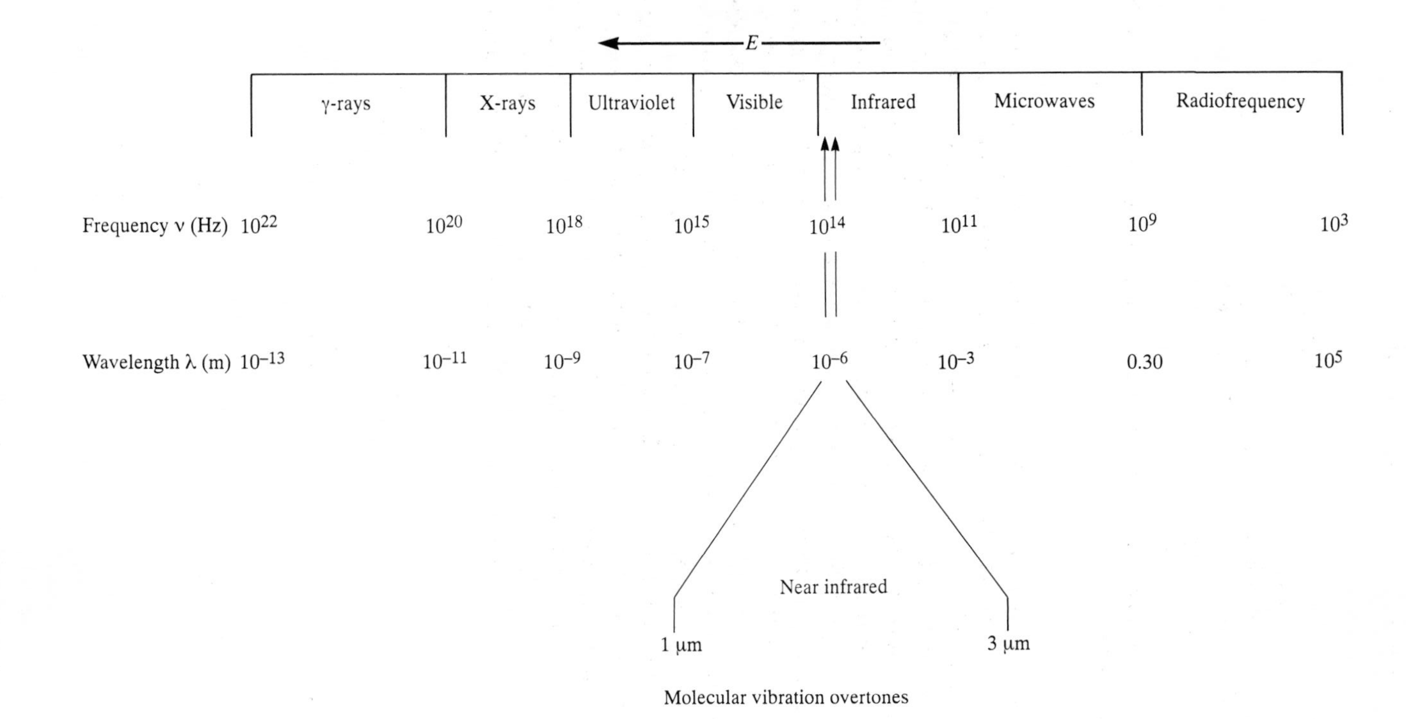

***Figure 5.1** Electromagnetic spectrum*

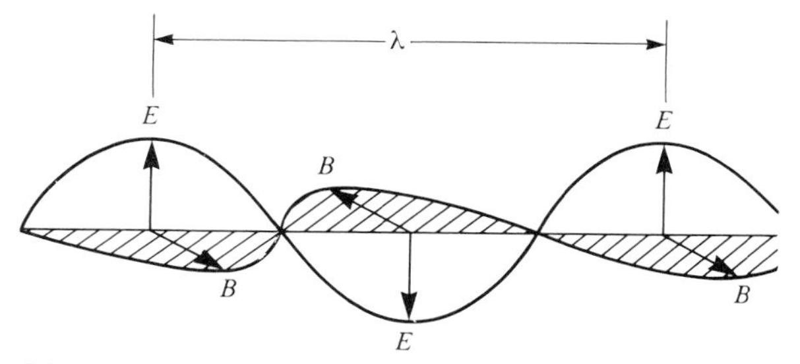

Figure 5.2 *Electromagnetic wave:* B *magnetic component,* E *electric component,* λ *wavelength*

electromagnetic radiation is related to the frequency ν (Hz) by $c = \nu\lambda$, where $c \approx 3 \times 10^8 \mathrm{m\,s^{-1}}$ is the velocity of light.) The water molecule is one of the best known examples of an NIR absorber and is capable of undergoing several vibrational modes, shown schematically in Figure 5.3. The frequency of oscillation of any mode is dependent upon the atomic masses and bond strengths of the –OH group. Provided that the vibration in question results in a change in dipole moment at its extremes of movement, NIR radiation at the appropriate frequency will be absorbed by the group, causing it to vibrate at this same frequency. Energy is thus taken from the radiation field and is dissipated as heat by frictional or collision damping. The mechanism described is known as infrared absorption and is covered comprehensively in other texts (Banwell 1983; Herzberg 1945).

These fundamental or natural vibrations are the basis of the familiar mid infrared spectroscopy commonly used by chemists in qualitative analysis. The mid infrared is, however, generally unsuitable for on-line compositional analysis for a number of reasons. In particular, the extremely low reflectivity of most solids above 2.5 μm results in unusable signal to noise levels for meaningful interpretation. While Fourier transform infrared

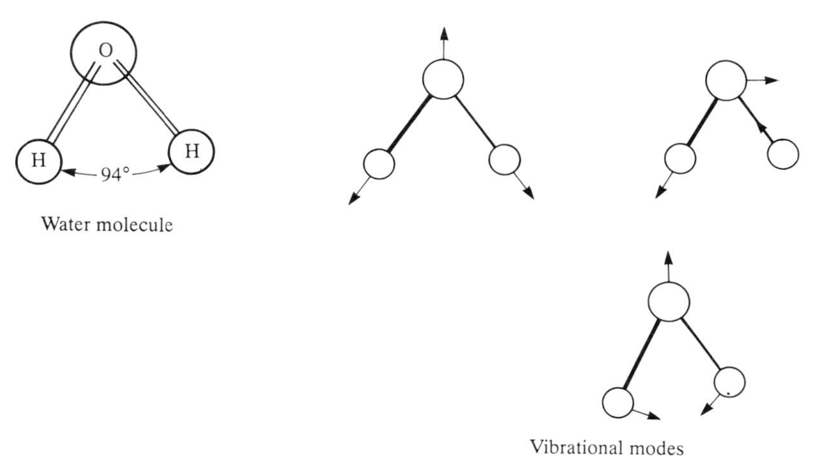

Figure 5.3 *Schematic diagram of the vibrational modes of the water molecule*

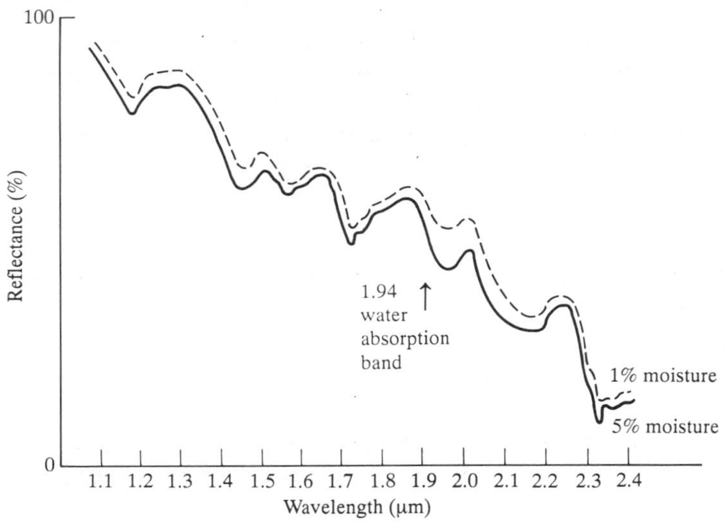

Figure 5.4 *Near infrared reflectance spectra of whole milk powder at two moisture contents (schematic representation)*

(FTIR) analysis in the mid IR has been demonstrated to be a useful technique, this method does not lend itself to on-line analysis (Benson et al. 1988). Secondly, the absorption bands in the mid IR are extremely intense and often display considerable fine structure, both features that do not lend themselves to quantitative analysis. The NIR spectral region, on the other hand, deals with absorptions which are a result of vibrations at harmonics or combination frequencies of the fundamental absorptions. The NIR region is therefore often referred to as the overtone region. The types of absorption that dominate the NIR are hydrogenic absorptions such as –OH, –NH and –CH vibrations. These types of absorption are displayed by moisture and virtually all other major constituents of foodstuffs.

Since these different molecular groups are made up of atoms of different masses and bond strengths, each one exhibits a unique set of absorptions at specific vibrational frequencies or wavelengths. This absorption specificity of the technique is one of the most important features in increasing the scope of its application in the food industry.

The amount of infrared absorption due to a particular absorber relates in a well defined way to the quantity of absorber present. The change in absorption with an increase in moisture within a material is illustrated in Figure 5.4, which shows diffuse reflectance spectra for milk powder at two different moisture levels. Such spectra for solid food products are conveniently obtained using a diffuse reflectance integrating sphere attachment fitted to an appropriate NIR scanning spectrophotometer, shown schematically in Figure 5.5. The spectral information is represented in Figure 5.4 as percentage reflectance plotted as a function of wavelength. Water shows three principal absorptions in the near infrared at 1.45, 1.94 and 2.95 μm. The most commonly used water absorption band at 1.94 μm is evident in Figure 5.4, and as the moisture content increases there is a corresponding deepening of the band, that is a reduction in reflectivity consistent with increased infrared absorption.

The absorption of infrared energy by solids approximates to the Beer-Lambert exponential law for transmission, of the form

$$I_t = I_0 \exp(-kx)$$

where x is the path length of the radiation, k is the absorption coefficient, I_0 is the incident infrared energy and I_t is the transmitted infrared energy. This can be written as

$$\log\left(\frac{I_0}{I_t}\right) = kx \qquad (1)$$

Therefore there is a simple linear relationship between the amount of absorber and the logarithm of the ratio of incident and transmitted energies.

To make use of this principle requires the design of instrumentation capable of accurately measuring absorption changes at a number of wavelengths in the NIR corresponding to the absorber to be measured. Numerous techniques have been devised over the years to achieve this in both hardware design and the subsequent mathematical evaluation of the signal.

For reflectance measurements, while potentially complicated by light scattering effects, a similar logarithmic relationship can be applied successfully in many instances. With some applications where light scattering plays a significant role, it will be shown that a more sophisticated approach is necessary to compensate for the effects that changes in the scattering characteristics of a product can have upon a measurement.

The reflectance spectra of milk powder in Figure 5.4, in addition to the

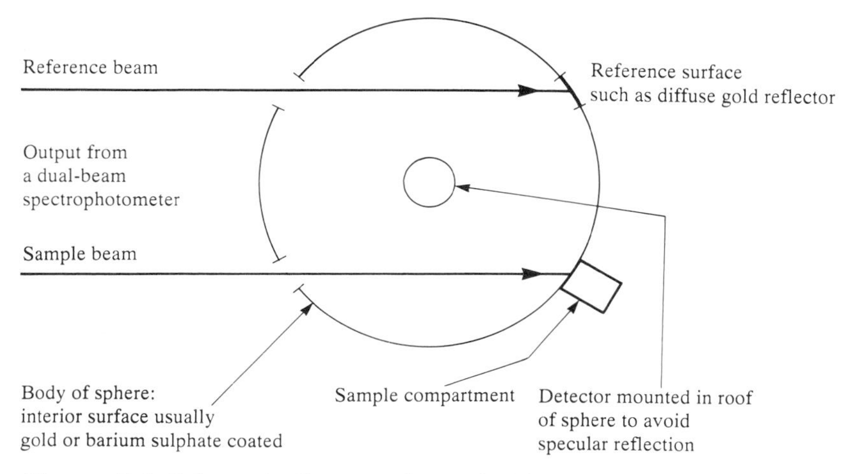

Figure 5.5 *Schematic diagram of a typical integrating sphere*

absorption bands due to water at 1.45 and 1.94 μm, show the presence of many other overlapping absorption bands. These are due to the various –OH, –NH and –CH stretches associated with the lactose, casein and fats present in typical milk powders. Although these can complicate a measurement of water content, requiring correction for the effects that their variation may have on the moisture measurement, they also provide an opportunity to measure these other constituents in addition to moisture.

3 Instrumentation

3.1 On-line Near Infrared Instrumentation

For an on-line NIR measurement to be acceptable in most industrial processes, severe constraints are imposed on the instrument design. For example, while most off-line analysers require the product to be carefully presented, pressed against a glass cell window, this is not practical in an on-line situation where a non-contacting system is much more appropriate. An on-line gauge must also be able to tolerate variations in operating conditions. Ambient temperature, relative humidity and the level of factory lighting are all factors that will change with time, yet they should not influence a measurement. Additionally, while certain processes may be impeccably clean and dust free, this is not usually the case; therefore instrument design must take into account the need to be able to operate in a mixture of dusty, damp, steamy and oily conditions!

Figure 5.6 shows a schematic drawing of the MM55 back scatter on-line NIR analyser (Infrared Engineering Ltd), which can be used in a diversity of applications in food moisture measurement. A broad spectrum of visible and infrared radiation is provided by a quartz halogen lamp. The source is underrun to prolong its life, and is optically pre-aligned in a lamp holder to ensure maximum energy throughput. The light from the lamp is passed via a series of focusing lenses and mirrors on to the product to be measured, where a well defined light patch is imaged. Interrupting the optical path is a rotating wheel, which contains optical interference filters (McCloud 1986). The wheel also contains a simple, visible transmitting glass to allow the beam patch of the instrument to be located. The interference filters are designed to transmit infrared energy at the wavelengths chosen for the measurement.

Typically, filters with bandwidths of between 1 and 5 per cent of the centre wavelength may be used. The number of wavelengths needed for a measurement is dependent upon the application, as will be discussed later. These 'colour' filters take the form of thin glass disks, about 25 mm in diameter, upon which multi-layer dielectric coatings have been vacuum deposited. The major benefits of using these devices for wavelength isolation are their compactness and efficiency with respect to energy

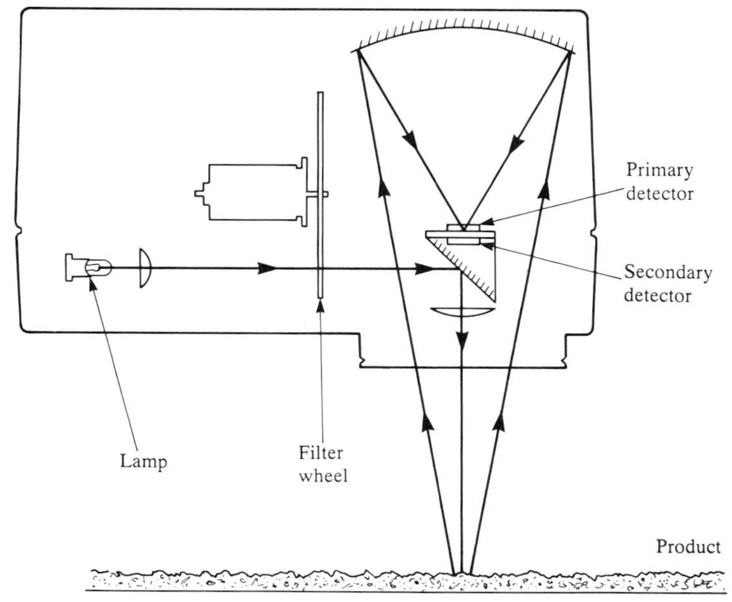

Figure 5.6 *MM55 optical sensing head (Infrared Engineering Ltd)*

throughput at the chosen wavelength. They are also robust elements and offer extremely good long term stability, both mechanically and in terms of their transmission, centre wavelength and bandwidth characteristics. The rotation rate of the filter wheel depends upon instrument design but is usually between 20 and 50 Hz. Therefore pulses of energy at the measurement wavelengths are arriving at the product many times a second, providing the basis for a fast continuous measurement. Immediately prior to the filtered light leaving the gauge, a small percentage of it is sampled by a secondary lead sulphide detector located behind a tiny aperture in the final mirror. The secondary detector effectively monitors the optical characteristics of the instrument, in particular those of the lamp and interference filters. The lamp emission decreases with time while interference filters show changes in transmission with temperature variation. Both changes will be detected by the secondary detector, thus providing for subsequent correction. The importance of this concept becomes clearer shortly.

Some of the radiation is absorbed by the product while the remainder is scattered. The concave collecting mirror in the optical sensing head is arranged so as to collect a portion of the back scattered light and to focus it on to the primary lead sulphide detector. The lead sulphide photoconductive device is ideal for NIR applications since its detectivity peaks at wavelengths of about 2 μm at room temperature, and it is a fast responsive element offering good signal to noise performance.

Because the primary and secondary detectors are matched in sensitivity and kept at the same temperature, when the two sets of detector signals are ratioed an instrument output will be obtained that is self-compensated for the effects of temperature on the sensing head optical components and for the ageing (colour temperature change with time) of the lamp. For example, a drop in lamp emission at one of the measuring wavelengths would be detected by both primary and secondary detectors and seen as the same percentage change in signal level, cancelled by ratioing the detector outputs. The ability to design an instrument that is unaffected by temperature variation is vital because in practical situations temperature changes occur with daytime and seasons.

To date the sensor design described above has been the most successful all round approach for meeting the requirements of an on-line NIR process analyser. However, other techniques have been developed which, if advances in technology permit, are likely to make an impact on basic gauge design.

For example, a solid state device designed by McFarlane (Beaconsfield Instrument Company, Beaconsfield, UK; linked with Hunter Associates) employs infrared light emitting diodes (IREDs) to generate the relevant NIR wavelengths for the measurement. These IREDs, however, emit a broad band of NIR energy and therefore fixed interference filters are still necessary to produce narrow bands of infrared light. The physical displacement between the devices (1 for each wavelength) also makes even illumination of the sample under analysis optically difficult. A more serious drawback for IREDs is that their spectral outputs change with temperature, which clearly, without sophisticated compensation, would have dire effects upon calibration stability. The wavelength range of operation of IREDs is perhaps the most severe limitation at the moment, since they are only economically available up to the 1.45 μm water absorption, thus missing the principal 1.94 μm band. In their favour, IREDs can be rapidly switched, thus providing a high speed measurement in a modulated form without any moving parts.

Solid state laser diodes offer another potentially interesting means of wavelength selection with a narrower band of emission of NIR energy. However, they are still affected by some of the limitations described above.

Another concept of potentially great interest is the use of acousto-optical tunable filters (AOTFs) to generate the relevant NIR wavelengths. The technique relies upon polarized light being diffracted by optically active materials under the influence of high frequency acoustic waves. The radiation transmitted through such a material is preferentially polarized at a certain wavelength dependent upon the acoustic frequency. Thus, by using a polarizer on the outgoing light beam, it is possible to select any number of wavelengths very quickly and without any moving parts. The technology is currently expensive and the concept is complex to physically implement, especially on-line, but it is likely to become important.

Consideration of equation (1) suggests that, irrespective of the optical system, to make use of NIR absorption requires a knowledge of the incident and reflected energies at the measuring wavelength. In practice it is not possible to measure how much energy falls on to the measured sample, that is the incident radiation. Therefore for on-line measurement, in the simplest case, this is approximated by making a two-wavelength measurement. Figure 5.7 shows a schematic diagram of the technique and

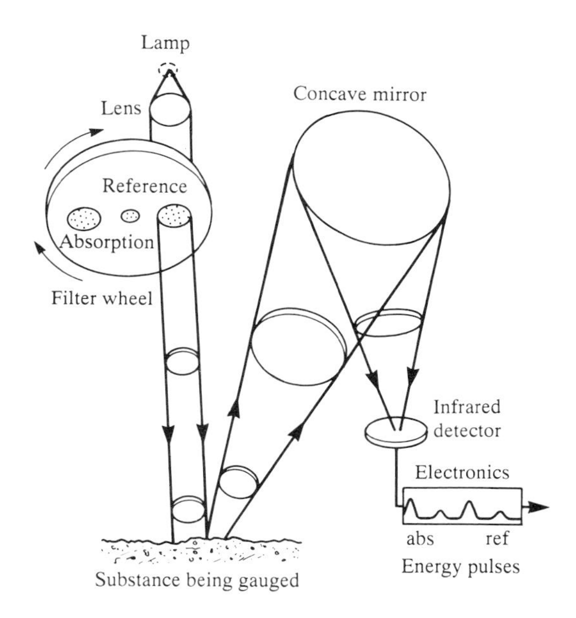

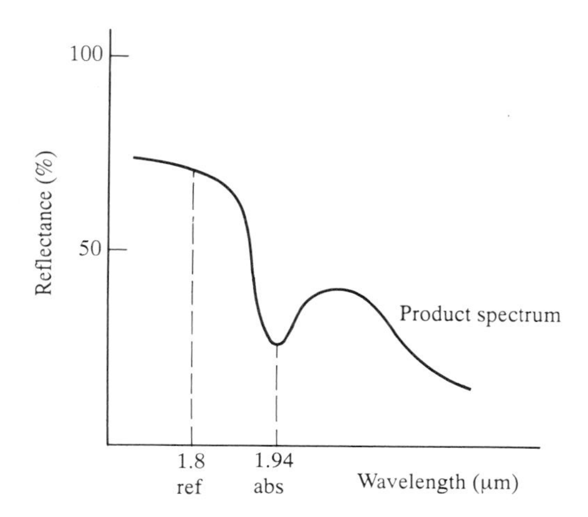

***Figure** 5.7 Schematic diagram of a two-wavelength moisture gauge with reference to the product NIR spectrum*

takes moisture as an example. One wavelength would be centred on the 1.94 μm absorption band of moisture. The other would be chosen from a part of the spectrum where there is little or no absorption due to moisture; commonly, a wavelength close to 1.80 μm is chosen as this reference.

The ratio of the reflected signals at the two wavelengths provides a measure of the peak height of the absorption. The logarithm of the ratio will relate linearly to moisture content as follows:

$$\log\left[\frac{\mathrm{sig}\,(\lambda_{\mathrm{ref}})}{\mathrm{sig}\,(\lambda_{\mathrm{abs}})}\right] \propto [H_2O] \qquad (2)$$

where sig (λ_{ref}) is the signal at the reference wavelength, sig (λ_{abs}) is the signal at the absorption wavelength and [H_2O] is the moisture content. Clearly, as moisture content increases the absorption increases, there is a corresponding reduction in sig (λ_{abs}), and thus the overall term increases.

The infrared detector generates signals which are proportional to the reflected intensities at each wavelength in an alternating current (AC) form. These are amplified and transmitted to an electronic control unit where they are integrated. The ratio calculation is followed by scaling with gain and offset (instrument span and zero) which can be selected by the user to calibrate the instrument in units of his or her choice.

The final gauge output of a simple two-wavelength measurement would be of the form

$$\mathrm{output} = a + b \log\left[\frac{\mathrm{sig}\,(\lambda_{\mathrm{ref}})}{\mathrm{sig}\,(\lambda_{\mathrm{abs}})}\right]$$

where a is the offset or zero adjustment and b is the gain or span adjustment. The modern infragauge will also provide proportional voltage (0–10 V) and current (4–20 mA) outputs and can be interrogated via its serial port to allow automatic feedback control to a dryer or a moisture conditioning process, for example.

The ratio calculation fulfils another fundamental requirement for on-line measurement. Many factors such as product-to-sensor distance (pass height variation) or dust build-up on external optical surfaces cannot be controlled. Changes in such factors would affect the size of the detected signals. However, both reference and absorption signals will be affected equally, and therefore the integrity of the ratio is maintained and the measurement is unaffected. The preceding description covers the main instrumental features of a simple on-line infragauge. Early instruments centred around analogue electronic design which limited the degree of complexity of computation (algorithm) that could be performed upon the detected signals. Furthermore, the available photometric precision (that is, the smallest incremental change in absorption that a sensor can reliably detect) limited the range of applications to the measurement of strong infrared absorbers such as moisture or relatively straightforward thickness measurements of organic coatings or plastic films.

However, recent developments in NIR on-line technology have seen radical changes in the instrument capabilities. The principal components of an infrared sensor remain the optical interference filters and the lead sulphide detectors. The key to improving photometric precision is a high energy throughput in the optical system together with highly sensitive, very low noise detectors. Lead sulphide detectors produced by a vacuum deposition process meet this requirement better than those made by the more usual chemical deposition technique, typically offering a fivefold improvement in signal to noise performance.

One of the consequences of tackling more complex measurements such as nicotine levels in tobacco and of improving the quality of existing moisture applications is that, inevitably, more infrared wavelengths are required. Microprocessors can readily handle this higher complexity of signal processing and are now commonplace in modern instrumentation. The modern infrared gauge can therefore combine complex multi-wavelength capability with user-friendly operation.

With careful design of the operating program it is still possible to produce an instrument for which the operator requires little prior skill or knowledge.

3.2 Laboratory NIR Instrumentation

The principles and functions of a laboratory analyser will clearly be similar to those of on-line devices, albeit with less emphasis being placed upon the tolerance to varied sample presentation and operating conditions, as discussed in the introduction. However, the techniques that have been and are being used for wavelength selection, detection and subsequent signal handling can differ. Benson et al. (1988) discuss the various techniques that have been employed for both off-line and on-line analysers, describing the relative merits of each approach.

For laboratory based analysers, optical filter technology is also used; however, clearly this restricts the choice of wavelengths, especially for a device that may be used for research and development. Full spectrum analysers are commonly used; these are based upon scanning monochromators or tilting filter systems, where the peak wavelength transmitted through a filter is a function of the angle of the incident radiation upon the filter. In virtually all instances the measurement is based on reflectance. However, devices do exist which operate in transmission mode at wavelengths short of 1.5 μm; these have been principally used for analysing whole cereal grains. They incorporate silicon detectors for greater sensitivity at the extremely short wavelengths (0.8–1.0 μm) necessary to achieve penetration (or forward scatter) through the sample. The schematic diagram of a typical laboratory reflectance analyser in Figure 5.8 shows the key

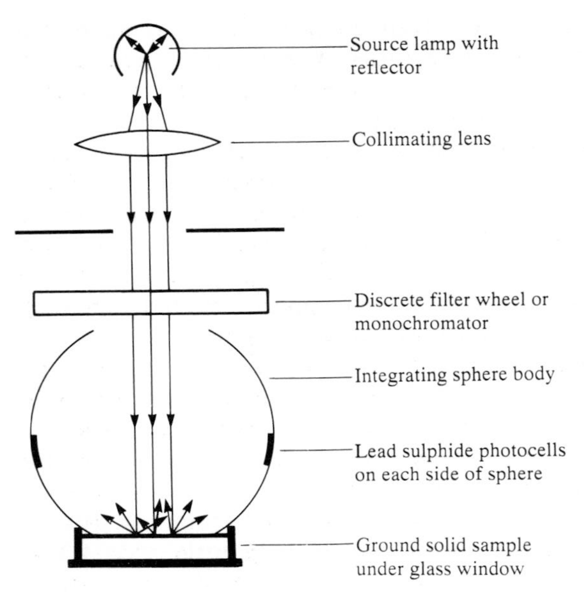

Figure 5.8 *Schematic diagram of a typical laboratory NIR analyser*

elements of the device and the difference in the collection of reflected radiation.

The signal processing methodology after data collection has been the area where the greatest energies have been devoted over the years. While on-line analysers tend to be based upon log(reflectance) mathematics or variants approximating more closely to Kubelka-Munk scattering equations (Kubelka and Munk 1931; Kubelka 1948), the off-line analysers have also been configured to perform first- and second-derivative spectral analysis; more recently, Fourier transformation has been proposed as a useful way of analysing the spectral data collected by a scanning instrument (Giesbrecht et al. 1981; McClure et al. 1984). The latest areas of interest for data analysis include principal component analysis (Cowe and McNicol 1985), Mahalonobis distances (Mark 1986) and partial least squares analysis (Martens and Martens 1986). The concepts behind these ideas are discussed in a short review by Davies (1987).

With all of the above instrumentation, the biggest criticism is the need, in most cases, for careful sample preparation and presentation, especially for materials which are not in a powdered form. This inevitably leads to the result potentially showing a high degree of operator dependence, since it is unlikely that different individuals will carry out the test in exactly the same way. For certain materials, especially those at higher moistures such as tobacco during its primary processing (20–45 per cent moisture), this would prevent a measurement because of the impracticalities of grinding damp materials. Also the size of the sample analysed is commonly only a few grams, and this must raise the question of how representative this would be when a process producing many tonnes of product per hour is monitored for quality control purposes.

A novel approach to overcome these shortcomings has been introduced in the form of a laboratory instrument which incorporates many of the lessons learnt from an on-line analytical background. One such device is the Infralab TM5000 (Infrared Engineering Ltd), which is a filter based instrument built exactly according to the technology described in the previous section. Importantly, the instrument uses a large sample tray (140 mm × 140 mm square) containing 50 grams or more of sample material, which the instrument then scans automatically, taking many readings and providing an average value for the variable being measured. As a consequence, sample presentation is no longer critical; the height of the product in the tray need not be controlled, and the sample does not require any specific preparation. The speed of measurement of the on-line sensor technology used in this instrument means that multiple measurement can be carried out rapidly, providing results, even for a large sample area, within 15 seconds. Since the instrument scans a large sample area, it effectively provides statistical information (mean and standard deviation of the multiple measurements) about that sample. It thus helps in judging its quality and reliability as a representative sample from the product flow, and minimizes the need for replicate analyses.

Digital multi-wavelength instrumentation, as for on-line analyses, simplifies the operation of off-line instrumentation. However, principally because of the better control of off-line measurement conditions (temperature, humidity) and the length of time available for a measurement, complex multi-component analysers have been in use for many years.

Finally, and very important conceptually, an approach in which the sample is automatically scanned provides a measurement which is no longer operator dependent.

3.3 NIR Measurement Characteristics

When considering the possibility of using the NIR technique to solve a particular measurement problem, it is necessary to appreciate both its scope and its limitations.

As will have been suggested by the description of on-line instrumentation, the technique is non-contact. This characteristic is very desirable since the measurement will not interfere with product flow – especially important with high moisture content materials! Also, for food processing, non-contact measurements are favoured by hygiene considerations. The non-destructive nature of the test minimizes product wastage associated with conventional off-line oven tests or wet chemical analytical methods.

For some high cost products, this latter feature alone can justify the use of an instrument if it can replace the majority of quality assurance testing.

The rotating filter wheel in the optical sensor allows many measurements per second and therefore provides an effectively continuous analysis of the moving product. This is ideal for control purposes because all the product viewed by the sensor is sampled and the average moisture (for example) derived can be representative of the product stream.

NIR measurement is unaffected by changes in the electrical properties of foodstuffs, for example conductivity or low frequency dielectric behaviour; such parameters can easily change if the salt or other ionic material content varies. Some alternative methods of on-line moisture measurement are based upon monitoring the electrical properties of the product, such as capacitance or conductance which can be related to moisture content. However, it is clear that these techniques will be susceptible to interference in instances where electrical properties change for other reasons. A foodstuff containing variable levels of salt would be an example of an unsuitable application because of the effect of the ionic material on resistance. The product density does not generally influence an NIR back scatter measurement but will certainly affect techniques based upon monitoring electrical capacitance properties. Similarly, product temperature will not influence NIR measurement whereas electrical techniques are temperature sensitive. Although there will be an increase in infrared emission from the product with temperature, this will be a relatively small change at the operational wavelengths and a continuous emission of DC form. It will therefore be ignored by a properly designed AC coupled detection system in an NIR instrument.

Whether an infrared gauge is sensitive to changes in ambient lighting levels is critically dependent on the instrument design. The form of the blocking filter, which shields a detector from visible and irrelevant parts of the infrared spectrum, plays an important role in reducing the influence of ambient lighting.

The rate of product throughput or the process line speed does not influence an infrared measurement, since the high speed of measurement more than adequately samples the material passing under the sensor, providing a representative time-averaged output.

In the application of NIR to moisture determination, the response to free or associated moisture and bound water (water of crystallization) should be appreciated. In most materials, the difference in wavelength between the absorption bands for these two forms of moisture is very small and therefore they cannot usually be treated separately. Whether this is an advantage or a drawback depends on the requirements for the individual application. For example, with whey powders the moisture content will be distributed between free and bound moisture, the latter being present as water of crystallization in the alpha-lactose. However, the storage stability of whey is only influenced by the free moisture and therefore an NIR measurement is of limited value. For detergent powders containing crystallized tripolyphosphates, the total moisture content is required and therefore NIR is suitable in this instance.

While on the subject of various moisture forms, the question of atmospheric moisture or relative humidity (RH) is inevitably raised and its effect on a moisture measurement queried. Inspection of Figure 5.9, which shows the transmission spectra of liquid water and water vapour (low resolution), indicates that, although the bands overlap, they do absorb at different wavelengths. Any overlap can be compensated for by careful wavelength selection and appropriate calculation.

The major limitation of the NIR measurement is the limited penetration of the infrared radiation into the product. Although this is dependent upon the water absorption band chosen (shorter wavelength radiation is more penetrating), it usually amounts to no more than a few tenths of a millimetre into the material. The scattering characteristics of the product

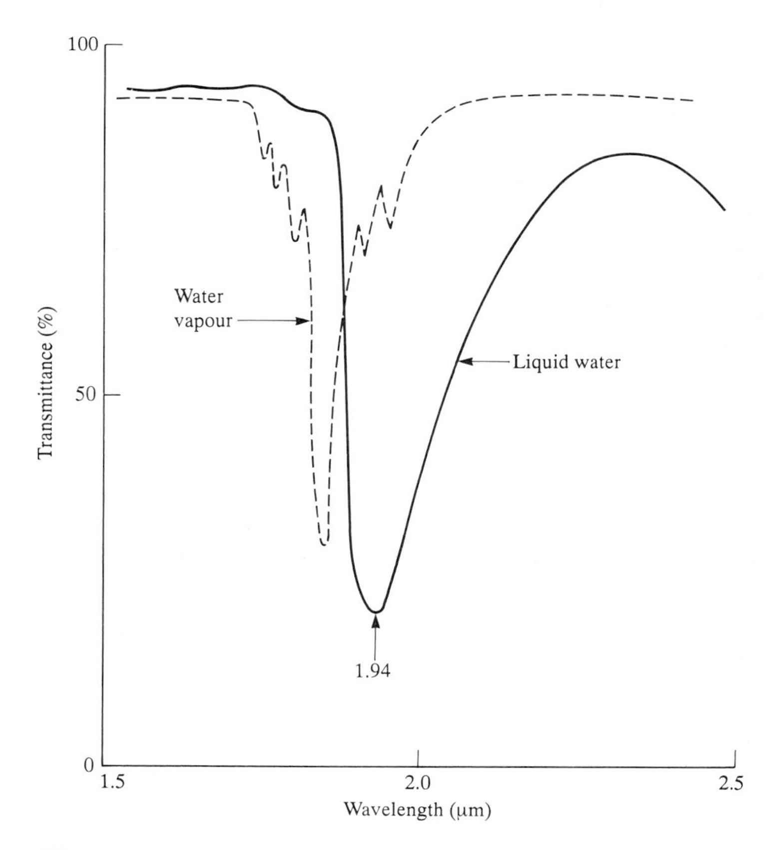

Figure 5.9 *Comparison of the relative absorption positions of liquid and atmospheric water (schematic representation)*

primarily determine the level of light penetration in a given wavelength region. Fortunately in many practical applications this does not present a problem since the bulk of the foodstuff, especially if made up of fine particles (powders, granulates etc.), is well represented by its surface layer. For example, in milk powder with particle sizes ranging from below 50 μm to over 200 μm, the individual particles are adequately penetrated by the radiation to give a meaningful moisture measurement. However, it is important that the surface of the product bed viewed by the sensor is representative of the bulk of the material which will not be seen. If the surface material tends to dry off relative to the bulk, because of long periods of exposure to the air or from having emerged hot from a dryer, precautions must be taken. The particulate product must be turned over prior to measurement, by installing the sensor at a point where a change in conveyors occurs, or even by ploughing the surface, this latter method being particularly suitable for fine powders.

The concern for materials with large particle size, especially those that have recently been dried or steam conditioned, is whether the surface moisture represents the internal condition. Indeed, it is frequently the case in drying processes that the surface and internal moisture levels of bulky materials differ. In many situations, this potential difficulty is not a problem because there is some form of relationship between the surface and total moisture content which can be exploited to provide a measurement. It is often the relatively constant conditions employed to dry a foodstuff that ensure this assumption is valid! The classic example of this is the on-line measurement of biscuit moisture. After oven drying, the surface moisture content of biscuits is very low, and will even vary across the surface, while the internal status may be of almost dough-like consistency at a high moisture level. However, the oven drying characteristics are very consistent and therefore the on-line measurement of surface moisture does provide the biscuit producer with useful information for a process control system.

NIR measurements specific to foods will be discussed in Section 4.1. Microwave measurement can be a useful alternative technique when the limited penetration of an NIR measurement is a problem. However, apart from the difficulties of engineering a microwave emitter/receiver on-line, often in intimate contact with the product, the technique is still susceptible to product temperature and density variations unless appropriate signal correction is applied. Sophisticated instruments are available which take these difficulties into account. The range of moistures over which a microwave system operates can be very limited owing to excessive attenuation of the microwave radiation by water. However, new techniques utilizing stripline or coaxial sensors can in part overcome these shortcomings. See Chapter 7 by M. Kent in this book for a detailed discussion of microwave measurement.

The measurement range of the NIR is well suited to the needs of food

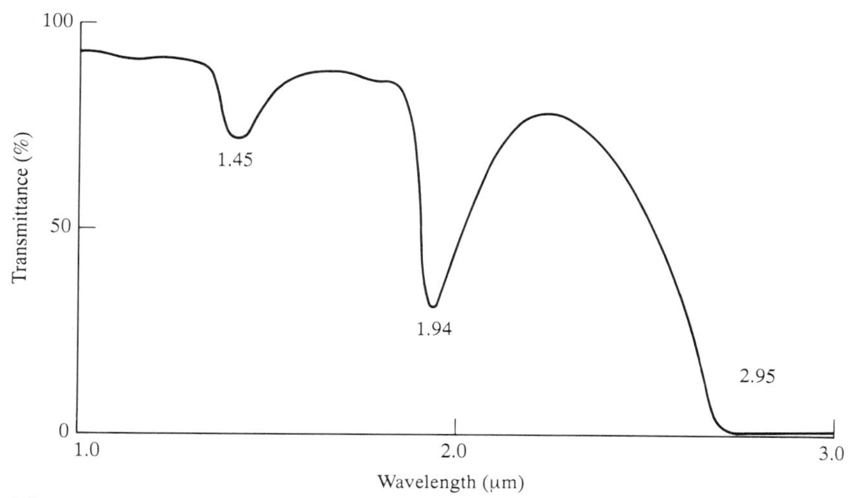

Figure 5.10 *Near infrared transmission spectrum of water*

processing since a wide range of moistures, fat and protein levels can be encountered. In the NIR the most important prominent absorptions are due to the –OH, –NH and –CH groups. Each of these groups characteristically exhibits three main absorption bands in this spectral region. For water, these absorptions occur at 1.45, 1.94 and 2.95 μm as shown in Figure 5.10. The relative intensities for the three bands are 1:3:90 respectively. To generate the spectrum in Figure 5.10 a water path length of 150 μm was used. Clearly with this quantity of water the absorption at 2.95 μm is of no use in practice because it exhibits total absorption, whereas both absorption bands at the shorter wavelengths would provide usable changes if the cell path length was changed. This means that a wide range of moistures can be accommodated merely by selecting the appropriate absorption band which shows the greatest level of change at the product moisture level.

It should be noted that in the mid infrared, apart from the extremely low reflectivity that most solid materials exhibit in this region (which reduces the signal to noise ratio in reflectance mode), the absorption bands are extremely intense and would be difficult to exploit, particularly for on-line quantitative analysis. The reason for this is that absorption changes with component level variation would be extremely small, thus putting impossible demands upon the photometry of the instrument. A secondary complication, although academic in the light of the above, is that the mid IR spectra are remarkably complex with highly resolved absorption features which are ideal for qualitative use but render the region problematical for quantitative analysis.

The NIR effectively provides a huge range of absorption sensitivities which can be selected to meet the application requirement. For example, with products at extremely low moistures, perhaps less than 0.5 per cent moisture, use might be made of the highly sensitive 2.95 μm absorption. On the other hand for a 'wet' material of around 80 per cent moisture, the weak 1.45 μm absorption would be more appropriate, to provide an adequate absorption change as a function of water content at these high moisture levels.

This is a real strength of the technique, enabling an instrument to be tailored for the application. Exactly the same characteristic holds true for –CH or –NH absorptions of materials such as fats or proteins: namely they each exhibit at least three regions of absorption in the NIR of different intensities and, importantly, at wavelengths different from the –OH absorptions of water.

The choice of the absorption band to use for a particular application depends upon many considerations. The degree of penetration into the product and the way in which it is scattered will affect the 'path length' of the radiation in the material and hence the amount of absorption. For the determination of water contents between the levels discussed above, it is usually a matter of selecting the one which shows the greatest change with a given change in moisture content. The absorption of NIR energy follows an exponential characteristic and therefore the maximum rate of change of absorption occurs at reflectance levels of around 60 per cent. This means that there is a point for a given moisture level above which greater sensitivity would be obtained by using a weaker absorption band. For food applications, the most commonly used water absorption is the 1.94 μm band. This typically can be used for moisture levels between 1 and 30 per cent.

3.4 Instrumentation and Installation in the Food Industry

The majority of on-line NIR measurements will be based upon a back scatter (reflectance) configuration of the sensor, this being the most suitable approach to the analysis of powdered or non-particulate solid materials. In this form the measurement is non-contact, the sensor being mounted some 200 mm away from the product surface. Careful optical design and the characteristic of the ratio measurement mean that quite large variations in the product pass height can be tolerated. Variations in height of ±50 mm will not cause problems and in fact larger ranges can be acceptable, especially if they are random rather than systematic with time. Installation is therefore simple and only requires a continuous flow of product so that the sensor does not 'see' the conveyor belt. In instances where this is not possible, such as in biscuit production, a special version of the on-line gauge with electronic gating is available.

Since the diffuse reflectance technique depends upon the sensor measuring *back scattered* light (scattered light has interacted with the product and has been partially absorbed) the measurement of shiny materials needs special consideration, especially when the amount of reflected light reaching the sensor is changing appreciably. In these instances the sensor must be arranged to avoid seeing the large surface reflected light component, which of course contains no absorption information.

A location meeting these relatively undemanding requirements can be found on most plants. On occasions it is necessary to monitor the product through a window in a duct. The use of toughened glass allows this and, provided it is kept clean and a moving flow of product is maintained against the window, the glass will not affect the measurement. In instances where glass may not be an acceptable window material it is possible to use plastics such as polyester or Aclar, but in these cases product contact with the window must be eliminated owing to the abrasive quality of most foodstuffs. Maize grits and meals, soya meals, coffee powders and milk powder are examples of products which have been successfully measured through glass (usually) windows. For high moisture or sticky products, on the other hand, a glass window is unlikely to succeed. In such cases, industrial sapphire which is hard, food-plant-acceptable and has a low coefficient of friction is an excellent but expensive alternative.

Products travelling in a screw conveyor may be analysed by measuring off-centre of the shaft of the screw and removing a small portion of the conveyor's screw blade or flight. This usually gives a product presentation with minimal height variation and no interference from the metal parts of the screw.

It is worth restating the need for the sensor to be viewing a surface which is truly representative of the bulk moisture content of the product. This consideration should also be extended to include the need for careful positioning of a static sensor on a wide band of food product. The chosen point for such an application must be where the gauge provides an output indicative of the average moisture content irrespective of any profile that may exist across the width of product. Alternatively the sensor may be scanned automatically. Whatever location is chosen, the point must always enable access for sample collection close by and immediately downstream from the sensor for the inevitable cross-checking of the instrument with the laboratory test. This will be particularly important in the early commissioning stages after installation. Calibration is discussed in Section 5.1.

In many cases, the food industry demands certain hygienic standards to be met; given the often dusty and high humidity environments or even the aggressive nature of some food products, the sensors must be designed accordingly. Since the optical sensor is a self-contained device connected only by a single cable to its control unit, this is relatively simple to achieve. Sensors are available in stainless steel with flush fitting joints to avoid harbouring dirt. For high temperature operations these units can be water cooled. As far as protection against dust or moisture ingress is concerned, an on-line sensor is usually built to the recognized IP65 (in accordance with BS5490) rating. In steamy or dusty environments it is necessary to keep the optical windows clean, and this can be effected by using an air purge system. The air purge device is attached to the sensor window. The purge is made up of two concentric tubes, the inner one being porous. Compressed air is passed through an aperture in the outer tube and this creates a steady flow of air through the inner tube, keeping the optical window clean.

One of the attractions of the on-line NIR analyser is the variety of seemingly adverse environments it can tolerate. This is largely attributable to the ruggedness of the design and the fact that the optics can be built from materials such as glass or quartz rather than the more fragile components usually associated with the longer wavelength mid infrared region. It is only necessary to ensure that the instrument is not subject to excessive vibration, as with any optical system.

In instances where the use of glass as a sensor optical window is unacceptable, it is possible to use either an Aclar (PTFE polymer) window or one where a polyester/glass laminate is incorporated which will prevent food contamination in the event of breakage.

NIR measurement within the food industry is not restricted to reflectance. In some cases a transmission measurement is more appropriate, such as in the measurement of alcohol in a beverage stream or even the water content of viscous syrups or caramels. The same type of sensor as for reflectance can be readily adapted to provide a transmission instrument appropriate for these applications.

4 Applications in the Food Industry

4.1 Moisture Measurements in Foodstuffs

The first NIR on-line analysers applied in the food industry used two wavelengths and were moderately successful in a number of applications. However, the early success was soon followed by criticism of the sensitivity of the technique to the product shape, the particle size or perhaps the chemical composition. Frequently, the last mentioned variable would be associated with colour changes, although these could be induced by changes in drying conditions alone. Infrared Engineering Ltd has used the concept of a balanced reference measurement which has been successfully applied to solve such problems.

To understand the reasons for the shortcomings of the two-wavelength instrument, it is necessary to consider the influence such variables have on a product's reflectance spectrum. Figure 5.11 shows schematically the reflectance spectra of milk powder at two different particle sizes but at the same moisture content. The spectral differences are exaggerated for clarity, but it is clear that the regions on each side of the water band suffer reflectance changes with particle size.

With a two-wavelength gauge, the reference wavelength signal is no longer stable and cannot be reliably used to provide a measurement of the absorption band peak height. The balanced reference method is depicted in Figure 5.12, where a third wavelength is used as an additional reference point in the product spectrum. Interpolation between the two references is then required to establish the spectral background characteristic at the water band and so provide a measurement that is independent of these slope changes.

The success of this technique is clearly demonstrated in Figure 5.13, which shows the calibration characteristics for skim milk powder for both a two-wavelength and a three-wavelength, balanced reference gauge. The single calibration for all milk powder particle sizes provided a measurement accuracy of ± 0.1 per cent moisture (2 standard deviations).

The particle size variation cited in this example typically ranged from returns from the cyclone separators to the agglomerated materials produced after a multi-stage spray and fluid bed drying process. Interestingly, this particular problem only arose with advances in the drying technology used for milk powders, which introduced fluctuation in particle size. As Figure 5.13 illustrates, it is possible to completely eliminate the problem by the use of additional, and appropriate, infrared wavelengths. The changes in the reflectance spectra can be accounted for by the way in which the scattering properties of the milk powder have been modified by a particle size change. The simple Beer-Lambert exponential law describing the

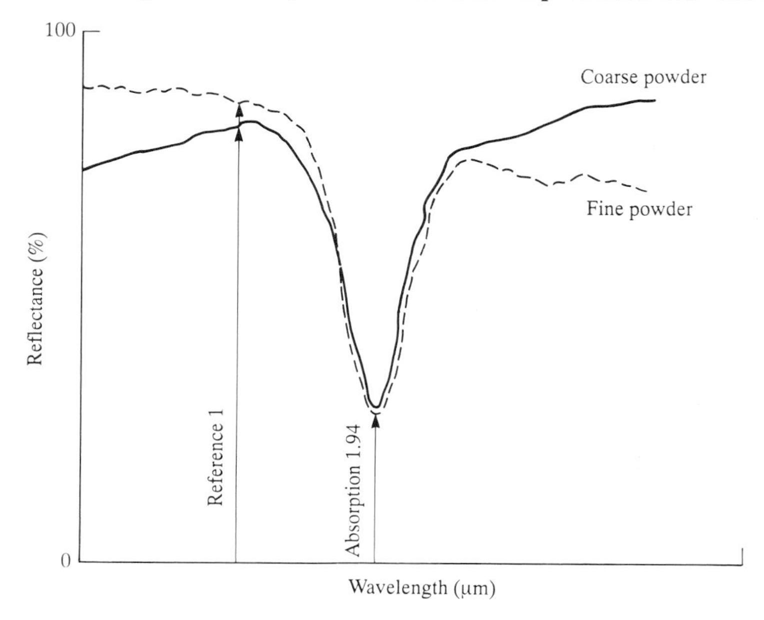

Figure 5.11 *Influence of milk powder particle size on the NIR reflectance spectrum: adverse effect upon a two-wavelength gauge*

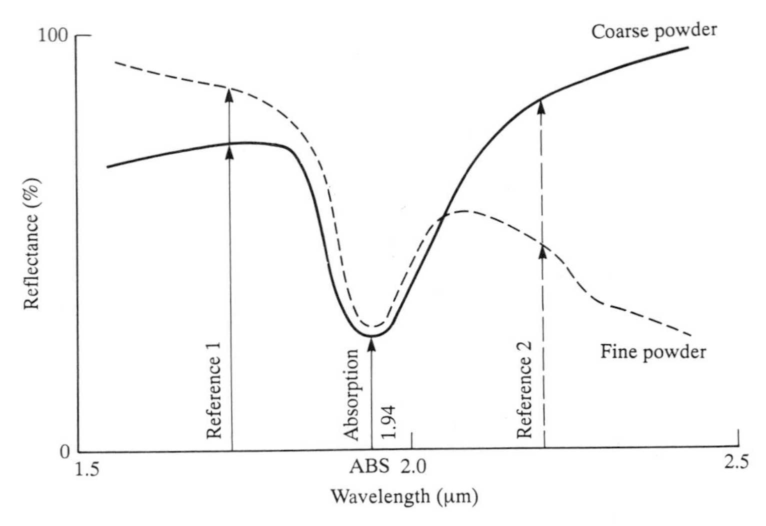

Figure 5.12 *Effectiveness of a second reference wavelength in compensating for the effect of powder particle size changes*

absorption of energy (equation (1)) does not take scattering into account, and therefore new models are required so that the spectral characteristics of the product can be described accurately enough for the moisture measurement to be of practical use. In most cases it is unusual to need more than three wavelengths for moisture measurements. Another example of the success of this approach is the measurement for varying grades and types of dried teas which is shown in Figure 5.14.

In instances where calibration non-linearity is evident, this again is due to the failure of the simple logarithmic law and can be overcome by more sophisticated algorithm designs. While non-linearity is not a problem in an on-line process, when operating over a narrow moisture range to a specified target it does complicate calibration. It is therefore good practice to use linear relationships, since most calibration procedures assume their existence!

In certain products, natural variation of another constituent can have an adverse effect on the moisture measurement. Milk powder again provides a good example of this, where the fat level or even the fat type (vegetable or animal) can vary depending on the product type. This can alter the optical properties of the powder, and hence the moisture calibration may be affected. However, by monitoring the fat absorption characteristics with at least one other wavelength, it is possible to compensate for fat contents of up to at least 30 per cent and provide a fat independent moisture calibration. The ability to monitor fat content implies that this constituent can be measured, and this will be discussed in the following sections.

A large and diverse range of moisture applications in the food industry

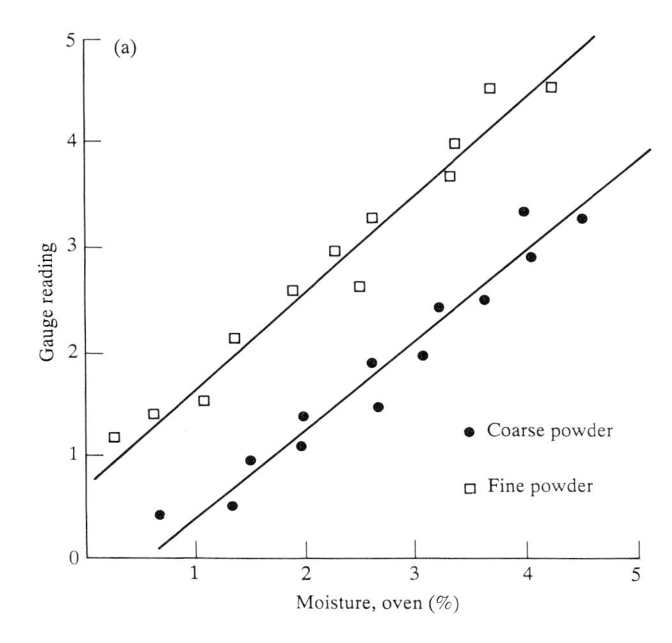

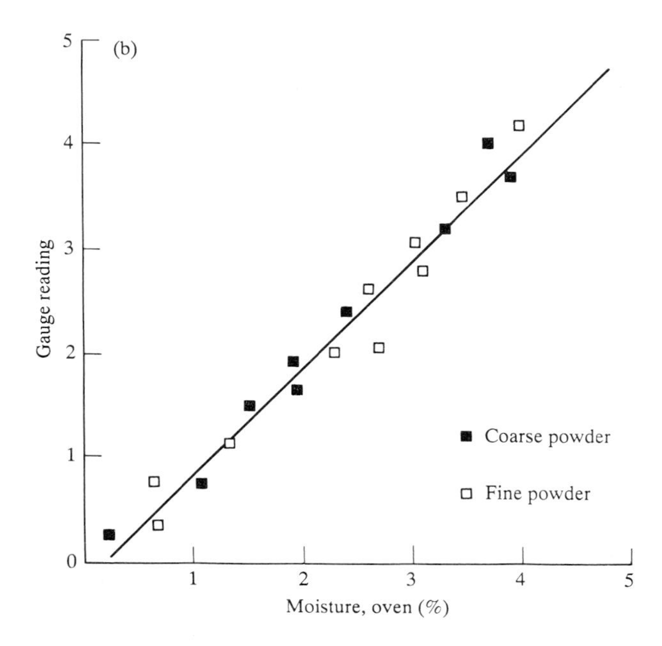

***Figure 5.13** Comparison of the calibration sensitivity of (a) two-wavelength and (b) three-wavelength infragauges to milk powder particle size variation*

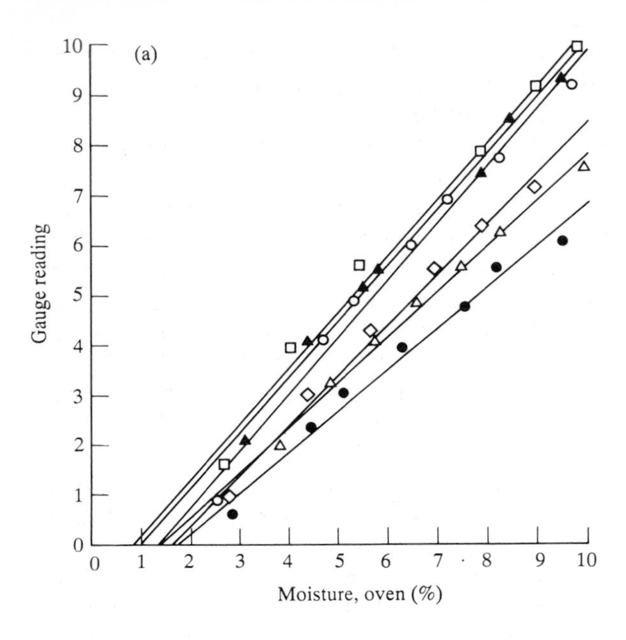

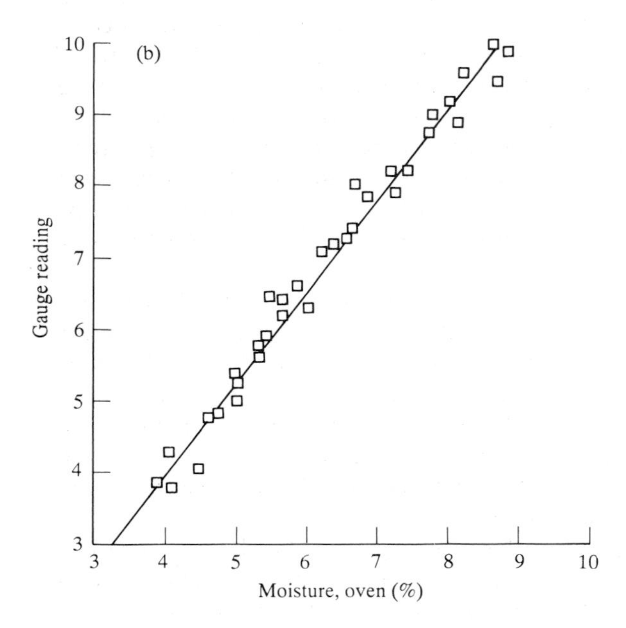

Figure 5.14 *Illustration of the benefits of additional wavelengths in eliminating calibration sensitivity for six different tea blends: (a) two-wavelength (b) three-wavelength infragauges*

has been successfully solved using on-line NIR measurement. Generally, the common denominator is that the material must be of powder, crystalline, fibrous or flaky form, although pelletized or bulky products of a wide range of particle sizes are amenable to this method providing their surface is representative of the material as a whole.

The acceptable moisture range is wide, this flexibility being a consequence of the range of absorption sensitivities available in the NIR. For example, at the low moisture end, materials such as citric acid crystals with moisture contents below 1 per cent can be analysed using the 1.94 μm band, whereas cottage cheeses and spent wash syrups from distillery wastes, both of around 80 per cent moisture, can be tackled using the weak 1.45 μm water band. With products at very high moistures there can be a tendency for the water to settle out, and therefore care must be taken in obtaining a presentation on-line where the foodstuff is well mixed.

Table 5.1 provides a representative list of well proven applications suitable for NIR measurement, and the typical accuracies obtainable are also given.

Of course, there are instances where NIR measurement may not be successful. By way of guidelines it is sensible to question whether very large particle size materials are suitable because of the almost certain difference in surface characteristics from the interior condition. Block or slab materials such as finished chocolate or even freshly baked bread provide extremely doubtful examples. Since, as stated earlier, for back scatter measurements it is important to avoid observing specularly reflected light from the surface of shiny materials, these types of products can be problematical. Caramels, syrups and fondants fall into this category. In these instances it is possible to consider a special back scatter optical arrangement for the sensor to ignore specular light, or even to use a transmission configuration.

Crystalline materials such as sugar can be measured because the percentage of specularly reflected light from the crystal faces, given a random presentation to the sensor, will be effectively constant. Foodstuffs which appear optically 'dark', that is they exhibit very low reflectance levels at the measurement wavelengths, usually present frustrating problems. For example, durum semolina powder can be easily measured; however, once it has been transformed into extruded dried pastas, an NIR measurement on these materials is impossible (to date!).

Certain products emerging from a dryer may in fact be virtually bone dry for the first few hundred micrometers of depth below their surface, and in these cases there can be no question of a reliable surface-to-volume moisture relationship existing. Some extruded confectionery products fall into this category. By contrast as mentioned before, biscuits can be quite readily measured after they have emerged from the oven. Such a measurement is improved by measuring an appropriate distance down the line to allow some moisture equilibration to take place. Apart from biscuits,

Table 5.1 *A selection of on-line moisture measurement applications*

Application	*Moisture range (%)*	*Accuracy (% absolute)*
Biscuits of all types, human food and animal feeds	0–5	±0.1
Cereals, breakfast types	0–6	±0.1 to ±0.2
Chocolate crumb	0–2	> ±0.1
Cocoa powders	0–5	±0.1
Coffee: instant	0–4	±0.1
ground	3–10	±0.1
roast beans	5–10	±0.3
green beans	9–14	±0.3
Crisps and other snack products	0–3	±0.1
Fish meal	5–15	±0.25
Flour	10–15	±0.2
Herbs, dried	5–15	±0.25
Grain products, wheat, barley, oats, spent grain (distillery waste)	5–15	±0.2
Maize products: flakes, powders, grits	5–15	±0.2
Milk powders: skim milk	0–5	±0.1
fat filled	0–5	±0.15
whole milk	0–5	±0.15
casein powder	8–14	±0.2
Non-dairy creamers	0–6	±0.1
Soya meals, flakes	8–18	±0.2
Starch powders	4–8	±0.2
Sugar, granulated (after centrifuging)	0–3	±0.1
Tea: instant	0–5	±0.1
finished cut leaf	0–15	±0.2
Yeast, powders or strips	5–20	±0.3

pelletized animal feeds emerging from an oven, or even leaf products such as tobacco which have recently been steam conditioned, all respond to measurement if installation is as far down the line as possible. The time delay for feedback control purposes from the conditioner to the point of measurement is not usually significant compared with the time taken for a change in drying conditions to affect the process, that is the process response time.

The measurement of moisture in liquid streams is occasionally required. The main provisos here are that (a) the liquid has an infrared 'window' at the water absorption band, and (b) the moisture is homogeneously distributed through the liquid, or miscible with it. For instance, measurement of residual moisture in polar solvents such as ketones or alcohols would be possible, while cooking oils present an example where heterogeneous moisture distribution would cause problems.

An interesting recent example of the use of a transmission NIR instrument is the measurement of the casing material used for enclosing sausages, including salami-like meat products. The cellulose or collagen based tube is flattened at certain stages of the process and provides an ideal presentation for measurement.

In the author's experience, new moisture measuring opportunities continue to arise in new and familiar areas, presenting a challenge to the technology. It is the quality of the applications engineering that decides whether a (new) foodstuff can be analysed, but the above examples should give the reader guidelines for an initial judgment.

4.2 Multi-component Analysis of Food Products

Current trends in legislation regarding food nutritional contents, the need to meet existing government specifications or the consumers' demand for low fat or high fibre products now compel the food manufacturer to consider on-line measurement of many components other than moisture. Equally important is the high cost of materials such as cocoa butter used in chocolate manufacture, where there is an incentive to control the amount added in a process stream. Until quite recently, control of these parameters depended upon historical tests derived from off-line sampling and analysis, using the traditional wet chemical technique or perhaps even an off-line NIR analyser.

The on-line measurement of such components puts severe demands upon the photometric performance of a sensor. The absorption bands due to fat, proteins, sugars or starches are considerably weaker than those of moisture. Figure 5.15 typifies this for fat, where the spectra of milk powders with a 30 per cent difference in fat content are illustrated. Calculations show the absorption band for fat to be some three times weaker than the equivalent band for moisture. For protein, the absorption is typically five times weaker, such that two reflectance spectra of a foodstuff with differing protein levels may show no obvious differences. The fact that these differences are not visually apparent puts into perspective the relative photometric performances of current scanning spectrophotometers and a dedicated multi-filter infragauge. From these considerations, an instrument would require a photometric precision of ± 0.05 per cent absorption. By way of comparison, scanning spectrophotometers rarely achieve better than ± 0.5 per cent photometry. The weak nature of these absorptions also means that any potential variation in the

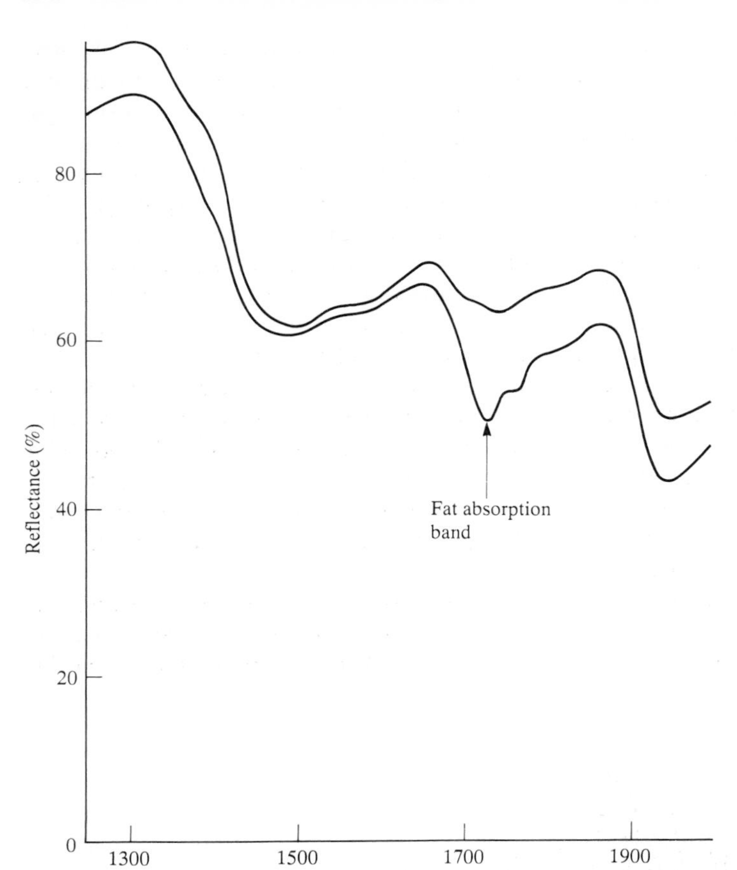

Figure 5.15 *Diffuse reflectance infrared spectrum of milk powder at two fat levels*

infrared characteristics of the product must be taken into account so that a measurement is not perturbed by other constitutents which could be varying in an uncontrolled way. It has already been shown that fat variation can influence a simple moisture measurement, and moisture alone, unless corrected for, will have an even larger influence on the fat or protein measurement.

Compensation for these changes is achieved by the use of additional wavelengths. The twelve-wavelength digital gauge (Infrared Engineering Ltd) shown in Figure 5.16 is an example of such an instrument which can meet these demanding photometric and multi-wavelength requirements. NIR measurements which have for many years been considered to be restricted to the realm of the laboratory analysers can now be performed on-line with this type of infrared gauge with exactly the same technology as described for moisture, albeit with more sophisticated instrument design. While conventional laboratory NIR analysers usually require samples to be highly consistent in both presentation and form (particle size), the on-line multi-component analyser still retains the industrial characteristics and robustness of the simpler gauges described earlier. Importantly, despite the greater degree of precision with which these instruments are required to work, there must be no compromise in their ability to tolerate changes in product appearance, variation in pass height or other process conditions.

The development of measurements of a food component that is a weak infrared absorber is considerably more demanding both technically and in terms of measurement time. Since an application is usually dealing with a product made up of many constituents, it is important to know how these may vary and to be able to take these into account in the solution. The problem is therefore a multi-variate one, requiring perhaps up to six wavelengths for a single component. The optimization of the algorithms for a given measurement usually requires the use of multiple regression statistical techniques.

This immediately raises a philosophical question over the correct way to approach these problems. For example, at one extreme, the development of a solution could be carried out in a very mechanical way by measuring the

Figure 5.16 *Digital twelve-wavelength NIR on-line analyser: provides simultaneous measurements of three components within a food (Infrared Engineering Ltd)*

reflectance characteristics throughout the NIR of a chosen group of samples which represents, as widely as possible, the variations expected in the product including the component to be measured. Subsequent data analysis would involve testing all possible combinations of a chosen number of wavelengths to derive a 'solution' based upon an algorithm which gives the statistically best measurement accuracy with little regard for physical meaning. Alternatively, a better approach would be to make use of statistics to help with the optimization of an algorithm using predetermined wavelengths which have been chosen based upon a sound appreciation of the chemical/physical characteristics of the product. The subtleties of measurement development are outside the domain of this discussion and are handled in other articles (Honigs 1985), but a broad understanding of the topic is important to those considering the use of NIR multi-component instrumentation in a new application. The quality and robustness of a calibration, that is the long term dependability of the measurement, will be strongly related to the soundness of this initial approach.

The author and his colleagues strongly favour an approach where the measurement is based upon sound chemical/physical principles. There are instances described in the literature where NIR measurement of a component, not existent in the material, is apparently being made, probably because of a correlation with some material in the natural product within the sample set collected for the analysis. The measurement of the ash content of wheat, the tar content of cigarette tobacco, the bakeability of bread (Hagburg factor) and the sensory qualities of peas are such examples. These measurements would be unlikely to succeed as on-line applications in the long term, and frequent recalibration of an instrument would be symptomatic of this lack of robustness.

By way of an example of the approach favoured by the author, the case of a protein measurement is now considered. The procedure described has been used to study various products and has resulted in accurate protein measurements being developed for wheat flours, soya products and milk powders. The description below relates to the work carried out on milk powder.

Milk powder, being derived from a natural product, can show a wide level of compositional variation. The basic components are casein, lactose and butter fats, together with moisture and some trace elements. The exact chemical compositon is dependent upon many things, but animal feed patterns probably account for the major seasonal changes. Since coincidental compositional changes will affect the NIR spectrum of a product, these changes must be included in the so-called training sample set.

Obviously, samples are also required that display as wide a range of proteins as may be encountered in subsequent process control measurements. While moisture principally absorbs in wavelength regions away from organic material, such as protein, it is a relatively strong absorber and there is considerable overlap of absorption so that it does influence the entire infrared spectrum. Therefore samples of the milk powder at a range of moistures must be included in the analysis so that the effect of moisture on the protein measurement can be quantified and corrected for.

Finally, the physical characteristics of the powder must be considered, such as particle size variation and colour, and therefore samples displaying the extremes of these variations must be examined.

This normally will amount of some 100–200 samples depending upon the chemical complexity and variability of the product. All samples should have been analysed for protein by the wet chemistry reference method. It would also be prudent to analyse for any potentially interfering components, so that the effect of these can be unequivocably quantified with specific checks. Wavelength selection is based upon interpretation of NIR spectra of the milk powder and its components. Reflectance values at the chosen wavelengths are collected using an infragauge to ensure that the information is recorded with adequate photometric accuracy.

Data analysis is based upon examining appropriate wavelength combinations which make chemical sense; for instance, an absorption band for protein should be included. Additional wavelengths are used to provide the necessary cross-compensation for any interference from the other constituents.

The coefficients applied to the signal for each wavelength are optimized using an appropriate form of multiple regression analysis. The net result is that an algorithm can be derived which is of the form

$$\text{output} = a_0 + g(a_1 \log S_1 + a_2 \log S_2 + a_3 \log S_3 + \cdots) \qquad (3)$$

where a_0 is the offset, g is the gain factor, a_i are coefficients and s_i are signal intensities at given wavelengths.

To establish the robustness of the measurement, it is imperative that proving samples not previously included in the training are measured with the gauge. The more thoroughly these two stages are carried out, the better will be the calibration.

The main benefit of this rigorous approach is that it is possible to deliver an instrument in a pre-calibrated condition. This is a great advantage with a process instrument since it can be commissioned in the shortest possible time and requires only minimal input from the user's laboratory facility. Experience to date has shown that such calibrations are very reliable with time and can accommodate the year to year changes of natural products with little effect on the measurement.

Examples of the multi-component applications tackled to date, apart from protein, include fat levels in a variety of snack products, milk powder, non-dairy creamers, chocolate based products or precursor materials to chocolate. These measurements are invariably performed simultaneously with that of moisture. Examples of the quality of the fat calibrations on powdered chocolate crumb and non-dairy creamer are shown in Figure

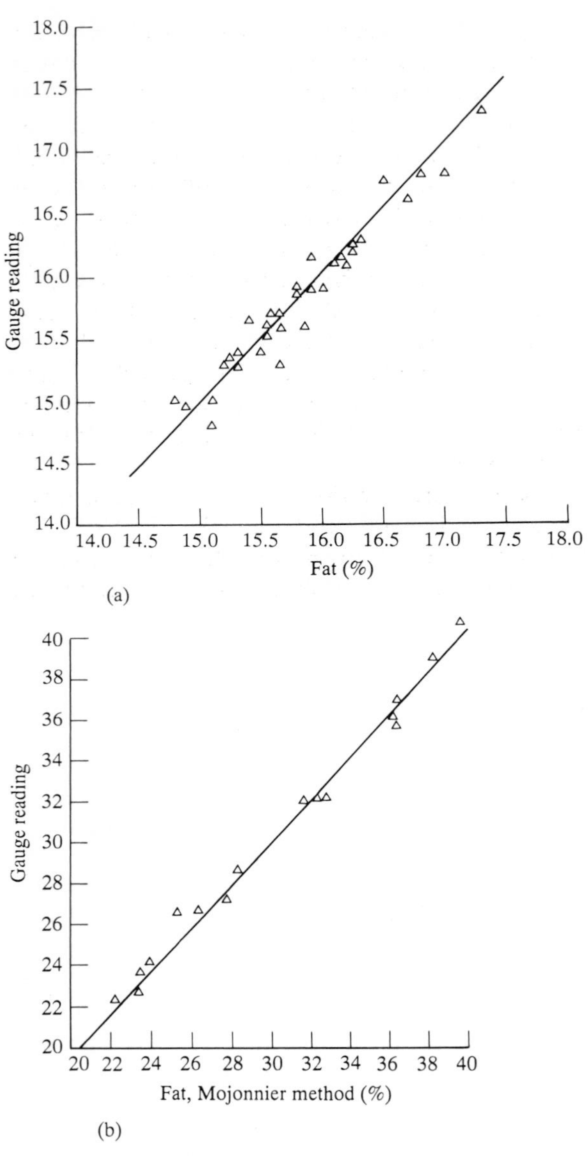

Figure 5.17 *Calibration graphs representative of on-line NIR fat measurements in (a) chocolate crumb (b) non-dairy creamer*

5.17. Measurement of the sugar content of these same products has also proved successful.

A recent on-line application which highlights the capability of NIR analysis was the measurement of caffeine levels in coffee. NIR does not have the sensitivity to reliably measure trace components at levels below say 0.5 per cent w/w, and hence could not be used on-line to indicate whether the decaffeination process had resulted in extraction of caffeine to levels below 0.1 per cent. However, the input to the process could be monitored since typically the caffeine level is between 1 and 2 per cent. This concentration range can be monitored accurately and, using a feed-forward control system, the end point of the decaffeination process could still be much more accurately determined than would be possible without a continuous measurement. Figure 5.18 shows the quality of the calibration, which yielded measurement accuracies better than ±0.15 per cent caffeine when compared with the wet chemical analysis. Although outside the scope of this subject, the application of NIR to provide on-line measurement of nicotine and total reducing sugars at concentrations of 1 to 5 per cent and 5 to 30 per cent respectively in tobaccos helps to further highlight the scope of this technique.

There are numerous other opportunities in this relatively new area of on-line NIR measurement. To help a potential user decide whether an on-line NIR gauge for his product may be appropriate, the guidelines discussed for moisture are generally useful. However, one advantage in the non-moisture

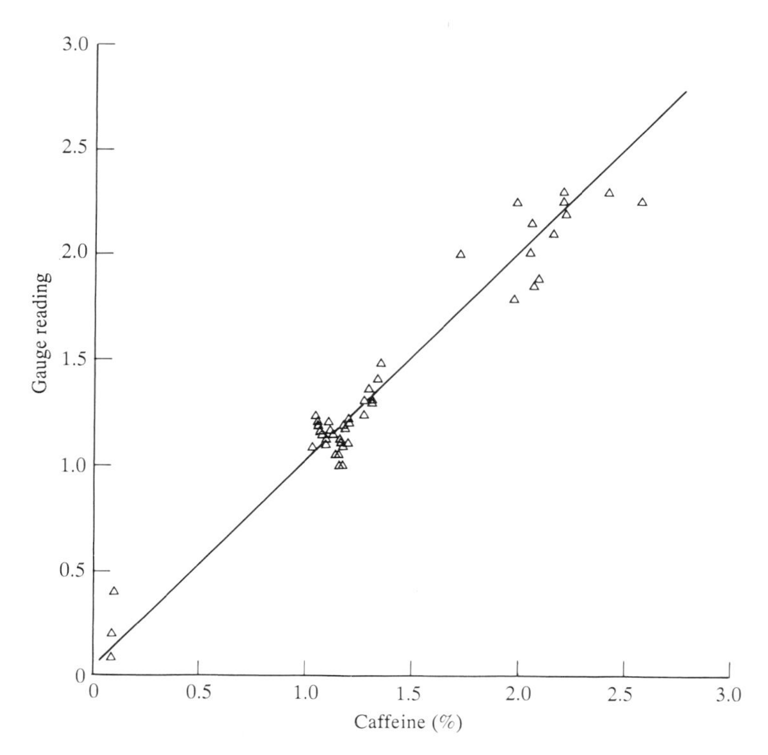

Figure 5.18 *Measurement of caffeine in ground green coffee beans*

area is that for a given product the constituent in question is quite likely to be uniformly distributed and should not change with time, and thus the installation requirements can be considerably simpler than for moisture where heterogeneous distribution can be a problem. Naturally there are exceptions; for example, with a milk powder fat measurement, care must be taken when siting the gauge because emulsifying agents such as lecithin are often sprayed on to the product, thus giving surface biased characteristics.

Measurements currently being performed with a laboratory NIR instrument are likely to be amenable to on-line analysis with the new instrumentation, but the comments made earlier in this section about the suitability of an application to NIR analysis should be borne in mind.

4.3 Quality Control for Food Packaging Materials

A number of the food industry's activities overlap with operations in other industries which provide materials for use by food producers. Many of these associated industries may already be making NIR measurements of their products to control the quality; however, the food technologist may be unaware of this occurring or even of the availability of such means to test these product specifications. The most important area must be that of general packaging materials. This encompasses a huge range of products including paper, board, plastic packaging films and metal foils. Many of the associated industries are actively using on-line NIR measurement to continuously monitor and control the manufactured product.

Many food products are packaged with one or more of these materials, and they can influence taste, shelf life or appearance of the foodstuff. For example, board is laminated with polyethylene (PE) to provide a moisture barrier for the packaging of milk or other liquids, while board/aluminium-foil/PE laminates are frequently used to package fruit juices. Similarly, wax coated papers are used to package anything from Camembert cheese to the ubiquitous hamburger, providing again a means of keeping in moisture but also avoiding grease migration. Aluminium foils are coated with thin layers of vinylic, PVC, or epoxy and other lacquers a few micrometres thick and then used to provide tops for yoghurts, wraps for cheese and a host of other similar applications.

All of these materials involve a coating operation, and the thickness of the deposited layer can be readily measured by on-line NIR instrumentation. It would equally be possible to use NIR in an off-line configuration to assess the quality of incoming packaging, in terms of both moisture content and coat weight. The coat weight measurement relies upon the identification of a characteristic NIR absorption due to an organic polymer –CH group which can then be monitored in much the same way as for the –OH group of moisture with a reflectance instrument.

Perhaps the technologically most exciting area is that of polymer film packaging. Most such materials are nowadays not single-layer films but rather complex multi-layer coextrudates made up of different polymers. Many modern food packages use combinations of these materials to protect the foodstuff. Bacon and other meats are often packaged in nylon/polyethylene structures, while very oxygen sensitive foodstuffs such as fish and fresh meats may be kept in a PE/EVOH/PE type composite (where EVOH stands for ethylene vinyl alcohol and PE for polyethylene).

The objective here is to exploit the differing barrier properties of the many polymers in existence to achieve a packaging medium with the desired specification to maximize food shelf life. For instance, polyethylene is a good moisture barrier but a rather poor oxygen barrier. Polymers such as nylon and EVOH are good and excellent oxygen barriers respectively but both are permeable to moisture.

Since the NIR technique is very selective it is possible to individually measure the critical thicknesses of these layers by identifying the differences in absorption and including the appropriate infrared filters for each polymer in a transmission instrument. For example, PE will show only –CH absorptions while EVOH will show both –CH and –OH absorptions, thus providing a means of their separation.

While these applications may at first sight seem rather remote from the food technologist's needs, it could be important to realize their existence either for the setting up of an incoming product screening method, or perhaps to insist that a supplier undertakes these measurements for greater security of product quality in such a vital area of health concern.

5 Practical Aspects of the Calibration of On-line NIR Instruments

The general principles behind calibration of any instrument and the objective of calibration are very simple concepts, but ones which can cause great anxiety amongst users. The reasons behind this may in some cases be a poor understanding of the operation of the instrument, but more commonly it is the practical difficulty of accurately assessing the instrument once it is installed on-line. For this reason it is convenient to divide the calibration of an NIR instrument into two stages:

(a) Obtaining as accurate a calibration of the instrument as possible off-line using pre-conditioned or pre-prepared samples.
(b) Installing the instrument on-line and deciding on an appropriate robust and realistic strategy to cross-check its performance with the conventional laboratory analysis.

Step (a) can often follow a clearly defined, unambiguous procedure whereas

(b) very much depends upon the application, the degree of flexibility in the particular process for allowing plant changes, and a good understanding of the *actual* short or medium term variation of moisture or another component in the process, since these will obviously influence the cross-checking strategy. This latter point is the most important aspect within (b), and the one that is the potential stumbling block, since in comparative tests the laboratory samples must be representative of that material 'seen' by the instrument. This subject of sampling, as it is known, will be discussed after calibration principles have been described.

5.1 Calibration Methods

In many non-moisture applications a gauge should be delivered to the user in a pre-calibrated condition, and therefore only requires the on-line comparative checking and subsequent minor adjustments that will be described in the next section. Therefore the following description will relate solely to moisture calibration.

Calibration is the process of scaling the output of an instrument that already shows a response which highly *correlates* with moisture variation, but does not concur with the absolute values as obtained from the accepted reference method for analysis. The important point is that an uncalibrated gauge can still show *relative* changes in moisture content very accurately and repeatably because the instrument algorithm should be designed to provide a linear response to changes in moisture content. If, for the NIR analysis of solids, we were dealing with absorption phenomena alone, it would be possible to produce an instrument which would measure moisture irrespective of product. Kubelka-Munk mathematics describing the interaction of light with solids shows that the apparent absorption coefficient k is equal to $\sqrt{2k/s}$, where k is the component absorption coefficient and s is the scattering coefficient. In reality there are large differences in the scattering characteristics of most products, which clearly makes it necessary to calibrate an instrument to suit that particular material.

It may be adequate in some instances to omit calibrating the gauge, and for process control purposes merely to ensure that the product is always dried or conditioned to a set point or target value. Provided the user has knowledge that, for example, when the gauge reads y units then the product moisture content is 4.5 per cent w/w, this may be adequate. However, the majority of users prefer the gauge to read accurately in familiar, scaled units such as percentage moisture. The difference in the two approaches is illustrated by Figure 5.19, where uncalibrated and calibrated gauge responses are shown in graphical form.

The conventional names of the two instrument controls which allow adjustment of the sensitivity and the intercept value are 'span' and 'zero', respectively, although 'gain' and 'offset' may also be found.

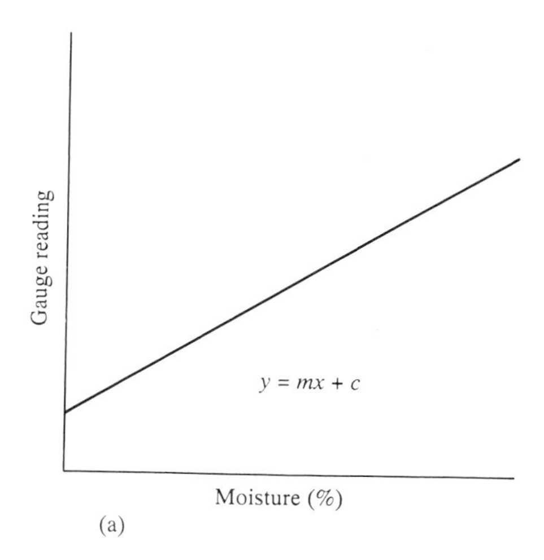

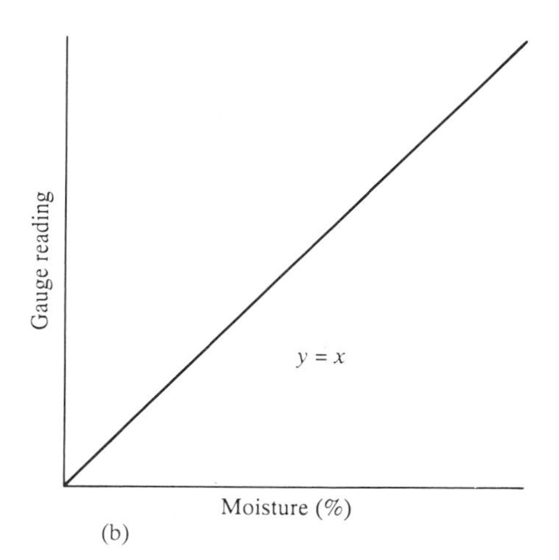

Figure 5.19 *Two modes of gauge calibration: relative changes accurately observed (a) with an arbitrary scaling (b) in absolute units (moisture content)*

To accurately calibrate a gauge requires samples over as wide a *range* of moisture contents as possible. The need for this should be clear since, if the gauge were operating over a very narrow range, the determination of its sensitivity (the slope of the calibration line) would be very inaccurate. Such samples can in many cases be prepared by conditioning a bulk of production material by adding or removing moisture from subsamples of the material. Although this procedure has received criticism in the past for the way it can physically change a product, or even because the moisture is combined differently in conditioned samples, it does work extremely well as far as the *NIR calibration* is concerned, especially if the precautions as described have been taken to desensitize the instrument to product appearance. The conditioned samples should be left to equilibrate before use; the period depends upon the particle size and the permeability of the particles, but typically 48 hours is more than enough for most food materials.

Alternatively in the unusual event that the production process can be altered (and agreement obtained to alter it!), a range of samples at different moistures can be collected. Another trick is to collect samples before and after the drying process, provided it has not radically altered the product. For example, it would not be sensible to use biscuits entering the oven as a dough and emerging cooked, because significant physical/chemical changes occur in the product during the cooking process, which will affect its NIR reflectance/scattering properties. In this instance, the cooked product can be easily conditioned to provide a set of samples at different moistures.

The calibration procedure should involve presenting the product in a form similar to that observed on-line. A rotating turntable is a useful substitute for a moving conveyor belt, but care must be taken that the product does not dry down too quickly, or in some cases pick up moisture while exposed to the air, bearing in mind the surface bias nature of NIR.

Generally, an instrument will be set up at least on a low and high moisture sample to get the sensitivity (calibration slope) approximately right. Subsequently, samples at a range of moistures would be presented to the gauge, and perhaps five average readings would be taken on each. Duplicate testing would then be carried out on each sample in the laboratory, using either a gravimetric air or vacuum oven test or a chemical analysis such as Karl Fischer titration. The data pairs can be plotted graphically or directly entered into a regression program; more commonly nowadays, the microprocessor built into the instrument will automatically analyse the data. Graphically plotting the data is always an extremely sensible precaution for visually assessing the quality of the results, thus preventing the calculation of calibration constants that could have been disproportionately affected by one or two outliers. In all instances the calculations derive the best fit regression line which provides a value for the slope and intercept of the calibration line so that any corrections can be made to give the ideal calibration $y=x$.

The number of samples required for this operation depends upon the type of product and its particle size, both of which may affect the variability of the reading obtained on the samples. For a fine powder, such as coffee or milk powder, very good calibrations can be achieved with just five samples, whereas with large flaky materials fifteen to twenty samples may be needed. Generally this is an area that *should not* be skimped, and therefore as many good quality samples as is practical (over a wide moisture range) should be used in the initial calibration. This will result in the derivation of a calibration which can be used with greater confidence.

5.2 On-line Installation and Sampling Procedures for Cross-checking

In the ideal world an instrument calibrated off-line should be operational on-line immediately. However, this is not often the case except on some fine powdered products. The reasons for this are many, and include the fact that some products may have emerged from a dryer and are not moisture equilibrated. On-line the gauge may have to look through a glass window or even operate at a much greater distance from the product than was used off-line.

The most sensible assumption to make at this stage is that any calibration correction needed will only be an intercept adjustment. Even if this is not strictly correct and there is a small error in the slope, this will not adversely affect an on-line calibration if a product is being made to a target moisture. Figure 5.20 illustrates this.

To correct for any intercept error in the calibration requires the collection of some ten samples from the line, and the recording of each gauge reading for comparison with a subsequent set of reference tests. The mean difference in reading can then be used with confidence to correct the on-line calibration.

The importance of correct sampling methods cannot be over-emphasized, since only when they are carried out properly will the user have any chance of assessing the gauge's performance.

The moisture in a product stream can vary enormously in the short term about a reasonably consistent average value. This is, for instance, particularly true for spray drying operations. Superimposed on this short term variation will be the longer term overall changes caused perhaps by a real change in moisture of the material input to the dryer. It is the longer term trends that can be corrected for by process control, but it is important for the user to be aware of the short term variations because of the effect they will have when comparing the gauge and laboratory methods if they are not taken into account. Typically the laboratory analysis will be based upon very small samples, about 5–10 g of product, while the gauge will be providing an average value of many kilograms or even tonnes of product!

The instrument, dependent upon the response time chosen, will take these short term variations into account but will not show the troughs or peaks of every short term deviation from the mean moisture value. In these instances, a single laboratory sample is very unlikely to agree with the average value shown by the gauge at any one time.

The response time of the gauge is usually an exponential time constant, effectively providing a running average of the moisture content of the product stream. By definition, this type of response time is the interval during which the gauge will have made a 63 per cent change towards the new value following a step change in moisture. The longer the chosen value, the more slowly the gauge will respond to a sudden change in moisture. The chosen response time therefore depends on how variable is the product moisture, and is usually derived empirically by adjustment until a sufficiently stable output is obtained for practical use. For the control of a drier, where its response time to provide a change in drying conditions may be of the order of minutes, a gauge response time of perhaps 10 or 20 seconds would be appropriate. However, on a potato crisp production line, where it can be important to quickly identify 'wet' batches of product so that they can

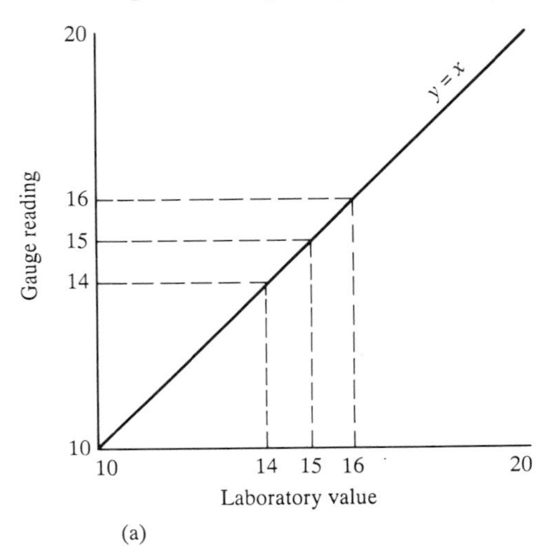

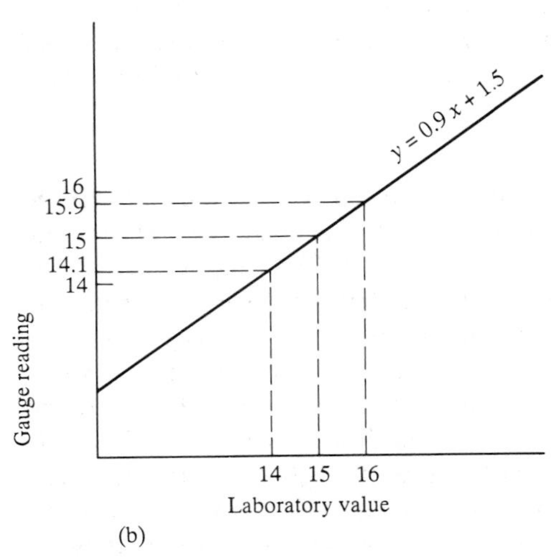

Figure 5.20 *Illustration of the minor effect of an error in calibration span when an instrument is accurately set at the product target moisture: (a) $y=x$ (b) $y=0.9x+1.5$. In the vicinity of the critical target value (15), both calibrations provide high accuracy*

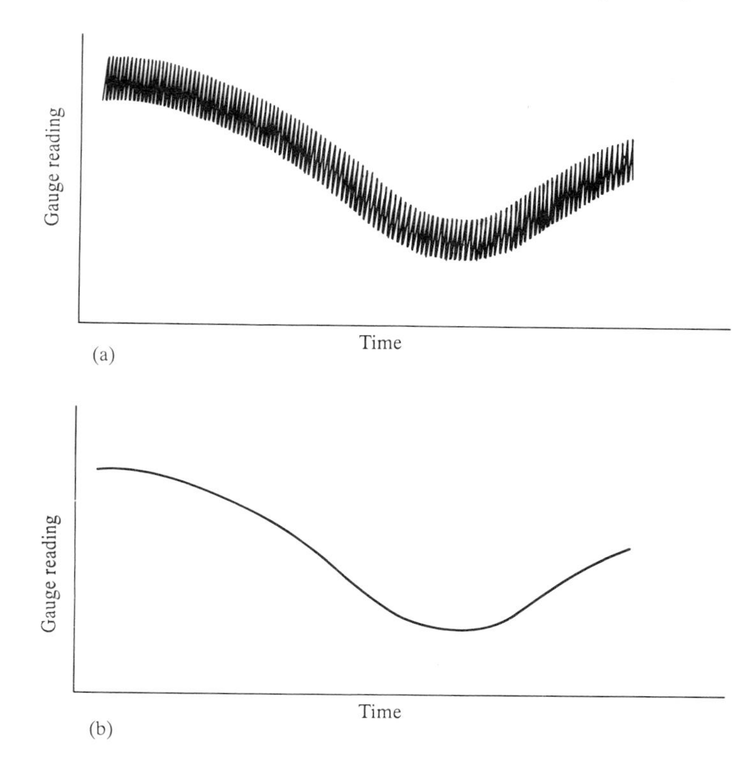

Figure 5.21 *Effect of instrument response time upon the measurement characteristic of the infragauge: (a) response time 0.2 s, output displays short and longer term process variation (b) response time 10 s, output displays longer term variation only*

be immediately diverted, a shorter response time of 2 to 5 seconds might be used to maximize the ability to detect the out-of-specification product. The ideas expressed above are schematically shown in Figure 5.21.

The sampling regime must therefore be devised to ensure that the samples collected are likely to be representative of the average moisture content. This can be achieved either by taking a large bulk of sample over a period and immediately subdividing it for multiple testing, or by collecting many small samples over the same period. Both ways will show the short term moisture variation and will enable a meaningful average moisture value to be calculated. If the standard deviation associated with the variation within these samples is derived, this will provide an estimate of the likely sampling error that can be expected in the particular process. When one is assessing the actual performance of the infrared gauge, this error must be taken into account. The period chosen for sampling should be between two and three times the chosen response time for the instrument, and samples should only be collected while the gauge output is steady with no obvious large changes in product moisture occurring.

These guidelines should assist in setting up procedures for comparative testing, not only for moisture but for any other components being measured by this technique.

Before concluding this important topic, it is worth stating the need for the user to be aware of the accuracy of the laboratory test against which the NIR sensor will be compared and the potential errors that can arise if the test is badly performed. Having this information in advance will allow a realistic measure of the accuracy of the gauge to be derived. The accuracy of a laboratory test is best assessed by multiple testing of subsamples taken from a bulk of equilibrated product and by calculating the average moisture content and associated standard deviation. The reliability of this figure will be improved if the test is carried out blind, that is without the knowledge of the operator. If it is large, this error could once again easily obscure the true performance of the infrared instrument.

6 Conclusion and What the Future Holds

The main objective of this chapter has been to familiarize the reader with the principles of NIR measurement and the capabilities of modern on-line instrumentation with regard to the food industry's requirements.

The practical considerations of how to use NIR on-line have been emphasized because this is often an area that gives a misleading impression about the quality of a given measurement. Laboratory NIR analysis has not been covered to any great extent, but it should be remembered that laboratory measurements using the same technology *derived from an on-line measurement philosophy* will probably give better results off-line because presentation conditions and sample handling can be far better controlled.

While moisture measurement features strongly, non-moisture measurements have been introduced since, with the new digital instrumentation, this will continue to be a major growth area for on-line NIR measurements. The use of NIR in the industries closely associated with the food industry have also been covered, since these may play an indirect role in determining food quality.

Given these facts, it is hoped that the reader will be well equipped to decide if NIR is a suitable technique for his or her particular measurement problem, and furthermore will be inspired to provide the instrument manufacturer with a whole series of new challenges.

There is currently considerable interest in the use of NIR as a means of identifying products, or in the pharmaceutical industry for the coarse assaying of preparations using the spectral characteristics of a product at a number of wavelengths to define it uniquely. In the realm of composition analysis, a huge diversity of requirements continue to arise; the measurement of free fatty acid content in frying oils for snack products and the sugar content of raw potatoes are two extreme examples which illustrate this.

The on-line NIR instrumentation clearly opens up a huge new range of opportunities to analyse components which perhaps would previously have been of limited interest for process control when only a historical off-line sampling analysis was available.

It is hoped that, through the development of a better general awareness and acceptance of NIR analysis, this powerful technique will continue to grow rapidly in the foreseeable future, and to diversify into areas of which even the instrument manufacturer is unlikely to be aware!

References

Banwell, C. N. (1983) *Molecular Spectroscopy (3rd edition)*, McGraw Hill

Benson, I. B. (1989) Industrial applications of NIR for the measurement of moisture, *Measurement & Control*, **22,** 45

Benson, I. B., Edgar, R. F., and Federle, H. (1988) On-line multi-component analysis of solid materials using NIR instruments. In *Process Optical Measurements*, SPIE, Vol. **1012,** 155-162

Biggs, D. A. (1979) Performance specifications for infrared milk analysis, *Journal of the Association of Official Analytical Chemists*, **62,** 1211-1214

Bruton, D. C. (1970) Measurement of moisture and substance by infrared radiation, *Measurement and Control*, **3**

Cowe, I. A., and McNicol, W. H. (1985) Principal component analysis, *Applied Spectroscopy*, **39,** 257

Davies, A. M. C. (1987) Near infrared spectroscopy: time for the giant to wake up, *European Spectroscopy News 73*

Edgar, R. F., Hindle, P. H. (1971) Infrared gauging, *Optics and Laser Technology*, February, 5

Giesbrecht, F. G., McClure, W. F., and Hamid, A. (1981) Fourier analysis of NIR spectra, *Applied Spectroscopy*, **35,** 210

Herzberg, G. (1945) *Infrared and Raman Spectra*, D. Van Nostrand, New York

Honigs, D. E. (1985) Near infrared analysis, *Analytical Instrumentation*, **14,** (1), 1-62

Hunt, W. H., Funk, D. W., Elder, B., and Norris, K. H. (1977) Collaborative study on infrared reflectance devices for determination of protein in hard red winter wheat and for protein and oil in soybeans, *Cereal Foods World*, **22,** 534-538

Kaye, W. (1954) Spectral identification and analytical applications, *Spectrochimica Acta*, **6,** 257

Kubelka, P., and Munk, F. (1931) Ein Beitrag zur Optik der Farbanstriche (A contribution to the optics of paints), *Zeitschrift für Technische Physik*, **12,** 593

Kubelka, P. (1948) New contributions to the optics of intensely scattering materials, *Journal of the Optical Society of America*, Part 1, **38,** 448

Mark, H. (1986) Qualitative NIR reflectance analysis using Mahalonobis distances, Analytical Chemistry, **58,** 379

Martens, M. and Martens, H. (1986) NIR Reflectance determination of the sensory quality of peas, Applied Spectroscopy, **40,** 303

McCallum, J. D. (1961) Limitations and design philosophy in ultraviolet, visible and near-infrared spectrophotometers, *Trans. New York Academy of Sciences*, **24,** 140–157

McCloud, H. A. (1986) *Thin Film Optical Filters*, 2nd edition, Adam Hilger

McClure, W. F., Hamid, A., Giesbrecht, F. G., and Weeks, W. W. (1984) Fourier analysis enhances NIR diffuse reflectance spectroscopy, *Applied Spectroscopy*, **38,** (3), 322

Miller, W. C., Hare, G., Strain, D. C., George, K. P., Stickney, M. E., and Beckman, A. O. (1949) A new spectrophotometer employing a glass Fery prism, *Journal of the Optical Society of America*, **39,** 377–388

Norris, K. H., and Hart, J. R. (1965) *Principles and Methods of Measuring Moisture in Liquids and Solids*, Vol 4, Reinhold, New York

Norris, K., and Williams, P. (1987) Near Infrared Technology in the Agricultural and Food Industries, American Association of Cereal Chemists Inc, St Paul, Minnesota, USA

Osborne, B. G., and Fearn, T. (1986) *Near Infrared Spectroscopy in Food Analysis*, Longman Scientific and Technical, Harlow

Star, C., Morgan, A. G., and Smith, D. B. (1981) An evaluation of near infrared reflectance analysis in some plant breeding programmes, *Journal of Agricultural Science*, Camb., **97,** 107–118

Willis, H. A. (1979) Industrial plant applications of infrared and Raman spectroscopy, *Advances in Infrared and Raman Spectroscopy*, **2,** 81–139

6 *Practical Aspects of Infrared Thermometry*

G. Beynon

Contents

1 Introduction 167

2 Radiation Thermometers 168
2.1 Collection 168
2.2 Detectors 170
2.3 Processing Electronics 172
2.4 Housings and Configurations 173
2.5 Line Scans and Thermal Imagers 175

3 Measurement Principles 178
3.1 The Black Body 178
3.2 Non-black Target 179
3.3 Background Temperature 179
3.4 Emissivity 181
3.5 Examples 182

4 Miscellaneous Techniques 184
4.1 Enhancing Emissivity 185
4.2 Avoiding the Need to Know Emissivity 185
4.3 Inaccessible Targets 186

References 188

1 Introduction

All objects emit infrared radiation. The emitted energy and its distribution in wavelength are strong functions of temperature. Infrared thermometers are fairly uncomplicated devices which detect this infrared radiation and deduce a temperature. They are widely used throughout process industry, both in hand-held 'camera' form for spot checks and in installed 'fixed system' form for on-line monitoring and control.

In the right circumstances the measurement can be very accurate indeed. Further, it is measurement at a distance – non-contact, non-invasive and non-contaminating. Measurements inside closed vessels are possible with suitable windowing. Fibre-optic cables can in some situations be used to help access products in hostile conditions. There is a misapprehension that infrared thermometry is applicable only at high temperatures; in reality, with modern equipment, measurements below 0°C are quite practicable.

The uptake of infrared technology by the food industry has been very limited. The reasons usually cited are:

- high costs
- insufficient accuracy in practical situations
- little really suitable equipment available.

All three criticisms have some justification, but new instrument developments are improving the situation.

Certainly developments within the industry provide strong incentives for adoption of infrared methods. Specific examples are microwave heating, which has associated problems in the use of conventional thermometers, and the widespread use of cool storage, with the associated need for rapid, non-contaminating temperature checks. More generally, progress from batch methods to continuous processing favours non-contact techniques and, in well controlled processes, the thermometry requirement is invariably repeatability, which infrared can often deliver in situations where absolute accuracy is difficult.

There is little published on the practicalities of infrared thermometry, particularly at lower temperatures. It is hoped that this chapter goes some way to filling that gap.

2 Radiation Thermometers

A radiation thermometer (Figure 6.1) comprises three essential elements:

- an infrared collection system
- an infrared detector
- some electronic signal processing.

Other topics in this section are housings and configurations, line scans and thermal imagers.

2.1 Collection

The infrared wavelengths of interest in radiation thermometry (1–15 μm) are not so very much longer than the wavelengths of visible light

(0.4–0.7 μm). As a result, infrared thermometers look like optical instruments, rather than plumbing or antennae, and operate according to simple laws of ray optics.

A well designed instrument collects radiation only from within a precisely defined field of view. Figure 6.2 shows a typical example. The field is axially symmetrical and necks down from 22 mm diameter at the thermometer to 10 mm at the focus point, diverging thereafter. The target spot, whose temperature is measured, is the intersection of this field of view with the target surface. For a well designed instrument the measurement is distance independent; in particular, the focus point has no significance to the user beyond being the point at which the target spot is smallest. The target surface must be large enough to fill the field of view, otherwise an average of the target and background temperature is returned. The measurement is insensitive to obliquity of the target surface, typically for angles up to 50° off normal.

The vast majority of target materials are highly opaque in the infrared. The emitted energy thus orginates only just below the target surface. Penetration depths are typically reckoned in micrometres and the temperature measured is, to all intents and purposes, a surface temperature.

Many instruments have adjustable focus. Typically the field angle is preserved, at least approximately, while the focus distance changes. For example, the instrument of Figure 6.2 might be adjustable to a 5 mm target spot at 600 mm or a 2.5 mm spot at 300 mm. Focus adjustment is usually effected by rotating a lens housing or fitting an auxiliary lens in front of the main optical system.

Most instruments incorporate some form of aiming system. An insertable telescope is a convenient approach for fixed instruments which are aimed just once, during installation. Portable instruments benefit from a more elaborate through-the-lens reflex arrangement.

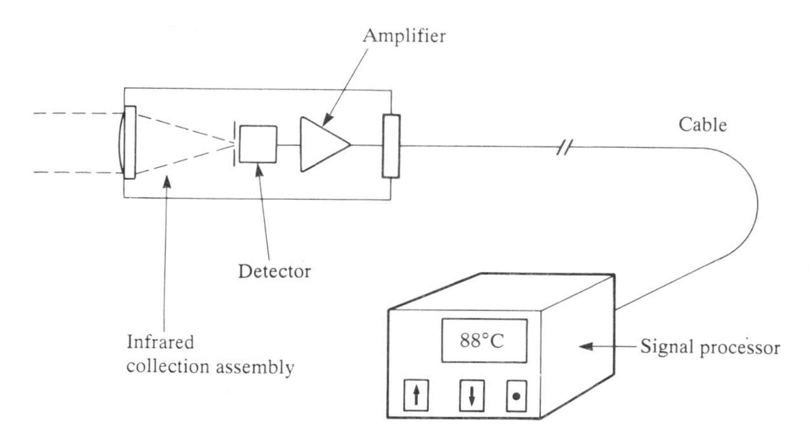

Figure 6.1 *Schematic diagram of infrared thermometer*

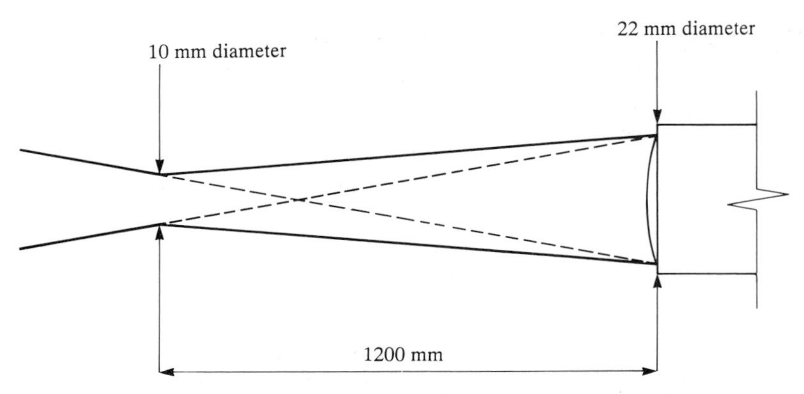

Figure 6.2 *Schematic diagram of thermometer field of view*

2.2 Detectors

The detector transduces the collected infrared energy to an electrical signal. Typically this signal is very small and the detector is combined with an amplifier to achieve a more usable level. The detector usually incorporates a spectral filter, which determines the range of infrared wavelengths sensed. We shall see in the next section that the operating waveband has a critical influence on instrument accuracy.

Details of detector operation need not concern the instrument user, but some appreciation of the available types and their performance limitations is valuable. Table 6.1 lists the operating waveband and detector combinations most commonly used in industrial general purpose thermometers, together with the target temperature ranges over which they are most useful. Note that emitted radiation shifts to shorter wavelengths as temperature increases.

The listed wavebands correspond to atmospheric transmission 'windows' (Figure 6.3), i.e. regions of the infrared spectrum where the

Table 6.1 *Operating wavebands and detector types for some commonly available infrared thermometers*

Waveband (μm)	Detector types	Useful range (°C)
0.7 to 1.1	Silicon	400+
1.0 to 1.8	Germanium	200 to 800
2.0 to 2.4	Lead Sulphide	50 to 600
8 to 14	Thermopile, pyroelectric	– 50 to 300

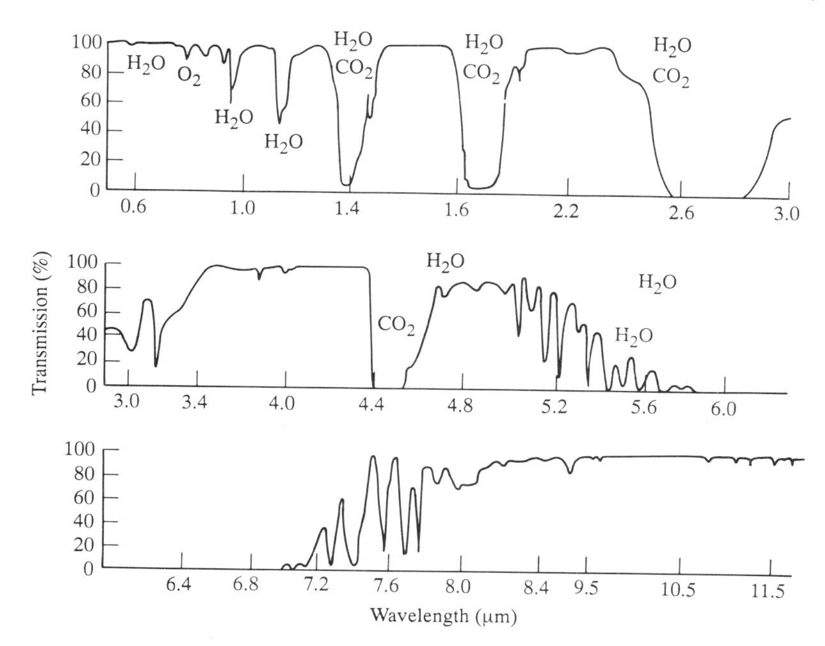

Figure 6.3 *Spectral transmission of the atmosphere: 300 m path length at sea level with 5.7 mm precipitable water at 26°C (Wolfe and Zissis 1978)*

atmosphere is highly transparent. Atmospheric absorption primarily derives from carbon dioxide and water vapour; symmetrical molecules (O_2, N_2 etc.) have no absorption bands in the infrared. A thermometer with a poorly chosen waveband will give readings which depend on target distance and also on humidity. The potentially very useful 3–5 μm window is not commonly utilized in industrial single-spot thermometers for reasons concerned with the cost and performance of available detectors; it is widely used in line scanners and thermal imagers (see Section 2.5).

The silicon photovoltaic diode is a superb detector of infrared near 1 μm wavelength. The detection mechanism is electron-hole pair generation by photons absorbed in a p–n junction. The signal current is linear with incident energy to better than 1 per cent over four to five decades. With careful amplifier design, the zero drift is negligible. The effective gain varies with ambient temperature, but in a way which is predictable and can be accurately compensated. There are no significant ageing mechanisms. Fast (microsecond) response times are achievable.

Silicon's usefulness is limited to target temperatures above about 400°C; the amount of 1 μm energy available for detection falls away very rapidly at lower temperatures. Efforts to squeeze lower temperature measurements by use of elaborate amplification schemes run into problems with reflected daylight.

Germanium operates in a similar fashion to silicon, but has a smaller bandgap, allowing absorption of lower energy, longer wavelength photons. It has all of silicon's attributes provided it is kept reasonably cool. Industrial germanium thermometers are typically specified to 50°C ambient but, in practice, benefit from the use of a water cooling jacket. Germanium is useful for targets above about 200°C.

Lead sulphide is a photoconductor, the detection mechanism being photon-induced resistance change in the bulk semiconductor. It is very sensitive but unstable. It is a practical necessity to modulate the incoming infrared and amplify the resultant AC signal. Modulation is invariably accomplished with some form of mechanical chopper, introducing issues of reliability and service life. The detector gain is subject to ageing and hysteresis effects.

Good thermometer designs incorporate lead sulphide only as a sensitive element and not as a calibrated device. Regrettably many instruments have been marketed which ignore this principle; they exhibit serious calibration drifts and have a tendency to 'remember' past illumination. Lead sulphide is useful down to about 50°C target. With good design it is the basis of the most accurate thermometers at these temperatures.

Thermopiles and pyroelectrics are thermal detectors whose sensitivity extends over all wavelengths. Thermopiles are easy to use and form the basis of a wide variety of simple and generally trouble-free thermometers. The response times are typically limited to about 50 ms. Pyroelectrics are somewhat more sensitive and faster (a few milliseconds) but a chopped radiation input is fundamental to the detection mechanism. Pyroelectrics are common in portable (i.e. intermittent use) thermometers, but the absence of moving parts is a powerful argument in favour of thermopiles for installed systems.

Used with an 8–14 μm filter, thermal detectors allow measurement down to room temperature and beyond.

2.3 Processing Electronics

The processing electronics performs an algorithm of the general form

$$T = f(\varepsilon, T_i, L)$$

where the inputs ε, T_i, L are, respectively, a user-set 'emissivity' factor, the thermometer internal temperature, and the infrared signal. The target temperature T is typically output as an analogue signal (e.g. 4–20 mA) plus a digitally displayed value.

Microprocessor-based processors conventionally take the form of a panel-mounted unit with digital temperature display and user keypad. These units are very flexible, and many processing functions may be available beyond the basic derivation of target temperature. For example, the user may be able to select peak or valley picking of the output signal with adjustable decay times; sample/hold actuated by an external trigger signal; high/low alarm activation with adjustable limits; on/off, PID (proportional integral differential) or more complex forms of control with adjustable set-point and other parameters. Mercifully the gee-whiz period is over and the focus is now on providing useful levels of flexibility with maximum user-friendliness. The microprocessors typically used are 8 bit single-chip microcomputers with the program stored in EPROM and keypad set parameters stored in NOVRAM. A very high level of noise immunity is an essential feature for industrial use, particularly in heavy plant, requiring careful attention to both hardware design and software self-checking.

Analogue processors continue to play an important role and there are good reasons for this in terms of both performance and cost-effectiveness. Digital processors are too expensive, power hungry, bulky and slow for some applications. It should be noted that, at short wavelengths, 20+ bit conversions are typically needed to accommodate the extreme non-linear radiance/temperature relation. The best of the analogue systems display an elegance bordering on genius and accommodate, with just a few components, enormous dynamic ranges in radiance with response times down to a few microseconds or power requirements down to a few hundred microwatts. Drift-free performance is achievable with good design. However, analogue cannot match microprocessors for flexibility.

2.4 Housings and Configurations

A first breakdown is between portable and fixed instruments.

Figure 6.4 shows a high specification portable radiation thermometer (Cyclops 33). This instrument covers the range -50 to $+1000$°C. It has an 8–14 μm operating waveband and uses a pyroelectric detector. The optics

Figure 6.4 *High specification portable radiation thermometer Cyclops 33. (Courtesy of Land Infrared Ltd)*

Figure 6.5 *Low cost portable radiation thermometer Compac 3. (Courtesy of Land Infrared Ltd)*

defines a 1 degree field of view, with focus adjustable from 750 mm to infinity. Reflex visual sighting is included. The measured temperature is displayed in the viewfinder and is also available as a digital output. Power is from a single PP3 9 V battery. A host of accessories is available, including data logger, printer and DA converter.

The instrument shown is one of a range offering various temperature spans and operating wavelengths within a common mechanical configuration. This range actually derives from a manufacturing agreement between a company specializing in infrared (Land) and a household-name camera manufacturer (Minolta). The success of this arrangement has, in recent years, spawned several such collaborations and something of a revolution in the cost and quality of portable radiation thermometers. Figure 6.5 shows a low-cost portable instrument (Compac 3) which is attracting interest in the food industry. It offers most of the features of the first instrument at a quarter of the cost.

Fixed instruments can be broadly divided into stand-alone and thermometer plus processor configurations. Figure 6.6 shows a thermometer plus processor. The optical collection, detector and associated amplification are housed in a cylindrical metal body connected, via a cable, to a remote processor and display unit. The instrument shown is actually a lead sulphide type covering 50–150°C. However, it is again one of a range (System 3) which includes various wavebands and temperature spans in a common format. The processor is a flexible, microprocessor-based unit which offers a variety of processing functions including three-term control.

Set-up parameters are input via touch keys on the front panel. Broadly similar equipment is available from several manufacturers.

Figure 6.7 shows a stand-alone thermometer (SOLO 3). Here the collection, detection and processing are contained within a single housing. The instrument shown is an 8–14 μm thermopile type measuring 0–250°C. Again it is one of a range.

Stand-alone instruments are typically available in two-wire and four-wire configurations. In four-wire types there are two connections for power-input and two for signal output. In the increasingly popular two-wire loop arrangement there are just two connections and the signal current is impressed on the current drawn from the power supply. Usually it is arranged for the loop current to be 4 mA at the bottom of the measurement range and 20 mA at the top, with linear interpolation. Clearly the whole of the thermometer electronics must operate off 4 mA, and this results in performance limitations.

For special applications, infrared thermometers can be built in a multitude of shapes and sizes. For example, Figure 6.8 shows a developmental miniature unit for use in commercial microwave ovens.

An essential part of any range of infrared thermometers is a selection of mounting fixtures, water and air cooling jackets, and air purge attachments. The latter in particular have reached a high state of development and are capable of maintaining clean thermometer optics over very extended periods under the most appalling conditions in heavy industrial plant.

2.5 Line Scans and Thermal Imagers

A line scan measures a temperature profile across a target surface. It is essentially an infrared thermometer whose field of view is scanned, in one

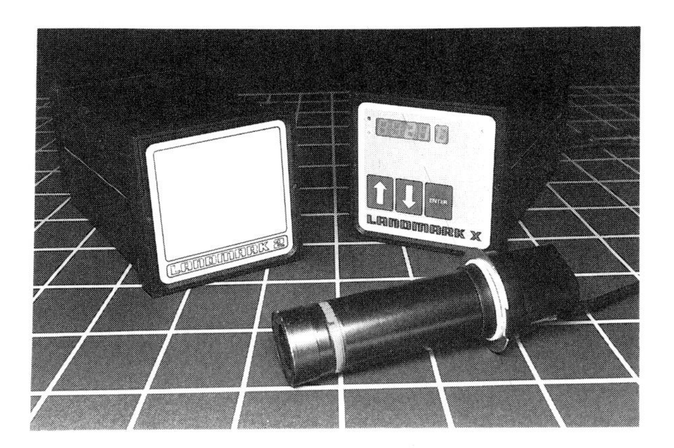

Figure 6.6 *Infrared thermometer with choice of remote processor units. (Courtesy of Land Infrared Ltd)*

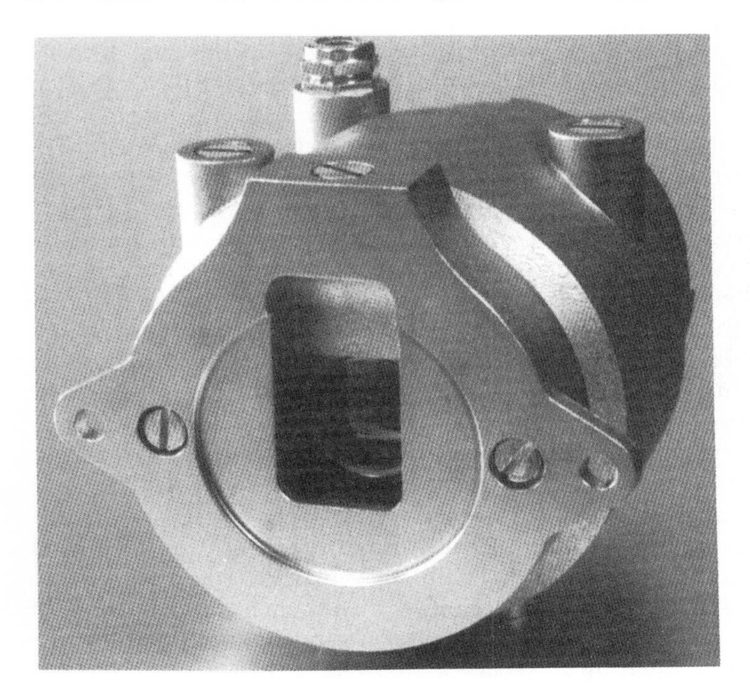

Figure 6.7 *Stand-alone bolt-on infrared thermometer SOLO 3. (Courtesy of Land Infrared Ltd)*

dimension, by some rotating mirror arrangement. In a thermal imager the field of view is scanned in a two-dimensional raster. In both cases the signal is usually processed by a computer; for a line scan a temperature/position trace is generated, and for a thermal imager a grey scale or false colour temperature contour map. Very elaborate image processing schemes are available, for example allowing difference detection via image subtraction.

Line scans measuring down to room temperature operate at 3–5 μm using thermoelectrically cooled photoconductors – lead selenide or mercury cadmium telluride. A typical specification might be 10 scans/second with a spatial resolution of 1/300 of the separation of the instrument from the target. In terms of accuracy of temperature measurement, exactly the same factors apply as for a single-spot thermometer.

Modern industrial thermal imagers operate at 3–5 μm using thermoelectrically cooled mercury cadmium telluride. Good quality (100+ lines) images are generated at 25 frames/second with minimum resolvable temperature difference a few tenths of a degree Celsius. Integration of successive frames allows temperature differences below 0.1°C to be resolved. Devices used in research and medicine and by the military achieve superior performance using detectors cooled with liquid nitrogen; both 3–5 μm and 8–14μm wavebands are used.

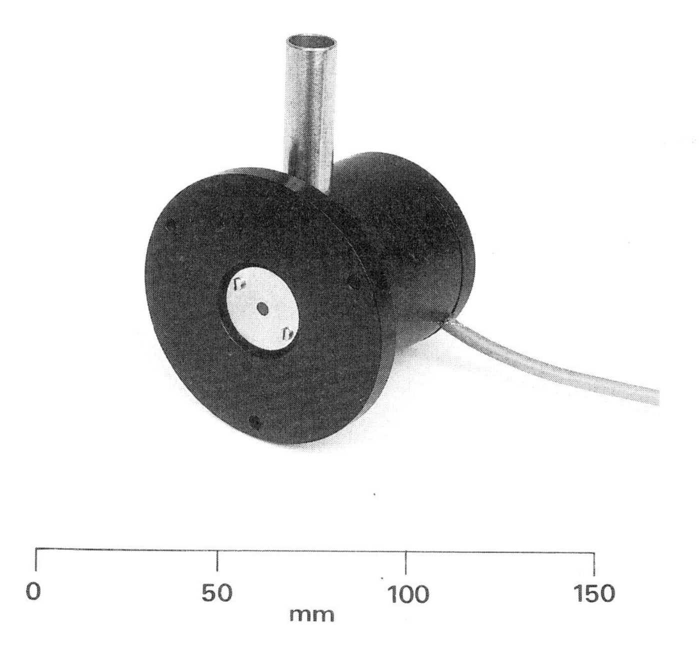

Figure 6.8 *Developmental miniature infrared thermometer for microwave ovens, and same fitted to oven ceiling: the spigot allows purge air derived from the oven fan to enter the probe. (Courtesy of Land Infrared Ltd)*

Historically, imager development has focused on spatial and temperature resolution, to the detriment of the absolute radiance calibration which is a prerequisite for accurate temperature measurement. The most recent devices claim a temperature accuracy comparable with that of single-spot thermometers of similar operating waveband.

3 Measurement Principles

An infrared thermometer directly measures infrared energy, specifically target radiance. From this radiance measurement a target temperature is deduced, implying a model linking temperature and radiance. Formulation of a satisfactory model can be trivial or practically impossible, dependent on often quite subtle aspects of the measurement situation. Herein lies much of the fascination of infrared thermometry and the source of its reputation as something of a black art.

One should recognize that the need for an interpretative model is not unique to infrared. Consider a mercury-in-glass thermometer: if the target is a stirred liquid then there is little problem in interpreting the measured height of the mercury column; but if the bulb is pressed against a solid target of indifferent conductivity, perhaps in a draught and with some incident radiation, then things are not so simple.

As the Holy Grail of contact thermometry is thermalization, so that of infrared thermometry is the black body radiator. One can view a 'black' target as exhibiting perfect radiative coupling to the thermometer; the thermometer signal is then determined solely by target temperature. A non-black target exhibits imperfect coupling and the thermometer couples partly to the target and partly to the surroundings. In this latter case, unravelling the target temperature from the thermometer signal requires a knowledge of the coupling factor (emissivity) and background radiances (conveniently expressed as a mean background temperature $\tilde{T}$).

3.1 The Black Body

The radiance of a black body is precisely described by the Planck equation and, at a given wavelength, is a function only of temperature. An infrared thermometer viewing a black target receives radiant energy L and produces a proportional detector signal

$$L = f(T) \qquad (1)$$

where the calibration function $f(T)$ is just the Planck function integrated over the thermometer operating waveband. If, for simplicity, we consider a thermometer with a narrow waveband near wavelength λ, then

$$f(T)=\frac{c_1}{\lambda^5}\frac{1}{\exp(c_2/\lambda T)-1} \tag{2}$$

where $c_1=1.191\times10^{-16}$ and $c_2=1.438\times10^{-2}$ in SI units, and T is in kelvins. Table 6.2 give $f(T)$ for narrowband thermometers operating at 1, 1.6, 2.2 and 10 μm. The task of the thermometer processing electronics is now essentially trivial; it is to invert equation (2) so as to recover T from measured L.

Laboratory targets are readily constructed which are perfectly black so far as industrial accuracy measurements are concerned. Many common materials – for example, anything containing enough water – are very close to black in certain parts of the infrared spectrum.

Non-black Target

For a non-black target, equation (1) becomes

$$L=\varepsilon f(T)+(1-\varepsilon)f(\tilde{T}) \tag{3}$$

where the first term represents emission from the target and the second represents radiation orginating in the surroundings and reflecting off the target spot into the thermometer. ε is the well known emissivity factor and has a value between zero and unity. $\tilde{T}$ is an effective background temperature, a formalism for describing a potentially very complex reflection situation.

To invert equation (3) it is necessary to ascribe values to ε and $\tilde{T}$. Most thermometers incorporate a means to input an emissivity value based on *a priori* knowledge. Some thermometers allow a background temperature value $\tilde{T}$ to be 'dialled' in, but a more common, if slightly dubious, practice is for $\tilde{T}$ to be taken equal to the thermometer internal temperature. Certain two-sensor systems allow $\tilde{T}$ to be input on-line from an auxiliary sensor, e.g. a thermocouple.

3.3 Background Temperature

Radiation incident on the target spot from any given direction has a certain probability of scattering towards the thermometer. For a mirror-like surface this probability function is sharply peaked at the specular reflection angle; for a diffuse reflector it may be close to direction independent; in either case the probability integrates, over all directions of incidence, to $1-\varepsilon$. Thus the formalism of equation (3) amounts to replacing the actual surroundings with an equivalent enclosure at uniform temperature $\tilde{T}$.

Our estimate (or the thermometer's assumed value) for $\tilde{T}$ is unlikely to be exactly correct, so we need to look at means to minimize the measure-

Table 6.2 *Calibration function f(T) (arbitrary units) and rate of change of f(T) (expressed as percentage change of f(T) per kelvin change in T for narrowband thermometers at 1.0, 1.6, 2.2 and 10 μm*

Temperature	f(T)				% per K			
(°C)	1.0 μm	1.6 μm	2.2 μm	10 μm	1.0 μm	1.6 μm	2.2 μm	10 μm
20	–	–	4.9×10^2	8.89×10^6	–	–	7.9	1.7
40	–	–	1.99×10^3	1.22×10^7	–	–	6.9	1.5
60	–	–	6.97×10^3	1.61×10^7	–	–	6.1	1.3
80	–	–	2.12×10^4	2.07×10^7	–	–	5.4	1.2
100	–	3.94×10^2	5.71×10^4	2.58×10^7	–	6.7	4.8	1.1
200	7.53	6.40×10^4	2.31×10^6	5.99×10^7	6.6	4.1	3.0	0.7
300	1.51×10^3	1.76×10^6	2.58×10^7	1.05×10^8	4.5	2.8	2.0	0.5
400	6.29×10^4	1.81×10^7	1.40×10^8	1.59×10^8	3.2	2.0	1.5	0.4
600	8.36×10^6	3.85×10^8	1.30×10^9	–	1.9	1.2	0.9	–
800	1.80×10^8	2.62×10^9	5.24×10^9	–	1.3	0.8	0.6	–
1000	1.48×10^9	9.77×10^9	–	–	0.9	0.6	–	–

ment error resulting from an incorrect $\tilde{T}$ assignment. For a small error $\Delta\tilde{T}$ in $\tilde{T}$ one has, to first order, a measurement error ΔT given by

$$\Delta T = \frac{1-\varepsilon}{\varepsilon} \cdot \frac{f(\tilde{T})}{f(T)} \cdot \Delta\tilde{T} \tag{4}$$

The first and second terms, respectively, show reflection to be little problem for near-black targets or where the surroundings are much colder than the target. But the second term is also extremely sensitive to operating wavelength; for example, reference to Table 6.2 shows that, for a 100°C target in 20°C surroundings, it has a value of 0.008 at 2.2 μm, compared with 0.34 at 10 μm. Thus, when the surroundings are cooler than the target, the system is rendered tolerant to quite large errors in background temperature assignment by use of the shortest possible operating wavelength. Conversely, when the surroundings are hotter than the target, the second term in equation (4) exceeds unity, and the situation is actually exacerbated by use of a short operating wavelength. Measurements in hot (i.e. hotter than the target) surroundings are difficult and require a very good estimate of $\tilde{T}$; by implication, high accuracies are obtainable only where the surroundings temperature is very uniform.

If the surroundings temperature is very close to that of the target, a different approach can be taken. For $\tilde{T} = T$ equation (3) reduces to

$$L = \varepsilon f(T) + (1-\varepsilon) f(T) = f(T)$$

i.e. the black body condition is recovered. If the thermometer emissivity control is now set to unity, a correct temperature reading is obtained irrespective of the actual emissivity or of the background temperature adopted. Techniques exist to artificially simulate this 'isothermal cavity' condition (see Section 4.1).

If the isothermality is imperfect by $\Delta\tilde{T}$ then the measurement error ΔT is of the order

$$\Delta T \sim \frac{1-\varepsilon}{\varepsilon} \Delta\tilde{T}$$

3.4 Emissivity

Emissivity is primarily a function of target material and surface condition, but also depends on wavelength and, to some extent, on temperature. In high accuracy thermometry it must be regarded as an empirically determined parameter. Reputable suppliers of infrared thermometers offer emissivity determination as a service at modest cost. Given the repetitive nature of industrial processes and the (usual) emphasis of measurement repeatability over absolute accuracy, it is variability of emissivity which needs to be carefully assessed.

Various tabulations of emissivity values exist, but need to be treated with great care. Sometimes the values quoted are wideband, i.e. integrated over all wavelengths, appropriate to heat transfer calculations but useless for thermometry. Sometimes values relate to laboratory-pure specimens bearing little resemblance to target surfaces found in industry.

Some generic examples appropriate to 10 μm and low temperatures are as follows. Water is extremely black near 10 μm, and anything with a wet surface film – slurries, pastes, flesh – may be expected to have an emissivity approaching 0.99. Many fat-based products – e.g. chocolate – have high emissivities of about 0.85. A vast number of organic or fibrous materials have emissivities of about 0.8. Most paints, varnishes etc. have an emissivity of about 0.8, irrespective of visible coloration. Opaque plastics typically have an emissivity of 0.85; rubber 0.9; glass 0.85; wood 0.85; asbestos 0.9; most building materials, i.e. brick, cement etc., 0.7–0.9; paper and cardboard 0.8–0.95; cloth 0.75. Black oxidized metals typically have high stable emissivities, e.g. steel 0.85, copper 0.8, brass 0.7. Carbon, soot etc. typically have an emissivity of 0.85–0.95. There is a (fallible) rule of thumb that anything that is dirty enough has an emissivity of 0.8.

By far the most difficult target materials are bright metals. Here emissivities are very low (≤ 0.1] and vary wildly with small amounts of surface oxidation. High accuracy thermometry on bright metals is rarely practicable.

In cold background conditions the system is rendered insensitive to moderate errors in emissivity setting by the choice of the shortest possible operating wavelength. For a given percentage error x in emissivity setting, the temperature measurement error is set by the per cent per degree rate of change of $f(T)$ with target temperature; this per cent per degree is much improved at short wavelengths, as is apparent from Table 6.2. A good per cent per degree is a panacea for other, e.g. constructional, deficiencies in thermometers since it defines how the basic radiance measurement precision translates to precision in temperature.

In isothermal conditions (see Section 3.3) the target emissivity becomes irrelevant.

3.5 Examples

To put the somewhat abstract discussion above into a practical context we consider two examples.

Firstly, we consider a target at 80°C with the background at a lower temperature, for example a food product emerging from an oven or being heated in cold wall (e.g. microwave) conditions. Table 6.3 gives measurement errors as a function of target emissivity, background temperature and thermometer operating wavelength. For each value of emissivity the centre column applies when the emissivity 'control' is correctly set; adjacent columns apply to $\pm$ 0.03 mis-settings. The calculations are based on a

Table 6.3 *Measurement errors (Celsius) for 80°C target in cold-wall conditions as function of background temperature, emissivity and thermometer operating wavelength*

Operating wavelength (μm)	Background temperature (°C)	Actual emissivity ε / Emissivity setting ε_{set} on the instrument								
			$\varepsilon=0.9$			$\varepsilon=0.8$			$\varepsilon=0.7$	
		0.87	0.90	0.93	0.77	0.80	0.83	0.67	0.70	0.73
2.2	0	0.6	−0.04	−0.6	0.6	−0.09	−0.8	0.7	−0.1	−0.9
	10	0.6	−0.03	−0.6	0.7	−0.06	−0.7	0.7	−0.1	−0.9
	20	0.6	0.0	−0.6	0.7	0.0	−0.7	0.8	0.0	−0.8
	30	0.7	0.05	−0.6	0.8	0.1	−0.6	1.0	0.2	−0.6
	40	0.8	0.15	−0.5	1.1	0.3	−0.3	1.4	0.6	−0.2
	50	1.0	0.3	−0.3	1.5	0.7	0.05	2.1	1.2	0.5
10	0	0.4	−1.2	−2.8	−1.0	−2.8	−4.5	−2.8	−4.8	−6.7
	10	1.0	−0.7	−2.2	0.4	−1.5	−3.2	−0.5	−2.5	−4.5
	20	1.7	0.0	−1.6	1.9	0.0	−1.8	2.2	0.0	−2.0
	30	2.4	−0.7	−0.9	3.5	1.6	−0.2	5.0	2.8	0.7
	40	3.2	1.5	−0.1	5.3	3.4	1.5	8.1	5.7	3.6
	50	4.1	2.4	0.7	7.3	5.3	3.4	11.4	8.9	6.7

Table 6.4 *Measurement errors (kelvin) for 80°C target in near isothermal conditions as function of background temperature and target emissivity*

Operating wavelength (μm)	Background temperature (°C)	Emissivity ε		
		0.9	0.8	0.7
2.2	70	−0.8	−1.7	−2.5
	75	−0.4	−0.9	−1.4
	80	0.0	0.0	0
	85	0.6	1.1	1.6
	90	1.2	2.4	3.5
10	70	−1.0	−2.0	−2.9
	75	−0.5	−1.0	−1.5
	80	0	0	0
	85	0.5	1.0	0.5
	90	1.0	2.0	3.1

monochromatic thermometer at 20°C which adopts its own temperature as the background temperature; they give a good indication of the errors expected from commercial lead sulphide (2.2 μm) and thermopile or pyroelectric (10 μm) instruments. The advantage of short wavelength operation is very apparent, as is the loss of precision with decreasing target emissivity (even if correctly set).

Secondly, we consider a target at 80°C in near isothermal conditions, for example in a well insulated convection oven. The thermometer emissivity 'control' is now properly set to unity, as discussed above. Table 6.4 gives errors as a function of target emissivity and true background temperature. The choice of operating wavelength is seen to be arbitrary in terms of precision, and may reasonably be made on cost and reliability criteria. The basis on which the thermometer adopts background temperature is irrelevant. High target emissivity gives tolerance to deviations from strict isothermality.

4 Miscellaneous Techniques

In this section we look firstly at simple means of enhancing emissivity; then at some techniques which avoid the need to know emissivity in specific conditions; and lastly at ways of measuring inaccessible targets.

4.1 Enhancing Emissivity

A simple way to improve the emissivity of say a bright metal target is to paint it. Paints with high (~ 0.9) stable emissivities are available, permitting accurate temperature measurement in virtually any background conditions.

An alternative approach is to create an isothermal cavity, for example by drilling a hole. A length to diameter ratio of about six will produce an effective emissivity close to unity, especially if the walls are roughened and the bottom is conical rather than square to the axis. A crack or groove can achieve the same effect, which can be useful in measuring a product with cracks or crevices on a conveyor belt; a small target spot is used and the thermometer signal is peak picked.

The isothermal cavity effect can be simulated by surrounding the target spot with a high reflectance enclosure, for example a gold plated hemisphere with a small hole through which the measurement is made. The separation of the hemisphere periphery from the target surface must be small compared with its diameter. Similar approaches include the use of a reflector plate to form a wedge with the target surface, or the use of naturally occurring wedges, for example where sheet target material spools off a bright roller. In all cases the precise reflectance value of the reflector becomes non-critical as the temperature of the reflector approaches that of the target.

4.2 Avoiding the Need to Know Emissivity

A number of techniques avoid the need for known emissivity, but at the expense of some assumption about the target surface. They have value, but tend to work only in very specific conditions. The image of infrared thermometry has been tarnished by overly aggressive marketing of some of the following methods.

A definitive instance is ratio (two-colour) thermometry. Here radiance is sensed in two separate wavebands and a signal is derived which is proportional to the *ratio* of the respective energies. The specious argument is that, while the energy in each waveband depends on emissivity, this cancels in the ratio, rendering the ratio signal a function only of temperature. This is true only for a grey target, whose emissivity is independent of wavelength. In general the ratio signal depends on the ratio of the respective emissivities in the two wavebands. Moreover, a ratio thermometer is typically several times more sensitive to a mis-setting of emissivity ratio than is a conventional thermometer to a mis-setting of emissivity. Ratio thermometers are actually more useful for coping with a partially obstructed or unfilled field of view, the ratio signal being unaffected as long as attenuation is identical in both wavebands.

Certain two-colour thermometers derive calibrated radiances in two wavebands in addition to the ratio. This additional information can, in certain very specific circumstances, be used to deduce emissivity on-line.

A variety of reflection methods exist which derive emissivity on-line. Looking back to equation (3), if the background temperature $\tilde{T}$ can be oscillated between two known values then the value for $1-\varepsilon$ and hence emissivity can be deduced from the AC ripple on the thermometer signal. This is practicable only for highly specular targets, where a reasonably small variable temperature source at the specular angle suffices.

If the background source is varied slowly and a polarizing filter is rotated rapidly in front of the thermometer, an AC ripple is produced (due to variation of emissivity with polarization) which falls to zero when the background temperature equals the target temperature. Simultaneous monitoring of background source temperature and thermometer output ripple thus permits an emissivity-free measurement. The requirement for high specularity is eased if one can assume that non-specular reflection is unpolarized.

Recently thermometers have been promoted which incorporate a laser reflectometer. A laser beam illuminates the target, and the fraction of beam power reflected back into the instrument is monitored. If this can be related to the *total* beam power reflected – i.e. over all angles – then emissivity is derivable from the relationship: emissivity equals one minus total reflectance. Clearly this entails a *model* relating reflectance into the solid angle subtended by the instrument's collecting aperture to total (i.e. 2π solid angle) reflectance – a point not overemphasized by some vendors of the equipment. In practice the assumption made is pure diffuse reflection. The scope of validity of this assumption is not yet well documented.

A two-wavelength version allows the emissivity ratio to be derived given only that the collected to total reflectance ratio is the same at both wavelengths.

4.3 Inaccessible Targets

An infrared transmitting window may be used to access the product within (say) a mixer or flowing in a pipe. With suitable design, high pressures (200 atmospheres) can be contained, permitting measurements in extruders.

The majority of window materials conventionally used are of doubtful applicability in food processes on grounds of toxicity, solubility or propensity to shatter into sharp fragments. Probably the most suitable material, by a considerable margin, is synthetic sapphire (single-crystal alumina). The relevant properties of sapphire are given in Table 6.5. It is a relatively inexpensive material in window diameters up to 25 mm or so. Fixture is generally via a suitable gasket or O-ring, but sapphire can also be brazed to Nilo or titanium.

Sapphire is highly transparent from the visible out to about 5 μm. Windowing 8–14 μm is much more problematical. (Please see Table 6.1 for the corresponding temperature ranges.) The usual materials – germanium,

Figure 6.9 *Infrared thermometer for hostile environments using fibre-optic transmission. (Courtesy of Land Infrared Ltd)*

Table 6.5 *Some properties of sapphire*

Solubility	10^{-4} g in 100 ml water
Hardness	1700 knoop*
Strength (rupture)	5×10^{8} N/m^{2}
Melting point	2015°C
Thermal conductivity	40 W/m K
Expansion	5×10^{-6}/K
Transmission	Visible to 5.5 μm

*Knoop: unit for empirical hardness scale. (A Knoop value of 1700 corresponds to a Mohs value of just under 9.)

zinc sulphide, barium fluoride – are very doubtful for the reasons given above. Polyethylene transmits reasonably well, but only in thin films (0.1 mm).

Targets which are inaccessible due to high ambient temperature, vibration or RF fields can sometimes be measured using fibre optics. The infrared collection assembly is separated from the detector unit with a flexible optic fibre. This allows the sensitive detector unit to be fixed in a benign environment. Collection assemblies can be built very tough, for example to withstand ambient temperatures over 400°C. Figure 6.9 shows a typical instrument.

The use of fibre optics, with commonly available glass and silica fibre, is limited to short wavelength (silicon and germanium) thermometers, and hence to high temperatures, owing to the transmission characteristics of the fibre. Typically fibre lengths of a few metres are supplied, but much longer lengths are technically possible. With selected fibre and in short lengths (≤ 1 m), measurement can be extended down to about 80°C using a filtered lead sulphide detector. Recently available zirconium fluoride fibre allows measurement to room temperature, but the cost is prohibitively high at the time of writing.

References

Burnay, S. G., Williams, T. L., Jones, C. H. (eds) (1988) *Applications of Thermal Imaging*. Adam Hilger, Bristol

Dewitt, D P. and Nutter, G. D. (eds) (1988) *The Theory and Practice of Radiation Thermometry*. Wiley, New York

Touloukian, Y. S., (Series ed) Ho, C. Y. (Series Technical ed) *Thermophysical Properties of Matter* Touloukian, Y. S., DeWitt, D. P. (1970) Vol 7. Thermal Radiative Properties – Metallic Elements and Alloys. Touloukian, Y. S., DeWitt, D. P. (1972) Vol 8. Thermal Radiative Properties – Nonmetallic Solids. Touloukian, Y. S., DeWitt, D. P., Hernicz, R. S. (1972) Vol 9. Thermal Radiative Properties – Coatings. IFI Plenum, New York

Wolfe, W. L. and Zissis, G. J. (eds) (1978) *The Infra-Red Handbook*. Office of Naval Research, Department of Navy, Arlington, VA

7 Microwave Measurements of Product Variables

M. Kent

Contents

7.3 Simultaneous Phase and Attenuation 221
7.4 Resonant Systems 223
7.5 Reflectance 223

8 Sensors 224
8.1 Horn Antennae 224
8.2 Stripline Sensors 227
8.3 Stripline Antennae 228
8.4 Open Ended Transmission Lines 229

9 Areas for Development 230

Appendix 1: Some Manufacturers of Microwave Moisture Measurement Instruments 231
Appendix 2: List of Symbols 231

References 232

1 Introduction

Microwave electromagnetic radiation has been used in the determination of the water content of various materials for at least three decades. This followed the germinal ideas of a number of workers in different parts of the world (Freymann 1954; Nedzvecki 1955; Watson 1957; Walker 1958). In spite of its successful use in that period there also have been frequent failures which of course remain unpublished. Many of the pitfalls which caused them arose from a lack of understanding of the nature of the method and might have been avoided. Part of the object of this chapter is to help provide that understanding. As a first step one should be aware of what is meant by 'microwave'. By definition such radiation lies within a broad frequency range from 10^8 Hz up to 10^{11} Hz. It is also always used in this application at powers well below any hazard level, typically a few milliwatts. Most commercial instruments have operated in the so-called S and X bands which cover the approximate frequency ranges 2–4 GHz and 8–12 GHz respectively, or wavelength ranges roughly 7–15cm and 3–4 cm.

At microwave frequencies the wavelengths of the radiation lie within the 0.3–300 cm range, and so we are always dealing with potentially serious scattering and interference problems arising from the similarity of the wavelengths both to the sample dimensions and to the dimensions of discontinuities within the samples. Too often this has been ignored and the power transmission regarded as amenable to interpretation in terms of scalar quantities or linear optics, in direct comparison with optical density measurements. In many cases such an approach is only useful as a first approximation and can sometimes be a very poor approximation indeed. Despite these potential problems, however, much of the earliest successful work (see for example Ince and Turner 1965) used the simple property of absorption of a beam of microwave power passing through a sample, relating this to the water content. Many of the subsequent applications were similar but, as this chapter will show, this is far from the limit of the technique's potential.

Reference to the review and bibliography published by Kraszewski (1980a; 1980b) shows how few applications of microwave measurement techniques which have appeared in the food industry have been published. Out of 400 publications recorded up to 1980 a mere handful referred to foods specifically. The last decade has barely changed that situation. Successful applications have been less publicly reported, however, for as diverse foodstuffs as grain, butter, margarine, cakes, nuts, breakfast cereals, extrusion-cooked products, meat products, starch, pasta, sugar beet products, chips, cheese, confectionery and many others.

Before discussing the applications in more detail it is perhaps a good idea to know a little more about the actual mechanisms involved in microwave power absorption and in microwave transmission in general, since it is in this area that many of the pitfalls lie. For example, the choice of measurement frequency is often crucial in avoiding the perturbing effects of dissolved ions. Equally the wrong choice of frequency could result in the measured microwave transmission loss decreasing with increasing water content where the reverse was expected. To avoid these and other unexpected problems it is useful to understand how process variables such as temperature, composition etc. for a particular foodstuff affect the dielectric properties over a wide frequency range. Literature values can be sought using the bibliography published by the author (Kent 1988) or, better still, laboratory measurements can be made over a wider range of variables than might be experienced in practice.

That said, however, it must be stressed that for many materials, particularly in the low moisture content range, say below 30 per cent w/w, the problems encountered will be slight and easily tackled. The most usual, as will be seen, are variation of calibration with temperature and, for particulate materials, with bulk density. Before delving into the background science of the method, examine first the advantages of its use and how they outweigh the disadvantages.

2 Advantages and Disadvantages of Microwave Techniques

2.1 Advantages

If problems exist with the applications of microwave dielectric measurements, the question must be asked, 'What are its advantages?' Consider the following:

(a) Unlike infrared measurements, the microwave technique can measure the bulk properties of the material and not just the surface. This is particularly important for non-homogeneous materials.
(b) It is non-hazardous, using very low power levels of microwave radiation of the order of a few milliwatts.
(c) As will be seen, many of the variables interfering with moisture measurement for example, which normally would be regarded as disadvantages, could in fact be determined rather than eliminated.
(d) It is non-destructive.
(e) The technique is non-invasive and so presents no great problem to hygienic design.
(f) Also because of its non-invasive nature, cleaning in place is not usually a problem.
(g) The measurement is not generally interfered with by the presence of dust or vapours, although condensation must be avoided as must excessive accretion of dust deposits.
(h) The accuracy of the technique can be as good as or better than other conventional methods, though obviously it can never be better than the method used to calibrate it. Typically figures of 0.25 per cent water content at levels of 10 per cent are quoted for simple attenuation measurements. With the possibility of measuring or at least eliminating some of the perturbing variables which in part give rise to this level of error, even more precise measurement may be achieved. Kress-Rogers and Kent (1987) estimated that for a coffee powder with a moisture content around 5 per cent and with bulk density variations occurring of up to 30 per cent, an error of 1.95 per cent moisture content would occur if the measurement was made using the loss factor alone. This represents a relative error of 39 per cent. Using the methods described in Section 5.1 would reduce this to 7 per cent and the first corrective iteration would take it down even further to less than 1 per cent, that is less than 0.05 per cent moisture content. The corresponding relative error in the density determination would be 0.3 per cent. The real limit would be higher than this, however, owing to the instrument measurement errors. Typically attenuation can be measured to 0.1 dB and phase to within a degree or two.
(i) It is not a problem to multiplex measurement heads and use a single instrument to monitor them.
(j) The sensing heads are robust and the electronics and microwave circuitry can be installed in some remoter, less hazardous spot if required.

2.2 Disadvantages

(a) It is expensive compared to low frequency methods (but comparable to NIR analysers), especially if all the features to be discussed are incorporated. These require, in addition to the microwave instrument, a microcomputer for data manipulation and storage of calibration data, although these are now common in most modern instrumentation.

(b) Applications often require careful design of sensing heads, which can be very dependent on the material and its other properties. For example, the simplest form of sensor available is that using transmitting and receiving horn antennae (Section 8.1), but care is needed in choosing the dimensions of the measuring cell to avoid errors.

(c) The method is not an absolute method and so requires calibration against some known standard method, e.g. oven drying, Karl Fischer etc. This in itself can introduce systematic errors, as Kress-Rogers and Kent (1987) have discussed. This is true of course of other common in-line methods for moisture measurements.

3 Dielectric Properties and their Parameters

3.1 Polarization

As we have all been taught, the atoms and molecules of matter comprise negatively charged particles (electrons) orbiting more massive, positively charged nuclei (consisting of protons and neutrons). When a static electric field is applied to a material, the electric charges in that material tend to separate, the negative charges moving towards the positive electrode and vice versa. In a conductor the applied field would cause a current to flow, the charges being free to move. In the case of a perfect dielectric material, where the charges are not mobile in this way, such a current occurs only until equilibrium is established within the material. This redistribution of the charge is called polarization. For the simple case of a dielectric whose charge is evenly distributed on each molecule, this polarization merely results from the distortion of the charge distribution, and takes place almost instantaneously on applying the field. Such dielectrics are termed 'non-polar'.

In 'polar' materials, positive and negative charges are distributed in such a way as to effectively form a small permanent electric dipole, analogous to a small bar magnet. Effective dipoles such as this can exist when the centres of mass (and hence the centres of charge) of the two charge distributions in the molecule (the orbiting electrons and the charged nuclei) do not coincide. In the presence of an applied electric field, the movement of the charge occurs through rotation of the dipole to a position of minimum potential in the field, i.e. aligned with the field, as well as by the previously mentioned distortion. This additional polarization is distinguished from distortional polarization by being termed 'orientational'.

3.2 Dielectric Dispersion

The orientation in the field is not instantaneous and is characterized by a time interval, referred to as the relaxation time τ. (For a summary of symbols see Appendix 2.) It is better understood from the reverse case, when the steady applied field is removed. The relaxation time is a measure of the time taken for the dipoles to return to a random distribution of orientation through collisions with other molecules, involving exchange of angular momentum.

In the static field case, we can assign a property to the material known loosely as the dielectric constant ε_s, and this is a measure of how well the material can polarize. If the field is made time-varying, then a different situation obtains. The dielectric, which in the static field case could have been considered perfect, i.e. had no losses, now exhibits a form of conduction loss. In an AC circuit, a capacitor filled with such a dielectric would be effectively a capacitive reactance with a parallel resistance.

For a sinusoidal variation of the field, at very low frequencies of oscillation, the dipoles may be able to align and realign exactly in phase with the field. In this case the capacitor still has a purely reactive impedance. As the frequency of the field oscillations is increased and the period approaches the relaxation time, then it becomes increasingly difficult for the polarization to become complete. As this occurs, the apparent dielectric constant begins to fall. At periods even shorter than the relaxation time, only the distortion polarization contributes to the dielectric properties and the dielectric becomes effectively non-polar. The net result of all these effects is a high dielectric constant ε_s at low frequencies, and a low dielectric constant ε_∞ at high frequencies. This behaviour, which is characteristic of polar materials, is known as dispersion. For reference later it should also be noted that, at any frequency, there is a relationship between the dielectric constant or permittivity ε and the more commonly known refractive index n:

$$n = \sqrt{\varepsilon} \tag{1}$$

At high frequencies, especially for non-polar materials, the optical refractive index is virtually identical to $\sqrt{\varepsilon_\infty}$.

It is found that the nature of polar materials is best understood in terms of a complex permittivity, that is a permittivity having both real and imaginary terms. This is expressed as

$$\varepsilon^\star = \varepsilon' - j\varepsilon''$$

where the real part or permittivity ε' is the component of the dielectric properties which represents the ability of the dielectric to store energy, and the imaginary part or loss factor ε'' is that part which relates to its ability to dissipate energy. The ratio of $\varepsilon''/\varepsilon'$ is called $\tan\delta$ or the loss tangent, and power dissipation in a material is proportional to this. The loss factor behaves in a somewhat different manner to the permittivity ε', in that it passes through a peak value as the frequency is swept. This peak occurs at the relaxation or critical frequency f_c, given by

$$f_c = \frac{1}{2\pi\tau} \tag{2}$$

The dependence of ε' and ε'' on frequency are shown in Figure 7.1 for a typical polar dielectric. As will be shown in Section 4.1, the microwave transmission variables of attenuation and phase are directly related to the complex permittivity. Water is a very good example of a polar molecule. The spatial configuration of hydrogen atoms with each oxygen atom is such that the orbiting negative electrons have a centre of mass slightly removed from the centre of mass of the positively charged nuclei, with the net result that a permanent electric dipole is formed. In fact the strength of this dipole, for the size of the molecule, is exceptional, and the dielectric dispersion is of quite a large magnitude: at room temperature ε_s is 80 and ε_∞ is 5.0. The peak loss factor of ~37 occurs at around 17 GHz at room temperature. It is all these facts which make water so easily detectable at microwave frequencies. This statement should only be taken as a general comment, since water in foods at a low concentration can be expected to

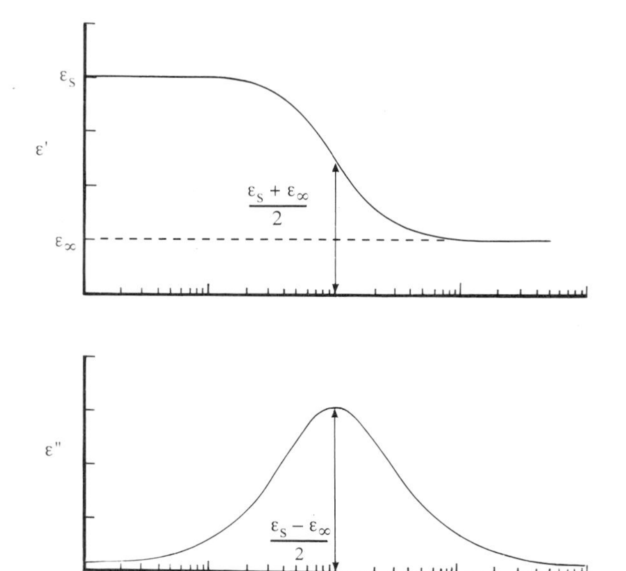

Figure 7.1 *Typical Debye dispersion showing frequency dependence of permittivity and loss factor: the frequency is shown as a ratio of frequency to critical frequency*

Table 7.1 *Dielectric data for some polar liquids (NBS 1958)*

	ε_s	ε_∞	f_c (GHz)	T (°C)
Water	80.4	5.2	16.9	20
Methanol	33.6	5.7	3.0	20
Ethanol	25.07	4.2	1.1	20
1-propanol	20.8	2.65	0.38	20
1-butanol	17.9	3.15	0.3	19

behave differently to a greater or lesser degree (see Section 3.3 and observe the non-linearity in Figure 7.4).

However, water is not the only polar molecule of interest to the food industry, and other important food constituents may display dispersion with the loss factor peaking at their particular critical frequencies. This not only allows the measurement of the concentration of, for example, fat in a non-polar matrix, but also raises the question of whether one could identify and measure compounds in materials containing a range of polar compounds by dielectric spectroscopy or multi-wavelength measurements as used in near infrared reflectance analysis. The alcohols, for example, as can be seen in Table 7.1, have relaxation frequencies ranging from megahertz to gigahertz. Unfortunately, mixtures of polar materials often do not have well resolved spectra, the dispersion and absorption spectrum of each covering a decade or two of frequency and overlapping considerably.

Very wide band dielectric spectroscopy in the microwave frequency range is barely possible when only three decades of frequency are amenable to microwave techniques. Although microwave spectroscopy is therefore not indicated for composition analysis, it may be possible to adopt other methods of distinguishing such components; we shall return to this subject in Section 5.

3.3 Dielectric Dispersion of Water in Foodstuffs

Water molecules hindered in their rotation, by solutes or proteins for example, have a lower relaxation frequency than pure water, and may even have a distribution of relaxation times. This tends to broaden even further the frequency range of the dispersion. Relaxation times are influenced by a number of factors: viscosity, binding at specific sites on other molecules, temperature variation, or change of state. Figure 7.2 shows the effect of concentration on the dielectric spectrum of sucrose solutions at 50°C. The

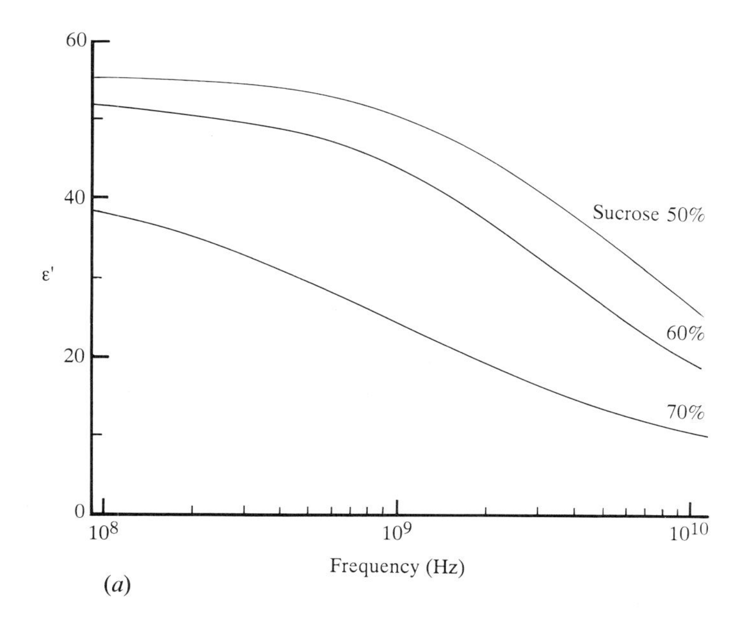

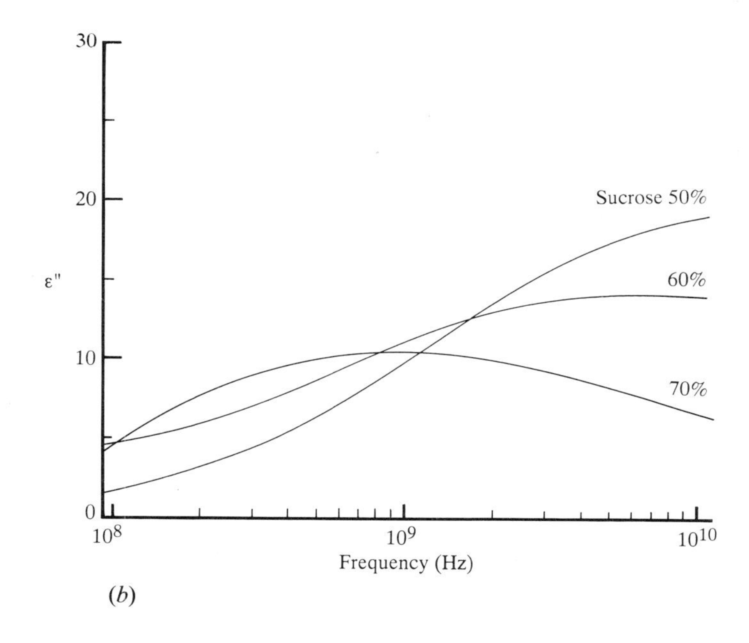

Figure 7.2 *Effects of concentration of sucrose solution at 50°C on (a) its permittivity relaxation spectrum (b) its loss factor*

changes are directly attributable to changes in viscosity. Similarly, change in temperature shifts the relaxation frequency, as Figure 7.3 shows for the case of a 70 per cent sucrose solution.

Any rotational hindering of the dipoles will increase the relaxation time. Ice at microwave frequencies is effectively a non-polar material, its relaxation frequency being in the audio frequency range. The consequences for other applications such as microwave heating/thawing are profound. In thawing of foodstuffs by microwaves, all the power dissipation is in the unfrozen fraction of water. Most foodstuffs, if not all, even at temperatures well below the freezing point, contain such a fraction. It is partly the relaxation properties of this water which enables the method to succeed. Ice is essentially transparent to microwaves: hence the well known inverse Baked Alaska or Norwegian Omelette. On the other hand the presence of liquid water on the surface, or a thawed layer in the case of a foodstuff which has been allowed to take up heat from warmer surroundings, completely alters this and can screen the interior of the material from the microwave energy.

Although we know that water strongly influences the dielectric properties of foods, in fact room temperature studies of protein powders at low levels of water content show very little evidence of any dispersion centred on the water relaxation frequency at 17 GHz (Kent and Meyer 1984). Elevation of both temperature and moisture content, however, does introduce a high frequency relaxation, but this is still very different in many aspects from the normal unhindered water relaxation. For heterogeneous systems the wide distribution of relaxation times smears out the dielectric spectrum. The consequence of this is to the advantage of the user of microwave methods for water content determination, since no effects are then visible such as were seen in the temperature dependence of sucrose solutions (Figure 7.3). In that case, for certain frequencies of measurement the loss factor could increase or decrease as the concentration or the temperature increased.

3.4 Temperature Effects

Some effect of all the perturbing influences may be seen in the temperature coefficients of the dielectric loss factor of many water bearing materials, particularly organic materials such as foods. For any polar system, increasing the temperature almost inevitably increases the relaxation frequency as can be seen in Figure 7.3. This is a consequence of the fact that relaxation is assisted by molecular collisions and obviously these increase in rate as the temperature rises. For a frequency of measurement lying below the relaxation frequency, the measured loss factor will thus decrease as the temperature increases, and the dispersion moves up in frequency further from the measurement point. This would be the case, for example, with

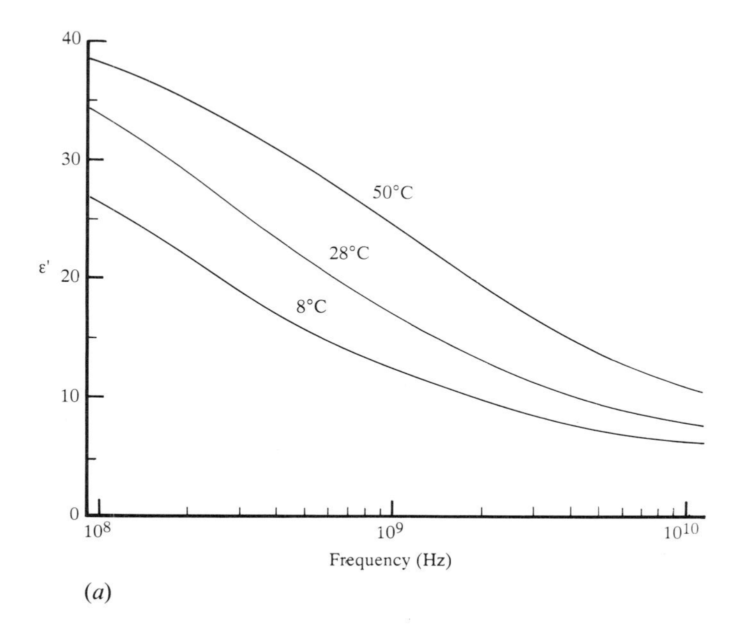

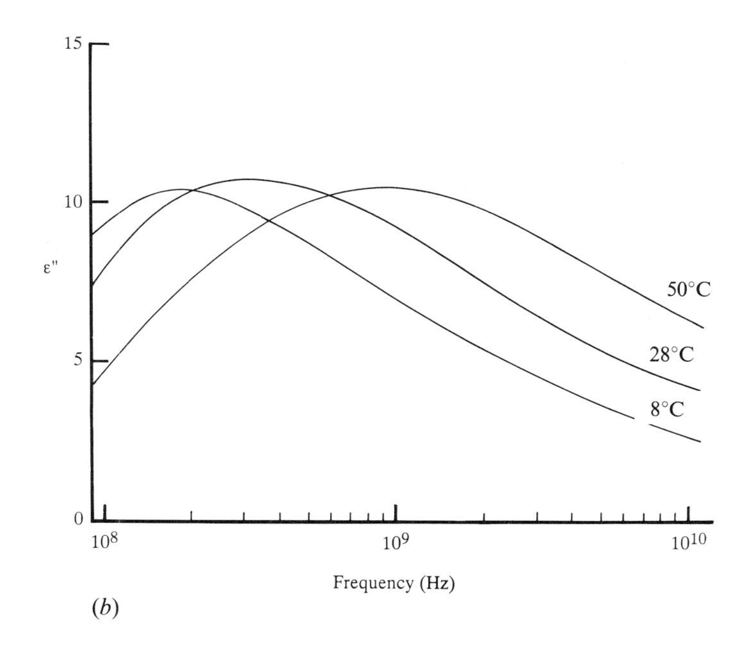

***Figure** 7.3 Effects of temperature of a 70% sucrose solution on (a) its permittivity relaxation spectrum (b) its loss factor*

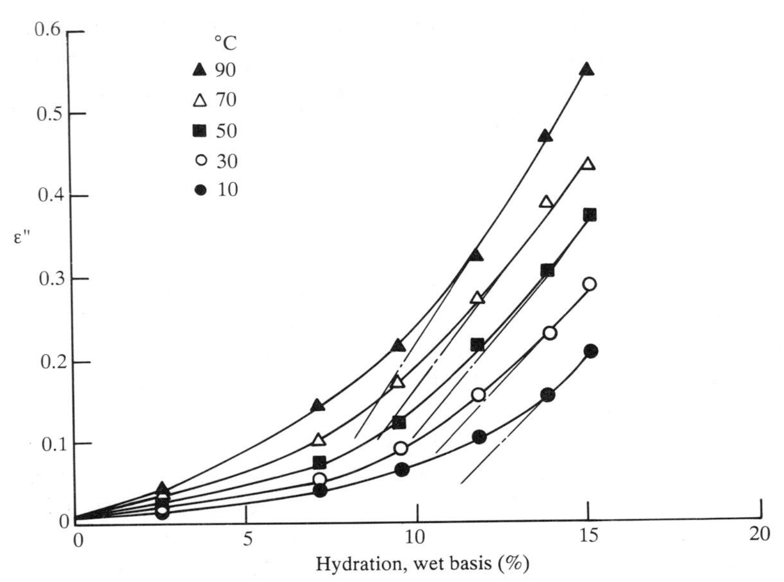

Figure 7.4 *Loss factor ε″ for fishmeal at 9.4 GHz as a function of temperature and hydration (from Kent 1970)*

pure water if measurements were being made at 10 GHz. For materials where the measurement frequency is above the relaxation frequency, in contradistinction, the temperature coefficient is positive until the relaxation frequency passes through the measurement frequency. A very wide variety of materials fall into this category at the measurement frequencies used for many microwave applications. A typical example is shown in Figure 7.4. This concerns data obtained at 9.4 GHz on fishmeal, which being proteinaceous contains many sites at which water may be bound. The strength of binding depends on the site, and the final result is a distribution of relaxation times with greatly hindered rotation of water molecules. There are of course alternative views of the reasons for this hindering, which both broadens and shifts the relaxation spectrum to a lower frequency range. Whatever the explanation, however, it is inescapable that water dipoles are hindered in their rotation, and at this microwave frequency there result positive temperature coefficients for ε''.

3.5 Bulk Density

Apart from temperature, which is relatively easy to correct, the first users of microwave attenuation methods found that measured power losses depended very much on bulk density. This is not at all surprising when it is considered that the dielectric properties are dependent on the number of dipoles within a given volume. Assuming no interactions therefore between dipoles, it could be assumed that the losses of energy were directly proportional to density. The author has observed (Kent 1977) that the dependences of the dielectric properties on bulk density follow square laws of the form

$$\varepsilon' = A'\rho^2 + B'\rho + 1 \tag{3}$$

$$\varepsilon'' = A\rho^2 + B\rho \tag{4}$$

where A, A' B and B' are constants and ρ is the bulk density. It has to be admitted that this is also a consequence of relationships derived by Beer (1853) and Gladstone and Dale (1863) based on observations of the dependence on density of the absorption of light and the refractive index of various materials (cf. equation (1)).

Much work on the bulk density problem has followed, principally by Nelson in the USA (Nelson 1983) and Kraszewski (1978), and dealing largely with the dielectric properties of grain and seeds. Various empirical relationships between complex permittivity and density have been explored. These have involved not only quadratic equations but also cubic relationships. It is beyond the scope of this chapter to describe them in detail here, and the reader is referred to Nelson (1983) for a full exposition.

3.6 Conductivity

In addition to the dipolar loss in a material, some contribution to the loss factor also arises from any DC conductivity possessed by the material. The effective loss factor introduced by the conductivity is inversely proportional to frequency and may be written

$$\varepsilon''_\sigma = \frac{\sigma}{2\pi f \varepsilon_0} \tag{5}$$

where σ is the DC conductivity in siemens per metre, f is the frequency and ε_0 is the permittivity of free space in farads per metre.

If ionic salts are present in a material the effective loss factor due to conductivity may mask any dipolar contribution from water, especially at low frequencies. This is also another reason that microwave ovens are able to heat foodstuffs. At one typical operating frequency of ovens (915 MHz) the dipolar loss of pure water is, at room temperature, very small (tan δ is 0.05). Only dissolved salts and the rotational hindering of some of the water dipoles (which shifts their relaxation peak down into the vicinity of the operating frequency) enable power to be rapidly absorbed by a foodstuff. For typical high moisture foods at 915 MHz, tan δ is 0.4 at room temperature, an order of magnitude higher than for pure water. At the higher oven frequency of 2.45 GHz, the difference between conductance and dipolar loss is less but still important. As the temperature rises, of course, the difference becomes even greater, the conductivity having a positive temperature coefficient and that for the dipolar loss of pure water at these frequencies being negative.

Returning to the problems of analysis, however, it is well known that salt content can be estimated from the DC conductivity of materials. What should now be noted is that this conductivity can be obtained from the frequency dependence of the dielectric properties, adding a further dimension to the potential of dielectric measurements. The advantage would be to have one instrument capable of simultaneous determinations of several composition related electrical variables, using non-contacting and non-invasive methods.

4. Methods for Measurement of Dielectric Properties

4.1 Attenuation Measurements

Reference has now frequently been made to 'attenuation' of microwaves; but to what exactly does this refer?

The dielectric loss in a material can be measured from the attenuation and velocity of a plane wave passing through the material. The attenuation is a direct consequence of Beer's law (Beer 1853), which is expressed as

$$I = I_0 e^{-\alpha x} \tag{6}$$

where I_0 is the power incident on a slab of thickness x, I is the power transmitted through it and α is a constant referred to as the attenuation per unit length. Note that this equation takes no account of the wave nature of the radiation.

The expression for α in terms of the dielectric properties is

$$\alpha = \frac{2\pi}{\lambda}\left(\frac{\varepsilon'}{2}\left\{\sqrt{[1+(\varepsilon''/\varepsilon')^2]}-1\right\}\right)^{1/2} \quad \mathrm{Np\,m^{-1}} \tag{7}$$

where λ is the free space wavelength in metres (on Np, see Appendix B of this book).

For values of $\varepsilon''/\varepsilon' \ll 1$, as is the case with many low moisture materials, this reduces to

$$\alpha = \frac{\pi}{\lambda}(\sqrt{\varepsilon'}) \tan \delta$$

At the same time, because of the difference in the refractive index of the material compared with free space and therefore as a consequence of the reduction of velocity of the wave in the material, we can write

$$\phi=\frac{2\pi}{\lambda}\left[\left(\frac{\varepsilon'}{2}\left\{\sqrt{[1+(\varepsilon''/\varepsilon')^2]}+1\right\}\right)^{1/2}-1\right]\ \mathrm{rad\,m^{-1}} \tag{8}$$

which again when $\varepsilon''/\varepsilon' \ll 1$ becomes

$$\phi=\frac{2\pi}{\lambda}[(\sqrt{\varepsilon'})-1]$$

This is the phase shift that such a wave undergoes in passing through the slab relative to that it would have in passing through the same distance of free space. Such equations, while being derived from wave equations, nevertheless refer only to internal transmission. They do not take into account the reflections at boundaries or the standing wave effects created by those reflections. Equation (7), however, was the basis, albeit implicit, for all the microwave transmission applications until 1973.

For medium to high water contents, say above 30 per cent w/w, the value of α becomes very large for even short path lengths in the material. At 10 GHz, 50 dB of attenuation is readily achieved by a few millimetres of pure water. This makes the presentation of the material to the sensing head extremely difficult, since small thicknesses are required to keep the attenuation measurable. Typically 30 dB should be considered maximal.

4.2 Resonant Methods

Another set of variables that might be considered as suitable for application are those which characterize a resonant system. For the problem of measuring the dielectric properties of food materials, Risman and Bengtsson (1971) have used microwave resonant cavities at various frequencies.

Others have in fact used such devices for dielectric measurement (e.g. Birnbaum and Fremeau 1949; Kaatze 1973; Bilbrough 1968; Henry and Berteaud 1980) both off-line and on-line. Commercial devices exist and have been used for materials such as butter. The usual variables that are measured are the resonant frequency and the Q factor (see Figure 7.5). The latter is a ratio of energy stored to energy dissipated, and as such is obviously conceptually similar to and in fact is inversely related to the ratio of the dielectric properties (tan δ). By careful design of the system, the resonant frequency is usually arranged to be dependent almost entirely on the real part of the permittivity:

$$\frac{f-f_0}{f_0}=-\frac{1}{2F}(\varepsilon'-1) \tag{9}$$

$$\frac{1}{Q}-\frac{1}{Q_0}=\frac{\varepsilon''}{F} \tag{10}$$

where F is a filling factor derived from sample geometry and the field distribution in the resonator cavity, f is the resonant frequency of the resonator, and Q is of course the Q factor. The subscripts 0 refer to the empty resonator case. Although resonant methods have been used for high moisture foodstuffs, the same problem is encountered as in the attenuation methods (Section 4.1), though in a different guise. In this case, too large a sample with too high a loss factor can degrade the Q so much that it is impossible to measure either resonant frequency or Q factor. For practical in-line applications rather than laboratory measurements it must therefore be limited to the same range of water contents.

4.3 Reflectance Measurements

When a transmission line, be it waveguide, coaxial line or whatever, is terminated with a dielectric material then, because of impedance mismatch at the interface, partial reflection of the incident wave occurs. The combination of the reflected and incident waves creates a standing wave. Either by measuring the position and magnitude of this standing wave or by measuring the intensity and phase of the reflected wave, we can compute the wave impedance of the dielectric and from this its dielectric properties. Thus once again we have a two-variable method with which density and water content could be determined.

A very useful reflection sensor can be constructed from an open ended coaxial line as described by Burdette et al. (1990) and many others. Knöchel and Meyer (1981) proposed the use of such a sensor for the measurement of the water content of slurries, and Chouikhi and Wilde

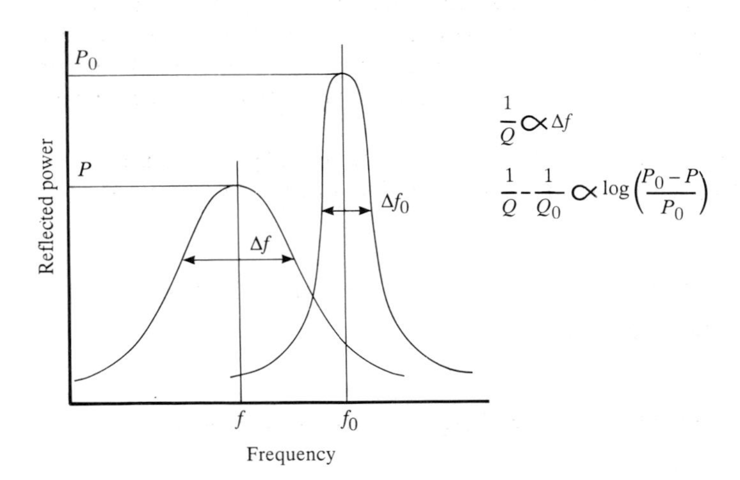

Figure 7.5 *Definition of* Q *factor and resonant frequency* f *with reference to the frequency response of a resonant microwave cavity impedance*

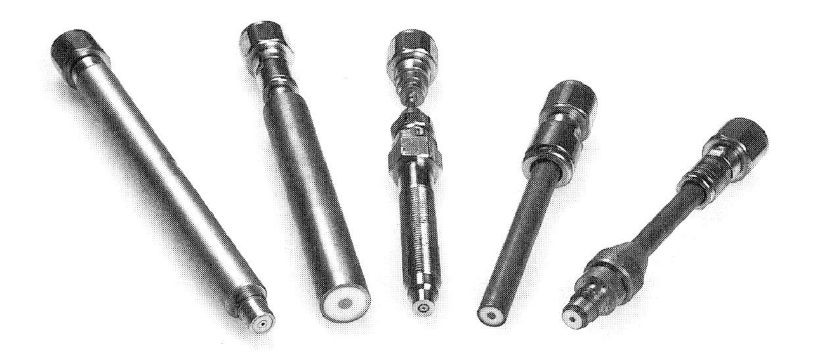

Figure 7.6 *Open ended coaxial sensors for reflectance measurements (courtesy of S. Chouikhi and the Institute of Food Research, Norwich)*

(1986) have adopted this for the moisture measurement in extrusion cookers (Figure 7.6). This type of sensor is discussed further in Section 8.4.

Because the measurement is made on one face of the material and not through it, reflection methods do not have the same limitations as transmission methods in respect of high water contents or high loss and permittivity samples. In this case the reflected power increases as the water content increases but other problems present themselves. Clearly the limiting condition is total reflection of the power when the reflection coefficient has unity magnitude. As the permittivity of the sample increases this limit is approached asymptotically. In addition, the limiting phase shift of the reflected wave compared with the incident is π, so for high water content samples small changes in moisture (permittivity) cause equally small (difficult to measure), changes in both magnitude and phase of the reflected wave.

The general conclusion is that reflection methods may be useful in a low to medium water content range, say up to 60 per cent w/w water content.

5 Dielectric Properties and Measurement of Bulk Density and Composition

5.1 Density Measurement and Compensation in Moisture Measurement

It was always recognized that some form of compensation for the bulk density of the material was necessary when attempting to measure moisture content by dielectric means in particulate materials (see Section 3.5). The usual solution was to incorporate some subsidiary means of density measurement, such as gamma-ray absorption (Mladek 1973; Zehnder 1967). This was an effective if somewhat expensive method, which was often considered unsuitable for the food industry because of its employment of radioactive sources. It is still widely used, however, and several commercially available systems incorporate it for density compensation (see Appendix 1 and Figure 7.7).

For many applications the density problem was ignored, and installations ran on the assumption that a reasonably constant density was maintained. Any variation in density likely to be encountered was considered to be tolerable and probably averaged to the same mean for a given water content. Another solution adopted, in particular for building materials, involved compressing the material into a solid block to ensure minimum density variation from sample to sample (Goodnig and Bird 1963).

The Polish workers Stuchly and Kraszewski (1965) were the first to recognize the full potential of the microwave technique for density independent moisture determination. Later, developing the idea further, Kras-

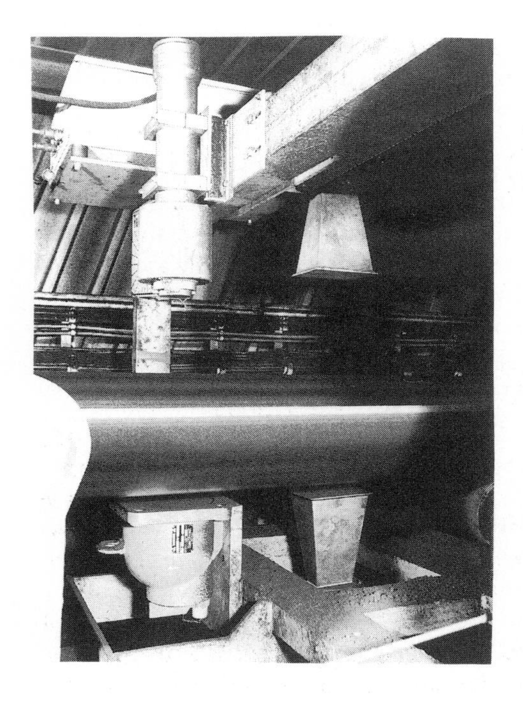

***Figure** 7.7 Installation of attenuation monitor and γ absorber gauge on a production line: the horn antennae are seen to the right (courtesy of Laboratorium Prof. Berthold)*

zewski and Kulinski (1976) devised a system which by measuring both phase and attenuation enabled density to be eliminated as a variable. They achieved this by using simultaneous linear calibration equations which required only a knowledge of their parameters for solution:

$$\alpha = x\frac{M_w}{V}a_1 + \frac{M_0}{V}a_2 \quad \text{nepers or decibels per unit length} \tag{11}$$

$$\phi = x\frac{M_w}{V}a_3 + \frac{M_0}{V}a_4 \quad \text{radians or degrees per unit length} \tag{12}$$

where M_w is the weight of water present, M_0 is the weight of dry material, V is the volume of the wet material, x is its thickness, and a_1-a_4 are constants found by calibration. Obviously in the same notation the bulk density ρ is given by

$$\rho = \frac{M_w + M_0}{V} \quad \text{mass per unit volume} \tag{13}$$

The moisture content fraction ψ expressed as $M_w/(M_0 + M_w)$ is readily shown to be given by

$$\psi = \frac{\alpha a_4 - \phi a_2}{\phi(a_1 - a_2) - \alpha(a_3 - a_4)} \tag{14}$$

Unfortunately this approach was ahead of its time. If microcomputers had been available then as they are now, nothing would have stood in the way of its implementation, but at that time the computing systems available for on-line application plus the cumbersome electromechanically operated microwave variable components militated against proper commercialization of the technique. It is also true to say that the assumed linearity in general is applicable only over a narrow range of water content. The dielectric properties of most food materials show distinct non-linear dependence on hydration (cf. Figure 7.4).

It was some years before further advances were made in this problem, but the work of Jacobsen et al (1980), Meyer and Schilz (1980; 1981) opened up new possibilities. With a more sophisticated microwave device to measure phase and attenuation, the proposition was made that a simple ratio R of attenuation to phase would provide adequate density compensation for many materials. The method was proposed for a variety of materials, and actual experimental trials were carried out on tobacco. Later an examination was made by Kent and Meyer (1982) of the application of the method to a very heterogeneous material, namely fishmeal. Despite the wide range of density, particle size and shape (see Figure 7.8) this technique promised great advances. Not only was the ratio R reasonably density independent but also it seemed independent of the structure of the material.

Figure 7.8 *Samples of fishmeal studied by Kent and Meyer (1982) showing the variety of structure which was covered by a single calibration: CU2 redfish; HL2 blue whiting; OFM1 whitefish offal; CU1 crab*

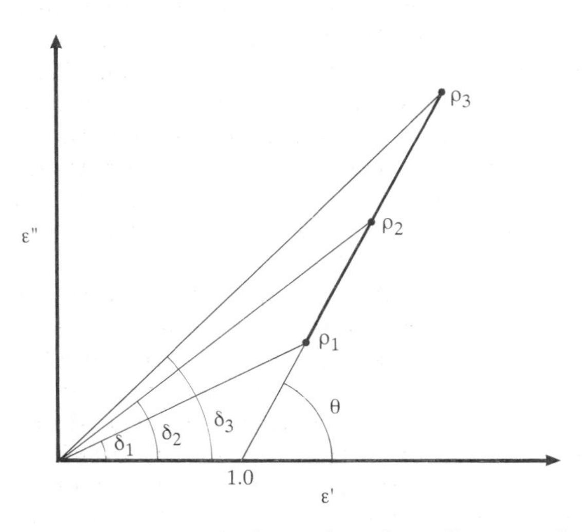

Figure 7.9 *Typical complex plane diagram of variation of complex permittivity with density at a constant temperature and water content: the loss angles δ_1, δ_2, δ_3 are shown for corresponding densities ρ_1, ρ_2, ρ_3 and the density independence of θ is clearly seen (after Kent and Meyer 1982)*

Part of the reason for its success in eliminating density dependence can be seen in the figurative diagram in Figure 7.9. When the dielectric properties of the material are plotted in the complex plane, it is seen that for many materials the locus generated by variation of the density parameter is a straight line. For different moisture contents these loci all intersect the abscissa at the point (1,0). This should not surprise anybody since these are the dielectric properties of free space, and zero density of any material must be precisely that. The more important property is that for some materials the angle created by these straight lines and the abscissa can be density invariant and is given by

$$\theta = \tan^{-1} \frac{\varepsilon''}{\varepsilon' - 1} \tag{15}$$

It was a relatively simple matter using equations (7) and (8) to demonstrate that for low loss materials this angle was very close to the simple ratio of attenuation to phase.

Thus, providing that the dielectric properties behaved in this very useful manner, a density independent, moisture dependent variable could be obtained either from direct use of the transmission variables of phase and attenuation, or from the computed dielectric properties. The latter open up the possibility for the similar exploitation of other measurement techniques which yield the dielectric properties, and these will be dealt with later.

Powell, et al. (1987) found that for wheat the simple relationship of equation (15) was far from independent of density. The work of Nelson (1983) has already been mentioned in relation to the development of models for the dependence of the dielectric properties on density. He had shown that $(\varepsilon')^{1/3}$ was a slightly better linear function of density than equation (3) and that $(\varepsilon'' + e)^{1/3}$, where e is a small constant error term, seemed also to be linearly dependent on density. Powell et al. (1987) therefore arbitrarily replaced ε' and ε'' in equation (13) by these alternative functions and improved matters slightly, reducing the overall variation in the ratio from around ± 15 per cent to ± 10 per cent for a density variation of ± 40 per cent. More interesting, however, was the homodyne microwave circuit that they devised for the measurement of phase and attenuation. This will be discussed in Section 7.3.

When Kress-Rogers and Kent (1987) looked at the method of Meyer and Schilz (1980; 1981) in relation to coffee and milk powders, they too found slight but significant deviation from the simple result of equation (15). They also noted that for these fairly dry materials the phase measurement was virtually moisture independent while retaining density dependence. Such dependence on water content as appeared to exist was largely in respect of the fact that the density itself depended on the moisture content. This almost clear-cut separation of the two unknown variables led to the proposition that by some suitable calibrations against both density and water content, iteration between the two would not only refine the accuracy of the water determination but would also yield a result for the density.

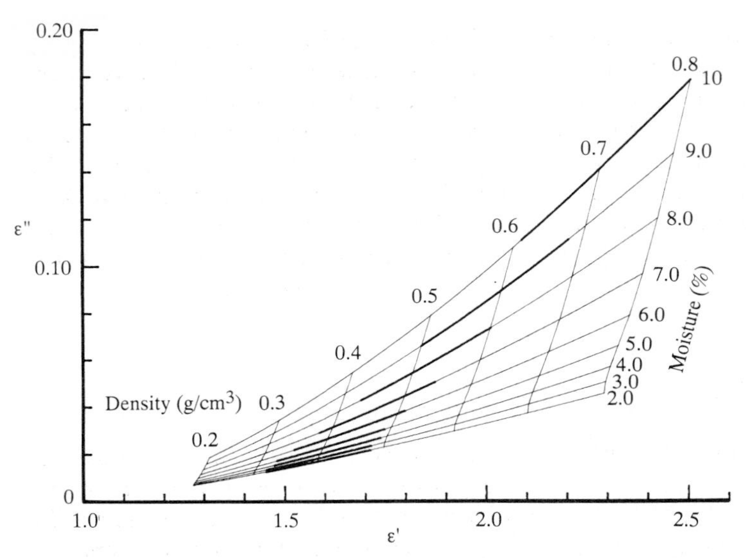

Figure 7.10 *Dielectric properties of a milk powder at 25°C and 10 GHz as a function of moisture content and density: the heavy lines indicate the range of experimental data, while the thinner lines show the extrapolated range based on equations (3) and (4) (after Kent 1989; Kress-Rogers and Kent 1987)*

As could be seen in some of the author's earlier work (Kent 1977) the complex dielectric properties occupy a regular space when plotted in the complex plane with moisture and density as parameters, with no crossing or folding of the curve within each family. This is shown in Figure 7.10 using results obtained on milk powder taken from Kress-Rogers and Kent (1987). It is obvious that any measurement of ε' and ε'' represents a unique solution for a given pair of variables, such as in this case moisture content and density.

In practice the complex permittivity is never measured directly; rather other variables are measured which are related to this complex quantity. Such variables are also complex and a choice exists as to which to use. For example, the use of intrinsic attenuation and phase shift has already been discussed as described in the work of Meyer and Schilz (1980). The density and the moisture content also occupy a very uniform space in the plane of the ratio R and phase, as can be seen in Figure 7.11. The mesh of calibration lines can be made to appear even more uniform by raising the denominator in the ratio (phase) to an arbitrary power m, but whether any advantage could be gained from this is doubtful. Bearing in mind the effects of errors in both phase and attenuation measurements it seems unnecessary to take the step of calculating R, which must be of greater relative uncertainty than attenuation alone.

As already discussed in Section 4.1, all direct attenuation methods are limited in use to moisture contents below about 30 per cent. This is simply because as water content increases then so does the measured attenuation, which means that the power levels being detected become impractically small. To obviate this problem a thinner sample may be used, but this is often impractical for mechanical reasons. In addition, the thinner the sample the more precisely must the thickness be known, since the variables are also dependent on this dimension. The problem can further be reduced by the choice of a lower operating frequency (longer wavelength) if that is indicated as a possibility from all the other considerations.

5.2 Fat and Water

Ohlsson et al. (1974) have published a great deal of data obtained at 2.8 GHz concerning the dielectric properties of ternary mixtures of fat, water and meat protein. They noted the possibility of determining the composition of such mixtures from complex permittivity measurements. In all the work described in this chapter, whether it be density and water determination or this ternary system work, one fundamental principle underlies the measurements. That is, that to measure variables with n degrees of freedom, one needs at least n independent measurements. This condition is satisfied by phase and attenuation, resonant frequency and Q factor, reflectance coefficient phase and magnitude, and any other pair of

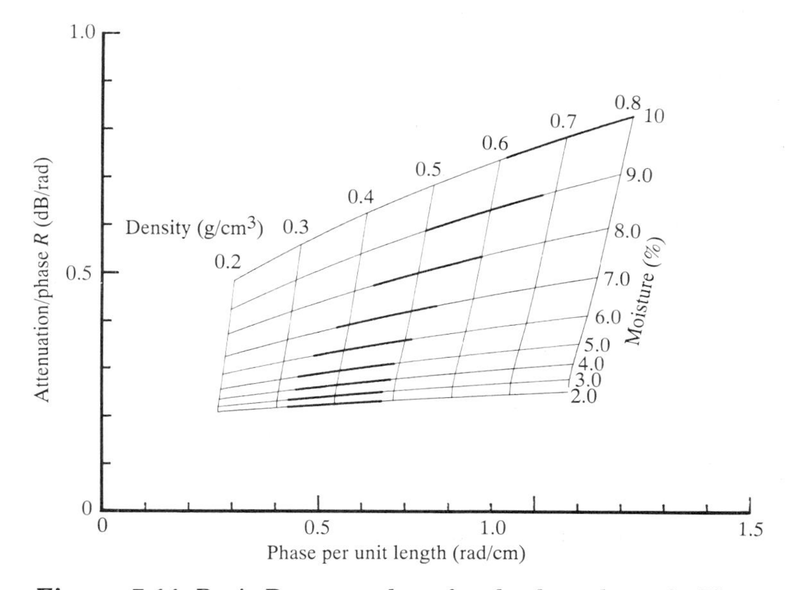

Figure 7.11 *Ratio* R *versus phase for the data shown in Figure 7.4 (after Kent 1989)*

variables related to the dielectric properties when dealing with systems of two degrees of freedom such as ternary compositions or the problem of density and water. (Remember that in ternary systems, knowledge of the proportions of two of the components determines the third.) The problem of density and moisture can itself be regarded as one of ternary mixtures, the components being air, dry matter and water. In their work on meat emulsions Ohlsson et al. (1974) showed triangular composition diagrams with the dielectric properties superimposed as contours of constant value. In some areas, loci of constant ε' and ε'' were parallel or very nearly so, and this meant that without any intersection there could be no solution in terms of the composition. These data can be replotted in Cartesian form to reveal more of the dependence of the complex permittivity on the composition (Figure 7.12). In this figure it can be seen that over quite a large region of the graph a unique solution obtains for any pair of permittivity values. Unfortunately, there is an area of confusion where the curves overlap, occurring when both fat and water content are low. In regions of composition other than this there could be usefulness in this approach. The addition of a fourth major component such as carbohydrates would cause some problems however.

As we have already noted, it is not the complex permittivity that is usually measured directly. It can be instructive to examine the relationship between the composition and other related variables that are measureable.

The use of the simple transmission loss (attenuation) and phase is precluded here, since for high values of permittivity it is extremely difficult to minimize the effects of reflected power and standing waves by the usual expedient of impedance matching. Even if matching were achieved for a particular sample composition, any deviation from this would produce sufficiently large changes in the permittivity for the impedance no longer to be matched. For the case of low moisture content powders this is not a problem because the permittivity of those powders usually never deviates far from ~ 2.0, matching is easily maintained, and in any case the reflection losses are slight (Brodwin and Benway 1980). This is even more so if the measurement system is incorporated into a control loop controlling the composition.

If transmission measurements are attempted without matching then, as briefly discussed in Section 4.1, the simple expressions for intrinsic phase shift and attenuation (equation (7) and (8)) do not apply. What is measured in fact is the transmission scattering parameter. It is not necessary to dig deeply into the theory of scattering networks to explain this new variable. Rather, suffice it to say that any device handling electromagnetic signals can be represented by a matrix of so-called scattering parameters S_{11}, S_{12}, S_{22}, S_{23} etc., which describe the signal at each input and output port of the device. A two-port system such as the dielectric sample in a transmission line can be described by four such S parameters, which completely describe the reflected and transmitted signals at each of the two faces.

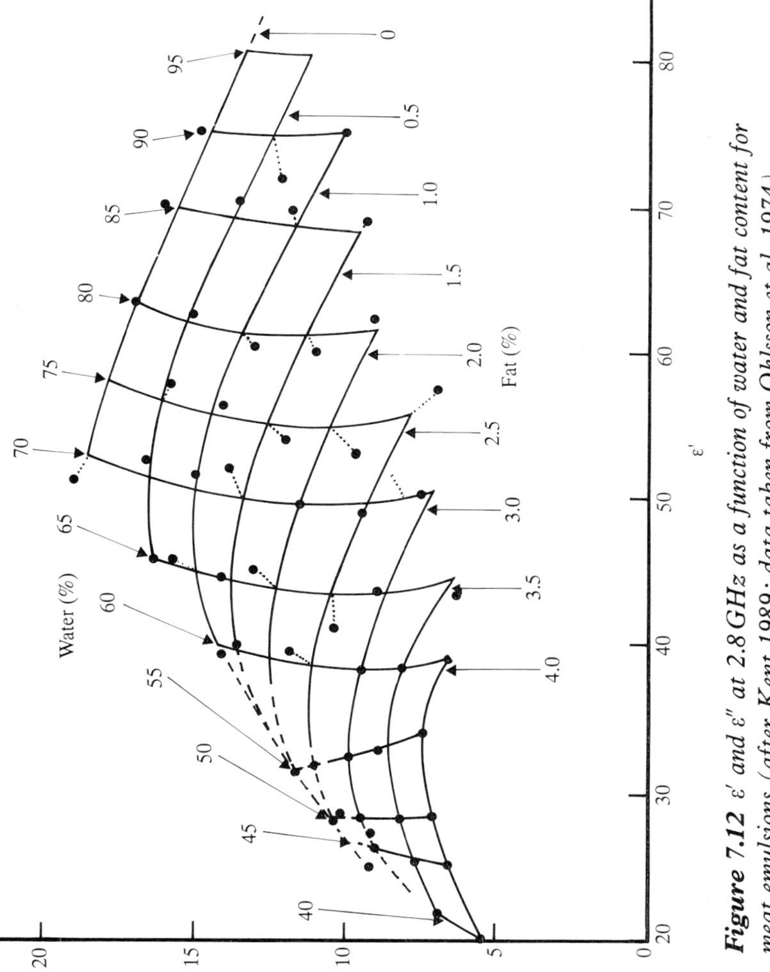

Figure 7.12 *ε' and ε'' at 2.8 GHz as a function of water and fat content for meat emulsions (after Kent 1989; data taken from Ohlsson et al. 1974)*

Symmetry reduces these further to two parameters, S_{12} for the transmission and S_{11} for the reflection. It is S_{12} that interests us here, and this may be written

$$S_{12}=\frac{(1-r^2)\exp[p(z-1)]}{1-r^2\exp(2pz)} \qquad (16)$$

where $r=(1-z)/(1+z)$, $p=\mathrm{j}2\pi d/\lambda$, $z=\sqrt{\varepsilon}$, d is the sample length and λ is the wavelength.

In practical situations, the microwave power is usually transmitted through dielectric windows (see Figure 7.13) which, unless arranged to be completely transparent at the wavelength of measurement (achieved by making them a multiple of half wavelengths thick), introduce their own reflections of the power. This modifies S_{12} significantly, and it now has to be written

$$S_{12}=\frac{(1-r_2^2)\exp[p_2(z_2-1)]\,D}{1-r_2^2\exp(2p_2z_2)-E} \qquad (17)$$

where D and E are complex factors incorporating the thicknesses and dielectric properties of the windows and the sample ε_2 and ε_1 respectively (with corresponding z_1 and z_2) and $r_2=(z_1-z_2)/(z_1+z_2)$.

Since this configuration is more realistic, S_{12} has been calculated using the data for the meat mixtures and assuming that the windows are constructed of polytetrafluoroethylene (PTFE) 20 mm in thickness. The sample is also assumed to be 20 mm thick. The results of these calculations are shown in Figure 7.14, where it can be seen that useful separation of the variables is obtained. The phase angle here requires some adjustment and $m\pi$ has been subtracted, where $m=0$, 1 etc. according to how many wavelengths were reckoned to be contained within the sample thickness. Phase of course can only be measured and distinguished within the range $-\pi$ to $+\pi$. Larger phase shifts must be inferred from a knowledge or estimate of the dielectric properties of the material and consequent modification of the observed phase shift.

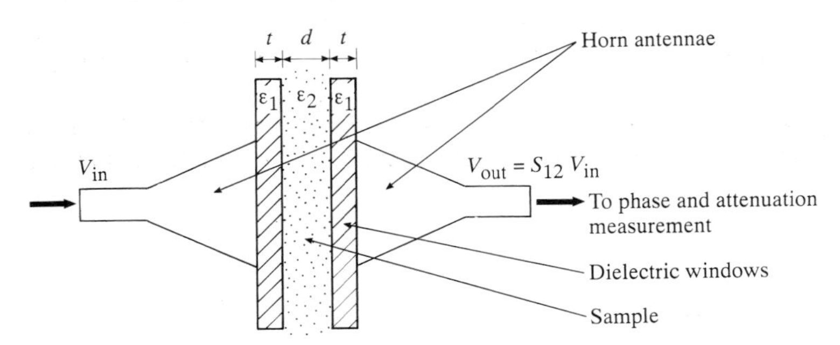

Figure 7.13 *Practical arrangement for microwave transmission measurements*

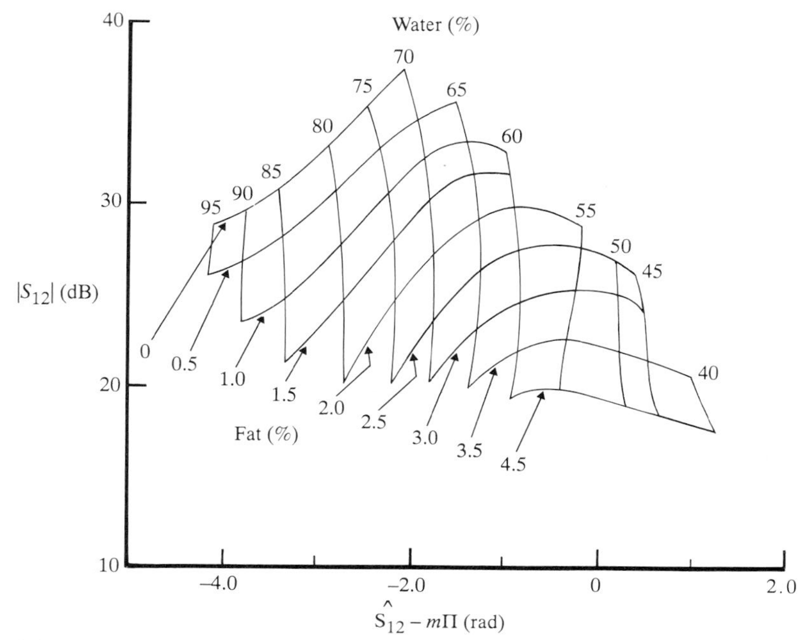

Figure 7.14 *Computed transmission variable S_{12} at 2.8 GHz for meat emulsions assuming plane wave propagation*

The magnitude of S_{12} is considerably greater than would be the intrinsic transmission loss derived from equation (7). This is because in this mismatched situation a very large proportion of the power is lost by reflection at the boundaries of sample and windows. In fact 10 dB or more of the observed power loss is by this mechanism. The problems of matching the impedance of the antenna to the sample have already been mentioned. Rough matching could be achieved by making the windows a quarter wavelength thick and of a dielectric whose permittivity was the geometric mean of the material and that of air, but this would only achieve the desired result for a very limited range of composition, as already explained. Clearly, as Figure 7.14 demonstrates, even without matching a useful calibration may be obtained.

The original data collected by Ohlsson et al. (1974) were obtained using a resonance technique for which the same problems of impedance matching do not apply.

5.3 Sugar Solutions

Having introduced the possibility of using microwave methods for more than simple moisture determination (see also Kent 1989), it is now necessary to explore these other regions of potential usefulness. One such could be in the on-line monitoring of the concentrations of hot sugar solutions. Dielectric data relating to this problem were obtained by a collaborative research programme operating in Europe (COST90bis: see Kent and Kress-Rogers 1987). Some of these data will now be used here to demonstrate this potential application. The data concern aqueous sucrose solutions over a concentration range from 20 to 60 per cent by weight and at temperatures from 30 to 90°C. The same arrangement has been adopted as for the meat mixtures, but in this case a sample thickness of only 10 mm was assumed. The results are plotted in Figure 7.15 in the plane of transmission loss $|S_{12}|$ versus phase $\hat{S}_{12}$.

In this example, because oı the smaller thickness of the sample, no further adjustment of phase was necessary except for the convenience of plotting. It is immediately clear that in this case the two unknown variables that can be obtained are concentration and temperature. Given that temperature can readily be measured by other means, then in fact only phase needs to be measured to yield the concentration. Use of the transmission loss alone would give ambiguous results owing to the curvature of the isotherms yielding two values of concentration for a given temperature and loss. This was the root of a problem encountered by the industry in trying to apply transmission methods to this system.

This type of transmission measurement may be entirely unsuitable for this particular problem since sugar boiling is a batch process. In those

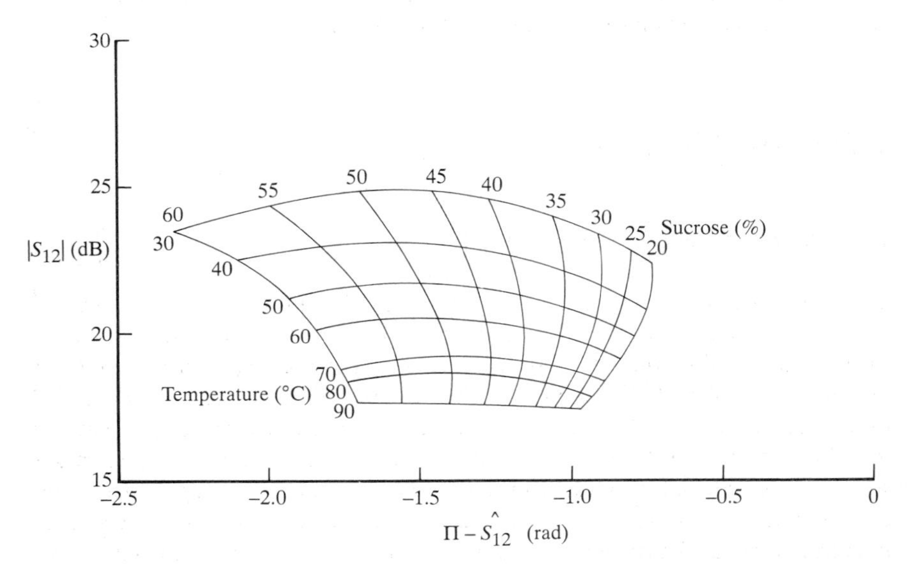

Figure 7.15 *Computed transmission phase and loss at 3.05 GHz for sucrose solution as a function of concentration and temperature: sample is 10 mm thick between 20 mm PTFE windows*

circumstances the use of some reflection technique is probably preferable and the coaxial probe already introduced could be the answer.

5.4 Other Mixtures of Materials

So far all the materials discussed have been mixtures containing water as the polar constituent, but this is not a necessary condition. As already alluded to, the method is applicable to any mixtures of polar and non-polar materials. Several publications have appeared which contain data for mixtures of various alcohols with or without water (Mudgett et al. 1974; Dutuit 1980). These data show that composition measurement by the methods described here is feasible. No data are known concerning ternary mixtures of alcohols, but they are certainly worthy of investigation considering the problems encountered in recent years with adulterated wines.

5.5 Salt Concentration

It has already been explained how dissolved ions affect the dielectric loss. Salt concentration may readily be correlated with the DC conductivity, which itself may be found from the frequency dependence of the loss factor. In the frequency regions where only conductivity contributes to the loss, i.e. outside dispersion regions, equation (5) is rewritten as $\sigma = 2\pi f \varepsilon'' \varepsilon_0$ and the loss can be calibrated against salt content.

6 Material Structure

6.1 Particle Shape and Distribution

The description of dielectric properties and their interaction with electromagnetic radiation in terms of composition may be inadequate in some cases. One of the most significant material properties, other than bulk density, to affect the absorption of microwave power is the shape of individual particles. This fact is demonstrated by a simple experiment (Chaloupka et al. 1980) in which water droplets are formed on a glass plate which is then placed in an electromagnetic field with its plane parallel to the electric field. The attenuation at a frequency of 16 GHz of this arrangement is found to be of the order of 0.5 dB. If another glass plate is brought into contact with the droplets, forming a sandwich, and the droplets are squeezed out into a continuous or semi-continuous film, then this attenuation leaps to around 20 dB. This is a consequence only of the change in structure of the water layer and is not due to any other influence. The same effect is seen in measurements on wet polyfoam as small quantities of detergent are added. The surface tension of the water decreases and the film formed on the foam surface becomes thinner.

The principal factor governing this effect in particulate materials can be described in terms of ellipsoids of revolution and a shape dependent depolarization factor. If we consider uniform ellipsoids of water held in some dry material of permittivity ε_m with a quite random distribution of orientation in space, then the effective permittivity is given by van Beek (1967) as

$$\varepsilon_{eff} = \varepsilon_m + \tfrac{1}{3} V_w(\varepsilon_w - \varepsilon_m) \sum_{r=a,b,c} \frac{1}{1 + (A_r/\varepsilon_m)(\varepsilon_w - \varepsilon_m)} \tag{18}$$

where ε_w is the permittivity of pure water, V_w is its volume fraction and A_r is the shape dependent depolarization factor. A_r takes values from 0 for needles to 1 for disks, with spherical particles having a value of $\frac{1}{3}$. The ellipsoidal principal axes are represented by $r = a$, b and c.

Clearly this example, with all the particles of equal shape, represents an idealized situation, but Chaloupka et al. (1980) have dealt with the more general case where the particles are of different shapes and have an arbitrary distribution. The outcome of their study is the description of a structure independent variable derived from linear combinations of real and imaginary parts of the dielectric constant at different frequencies, and also the determination of a mean depolarization factor A_0. This is very interesting, especially for materials which do not absorb or bind the water and for which the water is essentially in its pure form. However, this is not generally the case with foods and, as has already been noted, binding of water significantly alters its relaxation properties. Thus in equation (18) the value of ε_w required may be much closer to ε_m, in which case the effects of A_r become very much less significant. That this is so may be seen in the results of Kent and Meyer (1982) from which Figure 7.8 is taken. A greater distribution and range of particle shapes could not be imagined, yet it appears to have had little effect on the results.

Kraszewski (1989) has recently obtained similar results for wheat and grain where calibration seems independent of grain type.

6.2 Particle Size

Particle size may be important as well, though Kress-Rogers and Kent (1987) found very little difference between ground and unground milk and coffee powders once compensated for density changes. Such effects as were observed were attributed to thermal damage during grinding.

However, finer particles of non-food materials have been observed to differ in properties (Dube 1970). Where DC conductivity is present, i.e.

where a continuous conducting path exists, the increase in density of inter-particle contacts as the particle size is reduced can account for some increase in observed loss factor.

7 Apparatus for Microwave Measurement

7.1 Attenuation

Having discussed at some length the dependence of various microwave variables on dielectric properties, now we need to know how these variables are measured. Commercial instruments for the measurement of microwave variables exist and are usually applied to the problem of water determination alone (see Appendix 1 for list of manufacturers).

Microwave circuits for the measurement of attenuation can be fairly simple in operation. The most basic is as shown in Figure 7.16a. Here a microwave source feeds power through an isolator and then via a wave-guide or a coaxial line to a horn antenna which irradiates the sample. The function of the isolator is to prevent reflected power reaching the source and interfering with its operation. On the other side of the sample a horn antenna receives the unabsorbed power which is then detected by a crystal detector. The signal at the detector clearly diminishes as the water content in the material increases and more power is absorbed. The signal is calibrated against the water content. There are some disadvantages with this type of system. Firstly, fluctuations in the power output of the source are wrongly interpreted as moisture variations. This can be avoided by signal levelling, that is adding an additional detector and feedback circuit to control by some means the power output, as shown in Figure 7.16b. Secondly, the response is not linear, and for large values of attenuation changes in the water content have considerably less effect than the same absolute change would have at a lower attenuation. This could be avoided by displaying a log function of the detected current.

A better solution is shown in Figure 7.16c. This is a dual-path system where the power from the source is divided by a power splitter or directional coupler into a reference arm and a measurement arm. The signals in each arm are detected separately and the log ratio is taken electronically. In this case, fluctuations in the source output appear equally in both arms and the ratio represents a true measurement of the transmitted power fraction. By using a logarithm of this ratio the results can be expressed directly in decibels. The slight drawback with this system is that the detectors must always be working within a square law such that the output *voltage* is proportional to the detected *power* level. This is not a problem if results from different instruments are not to be compared, but

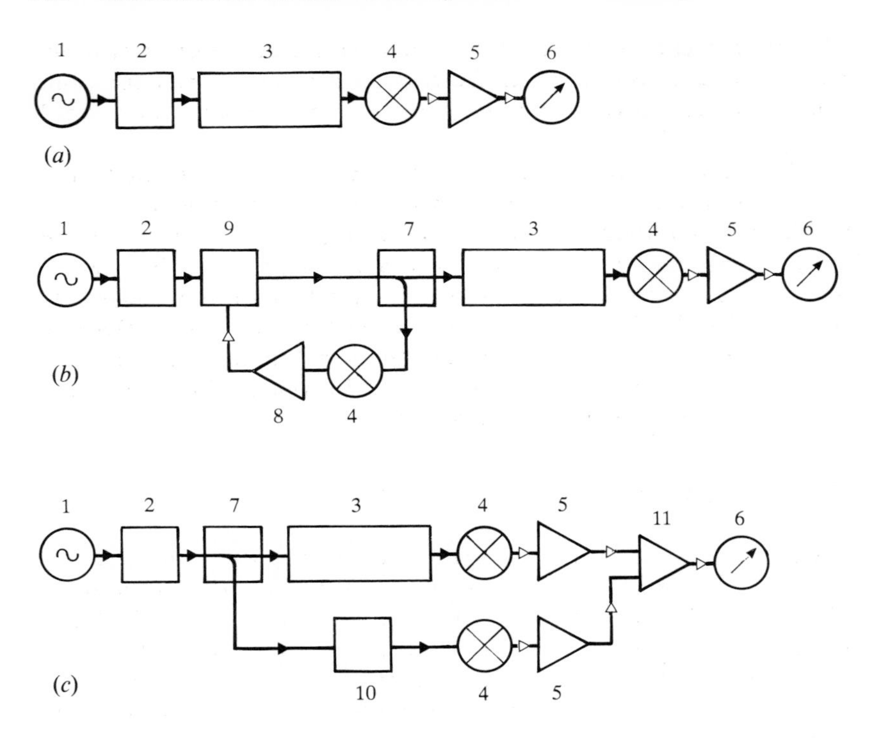

Figure 7.16 *Methods for transmission measurement (attenuation): 1 microwave source; 2 isolator; 3 sample head and sample; 4 detector; 5 log amplifier; 6 Output; 7 directional coupler; 8 feedback amplifier; 9 amplitude modulation and control; 10 reference attenuator; 11 difference amplifier*

for transferable calibration it is essential that the detectors operate with the same response. Too high a power at the detector will take it out of its square law response into a linear region where the output is proportional to the field strength and not the power.

All the systems described have involved straightforward DC detection and amplification. Some devices have been constructed with the addition of amplitude modulation of the source power at a few kilohertz, followed by detection and amplification at this frequency. The advantages are simply that spurious signals or drift currents in the detectors and electronic circuits are not interpreted as fluctuations due to the sample.

7.2 Phase

Measurement of phase alone has not been explored enough owing to lack of availability of commercial devices. There are situations, however, where it would seem eminently suitable. One example would be the determination of sugar concentration discussed in Section 5.3, where coupled with temperature measurement it would provide a more readily interpreted result than attenuation.

It has been used, however, for the measurement of the water content of tobacco (Ozamiz and Hewitt 1979) and the water content of coal (Klein 1981). The circuit devised by Ozamiz and Hewitt (1979) is shown in Figure 7.17. This is a heterodyne system in which the power in the reference and sample arms is down-converted from 1.25 GHz to an intermediate frequency of 160 MHz. It is easily demonstrated that the phase difference between the two intermediate frequency (IF) signals is the same as in the microwave case. These signals are amplified and clipped to form a pseudo-square wave and then fed to a phase detector.

7.3 Simultaneous Phase and Attenuation

It is relatively simple to build a manually operated bridge for the determination of both phase and attenuation, and the early device of Kraszewski and Kulinski (1976) was basically a mechanical version of that. Kalinski (1976) has devised several variations of a homodyne circuit which proved useful for phase compensation in glass water content measurements.

The circuit used by Meyer and Schilz (1980) is shown in Figure 7.18a, and its comparative complexity is obvious. Based on heterodyne principles, it functions by first down-converting the 9 GHz signal to 10 MHz. The required 10 MHz difference between the local oscillator at 8.99 GHz and

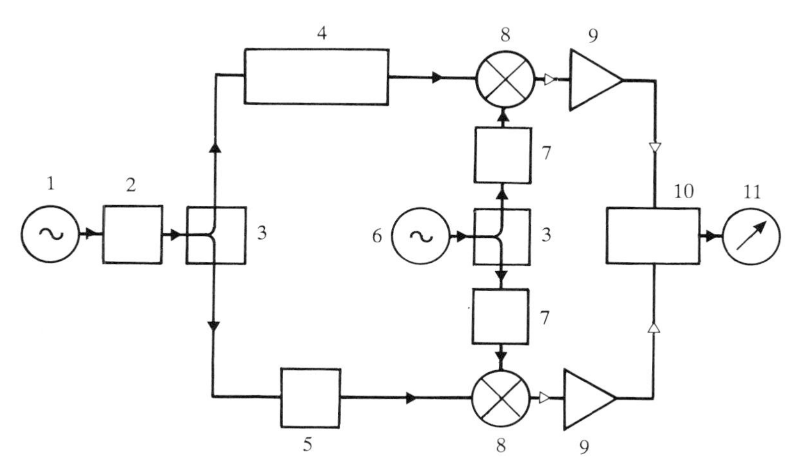

Figure 7.17 *Method of phase measurement (after Ozamiz and Hewitt 1979): 1 microwave source; 2 isolator; 3 power splitter; 4 sample head and sample; 5 phase shifter; 6 local microwave oscillator; 7 attenuator; 8 mixer detector; 9 amplifier; 10 phase detector; 11 output*

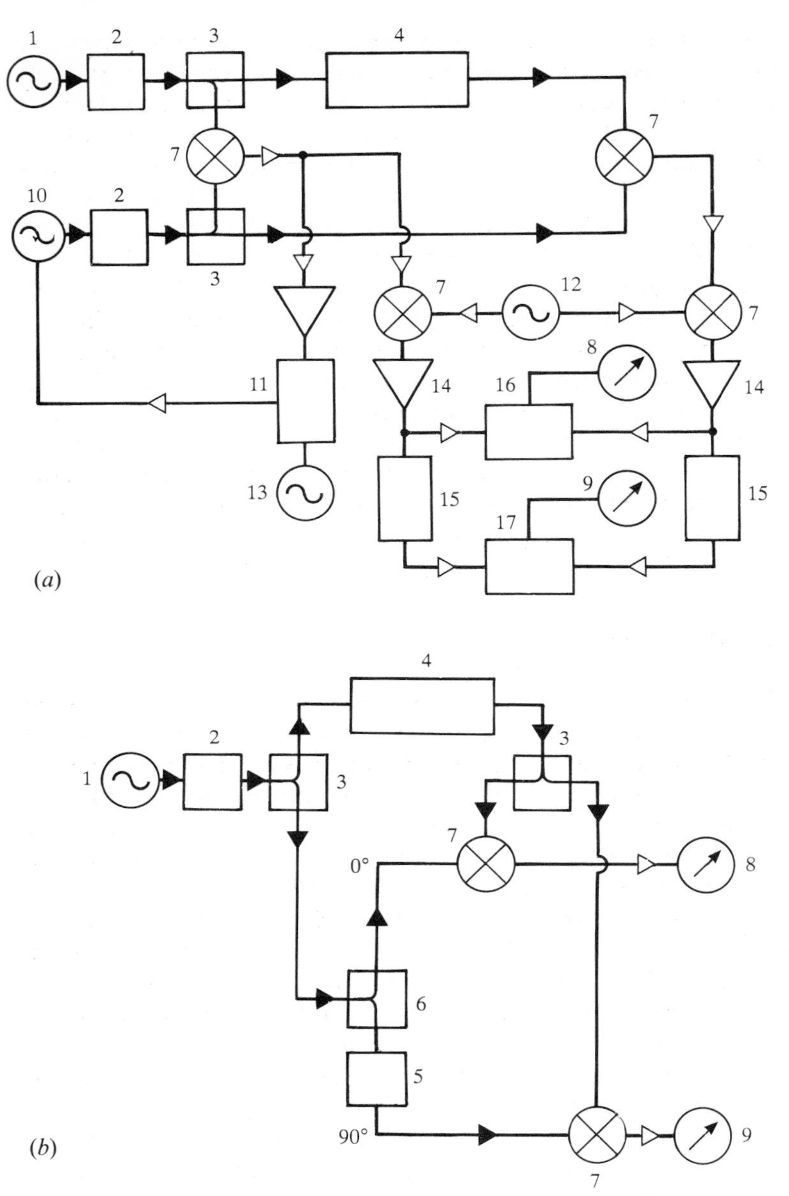

Figure 7.18 *Methods for phase and attenuation measurement after (a) Meyer and Schilz (1980) (b) Powell et al. (1987): 1 microwave source; 2 isolator; 3 power splitter or directional coupler; 4 sample head and sample; 5 phase shifter; 6 hybrid splitter; 7 mixer detector; 8 attenuation output; 9 phase output; 10 local microwave oscillator; 11 phase locked loop; 12 local IF oscillator (9.99 MHz); 13 second local IF oscillator (10 MHz); 14 tuned IF amplifier; 15 squarer; 16 log difference amplifier; 17 phase difference*

the source oscillator is maintained by a quartz crystal controlled 10 MHz oscillator and a phase locked loop. The 10 MHz signal is further mixed with 9.99 MHz from another quartz oscillator to produce a final IF of 10 kHz. The amplitudes of the signals at this frequency from both reference and source are compared to yield the attenuation, and after the waveform is squared the phase shift is obtained from the difference in time of the zero crossing of the two waveforms.

Finally, Powell et al. (1987) have described probably the simplest so far of all devices for phase and attenuation measurements (Figure 7.18b). Microwave power is divided in the standard way into two arms, one feeding through the sample. After the sample the signal is further split, one half being mixed in the first mixer with half of the reference signal from a further quadrature splitter. The other half is mixed with the other phase of the reference signal in the second mixer. The results of these various divisions and mixing is to produce outputs from the two mixers proportional in the one case to the attenuation, and in the other to the phase. Departures from true quadrature power splitting need to be corrected by incorporation of phase shifters at the appropriate places.

7.4 Resonant Systems

The types of resonators most commonly used have been (a) resonant waveguide cavities and (b) open ended transmission lines coupled in such a way as to enable resonance to occur. For the determination of the resonator variables many of the principles already employed for transmission measurements can be used. The change in transmitted power at resonance is a direct measure of change in Q factor and tan δ. The frequency shift can be measured readily by a feedback circuit designed to maintain the source frequency at the resonant frequency of the loaded cavity. Alternatively, for small deviations of resonant frequency the source can be at a fixed frequency and the phase of the transmitted power can be monitored. For a narrow range around resonance this can be a reasonably linear function of permittivity (Bosisio et al. 1970). The variety of methods to achieve these measurements is too great to consider here, but since all are designed to measure the effects of equations (14) and (15) we will generalize and show only a basic type of system (Figure 7.19). Here the transmission variables are measured either as a function of frequency or at a fixed frequency. A further variation is to measure at the resonance frequency. All possibilities enable the important variables to be measured.

7.5 Reflectance

Little can be said concerning reflectance measurement since implementation of such methods is not widespread and is in the experimental stage.

Suffice it to say that commercially available impedance meters have been used, particularly in the work of Chouikhi and Wilde (1986) on extrusion cooking. A basic reflectometer set-up would be as shown in Figure 7.20. Here power is fed to the sensor in the normal way and the reflected power is detected by means of a directional coupler or circulator. Thereafter techniques similar to transmission methods may be used to obtain the required data. The use of vector voltmeters or impedance meters allows for immediate acquisition of two variables, i.e. magnitude and phase of reflected power.

The reflection sensor incorporated in this circuit could just as easily be a resonator, as discussed in the previous section.

8 Sensors

8.1 Horn Antennae

Now that both the microwave measurement techniques available and the basic construction of the microwave circuits involved have been described, a few words are called for on the subject of sensors. The term 'sensor' in this context means the interface between the microwave apparatus and the material being measured. Two such interfaces have already been met, the horn antenna and the coaxial probe. There are other possibilities, however, and these will now be considered. The horn type of antenna itself requires a little further discussion in order to comprehend both its use and its misuse. Basically, a horn is an attempt to match the impedance of a waveguide to that of free space, at the same time increasing the directivity of the

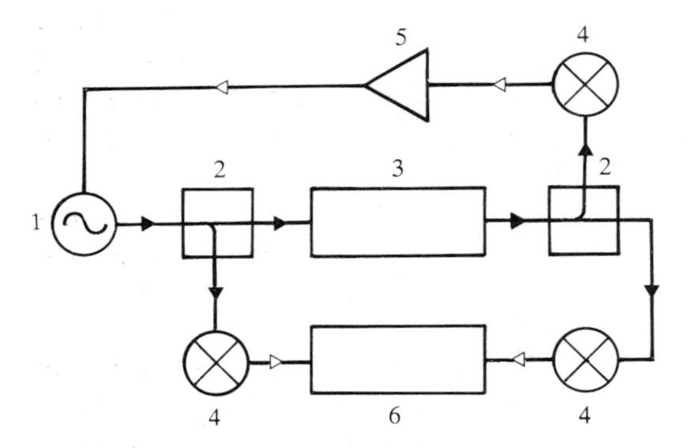

Figure 7.19 *General system for measurement of resonator variables* Q *and* f*: 1 microwave oscillator (swept frequency, frequency modulated or fixed frequency); 2 directional coupler; 3 resonator; 4 detectors; 5 frequency control loop amplifier if required; 6 vector voltmeter*

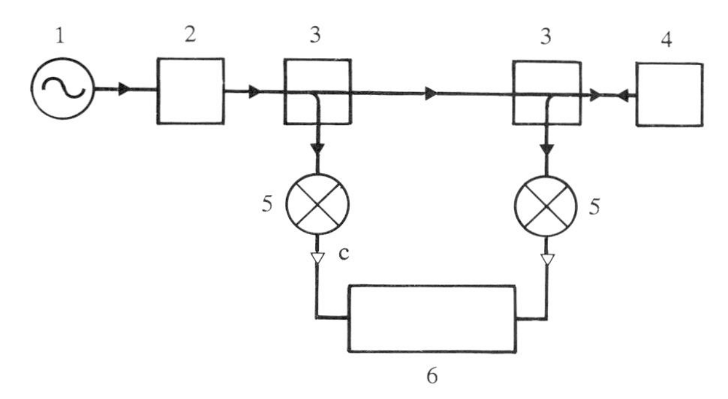

Figure 7.20 *Method for reflection measurement: 1 microwave oscillator; 2 isolator; 3 directional coupler or circulator; 4 reflection sensor; 5 vector voltmeter or impedance meter*

radiation. A sudden transition between a waveguide and free space, or indeed any material, results in a large reflection of power back into the waveguide owing to the impedance mismatch that occurs. By flaring the waveguide into the shape of a horn, this mismatch is reduced and a greater efficiency of transmission of power occurs.

Antennae are distinguished by certain interesting and relevant variables. The first of these, referred to as the 'gain', is a measure of the efficiency of such a device in directing power in a certain direction with reference to a perfect isotropic radiator, and is related to both its aperture size and the spread of the beam. In simple terms the beamwidth can be visualized as the cone of radiation emitted or received by such a device. Thus a short horn with a given aperture will have a much wider beam than a longer horn of the same aperture. Figure 7.21 shows, for an optimally designed rectangular horn, the relationship between the various dimensions. The beamwidths are different for the two perpendicular planes of the electric and magnetic components of the wave. A typical horn 4 wavelengths long with an aperture 3 wavelengths square would have beamwidths between 15° and 20° in the respective electric and magnetic planes.

As we move away from the central axis the field strength diminishes continously, the pattern of field strength versus angle being an elongated pear shape or lobe with several side lobes. The side lobes result from the effects of interference and diffraction of the wave.

The existence of these lobes means that it is possible for the antenna to pick up signals, albeit of low response, from places far off the axis. This is not a great problem in the measurements described here unless the criteria discussed below are not adhered to.

Mladek and Beran (1980) have shown that certain critical dimensions must be achieved for a successful application of even such a simple sensor. These dimensions depend very much on the material properties and are chosen so as to minimize errors from standing waves, reflections at the side boundaries of the material, and diffraction of waves around the sample to the receiver rather than through it. Thus for wheat, say with a moisture content of 5 per cent at a measurement frequency of 10 GHz, a layer at least 15 cm thick is required. To avoid reflections from the boundaries of the sample it should then be at least 1.5 times wider than this thickness.

The dimensions of the horns are also critical in the reduction of reflections. It has been calculated that, when measuring wheat at this frequency, a 14 cm thick sample should be at least 28 cm wide with rectangular horn antennae apertures of 4.5×7.0 cm. In this way, the anticipated error resulting from the shape and size of the sample could be reduced to less than 1 per cent.

Many improvements can be made to the simple horn antenna, all of which add to its expense. For example, Fresnel lenses may be fitted to more clearly collimate the beam, i.e. narrow its width. Perfect pencil beams, however, are impossible to achieve with apertures that are only a few wavelengths in size; interference effects always operate to produce side lobes. Even longer horns cannot achieve much better than roughly 10° beamwidth within the limits of size dictated by practicality. This fact should never be ignored.

The propagation from horns is usually considered to be a plane wave. This is not actually the case, especially close to the antenna, and small errors occur in the so-called near-field region if equations such as (17) which are derived assuming plane wave propagation are assumed to apply exactly. This is not important, however, and merely means that real

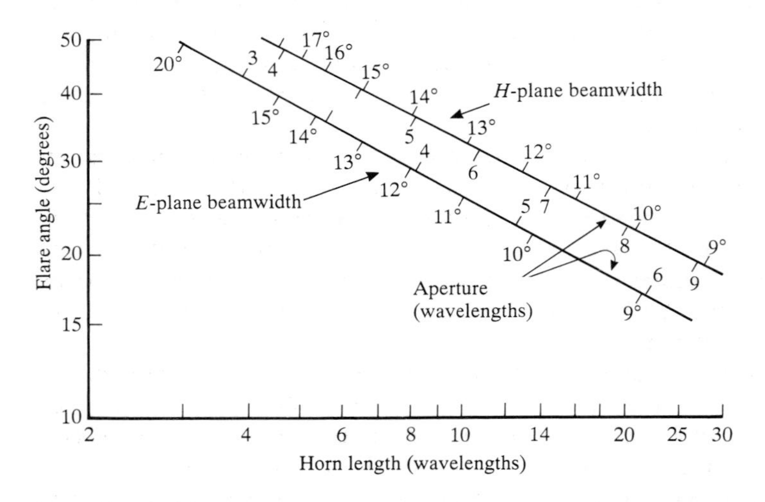

***Figure 7.21** Relationship between length, aperture and beamwidth for an optimally designed rectangular horn antenna*

calibrations will yield slightly different results from the theoretical ones discussed in Sections 4 and 5.

8.2 Stripline Sensors

The author has described in several publications (Kent 1972; Kent and Price 1979) the use of a form of transmission line known as stripline or microstrip as a sensor for transmission microwave measurements. Applications to both rice and grain (Miyai 1978) have since been described, as also have measurements on other foodstuffs such as sugar beet (Steele and Kent 1978). It is known that it has also been used successfully in soap.

The device can best be visualized by beginning with a coaxial line which is then split lengthwise, opened out and flattened. The result is a dielectric sheet or slab with a conducting ground plane on one side, and a conducting strip on the other (Figure 7.22). When another dielectric material is placed on top of the strip, the measured transmission loss of the line increases. It seems possible that this is because power is coupled out of the strip and radiated into the second dielectric. Provided this dielectric is of great enough extent, no power is reflected back to the strip from the dielectric boundary. The measured loss in the sensor can be calibrated against permittivity or water content, whatever is required. The phase shift is also found to be dependent on the properties of the second material and so can also be used for composition measurements, as Ozamiz and Hewitt (1979) have done with tobacco. Kent and Meyer (1981) have also used this type of sensor in a two variable technique in an attempt to compensate for the density variations of loose tobacco. Using the simple ratio of attenuation to phase proved to be not entirely successful, but as is now realized the density need not be eliminated but can also be measured by suitable calibration (see Section 5.1).

The advantage of this type of sensor over normal transmission measurements is that higher water contents can be measured. In this case of simple transmission, as we have seen in Section 4, the losses measured in high water content foods are high even for very short path lengths in the material. The stripline sensor is a much less sensitive device and can cope with a much larger bulk of material. Its low sensitivity, however, is only an advantage in that respect.

Because of its structure it can be constructed to dip into liquids and so is useful for batch processes. Unfortunately the sensitivity decreases markedly at very high water contents. A large number of different variants have been proposed but few have as yet found application. Figure 7.23 shows an experimental installation in milk processing. One of its problems is that in contrast to the normal transmission measurements this sensor requires contact with the material of interest. Given the highly corrosive cleaning

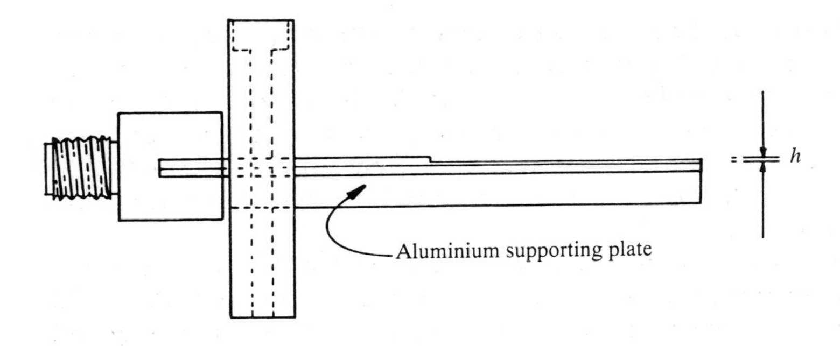

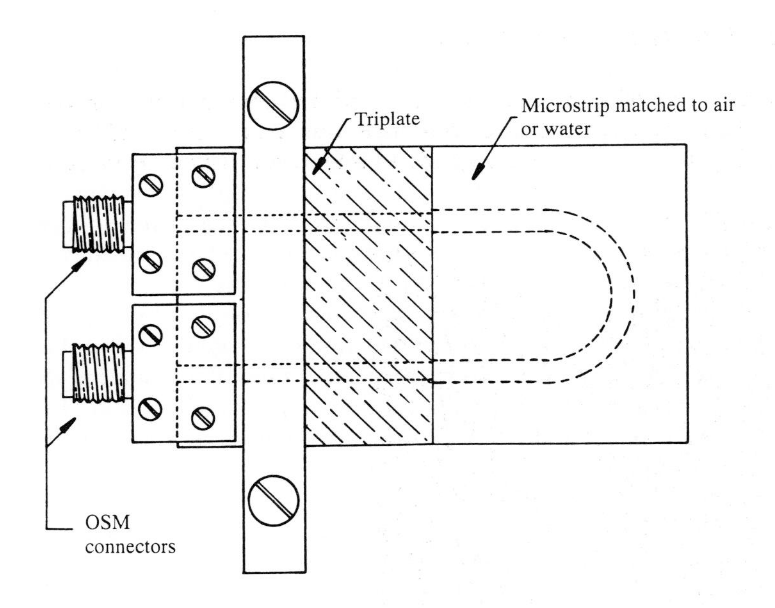

***Figure** 7.22 Stripline sensor*

agents that are used in modern cleaning-in-place practices, for example, this places severe design constraints on the materials of the sensor.

8.3 Stripline Antennae

The use of stripline as a microwave sensor has just been described, and its operation was discussed briefly in terms of radiative power loss from the strip in the presence of a second dielectric above the strip. Stripline can be constructed, however, to radiate more efficiently into free space, and such stripline antennae could be very useful in the kinds of applications that we are discussing here. Basically such an antenna consists of a very broad strip or patch or even of a hole in a patch. The instantaneous distribution of charge on the patch and its movement excite radiation. Numerous variations can be devised.

The crucial aspect for us, however, is that they would be far cheaper to manufacture than even the simplest horn antenna. The performance could not be expected to be as good as the horn, with poorer directivity and radiation pattern, but these are problems that could be reduced by careful use. Horns are expensive items to construct, whether they be electroformed or constructed piecemeal. The use of a stripline antenna offers a very cheap alternative.

8.4 Open Ended Transmission Lines

For the measurement of reflectance a number of different sensors can be considered. The impedance mismatch that is minimized in the use of horn antennae can be put to good advantage if the reflected power is used as a

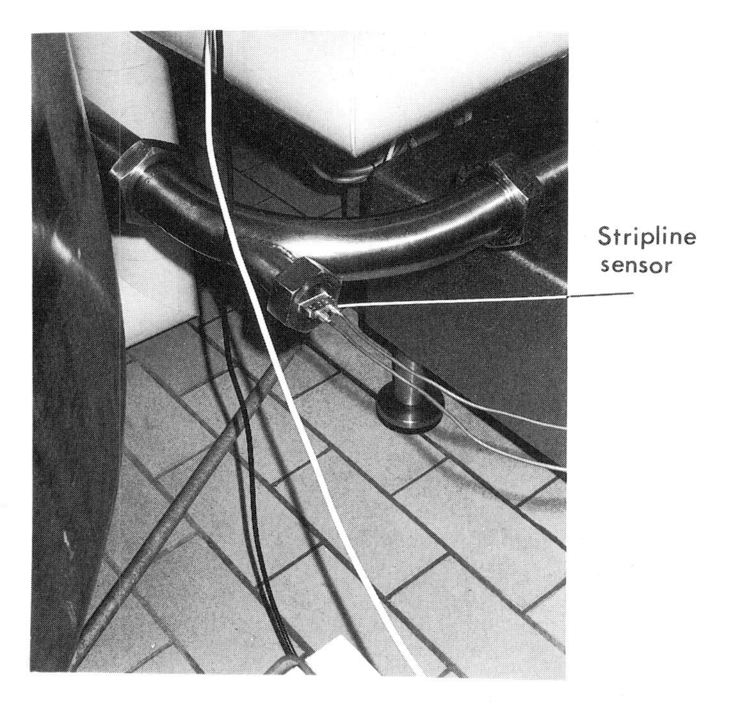

***Figure** 7.23 Stripline sensor installed in a milk concentrate flow*

measure of the material's properties. Similarly, open ended coaxial sensors have been described in Section 4.3 for use as reflection sensors. This type of configuration can be considered in terms of the fringe capacitance that exists at the open end of the line. This capacitance is changed by the presence of a dielectric material and so the measured reflectance changes.

This simple structure can be modified by extending the centre conductor into the material, when it acts as a monopole antenna radiating power into the sample. The antenna impedance is modelled in effect by the following equation:

$$\frac{Z(\omega, \varepsilon^*)}{\eta} = \frac{Z(\omega\sqrt{\varepsilon^*}, \varepsilon_0)}{\eta_0} \tag{19}$$

where Z is the impedance, ω is the angular frequency ($2\pi f$), ε^* is the permittivity, ε_0 is the relative permittivity of free space, η is the intrinsic impedance of the material and η_0 is the impedance of free space.

Such antennae have been widely used for measurements of biological tissues (Burdette et al. 1980). The special case when the monopole is of zero length – otherwise recognized as the open ended coaxial line probe – has been used in food applications (Knöchel and Meyer 1981; Chouikhi and Wilde 1986).

Since preparing this chapter, commercial equipment has become available for swept frequency dielectric spectroscopy measurement using the open-ended coaxial line as a sensor (Hewlett Packard dielectric probe and automatic network analyser). This opens up the possibilities of wider use of the dielectric spectrum.

9 Areas for Development

During the 10th European Microwave Conference held in Warsaw in 1980, some of the few individuals working in the field that has been called 'microwave aquametry' were able to meet and discuss its development. It is instructive to see both the transcript of that discussion and the answers to a questionnaire circulated before the meeting. They can be found in a special issue of the *Journal of Microwave Power* edited by Andrzej Kraszewski (see Kraszewski 1980a; 1980b). Many of the questions asked have yet to be answered adequately. These refer not only to technical problems but economic ones as well. Other problems are only now at the point of solution and, as can be seen from this chapter, the field is still one of active investigation.

The most important requirement now is that more instruments become available using multi-variable measurements for the implementation of the kind of methods of composition analysis described above. As has been shown, the technique need not stop at two variables alone. For example, it should be possible to include DC conductivity in a measurement and to extend the method to the simultaneous determination of fat, water, protein and ash. Other suitable independent variables can be measured, which enable the dimensions of the technique to expand even further.

The advent of intelligent instruments renders calibration a simpler task, and techniques of continuous recalibration can improve the accuracy of the output.

Appendix 1: Some Manufacturers of Microwave Moisture Measurement Instruments

Advanced Moisture Technology Inc., Buffalo Grove, Illinois, USA
Hamworthy Engineering Ltd, Poole, Dorset, UK
Hydronix Ltd, Cranleigh, Surrey, UK
Infrared Engineering Ltd (Moistrex), Maldon, Essex, UK
Kay-Ray Inc., Wheeling, Illinois, USA
Laboratorium Prof. Berthold, Wildbad, Germany
Pleva GmbH, Horb-Bildechingen, Germany
Wilmer Instruments and Measurement, Warsaw, Poland

Appendix 2: List of Symbols

A_r	depolarization factor
A, B, A', B'	constants
a_1, a_2, a_3, a_4	constants
d, x	distance or thickness
F	resonator filling factor
f	frequency
f_c	critical frequency or relaxation frequency of dispersion
f_0	resonant frequency of a resonator
I	transmitted power
I_0	incident power
j	$\sqrt{-1}$
M_0	weight of dry material
M_w	weight of water
n	refractive index
Q_0	unloaded Q factor of resonator
Q	loaded Q factor
R	ratio of attenuation to phase

r, r_1, r_2	reflection coefficients
r, a, b, c	ellipsoid axes
S_{11}	reflection S parameter
S_{12}	transmission S parameter
V	volume of wet material
V_w	volume fraction of water
Z	impedance
z	square root of complex permittivity
α	attenuation
δ	loss angle $\tan^{-1}(\varepsilon''/\varepsilon')$
ε_0	permittivity of free space (vacuum)
ε_s	static permittivity or dielectric constant relative to ε_0
ε_∞	permittivity at very high frequency relative to ε_0
$\varepsilon, \varepsilon^*$	complex permittivity relative to ε_0
ε'	real part of ε^*
ε''	imaginary part of ε^*
ε_{eff}	effective complex permittivity of a mixture
ε_w	complex permittivity of water
ε_m	complex permittivity of dry material
ε_0''	loss factor at a given frequency due to DC conductivity
η	impedance of dielectric
η_0	impedance of free space (vacuum)
θ	reduced loss angle $\tan^{-1}[\varepsilon''/\varepsilon'-1)]$
λ	wavelength
ρ	bulk density
σ	DC conductivity
τ	relaxation time of dispersion
ϕ	phase shift
ψ	moisture content fraction

References

Beer, A. (1853) *Einleitung in die höhere Optik*, 1st edn. Brunswick

Bilbrough, J. (1968) Microwave apparatus for measuring moisture content. UK Patent 1, 111, 384

Birnbaum, G. and Fremeau, J. (1949) Measurement of the dielectric constant and loss of liquids and solids by a cavity perturbation method. *Journal of Applied Physics*, **20**, 817–818

Bosisio, R. G., Giroux, M. and Coudere, D. (1970) Paper sheet moisture measurements by microwave phase perturbation techniques. *Journal of Microwave Power*, **5**, 25–34

Brodwin, M. and Benway, J. (1980) Experimental evaluation of a microwave transmission moisture sensor. *Journal of Microwave Power*, **15**, 261–265

Burdette, E. C., Cain, F. and Seals, J. (1980) In-vivo probe measurement technique for determining dielectric properties at VHF through microwave frequencies. *IEEE Transactions*, **MTT-28**, 414–427

Chaloupka, H., Ostwald, O. and Schiek, B. (1980) Structure independent microwave moisture-measurements. *Journal of Microwave Power*, **15**, 221–231

Chouikhi, S. M. and Wilde, P. J. (1986) Reflection of an open-ended coaxial line and application to moisture content measurement. *Proceedings of the International Measurement Conference on Tests and Transducers*, vol. 2, 251–264

Dube, D. C. (1970) Study of Landau-Lifshitz-Looyenga's formula for dielectric correlation between powder and bulk. *Journal of Physics D, Applied Physics* **3**, 1648–1652

Dutuit, Y. (1980) Construction of a time domain spectrometer. A contribution to the improvement of methods of analysis of dielectric relaxation spectra and application. PhD thesis, University of Bordeaux I

Freymann, R. (1954) Procedure and apparatus for determining the water content and the properties of free and bound water contained in various materials. French Patent 1,102,199

Gladstone, J. H. and Dale, T. P. (1863) Researches on the refraction, dispersion and sensitiveness of liquids. *Philosophical Transactions*, **153**, 317–343

Gooding, R. G. and Bird, D. (1963) An apparatus for continuous measurement of water content of foundry sand. *BICRA Journal*, **11**, 641–661

Henry, F. and Berteaud, A. J. (1980) New measurement technique for the dielectric study of solutions and suspensions. *Journal of Microwave Power*, **15**, 233–242

Ince, A. D. and Turner, A. (1965) The determination of moisture in plain cakes by a microwave attenuation technique. *The Analyst*, **90**, 692–696

Jacobsen, R., Meyer, W. and Schrage, B. (1980) Density independent moisture meter at X-band. *Proceedings of the 10th European Microwave Conference*, Warsaw, 526

Kaatze, V. (1973) A cavity resonator method for measuring the complex permittivity of small amounts of liquids in the frequency range 0.7–25 GHz. *Applied Physics*, **2**, 241–246

Kalinski, J. (1976) Automatic phase control system for microwave industrial on-line moisture-attenuation-voltage (MAV) convertor. *IEEE Transactions*, **IECI-25**, 425–427

Kent, M. (1972) The use of strip-line configuration in microwave moisture measurement. *Journal of Microwave Power*, **7**, 185–193

Kent, M. (1977) Complex permittivity of fishmeal; a general discussion of temperature, density and moisture dependence. *Journal of Microwave Power*, **12**, 341–345

Kent, M. (1988) *Dielectric and Electrical Properties of Foodstuffs*. Science and Technology Publishers, London

Kent, M. (1989) Application of two-variable microwave techniques to composition analysis problems. *Transactions of the Institute of Measurement and Control*, **11**, 58–62

Kent, M. and Kress-Rogers, E. (1987) The COST90bis collaborative work on the dielectric properties of foods. In *Physical Properties of Foods*, vol. 2, eds R. Jowitt, F. Escher, M. Kent, B. McKenna and M. Roques, 171–197, Applied Science Publishers, Elsevier, London

Kent, M. and Meyer, W. (1981) Density independent microwave moisture measurements using stripline sensors. *Proceedings of the 1981 IMPI Symposium on Microwave Power*, 6A.1, Toronto

Kent, M. and Meyer, W. (1982) A density-independent microwave moisture meter for heterogeneous foodstuffs. *Journal of Food Engineering*, **1**, 31–42

Kent, M. and Meyer, W. (1984) Complex permittivity of protein powders as a function of temperature and hydration. *Journal of Physics D, Applied Physics*, **17**, 1687–1698

Kent, M. and Price, T. E. (1979) Compact microstrip sensor for high moisture content measurement. *Journal of Microwave Power*, **14**, 363–365

Knöchel, R. and Meyer, W. (1981) Continuous moisture determination in fluids and slurries. *Proceedings of the 1981 IMPI Symposium on Microwave Power*, Toronto, 193–195

Kraszewski, A. (1978) A model of the dielectric properties of wheat at 9.4 GHz, *Journal of Microwave Power*, **13**, 293–296

Kraszewski, A. (1980a) Microwave Aquametry – a review. *Journal of Microwave Power*, **15**, 209–220

Kraszewski, A. (1980b) Microwave aquametry – a bibliography 1955–1979. *Journal of Microwave Power*, **15**, 298–310

Kraszewski, A. (1989) Moisture monitoring of moisture content in grain – further considerations. *Journal of Microwave Power*, **23**, 236–246

Kraszewski, A. and Kulinski, S. (1976) An improved microwave method of moisture content measurement and control. *IEEE Transaction*, **IECI-23**, 364–370

Kress-Rogers, E. and Kent, M. (1987) Microwave measurement of powder moisture and density. *Journal of Food Engineering*, **6**, 345–376

Meyer, W. and Schilz, W. (1980) A microwave method for density independent determination of the moisture content of solids. *Journal of Physics D, Applied Physics*, **13**, 1823–1830

Meyer, W. and Schilz, W. (1981) Feasibility study of density-independent moisture measurement with microwaves. *IEEE Transactions*, **MTT-29**, 732–739

Miyai, Y. (1978) A new microwave moisture meter for grains. *Journal of Microwave Power*, **13**, 213–218

Mladek, J. (1973) Determination of the moisture content of granular materials by a microwave method. *Zemedelska Technika*, **19**, 453–458

Mladek, J. and Beran, Z. (1980) Sample geometry, temperature and density factors in the microwave measurement of moisture. *Journal of Microwave Power*, **15**, 243–250

Mudgett, R. E., Wang, D. I. C. and Goldblith, S. A. (1974) Prediction of dielectric properties in oil-water mixtures at 3000 MHz, 25°C based on pure component properties. *Journal of Food Science*, **39**, 632–635

NBS (1958) Tables of dielectric dispersion data for pure liquids and dilute solutions. National Bureau of Standards circular 589, US Department of Commerce, Washington DC

Nedzvecki, Y. E. (1955) Electrical determination of moisture content in nonconducting materials and arrangement for its realisation. USSR Patent 107,977

Nelson, S. O. (1983) Observations on the density dependence of dielectric properties of particulate materials. *Journal of Microwave Power*, **18**, 143–152

Ohlsson, T., Henriques, M. and Bengtsson, N. E. (1974) Dielectric properties of model meat emulsions at 900 and 2800 MHz in relation to their composition. *Journal of Food Science*, **39**, 1153–1156

Ozamiz, J. M. and Hewitt, S. J. (1979) Microwave moisture measurement system. *Proceedings of the 9th European Microwave Conference*, Brighton, UK, MS2, Microwave Exhibition and Publishers, UK

Powell, S. D., McLendon, B. D., Nelson, S. O. and Allison, J. M. (1987) Use of a density independent function and microwave measurements system for grain moisture measurements. Paper 87-3053, Meeting of ASEA, Baltimore

Risman, P. O. and Bengtsson, N. E. (1971) Dielectric properties of foods at 3 GHz as determined by a cavity perturbation technique. *Journal of Microwave Power*, **6**, 107–123

Steele, D. J. and Kent, M. (1978) Microwave stripline techniques applied to moisture measurement in food materials. *Proceedings of the 1978 IMPI Symposium on Microwave Power*, Ottawa, 31–36

Stuchly, S. and Kraszewski, A. (1965) Method for the determination of water content in solids, liquids and gases by means of microwaves and arrangement for application of this method. Polish Patent 51,731

Van Beek, L. K. H. (1967) Dielectric behaviour of heterogeneous systems. *Progress in Dielectrics*, **7**, 69–114

Walker, C. W. (1958) Apparatus and method for measurement of moisture content. US Patent 3,079,551

Watson, A. (1957) Improvements in and relating to the determination of moisture content. UK Patent 897,956

Zehnder, C. B. (1967) Application of the combination microwave-gamma ray gauge to wood chip weight and moisture measurement. *Pulp and Paper Magazine of Canada*, no. 10, 678–688

8 *Ultrasound Propagation in Foods and Ambient Gases: Principles and Applications*

Erika Kress-Rogers

Contents

5.1 Sources of Attenuation 270
5.2 Relaxation 271
5.3 Scattering 273
5.3.1 Hindrance or Help? 273
5.3.2 Concentration of Suspensions 277
5.3.3 Juice Stability and Produce Defects 277
5.3.4 Flow Metering 277
5.3.5 Emulsions – the Three Contributions to Scattering Losses 278

6 Conclusions 280

Appendix 1: Ultrasound Measurement Applications in and for the Food Industry 281

Appendix 2: List of Symbols 283

References 284

1 Introduction

Ultrasound instrumentation is already established for a number of process control tasks in the food industry. These applications include the measurement of level, flow and suspended solids concentration and the detection of interfaces between liquids. The concentration of certain simple solutions is also being monitored with current instrumentation. Automatic qualitative distinction between food and cleaning liquids is carried out in other situations. Details on the established systems are given in Chapter 9, 'Ultrasonic instrumentation in the food industry', by Nicholas Denbow.

In this complementary chapter, the underlying principles are described and illustrated with data for the relevant properties of foodstuffs, ambient gases and container materials. Techniques developed recently for a wider range of applications in the food industry, and awaiting commercial implementation, are discussed within the context of the principles involved. The new applications include, for example, the measurement of solid/liquid ratios for fats or of the original gravity of beer. Techniques for the determination of the size and the size distribution of droplets in emulsions are under development. The potential for a much wider range of applications with good reliability is based on advances in the understanding of the ultrasonic properties of foods and in the electronics for measurement and data processing.

An overview of ultrasound measurement applications for the food industry, either implemented or at the research stage, is presented in Appendix 1.

This chapter is focused on the propagation of ultrasound in food liquids. The ultrasonic properties of ambient gases and solid foods are discussed more briefly. A few data for container and transducer materials are given in the tables in this chapter. The properties of the materials of processing vessels and pipes in the food industry are quite similar to those in other industries, and a description of these can be found in general ultrasonics engineering guides. (One container material unique to the food industry is discussed very briefly, namely egg shells.) The overview of the principles underlying ultrasonic techniques follows the theoretical treatment presented in the textbooks by Blitz (1963), Cracknell (1980), Crawford (1968) and Lynnworth (1989) or (mainly for Section 5.3) in recent articles. Measured values and food application details are from articles and handbooks as indicated in the tables, figures and text. Experimental studies carried out at the LFRA have also been taken into account.

2 Overview of Ultrasound Applications

2.1 What is Measured?

Longitudinal sound waves consist of periodic displacements of the particles within the medium, setting up regions of compression and rarefaction propagating through the medium. There are many parallels to the propagation of mechanical waves on coupled bead-loaded springs (Section 3.1). However, for sound waves, the regions of compression and rarefaction are associated with temperatures above and below the equilibrium, respectively, which in turn influence the excess pressure within these regions. The thermodynamic (as well as the mechanical) properties of the medium thereby influence wave propagation.

Sound waves with frequencies above the audible range are called ultrasound waves. A number of variables characterizing the propagation of an ultrasound signal through a medium can be measured. These include:

- the transit time for the transmission of a pulse through the sample (yielding the speed of sound for a known sample depth or vice versa)
- the attenuation of a signal transmitted through the sample
- the reflected and transmitted proportions of a signal meeting an interface between the sample and a medium with known properties (yielding the acoustic impedance of the sample)
- the amplitude, phase and frequency of a scattered ultrasound signal at specified angles.

In addition, such ultrasound measurements can be performed at several frequencies. This range of measurement types, together with the relationships of these measured variables with the physical and chemical properties of gases, liquids and solids, makes a wide range of applications possible.

New measurement techniques for the food industry are often the result of technology transfer from applications in other industrial sectors. Applications both within and outside the food industry are therefore briefly reviewed in this section. Later sections discuss the ultrasonic properties of materials relevant to the food industry in more detail and describe applications.

Only those aspects of measurement techniques that are specific to food applications are considered in this chapter. An overview on general experimental methods in ultrasound measurement (with an emphasis on food applications) has been given by Povey and McClements (1988, pp. 233–241). They refer to Breazeale et al. (1981) and Papadakis (1976) for more general reviews of ultrasonic techniques in the laboratory and in industry, respectively. Further information can be found in Lynnworth (1989). Ultrasound techniques used in the measurement of level, flow rate and suspended solids content in the food industry are treated in Chapter 9.

2.2 Communication, Detection and Location

For a given target variable, it is often possible to choose between competing technologies, one using ultrasound waves and the other electromagnetic waves. The characteristics of their propagation and the implications for measurement applications are therefore compared here for the two wave types.

In common with electromagnetic waves, sound waves can propagate through a gas, a liquid or a solid. Sound waves cannot propagate through a high vacuum, however. Their propagation characteristics, such as velocity or attenuation as a function of frequency, differ from those of electromagnetic waves, thus defining their respective application areas and associated frequency ranges.

Figure 8.1 shows the spectrum of sound waves in comparison with the electromagnetic spectrum. The audiosound range that we use for speech communication extends from about 15 Hz to 20 kHz, and wavelengths are in the metre and centimetre range. Similar wavelengths are encountered in radio, television and radar communication by freely propagating electromagnetic waves. This is the spectrum from the VHF radio range over the UHF range to the lower (by frequency) microwave range. The velocity of propagation in a gas is six orders of magnitude lower for sound than for electromagnetic waves. The frequency ranges for sound and radio communication are displaced correspondingly.

Moving up slightly in frequency and down to wavelengths of approximately 1 cm, we arrive in the detection and location ranges within the

Frequency (Hz)		Sound waves: Wavelength in typical solid S1	liquid L1	gas G1		Electromagnetic waves			Wavelength
10^{15}						V			
									1 µm
10^{14}	100 THz					↑	NIR	●	
							MIR		10 µm
10^{13}	10 THz					IR		■	
					↑				100 µm
10^{12}	1 THz	6 nm				↓	FIR		
									1 mm
10^{11}	100 GHz		15 nm			↑	EHF		
					MS	MW			1 cm
10^{10}	10 GHz			30 nm			SHF		
						↓			10 cm
10^{9}	1 GHz				↓		UHF		
						↑			1 m
			●		↑				
10^{8}	100 MHz	60 µm					VHF		
			■						
10^{7}	10 MHz	0.6 mm	150 µm				HF		
					US	RW			100 m
10^{6}	1 MHz		1.5 mm				MF		
10^{5}	100 kHz			3 mm	↓		LF		
10^{4}	10 kHz	60 cm	15 cm	3 cm	↑	↓			
									100 km
10^{3}	1 kHz				AS				
10^{2}	100 Hz	60 m	15 m	3 m	↓				
10^{1}	10 Hz				↑				
					IS				
10^{0}	1 Hz				↓				

Acoustic media S1/L1/G1 have sound velocities 6000/1500/300 m/s^{-1}.
Examples for particulates (and distance) of dimensions corresponding to the wavelength are entered for illustration of scattering characteristics (Section 5.3) as follows:
■ suspended solids particle of 50 µm
● emulsion droplet of 5 µm
Sound wave ranges are: infrasound (IS), audiosound (AS), ultrasound (US) and microsound (MS).
Electromagnetic wave ranges are:
RW radiowaves, subdivided into earth-bound (MF, LF) and freely propagating (HF, VHF, UHF)
MW microwaves
IR infrared, subdivided into far infrared (FIR), mid infrared (MIR) and near infrared (NIR; 0.7–2.5µm)
V visible (approximately 400–700 nm)
Ultraviolet, X-rays and gamma-rays are at higher frequencies and are not shown.

Figure 8.1 *Spectra for sound and electromagnetic waves*

sound and the electromagnetic spectrum known as sonar and radar, respectively. Radar in the mid microwave range is employed in the detection of aircraft. It can operate over long distances in air, but is strongly attenuated by water. The reverse is true for sonar, where sound waves instead of electromagnetic waves are emitted. Using a scanning system, the reflections from the object can be visually displayed showing bearing and range, similar to radar detection systems. Sonar was initially developed for the detection and location of submarines and later for the measurement of the depth of the seabed or of lakes. Sonar operates in the ultrasound range, that is with frequencies in excess of about 20 kHz, beyond the response limit of the human ear.

With ultrasound pulse-echo systems developed subsequently, the search for shoals of fish or the inspection of off-shore installations became possible. With a high-power echo sounder it is even possible to detect a single fish of 35 cm length at a depth of 460 metres using 30 kHz (pulse length 500 μm, peak energy 8 kW). With higher frequencies it becomes possible to identify the size and species of fish, but at the expense of operating over shorter distances only, since ultrasound attenuation by sea water increases with frequency (Cracknell 1980).

2.3 Level and Flow Rate Measurement

The ultrasonic pulse-echo technique was later adapted to operate in other media, including gases and solids as well as liquids, and for a wide variety of applications, including level measurement and flaw detection. The tradeoff in the case of marine applications, namely better spatial resolution when the frequency is increased, at the expense of higher attenuation and thus shorter range, applies also to many of the industrial applications of ultrasound. However, the relationship between attenuation and frequency is often more complex in the media encountered here (see Sections 5.2 and 5.3), for example for transmission through gases and liquids carrying dispersed particles that scatter ultrasound.

It is also worth noting that resolution in ultrasound ranging is limited not only by long wavelengths but also by short pulse duration and low signal to noise ratios in the received echo (Cracknell 1980). Other factors influencing the choice of frequency are the transducer size and instrumentation costs. The ultrasound range extends to 500 MHz approximately. For process control measurements in the food industry, the frequency rarely exceeds 10 MHz.

The lowest part of the ultrasound range, up to 200 kHz approximately, is used for transmission through gases, for example in echo-ranging techniques for tank fill level measurement. In this frequency range, the ultrasound attenuation in most common gases is acceptably low so that a satisfactory reflected signal is received and the wavelength in gases is short enough to allow adequate spatial resolution. (Bats use much the same frequency range for their guidance). In some applications of echo ranging through gases, microwave techniques are more appropriate than ultrasound. This is the case where strong temperature variations along the path of the beam may occur, for example in the ullage space of ships' holds.

Liquids attenuate ultrasound less than gases, in general, so that higher frequencies can be employed. For process control applications based on ultrasound echo ranging in liquids, frequencies in the range 1 to 5 MHz are commonly used. These frequencies are higher than for sonar since transmission over long distances (as would be encountered in the location of fish shoals) is not required and demands on spatial resolution or on accuracy in velocity measurement are more stringent. The velocity of sound is higher in liquids than in gases so that the wavelengths encountered in echo ranging through liquids are of the same order as for gases, despite the frequency difference.

Details on ultrasonic level and flow measurement in the food industry are provided in Chapter 9. Some additional notes are given here in Section 5.3. Further information on non-invasive flow measurement is given in CPPM (1989) for clinical applications.

2.4 Non-destructive Testing

In non-destructive testing (NDT; see for example, Krautkrämer and Krautkrämer 1983), the ultrasonic flaw detector is widely used, with frequencies up to about 20 MHz for the examination of fine-grained metals and thin objects. Ultrasonic thickness gauges are used in the inspection of components such as turbine blades; a resolution of 0.15 mm can be achieved with a frequency of 20 MHz. Ultrasonic wavelengths in water and in many liquid and semi-solid foods at 5 MHz are similar to those in steel at 20 MHz, so that one could expect a similar resolution. However, in high-speed processing lines it is difficult to arrange for good acoustic contact between the transducer and a food coating layer, for example, without disturbing or damaging the layer. Also, resolution can be limited by weak echoes as they result from the high attenuation in many foods. (In laboratory studies, immersion tanks have been employed to measure the thickness of egg shells, and the use of soft-tipped probes has been proposed as an alternative. See Section 3.5).

Polycrystalline solids scatter sound in a manner similar to gases or liquids containing fine-grained particles. In the determination of metal grain size, scattering techniques using frequency scanning are now gaining ground despite the need for considerably more signal processing compared with the older single-frequency attenuation method. For example, scanning from 3 to 12 MHz is used to determine grain sizes in the range of 50 to 350 μm in steel (Lynnworth 1989). The rapid advances in the signal processing capabilities of industrial process control systems are also increasing the feasibility of new ultrasound applications in the food industry. Methods are being developed for the determination of droplet size in emulsions, for example, where several scattering mechanisms overlap (Section 5.3).

Ultrasonic imaging has found well-established clinical applications at 1–2 MHz (Crecraft 1983; Wells 1977; Newhouse 1988). Ultrasound is also used for the assessment of back fat in live agricultural animals (Section 3.5).

2.5 Concentration Measurement

Current ultrasound applications in the food industry are primarily level and flow measurement, but also the monitoring of suspended solids concentration and the detection of interfaces between immiscible liquids (see Chapter 9). In other industries, primarily the petrochemical and nuclear sectors, the relationship between the acoustic properties of a material and its chemical composition is being used to determine solute concentration or to identify liquids. The possibility of mounting ultrasound transducers non-invasively on the outside of vessels containing hazardous materials is particularly suitable for these industries (Asher 1983).

The possibility of non-contact application is also attractive to the hygiene-conscious food industry. Concentration measurements are already used for the monitoring of simple solutions, such as sodium hydroxide cleaning solutions, or pure sucrose or sodium chloride solutions, and in the control of coffee extract dilution and corn starch processing.

The idea of an ultrasonic 'solution analyser' (better termed a concentration meter) for alcoholic beverages and dairy liquids was proposed in the late 1940s. Interest in practical applications in the food industry received a new impetus in the late 1960s, when the implementation of ultrasonic concentration monitoring was successful in the chemical industry, where ultrasound applications soon expanded to other tasks (Zaccharias and Parnell 1972; Steele 1974). (The term 'moisture measurement' was sometimes used to describe the determination of the water content of aqueous solutions.) Applications in the in-line monitoring of food properties (rather than just process variables such as flow rate or fill level) have so far largely been restricted to a few simple process streams or else to a distinction of beverage from cleaning solution. Recent advances in the understanding of the ultrasound properties of food and the greater feasibility of multi-variable measurement in-line is now offering the potential of widening the range of applications in the food industry. The measurement of solute concentration or the water content of solutions is discussed in more detail in Section 3.4.

2.6 Passive Ultrasound Equipment: Acoustic Emission Monitoring

The ultrasonics chapters in this book describe measurements of food properties or of process variables, based on the interaction of an applied low-intensity ultrasound signal with the food, ambient gas and container material. Before describing these in detail, it is appropriate to include an alternative acoustic measurement technique in this overview section. In this technique, the equipment does not emit an acoustic signal but analyses audiosound and ultrasound created by the process itself. This passive technique is known as acoustic emission monitoring (AEM).

The AEM equipment listens to leaks or bubbles, mixing or grinding operations, solids flow or bearings. It detects deviations from the standard pattern, indicating a defect in a component or an out-of-specification processing operation. It can indicate the end point of an operation such as milling. The sound (in the audiosound or ultrasound range) that is detected in acoustic emission can result from stresses or impact created within the process that is monitored; or stresses can be induced to generate sound emission as a crack grows, for example. These techniques have evolved from the ancient methods of tapping objects and listening to deduce unseen properties such as the soundness of crocks or the ripeness of water melons. Applications are primarily in detecting flaws and anticipating failure in engineering structures (see for example Szilard 1982; and issues of the *Journal of Acoustic Emission*, published by the Acoustic Emission Group, University of California).

Although an early application of AEM to the assessment of the quality of sulphur samples was reported in the *Codex Germanicus* in AD 1350, the extent of AE from chemical reactions was realized much later (Betteridge et al. 1981). Now, the monitoring of hydration, titration, precipitation and electrolysis processes by AEM is being investigated, or an enzyme is immobilized on an AE sensor which then monitors effervescence from gas production (Wade et al. 1991).

Drying processes in biological tissues can cause AE. Under conditions of draught, suction pressures in plant conduits can reach −10 MPa. When these tensions are released by cavitation, AE occurs. This is being investigated for irrigation control. The drying of woody tissues can also be monitored to control drying rates. One can conjecture that it may be possible to develop AEM techniques for the observation of drying processes in foods with a fibrous structure or for the assessment of other water migration processes in foods.

Applications in the monitoring of food processing operations have been developed at the LFRA. These include the monitoring of food powder flow rates and of the impact of cleaning fluid sprays on vessel surfaces. An acoustic emission sensor mounted underneath a cooker extruder provided a signal related to the density of the final food product emerging from the extruder (Roberts and Wiltshire 1990a).

'AE can be simple to detect and difficult to interpret' (Wade et al. 1991). Neural network techniques (Jansson 1991) are being explored to help with the analysis of AE spectra (Zgonc and Grabec 1990).

2.7 High-power Ultrasound Equipment

Instrumentation for ultrasonic measurements in industrial process control uses low signal intensities, or short pulses of higher peak intensity separated by long 'dead' times. This low time-averaged signal intensity leaves the sample unchanged. The sample is also unaffected by AEM, where the applied ultrasound intensity is zero. At the other end of the power scale is the ultrasound equipment designed to take part in the food processing operation rather than monitor it. Many readers will have used ultrasonic high-power appliances in sample preparation in the laboratory; others already use ultrasonic food processing plant or are considering this option. The effects of high-power ultrasound are therefore briefly introduced in this section, although they do not occur with the instrumentation used in ultrasonic measurement.

High continuous power levels are used in equipment for sonoprocessing, such as ultrasonic cleaning baths, homogenizers, emulsifiers or (microbial) cell disrupters (see Ensminger 1973; Puskar 1982 for a general introduction to high-power ultrasonics). The effect of these high ultrasound power appliances on the medium is often due to cavitation, that is the sudden collapse of gas bubbles in the liquid and the associated high pressures. It is caused by rapidly alternating pressures of high amplitude. Frequencies of typically 20–50 kHz are used since the threshold intensity for cavitation rises rapidly for higher frequencies, and lower frequencies would be within the audio range (Cracknell 1980; Young 1989).

High-power ultrasound can also induce or promote chemical reactions, and this application has acquired the name sonochemistry (Henglein 1987; Heusinger 1987; Lorimer and Mason 1988). The reaction rate is higher owing to the high temperatures, particularly in the cavitation regions, and to the increased intermolecular contact in regions of high and fluctuating pressures.

Sajas and Gorbatow (1978) have studied the nature of the changes in meat and meat products during high-power ultrasound treatment and have recommended treatment procedures based on this. This work has included the following meat processing technologies: tenderizing of muscles, curing of meat, emulsifying sausage mix for scalded sausages, preparing emulsions from spice extracts and spray drying of meat products. They have carried out a detailed investigation of the effect of the properties of ultrasound treated meat and meat products, including the physical structure (the fibrous structure of meat and the droplet size distribution in emulsions), the biochemical properties (fatty acids composition, amino acids composition and the presence of enzymes), the nutritional properties (digestibility) and the microbial flora.

Many other high-power ultrasound applications exist in food processing (Roberts and Wiltshire 1990b). Sonoprocessing increases extraction rates, promotes crystallization with the formation of finer crystals and homogenization resulting in stable emulsions with fine droplets, thins polysaccharide solutions and helps to degas liquids and clean contaminated surfaces. Cavitation occurs preferentially in existing fine bubbles or small voids at liquid–solid interfaces (owing to imperfect wetting) and in high-temperature regions. This can be used to concentrate the applied energy at a contaminated surface or at a heat exchanger surface, for example. Standing wave techniques are used in separation processes.

Sonochemistry enhances oxidation processes (helpful in the accelerated ageing of alcoholic beverages, undesirable in the processing of fats). It promotes catalytic reaction by keeping catalyst surfaces clean and by increasing the area of interfaces between immiscible liquids and gases taking part in the reaction (for example the hydrogenation of oils).

One high-power technique not relying on cavitation is ultrasound-assisted drying. This is based on resonant absorption of bubbles (see Section 3.4) in capillaries and resulting pressure differentials and on increased convection induced by the passage of high-intensity sound waves in air.

None of these effects occurs at the time-averaged power intensities used in ultrasound measurement.

3 Speed of Sound

3.1 Velocity of Propagation for Bulk Longitudinal Waves

For a summary of the symbols used in this chapter, see Appendix 2.

The speed of sound is

$$c=(E_M/\rho)^{1/2}$$

where, for longitudinal (compressional) waves, the elastic modulus E_M is taken as

$$E_M=K_s=\gamma P \qquad \text{for gases}$$

$$E_M=K_s \qquad \text{for liquids}$$

$$E_M=Y+(4/3)G \qquad \text{for bulk solids}$$

The three elastic moduli are given by the ratio of applied stress to resulting strain:

K=(hydrostatic pressure)/(relative decrease in volume)
Y=(stretching tension per cross-sectional area)/(relative increase in length)
G=(shearing force per area)/(angle of shear)

They are a measure of the compressibility ($1/K$), the stretchability ($1/Y$) and the rigidity (G) of a material.

A similar, more familiar waveform is the mechanical wave propagating on coupled springs (see Figure 8.2).

3.2 Other Ultrasonic Wave Types

In solids, transverse (shear) ultrasonic waves, as well as longitudinal (compressional) waves, will propagate. Here the displacement is transverse to the direction of wave propagation rather than in the same direction. Shear waves can be induced by applying a transverse wave transducer or by mode conversion of longitudinal waves at an interface. The velocity of transverse waves is

$$c_{\text{shear}} = (G/\rho)^{1/2}$$

For animal bone, the shear wave velocity has been measured as 1970 m/s (Kaye and Laby 1986). Very few foods have such high rigidity, and shear wave velocities in food are therefore much lower.

Transverse (or shear) ultrasonic waves are not supported by gases or by most liquids. They are strongly attenuated even in viscoelastic liquids. In food gels (gelatine, pectin, agarose or sugar and gelatine in water), attenuation for shear waves was found to be too high for transmission through the sample. The shear velocity (0.5 m/s and below) was by three orders of magnitude lower than the compressional wave velocity and thus very difficult to measure. Even for an exceptionally stiff food gel (40 per cent gum arabic in water), the shear wave velocity (2.6 m/s) is quite small compared with the compressional wave velocity (Povey 1989).

Low velocities combine with high attenuation to render shear wave measurements in foods extremely difficult. For solid foods, where the intrinsic shear wave properties could be expected to be more favourable,

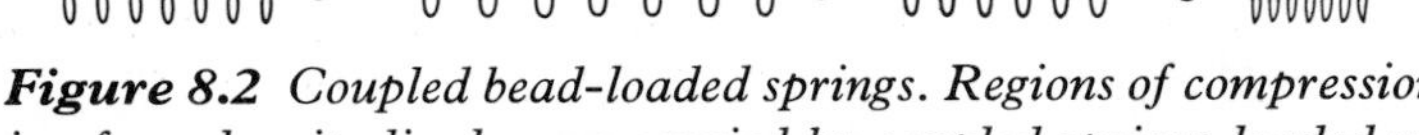

Figure 8.2 *Coupled bead-loaded springs. Regions of compression and stretching for a longitudinal wave carried by coupled springs loaded with beads. At equilibrium, the distance between the beads is a; the average mass per length is* $\rho_0 = M/a$*; the spring constant is* K_{sp}. *The wave propagates with velocity* $c_{sp} = (K_{sp}\, a/\rho_0)^{1/2}$. *The form of this equation is similar to that for the propagation of sound waves.*

other factors conspire against ultrasound transmission (scattering, coupling problems, etc.; see Section 3.5). The application of very-low-frequency shear waves in the audio range has been suggested but here the sample geometry can be a significant factor.

It is possible that transverse waves in foods could play a role in acoustic impedance techniques (Section 4.3) as these sample the interface properties and can be applied to highly attenuative acoustic media. However, no food applications that bear this out conclusively are known to the author.

Readers of the ultrasonics literature are likely to come across Rayleigh waves. These are surface waves and are among the principal wave types encountered in industrial measurements. Rayleigh waves have both a longitudinal and a transverse component and propagate in solids only. For food industry measurements, Rayleigh wave propagation in container materials and attached ultrasound waveguides can be relevant.

Rayleigh waves can be generated either by a special transducer or by mode conversion of bulk waves. They are characterized by a very low attenuation compared with bulk waves, and this is used in some sensing systems to transmit the signal for remote measurement applications in vessels where access for transducer maintenance is not possible. ('Remote' sensing in industrial applications usually implies distances of tens of metres, rather less than in the remote sensing from satellites.)

Lamb waves are also restricted to propagation in solids. Lamb showed that a solid plate can resonate at an infinite number of frequencies. Laminar defects just below the surface can be detected by observing the resonances in the plate between surface and lamination (Blitz 1963). Lynnworth (1989) describes the edge view of the resonating plate as resembling 'a straight snake swallowing a series of equally spaced eggs' (for symmetrical Lamb waves) or as 'a flag in a steady wind, with periodic waviness' (in the asymmetrical case). (See also Lynnworth 1989, Tables 3.1, 3.2 and 3.5 on ultrasound wave types.)

The leakage of Lamb waves in transducer or vessel walls into surrounding liquid can be used to detect the presence or absence of the liquid at a particular level. Level switches and level meters using this principle are described in Chapter 9, Sections 3.2 and 3.3.

A special acoustic device, the surface acoustic wave (SAW) device, is used in ultrasonic high-frequency applications. The propagation of the ultrasonic wave at the surface of the device is influenced by changes in the physical properties of the surface layer. A wide range of sensors for chemical and biochemical variables has been developed by coating the sensing surface with specifically absorbing or binding layers (see Chapter 17 for chemical sensors and biosensors).

Following these brief notes on transverse waves and on surface waves, the remainder of the chapter is concerned with bulk longitudinal ultrasonic waves.

3.3 Speed of Sound in Gases

In a gas, only longitudinal sound waves will persist, as shear stresses are not supported. The speed of sound is

$$c=(\gamma P/\rho)^{1/2}$$

For an ideal gas,

$$c=(\gamma RT/M)^{1/2}$$

The speed of sound for the ideal gas is thus pressure independent. This is not the case for non-ideal gases, for example organic vapours or gases under high pressures, where

$$c=[\gamma(RT+2BP)/M]^{1/2}$$

In both cases, the speed of sound increases with temperature and decreases for increasing molecular weight. The ratio γ of the principal specific heats is approximately 1.4 for diatomic gases such as nitrogen or oxygen, and 1.3 for carbon dioxide.

For the propagation of ultrasound in certain polyatomic gases such as carbon dioxide, velocity dispersion can occur (that is the speed of sound varies with the frequency), still within the frequency range for practical applications in gases. This velocity dispersion will be accompanied by strong attenuation, often precluding measurements (see Section 5.2) in such gases. As a consequence, the frequency dependence or the additional pressure and temperature dependence of the dispersive velocity will not usually be a concern in food industry applications.

3.4 Speed of Sound in Liquids

Influence of Solutes, Solvents and Temperature

In most liquids, in common with gases, only longitudinal waves will propagate. (As an exception, shear waves can be observed in highly viscous liquids when viscoelasticity occurs.)

The speed of sound is

$$c=(K_s/\rho)^{1/2}=[c_p(\gamma-1)/(\beta_p{}^2\,\mathrm{T})]^{1/2}$$

For transmission through liquids, as in the case of gases, compensation for temperature variations will normally be required for any instruments relying on a measurement of the speed of sound (unless temperature is the target variable).

Whereas for gases the velocity increases with temperature, the reverse is true for most pure liquids (away from the critical temperature). For distilled water, however, the speed of sound exhibits a maximum at about

74 °C (see for example Kaye and Laby 1986 for further data). The temperature dependence changes with the introduction of solutes or of miscible liquids. This is of interest when compensating measured velocities for temperature changes. For aqueous sugar solutions, for example, the shape of the curve for the temperature dependence of the speed of sound will vary considerably with the concentration and temperature range of the solution (see Figure 8.3).

Solute molecules influence the speed of sound by acting on the solvent as structure formers or structure breakers, and also by influencing the density of the solution. For sugar solutions, the increase of the sound velocity with concentration (Figures 8.3b, 8.4, 8.5) is well documented. The density increase accompanying the solute increase would be expected to lead to a decrease in the velocity, and it is therefore clear that the velocity change is, instead, dominated by a decrease in compressibility with increasing solute concentration. This has been examined in more detail for monosaccharides by Smith and Winder (1983).

The increase of sound velocity with concentration for aqueous NaCl solutions is shown in Figure 8.6. As for the sugar solutions, the change of velocity with concentration is dominated by the change in compressibility rather than that in density.

Even within the group of monosaccharides, the speed of sound varies (at the same concentration), and there are even greater differences between monosaccharide and disaccharide or polysaccharide solutions (see Figures 8.4 and 8.5). This would pose a problem, for example, in the monitoring of the concentration of sugar syrups of varying saccharide composition by measuring their speed of sound. Similarly, it would introduce an error in monitoring the temperature of such a syrup (at constant concentration) by the same measurement.

The non-contact application of ultrasonic time-of-flight measurement to concentration measurement is attractive owing to its inherent hygiene advantage and protection from fouling by the sample. It is, however, restricted to those solutions either where only one of the relevant sample properties (solute composition, solute concentration, sample temperature) varies, or where variations in the other two can be monitored and compensated for. (The latter may be complicated by the change in the dependence on solute type with temperature.)

Qualitative monitoring of liquids by ultrasound velocity measurement is more easily implemented. An example is the distinction between a food liquid, a caustic solution (used periodically for cleaning-in-place or CIP) and plain water (used for flushing after cleaning). Control systems using ultrasonic instrumentation for this distinction have already saved many hectolitres of beer from entering the waste pipe.

When monitoring the mixing of two liquids, of which one or both are associated (such as water or alcohol), the non-monotonic dependence of the speed of sound on the mixing ratio (over part of the range) is of interest.

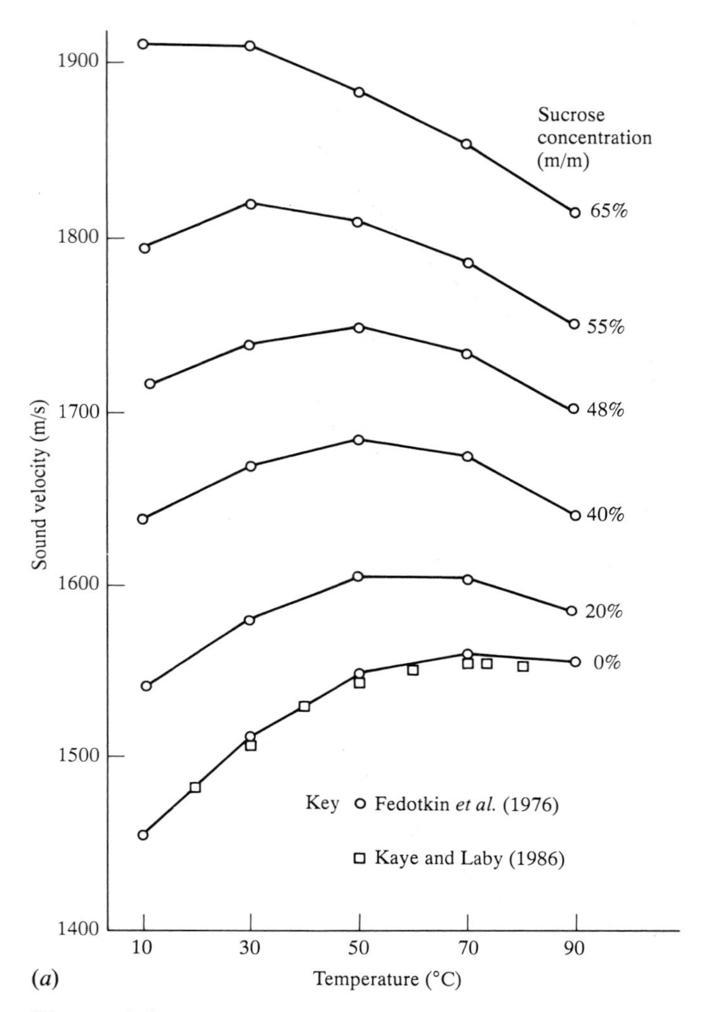

Figure 8.3

For low concentrations of ethyl alcohol in water (less than about 20 per cent, depending on temperature), velocity increases with alcohol content; for high concentrations (above about 30 per cent), velocity decreases with increasing alcohol content (Blitz 1963).

For low alcohol concentrations in aqueous sucrose solutions, there is a temperature for which the velocity remains constant during yeast fermentation as ethanol is formed and sucrose content decreases. The velocity at this temperature can be used as an indicator for the original gravity. Unfortunately, the temperature at which this occurs is above the process stream temperature for beer fermentation. For the determination of

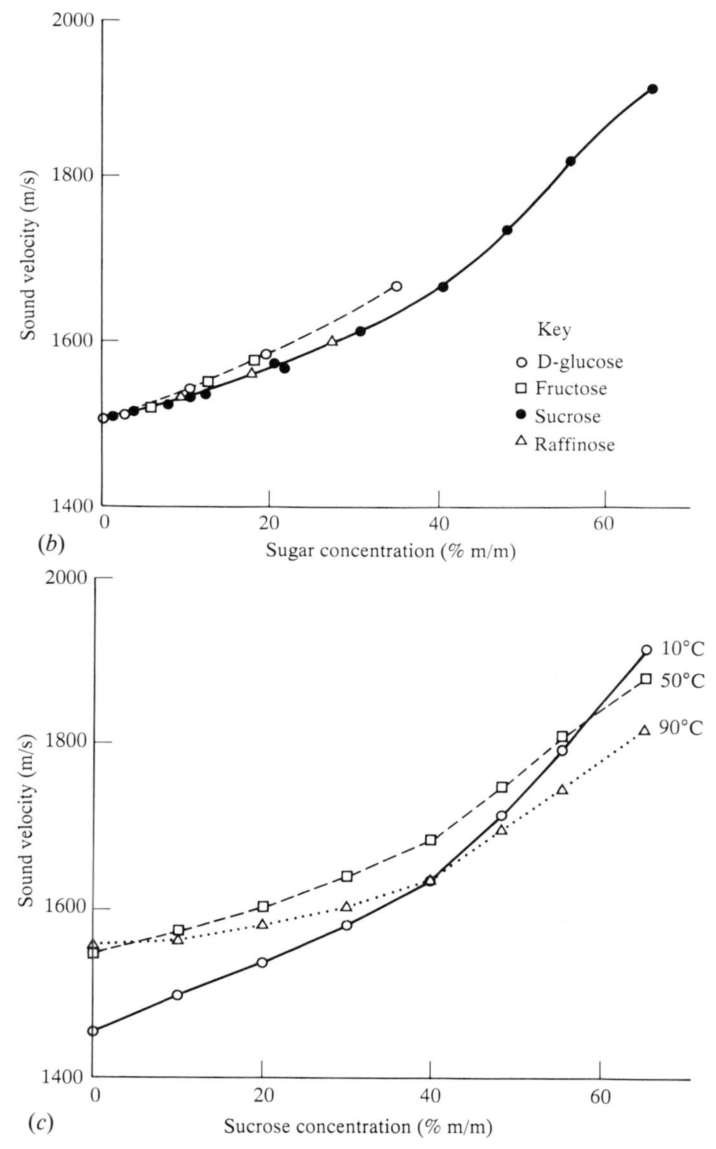

Figure 8.3 *(a) Sound velocity as a function of temperature for water and aqueous sucrose solutions. (b) Sound velocity as a function of sucrose concentration in aqueous solutions at 30 °C. (c) Change of relationship between sound velocity and sucrose concentration with temperature. Data in Figure 8.3b from Schaaffs (1963) (given there in tabular form), and in Figures 8.3a and c from Fedotkin et al. (1976) (as part of a graph there). To illustrate the error in reading data from a small graph, data for water from Kaye and Laby (1986) have been entered in Figure 8.3a for comparison.*

original gravity at process temperatures, a combination of sound velocity and refractive index measurement has been proposed instead. This has been tested successfully in the laboratory (Forrest 1987). In-line application would be more difficult in the presence of entrained gas or suspended yeast particles.

The ultrasonic properties of aqueous solutions of glycine, alanine and histidine have been examined by Wang and Feng (1990). In neutral solutions, the amino acid molecules exist in dipolar form and interact strongly with the surrounding water molecules to form a hydration area resulting in an increment to the sound velocity in pure water. For other pH values, a smaller velocity increment is observed. The hydration of polyamines in aqueous solution and its influence on the sound velocity have been investigated by Juszkiewicz and Figlerowicz (1990). Such studies have contributed to an understanding of both the ultrasonic properties and the molecular kinetics of biological media (as they are encountered in both clinical and food samples).

It has been suggested that thick egg white could be distinguished from thin egg white by a measurement of ultrasound velocity or attenuation. As eggs are stored, thin white becomes more prevalent as the ovomucin content decreases. These eggs are then less suitable for the manufacture of certain food products. It would therefore be helpful if they could be distinguished from fresh eggs without being broken open. A method for the measurement of attenuation of ultrasound transmitted through whole eggs

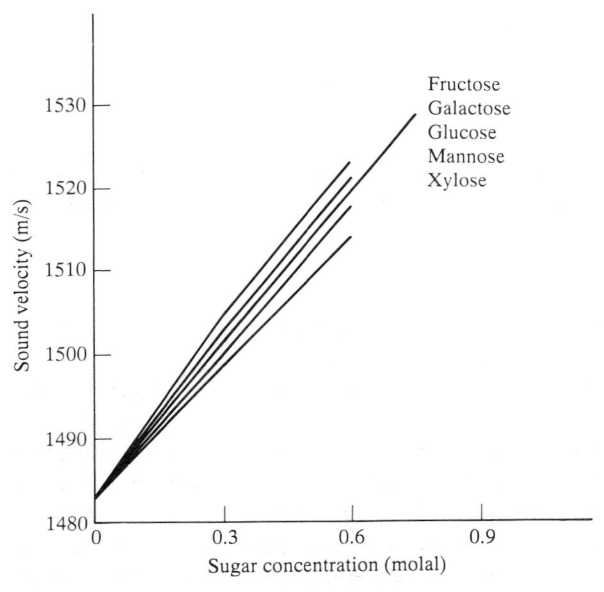

Figure 8.4 *Sound velocity as a function of sugar concentration in aqueous solution for five monosaccharides at 20 °C (Smith and Winder 1983).*

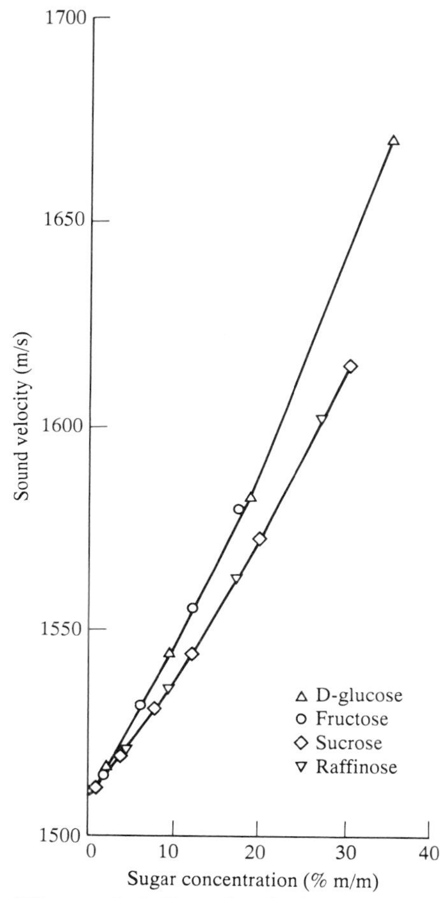

Figure 8.5 *Sound velocity as a function of sugar concentration in aqueous solution for two monosaccharides and two higher saccharides at 30 °C (Schaaffs 1963).*

has indeed been suggested (Section 3.5). However, literature data on the effect of ageing on the ultrasound characteristics of egg albumen are divergent (Povey and Wilkinson 1980; Javanaud et al. 1984; Choi et al. 1987).

The in-line measurement of ultrasound velocity in a foodstuff in order to derive its temperature has been suggested. This would be restricted to those applications where the process stream is predictable and constant in its other properties influencing the velocity. For liquid foods, this refers mainly to the precise chemical composition (including, for example, types and concentrations of dissolved saccharides or salts).

An alternative less liable to such cross-sensitivities is the use of a specially designed rod coupled to an ultrasonic transducer (see for example Gopalsami and Raptis 1983). The temperature of the rod material is then determined by measuring the speed of sound in that material. A particular section of the rod can be picked out for the temperature measurement by marker notches reflecting the signal. This rod section is immersed in the process stream. The method has been used in other industries, particularly in hostile environments, and has also been proposed for thermometry in the food industry (Povey 1989). It has been suggested that this method should be applied to semi-solid foods in microwave ovens. For plain water as the sample, this was shown to provide an accurate temperature measurement from approximately 30 to 55 °C (Richardson and Povey 1990).

For applications in semi-solid foods, however, such a probe would share a number of characteristics with other contact thermometers. The temperature of the food portion in good thermal contact with the probe tip would be measured. Temperatures in the remainder of the product would not necessarily be adequately represented. This would be particularly relevant in microwave ovens, where uneven heating can result from inhomogeneity in the dielectric properties of the foodstuff, from the shape of the food item and from the oven design. Potential advantages over the current fibre-optic thermometers for industrial microwave ovens require clarification.

Influence of Dispersed Particles, Droplets and Bubbles

Solid particles dispersed in a liquid will influence the velocity of sound propagating through the material. This is due to the velocity difference between the liquid and solid phases of a fat. For common frying oils, for

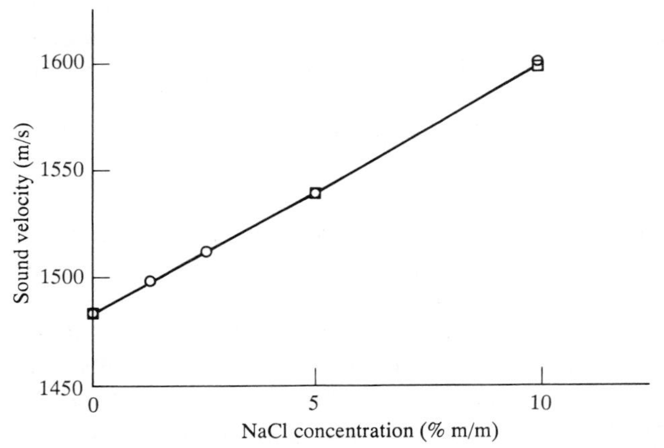

Figure 8.6 *Sound velocity as a function of concentration for aqueous NaCl solutions: × at 21 °C and 1 MHz (Kress-Rogers 1986a); • at 20 °C and 2 MHz (McClements and Fairley 1991).*

example, the difference in velocity between different oils at room temperature is of the order of 1 per cent (Table 8.1). The net increase (compensated for the temperature change) on solidification is of the order of 20 per cent (see figures in Miles et al. 1985). For frying fats such as lard or dripping, less than half of the volume is taken up by the solid phase at 20 °C. The velocity differences between such fats at room temperature are due to the variations in the solid/liquid fat ratio as much as to the composition of the fats.

This has been studied for the assessment of solid/liquid ratios in fats where fat crystal seeds are dispersed in the liquid fat phase (Miles et al. 1985). The authors conclude that this method can estimate the volume fraction of solid in a partially melted fat (see Section 5.3 in this chapter on the effect of particle size). They determined the second-phase content, that is the proportion of solid fat suspended in the liquid fat, by a measurement of the ultrasound velocity at 2.5 MHz.

McClements and Povey (1987) showed that a reproducibility of ± 0.2 per cent can be achieved in the determination of solid fat mass fractions (up to 20 per cent) dispersed in a liquid fat. It is worth noting that such a high reproducibility is dependent on a predictable composition of the fat. However, given that the difference between liquid and solid fats is much greater than that between common edible fats used for spreading or frying, this should not cause an undue problem. McClements and Povey conclude that the technique is easily automated, cheap, accurate, rapid, non-intrusive and based on sound physical principles, and should therefore prove useful both for in-line quality control monitoring and for food research applications.

Both groups of workers used Urick's (1947) equation, which describes the velocity as a function of the volume averages of compressibility and density. The model assumes that the dispersed particles are rigid, negligibly small compared with the wavelength, and present as a low mass fraction. Deviations from this model due to scattering are discussed in Section 5.3.

Another application of ultrasonic velocity measurement suggested by McClements and Povey (1987) is the determination of the solubility of triglycerides.

Applications for ultrasonic instrumentation for the oils and fat sector before 1980 were largely based on empirical relationships. A better understanding of the underlying principles, together with advances in electronics, now provides the basis for wider and more reliable applications. A commercial system, once developed, would compete with the currently most widely used technique for solid/liquid fat ratio determination, namely pulsed nuclear magnetic resonance (NMR) (Sleeter 1985). Equipment cost and response time characteristics could be in favour of the ultrasonic method, but problems might be encountered with aerated samples such as some low-fat spreads.

Table 8.1 *Sound velocity (measured) and impedance (calculated)*

(a) Solids: machinery, container and transducer materials; biological tissues; solid foods

Material	*Source of velocity*	*Velocity* * c (m/s^{-1})	at T (°C)	*Z(20 °C) calc.* † (10^6 $kg\ m^{-2}\ s^{-1}$)
Steel, stainless	K	5980	~20	46.9
Steel, stainless	R	5790	~20	45.7
Quartz, ×-cut	K	5720	~20	15.1
Glass, Pyrex	K	5640	~20	13.1
Pine, 0% moist	SM	5080		
Pine, 50% moist	SM	4170		
Lead zirconate titanate (PZT)	C	~4000	~20	30
Bony tissue	SG	3365	16	6.12
Bone	C			8#
	K	4000	~20	
Sugar	P			4
Ice (high density)	K	3840	−20	
Tin, rolled	R	3320	~20	24.2
Polythene	R	1950	~20	1.76
Milk chocolate:				
deaerated	AA	2050	20	
untreated	P	≤1750	25	
aerated	P	≥900	25	
Cardiac muscle, normal	O	1625		

Muscle, clinical	C			1.7#
Muscle:				
beef	SG	1572	18	1.63
veal	SG	1584	18	1.62
pork	SG	1589	18	1.63
mutton	SG	1588	18	1.64
Fish flesh	P	1550	22	1.6
Offal:				
liver	SG	1545	18	1.65
kidney	SG	1562	16	1.62
spleen	SG	1512	16	1.61
brains	SG	1503	16	1.52
Adipose tissue:				
beef	SG	1431	18	1.37
pork	SG	1452	18	1.37
pork	MF	1674	2	
Cheeses	P	1360–1650	10	
Rubber, gum‡	R	1550	~20	1.47
Teflon	AC	1340	~20	2.95
Potato flesh	P	~800	21	
Biscuits	PH	140–320	20	

Table 8.1 *(contd.)*
(b) Liquid and semi-solid foods: fats and oils; solutions, gels, emulsions

Material	*Source of velocity*	*Velocity* ★ c (m/s^{-1})	at T $(°C)$	*Temp. coeff.* dc/dT $(m/s^{-1}\ K^{-1})$	c(20 °C) (m/s^{-1})	Z(20 °C) calc.† $(10^6\ kg\ m^{-2}\ s^{-1})$
Dripping	MF	1570			1570	
Lard	MF	1517			1517	
Milk fat	MS	1500	25			
Cocoa butter	P					1.5
Corn oil	MF	1477			1477	
Sunflower oil	McP	1470	20		1470	1.33
Peanut oil	Z	1450	25	−3.3	1467	1.33
Olive oil	Z	1448	25	−3.4	1465	1.32
	MF				1464	
Light corn syrup	Z	1958	25	−1.2	1964	
NaCl solution, 10%	A	1600	21		1598	1.71
NaCl solution, 5%	A	1539	21		1537	1.59
Sucrose solution, 50%	§	1740	10		1753	
Sucrose solution, 25%	§	1560	10		1579	
Glucose solution, 35%	¶	1670	30		1655	1.90
Glucose solution, 19%	¶	1583	30		1563	1.67
Egg white	P	1560	20		1560	
Apple juice	Z	1544	25	+2.2	1533	
Salad cream	McP	1528	20		1528	
Skimmed milk	P	1522	28			
[illegible]	P	1500	20		1500	

(c) Simple liquids

Material	*Source of velocity*	*Velocity* * c (m/s^{-1})	at T $(°C)$	*Temp. coeff.* dc/dT $(m/s^{-1}\ K^{-1})$	$c(20\ °C)$ (m/s^{-1})	$Z(20\ °C)$ *calc.*† $(10^6\ kg\ m^{-2}\ s^{-1})$
Glycerol, pure	R	1904	25	−2.2	1915	2.42
Glycerine	K	1920	25	−1.9	1930	
Water, dist.	K	1447.3	10			
		1482.3	20		1482	1.48
		1509.1	30			
		1542.6	50			
		1554.8	70			
		1550.5	90			
Acetic acid	K	1173	20			
Ethyl alcohol	K‖	1145	25	−3.3	1162	0.94
Methyl alcohol	K‖	1103	25	−3.3	1120	0.88

Table 8.1 *(contd.)*

(d) Ambient gases

Material	*Source of velocity*	*Velocity* * c *(m/s⁻¹)*	*at T (°C)*	*Temp. coeff. dc/dT (m/s⁻¹ K⁻¹)*	*c(20 °C) (m/s⁻¹)*	*Z(20 °C) calc.† (10⁶ kg m⁻² s⁻¹)*
Air, dry	R**	331.45	0	0.59	343	0.0004
Nitrogen	K	354.4	29			
	R	334	0	0.6	346	
Oxygen	K	332.2	30			
	R	316	0	0.56	327	
Carbon dioxide	K	280 (v_h=293)	51			
	R	259	0	0.4	267	
Ethylene	K	327	20		327	
Water vapour	K	477.5	100			

* Velocity values are for compressional (longitudinal) waves in bulk material. Gases are at standard pressure unless stated otherwise.

Velocities will vary with the composition and processing history of the material. Values for metal alloys, glass, plastics or foodstuffs can therefore be regarded as typical only. Furthermore, some of the data were obtained many years ago with less sophisticated equipment. This also applies to some of the data cited in recent handbook editions or reviews.

The velocity at 20 °C is calculated, where necessary, with the help of the temperature coefficient or further velocity data in the source.

v_h is the high-frequency limit.

† Impedance Z calculated as c(20 °C) × ρ (20 °C) (unless marked by #). Velocity values from sources as indicated; density values largely from R, one from C.

‡ Not the chewable variety!

§ Sucrose solutions: data for 10 °C interpolated for 50 and 25 per cent concentration from Figure 8.3c. Temperature dependence from F.

¶ Glucose solutions: velocity data for 30 °C from S, temperature dependence from F.

|| Ethyl and methyl alcohol: similar values also from G, but velocity value for ethanol is given as 1207 m/s^{-1} at 25 °C in R. Further values are cited in AC.

Sources:

A	Author, data for NaCl at 21 °C, 1 MHz (see Figure 8.6). Part of the data was published in Kress-Rogers 1986a. Temperature dependence for NaCl solutions derived from data in ICT 1930 combined with own data
AA	Author, data for chocolate at 0.8 MHz. Related data were published in Kress-Rogers 1986b
AC	Anson and Chivers 1990
C	Crecraft 1983
F	Fedotkin et al. 1976
G	Gray 1963
K	Kaye and Laby 1986
McF	McClements and Fairley 1991
McP	McClements et al. 1990; salad cream values for typical UK formulation at 2.25 MHz
MF	Miles et al. 1985
MS	Miles et al. 1990
O	Okawai et al. 1990
P	Povey 1989
PH	Povey and Harden 1981
R	CRC Handbook 1976–77
S	Schaaffs 1963
SG	Sajas and Gorbatow 1978
SM	Sakai et al. 1990
Z	Zaccharias and Parnell 1972

Creaming profiles in oil-in-water emulsions can be obtained by relating the ultrasound velocity to the dispersed volume fraction at varying depths (Hibbard et al. 1986). In assessing a depth profile of creaming rather than trying to measure an absolute fat content, the technique is less sensitive to variations in dissolved and dispersed air content, the variable composition of the carrier liquid, and other factors influencing the velocity of ultrasound propagation.

For air bubbles dispersed in water, resonant scattering occurs when the frequency of measurement matches the resonant frequency of the bubbles (Lynnworth 1989). The latter amounts to 5.5 kHz/d [mm], for example 550 kHz for bubbles with a diameter of 10 μm or 55 kHz for 100 μm. The velocity of ultrasound propagation is that of water without air for frequencies above resonance, but is reduced for lower frequencies. Near resonance, velocity is highly frequency dependent. For very low frequencies well below the ultrasound range, the velocity can be reduced to as little as 22 m/s (for 50 per cent air volume fraction, for example with a bubble size of 100 μm at 1 kHz or below). Attenuation of ultrasound is strongly enhanced for frequencies near resonance. This phenomenon can interfere with measurements, but can also help to assess the volume fraction and size of entrained gas bubbles in liquids (see Section 5.3). It is also used in drying processes (see Section 2.7).

Comparison of Methods for Composition Monitoring

For the determination of water content and of other compositional variables, methods based on the physical interaction of the sample with ultrasonic, microwave and infrared waves are discussed in this book. It is therefore appropriate to compare the characteristics of these three approaches and consider their suitability for particular applications in food process control. All three are also applied to the measurement of food properties other than composition. The comparison here is for composition measurement, and particularly water content determination, which is a common application for all three methods.

The in-line measurement of the water content of a solid or semi-solid sample is usually referred to as moisture content, whereas the water content of liquids would often be expressed as the complementary (dissolved or dispersed) solids content. This differs from the terminology in the analytical laboratory, where moisture can imply both water and other volatiles driven off in an oven drying procedure.

Near infrared (NIR) composition measurement is derived from the highly specific mid infrared chemical analysis, but deliberately shifted to the near infrared range to reduce specificity. NIR thus allows the quantitative monitoring of groups of compounds such as fat, protein or starch as well as of water. This multi-component analysis is a valuable advantage of near infrared monitoring compared with microwave moisture or ultrasound solids content determination. With suitable reference wavelength selection and appropriate calibration procedures, reliable operation has been demonstrated for many industrial food applications. These are described in Chapter 5.

A disadvantage of NIR analysis (NIRA) in food process control applications is the generally low penetration depth. This necessitates a predictable relationship between the surface composition and the composition in the bulk of the sample. Problems are encountered when the surface is dried and hardened or covered with an oil film, for example.

Microwave methods offer an alternative when a bulk measurement of moisture content is required. This is then insensitive to variations in the distribution of moisture within the sample. Ultrasound is rarely applicable to the in-line monitoring of the composition of solid foods, but is applied to the determination of solute concentration. The following discussion centres on a comparison of water content measurement with microwave and ultrasound methods.

Microwave moisture measurement is characterized by the very distinct dielectric properties of free water (see Chapter 7). In contrast, ultrasound velocity in liquids is a function of compressibility and density. It depends quite sensitively on the composition of solids and liquids dissolved in the water. Accurate solute concentration measurement is thus confined to applications where the composition of the dissolved solids and the liquid are stable or defined by other simultaneous measurements.

Microwaves in the gigahertz range are strongly attenuated by water. They can penetrate only short distances into water or high-moisture materials, but they can readily be transmitted through deep layers of low-moisture materials. As the term 'microwave moisture' (as opposed to dissolved solids content) measurement implies, microwave techniques are used primarily for the determination of water content in solids, e.g. margarine, grains and powders. In such applications, microwaves can be transmitted through product streams travelling on conveyor belts or in pipelines. They will then provide a non-contact measurement sampling the entire process stream volume, yielding a bulk moisture value. Ultrasound cannot be applied to such solids–air systems owing to the high acoustic mismatch between the particulates and the air gaps. (NIRA is applicable provided that the surface is representative for the sample average.)

For the determination of solids content in aqueous solution, both microwaves (with stripline applicators) and ultrasound can be considered (as can NIRA). For homogeneous liquids with a low air content, ultrasound can often be transmitted through a substantial sample depth. For aerated liquids, on the other hand, ultrasonic time-of-flight techniques suffer from high signal attenuation. Acoustic impedance measurement can be considered but this is a measurement of the properties at the interface only and thus requires a representative sample layer there. Microwave measurements with stripline applicators can be used for both aerated and other liquids. For high water content combined with little aeration, they will have low penetration depth (but generally considerably deeper that with NIR reflection analysis).

A further aspect to be considered in the choice of a process control method is the windowing arrangement. Ultrasonic waves can be transmitted through a metal wall, although this can be frustrated by a high attenuation in the liquid inside the container, when most of the detected signal may have travelled around the vessel wall rather than through the liquid. For microwaves, free access over a conveyor belt, transmission through a Teflon window or the introduction of a thin stripline applicator can be considered. For near infrared applications, free access or transmission through Aclar (PTFE polymer) or polyester/glass laminate shatterproof windows is used (see Chapter 5). Glass windows are also used, but the possibility of glass fragments entering the food process stream must be eliminated. For remote infrared thermometry above 50 °C, sapphire windows have been suggested to provide the non-toxicity, non-solubility and strength that are important for safe applications in food process control. Fibre-optic applications have also been considered (see Chapter 6).

3.5 Speed of Sound in Solids

Of the applications of ultrasound measurement to the determination of the composition of solid food, the measurement of the ratio of fat to lean meat on carcasses is one of the most widely accepted. Ultrasonic pulse-echo reflection analysis is used in the assessment of back fat in live agricultural animals using paraffin as a couplant (Miles et al. 1983; Miles et al. 1984; Alliston et al. 1982). Further studies by Miles and coworkers have investigated the assessment of the proportion of ice in partially frozen food such as beef carcasses by ultrasound velocity measurement (Miles and Cutting 1974).

In other sectors of the food industry, applications to a number of solid foods have been explored but are much further from industrial application. Techniques for the measurement of solid/liquid fat ratios in semi-solid fats are well advanced (see Section 3.4).

The characterization of solid foods such as fruit, vegetables (to assess conditioning or ripeness) and biscuits (to assess crispness) by ultrasonics has been studied. Industrial implementation would be complicated by the problem of providing good acoustic coupling without damage to objects passing at a high throughput rate, and by the high attenuation of some of these materials for ultrasound. There are also problems in interpreting the ultrasound characteristics of such cellular structures with models that work for steel or quartz. Crispness, together with other texture attributes, forms a complex food property that is difficult to correlate with instrumental measurements (see Chapter 11).

In laboratory studies, immersion tanks have been used to provide acoustic coupling. Under these conditions, slices of potatoes, for example, changed their properties, probably as a result of water absorption (Sarkar and Wolfe 1983). Povey (1989) reported ultrasonic measurements at 0.5 MHz on a wide range of fruit and vegetables. Slices of the produce rather than whole fruit had to be used to obtain a transmitted signal, so that a non-destructive measurement does not, as yet, appear to be feasible.

Soft-tipped probes, held on to the sample by spring-loaded clamps, have been used for biscuits to obtain ultrasound data (Povey and Harden 1981). Such probes have also been proposed for whole eggs (Povey and Wilkinson 1980) to measure both the egg albumen properties (see Section 3.4) and the egg shell thickness (Gould 1972; Voisey and Hamilton 1976). (Break strength would be of interest but cannot be obtained with low-intensity ultrasonic measurements). Should a reliable measurement technique and a dependable correlation of ultrasonic data to industrially important food properties be confirmed for such fragile solid foods, the implementation of the proposed techniques in high-speed lines would present a further challenge.

For cheese, ultrasonic monitoring of maturation has been proposed. Povey (1989) points out that this would require a measurement indicative of rigidity and breaking strength. The rigidity of cheese is low compared with its non-compressibility and is therefore difficult to measure with the proposed technique. The breaking strength is not accessible with low-intensity ultrasound measurements.

For food gels, the compressional wave velocity is usually indistinguishable from that in the equivalent sols prior to setting. The shear wave velocity is quite small (Section 3.2). In the measurement of speed of sound, the gel thus behaves like a liquid rather than a solid. Attenuation increases on setting for certain gels (see Section 5.2).

4 Acoustic Impedance

4.1 Definition

When an ultrasonic wave meets the interface between two media, the reflection and transmission rates are dependent on the ratio of the acoustic impedance of the two media. The acoustic impedance, in analogy to the impedance of a dielectric medium for electromagnetic waves such as microwaves (see Chapter 7), is governed by the ratio

$$(\text{excess pressure})/(\text{particle displacement}) = E_M k$$

The characteristic impedance is given by

$$Z = E_M k/\omega = E_M/c = c\rho = (E_M \rho)^{1/2}$$

where again $E_M = \gamma p$ for gases and $E_M = K_s$ for liquids. This is analogous to the impedance for the imposition of a longitudinal wave on bead-loaded coupled springs, where $Z_{sp} = (K_{sp} a \rho_0)^{1/2} = c_{sp} \rho_0$ (with symbols as given in Section 3.1). Z_{sp} is the constant of proportionality in the response (that is the force opposing the motion) of the spring to the imposed amplitude variation with time.

The specific acoustic impedance is usually given in the form

$$Z = \rho c$$

If follows from this relationship that the acoustic impedance will be smallest for gases (having both lower velocities c and much lower densities ρ at standard temperature and pressure (STP) compared with liquids) and highest for solids (see Table 8.1).

4.2 Implications for Velocity Measurement

When two adjacent media have impedances of the same order of magnitude, there is good coupling. This is the case for the interface between two metals or the interface between a soft solid such as nylon and a liquid. Impedances differing by less than two orders of magnitude are still reasonably well matched. This generally applies for a metal immersed in a liquid or a metal coupled to a less rigid solid such as Perspex.

The rate of transmission across a boundary is calculated from the coefficient of transmission

$$\alpha_T = 4Z_1 Z_2 / (Z_1 + Z_2)^2$$

For a solid–liquid interface with a ratio Z_l/Z_s of the order of 0.1, the coefficient is about 0.3, leading to a loss of approximately 5 dB. For a solid–gas interface with a ratio of the order of 10^{-5}, the coefficient is about 4×10^{-5} and the loss amounts to about 44 dB. For a liquid–gas interface with impedances differing by four orders of magnitude, the loss would be about 34 dB.

For a liquid or semi-solid food sample (with modest attenuation for ultrasound) confined between solid parallel walls, the impedance mismatch at the two food/wall interfaces gives rise to an echo pattern from an ultrasonic pulse reverberating within the food sample and losing a proportion of the intensity at each reflection (Figure 8.7). The distance between the pulse echoes can be used to measure the velocity of propagation within the food. The amplitudes in a multiple echo pattern can help to estimate attenuation in the food sample.

Only a small fraction of the acoustic energy is transmitted into a gas at the interface to a solid or a liquid. It is therefore necessary to fill even the minute air gap at the interface between two solids (resulting from the surface roughness) with a couplant. Immersion of the sample in water or the application of gels and greases on the sample surface are often employed

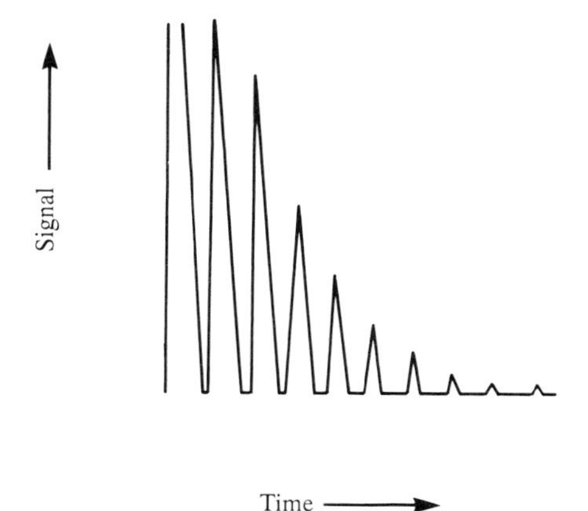

Figure 8.7 *Ultrasound pulse-echo pattern in a slice of redcurrant jelly (at ~20 °C, 1 MHz; measured speed of sound* c = *1.6 km* s^{-1}*)*

in laboratory measurements to avoid air gaps. For long-term mounting, the transducer can be cemented on to a solid sample or container. The requirement for good acoustic contact (either directly with the sample or via a container wall in good contact with the sample) is a major hindrance for the application of ultrasonic techniques to in-line measurements on solid foods (see Section 3.5).

4.3 Measuring Impedance Instead of Velocity

The measurement of the acoustic impedance of a sample can be used in place of a time-of-flight technique when the attenuation of the sample is too high for transmission. Clearly, this requires that the sample interface with the probe is representative for the bulk of the sample. In some applications, it will therefore be necessary to arrange for continuous wiping of the probe face. Reproducible coupling at the interface is also required for quantitative measurements of the sample impedance, and this will generally restrict the applications to liquids or samples that solidify before or during the course of the measurement without detaching from the probe. Further, temperature compensation will be required for such quantitative measurements just as in the case of time-of-flight measurements. Stable coupling only is required when the time dependence of a transition process is monitored, for example melting or setting behaviour or tempering characteristics in fat crystallization.

Randall et al (1987) have used an impedance mismatch probe consisting of a buffer rod coupled to an ultrasonic transducer. The buffer rod, which is partly immersed in a liquid food sample, ensures the resolution of the reflected pulse from the transmitted pulse so that the ratio of their amplitudes can be monitored. The transducer itself remains outside the process vessel.

A similar arrangement is used by McClements and Fairley (1991) as part of an instrument for the determination of ultrasonic velocity, attenuation coefficient and characteristic impedance of a liquid sample. This 'ultrasonic pulse-echo reflectometer' is intended for adaptation to the automatic determination of a number of physical properties of liquid samples in tanks or pipes. Most readily determined is the density as $\rho = Z/c$. For the monitoring of the concentration of dissolved solids, impedance Z can be measured as an alternative to velocity by time of flight for materials with high attenuation. As in the case of direct velocity measurement, the calibration for the application in question needs to be established first, and variations in solute or solvent composition or in the temperature will affect the readings.

5 Attenuation

5.1 Sources of Attenuation

The attenuation of ultrasound in gases is relatively high and increases with frequency. Also, their acoustic impedance is low compared with that for solids so that only a small fraction of the energy from the transducer is transmitted into the medium. Measurements at frequencies in the megahertz range are therefore impractical for gases.

Classical theory predicts an increase of the coefficient of sound attenuation with the square of frequency, and this holds for most common gases (but not for carbon dioxide; see the next section) in the ultrasonic range for process control measurements (up to 100 kHz). This frequency dependence also applies to most common liquids within the ultrasonic range considered here (up to about 10 MHz), but the attenuation is often much higher than predicted by classical theory, even for pure liquids without entrained bubbles or suspended solids.

Attenuation can arise from

- a relaxational type of absorption, or
- deviation of energy from the parallel beam.

In the first case, an energy mode with a time lag to establish equilibrium conditions (for example, thermal conduction, viscous flow, vibrational motion of molecules) is coupled to an energy mode with instantaneous transfer (such as translational motion). With increasing frequency, the phase lag increases (thus increasing absorption) but the probability of energy transfer decreases (thus decreasing absorption). As a result, a broad maximum in the loss per cycle is observed at the relaxation frequency ω_0.

In the second case, the beam is scattered or diffracted by inclusions of a differing acoustic impedance, or reflected and refracted at interfaces (boundaries) between different acoustic media.

For monatomic gases, absorption is due to viscosity and thermal conduction. Both can be described as relaxational phenomena, but with relaxation frequencies well above the ultrasonic range for industrial process control measurements. Within this range, absorption is therefore approximated by classical theory. The loss per cycle is proportional to the frequency, the coefficient of absorption increases with the square of the frequency, and the velocity of propagation is independent of frequency. For diatomic gases at STP, the classical description is still adequate within the ultrasonic range considered here. Considerable discrepancies can be observed for polyatomic gases (see Section 5.2).

For most liquids, absorption is dominated by viscous and vibrational relaxation, and for associated liquids such as water or alcohol also by structural relaxation. Only the contribution from viscous relaxation is included in the classical description. This accounts for the attenuation well in excess of the classical description. For rotational isomers, for example the ester ethyl acetate, a relaxation frequency just within the ultrasonic range considered here can also occur.

Food liquids often contain entrained gas bubbles or suspended solid particles, and these give rise to high attenuation (see Section 5.3).

5.2 Relaxation

For certain media, an unusually high attenuation not following the frequency dependence expected from classical absorption theory can be observed within a part of the ultrasound range. This is, for example, the case for carbon dioxide, where the loss per cycle exhibits a broad maximum within the ultrasound frequency range used for transmission through gases (shifting to higher frequencies for carbon dioxide with water vapour inclusions).

Similar characteristics are observed for other gases composed of linear triatomic molecules. This broad peak in attenuation at a characteristic frequency is due to thermal relaxation associated with the exchange of acoustic energy with an internal vibrational mode of the CO_2 molecule. It is accompanied by velocity dispersion, that is the velocity is now frequency dependent.

A practical consequence is the restriction of ranging applications (for level measurement) to headspace gases with low carbon dioxide content (see Chapter 9, Section 2.1).

For attenuation due to relaxation, characterized by a relaxation time τ, and for a dispersive sound velocity $v(f)$ with a low-frequency limit c, the coefficient of attenuation is

$$\alpha = \frac{\omega^2 \tau v}{2c^2(1+\omega^2\tau^2)} = \frac{v/\tau}{2c^2}\frac{\omega^2\tau^2}{1+\omega^2\tau^2}$$

The loss per cycle is

$$Q^{-1} = \frac{\alpha\lambda}{\pi} = \frac{2\alpha}{k} = \frac{v^2}{c^2}\frac{\omega\tau}{1+\omega^2\tau^2}$$

where ω is the angular frequency ($2\pi f$), and a broad maximum of Q^{-1} occurs at $\omega = 1/\tau = \omega_0$.

For frequencies well below the relaxation frequency, that is for $\omega \ll \omega_0 = \tau^{-1}$, these equations are simplified to

$$\alpha_{\text{lf}} = \frac{v^2}{2c^2\tau}\omega^2\tau^2 = \frac{v\tau}{2c^2}\omega^2 = \frac{\tau\omega^2}{2c}$$

$$Q_{\text{lf}}^{-1} = \frac{v^2\tau}{c^2}\omega = \tau\omega$$

The coefficient of attenuation now increases with f^2 and the loss per cycle is directly proportional to frequency f.

For most common gases, absorption is adequately described by the contributions from viscosity and thermal conduction, where

$$\tau_{\text{vis}} = \frac{4}{3}\frac{\eta}{\rho c^2}$$

$$\tau_{\text{thc}} = \frac{\kappa}{\rho c^2 c_v}\frac{\gamma - 1}{\gamma}$$

In the low-frequency limit ($\omega \ll \omega_0$), the corresponding attenuation coefficients are

$$\alpha_{\text{vis, lf}} = \frac{8\pi^2}{3}\frac{\eta}{\rho c^3} f^2$$

$$\alpha_{\text{thc, lf}} = \frac{2\pi^2\kappa}{\rho c^3 c_v}\frac{\gamma - 1}{\gamma} f^2$$

The classical prediction is equal to the sum of these low-frequency limits of the relaxation model:

$$\alpha_{\text{class}} = \alpha_{\text{vis, lf}} + \alpha_{\text{thc, lf}}$$

The attenuation is inversely proportional to gas density, accounting for the difficulty of transmitting ultrasound through a gas at low pressure.

Values for relaxation times in gases and liquids are given in Table 8.2a; they are put into the context of the ultrasonic spectrum in Figure 8.8. Calculated classical absorption coefficients are given in Table 8.2b, and measured absorption coefficients in Table 8.2c.

Relaxation studies for gels have been described by Wyn-Jones et al.

(1982) and by Ross-Murphy (1984). For random-coil polymers, a substantial attenuation was observed in the range 1–100 MHz. This feature was found to be less pronounced for conformationally ordered biopolymers.

For the gels of the polysaccharide agarose, the increase in attenuation compared with the sol (prior to setting) was found to be a function of concentration. For 3 per cent w/v agarose in water, the attenuation coefficient at 1 MHz was by a factor of about 10 higher in the gel than in the sol. The difference decreased with increasing frequency (Wyn-Jones et al. 1982). Relaxation phenomena have already been described in Chapter 7 in the context of composition measurements based on the interaction of microwaves with the sample. For the propagation of electromagnetic waves in dielectric media, relaxation was shown to be characterized by a broad maximum in dielectric loss and a decrease in permittivity as frequency increases (Figure 7.1). The absorption of microwaves by water in the relaxation regime is the basis of microwave cooking.

In the case of microwave moisture measurement, for example, the engineer testing a microwave attenuation monitor for a new group of materials may find that when the sample moisture rises beyond a certain value, the instrument output is no longer increasing gradually with the moisture content, but instead reaches a maximum and then falls. Having identified the relaxation characteristics, one can then try a more appropriate frequency range for the attenuation measurement (Figure 7.2b), or switch to microwave phase shift (Figure 7.2a) as the measured variable.

In the application of ultrasonic techniques in food liquids and headspace gases, relaxational phenomena can irritate the engineer with unexpectedly high attenuation levels for certain media. For high carbon dioxide contents as they occur in the headspace of fermenting or carbonated liquids, strong attenuation can preclude measurements involving ultrasound transmission through the headspace.

For most food liquids, the relaxation frequencies will be above the ultrasonic frequency used in process measurement (except perhaps in the case of esters). However, for many food liquids, the measured attenuation will be in excess of the values calculated from classical theory (even without the inclusion of scatterers).

It is usually this increased attenuation, rather than the frequency dependence or the additional pressure and temperature dependence of the dispersive velocity, that is likely to present a problem in process measurement in the food industry. A detailed discussion of relaxation phenomena in ultrasonics can be found in Blitz (1963) and in Kinsler et al. (1982).

5.3 Scattering

Hindrance or Help?

For liquids, attenuation at high frequencies can be very high where

Table 8.2 *Relaxation times and absorption coefficients*
(a) Relaxation times (seconds) for gases and liquids at 20 °C and STP (Blitz 1963, pp. 87–105, 117–123)

*Relaxation times**	*Gas*	*Liquid*
τ_{vis}	$\sim 10^{-10}$ 1.6×10^{-10} air	$\sim 10^{-12}$ 0.6×10^{-12} water 2×10^{-12} ethanol§ 60×10^{-12} olive oil 400×10^{-12} glycerol¶
τ_{thc}	$\sim 10^{-10}$ 0.7×10^{-10} air	$\sim 10^{-17}$
τ_{rot}	$\sim 10^{-9}$ N_2,O_2	
τ_{vib}†	$\sim 10^{-5}$ CO_2 $\sim 10^{-7}$ SO_2 $\sim 10^{-8}$ H_2O vapour	$\sim 10^{-10}$
τ_{str}		$\sim 10^{-10}$ glycerol $\sim 10^{-12}$ water $\sim 10^{-14}$ *n*-propyl alcohol
τ_{iso}		$\sim 10^{-7}$ ethyl acetate (ester)
$1/(2\pi f_{max})$‡	$\sim 10^{-6}$	$\sim 10^{-8}$

*vis viscosity
thc thermal conduction
rot molecular rotation
vib molecular vibration
str structural
iso rotational isomers
† τ_{vib} is the longest quoted if several present.
‡ f_{max} is the highest frequency used in the industrial monitoring applications described in Chapter 9.
§ Ethyl alcohol is the better label but is less commonly used.
¶ Pure glycerol, known as glycerine in extracts of varying impurity.

(b) Calculated absorption coefficients for classical model, α_{class}/f^2(Np m^{-1} Hz^{-2})*, at 20 °C and STP (Blitz 1963, pp. 88, 117)

Gas	*Liquid*
$\sim 10^{-11}$	$\sim 10^{-14}$
1.4×10^{-11} air	0.8×10^{-14} water
1.3×10^{-11} CO_2†	3×10^{-14} ethanol‡
	80×10^{-14} olive oil
	600×10^{-14} glycerol§

* Neper and decibel: 1 Np = 8.686 dB (where 8.686 = 20 lg e = 20/ln 10). Expression of intensity ratios: 10 lg (I_2/I_1) dB or 0.5 ln (I_2/I_1) Np.
† For CO_2 the calculated attenuation is particularly unrealistic, as τ_{vib} is longer than the period of the measurement frequency.
‡ Ethyl alcohol is the better label but is less commonly used.
§ Pure glycerol, known as glycerine in extracts of varying impurity.

(c) Measured absorption coefficients, α_{meas}/f^2($\times$ 10^{-14} Np m^{-1} Hz^{-2}) (Kaye and Laby 1986; Gray 1963; Anson and Chivers 1990)

*Substance**	*At 25 °C*	*At 20 °C*
Water	2.2	2.6
Ethanol†	5	9.0
Glycerine‡	300	
Olive oil		135.0
Steel (tool)		4.9 (10 MHz)
Quartz (x-cut)		0.013 (10 MHz)
Glass (crown)		2.0 (10 MHz)
Polythene		5 400 (1 MHz)
Rubber		12 000 (0.35 MHz)

* For solids, the measurement frequency is given in brackets.
† Ethyl alcohol is the better label but is less commonly used.
‡ Glycerol in extracts of varying impurity.

Gases	Liquids	Frequency f_o (Hz)
	τ_{str} water τ_{vis} water, ethanol	10^{11}
		10^{10}
τ_{the} gas τ_{vis} gas	τ_{str} glycerol τ_{vib} olive oil τ_{vis} glycerol	10^9
τ_{rot} N_2, O_2		10^8
τ_{vib} H_2O vapour		10^7 L
τ_{vib} SO_2	τ_{iso} ethyl acetate	10^6 L
		10^5 G
τ_{vib} CO_2		10^4

Figure 8.8 *Relaxation frequencies within the sound spectrum (see Table 8.2a). G (gases), L (liquids) indicate typical frequency ranges for the industrial process control measurement described in Chapter 9. Figure shows that for ultrasound transmission through CO_2 or ethyl acetate, severe attenuation and a dispersive propagation velocity have to be expected.*

entrained air bubbles or suspended solid particles are present, as is often the case in food liquids and semi-solids. During food processing operations, pumping and mixing can lead to substantial air entrainment and heavy frothing, and this can spoil an ultrasound velocity measurement that was possible in the tranquil laboratory sample.

Attenuation is also high for foods such as chocolate where the liquid fat phase carries not only suspended solid fat crystal seeds, but also solid sugar particles and air inclusions. The sugar particles have an acoustic impedance much higher than that of the cocoa butter (by a factor of 3; see Table 8.1a, b) and are therefore a cause of considerable scattering.

In such two-phase (or three-phase) systems, a lower frequency for ultrasound transmission may be chosen to reduce attenuation (Kress-Rogers 1986b) where Rayleigh-type scattering (see later in Section 5.3) is dominant. Frequencies of 0.5 to 0.8 MHz have been used for measurements on chocolate (see Table 8.1a). Alternatively, a different ultrasound technique (based on the measurement of acoustic impedance, for example; see Section 4.3) can be considered.

A hindrance in some applications, the attenuation due to scattering is the basis of others, as described below.

Concentration of Suspensions

The increase in attenuation with suspended solids inclusion is the basis of a technique used in the food industry for the determination of the solids content of slurries up to 50 per cent w/w (see Chapter 9, Section 3.4). Also, small air bubbles can be detected even at low volume fractions under resonant scattering conditions (Section 3.4 in this chapter). For concentrated suspensions such as foams or slurries with high dispersed solids contents, a full theoretical description does not exist at present owing to the complexity of the scattering properties of such systems (Povey and McClements 1988).

Applications in other industries have demonstrated that it is possible to detect dispersed air bubbles, oil droplets and solid particles and distinguish them from each other by ultrasound measurements (including attenuation, forward and backward scattering). This has proved applicable to high flow rates of the process stream. Examples include air (as bubbles, in foam or slugs) in blood during processing, oil droplets in recirculated water during secondary oil recovery, wear particles in engine oil, and small suspended air bubbles (as distinct from large loose bubbles) in paper pulp suspensions (Lynnworth 1989).

Juice Stability and Product Defects

The scattering of ultrasound by colloidal particles has been studied by Sarkar and Wolfe (1983) in monitoring the separation of reconstituted orange juice. They measured the attenuation of ultrasound at 1 MHz transmitted through the juice. The separation level of the settling colloidal particles could be observed in the attenuation as a function of liquid depth. This technique is intended for the non-destructive inspection of reconstituted juice in cartons.

Back scatter of ultrasound at 5 MHz has been studied by the same authors for characterizing the smoothness of orange skin, detecting cracks in tomato skin and distinguishing between normal and partially filled or shrivelled cobs of husked sweet corn (Dull 1986; Sarkar and Wolfe 1983). The produce was mounted in an immersion tank. (Optical sorters would probably be easier to apply, as long as there was a visible defect; see Chapter 4.) For the assessment of ripeness or conditioning of produce, attenuation of submerged slices was measured (see Section 3.5).

Flow Metering

Scattering from suspended solid particles or droplets (of a second immiscible liquid) or of gas bubbles in a liquid is used in the Doppler flow meter (see Chapter 9, Section 3.1). This is not applicable to molten chocolate, unfortunately, despite the abundance of scatterers there. The attenuation here is too high and the signal, if any, would originate from the outer layer of the flow profile, possibly a stagnant part of the sample stream. Doppler flow meters can, however, be used for the monitoring of some components used in chocolate manufacture (McFarlane 1987).

Both transit-time and Doppler flow meters are based on a measurement of flow velocity. This has to be combined with an assumption on the flow profile and the part of this that is actually sampled by the flow meter in order to derive the mass flow rate. Food liquids often exhibit non-parabolic flow profiles and, unless their flow rate profile is known and taken into account, the flow rate reading will be unreliable. Time-gated Doppler flow meters can give information on the flow velocity profile.

For larger particulates, the flow rate can differ from that of the carrier liquid. Also, their distribution across the flow profile can be uneven. For such sample streams, special care is needed in the interpretation of flow measurements.

Ultrasonic flow meters based on the Doppler shift or on transit time are discussed for food applications in Chapter 9, Section 3.1, and for general applications by Lynnworth (1989). The latter also describes vortex shedding flow meters, which are widely used in other industries. In the food industry, the presence of the bluff body in the process stream (required for this technique) is a disadvantage from the hygiene point of view.

Cross-correlation techniques (Beck and Plaskowski 1987) have been suggested as a further non-invasive alternative. With advances in computer technology, the real-time industrial application of correlation techniques has become feasible. Commercial equipment has become available in recent years; it is still too early to judge how widely this technique will be used in industry. In cross-correlation flow metering, the transit time of random fluctuations in the scattered signal (its amplitude, phase and frequency) is measured. These can be due to turbulent eddies or to density fluctuations in two-phase liquids, for example.

Emulsions: the Three Contributions to Scattering Losses

Scattering is weaker for oil-in-water emulsions where a liquid phase is dispersed within a second liquid phase of comparable compressibility $1/K$ and density ρ. In an ideal non-scattering two-phase system, the velocity is calculated from the volume averages of the respective values of K and ρ for the component phases (the velocities themselves are not averaged); attenuation is calculated as a volume average. In real food emulsions, however, both the velocity and the attenuation depend on the average size and the size distribution of the droplets as well as on the volume proportion of the second liquid phase. This has been demonstrated for salad creams with ultrasound measurements at 1–6 MHz (McClements et al. 1990). The authors suggest a measurement at two frequencies for a simultaneous in-line determination of droplet sizes and fat content. Ultrasonic particle sizing offers the advantages of non-invasive *in situ* measurement without prior dilution over particle sizers on the market now.

The excess attenuation (additional to the volume average) is attributed to three contributions: simple scattering (Rayleigh type in the long-wavelength limit); scattering due to viscous loss; and scattering due to thermal loss. The latter two contributions are included under the heading of 'scattering' by some authors; others reserve this term for the simple scattering. For each of the three contributions, the increase in the coefficient of attenuation is proportional to the volume fraction of the suspended particles but depends also on other sample properties.

The three contributions have been discussed for a single particle within a medium in Landau and Lifschitz (1975, p. 297). A theoretical description for emulsions and suspensions including the three contributions was later given by Allegra and Hawley (1972). The thermal properties of the fluid and suspended second phase have only recently been included in the prediction of ultrasound propagation in food emulsions. McClements and Povey (1989) have examined simple oil-in-water emulsions; McClements et al. (1990) have studied salad creams (see above). Miles et al. (1990) have investigated milks and creams.

In the following description, a non-dispersive velocity regime (see Section 5.2) is assumed, and this will apply in most food industry applications of ultrasound.

In the simplest analysis of ultrasound propagation in dilute suspensions, lossless spherical particles in a lossless inviscid fluid are taken to be the model. When the particle radius r_p is small compared with the wavelength, the coefficient of absorption increases with $(f/c)^4$, ${r_p}^3$ and with the difference in compressibility and density between the liquid and the suspended particles. For milks and creams at 2–7 MHz, this contribution is negligible.

The second contribution is due to the effect of the fluid viscosity on the motion of the particles. This relative motion must occur when the density of the particles differs from that of the carrier liquid. This contribution is a complex function of frequency, density ratio, particle radius and viscosity.

The third contribution is due to the conduction of heat between the fluid and the particles. The pressure variations travelling with the ultrasonic wave are accompanied by temperature variations, and the differing thermal properties lead to a temperature gradient at the particle surfaces and thus to heat flow between the components of the emulsion. For milks and creams at 2–7 MHz, this contribution is dominant and decreases with increasing particle size. The same was true for the salad creams at 1 MHz.

The relative importance of the thermal contribution to the attenuation in liquids containing suspended solid particles has been studied for a wide range of liquids by Anson and Chivers (1990). They conclude that the thermal contribution must be included in virtually all systems for $ka \leqslant 0.5$ (where $k = 2\pi/\lambda$). The relative importance of the other contributions depends critically on the values of the relevant parameters (see above) of the two phases. For $ka \geqslant 1$, the thermal contribution needs to be considered only for liquids containing a second liquid phase in suspension and where these two liquids are similar in all but the thermal properties. Otherwise, simple scattering and viscoinertial scattering dominate in this regime. For solids suspended in liquids, the simple scattering dominates.

For the milks, creams and salad creams (see above), the mean fat droplet size (about 0.5–10 μm diameter, depending on homogenization treatment) was much smaller than the wavelength at 1 MHz (about 1.5 mm). The excess attenuation was dominated by the thermal conduction between fat globules and fluid phase. The simple, Rayleigh-type scattering contribution was negligible here and the attenuation increased with homogenization treatment, that is with decreasing particle size. For milks (with fat contents up to 4 per cent), the attenuation increased in proportion with the fat content for a given particle size. The measurement of both the degree of homogenization and of the fat content of milk is of industrial interest.

6 Conclusions

A range of applications of low-intensity ultrasound are well established for process control in the food industry. Liquid level, solids level and liquid flow as well as suspended solids concentration (for constant particle size) are routinely being determined.

New applications are emerging based on developments in ultrasonic instrumentation and in the characterization of the ultrasonic properties of foodstuffs. Solute concentration is already being monitored in a few well-defined applications. Further sample characteristics are now being inferred, for example droplet size in oil-in-water emulsions, crystallization characteristics in fats or setting behaviour in gels.

The complex relationships between the propagation characteristics of ultrasound and the physical and chemical properties of foodstuffs and their environment are the basis of this wide range of application areas, but they are also a source of cross-sensitivity. With due care in the selection of new applications, the scope of industrial ultrasound techniques can be expanded to probe a range of food properties in addition to the established process variables.

The simultaneous measurement of several variables is becoming more widely applicable as an option in process measurement with the advances in the electronics employed in measurement and signal processing. For example, a combination of ultrasound velocity, impedance and/or attenuation can be used, or an in-line measurement of ultrasound velocity can be combined with that of refractive index (optical). Such multi-variable measurements provide a more reliable performance for many existing applications and allow new industrial applications to be introduced.

Appendix 1: Ultrasound Measurement Applications in and for the Food Industry

As this chapter describes the principles underlying ultrasound measurement in the food industry, the sequence here is determined by the measured ultrasound variable such as speed of sound or acoustic impedance. In Chapter 9, applications of ultrasonics currently established in the food industry are discussed and the sequence there is by instrumentation type.

In addition to giving an overview of ultrasound applications in and for the food industry, this appendix is intended to help locate the section describing the measurement of a particular target variable of interest to the reader, such as flow rate or solute concentration. The sections are keyed as follows:

8/*n.n* Application, either implemented or at the research stage, discussed in Chapter 8, Section *n.n*, in the context of the underlying principle.

9/*n.n* Application details for current instrumentation discussed in Chapter 9, Section *n.n*.

For data on the ultrasound properties of specific foods and literature sources, see Tables 8.1, 8.2 and Figures 8.3–8.6.

Measurement of Process Variables

Fill level indication:	
liquids	9/2.1, 9/3.2, 9/3.3
solids	9/2.1
Flow rate metering, liquids	9/3.1, 8/5.3
Interface detection and location:	
liquid to liquid	9/3.4
emulsion to single phase	9/3.4
Counting of items passing	9/2.2

Measurement by Acoustic Emission Monitoring

Powder flow	8/2.6
Cleaning fluid impact	8/2.6
Cooker extruder performance	8/2.6

Measurement of the Properties of Liquid Foods

Distinction between:	
foam and liquid	9/3.2

cream and milk	9/3.2
beer and cleaning solution	8/3.4
Concentration of solutes and solvents in aqueous solution:	
NaOH (in cleaning solution)	9/3.5
NaCl	8/3.4, 9/3.5
monosaccharides (fructose, galactose, D-glucose,	8/3.4
mannose, xylose)	8/3.4
disaccharide (sucrose)	8/3.4, 9/3.5
polysaccharide (raffinose)	8/3.4
coffee extract	9/3.5
modified corn starch	9/3.5
ethyl alcohol	8/3.4
Original gravity of beer before, during and after fermentation:	
measure sound velocity + optical refractive index	8/3.4
Hydration:	
amino acids	8/3.4
polyamines	8/3.4
Mass or volume fraction in liquid carrier:	
suspended solids (slurry)	8/5.3, 9/3.4
suspended solids (fat crystals in liquid fat)	8/3.4
dispersed droplets (in milk, cream, salad cream)	8/5.3
dispersed bubbles	8/3.4
Size and size distribution:	
small air bubbles in liquid	8/3.4
droplets in emulsion (milk, cream, salad cream)	8/5.3
Creaming:	
oil-in-water emulsions	8/3.4
Settling (separation, sedimentation):	
juice colloids	8/5.3
Temperature	8/3.4
Density	8/4.3

Note See also Table 8.1b for references to specific liquid foods.

Measurement of the Properties of Solid and Semi-Solid Foods and of Animals

Back fat, live animals	8/3.5
Fat/lean ratio, carcasses	8/3.5
Frozen/liquid water ratio in meat	8/3.5
Solid/liquid ratio (fat)	8/3.4
Melting, solidification	8/4.3
Crystallization, tempering	8/4.3
Solubility (triglycerides)	8/3.4

Setting (gels)	8/4.3, 8/5.2
Temperature	8/3.4

Note See also Tables 8.1a, b for references to specific solid and semi-solid foods.

Potentially more Difficult Measurement Applications

Freshness of eggs (thick/thin egg white ratio)	8/3.4
Maturing of cheese	8/3.5
Crispness, biscuits	8/3.5
Ripeness, conditioning (produce)	8/3.5, 8/5.3
Smoothness, cracks, shrivelling, missing kernels (produce)	8/5.3
Thickness, egg shells	8/3.5

Note See also Tables 8.1a, b for references to specific solid and semi-solid foods.

Applications Not Feasible

Determination of breaking strength	8/3.5
Echo ranging through air containing high carbon dioxide levels	8/5.2, 9/2.1

Appendix 2: List of Symbols

c speed of sound
v dispersive ultrasound phase velocity
Z specific acoustic impedance
α coefficient of ultrasound attenuation
α_T coefficient of transmission through boundary

f frequency
λ wavelength
ω angular frequency; $\omega = 2\pi f$
k wave vector; $k = 2\pi/\lambda$
τ relaxation time

ρ density
P absolute pressure
M molecular weight
η coefficient of viscosity
r_P radius of suspended spherical particle

V volume
T absolute temperature

E_M elastic modulus (as appropriate)
K bulk modulus
Y Young's modulus
G shear modulus

K_S adiabatic bulk modulus
β_P coefficient of expansion at constant pressure

γ ratio of principal specific heats c_P/c_V
c_P, c_V specific heats at constant pressure, volume
κ coefficient of thermal conductivity

R universal gas constant
B second virial coefficients
STP standard temperature & pressure

References

Acoustic Emission Group, University of California, Los Angeles. *Journal of Acoustic Emission*, from 1981

Allegra, J. R. and Hawley, S. A. (1972) Attenuation of sound in suspensions and emulsions: theory and experiments. *Journal of the Acoustical Society of America*, **51,** 1545–1563

Alliston, J. C., Kempster, A. J., Owen, M. G. and Ellis, M. (1982) An evaluation of three ultrasonic machines for predicting the body composition of live pigs of the same breed, sex and live weight. *Animal Production*, **35,** 165–169

Anson, L. W. and Chivers, R. C. (1990) Thermal effects in the attenuation of ultrasound in dilute suspensions for low values of acoustic radius. *Ultrasonics*, **28,** 16–26

Asher, R. C. (1983) Ultrasonic sensors in the chemical and process industries. *Journal of Physics E: Scientific Instruments*, **16,** 959–963

Beck, M. S. and Plaskowski, A. (1987) *Cross-Correlation Flowmeters: Their Design and Applications*. Hilger, Bristol

Betteridge, D., Joslin, M. T. and Lilley, T. (1981) Acoustic emissions from chemical reactions. *Analytical Chemistry*, **53,** 1064–1073

Blitz, J. (1963) *Fundamentals of Ultrasonics*. Butterworths, London (out of print, but still in many libraries; alternatives available: ask your librarian)

Breazale, M. A., Cantrell, J. H. and Heymann, J. S. (1981) Ultrasonic wave velocity and attenuation measurements. In *Methods of Experimental*

Physics, **19,** Ultrasonics, P. D. Edmonds (ed), Academic Press, New York, pp. 67–133

Choi, P. K., Bae, J. R. and Takagi, K. (1987) Frequency dependence of ultrasonic absorption in egg white. *Journal of the Acoustical Society of America*

CPPM (1989) Ultrasound in the Diagnosis and Management of Cerebro-Vascular Disease. *Clinical Physics and Physiological Measurement*, **10,** supplement A, February

Cracknell, A. P. (1980) *Ultrasonics*. Wykeham, London

Crawford, F. S. (1968) *Waves. Berkeley Physics Course*, vol. 3, McGraw-Hill, New York (out of print)

CRC (1976–77) *Handbook of Chemistry and Physics*, 57th edn R. C. Weast (ed). CRC Press, Cleveland

Crecraft, D. I. (1983) Ultrasonic instrumentation: principles, methods and applications. *Journal of Physics E: Scientific Instruments*, **16,** 181–189

Dull, G. G. (1986) Non-destructive evaluation of quality of stored fruits and vegetables. *Food Technology*, May, 106–110

Ensminger, D. (1973) *Ultrasonics: The Low and High Power Applications*. Marcel Dekker, New York

Fedotkin, I. M., Klimenko, M. I., Tchepurnoj, M. N. and Schneider, W. E. (1976) Velocity of ultrasound propagation in the production of sugar from beet juice. *Izvestiya Vysschych Utschebnych Zawedeny – Pischtschewaya Technologya*, no. 3, 105–107

Forrest, I. S. (1987) Methods of in-line OG measurement. *Brewers' Guardian*, **116,** 8, 9, 26

Gopalsami, N. and Raptis, A. C. (1983) Simultaneous measurement of ultrasonic velocity and attenuation in thin rods with application to temperature profiling. IEEE Ultrasonics Symposium, 856–860

Gould, R. W. (1972) Non-destructive egg shell thickness measurements using ultrasonic energy. *Poultry Science*, **51,** 1442–1446

Gray, D. E. (1963) *American Institute of Physics Handbook*, 2nd edn. McGraw-Hill

Henglein, A. (1987) Sonochemistry. *Chemical Society Reviews*, **16,** 239–311

Heusinger, H. (1987) A comparison of the product formation induced by ultrasonic waves and γ-rays in aqueous D-glucose solution. *Zur Lebensmitteluntersuchung und Forschung*, **185,** 106–110

Hibbard, D. J., Howe, A. M., Makie, A. R., Purdey, P. W. and Robbins, M. M. (1986) Measurement of creaming profiles in oil-in-water emulsions. In *Food Emulsions and Foams*, E. Dickinson (ed), Royal Society of Chemistry, London

ICT (1930) *International Critical Tables of Numerical Data, Physics, Chemistry and Technology*. National Research Council of the USA, McGraw-Hill, New York

Jansson, P. A. (1991) Neural networks: an overview. *Analytical Chemistry*, **63,** 357A–362A

Javanaud, C., Rahalkar, R. R. and Richmond, P. (1984) Measurement of speed and attenuation of ultrasound in egg white and egg yolk. *Journal of the Acoustical Society of America*, **76,** 670–675

Juszkiewicz, A. and Figlerowicz, M. (1990) Ultrasonic velocity hydration numbers of polyamines. *Ultrasonics*, **28,** 391–393

Kaye, G. W. C. and Laby, T. H. (1986) *Tables of Physical and Chemical Constants – and some Mathematical Functions*. Longman, London and New York

Kinsler, L. E., Frey, A. R., Coppens, A. B. and Sanders, J. V. (1982) *Fundamentals of Acoustics*. Wiley, New York

Krautkrämer, J. and Krautkrämer, H. (1983) *Ultrasonic Testing of Materials*, 3rd edn. Springer, New York

Kress-Rogers, E. (1986a) Instrumentation in the food industry. I: Chemical, biochemical and immunological determinands. *Journal of Physics E: Scientific Instruments*, **19,** 13–21

Kress-Rogers, E. (1986b) Instrumentation in the food industry. II: Physical determinands in quality and process control. *Journal of Physics E: Scientific Instruments*, **19,** 105–109

Landau, L. D. and Lifschitz, E. M. (1975) *Fluid Mechanics. Course of Theoretical Physics*, vol. 6 (first published 1959). Pergamon, Oxford

Lorimer, J. T. and Mason, P. J. (1988) *Sonochemistry: Theory, Applications and Uses of Ultrasound in Chemistry*. Ellis Horwood, Wiley, New York

Lynnworth, L. C. (1989) *Ultrasonic Measurements for Process Control: Theory, Techniques, Applications*. Academic Press, New York

McClements, D. J. and Fairley, P. (1991) Ultrasonic pulse echo reflectometer. *Ultrasonics*, **29,** 58–62

McClements, D. J. and Povey, M. J. W. (1987) Solid fat content determination using ultrasonic velocity measurements. *International Journal of Food Science and Technology*, **22,** 491–499

McClements, D. J. and Povey, M. J. W. (1989) Scattering of ultrasound by emulsions. *Journal of Physics D: Applied Physics*, **22,** 38–47

McClements, D. J., Povey, M. J. W., Jury, M. and Betsanis, E. (1990) Ultrasonic characterisation of a food emulsion. *Ultrasonics*, **28,** 266–272

McFarlane, I. (1987) Instrumentation. In *Industrial Chocolate Manufacture and Use* S. T. Beckett (ed), Blackie, London and AVI-Van Nostrand Reinhold, New York, 277–304

Miles, C. A. and Cutting, C. L. (1974) Changes in the velocity of ultrasound in meat during freezing. *Journal of Food Technology*, **9,** 119–122

Miles, C. A., Fursey, G. A. J. and Jones, R. C. D. (1985) Ultrasonic estimation of solid/liquid ratios in fats, oils and adipose tissue. *Journal of the Science of Food and Agriculture*, **36,** 215–228

Miles, C. A., Fursey, G. A. J. and Pomeroy, R. W. (1983) Ultrasonic evaluation of cattle. *Animal Products*, **36,** 363–370

Miles, C. A., Fursey, G. A. J. and York, R. W. R. (1984) New equipment for measuring the speed of ultrasound and its application in the estimation of body composition of farm livestock. In *In-Vivo Measurement of Body Composition in Meat Animals* D. Lister (ed), Elsevier, Amsterdam, 93–105

Miles, C. A., Shore, D. and Langley, K. R. (1990) Attenuation of ultrasound in milks and creams. *Ultrasonics*, **28,** 394–399

Newhouse, V. L. (1988) *Progress in Medical Imaging*. Springer, Heidelberg

Okawai, H., Tanaka, M. and Dunn, F. (1990) Non-contact acoustic method for the simultaneous measurement of thickness and acoustic properties of biological tissues. *Ultrasonics*, **28,** 401–410

Papadakis, E. P. (1976) Ultrasonic velocity and attenuation: measurement methods with scientific and industrial application. In *Physical Acoustics*, vol. XII, W. P. Mason and R. N. Thurston (eds), Academic Press, New York, 277–374

Povey, M. J. W. (1989) Ultrasonics in food engineering. Part II: Applications. *Journal of Food Engineering*, **9,** 1–20

Povey, M. J. W. and Harden, C. A. (1981) An application of the ultrasonic pulse-echo technique to the measurement of crispness of biscuits. *Journal of Food Technology*, **16,** 167–175

Povey, M. J. W. and McClements, D. J. (1988) Ultrasonics in food engineering. Part I: Introduction and experimental methods. *Journal of Food Engineering*, **8,** 217–245

Povey, M. J. W. and Wilkinson, J. M. (1980) Applications of ultrasonic pulse-echo techniques to egg albumen quality testing: a preliminary report. *British Poultry Science*, **21,** 489–495

Puskar, A. (1982) *The Use of High Intensity Ultrasonics*. Elsevier, Amsterdam

Randall, N., Wilkinson, V. M. and Roberts, R. T. (1987) Possible applications for an ultrasonic acoustic impedance probe. Leatherhead Food Research Association research report 588. This report is available to LFRA members only. However, a very brief description of the probe is given in Roberts and Wiltshire (1990a, p. 114). A similar probe is described in more detail in McClements and Fairley (1991)

Richardson, P. S. and Povey, M. J. W. (1990) Ultrasonic temperature measurement and its potential for food processing systems. *Food Control*, **1,** no. 1, 54–57

Roberts, R. T. and Wiltshire, M. P. (1990a) Sensing with sound waves. *Food Technology International Europe 1990*, Sterling Publications, 109–115

Roberts, R. T. and Wiltshire, M. P. (1990b) High intensity ultrasound in food processing. *Food Technology International Europe 1990*, Sterling Publications, 83–87

Ross-Murphy, S. B. (1984) Rheological methods. In *Biophysical Methods in Food Research. Critical Reports on Applied Chemistry*, vol. 5, H. W. S. Chan (ed), Blackwell Scientific, Oxford

Sajas, J. F. and Gorbatow, W. M. (1978) Anwendung von Ultraschallschwingungen in den technologischen Vorgängen der Fleischindustrie. Parts I, II, III. *Fleischwirtschaft*, **58,** 1009–1021 (English summary on 997), 1143–1152, 1325–1332

Sakai, H., Minamisawi, A. and Takagi, K. (1990) Effect of moisture content on ultrasonic velocity and attenuation in woods. *Ultrasonics*, **28,** 382–385

Sarkar, N. and Wolfe, R. R. (1983) Potential of ultrasonic measurements in food quality evaluation. *Transactions of the American Society of Agricultural Engineering*, **26,** 624

Schaaffs, W. (1963) *Molekulare Akustik. Eine Einführung in die Zusammenhänge zwischen Ultraschall und Molekularstruktur in Flüssigkeiten und Gasen*. Springer, Berlin (out of print)

Sleeter, R. T. (1985) Instrumental analysis for quality control and quality assurance. In *Bailey's Industrial Oil and Fat Products*, vol. 3 T. H. Applewhite (ed), Wiley, New York, 203–242

Smith, D. E. and Winder, W. C. (1983) Effects of temperature, concentration and solute structure on the acoustic properties of monosaccharide solutions. *Journal of Food Science*, **48,** 1822–1825

Steele, D. J. (1974) Ultrasonics to measure the moisture content of food products. *British Journal of Non-Destructive Testing*, **16,** 169–173

Szilard, J. R. (1982) *Acoustic Emission: a Diagnostic Tool in Non-Destructive Testing*. Wiley, New York

Urick, R. J. (1947) A sound velocity method for determining the compressibility of finely divided substances. *Journal of Applied Physics*, **18,** 983–987

Volsey, P. W. and Hamilton, R. M. C. (1976) Ultrasonic measurement of egg shell thickness. *Poultry Science*, **55,** 1319–1324

Wade, A. P., Sibbald, D. B., Bailey, M. N., Belchamber, R. M., Bittman, S., McLean, J. A. and Wentzell, P. D. (1991) An analytical perspective on acoustic emission. *Analytical Chemistry*, **63,** 497A–507A

Wang Jin and Feng Ruo (1990) Ultrasonic velocity of aqueous solutions of amino acids. *Ultrasonics*, **28,** 37–39

Wells, P. N. T. (1977) *Biomedical Ultrasonics*. Academic Press, London

Wyn-Jones, E., Pereira, M. C. and Morris, E. R. (1982) Ultrasonic relaxation studies in sols and gels. *Progress in Food and Nutritional Science*, **6,** 21–31

Young, R. F. (1989) *Cavitation*. McGraw-Hill

Zaccharias, E. M. and Parnell, R. A. (1972) Measuring the solids content of foods by sound velocimetry. *Food Technology*, **26,** no. 4, 160–166

Zgonc, K. and Grabec, I. (1990) A neural-like system applied to acoustic emission analysis. International Neural Network Conference, IEEE/International Neural Network Society, Paris, July

9 *Ultrasonic Instrumentation in the Food Industry*

Nicholas Denbow

Contents

1 Introduction

The commonly accepted definition of 'ultrasound' is a pressure wave that is oscillating at a frequency above that of human hearing. Whilst the actual highest frequency audible can vary from person to person, ultrasonic waves can have any frequency above 20 kHz up to a practical limit of around 10 MHz for the applications to be discussed in this chapter. Within this band, the measurement of the speed of propagation, the attenuation, or the reflection of ultrasonic signals can be the basis of instrumentation systems for the food industry. Often these systems have distinct advantages for on-line food applications, as the ultrasonic sensors normally have no moving parts, can usually be sealed behind stainless steel diaphragms, and occasionally can function as non-contact devices, either above the food product or attached to the outside of the containment vessel or pipe.

Within the ultrasound range, the choice of frequency is dependent on the attenuation characteristics of the medium in which it is to propagate. In gases, the attenuation for ultrasound is high and increases with frequency. Process measurements requiring ultrasound transmission through gases are therefore generally impractical for frequencies much above 100 kHz. (See Chapter 8 for an overview on the characteristics of ultrasound propagation in foodstuffs and ambient gases.)

In applications requiring good spatial resolution, short wavelengths are used in preference. For ranging tasks, they are generally in the millimetre range. The corresponding ultrasound frequencies are of the order of 50 kHz for gases and 1 MHz for liquids and solids (see Table 9.1).

Further considerations in the choice of the measurement frequency are the sizes and mechanical properties of the transducer and the cost of the associated instrumentation.

Transducers for ultrasonic equipment are mostly based on piezoelectric

Table 9.1 *Ultrasonic velocity and wavelength examples for typical ultrasound ranging applications*

Medium	*Usable frequency range*	*Speed of propagation*	*Wavelength (at typical frequency)*
Gas: air	20 kHz to 100 kHz	332 m/s	6.6 mm (50 kHz)
Liquid: water	200 kHz to 5 MHz	1500 m/s	1.5 mm (1 MHz)
Solid: epoxy	500 kHz to 10 MHz	2500 m/s	2.5 mm (1 MHz)
steel	500 kHz to 10 MHz	6000 m/s	0.6 mm (10 MHz)

materials; most popular are heavy man-made lead zirconate titanate (PZT) (ceramic) slices with silver-coated top and bottom faces to attach electrical connections. Quartz is a naturally occurring piezoelectric material, but is generally less efficient as a transducer and more expensive. The PZT disks are of various thicknesses to make them resonant at the required frequencies; a 1 MHz crystal is approximately 1.5 mm thick, and a 50 kHz crystal 12 mm thick. To achieve a directional transducer, the disk diameter needs to be around ten times the ultrasonic wavelength being used; this means that low-frequency transducers are usually large.

At the other end of the scale, flaw detection equipment based on ultrasonics is widely used industrially to detect weld defects or cracks in structures. To detect small defects, non-destructive testing (NDT) equipment operates at a very high frequency, and the probes can be very small. A practical limitation in the construction of these transducers is that for frequencies above 5 MHz, the PZT disks are so thin that they become difficult to mount securely against a transducer face.

An industrial or food processing plant is full of sources of ultrasonic noise. All the normal sources of plant vibration or noise also produce inaudible ultrasonic noise. Additionally, there are other sources such as microcracks and temperature changes at joints. The ultrasonic transducer-receiver collects these noise inputs, plus electrical noise on the transducer cables in all frequency bands, and then has to pick out the required signals. In general, this is achieved with the self-tuning of the resonant piezoelectric material, receiver-amplifier tuning and microprocessor-based signal recognition techniques. The availability of microprocessors has allowed a significant increase in the food industry's use of ultrasonic instrumentation.

2 Low-frequency Techniques

Instruments using low ultrasonic frequencies, between 10 kHz and 100 kHz, are designed to transmit the ultrasonic beam through air. Such instruments form the basis of two measurement systems: pulsed signal echo ranging and continuous signal echo detection. Echo ranging is used for liquid or solids level measurement, and echo detection provides beam-break detectors or counting systems.

2.1 Level Measurement by Echo Ranging

Instrumentation for ultrasonic level measurement, for example for liquids, is based on pulsed signal transmission followed by detection of the time-delayed echo. A transducer above the surface of the liquid to be monitored transmits a short (approximately 1 ms) burst of ultrasound, typically at 50 kHz. This travels through air (or other gas in the ullage space) vertically down to the liquid surface. Because of the ultrasonic impedance mismatch between the liquid and the air, most of the energy in the airborne beam is reflected vertically back along the same path. By the time the ultrasonic pulse reaches the transducer, this has stopped transmitting and switched over to being a receiver. The time delay between the start of the transmit pulse and detection of the leading edge of the received pulse is measured, and used to calculate the liquid level.

Inherent in this simple description are some major assumptions. The first is that the speed of sound in the air space above the liquid surface is known, or is predictable. The second is that the pulse transmission, reflection from the surface and reception are of high enough signal to noise quality that the same part of the pulse is detected in the receiver circuits as is used to trigger the clock on transmission.

2.1.1 Calculating or Measuring the Speed of Sound in the Headspace

The speed of sound c in gas is related to pressure P, density ρ and ratio of principal specific heats γ by the expression

$$c=\sqrt{(\gamma P/\rho)} \qquad (1)$$

Since for pressures at or near atmospheric values, air pressure is proportional to density, the speed of sound is pressure independent under such conditions:

$$c=\sqrt{(\gamma R)}\,\sqrt{(T/M)} \qquad (2)$$

where T is the absolute temperature and M is the molecular weight. The factor $\sqrt{(\gamma R)}$ includes the ratio of specific heats γ, typically 1.4 for diatomic gases, and the universal gas constant R.

The speed of sound, then, is proportional to the square root of the absolute temperature of an ideal gas (see Chapter 8 for other gases). This means that at ambient conditions the speed of sound increases by 0.2 per cent for every Kelvin increase in temperature (0.17%/K at 20°C), showing that for accurate level measurement, compensation for temperature variations is required. This is most simply achieved by including a temperature sensor in the ultrasonic transducer head and using the microprocessor in the level measurement electronics to compute the appropriate speed of sound. (On temperature units, see Appendix B3.)

Equation (2) shows that sound is not dependent on the pressure of an ideal gas in the ullage space. Over the operating range of ultrasonic systems, and for the gases normally encountered, this relationship is valid and no correction for pressure changes is needed. There are other limitations related to gas pressure. Ultrasound is a pressure wave and it

Table 9.2 *Speed of sound at 0°C in various gases. Values calculated from Kaye and Laby (1986). In this form, they are used to find the start value to be entered. The set value will subsequently be adjusted by the instruments' microprocessor in accordance with the prevailing temperature*

Gas or vapour	*Speed of sound at 0°C**	*Molecular weight*
Nitrogen	337	28
Oxygen	315	32
Air (dry)	331.45	29
Methane	435	16
Ammonia	418	17
Water (dry steam)	409	18
Ethylene	316	28
Hydrogen sulphide	296	34
Hydrogen chloride	296	37
Fluorine	283	38
Acetaldehyde	278	44
Freon 11	139	138

*Calculated set value. For gas mixtures, weighted averages are used.

needs gas molecules to interact with each other to propagate through a medium; it cannot travel through vacuum. In practice, most manufacturers quote a low-pressure limit of −25 kPa (−0.25 bar) gauge pressure, although there is a progressive increase in sound attenuation with decrease in pressure. The high-pressure limit is set by the transducer itself, that is the stress that its construction can withstand. Since the necessary diaphragm must be flexibly mounted in the transducer housing, the high-pressure limit is typically below 500 kPa (5 bar) excess pressure. The actual composition of the gas in the ullage space determines the speed of sound. Equation (2) shows the relationship to molecular weight, and this is illustrated with data in Table 9.2. Further data can be found from such reference books as Kaye and Laby (1986). In most food industry applications, the ullage space in tanks or vessels will be air with a high water vapour content (humidity). Practical experience has shown that the best accuracy is achieved by setting the *c* value assumed for air to 332 m/s. This dry-air setting is valid up to approximately 40°C, but above this temperature, at high humidities, the water vapour (molecular weight 18) increases the speed of sound considerably.

Where such high-humidity, high-temperature conditions prevail, or where high-volatility liquids (solvents or alcohols) are involved, the complex relationship between speed of sound and gas composition makes it impossible for the microprocessors to use data from the pre-programmed tables. In these cases, it is usually necessary to use an ultrasonic technique to measure the speed of sound. In practice, a 'reference reflector' is mounted above the maximum contents level of the tank, and the distance from the transducer to this reflector is measured. The receiver system microprocessor can then compute the operational speed of sound in the variable gas from the time delay of echoes from this known reference. It assumes that this reference path is representative of the whole tank ullage in both vapour composition and temperature, and determines the liquid level using the measured speed of sound. The microprocessor updates sound data at intervals of typically 10 to 30 s.

These alternative techniques are shown in Figures 9.1 and 9.2, with typical transducer constructions. For most food applications, the temperature correction technique is sufficient to allow for the speed of sound variations experienced in practice, and provides a simpler system. The reference pin system will only be used for the highest accuracy requirements or where the variations in ullage gas composition are significant.

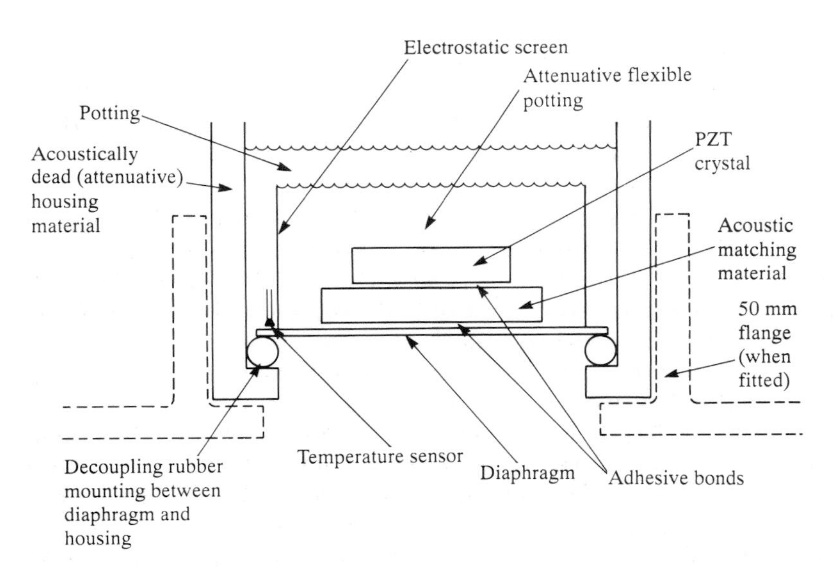

Figure 9.1 *Typical pulse-echo level measurement transducer construction with built-in temperature sensor*

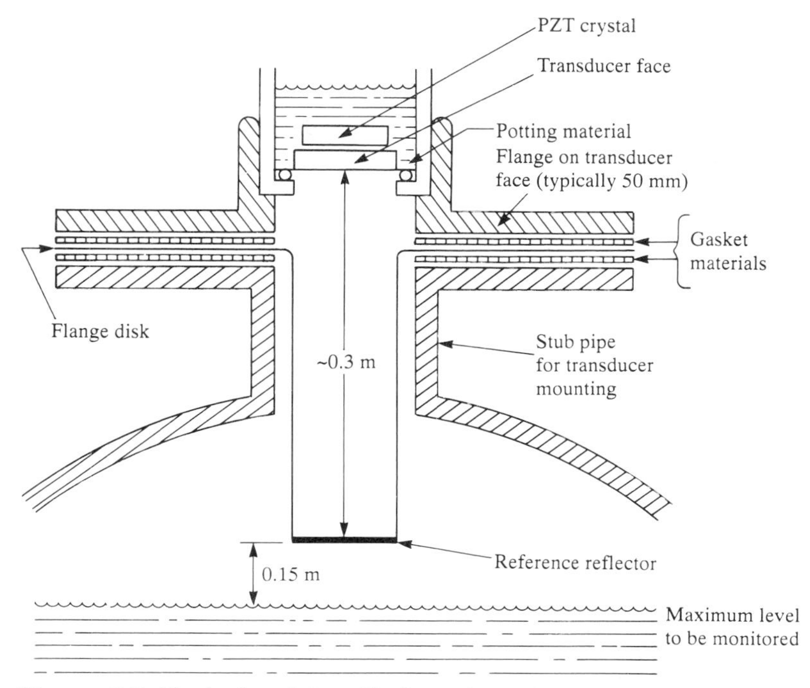

Figure 9.2 *Typical tank installation using reference pin reflector for speed of sound measurement*

2.1.2 Pulse Transmission, Reflection and Reception

The designer of transducers for airborne pulse-echo measurement systems faces the need to produce a transmit pulse with a sharp leading edge, so as to allow accurate timing, and to provide the maximum of sound energy in a resonant crystal material, which must then be switched off very quickly to act as a receiver. The diaphragm must be of sufficient diameter to produce a narrow beam and flexible enough to vibrate, but must be rigidly attached to the piezoelectric disk. It must then be strong enough to withstand the temperature and pressure ranges required.

In general, a 50 kHz system requires a 50 mm diameter transducer face. This will produce a beam that is within a $\pm 6°$ angle, but is still economical to fit to process vessels by flanges. For such a system, a twenty-cycle pulse train takes about 0.4 ms to transmit, but the transducer ringing-after effects last for a further 1.5 ms. This limits the closest detectable echo to a range of approximately 0.3 m from the transducer face.

The pulse train is now reflected from the liquid surface – from an illuminated area of radius 0.1 m for each metre of range from the transducer. Each ripple or surface discontinuity will give its own reflections, which arrive back at the receiver either with the main echo, or slightly delayed because of the longer path length. Only the first echoes will therefore be recognizable as a pulse train; the later echoes will have secondary reflections superimposed. The waveforms might be as shown in Figures 9.3 and 9.4.

The wavelength between successive peaks of this wave represents approximately 6 mm in distance. It is therefore important for the receiver to identify the correct pulse on the leading edge of the receiver waveform, to remove a potential 6 mm error (or multiples thereof). Using a microprocessor, this is readily achieved. The first technique is to reject echoes that are detected but that fall outside a range gate 'window' set either side of the range of most of the echoes. The second technique is to measure the height of the strongest pulses in the returned echo pulse train, and set a threshold level at, say, 25 per cent of this height. The first echo pulse is then identified as the first detected signal above this automatically adjusted threshold level.

2.1.3 Microprocessor-Based Electronics

The microprocessor is an essential tool in the signal processing and detection circuitry used in pulse-echo level measurement systems. With crystal-based timing circuits, the basic measurements can be made accurately; then the store, hold and compare functions available in the microprocessor allow rejection of invalid echoes on the basis of time delay, signal strength or the fact that they are not confirmed in successive pulses. In addition, the temperature or reference echo detection system can be interrogated on an intermittent basis, to amend the speed of sound value as necessary.

These are the direct measurement functions of the microprocessor. The indirect functions are there as support to improve the system reliability. For example, the cycle time between pulses is subject to a 5 per cent jitter at

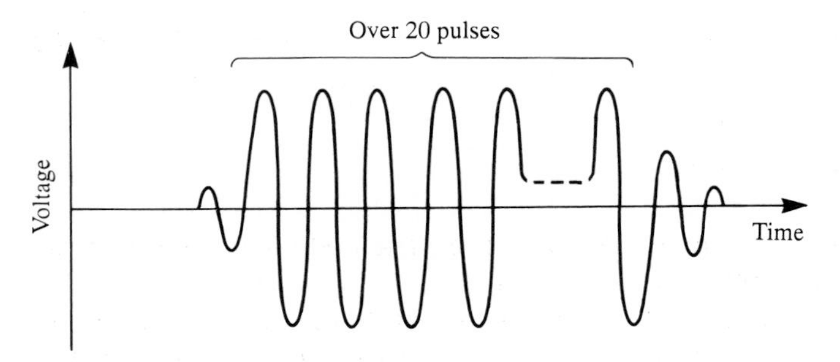

Figure 9.3 *Transmitted pulse train*

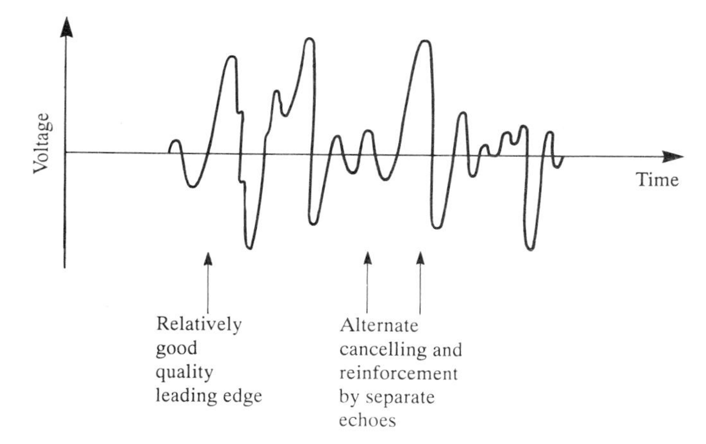

Figure 9.4 *Received pulse train*

random around a typical 200 ms cycle for a 10 m range system; this overcomes the effect of any low-frequency cyclical noise. On such a 200 ms cycle, the microprocessor looks for echoes for 60 ms. The remainder of the cycle is used to control the logic of pump or alarm control relays, and set a current output proportional to level, or proportional to tank contents, by using a look-up table. The microprocessor optimizes the ultrasonic pulse transmission frequency itself, by adjusting the frequency up or down by, say, 1 kHz. If the surface echo received is stronger at the new frequency, it will track to that frequency for the next 500 cycles. This takes account of temperature effects on the resonant frequency of the PZT crystal itself, which is significant. The typical operational band can vary from 45 to 60 kHz, over the full temperature range of typically −20°C to +70°C.

The last function of the microprocessor is to decide what to do after a predetermined period of no valid new data. Typically, this can be up to 90 s; after this time the microprocessor can set relays and current output to a safe state, indicate a problem status on the display and even de-energize a faulty signal relay. The most common cause of such a problem is the presence of foam on the liquid surface, which has attenuated the echo to such a degree that it cannot be detected; an understanding of this potential problem is essential to designing a fail-safe measurement system.

2.1.4 Liquids Applications

Ultrasonic pulse-echo techniques have been adopted significantly in the food industry since 1983. The first applications were in storage tanks for edible oils. The oil level was monitored with systems using reference pin speed of sound values together with the time-of-flight measurement. Rectangular tanks with approximately 4 m sides were monitored, as well as cylindrical tanks 7 m tall with a conical bottom section. Using the microprocessor to calculate the tank contents in litres allowed blending operations to be monitored and controlled.

A number of dairies have used ultrasonic techniques to monitor the contents of milk tanks, and also of tanks containing aqueous sugar solutions and other liquids on the site. Sugar refineries have used ultrasonics on sugar, molasses and other liquids in storage or process tanks. In all these cases, there is some degree of foaming on the liquid surface, which results in weak echo signals at longer range (for example 5 m). However, the highest degree of foaming occurs during filling, and it is towards the end of this process that a reliable signal is needed to prevent spillage. Because the transducer is top-mounted, the critical point for such high-level alarms is encountered when the liquid surface is around 1 m or less from the transducer. At this range, even the echo from thick foam is usually sufficiently strong to be detectable.

Applications in dairies require that the transducers withstand cleaning processes, even if they do not need to function properly during the cleaning cycle. It is relatively undemanding for an ultrasound transducer to withstand chemical and hot water cleaning, but steam cleaning can damage, if not blow apart, transducers made from plastic materials. There are now stainless-steel-faced transducers available, but the flexible resins and glues in these will be damaged by anything other than low-pressure steam cleaning.

The use of ultrasonics for monitoring the blending of oils, and successful applications in occasionally foaming sugar and milk, led to applications on yeast tanks. This time, ultrasonic pulse-echo systems were shown to be incapable of reliable operation. In yeast production, the fermentation produces a thick foam. This indeed reflects a sufficient echo signal, but the yeast fermentation produces carbon dioxide gas as well. This carbon dioxide has an unusual characteristic: it attenuates ultrasound severely at the frequencies used in ultrasound echo ranging. (See Chapter 8, Section 5.2 on the attenuation characteristics of CO_2.)

This attenuation prevents the use of ultrasonic pulse-echo techniques in any process that produces carbon dioxide in the ullage space, above concentrations of approximately 5 per cent. This not only applies to yeast fermentation vessels but has also been observed for storage tanks containing blackcurrant and similar juices, which have high carbon dioxide concentrations in the ullage space.

2.1.5 Solids Applications

The measurement of solids level in storage tanks suffers from two disadvantages compared with liquids. Firstly, the solids do not flow to achieve a flat surface, so that the returned echoes do not normally come from approximately the same range. Secondly, solids level applications tend to be for longer ranges than the liquid equivalents. Large grain silos are a typical example.

The longer range required for grain silo monitoring can be achieved with higher-power and physically larger transducers. It is also useful to reduce the operating frequency, so as to increase the wavelength and therefore reduce scattering by grain particles in the air. Silos for grain storage have been successfully monitored using such techniques, with reported ranges up to 50 m (Milltronics 1990).

The coring effects on such a silo would possibly result in a ± 1 m uncertainty as to where the level was, so that even a 1 per cent accurate electronics system giving an error of ± 0.5 m at this extreme range would be adequate.

For smaller silos or for solids level measurement applications on conveyor belts or similar positions, the accuracy and resolution resulting from a higher-frequency system would again be needed. Even so, the solid surface level uncertainty is likely to exceed the equipment inaccuracy. In this class, known applications of ultrasonic pulse-echo techniques on solids include monitoring chocolate crumb and peas on tanks feeding process lines, and also monitoring the total thickness of the layer of potato emerging stacked on a conveyor from a slicing machine for potato crisp manufacture. Such measurements of level on a conveyor can be used to give an approximate (total volume per second) flow rate.

Returning to the applications in storage or mixing tanks, ultrasonic non-contact level measurement is widely used for sticky semi-solid substances. Such mixtures occur in the food industry, as pastes and blends, which are then fed to individual containers. The microprocessor can usually be programmed to ignore occasional reflections from the blades or paddles of rotating stirrers, thus giving a true level measurement.

2.1.6 Accuracy

With pulse-echo ultrasonics there are many factors to be taken into account, and a sensibly engineered system will achieve a good result across all of them rather than seek to perfect the response to one or two.

Generally, the microprocessor timing is based on a high-frequency crystal oscillator (e.g. 12 MHz), which is specified across the operating temperature range as better than one part in 20,000, leading to better than 0.01 per cent time measurement accuracy. Timing resolution is typically 1 μs, which represents 0.16 mm steps of range in air.

The major source of error is the assumption about the speed of sound in the ullage space, and whether this can be stated more accurately than ± 0.2 per cent taking into account variation of ambient gas composition and vapour percentages. Once the composition is established, the temperature averaged over the sound path needs to be estimated. If a temperature sensor is used, how representative is the reading for the sound path average? Inevitably, temperature sensors in transducers cannot track variations in the temperature in the ullage space accurately, and even in steady-state conditions will also monitor the tank surface temperature.

This situation can be improved by positioning a temperature sensor separately into the tank, but again the relationship to the average value is questionable. In most cases a further ±1 K error is expected even in steady-state conditions, leading to a ±0.2 per cent error in speed of sound.

The result of these errors is a typical accuracy of ±0.5 per cent in the measurement of liquid surface range from the transducer. With a typical readout showing the tank contents, the range and contents are inversely related: long range means low tank contents. Therefore, as shown in Figure 9.5, when a 3 m high tank is 10 per cent full, the range error is ±0.5 per cent of 3.1 m (i.e. ±1.6 cm). When expressed as a percentage of the depth of the tank contents (0.3 m), this results in a reading inaccuracy calculated as ±1.6/30, that is ±5 per cent of the reading. Always, ultrasonic systems of this type have higher uncertainty when tanks are nearly empty, and good accuracy of tank contents when they are nearly full.

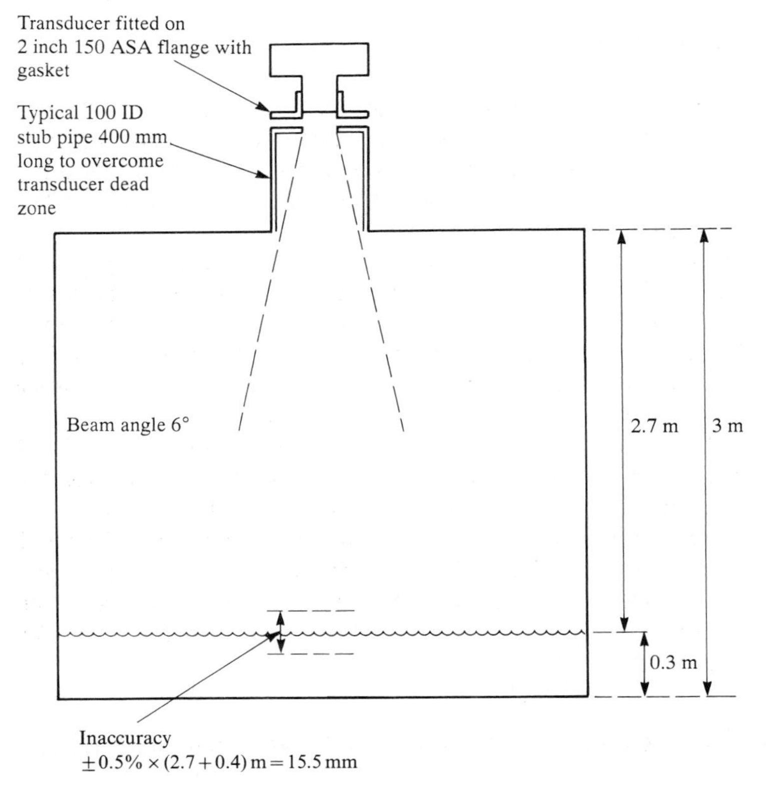

Figure 9.5 *Ultrasonic system level measurement installation and accuracy example*

In many applications, where for example a certain quantity of liquid has to be dispensed from a storage tank for blending, the repeatability and resolution are more relevant than absolute accuracy. In such applications, the vapour space over the liquid will remain constant in its temperature and composition, making the dispensed quantity itself measurable to an accuracy of ± 0.5 per cent. With the resolution of the timing measurement typically equivalent to ± 0.2 mm, the main limitation will be the resolution of the current output from the measurement electronics.

For an earlier microprocessor with 8 bit mathematics, 256 steps were used between empty and full, that is across the range of the current output. The smallest step change detectable was therefore 0.4 per cent of the span of this current output. A similar 256-step look-up table was used to relate liquid level to tank contents, where the microprocessor is used to derive tank volume; this used to result in approximately a 0.4 per cent (of volume) minimum step size on the electronics display more recent microprocessors use 12 bit mathematics resulting in 0.025% resolution, that is 4000 steps, (Platon 1992).

2.2 Beam-break Detectors as Counters

Since any relatively solid object placed in an ultrasonic beam will reflect or deflect the sound wave, systems have been developed to detect the presence of such solid objects using the interruption of an ultrasonic beam between transmitter and receiver. Working at 30–50 kHz, such systems are used as counters for packages on conveyor belts.

The advantage of ultrasonics over other techniques such as optical or infrared beam-break detection systems is that the ultrasonic beam is less severely attenuated by dust in the air, and more importantly is unaffected by dust build-up on the transducer faces themselves. This advantage arises because the transducer faces physically move to transmit the beam, rather than being purely a window as in the optical system.

Ultrasonic beam-break detection techniques were also used as level controls (employing transducers mounted on opposite sides of the tank) for many years, particularly on dusty materials such as grain or even coal in silos. Systems based on this technique are now no longer manufactured because the bulk of the simpler applications are served by infrared beam-break detection systems that are available at very low prices, and the ultrasonic systems are expensive when produced in small quantities for special applications.

2.3 Future Developments in Low-frequency Instrumentation

The main developments in low-frequency ultrasonic techniques will be in the design of transducers. Refinement to existing designs will be to increase the operating temperature rating to 130°C, but probably no higher; to increase the pressure rating to the economically viable maximum, which is likely to be 500 kPa (5 bar) excess pressure; and to make different materials of construction available in addition to the current materials, namely plastics, PTFE and stainless steel. In performance, the only real change will be in the closest allowable measurement distance, which might be reduced to 0.1 m.

There is no prospect of producing a non-penetration level measurement system using this technique, because of the difficulty of using the tank wall as a low-frequency transducer diaphragm. Techniques have been developed, notably by UKAEA, Harwell, for using a tube to guide the ultrasonic pulse down into a tank (Asher 1983). This could be of use in the food industry to avoid beam interference from major tank obstructions such as heating coils or multiple stirrer systems.

3 High-frequency Techniques

Instruments using higher ultrasonic frequencies, between 500 kHz and 5 MHz, are designed to transmit the beam through the liquid. Systems in this range are based on the measurement of one of the following:

- the attenuation of the beam (measuring the amplitude at a receiver)
- the propagation velocity of the beam in the liquid (measuring the delay of a pulse)
- the frequency of a received signal
- the loading of the liquid on a transmitted signal (monitoring the decay or reflection of an ultrasonic pulse resulting from the impedance mismatch between transducer surface and liquid).

Based on these techniques, liquid level, liquid flow, suspended solids and liquid composition monitoring systems applicable to the food industry have been commercially developed.

3.1 Flow Measurement Systems

Ultrasonic flow measurement systems have the major application advantage that, in most cases, the transducers can be mounted in or on the walls of the pipe carrying the flow to be monitored. They are therefore obstructionless, which is a major advantage, but can also be non-contact devices, and thus have the further major advantage of being hygienic.

In medicine, blood flow measurement pioneered the use of ultrasonic flow meters based on the Doppler effect. These were then applied indus-

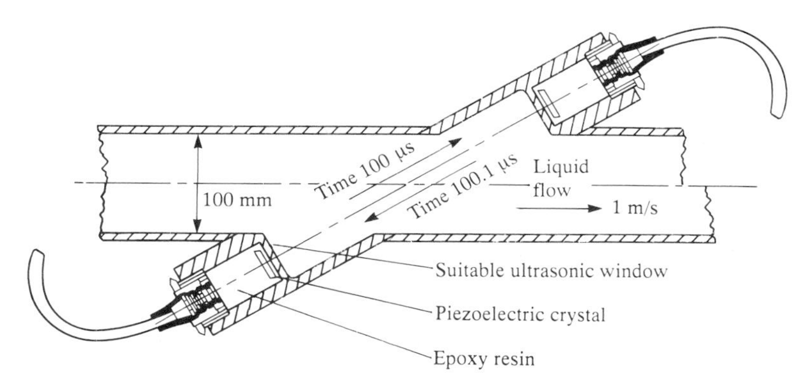

Figure 9.6 *Time-of-flight flow measurement system*

trially from 1977 onwards, with many applications in the food industry. Work at Harwell led to the time-of-flight ultrasonic flow measurement system at about the same time, but the two techniques are totally different in their underlying principles and performance characteristics, and lead to different applications.

3.1.1 Time-of-flight Flow Measurement

Two ultrasonic transducers are fitted into bosses welded on to a section of pipe. They are positioned so that an ultrasonic beam transmitted from one to the other travels at an angle of typically 60° to the pipe axis and therefore the liquid flow (Figure 9.6). A pulse of ultrasound is timed travelling through the liquid between transducer 1 and transducer 2; because the pulse is travelling downstream with the flow, the transit time is shorter than that for a similar pulse transmitted upstream from transducer 2 to transducer 1.

Electronics systems driving the transducers can be used to derive the difference between these two transit times. Typical values for a 100 mm bore pipe might be a transit time of 100 μs, and a difference in transit times of 100 ns for a 1 m/s flow velocity. Measuring this to a 1 per cent accuracy requires a stability of 1 ns in this time measurement, which is easily achievable.

Where the flowing liquid contains suspended solids, these solids will scatter the sound and attenuate the pulse as it travels between transducers. This will prevent reliable operation on liquids where the solids loading is too great. Temperature changes or dissolved solids that affect the speed of sound are automatically compensated for in the electronics, by measurement of the average transit time between transducers to derive the actual speed of sound in the liquid.

The major application for this technique is in the larger-diameter pipeline applications in the water supply industry. Here the ability to install transducers into bosses welded on to existing pipework, even without interrupting the flow, is a major advantage. Systems are installed on pipelines of up to 3 m diameter, monitoring flow with accuracies better than 0.5 per cent of reading. Such systems have been installed to measure wastewater flows in sewers, and on site for process control in both sewage treatment works and water purification plants.

Flow measurement applications in other parts of the food industry are less numerous, since the advantages of this technique are less significant there, and difficulties in installation occur in smaller-diameter pipelines. To provide sufficient ultrasonic path length for pipe diameters below 75 mm, the transducers are usually positioned to send the ultrasonic beam axially along part of the pipe, achieved by introducing a double bend in the pipework. Despite the flow profile distortion, very good linearity has been achieved on examples of such systems.

Future developments are in the area of clamp-on transducers for operation on existing pipework, possibly using a zigzag path between transducers to overcome the problems of small-diameter lines.

3.1.2 Doppler Flow Measurement

The Doppler effect is characterized by a change in frequency of a wave produced by or reflected by an object that is moving towards or away from the observer. In a Doppler ultrasonic flow meter, a piezoelectric crystal transducer produces an ultrasonic beam which is launched into a flowing liquid. Discontinuities in that liquid, typically suspended solids, reflect the beam, and if these solids are moving towards the transducer then the reflected beam has a higher frequency. The frequency shift, as a percentage of the transmitted frequency, is proportional to the speed of movement of the suspended solids, as a percentage of the speed of sound in the liquid.

The transducer is normally clamped on to the outside of the pipeline, using silicone grease or epoxy resin as an acoustic couplant between the transducer face and the pipe wall (Figure 9.7). Within the transducer housing there are usually two piezoelectric crystals, angled by typically 30° from perpendicular to the pipe surface. One crystal transmits a 1 MHz continuous wave signal into the pipe and the flowing liquid. The returned reflected signals are detected by the received crystal and mixed with direct coupled transmitted signals. Various electronic circuit detection systems are used to average the returned frequency shifted signal, to obtain a characteristic frequency quoted as the Doppler frequency. This is then processed to give flow rate.

Because of their advantages – clamp-on, obstructionless and hygienic – Doppler flow meters are widely used in the food industry, particularly on liquids containing particulates, for example soups or jams. Slurries also present excellent applications; for example, the sugar beet slurry produced in the first stages of the refining process is measured with a Doppler flow meter. In brewing, Doppler flow switches are used on the pipelines containing wort and yeast, and in sugar refining the flow control systems on lime milk are based on Doppler flow meters.

While the Doppler technique is ideal for slurries, or any liquid containing solid particles, its advantages of clamp-on configuration affording convenience and hygiene make it very attractive for other flow measurement duties, sometimes on clean liquids. Results are often reported to be good despite the lack of apparent reflectors. Whilst this is often ascribed to reflections from turbulence or eddies, the reflected signal is more likely to come from entrained air bubbles. It is very important to use care in the application of Doppler techniques to such flow application in clean liquids. The normal plant noise and vibration of the transducer on the pipe wall can feed an apparent 'Doppler' frequency signal into the detection circuits, giving a false flow measurement.

To increase confidence in the Doppler signal, most Doppler flow meters have a signal strength meter to measure the amplitude of the returned signal. This can be used to give a measure of the reliability of the flow reading. On relatively clear liquids there are many applications where good results can be obtained. Milk and orange or other fruit juices normally give good results. Some liquids such as oils and condensate in return lines give poor results, but such is the enthusiasm for the use of Doppler systems that at times air bubble injection systems have been used to improve the flow meter performance on these liquids. This is perhaps acceptable in plant commissioning or proving trials, where a portable Doppler flow meter might be used for troubleshooting, but is hardly recommended for permanent installation.

With most Doppler systems being clamped to the pipe wall, it is

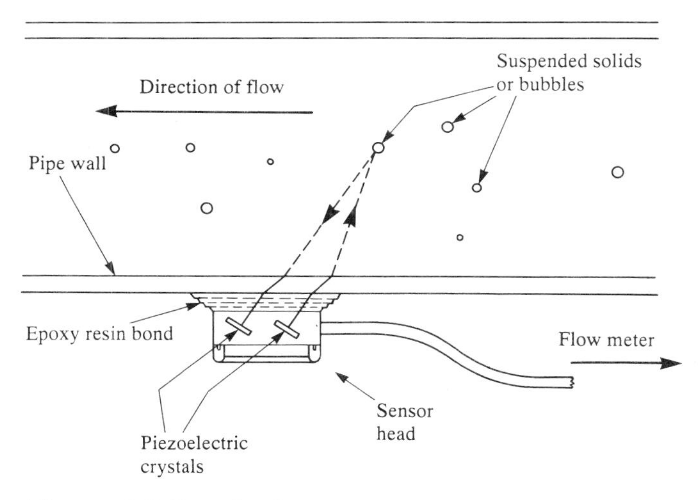

Figure 9.7 *Doppler flow meter sensor*

important to consider the effect of the pipe wall as part of the transducer. The inside surface must be smooth and parallel to the outside surface, and made of a single continuous material. This excludes lined pipes, where any gap between the lining and the pipe material will stop all transmission. It also excludes inhomogeneous materials like cement, cast iron or glass reinforced plastic (GRP). Any paint on the outside of the pipe should preferably be removed. The best types of pipe are plastic or glass, because they transmit the ultrasound efficiently through to the liquid and yet do not allow pipe noise or plant vibration to travel too far. While the speed of sound in the liquid will affect the angle of refraction of the ultrasonic beam, and this therefore varies with temperature and type of liquid, the frequency shift is dependent on the component of the liquid velocity along this beam angle. A mathematical evaluation of beam refraction theory shows that the Doppler shift is independent of liquid temperature and composition, provided that the pipe walls are parallel to one another and the pipe axis (Cousins 1978; Scott 1979).

Doppler flow measurement systems are most useful on pipe sizes from 20 to 200 mm, monitoring flows where the linear velocity along the pipe is normally above 0.3 m/s. In such cases the accuracy achieved will be better than 5 per cent, although many users will claim much better results than this. The main application is in flow measurement of slurries, or any liquid containing a second phase in suspension.

For liquids with a heavy suspended solids loading or a high density of entrained gas bubbles, the resulting strong attenuation of the ultrasound signal means that the flow velocity is measured close to the pipe wall only. This leads to an error in the derived value for the pipe overall velocity.

3.2 Liquid Level Switches

Liquid level sensors based on ultrasonic transmission have been manufactured commercially since 1973, and have evolved into various styles of sensor for high- or low-level alarm, either penetrating the tank or for use across a vessel or pipe as non-penetration systems. Advances in transducers and electronics have recently made pulse attenuation measurement systems viable, based on original work at Harwell (Asher 1983). This also leads to a device for level control without penetration of the vessel wall.

A series of level switches can be chosen as an alternative to the echo ranging systems for level measurement (Section 2.1). Level switches (point sensors) can offer the following advantages: steam cleanability, high-pressure tolerance, fail-safe features, reliability, low cost, and insensitivity to ullage gas composition and to stirring paddles. The echo ranging systems, on the other hand, are preferable when analogue output, one-tank entry and measurement over a continuous range of liquid levels are required.

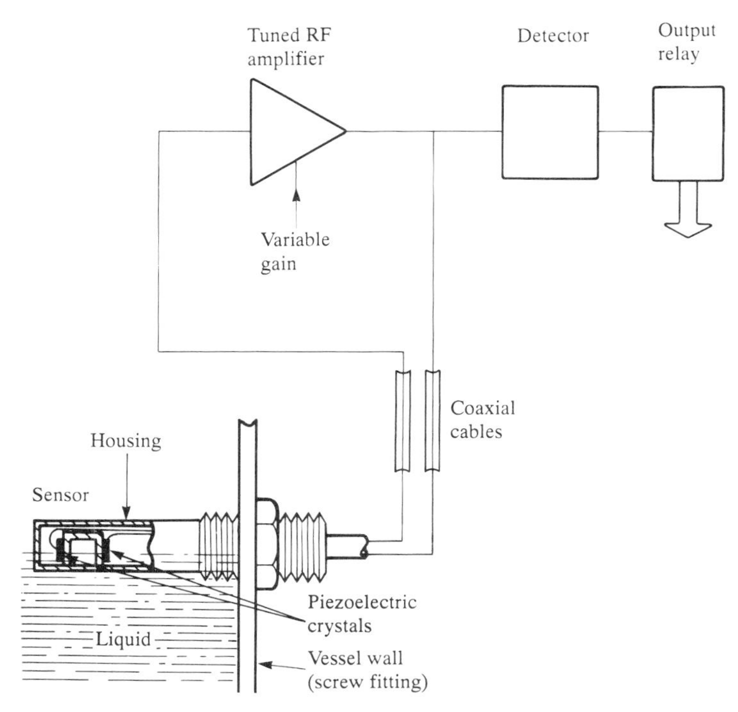

Figure 9.8 *Ultrasonic liquid level sensor circuit*

3.2.1 Ultrasonic-transmission-based Level Detectors

At an operational frequency of 1 MHz and above, the attenuation of ultrasound in air is very high. Transmission through liquids, on the other hand, is very good, so that when two transducers face one another, ultrasound will only pass from transmitter to receiver when the gap between them is filled with liquid. The technique is used in a liquid level sensor circuit as shown in Figure 9.8.

In practice, the transmitter crystal is driven from the output of a tuned RF amplifier, with the receiver linked directly to the input. The system sensitivity is adjusted using the gain potentiometer on the amplifier. With liquid in the sensor gap, the loop gain is greater than unity, and the system oscillates at its resonant frequency. This is detected on the output of the amplifier and an output relay is energized. When air only is present in the gap, the coupling between the amplifier output and the input is from the cross-talk between the two coaxial cables to the sensor, and from ultrasonic energy that travels from the transmitter crystal down the 'bridge' to the receiver crystal. Direct coupling through the air is negligible.

All the components in this feedback loop have resonant frequencies that can vary with temperature and liquid type. The stainless steel windows of the transducer itself must be machined to the correct thickness; but more surprising is the effect of the gap between the two transducer faces filled with liquid. Because of the mismatch between the acoustic impedances of steel and water, or any other liquid, standing waves are set up of multiple reflections backwards and forwards in this gap. By using a self-tuning self-seeking resonant feedback circuit, the oscillation settles on the frequency that gives the highest loop gain, and tracks this with temperature and variations of fluid composition which affect the speed of sound.

Different designs of sensor can be produced, based on this same principle, using bodies made from different materials, for example plastic or epoxy resins or PTFE. There are stainless steel sensors of different designs, where the 'bridge' is less significant in producing false dry signals. The main advantages for the user are that the sensor is small, has no crevices or moving parts and is suitable for both conductive and non-conductive liquids. It ignores foam and moisture droplets and is unaffected by most coatings. The disadvantages reside in the relatively low maximum temperature for the standard unit (150°C) and the effects of bubbles or suspended solids. Either of these attenuates the signal passing between transmitter and receiver because the second phase scatters the ultrasound. To overcome this it is necessary to increase the gain of the amplifier, and usually a different sensor design is used for such liquids where the 'bridge' is less significant, producing less coupling, so that the amplifer gain can safely be increased to a high level. Special sensors have therefore been developed specifically for aerated or carbonated liquids and for liquids containing suspended solids (Figure 9.9), using extended bridges. The use of epoxy resin or plastic sensor housings improves the acoustic match between sensor and liquid, therefore increasing the signal obtained in a liquid.

Applications for this type of sensor in the food industry abound, both for the standard sensor designs and for designs based on these principles for specific applications. General purpose high-level, low-level and pump control points are detected in all types of beverage, including carbonated drinks. The small size of the sensors is used to insert them into discharge lines from blending tanks, to ensure that the tank is empty. A specific example of this was in a whisky blending plant, where the requirement was that the sensor should not treat whisky foam the same as pure liquid. In fact, this type of sensor treats foam the same as air. HM Customs and Excise required the control system to indicate tank empty, but did not wish to insist that all the foam had been drawn off too. Similarly, sensors are used in delivery pipelines to alarm when the pipe is no longer full, when a flow meter reading will no longer be valid or when a coolant flow has failed.

Specific sensor developments have been made for cryogenic applications, to control the level of liquid nitrogen, for example, in vessels for freezing fresh fruit such as raspberries. For beer brewing systems the insensitivity to foam has been used to advantage, with a sensor on the end of a tape to allow hand dipping of the tanks. This replaced the standard HM Customs and Excise approach, using a dipstick, which is relatively difficult to apply in certain real ales. Another special sensor was developed for one customer based on a section of pipe, with the piezoelectric crystals potted into recesses machined in the outside of the pipe walls. The inside of the pipe was therefore of normal smooth bore, which achieved the very high standard of hygiene required on this plant. Transmitting through a pipe wall from the outside is relatively easy, although limited to pipe diameters between 20 and 75 mm. The standard approach uses two separate epoxy resin transmitter and receiver blocks. These are bonded to the outside of the pipe wall, while held accurately parallel. This technique is regularly used on sight glasses attached to beer vats, to give high- and low-level alarms for automatic control. It is particularly useful when the level is no longer visible to an operator because of discoloration of the sight glasses.

A recent development has been to turn the ultrasonic sensor 'inside out' by mounting the piezoelectric crystals on the inside of a cylinder (Figure 9.10). The ultrasonic beam travels around the cylinder as a plate or Lamb wave (see Chapter 8, Section 3.2), but can be thought of as reflecting backwards and forwards within the thick wall of the transducer. At each outside surface reflection, where the sensor is immersed in liquid, a proportion of the energy is transmitted into the liquid. This is lost in the surrounding volume, and the wave is attenuated compared with the direct coupling between the two transducers when the sensor is in air. Since this

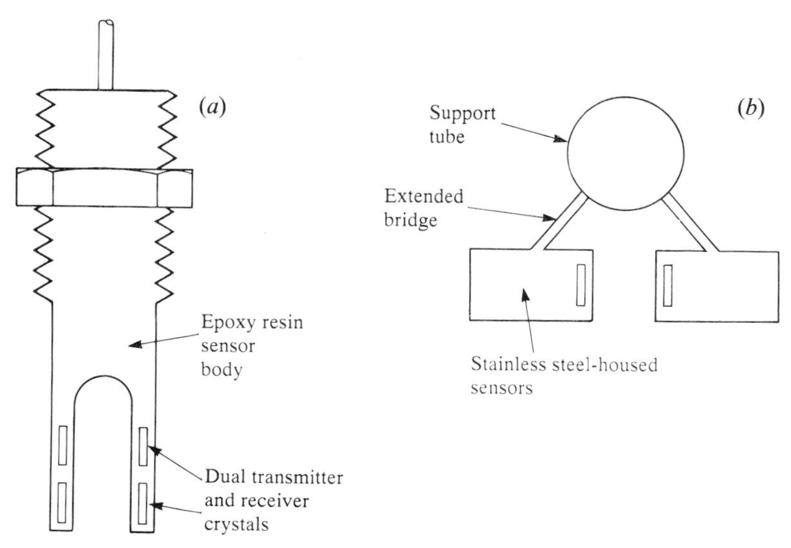

Figure 9.9 *Special sensor designs: (a) sensor for aerated liquids, (b) top view of sensor for liquids containing suspended solids*

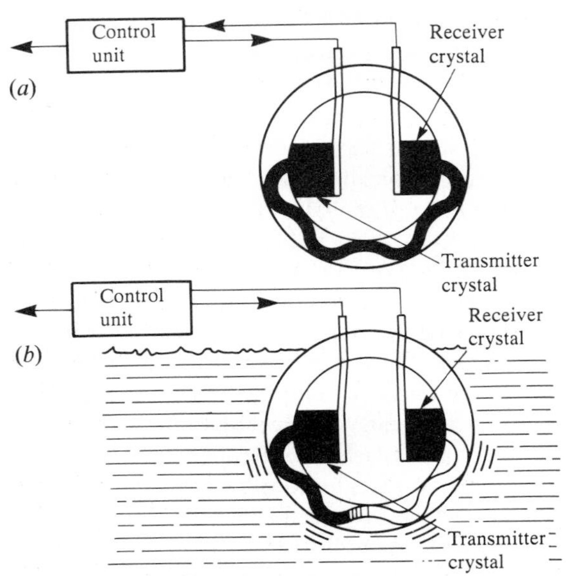

Figure 9.10 *Mobrey Hi-Sens sensor. (a) Sensor in air: ultrasonic signal reaches receiver crystal. (b) Sensor in liquid: ultrasonic signal is lost into the liquid and does not reach receiver crystal*

transducer is active when it is dry, it is suitable for a fail-safe high-level alarm, and this is its main duty. In the food industry, the main applications are in overfilling alarm systems on process vessels or cooking vessels.

A major limitation of the ultrasonic technique must be mentioned, which particularly relates to gap-type sensors. Because these sensors rely on transmitting the ultrasonic signal between the transducers, through the liquid, the assumption is that the liquid will transmit this sound relatively easily. Special sensors have been reported for use in liquids with suspended solids or for carbonated liquids. A more severe problem occurs in the dairy industry with milk containing a high proportion of cream. Either because this cream produces an emulsion layer, where the second-phase fat droplets act as scattering centres for the ultrasound beam, or because the highly viscous cream layers trap air bubbles, following agitation or splashing of the milk, the ultrasonic beam is highly attenuated. In many cases ultrasonic gap sensors have been found unsuitable for level detection in such dairy applications. The possibility of using this effect to detect the difference between cream layers and milk is a form of interface detection (see Section 3.4).

3.2.2 Pulse-attenuation-based Level Detectors

The old school of practical engineers will tap the outside of a tank and listen for a different 'dull' sound when the tapping is below the unseen liquid level. This technique is taken one step further using an ultrasonic transducer and 'near wall attenuation' measurements as originally proposed by UKAEA Harwell (Asher 1983). A high-frequency (2 MHz) pulse is coupled into the vessel wall from an externally bonded transducer. If the internal face of the wall is under the liquid, some of this pulse is transmitted into the liquid and lost in the volume of the vessel. If the internal face is in air, the ultrasound pulse is all reflected back towards the transducer. The proportion of the ultrasound reflected from this boundary is dependent on the acoustic impedance mismatch between the vessel wall and the liquid or air behind the wall. For any gas, the acoustic impedance (which is the product of sound velocity and density) is so low that the reflections within the metal wall are total. For a liquid, there is still a considerable mismatch, but the proportion reflected is lower than 100 per cent, and after many traverses of the vessel wall, the ultrasonic pulse amplitude is significantly reduced (Figure 9.11). By adjustment of the signal amplitude threshold for detection and the time delay before a measurement is taken, a non-penetration liquid presence detection system can be produced.

The advantages of this technique are that it is hygienic, it requires access to one outside surface of the vessel only, and it can be retrofitted by external clamping or bonding to existing tanks or pipework. A significant application in the food industry has been in fitting level control sensors to refrigerant liquid tanks on commercial cold cabinets and freezer installations. The disadvantages are those of on-site adjustment and set-up procedures which are an inevitable part of any system that involves the vessel wall as part of a transducer system.

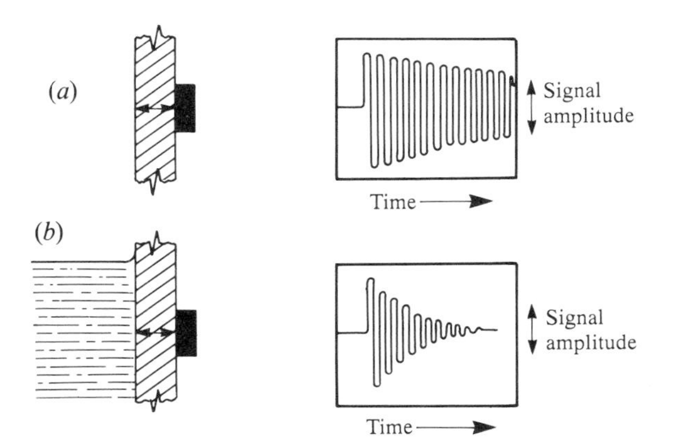

Figure 9.11 *Near wall attenuation level sensor: (a) vessel empty, (b) vessel full*

3.3 Liquid Level Measurement Systems

The investigation of liquid level measurement using high-frequency ultrasound either in the liquid itself or in the vessel wall (or equivalent) has produced many prototype techniques based on research but not many practical or commercial systems.

By monitoring the attenuation or reflection of plate (Lamb) waves (see Chapter 8, Section 3.2) in a metallic vessel wall or probe inserted into a vessel, the change occurring when this surface is submerged in a liquid can be measured. Analogue or time-based measurements can be used to deduce a liquid level. The technique has been used to monitor ultrasound pulses in a vertical rod inside the vessel, penetrating into the liquid. System accuracy was enhanced by having a magnet in a float on this rod and detecting the magnetic/ultrasonic discontinuity effects at the surface. Such techniques have been limited to measurements in the nuclear industry or for other high-specification applications, because of the high cost.

Ultrasonic pulse-echo techniques using a transducer in the liquid have been much more successful in the brewing industry in particular. In this system, the ultrasonic pulse from a bottom-mounted transducer is timed up to the liquid surface and back through the liquid. Because it is necessary to know the speed of sound in the liquid, a separately mounted transducer on the side of the vessel is used to measure the time delay across the vessel in a horizontal plane – a known distance. The microprocessor-based electronics then compute the required answer. This technique does not suffer from the surface foam problem that severely restricts airborne pulse-echo systems in this industry. Using a liquid as the transmission medium, the transducer is more efficient. The signal attentuation in a clear liquid is low, but becomes higher when any gas-producing fermentation occurs or when any solids are in suspension, because of the scattering produced by these second phases. In tanks where the sediment build-up is considerable, the transducers occasionally have to be positioned off the bottom of the tank. The technique has been used widely in breweries; usually the electronics are switched between tanks to provide level information at specific stages in the process.

3.4 Suspended Solids and Interface Detection Systems

The technique previously described for liquid level switches – the monitoring of ultrasonic transmission amplitude between transmitter and receiver crystals – can be used to detect different characteristics of the liquids being monitored. The most significant measurement of this type is of the concentration of suspended solids, which increase ultrasound attenuation because they act as scattering centres. Similar systems have also been used to detect interfaces between clear liquids, by use of either the different

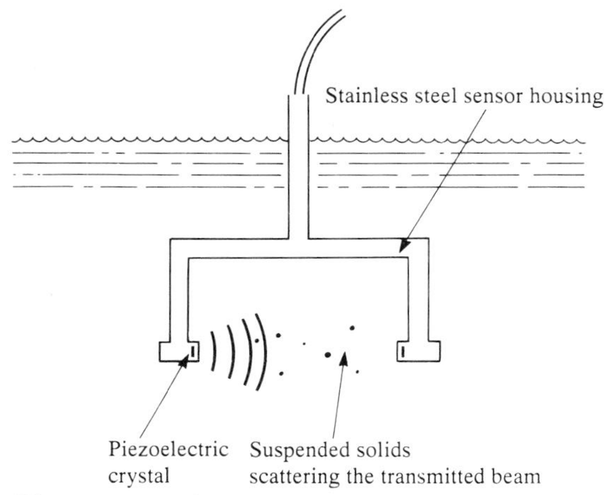

Figure 9.12 *Sensor for blanket detection in sedimentation tanks*

attenuation in each liquid or the reflections that occur at the interface. An emulsion contains two phases separated by many interfaces. The increased attenuation due to scattering provides a means of distinguishing the emulsion from the single-phase liquid without second-phase inclusions. A sensor used as an ultrasonic attenuation monitor is typically either suspended at a fixed point in a tank or used to monitor a liquid flowing down a pipeline. For either case, a gap between the sensor faces of between 100 and 300 mm is typical, to provide an adequate liquid sample length. For a suspended sensor in a settlement tank, a 150 mm gap device would be used, constructed as a fork arrangement (Figure 9.12). For pipeline use, the sensors would be mounted in bosses welded on to the side of the pipe, and would have their faces flush with the inside wall to minimize flow obstruction and silting up of the sensors (Figure 9.13).

3.4.1 Suspended Solids Measurement

In practice, it is found that the attenuation produced by suspended solids particles, when expressed in decibels (dB), is:

- proportional to the gap between the sensors over the range 20–600 mm
- proportional to the ultrasonic frequency over the range 1–4 MHz (wavelengths approximately 0.5–2 mm)
- proportional to the percentage of solids by weight in the solid–liquid mixture, over the range 0.5–50 per cent.

The proportionality of ultrasonic attenuation to weight percentage of solids applies to solids in suspension that have submillimetre particle sizes, that is radii smaller than the wavelength. The excess attenuation is a result of ultrasonic scattering rather than beam interruption or diffraction, which occurs when the wavelength is smaller than the particle size (Flammer 1962; Lenn 1987).

Different solids in suspension have different characteristics of attenuation, which relate to the size and density of the solid particles and the acoustic properties of the solid material. In general, practical experimentation is necessary to establish these characteristics to choose the correct combination of frequency and sensor gap so that the monitoring electronics give an output of percentage solids over the required range.

There are two basic control output requirements for suspended solids monitors, either a relay high/low output or an analogue output proportional to suspended solids. In the simpler relay output system, a gain control is used to set a fixed attenuation level, which defines the transition from high to low solids concentration. This is used in settling tanks typically as a blanket level alarm, to trigger or stop the desludging of a tank.

In the water industry, such ultrasonic sensors are used for water treatment floc blanket alarms, to prevent the floc from spilling over into the water supply system. Similarly, they are used for control of sand filter backwash systems, alarming when the bed is fluidized too much and in danger of being washed away. Similar techniques have been investigated by the Brewing Research Foundation for use in brewery fluidized bed backwash systems. More severe sludges are met in the sewage industry, where primary sedimentation tank control systems are normally based on ultrasonic attenuation sensors. Many food processing sludges can be treated in a similar way. For example, Australian sugar cane processing uses ultrasonic

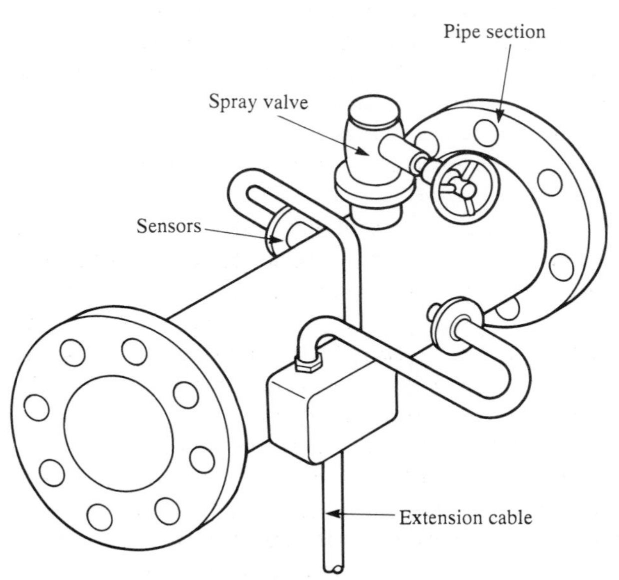

Figure 9.13 *Suspended solids measurement using pipe-mounted sensors*

blanket detectors in the subsiders, to separate juice and suspended impurities.

The analogue output system for measuring the percentage solids of a slurry has been used primarily on mineral slurries such as china clay and coal tailings, but has also been used in the food industry. In the UK sugar-beet industry, lime additions are monitored using these techniques. In the brewing industry, recirculation of the yeast in the sediments via membrane filters is controlled based on such measurements. Other applications have been found in peanut processing in the USA.

The electronics used to monitor the ultrasonic attenuation are typically designed to overcome the variations produced by changes in the temperature and in the concentration of dissolved salts or other solutes in the liquid (and thus changes in the speed of sound), and those caused by changes in the transducer window thickness and spacing. The technique will not be affected by flow or movement of the liquid. To a small degree, the acoustic properties of the carrier liquid will affect the measurement, so that particularly for low percentage solid applications, the zero solids adjustment should be made in the appropriate mother liquor. The major factor responsible for erroneous readings in most applications is another type of suspended scatterer such as small gas bubbles, or a reflector in the beam such as a large solid (e.g. a stone) or large air bubbles. The installation must be made remembering that the effect of air bubbles is severe; they also attenuate ultrasound. This means that the technique is not applicable to monitoring biologically active liquids, for example fermenting liquids or digested sewage, or liquids where gas bubbles are produced, as in cooking processes for marmalade or jams. Where the suspended solids themselves contain, or can trap, variable amounts of air, the slurry must be monitored with care, and not allowed to free-fall or be subject to violent agitation with possible air entrainment before it is monitored.

3.4.2 Liquid Identification and Interface Detection Systems

Ultrasonic attenuation measurements can, in theory, be used to identify the difference between two liquids by the different signal attenuation observed on placing a given ultrasonic sensor in the two liquids. This effect is a result of the different acoustic impedances (functions of bulk modulus and density) of the two liquids. These determine the proportion of the signal energy transmitted from the stainless steel transducer face into either of the liquids. Practical applications for this type of detection exist in distinguishing between crude oil or diesel fuel and sea water, and this is used extensively in offshore storage systems, for example in caissons in the Persian Gulf. Only when the two liquids are significantly different, as in this case of oil and water, can this technique work reliably, and no practical food industry applications in current use are known to the author.

If, however, the interface between two liquids itself is used to cause a

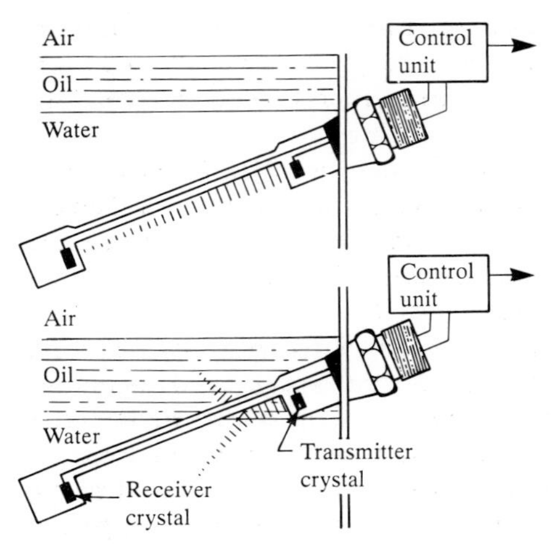

Figure 9.14 *Interface detection system by reflection*

discontinuity, which interrupts or diverts the ultrasonic beam, then a significant signal loss between transmitter and receiver can occur, which is the basis of a practical system. This is frequently used for interface detection on immiscible liquids where the sensor is angled at approximately 15° to the interface (Figure 9.14). If the interface lies across the beam, the ultrasonic signal is either reflected or refracted away from the receiver; no signal is detected. This is used in liquid/liquid separation systems to monitor interfaces, for example in the extraction of flavours and fragrances, or to control the thickness of acetone or similar organic liquids in biochemical culturing vessels.

In an emulsion produced by agitation or mixing of two immiscible liquids, the interfaces between these two liquids can affect an ultrasonic beam significantly, depending on the volume fraction and the size of second-phase globules, and on frequency (see Chapter 8, Section 5.3). As a result, the ultrasonic transmission can be highly attenuated. This is used for detection systems in the petrochemical industry, giving an alarm output indicating the presence of an oil/water emulsion in the discharge line from a separator, for example. Such systems are less applicable in the food industry, except perhaps in the dairy sector, where the effect is potentially usable for the detection of cream layers and suspended inhomogeneously distributed fat inclusions.

3.5 Concentration Measurement Systems

A pulse of ultrasound travels through a liquid at a velocity dependent on physical properties of the liquid, namely the bulk modulus of compressibility and the density. These are not typically the parameters used as inputs to a process control system. However, they vary with concentration changes of dissolved salts, or other solutes, and as a function of temperature.

A sensor for measurement of the speed of sound in a liquid is typically a flange-mounted unit fitted with an ultrasonic transducer and a reflector. The transducer acts as both a transmitter and a receiver monitoring the pulse echo from the reflector, allowing the speed of sound to be monitored over a path length of approximately 100 mm total.

Separately, a temperature sensor provides additional information to the microprocessor-based control electronics. The microprocessor uses pre-programmed look-up tables, to derive a value for the concentration of the known solute from the sound velocity and temperature measurements.

The major limitation of this technique is that solute composition must be known and that the characteristic curves must be available from suppliers' test data, or these curves must be experimentally established for the solvent-solute system involved. The results will then apply to that solution provided that no other solutes or solvents are present in variable amounts, since these would produce their own variations in the speed of sound. The major application is therefore in simple mixtures, that is one solute and one solvent, preferably over a limited temperature range.

Current food industry applications are in the measurement of sodium hydroxide concentration for making up cleaning solutions (Figure 9.15), monitoring the concentrations of sucrose solutions and sodium chloride solutions, and monitoring coffee extract and corn starch concentrations. Further applications are discussed in Chapter 8, Section 3.4. UKAEA

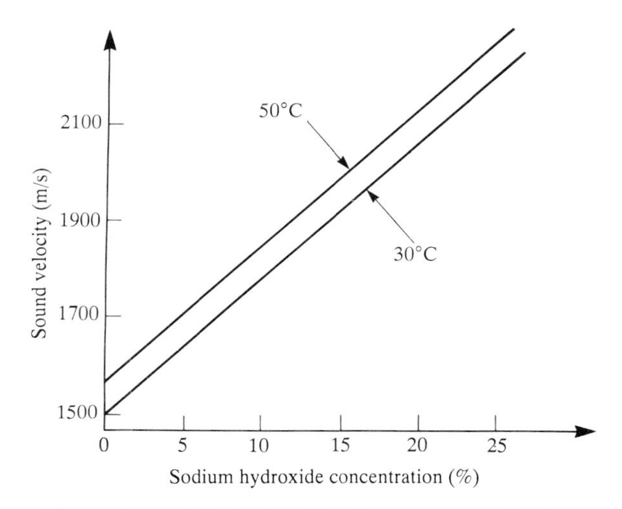

Figure 9.15 *Relationship between speed of sound and concentration for sodium hydroxide (schematic representation)*

Harwell has produced special measuring systems that either operate across a specially constructed section of pipework or can be mounted on existing pipework in some cases.

3.6 Future Developments in High-frequency Instrumentation

In high-frequency techniques, the main future developments will relate to the use of clamp-on transducers because of the hygiene advantage this offers. The variable nature of the bond between transducer, pipe or vessel wall, and liquid gives rise to signal amplitude variations. Techniques that use a time measurement will therefore be easier to apply than those that rely on an amplitude measurement. Already, the Doppler flow meter (which measures a frequency shift) is available as a clamp-on system. Time-of-flight ultrasonic flow meters will also be developed as clamp-on devices, probably with microprocessor-based electronics that allow a readout of flow velocity and possibly pipe wall thickness (to allow flow volume computation). Unfortunately, this development is likely to be aimed more at industries other than the food industry, and will probably remain at the high end of the flow metering price range.

The speed of sound measurement system for concentration measurement, for example, is applicable to the same design of clamp-on transducers, and could use the same signal processing electronics. However, again real development of the technique is limited by the relatively low demand for the resulting equipment, which will inevitably be in the higher price bracket.

Ultrasonic signal attenuation measurement systems offer several potentially useful applications in the food industry, for which the basic research work and application engineering necessary are being undertaken by various research laboratories. Through tripartite cooperation between these laboratories, instrument suppliers and process plant users, useful and viable developments could arise. However, some of the techniques described here have been developed originally for use in other industries, and then have found spinoff applications in the food industry after only slight modification. The ultrasonic point level sensor is an example of this, and developments can be expected in single-sided clamp-on level switches or special hygienic sections of pipework incorporating such sensors.

4 Contacts for Further Information

The science of ultrasound propagation applied to measurement or sensing is fascinating in any industry, but is particularly relevant to the food industry with the potential advantages described above. For practical applications of the standard techniques, the appropriate manufacturers can provide literature and give advice based on current experience (KDG Mobrey 1990: Nusonics 1985: Platon 1992). For research into specific problems and for advice on the more difficult industrial applications there are several centres of ultrasonic expertise. For the UK, these are associated with the nuclear industry (for example, Instrumentation and Applied Physics at Harwell, and the Northern Research Laboratories at Risley) or are based in academic institutions (Department of Fluid Engineering and Instrumentation, Cranfield, and departments at Bradford and Leeds Universities and Teesside Polytechnic). Food industry applications have been pioneered by groups working at the Leatherhead Food Research Association, the Brewing Research Foundation, the Institute of Food Research Laboratories at Norwich and formerly at Langford near Bristol. The work of the Langford group on ultrasonic measurements on meat has now moved to the University of Bristol (C. A. Miles). The Campden Food and Drink Research Association has also begun ultrasonic studies.

Ultrasonics today offers a significant range of useful instrumentation to the food industry. The standard products have been described, with information on their limitations and likely developments. More specialized products are also available, but need care in their applications at present. Once their valid applications are established by practical testing, these will become the hygienic, non-contact, non-penetration sensors for many of the required process parameters of tomorrow. Ultrasonic techniques are growing in importance to the food process engineer and can give significant advantages in this industry.

References

Asher, R. C. (1983) Ultrasonic sensors in the chemical and process industries. *Journal of Physics E: Scientific Instruments*, **16**, 959

Cousins, T. (1978) The Doppler ultrasonic flowmeter. In *Flow Measurement of Fluids*, H. H. Dijstelbergen and E. A. Spencer (eds) North-Holland, 513–518

Flammer, G. H. (1962) Ultrasonic measurement of suspended sediment: general geology. Geological Survey Bulletin 1141, US Government Printing Office

Kaye, G. W. C. and Laby, T. H. (1986) *Tables of Physical and Chemical Constants*. Longman, London and New York

KDG Mobrey (1990) Ultrasonics for level and flow measurement, suspended solids measurement, interface detection. KDG Mobrey Ltd, Slough

Lenn, C. P. (1987) *In situ* particle size measurement by scattering of ultrasound from acoustically hard spheres. In *Particle Size Analysis* (in Lloyd, P. J. (ed.) below) 647–659

Lloyd, P. J. (ed.) (1987) *Proceedings of 1985 Conference of the Analytical Division of the Royal Society of Chemistry*. Wiley

Milltronics (1990) Ultrasonic level measurement systems for grain silos. Milltronics Ltd, Ontario, USA

Nusonics (1985) Ultrasonics for concentration measurement. Nusonics Inc., Tulsa, Oklahoma

Platon (1992) Ultrasonics for liquid level measurement. Platon Instrumentation Ltd, Basingstoke, UK

Scott, M. J. (1979) The Doppler ultrasonic flowmeter. *Chartered Mechanical Engineer*, October

Part II

Instrumental Techniques in the Quality Control Laboratory

10 Rheological Measurements

Brian M. McKenna

Contents

1 Introduction

Food rheology is the study of the response of foods to an applied stress. Escher (1983) gives the following purposes for which such responses are important:

- plant design: pumps and pipe sizing and selection, heat and mass transfer calculations, filler designs
- quality control of both raw material and the product at different stages of the process
- quantitative measurement of consumer determined quality attributes by correlating rheology measurements with sensory data
- assessment of food structure (Hermannson 1980; 1983).

Food rheology is often confined to the behaviour of liquid foodstuffs. However, there is an increasing tendency to consider the response of both solid and liquid materials to applied stresses and strains as being two extremes of the same science. There are in fact some foods that will exhibit either behaviour depending on the stress applied; molten chocolate, fat based spreads, mashed potato and some salad dressings will exhibit a solid-like behaviour at low stresses and a liquid-like behaviour at high stresses (Mitchell 1984). This tendency is increasing as more food products are developed that would be classed by the consumer as being semi-solid or semi-liquid. A more exact definition would therefore be the study of both the elastic and the plastic properties of foods. It is proposed, however, to place the emphasis of this chapter on classical liquid rheology, leaving the detail of solid properties to Chapter 11 on texture measurement.

There are many substantial reviews of basic rheology (Prentice 1984; Sherman 1970; Rao 1977), and while the objective of this chapter is to review instrumentation and measurement techniques in food rheology, it is nevertheless necessary to briefly consider some of the fundamentals. It is also necessary to justify the need for measurement given the wealth of published data already available. For a selection of these see Steffe et al. (1986), Sherman (1970) and the bibliography of McKenna (1990). The reason for this is as stated by Prins and Bloksma (1983); 'Rheological measurements have to be made under the same conditions as those which exist in the system studied.' In other words, there is little use in carrying out measurements on a product or extracting values from the literature, if the stresses used and their rates of application during the measurement differ from those in the process or calculation for which the measurement is required.

2 Relevance of Rheological Properties of Foods

The relevance of food rheology has been summarized above into the four categories of plant design, quality control, sensory attributes, and the research and development of food structure.

2.1 The Consumer's Perception

Ultimately the food product must be eaten, so sensory attributes become the most important. However, *en route* from the farm to the mouth the product may have to be pumped, heated, stored or subjected to other processes, and must be amenable to flow when being placed in a container/package. Equally important is its ability to flow out of the container before consumption. Indeed, it is this ability (or the occasional lack of it) that first brings the consumer into a direct and sometimes frustrating contact with rheological principles. How often has the consumer experienced the dilemma of tomato ketchup refusing to flow from its bottle and found that the application of a sharp blow to the bottle base resulted in an excess amount being deposited on the plate? This provides an excellent example of a situation in which a product has a yield stress below which it will not flow, but flows perhaps too well once the consumer unknowingly provides the stimulus that exceeds it. Not only does this example illustrate yield stress, but it also shows the relationship between force and deformation and flow!

This simple example also gives emphasis to one of the basic rules of rheological measurements, namely that the product should be tested under a range of conditions of stress and shear rate that reflect those experienced during subsequent use, whether that use be tasting, pouring, shaking, stirring or any other action that requires movement of the material.

Of course, rheological relevance does not stop when a food reaches the plate but influences the sensory perception or 'mouthfeel' of the product. Matz (1962) defines mouthfeel as the mingled experience deriving from the sensations of the skin of the mouth after ingestion of a food or beverage. It relates to density, viscosity, surface tension and other physical properties of the material being sampled. These relationships between rheology and mouthfeel have been the subject of extensive research, recently reviewed in the author's bibliography on food rheology (McKenna 1990). It will, however, be obvious that a change in the manner in which a food may move or flow in the mouth and throat will influence our perception of it as a desirable food.

2.2 The Requirements of the Processor

Rheology assists the processor in several areas. At the plant design stage, pumps, pipes, heat exchangers, stirrers etc. need to be selected.

The rate at which a liquid will flow in a pipe is highly dependent on its rheological properties (see equations (7) to (13)). An alternative interpretation is that for a given flow rate of a given liquid, a particular pressure drop will be required along the length of the pipe. This, in turn, will influence the quantity delivered by the pump chosen for the system. The nature of the process itself may lead to further influences. For example, if the pipe system leads to or through a heat exchanger, the rheological properties may be changed by the heating effects and lead to changes in the flow system. In the extreme case of a large, heat induced reduction in viscosity, the velocity of flow might increase and give a product that has had too short a residence time in the system. More seriously, the rheological changes may lead to a change in the flow (velocity) profile of the liquid in the system, to a change in the residence time distribution, and again to an under-processed product. Of course, not all heat induced changes lead to 'thinning' of the liquid. Starch gelation and similar processes can induce the opposite effects, slow the liquid and increase the severity of the heating process.

There are, however, many other rheological problems in processing. Yield stress, as exhibited in the ketchup example above, may lead to more serious processing problems with significant economic relevance. Enrobed food products, which may range from chocolate enrobed confectionery to batter enrobed fish or meat products, all demand an enrobing material that exhibits a yield stress. If this yield stress is too low, the weight of enrobing liquid adhering to the sides of the product will induce a stress in excess of the yield stress, either on the vertical side of the product or on a plane parallel to this within the enrobing material, and will cause the material to flow off the product. Conversely, too high a yield stress will lead to excessive thickness of enrobing material – possibly attractive to the consumer of a chocolate bar, but with adverse economic consequences for the processor.

Quality control is also an area of rheological significance for the processor. While there is the obvious need to induce the desired characteristics into the product and to test the product for these attributes, rheology can provide other quality control information by drawing on the wealth of correlations between rheological and other data that have been developed over many years. For example, Sharma and Sherman (1966) have shown that for ice cream the rheological measurements correlate with fat droplet size, the amount of air incorporation (overrun), ice crystal size and product temperature. For chocolate, information on the hardness and consequently the fat composition of the major ingredient, cocoa butter, can be deduced (Lovegren et al. 1958), while many of the processing influences are detailed in the extensive series of monographs published on chocolate rheology (Sherman 1970; Elson 1971; Kleinert 1976; Tscheuschner and Wurische 1979).

The dairy industry also provides many examples of the use of rheological control techniques. While there are obvious textural related rheological attributes for both set and stirred yoghurts, the ever increasing range of dairy based spreads demands that the successful product should have the correct viscoelastic properties for spreadability. Soft and cream cheeses also have liquid properties that must be kept within chosen ranges and which are highly dependent on the ongoing microbiological activity, proteolysis and syneresis within the product. Less obvious is the need for rheological control of concentrated milk products during evaporation and drying; changes in the rheology will change the drop size range produced by the spray drier atomizers (McKenna 1967), which will in turn change the particle size distribution in the finished powder, not only altering its bulk density and ease of reconstitution but also leading to increases in powder losses in the final air–powder cyclone separators.

Probably the most extensively researched area of food rheology has been that of dough of various types. Typical of the many reviews of this topic are those of Sherman (1970), Rasper (1976) and Baird (1983). Dough rheology is important because of its influence in determining the texture of the bread crumb produced and also on the final volume of the baked product. Indeed, this area is increasing in importance because of the continuing developments in snack foods produced by extrusion cooking. In this situation, the expansion of the product as it passes through the extruder die is dependent on the viscoelastic properties of the dough as is the flow behaviour of the paste within the screws of the system. Such has been the interest in dough rheology that a series of specialized instruments have been developed over the years to monitor these properties. Unfortunately, while they are widely used, many of the properties measured are machine specific and are not the absolute properties defined in the next section.

3 Basic Rheology

As has been stated, rheology is the response of a liquid to an applied stress (normally a shear stress). Foods of differing internal structure and bonding will react in different manners to an applied stress. In the simplest case the shear stress developed in the fluid is directly proportional to the rate of deformation or the rate of strain. In such cases, the liquid is said to be *Newtonian* and obeys the relationship

$$\tau = \mu \dot{\gamma} \tag{1}$$

where τ is the shear stress and $\dot{\gamma}$ is the shear rate. (For a summary of symbols see the Appendix.) Such a relationship is shown by line (a) of Figure 10.1. In SI units, τ will normally be in pascals (Pa), $\dot{\gamma}$ in reciprocal seconds (s^{-1}) and μ in pascal seconds (Pa s). The constant of proportionality μ between the shear stress and the shear rate is termed the viscosity of the fluid, and from the 1663 definition of a fluid by Pascal can be viewed as a measure of its internal friction. The nature of this friction can be demonstrated by a hypothetical experiment shown in Figure 10.2, in which a fluid is placed between two infinite parallel plates (Leninger and Beverloo

1975). The bottom plate is stationary while the top plate moves in the x-direction at a fixed velocity. A force per unit area perpendicular to the y-axis is required to maintain movement of the top plate. This is termed the shear stress and is given the symbol τ. Force balances will demonstrate that this same shear stress is exerted between any two layers of fluid parallel to the plates. A velocity gradient or shear rate exists which is simply the difference in velocity of the two plates divided by the distance between them. In such a simple experiment it is assumed that the fluid at each plate has the same velocity as the plate, in other words that there is no slip occurring between the fluid and the surface.

One might be forgiven for thinking that inclusion of the early Pascal model of Figure 10.2 is using too early a concept in a chapter on modern instrumentation. However, while the modelling of fluid behaviour has progressed significantly since that time, the basic principles of many instruments still use the two-surface concept, one moving and one stationary, with the fluid being characterized by force measurements at one of the surfaces. It also highlights the significance of slippage between the fluid and the instrument surfaces.

It is now useful to return to Figure 10.1 and the concept of a Newtonian fluid in which there is a fixed proportionality between shear stress and the applied shear rate. Because of the relatively simple form of the flow curve, such liquids can be characterized by a single term, namely the constant of proportionality or the viscosity. More importantly, a single experiment such as the measurement of the shear stress at one surface at a *single* shear

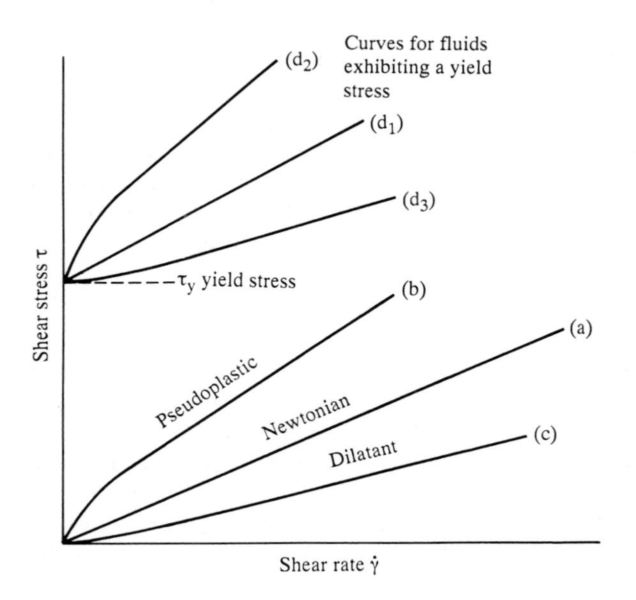

Figure 10.1 *Typical flow curves*

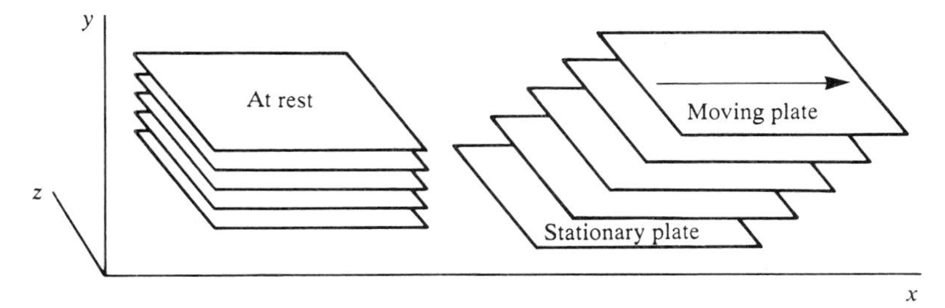

Figure 10.2 *Model system*

rate is sufficient to quantify the rheological characteristics of the fluid. However, only a few liquids follow this simple relationship (water, unconcentrated milk, vegetable oils, some dilute solutions); most foods may be classified as non-Newtonian and exhibit responses (termed flow curves) such as curves (b), (c) and (d). It will be shown that such fluids cannot be characterized by a measurement at a single shear rate as can the simple Newtonian fluid, and it is in ignoring this requirement that measurement errors are most often made in the food industry. Furthermore, there are many food liquids in which shear stress is not only determined by shear rate but is also time dependent, a factor which demands its own unique measurement system.

Many foods are termed *pseudoplastic* and their response to an applied deformation varies with the rate of application of the deformation. Typically, plots or flow curves such as curve (b) of Figure 10.1 represent such fluids. Because the slope of the curve decreases as shear rate increases, the term 'shear thinning' is often applied to such fluids. Of lesser importance in the food industry are foods with curves of type (c), which are 'shear thickening' or *dilatant*.

Rather than apply polynomial regression analysis to obtain equations for such behaviour, it has been found more convenient to plot the logarithm of shear stress against that of shear rate. For most pseudoplastic or dilatant fluids this results in a straight line and leads to the equation

$$\tau = k\dot{\gamma}^n \qquad (2)$$

which is normally termed the power law equation. In this equation, n is the power law exponent and k is the apparent viscosity or consistency index. While this equation is a useful mathematical model for most foods, there is a theoretical objection to its use, namely that the dimension of k is dependent on the value of n. A Newtonian fluid would of course have an n value of 1.0 and k would equal its viscosity. For pseudoplastic fluids, n will lie between 0 and 1.0, while for dilatant liquids the value will be greater than 1.

Though widely used, the power law model is not the only available, and in some cases its two-parameter equation represents an oversimplification (Launay and McKenna 1983). Ree and Eyring (1958) proposed a three-parameter model

$$\mu = \mu_0 + (\mu_0 - \mu_\infty)\frac{\sinh^{-1}\beta\dot{\gamma}}{\beta\dot{\gamma}} \tag{3}$$

where μ_0 and μ_∞ are the Newtonian viscosities at zero and infinite shear rate, while β is a characteristic relaxation time. Obviously, such a model facilitates consideration of time dependent behaviour. Cross (1965) proposed a four-parameter model

$$\mu = \mu_0 + (\mu_0 - \mu_\infty)[1 + (t\dot{\gamma})^{1-n}] \tag{4}$$

where t is another relaxation time. However, while equations (3) and (4) give more precise modelling of the flow curves of many foods, the widespread use of power law values in engineering equations make equation (2) the most useful if not the most exact model. Neither do three- or four-parameter models imply a better understanding of the structure of the food in question.

Finally, one must consider the family of curves marked (d) in Figure 10.1. Such liquids exhibit a yield stress τ_y which must be exceeded before any deformation or flow can occur. For certain food processes (e.g. chocolate, confectionery and other coatings) the existence of a yield stress in the food is essential for application of the technology. Indeed, in the absence of rapid crystallization or solidification of a coating, the magnitude of the yield stress will determine the thickness of the coating on a vertical surface. If the weight of coating divided by the vertical area (i.e. the shear stress exerted by the coating itself) exceeds the yield stress, then the coating will flow off the product. If not, it will neither flow nor deform and will remain to set on the product.

Equations which describe such products mathematically are those of Casson (1959) and Herschel-Buckley (see Charm 1971):

Casson: $$\tau^{0.5} = \tau_y^{0.5} + k'\dot{\gamma}^{0.5} \tag{5}$$

Herschel-Buckley: $$\tau = \tau_y + k''\dot{\gamma}^n \tag{6}$$

where τ_y is the yield stress and k' and k'' are constants. While the Casson equation is widely used (particularly in the chocolate industry), the Herschel-Buckley equation has the added attraction of merely adding a yield stress to the power law model.

Time dependent behaviour of liquid foods is not considered in detail in this chapter. This is not because such aspects are unimportant for many foods but because, in steady state flow in pipes or channels in a food processing operation, little or nothing of time dependent behaviour is observed. However, in storage of foods these properties becomes increasingly important as the onset of undesirable change may limit the effective shelf life of a product.

In concluding this section on basic rheological concepts, attention must be drawn to the dependence of such behaviour on temperature. As internal friction is a molecular phenomenon, anything that alters molecular movement will influence internal friction. Consequently, the rheology of most liquid foods is highly temperature dependent. In particular, the viscosity of Newtonian liquids exhibits such a dependency, as does the consistency index or apparent viscosity of power law fluids. The power law exponent is, however, relatively unaffected. No attempt will be made to quantify this phenomenon mathematically or to give a thermodynamic explanation for its existence. It is merely highlighted here to stress the importance of temperature control on the accuracy of *any* of the experimental rheological techniques detailed in later sections. For example, since the viscosity of water at 20°C (293 K) will change by 2.5 per cent per kelvin temperature change, an accuracy of 0.1 per cent in the measurement of this viscosity will demand temperature control to within 0.04 K. Many oils will change in viscosity by 10 per cent for each kelvin temperature change at 298 K (25°C), thus demanding temperature control to 0.1 K for a 1 per cent accuracy. It should be assumed that close temperature control is an essential feature of any of the measurement systems described in the following section.

4 Measurement Systems

Measurement systems can be classified into three groups: (1) capillary viscometers that make use of gravity flow in capillary tubes for the measurement process; (2) rotary viscometers in which the liquid is enclosed between rotating surfaces; and (3) other minor systems that may measure a rheologically affected phenomenon as a correlate for the desired variable.

4.1 Capillary Viscometers

4.1.1 Theory

Capillary viscometers are the simplest form of viscometer available from which it is possible to obtain absolute values of viscosity for Newtonian fluids and to obtain limited information on power law fluids. The basic measurement made is of the time t taken for a fixed volume V of the test fluid to pass through a length L of capillary tubing. Relative movement takes place between the axial part of the sample and that in contact with the tube walls. The driving force used is normally gravity as determined from the hydrostatic head difference between two liquid reservoirs in the viscometer (see Figure 10.3).

From first principles it is possible to derive an equation for the flow rate of fluid through such a tube or pipe. For Newtonian fluids, this equation is known as the Hagen-Poiseuille law (Hagen 1839; Poiseuille 1841) and relates the flow rate to the driving pressure for flow, with many of the variables of such a system incorporated into the constants of the equation:

$$\frac{Q}{d^3} = \frac{\pi\, d\, \Delta p}{128 \mu L} \qquad (7)$$

where Q is the flow rate through the tube (m^3/s), d is the tube diameter (m) and Δp is the pressure difference across the tube (N/m^2). For a given instrument d and L are fixed, so by measuring Q at a known Δp the coefficient of viscosity μ may be calculated. Indeed, since the volume processed in a given instrument is fixed at V, then Q may be replaced by V/t, where t is the time required for the flow. Furthermore, the driving force for flow will normally be the hydrostatic head within the system and will be equal to the product $\rho g h$, where ρ is the liquid density, g is the gravity constant and h is the difference in liquid levels between the reservoirs of the system. It is then possible to simplify equation (7) and write it in the form

$$\mu = K \rho t \qquad (8)$$

where ρ is the density of the fluid under test, t is the time taken for the fluid to flow through the capillary tube, and K is a constant for the instrument given by

$$K = \pi g h d^4 / 128 L V \qquad (9)$$

This value is often supplied by the viscometer manufacturer. However, a common alternative approach is to use such capillary viscometers for comparative measurements against standard fluids of known viscosity. If the pressure difference causing flow is the same while measuring both fluids (atmospheric pressure and gravity flow are usually applied) then the ratio of the viscosity of the food sample to that of the standard fluid will be equal to the ratio of the time required for equal volumes of the fluids to flow through the viscometer tube. Similarly, such standard fluids may be used to compute or to check the value of K given in equation (9).

The equations above have traditionally been used not only for viscometry but also to quantify the flow rate in a pipe system by monitoring the pressure drop along a section of the pipe. However, as the following section will demonstrate, this method should only be used as a rough estimate with food liquids as their generally non-Newtonian behaviour will demand that more complex relationships be used.

The flow of more complex fluids is governed by variations on the above equation. For laminar flow of power law fluids through a cylindrical tube under the influence of a pressure difference Δp, the following equation is obtained:

$$\frac{Q}{d^3} = \frac{\pi}{8(3+1/n)}\left(\frac{d\,\Delta p}{4kL}\right)^{1/n} \quad (10)$$

where n and k are the power law constants. At constant temperature, the apparent viscosity, k will be constant, so a plot of log Q versus log Δp will give a straight line of slope $1/n$, with the value of k being abstracted from the intercept value of the plot:

$$\log \frac{\pi d^{3+1/n}}{(4kL)^{1/n}[8(3+1/n)]} \quad (11)$$

However, with such simple equipment, facilities are seldom available to apply different pressure differences so as to obtain the points for such a plot. A more limited possibility is to take a range of viscometers of different capillary diameters but similar tube lengths and then to test the power law liquid in each using gravity flow. A plot of log Q versus log d should then give a straight line of slope $3+1/n$. Again, k could be abstracted from the intercept value:

$$\log \frac{\pi \Delta p^{1/n}}{(4kL)^{1/n}[8(3+1/n)]} \quad (12)$$

A food liquid that behaves as Newtonian once its yield stress value has been exceeded (curve (d_1) in Figure 10.1) will have a characteristic behaviour equation as follows:

$$\frac{Q}{d^3} = \frac{\pi d \Delta p}{128\mu_p L}\left(1 - \frac{16\tau_y L}{3d\Delta p} + \frac{256\tau_y L}{3d\Delta p}\right)^4 \quad (13)$$

where μ_p is the slope of the straight line plot of shear stress versus shear rate once the yield stress had been exceeded. For a more complete discussion of this equation see Leniger and Beverloo (1975); Prentice (1984) details the flow of Herschel-Buckley and Casson liquids in tubes or capillaries. As this is a more complex equation and post yield stress linear behaviour is seldom experienced with food products, these simple viscometers cannot be recommended for examination of such products.

They are, however, widely used, often in circumstances where their limitations are not fully understood. This is because they are relatively cheap and are easily available from most laboratory supply companies. Indeed, when one considers the equations involved and the multiple measurements required for all but simple Newtonian fluids, the use of such unsophisticated equipment presupposes a knowledge of the basic behaviour of the fluid under test. In other words, these viscometers should only be used for known Newtonian fluids. This would confine their use to dilute solutions and vegetable oils. For other foods, they can only provide rough quality control tests.

The sizes of the food sample and of the constituents within the sample are important with viscometers of this type. As they rely on measuring the time taken for a given volume of sample to flow through the capillary tube, it is important to ensure that a homogeneous sample of the volume required can be obtained from the food. Difficulty may be experienced with foods containing large amounts of suspended solids. Indeed, suspended solids will contribute to large errors in the measured times if they are of a size that is significant when compared with the diameter of the capillary tube. Further, particles that effect laminar streamline flow within the capillary will change the time measurement. Of course, these comments are equally relevant to droplets within an emulsion as they are to solid particles. Care must also be taken to ensure that suspended particles within a food do not settle during the duration of a test. Nor should any separation occur within a food emulsion.

Examples have already been given that place general emphasis on the need for exact temperature control during measurements with this as with any type of viscometer. Prentice (1984) quotes an instance where temperature variations of $\pm 0.12\,K$ will alter the linearity of the flow curves obtained.

4.1.2 Instruments

Figure 10.3a shows a typical capillary viscometer. Often referred to as the

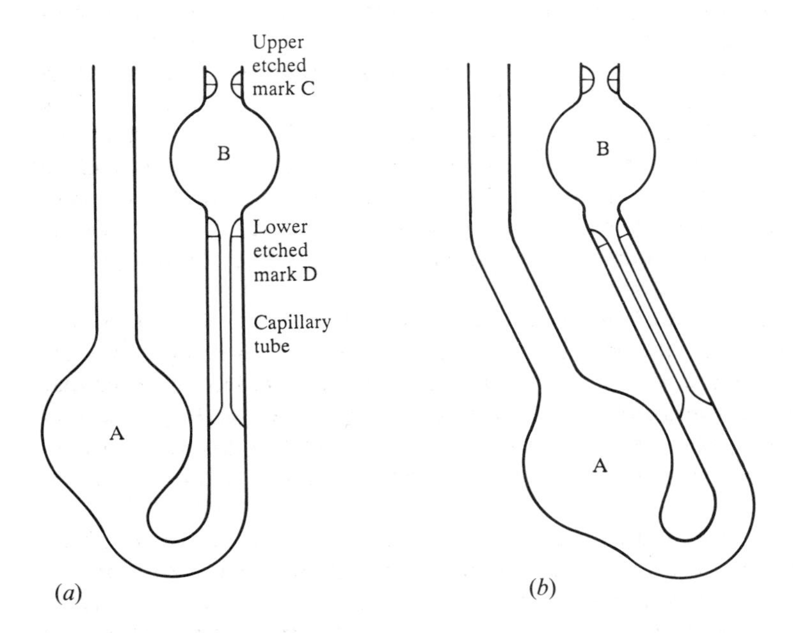

Figure 10.3 *Capillary viscometers: (a) normal (b) inclined tube*

Ostwald viscometer, it is the simplest available. However, there are many variations of this on the market, each of which would claim a special advantage and may have its own specific name applied by its manufacturers.

Operation is as simple as the design of the system. A standard volume of the test food liquid is entered into reservoir A of the viscometer and the U-tube below it. The instrument should preferably be held exactly vertical. If not, the support fixture should be such as to always hold the instrument at the same angle from the vertical. The instrument and the test liquid must now be equilibrated at the test temperature by immersing the viscometer in a controlled temperature water bath. Earlier sections have discussed the influence of the precision of this temperature on the accuracy of the results obtained. As these are related to the temperature sensitivity of the viscosity of the test liquid and this is often unknown before the measurements are undertaken, this author recommends that ± 0.1 K be taken as a target temperature variation. Equilibration may take up to 0.5 h, during which the earlier comments on sedimentation or separation become relevant. Suction is then used to raise the liquid through the capillary into reservoir B until the meniscus of the liquid is above the etched mark C. The liquid is now allowed to flow under gravity and the time is taken for the meniscus to pass between marks C and D.

During this process the hydrostatic head will fall as the liquid level falls on the right hand side in Figure 10.3a and rises in the left hand leg. However, because the geometry of the system is so arranged as to always have test liquid within reservoir A with its large cross-section, the rise in the level in reservoir A will be very small. Consequently, the variation in hydrostatic head will be minimized. In addition, the shape of reservoir B is such that most of the measured flow will occur with the level central in this chamber and further reduce the variation in head. A mean value will be quoted by the manufacturer. Examination of equations (7) and (10) shows that this variation has no effect on Newtonian fluid measurements, while its effect on power law fluids could be considerable if the power law exponent n varied significantly from 1.

As previously stated, variations in design of capillary viscometers are many in numbers. One common form involves bending both legs of the U-tube slightly so that the bulb of the lower reservoir A is directly below that of reservoir B (Figure 10.3b). Another variation is the use of light sensors to note the passing of the meniscus across the etched marks C and D coupled to electronic timing, thereby ensuring more accurate measurement.

As with most scientific instruments, corrections are necessary if a high level of accuracy is required. These include energy losses in the expansion section of the flow (where the liquid emerges from the capillary), changing meniscus at the entry and exit points, liquid adhering to the wall of the capillary and the opposite problem of slippage of the liquid at the tube wall.

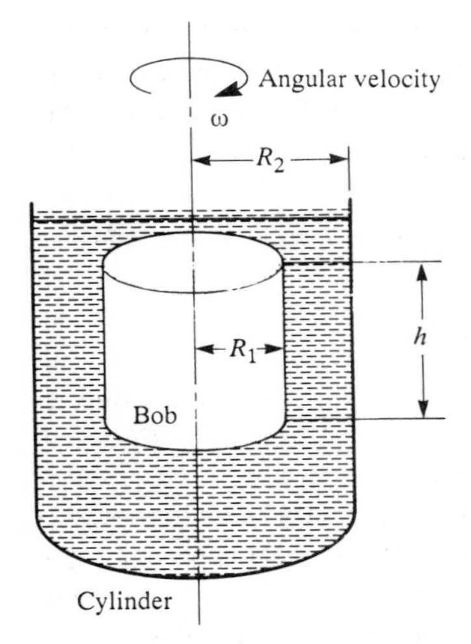

Figure 10.4 *Concentric cylinder viscometer*

Many authors cover these corrections in some detail and the reader is referred to Van Wazer et al. (1963) for a complete discussion.

4.2 Rotary Viscometers

There are two main types of rotary viscometers: concentric cylinder or Couette viscometers, and cone and plate viscometers.

4.2.1 Concentric Cylinder Viscometers

This type is shown schematically in Figure 10.4 and owes its development to the pioneering work of Couette (1890). These instruments consist of a cylindrical bob positioned concentrically in a hollow cylinder. Either the bob or the cylinder can be rotated, and the measured variable is either the torque transmitted through the liquid to the stationary cylinder or the torque required to keep the moving cylinder rotating at a given velocity. More advanced instruments apply a controlled shear stress. The shear-stress/shear-rate relationship is the same with each system of rotation. Continuous measurements may be made and time-dependent effects studied. Continuous or step variation over a wide range of velocities is normally available. Because of this, a range of shear rates may be readily obtained, thus permitting analysis of Newtonian or non-Newtonian behaviour. However, a major disadvantage is that the liquid is not subjected to a spatially uniform shear rate even if it is a simple Newtonian liquid.

These systems are, without doubt, the most widely used in rheological measurements, and fluid behaviour within the annular gaps of these instruments has been the subject of intensive investigation. Consequently, there is a wide range of analytical equations available for assessing their results and for modifying the readings obtained to correct for a wide variety of error sources.

For Newtonian fluids, the simplest relationship available is the Margules equation (1881):

$$\mu=(T/4\pi h'\omega)(1/R_1^2-1/R_2^2) \tag{14}$$

where μ is the viscosity, T is the torque on the cup or bob, ω is the angular velocity of the rotating cup or bob, h' is the height of the bob, R_1 is the radius of the bob and R_2 is the radius of the cup.

For non-Newtonian materials, Van Wazer et al. (1963) derive the general equations for flow in the annular space between the concentric cylinders and provide solutions for Newtonian fluids, power law fluids, power law fluids with a yield value (Herschel-Buckley), Eyring model fluids and several others. For simple power law fluids the following relationship is obtained for the mean shear rate $\dot{\gamma}$:

$$\dot{\gamma}=\omega(R_2^2+R_1^2)/(R_2^2-R_1^2)\cong\omega(R_2/\Delta R) \tag{15}$$

where ΔR is the width of the gap between the cylinders, and the shear stress at the bob τ_b is calculated from the measured torque as follows:

$$\tau_b=T/2\pi R_1^2 h' \tag{16}$$

Plotting the logarithms of the values derived from equations (15) and (16) should give a straight line of slope n and intercept k. It may, however, prove more convenient to calculate τ_b from equation (16) and plot ln ω versus ln τ_b, which should give a curve that fits the equation

$$\ln\omega=(1/n)\ln\tau_b+\ln\{(n/2)(\sqrt[n]{k})[1-(R_1/R_2)^{2/n}]\} \tag{17}$$

It is normally necessary to apply corrections to the readings from concentric cylinder viscometers. The most common of these is that for end effects. Many authors recommend that these effects be measured by taking torque readings at several different immersion depths of the cylinders in the test fluid. If T is then plotted against h, the resulting graph will intersect the h-axis at a negative value h_c that corresponds to the correction to be added to h in any of the above equations. Alternatively, this may be calculated from the following equation of Oka (1960):

$$h_c/R_1=(R_1/8e)[1-(R_1/R_2)^2]$$

$$\{1+(4e/R_1)\sum_{n=1}^{\infty}A_n I_2(n\pi R_1/e)+(8e/\pi R_1)\sum_{n=1}^{\infty}B_n[\sinh(K_n h)]/K_n R_1\} \tag{18}$$

where e is the distance between the bottom of the bob and the cup, I_2 is a modified second-order Bessel function, K_n is the nth positive root of a derivation of the Navier-Stokes equation for incompressible fluids, and A_n and B_n are functions of the variables R_1/R_2, h/R_2 and e/R_2. However, if the immersion system is such that the gap between the end of the bob and either the cup or the fluid container is large, then the end effects become negligible and the difficult application of equation (18) is avoided.

An equally important correction is that for shear rate variation across the sample. Equation (15) gives a mean value for shear rate. However, for power law fluids this can be corrected by the relationship

$$\dot{\gamma}_{\mathrm{eff}}/\dot{\gamma}_{\mathrm{meas}} = (1/n)(1/2^{1/n-1})[1-(R_1/R_2)^2] \\ [1+(R_1/R_2)^2]^{1/n-1}[1-(R_1/R_2)^{2/n}]^{-1} \quad (19)$$

where $\dot{\gamma}_{\mathrm{eff}}$ is the effective shear rate and $\dot{\gamma}_{\mathrm{meas}}$ is the measured value. The reader is referred to a correction table available in Prentice (1984) which obviates the need to carry out this detailed calculation.

There is a large range of concentric cylinder viscometers available from many different manufacturers. All use the same basic configuration, but they vary significantly in their degree of sophistication. Digital displays of rotational speed and torque are becoming standard, while many now have their own microprocessor incorporated. It is impossible to make specific recommendations in this general chapter other than to emphasize the guidelines of Prins and Bloksma (1983) to which reference has already been made in section 1. However, it is essential that when selecting an instrument, consideration be given to the range of shear rates required. The liquids must be subjected to the same shear rates as those in the application for which the rheological characteristics are required. In particular, processors of fluids such as chocolate which have a yield value must select an instrument capable of accurate measurement at very low shear rates.

Most instruments are now available with interfaces enabling data capture on a microcomputer. Software for standard calculations is available from the instrument manufacturers.

4.2.2 Cone and Plate Viscometers

A much recommended system for rotary measurement is the cone and plate viscometer (Figure 10.5). This consists of a cone of shallow angle, normally of less than 3 degrees and possibly with a truncated tip, that almost touches

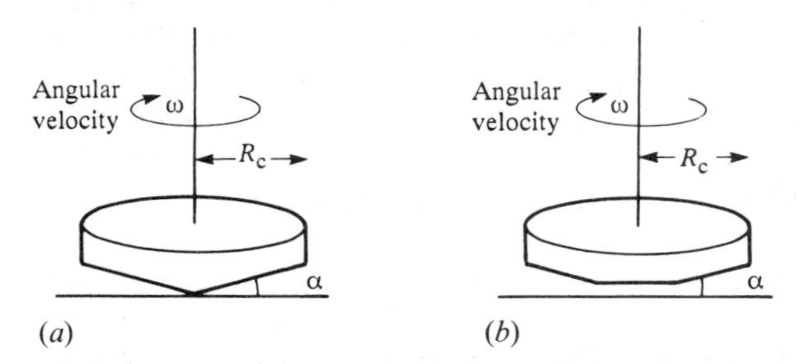

Figure 10.5 *Cone and plate viscometers: (a) normal (b) truncated cone*

a flat plate. The sample for assessment is placed in the intervening space and either the cone or the plate is rotated at a range of velocities. More recent instruments permit continuous variation of velocity over the range rather than the discrete step variations possible with older machines. While in theory it is possible to rotate either the cone or the plate and measure the torque transmitted through the intervening liquid, the normal procedure is to rotate the cone and measure either the transmitted torque on the plate or the torque required to rotate the cone at a constant angular velocity.

The major advantage of this measuring system is that the shear rate is constant at all points in the fluid. This feature is only true when small conical angles are used and makes the system particularly useful when characterizing non-Newtonian fluids, since the true rate of shear may be determined without the need for detailed corrections as in the concentric cylinder type. This constant shear rate may be determined from the relationship

$$\dot{\gamma} = \omega/\alpha \tag{20}$$

where ω is the angular velocity (rad/s) and α is the angle of the cone (rad).

The shear stress may be calculated from the following equation:

$$\tau = 3T/2\pi R_c^3 \tag{21}$$

For a Newtonian fluid these may be simply combined to give

$$\mu = 1.5T\alpha/\pi\omega R_c^3 \tag{22}$$

Indeed, even when measurements are performed on more complex systems, the analysis of results simply requires the substitution of the above equations for $\dot{\gamma}$ and τ into the relevant behaviour model for the fluid. Carrying out this substitution for power law (equation (2)), Casson (equation (5)) and Herschel-Buckley (equation (6)) fluids leads to the following set of relationships:

Power law: $$3T/2\pi R_c^3 = k(\omega/\alpha)^n \tag{23}$$

Casson: $$(3T/2\pi R_c^3)^{0.5} = \tau_y^{0.5} + k'(\omega/\alpha)^{0.5} \tag{24}$$

Herschel-Buckley: $$3T/2\pi R_c^3 = \tau_y + k''(\omega/\alpha)^n \tag{25}$$

While these instruments are much recommended, particularly for transient measurements, care must be exercised when treating any fluid containing suspended solids. The gap between the plate and the truncated cone is normally 50 μm or less. The problems and errors that would be encountered with particles of this size or larger need not be stressed. There is a general recommendation that particles should be at least ten times smaller than the size of the smallest gap between the cone and plate.

4.2.3 Other Rotary Viscometers

Many rotary viscometers have been developed in which cylinders, bars or agitator paddles rotate in the test fluid. Analysis of such systems is difficult because of the geometrical complexities of the system. Consequently, their use depends on the existence of empirical relationships which relate their measuring variables, normally the torque required to rotate the instrument at a known speed, to the rheological characteristics. However, it must be stressed that many such instruments provide correlations with Newtonian viscosity only and consequently may have limited use when one is considering the more complex fluids normally found in the food industry.

Two instruments are, however, worthy of mention because of their widespread use in the food industry. In the Brookfield Synchro-Lectric Viscometer a series of cylindrical spindles and horizontal disks are rotated at fixed speeds while the torque required to overcome the viscous drag of the fluid is recorded. Conversion tables are available to convert this into Newtonian viscosity. It is, however, very difficult to estimate accurately the shear rates being used and consequently its use for non-Newtonian fluids is limited. Some work has, however, been carried out (Mitschka 1982) to enable calculation of some of the basic power law (equation (2)) variables from Brookfield readings. In repetitive quality control use, many processors find its rugged simplicity useful and happily use its readings for comparative purposes. For more precise rheological evaluations, an optional attachment converts it into the more useful concentric cylinder geometry.

The Brabender Viscocorder measures the torque imparted to a paddle by the viscous drag of the test fluid in a rotating cup. Various forms of the instrument are available and a version capable of heating the test fluid in the rotating cup has found widespread use in the starch industry. Again it is difficult to relate the results obtained to the fundamental rheological properties, and the instrument, while widely used in quality control in the baking industry, does not see extensive use in other areas.

Details of both the Brookfield and the Brabender instruments are widely available in reviews including those of Matz (1962) and Sherman (1970).

5 On-line Measurement Systems

On-line systems are finding ever increasing applications in process control. An excellent comprehensive review of the instruments available was published by Cheng et al. (1984) of the Warren Springs Laboratory, UK, while a more recent review has been published by Davidson et al. (1989).

Many of the instruments available are modifications of those outlined in the previous sections of this chapter. In general, all share the same limitations and difficulty or ease of use as their laboratory counterparts. In particular, they must operate at a shear rate or range of shear rates that correspond to those experienced (at the relevant stage of processing or

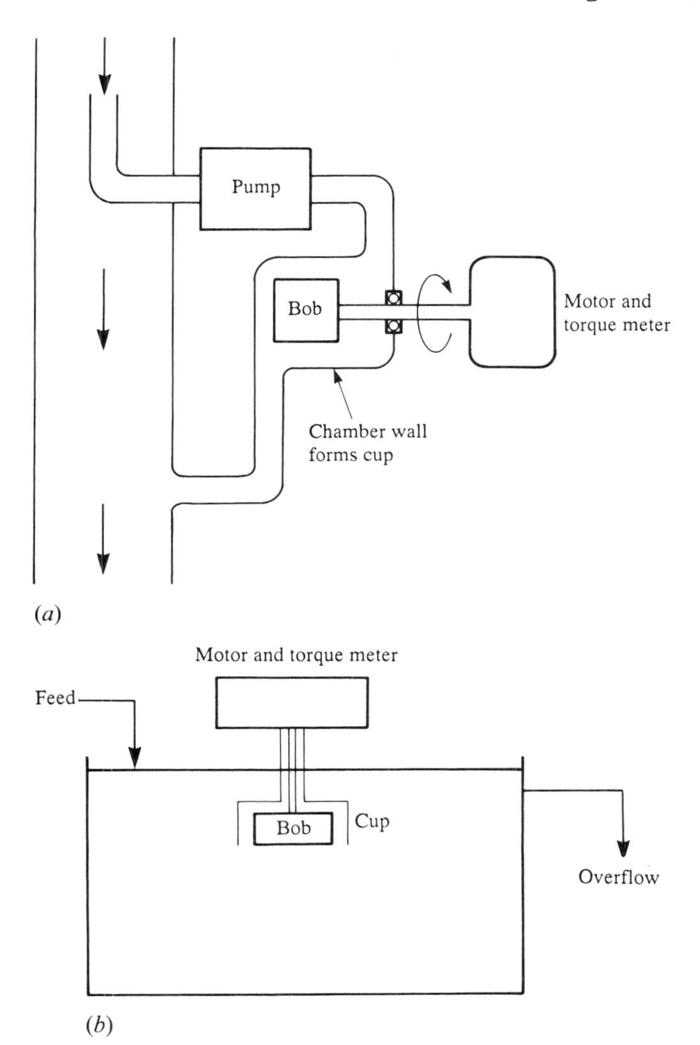

Figure 10.6 *Rotary viscometers: (a) on-line (b) tank mounted in-line*

consumption) in the system being controlled. They also have the additional requirements of hygienic design and suitability for cleaning in place.

Rotary systems may incorporate a chamber in the process line through which all or a bypass stream of the process fluid may flow. Within this chamber may be fitted the cylinder and cup of a concentric cylinder Couette viscometer. In some instruments the measuring chamber itself is also the outer cylinder within which a concentrically mounted bob rotates (Figure 10.6a). A second variation is the tank mounted system (Figure 10.6b). In addition to these geometrically well defined systems, there is a range of systems which continuously monitor the torque required to rotate blades and paddles at fixed speeds. As with the laboratory systems, these must rely on empirical correlations to convert the torque/speed readings into rheological parameters. Alternatively, process specific target and limit values are set for the torque/speed readings without conversion.

On-line capillary instruments are also available. As the entire process stream cannot be sent through a small capillary, these always use a bypass stream through which some of the process fluid is accurately metered (normally by a gear pump) through a capillary tube. Pressure sensors at each end provide the only measurements required for their operation. Capillary tubes may of course be changed for others of different diameter to give a variation in shear rate, but of course this is an intermittent rather than a continuous variation. Cleaning-in-place difficulties may also be experienced with these narrow bore capillaries. As they are subject to all the same limitations as their laboratory counterparts, their use tends to be limited to Newtonian liquids of low viscosity.

In addition to these adaptations of laboratory systems, other systems include vibrational techniques in which the power to keep spherical bobs or rods vibrating at fixed amplitudes within the liquid food is measured and correlated with rheological parameters. Other systems use falling cylinders within the fluid coupled to magnetic measuring systems. However, shear rates within such instruments are not readily available, thus limiting their use to quality control systems. Recently, hot wire methods have been suggested (Sato et al. 1990) using the temperatures in the fluid surrounding sheathed and heated wires as rheological correlates. This system requires development to overcome any potential hygiene problems and to ensure the degree of robustness necessary for continuous industrial use.

6 Instrument Selection

Guidance on the choice of correct instrument is a very subjective area in which authors will generally have a personal preference for a particular instrument or for a particular manufacturer. Therefore, the intention in this section is not to express such preferences but merely to give broad guidelines.

For occasional monitoring of the viscosity of a Newtonian food, a capillary viscometer is the cheapest and often quite a suitable selection. It must be emphasized, however, that such behaviour is rather rare in foodstuffs and is rarer still in food liquids that require frequent rheological checking. Rotary viscometers should therefore be regarded as the effective minimum equipment requirement. This is particularly true for all fluids exhibiting non-Newtonian behaviour.

Temperature control of the test sample is essential with fluid types. Launay and McKenna (1983) recommend a control limit of ± 0.1 K. It should also be noted that viscous heating of the test sample at high shear rates can lead to significant underestimation of the viscosity.

There are many different instrument manufacturers producing both concentric cylinder and cone and plate viscometers. They will differ in their sensitivity, degree of variation of shear rate avai'able, and consequently price. There is little published on comparative measurements using the various instruments available. However, the COST90 collaborative European study of food rheology as reported by Prentice and Huber (1983) is of interest in their presentation of results on the same samples from eight laboratories using many different viscometer types of different degrees of sophistication. On the test Newtonian liquids used, namely sucrose solutions, a cooking oil and mineral oils, agreement between the various instruments is remarkably close.

For non-Newtonian liquids the agreement is not as good. Indeed, at low shear rates many of the flow curves diverged. Consequently, it is important to select an instrument with shear rates covering the range of interest. Launay and McKenna (1983) stress this as, even with the simplest of non-Newtonian models, the power law fluids, this behaviour may not be exhibited over the entire range of shear rates.

A similar caution should be noted with liquids exhibiting a yield stress. The report by Prentice and Huber (1983) on the COST90 study cites values for a chocolate melt using the different viscometers. The measured yield stress (found by extrapolating the stress/shear-rate curve to zero shear rate) varied by more than a factor of three between the lowest and highest values, with a similar variation in the constant k' of the Casson equation (24). Only those instruments capable of applying very low shear rates gave good agreement between instruments. This again emphasizes the importance of choosing an instrument with an appropriate shear rate range.

Other foods tested exhibited shear thinning together with a yield stress. This was particularly true for an apple sauce, which prompts the question of how to measure the rheological properties of such a suspension which may be subject to settling or other separation during the shearing in the viscometer. Launay and McKenna (1983) recommend that for such a material with time dependent foods, care should be taken to ensure that standardized methods are employed, such as varying shear rate in the same manner in each test, allowing equilibrium to be established or standardizing the reading time, and so on.

Appendix: List of Symbols

A_n	function in equation (18)
B_n	function in equation (18)
d	capillary tube diameter
e	distance from bottom of bob to cup or container
g	gravitational constant
h	difference in levels between liquid in reservoirs of a capillary viscometer
h_c	end effects correction in concentric cylinder viscometry
h'	height of bob for a rotary viscometer
I_2	Bessel function
K	constant for a capillary viscometer
K_n	root of modified Navier-Stokes equation
k	consistency index for a power law fluid
k'	constant in Casson equation
k''	constant in Herschel-Buckley equation
L	length of capillary tube
n	power law exponent
Δp	pressure drop along a capillary tube
Q	flow rate
R_c	radius of cone
R_1	radius of bob in a concentric cylinder viscometer
R_2	radius of cup in a concentric cylinder viscometer
ΔR	$R_2 - R_1$
T	torque
t	relaxation time (equation (4)) or time (equation (8))
u_p	slope of plot of shear stress versus shear rate once yield stress has been exceeded
V	volume of liquid used in a capillary viscometer
α	conical angle
β	relaxation time
$\dot{\gamma}$	shear rate
$\dot{\gamma}_{eff}$	effective shear rate
$\dot{\gamma}_{meas}$	measured shear rate
μ	viscosity
μ_0	viscosity at zero shear rate
μ_∞	viscosity at infinite shear rate
ω	rotational speed
π	3.1417
ρ	density
τ	shear stress
τ_b	shear stress at the bob
τ_y	yield stress

References

Baird, D. G. (1983) Food dough rheology. In *Physical Properties of Foods*, M. Peleg and E. B. Bagley (eds), AVI, Westport, CT, 343–350

Charm, S. E. (1971) Fundamentals of Food Engineering. AVI, Westport, Conn.

Casson, N. (1959) In *Rheology of Disperse Systems*, C. C. Mill (ed), Pergamon, London, 84–104

Cheng, D. C.-H., Hunt, J. A. and Madhvi, P. (1984) Status report on process control viscometers: current applications and future needs. Warren Springs Laboratory, Stevenage, UK

Couette, M. M. (1890) Études sur le frottement des liquides. *Annales de Chimique et de Physique*, **21**, 433–510

Cross, M. M. (1965) Rheology of non-Newtonian fluids: a new flow equation for pseudoplastic systems. *Journal of Colloid Science*, **20**, 412–437

Davidson, V. J., White, J. and Hayward, G. (1989) On-line viscosity measurement in food systems. In *Proceedings of the 5th International Congress on Engineering and Food*, W. E. L. Spiess et al., (eds) Elsevier, Barking, UK, Vol 1 p 752–760

Elson, C. R. (1971) The flow behaviour of molten chocolate. BFMIRA research report 173, Leatherhead, UK

Escher, F. (1983) Relevance of rheological data in food processing. In *Physical Properties of Foods*, R. Jowitt, F. Escher, B. Hallstrom, H. F. Th. Meffert, W. E. L. Spiess and G. Vos (eds), Elsevier, Barking, UK, 103–110

Hagen, G. (1839) Über die Bewegung des Wassers in engen zylindrischen Röhren. *Pogg. Ann.*, **46**, 423

Hermannson, A. -M. (1980) Effect of processing conditions on microstructure and functional properties of comminuted meat batters. 26th European Meeting of Meat Research Workers, Colorado Springs

Hermannson, A. -M. (1983) Use of rheological data in structure analysis and food engineering. In *Physical Properties of Foods*, R. Jowitt, F. Escher, B. Hallstrom, H. F. Th. Meffert, W. E. L. Spiess and G. Vos (eds), Elsevier, Barking, UK, 111–121

Kleinert, J. (1976) Rheology of chocolate. In *Rheology and Texture in Food Quality*, J. M. de Man, P. W. Voisey, V. F. Rasper and D. W. Stanley (eds), AVI, Westport, CT, 445–473

Launay, B. and McKenna, B. M. (1983) Implications for the collection and use of rheological property data of experience from the collaborative study. In *Physical Properties of Foods*, R. Jowitt, F. Escher, B. Hallstrom, H. F. Th. Meffert, W. E. L. Spiess and G. Vos (eds), Elsevier, Barking, UK, 193–203

Leninger, H. A. and Beverloo, W. A. (1975) *Food Process Engineering*. Reidel, Dordrecht, The Netherlands

Lovegren, N. V. Guice, W. A. and Feuge, R. O. (1958) An instrument for measuring the hardness of fats and waxes. *Journal of the American Oil Chemistry Society*, **35**, 327

Margules, M. (1881) Über die Bestimmung des Reibungs- und Gleitungskoeffizienten aus ebenen Bewegungen einer Flüssigkeit. *Wien Sitzungsberger Abt.2A*, **83**, 588

Matz, S. A. (1962) *Food Texture*. AVI, Westport, CT

McKenna, B. M. (1967) The influence of rheology on the characteristics of atomization. MEngSc thesis, University College Dublin

McKenna, B. M. (1990) *The Liquid and Solid Properties of Foods – a Bibliography*. Food Science Publishers, London

Mitchell, J. R. (1984) Rheological techniques. In *Food Analysis – Principles and techniques*, vol. 1, D. W. Gruenwedel and J. R. Whitaker (eds), Marcel Dekker, New York, 151–220

Mitschka, P. (1982) Simple conversion of Brookfield R.V.T. readings into viscosity functions. *Rheologica Acta*, **21**, 207–209

Oka, S. (1960) The principles of rheometry. In *Rheology, Theory and Applications*, vol. 3, F. R. Eirich (ed), Academic, New York, 83–144

Poiseuille, J. L. M. (1841). Recherches experimentales sur le mouvement des liquides dans les tubes de très petits diametres. *Compte Rendu*, **112**

Prentice, J. H. (1984). *Measurements in the Rheology of Foodstuffs*. Elsevier, Barking

Prentice, J. H. and Huber, D. (1983) Results of the collaborative study on measuring rheological properties of foodstuffs. In *Physical Properties of Foods*, R. Jowitt, F. Escher, B. Hallstrom, H. F. Th. Meffert, W. E. L. Spiess and G. Vos (eds), Elsevier, Barking, 123–183

Prins, A. and Bloksma, A. H. (1983) Guidelines for the measurement of rheological properties and the use of existing data. In *Physical Properties of Foods*, R. Jowitt, F. Escher, B. Hallstrom, H. F. Th. Meffert, W. E. L. Spiess and G. Vos (eds), Elsevier, Barking, 185–191

Rao, M. A. (1977) Rheology of liquid foods – a review. *Journal of Texture Studies*, **8**, 135

Rasper, V. F. (1976) Texture of dough, pasta and baked goods. In *Rheology and Texture in Food Quality*, J. M. de Man, P. W. Voisey, V. F. Rasper and D. W. Stanley (eds), AVI, Westport, CT, 309–354

Ree, T. and Eyring, H. (1958) The relaxation theory of transport phenomena. In *Rheology, Theory and Applications*, vol. 2, F. R. Eirich (ed), Academic, New York, 83–144

Sato, Y., Miyawaki O., Yano, T., Ito, K. and Saelei, Y. (1990) Application of the hot-wire technique to monitoring viscosity of a fluid in a non-baffled agitated vessel. *Journal of Food Engineering*, **11**, 93–102

Sharma, F. and Sherman, P. (1966) The texture of ice cream. *Journal of Food Science*, **31**, 699–706

Sherman, P. (1970) *Industrial Rheology*. Academic, London
Shoemaker, C. F., Lewis, J. I. and Tamura, M. S. (1987) Instrumentation for rheological measurements of food. *Food Technology*, **41** (3), 80–84
Steffe, J. F., Mohamed, I. O. and Ford, E. W. (1986) Rheological properties of fluid foods; data compilation. In *Physical and Chemical Properties of Food*, M. R. Okos (ed), American Society of Agricultural Engineers, St Joseph, MI, 1–13
Tscheuschner, H. -D. and Wurische, D. (1979) Rheological properties of chocolate masses and the influence of some factors. In *Food Texture and Rheology*, P. Sherman (ed), Academic, London, 355–368
Van Wazer, J. R., Lyons, J. W., Kim, K. Y. and Colwell, R. E. (1963) *Viscosity and Flow Measurement: a Laboratory Handbook of Rheology*. Wiley, New York

11 *Modern Methods of Texture Measurement*

D. Kilcast and A. Eves

Contents

1 Introduction 349

2 Current Methods of Texture Measurement 350
  2.1 Sensory Texture Assessment 350
  2.2 Instrumental Texture Measurement 352
    2.2.1 Empirical Methods 352
    2.2.2 Imitative Methods 353
    2.2.3 Fundamental Methods 357

3 Physiological Aspects of Chewing 358
  3.1 Monitoring Human Subjects 358

4 Texture Measurement by Electromyography 359
  4.1 Instrumentation and Measurement Procedure 360
  4.2 Reproducibility 363
    4.2.1 Between Subjects 363
    4.2.2 Between Occasions 363
  4.3 Data Interpretation 364
    4.3.1 Relationship to Biting Force 364
    4.3.2 Relationship to Sensory Data 365

5 Future Developments in Texture Measurement 369

6 Conclusions 372

References 372

Introduction

Enjoyment of food is determined by the appearance, flavour and texture of that food. The relative importance of each of these sensory modalities is determined by the nature of the food and also by factors such as the eating situation. Products such as sugar confectionery and snack foods are often distinguished primarily by their textural characteristics. Major difficulties continue to be encountered, however, in designing and controlling specific textural characteristics in food. This results not only from a lack of fundamental knowledge of the food properties that manifest themselves as texture, but also from a lack of understanding of the physiology of texture perception. Successful design of texture relies on having a means of defining and quantifying the many distinct textural attributes that foods possess. The perception of texture is a physiological response to a tactile stimulus modified by subject psychology, and therefore a full description of texture can only be achieved through the use of human subjects. Salivation, for example, has a strong influence on perceived texture, and is influenced by both physiological and psychological factors.

The initial texture impression is visual, followed by the tactile reaction when the food is either cut or placed in the mouth (Peleg 1980). The physiological reactions to food in the mouth can trigger a psychological reaction. The complexity is emphasized by Szczesniak and Kahn (1971), who described perception in the mouth as 'a mixture of conscious and unconscious processes, the awareness being accentuated when visual expectations are violated'. The neurological basis of oral perception involves stimulation of at least two different sensory systems. Food is a tactile stimulus to the tongue, palate and pharyngeal regions, and chewing, through movement of both the jaw and the tongue, is the cause of muscular sensation.

Food texture research has been hindered by problems in agreeing a suitable terminology for describing textural characteristics and in understanding the physiological and psychological processes involved in the perceptual process. Szczesniak (1963a) linked texture to sensory, structural and physical parameters, and Sherman (1970), in a modification of Szczesniak's (1963b) definition, expressed texture as 'the composite of those properties arising from the structural elements, and the manner in which they register with the physiological senses'. Jowitt (1974) stated that the appreciation of texture involves the subtle interaction between both motor and sensory components of the masticatory and central nervous systems. In other words, the complex reactions occurring during the chewing of food are all integrated by the brain into the sensation perceived as texture.

2 Current Methods of Texture Measurement

2.1 Sensory Texture Assessment

The principles described by Szczesniak were incorporated into the first texture profile system to be adopted by the food industry (e.g. Brandt et al.

1963; Szczesniak et al. 1963; Civille and Szczesniak 1973). Standard scales were devised for rating various textural variables, such as hardness, chewiness and adhesiveness. Each scale point was identified by a reference food, which was specified by its descriptive name, brand or type, manufacturer's name, sample size and temperature of serving. Each parameter was given a careful definition. Other textural characteristics were evaluated less quantitatively. The order of appearance of the various characteristics was specified, and grouped into three stages: initial (first bite), masticatory and residual. Panellists received extensive training in the terminology and methodology of profiling. Initially they had foods that corresponded to points on the standard rating scales, and later they gained experience in preparing profiles for a range of other foods.

A major practical problem facing companies using this method arises from the lengthy training procedures; for example, for training twelve panellists, about 1900 man hours are required over a six-month period. As a consequence, although some modifications have been made to the original system (e.g. Munoz 1986), most of the industry has attempted to use profile methods that are less demanding on time and cost. The most commonly used methods are variants of the quantitative descriptive analysis (QDA) procedure, first described by Stone et al. (1974).

Unlike the previously described method, a QDA panel is normally used for the evaluation of one product type only. Potential assessors are first screened for their suitability for such work. The panel then, through discussion sessions, derives a list of descriptive terms that define the product range, using the products themselves. These terms may also include appearance and flavour terms. The panel is then trained to score the intensity of the attributes on a line scale, anchored by terms such as 'absent' and 'strong'. Training is continued until the panellists can score the attributes reproducibly and the results are presented both numerically and graphically. Although the method is more cost effective, it is still demanding on panellist time, and the performance of a panel needs to be maintained through ongoing training if it is used only irregularly.

An additional criticism often levelled at QDA and associated methods is that panellists are forced into a consensus on the use of terms and that as a consequence there is often confusion. The different use of terms can often be identified by statistical methods such as principal components analysis. However, an alternative method, called free choice profiling, has been described (Arnold and Williams 1986) in which each assessor constructs his/her own set of descriptors, which is then used for subsequent scoring sessions. The use of personal descriptor sets minimizes the need for training. Since there is generally considerable confusion in the use of textural terms, the method has great potential for texture profiling, but suffers from the problem that the large number of individual configurations that result requires statistical techniques such as generalized Procrustes analysis. The techniques are complex, demand substantial computing power and ideally require substantial statistical knowledge. The practical difficulties posed by these subjective methods have resulted in the food industry seeking simple and inexpensive objective measurements of texture.

2.2 Instrumental Texture Measurement

Instrumental methods have been classified into three main categories: empirical, imitative and fundamental (Scott-Blair 1958).

2.2.1 Empirical Methods

Empirical tests often measure ill-defined variables that are indicated by practical experience to be related to some aspect of textural quality. Devices have been developed within different sectors of the industry that are appropriate to specific product types. Even for the same product type, different food manufacturers have developed their own in-house devices. Fuller details of the devices described in this section are given in Bourne (1982).

Puncture or penetration devices measure either the force needed to push a probe into the food to a specified depth or the penetration distance achieved by application of a specified force. Examples include Magness-Taylor testers (for fruit), the Bloom Gelometer and the FIRA Jelly Tester (for gels), the cone penetrometer (for fats) and the Christel Texture Meter (for peas).

Shearing devices measure the force needed for one or more blades to shear through the food. The maximum force is often assumed to measure toughness, firmness or fibrousness. Instruments include the Warner-Bratzler Shear (for meat), the Kramer Shear Cell (general purpose) and the FMC Pea Tenderometer.

Compression devices measure the force needed to achieve a given compression or the compression achieved at a given force. Examples include the Baker Compressimeter (for bread) and the ball indenter (for fats).

In extrusion tests the food is forced through one or more orifices and the maximum force, average force or work done over a specified period is measured. The measured values are assumed to relate to firmness, toughness, consistency or spreadability. Examples include the FIRA-NIRD Extruder (for fats) and various cells used in conjunction with general purpose instruments.

Cutting devices use wires or blades (sometimes rotating) to cut through the food and measure the maximum force developed or the time needed to cut through a standard size of sample. Measurements are assumed to relate to fibrousness, firmness or hardness. The FMBRA Biscuit Texture Meter is a rotating blade device used to measure biscuit hardness.

Flow and mixing devices are used to give a measure of viscosity or consistency of liquid and semi-liquid foods. They often measure the extent to which samples flow or spread under specific geometric conditions, e.g. the Bostwick Consistometer and the Lyons Gel Flow Meter.

Although such empirical devices are often simple, inexpensive and portable, precision and reproducibility are generally poor and the measured parameters are poor measures of perceived texture. Extensive use is still made of them in industry, however, mainly for quality control purposes.

2.2.2 Imitative Methods

Imitative methods of measurement mimic the conditions to which the material is subjected in practice during eating. Volodkevich (1938) described the bite tenderometer, which attempts to mimic the action of teeth on food. It records the force of biting on a piece of food as a function of the deformation incurred. Two wedges with rounded points substitute for teeth, the lower being fixed to a frame. The upper wedge can be moved with a linear motion through the arc of a circle by a lever, thus squeezing a sample between the wedges.

The closest analogy to human chewing was attained by the denture tenderometer (Proctor et al. 1955), an adaptation of Volodkevich's apparatus utilizing a pair of human dentures to enable determination of food-crushing forces. The driving mechanism was able to impart both lateral and forward motion to the lower jaw. The amplified output from strain gauges placed on the articular arm measured the forces encountered. This served as the prototype for the General Foods (GF) Texturometer (Friedman et al. 1963), in which the dentures are replaced by a plunger. The location of the sensing element was moved from the articular arm to the sample area to eliminate gravity forces, and the oscilloscope was replaced by a chart recorder, enabling easy and permanent recording of any chosen number of consecutive chews. In this device, the driving mechanism no longer imparts a combined lateral and forward motion to the lower jaw, although it still drives the plunger through the arc of a circle. The linear speed of travel of the plunger varies sinusoidally with time (Brennan et al. 1975).

A schematic diagram of the GF Texturometer is shown in Figure 11.1. As a result of the complex dynamics of the instrument, interpretation of the data was of a semi-empirical nature. The force–time output was analysed in terms of a number of measured parameters, and these were related to specific sensory attributes using a texture profile method.

Although the GF Texturometer remains in use in North America and in Japan, the general purpose testing machines exemplified by those made by

Instron, Lloyd and Stevens are most commonly used in the UK food industry. The instruments differ in their mechanical construction and in their data acquisition and data analysis capabilities, but they have a number of important features in common (Figure 11.2). All have a crosshead containing a load cell, which is driven vertically at a range of constant speeds, and which can cycle over a fixed distance or load range. Probes can be attached to the crosshead for penetrating, shearing or crushing food, which can be held in a variety of cells. The load is recorded relative to time or to penetration/deformation distance, and displayed on a suitable recorder. Computer control of the instrument and computer analysis of the data are increasingly common. A typical modern instrument is shown in Figure 11.3.

An idealized force–distance relationship is shown in Figure 11.4, together with some interpretations that are commonly placed on the measured parameters. It should be noted, however, that any interpretation should only be made via correlation of the instrumental parameters with quantified sensory attributes. This is too seldom the case in published literature. Force–time curves from real foods are in general considerably more complex than the above. Figure 11.5 shows data from a gelatin gel penetrated using a flat-ended cylindrical probe. Measurements of such curves require considerable care in both execution and interpretation.

A major advantage of such instruments is that flexibility of design allows them to be used for a wide range of foods. This is particularly useful for companies that are handling or manufacturing a varied product range.

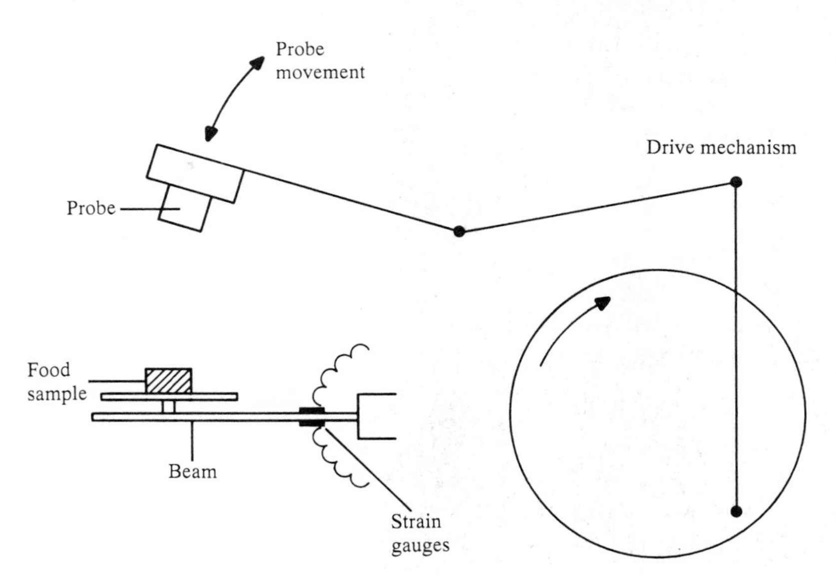

Figure 11.1 *Schematic diagram of General Foods Texturometer (after Brennan et al. 1975)*

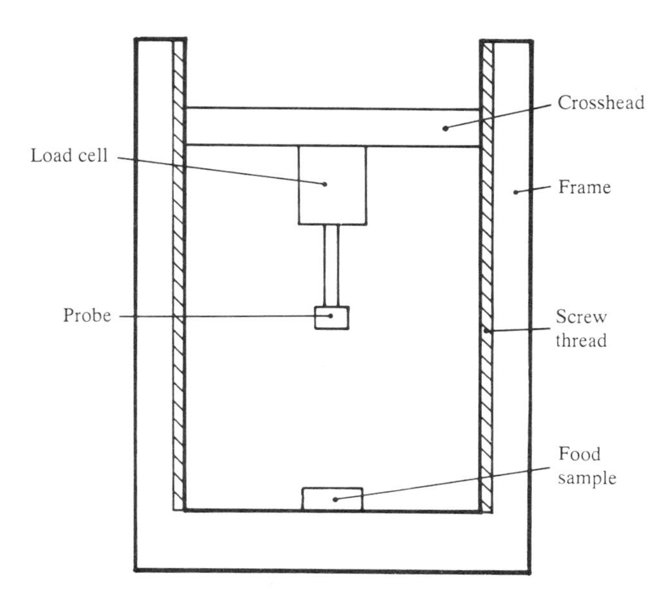

Figure 11.2 *Schematic diagram of general purpose testing machines*

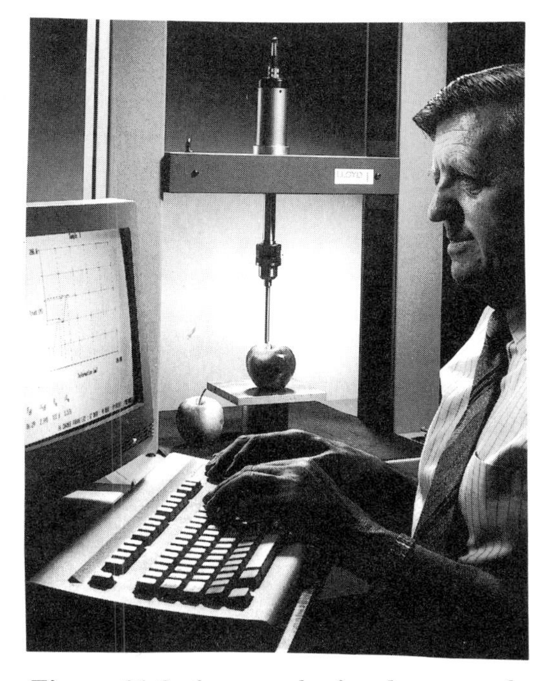

Figure 11.3 *An example of modern general purpose testing machine (courtesy of Instron Ltd, High Wycombe, UK)*

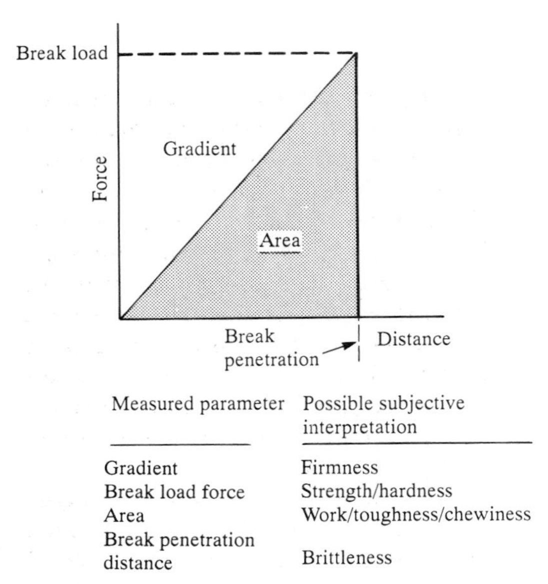

Measured parameter	Possible subjective interpretation
Gradient	Firmness
Break load force	Strength/hardness
Area	Work/toughness/chewiness
Break penetration distance	Brittleness

Figure 11.4 *Idealized force–distance relationship: the terms used to describe subjective interpretation do not necessarily have the same meanings as in classical physics*

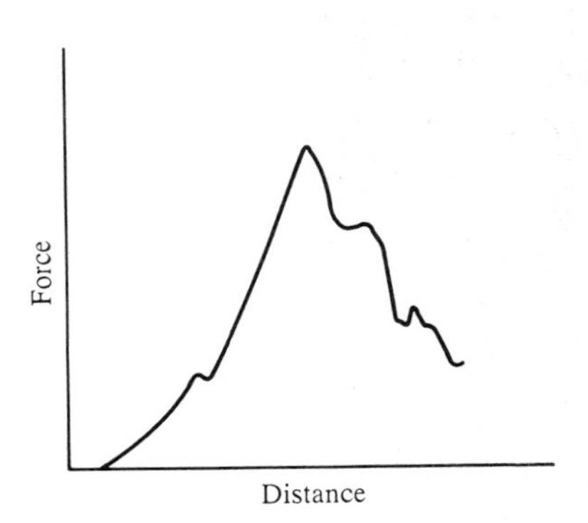

Figure 11.5 *Actual force–distance relationship: gelatin gel penetrated by a flat-ended cylindrical probe*

Load cells can be changed to give a high level of accuracy for relatively soft foods through to very hard foods. Probes and sample holders can easily be changed to accommodate measurements on different product types. An additional advantage is that such instruments can often be adapted for fundamental texture measurement.

2.2.3 Fundamental Methods

Fundamental methods involve measuring well-defined physical properties of food, which, if measured properly, are independent of the method of measurement. The most common fundamental parameters are Young's modulus, shear modulus, bulk modulus and Poisson's ratio (for solids) and viscosity (for liquids). Fundamental parameters for solids can be measured on general purpose testing machines, but such measurements require a carefully designed experimental set-up and are consequently slow. In addition, foods are generally heterogeneous and do not exhibit simple elastic behaviour. Fundamental parameters can be measured on liquids using suitable instrumentation, for example the Weissenberg Rheogoniometer and the Carri-Med Rheometer. Again, however, liquid foods rarely exhibit simple viscosity behaviour. A typical modern rheometer is shown in Figure 11.6 (see Chapter 10 on the rheology of liquids).

Such fundamental measurements are valuable in investigating the physical properties of food, but are too complex for routine use and, in common with other instrumental measurements, rarely correlate well with perceived texture. Some reasons for this can be identified on examining some of the physiological factors associated with chewing.

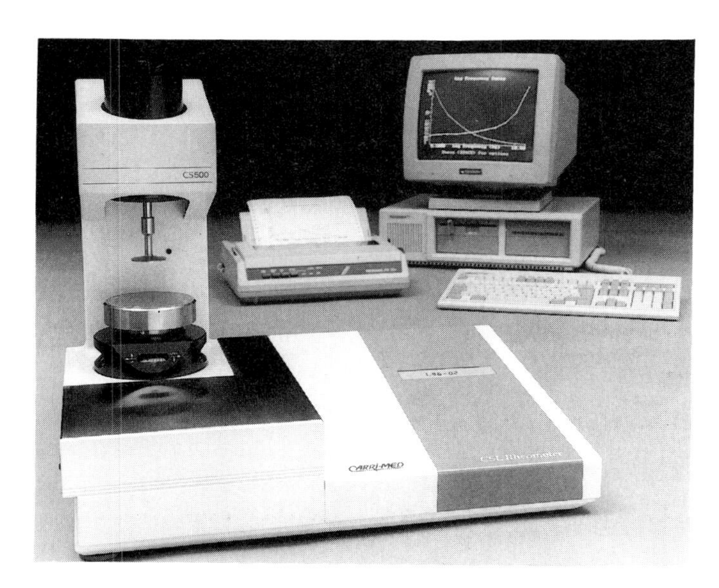

Figure 11.6 *An example of a modern rheometer (Carrimed CS500: courtesy of Carrimed Ltd, Dorking, Surrey, UK)*

3 Physiological Aspects of Chewing

Once food is bitten with the incisors or placed in the mouth, a complex series of events occurs. Firstly, the food may be positioned in an appropriate part of the mouth using tongue movement. Pressure from the tongue may press the food against the hard palate, giving information on the response of the food to such pressure and on the surface textural characteristics. Biting using the lateral sets of teeth occurs in two stages, the first at low loads with a high rate of movement (up to 400 cm/min for whole meat: Tornberg et al. 1985) and the second at high loads with low speeds (Brennan 1988). The motion itself can be composed of complex combinations of vertical and horizontal movements, and can depend on the state of occlusion of the teeth (Boyar and Kilcast 1986). These movements lead to the foods being subjected to a complex combination of compressive, shearing and tensile forces.

During this process, saliva is being introduced to the food, acting as a lubricant and helping in the formation of a bolus with the correct physical properties needed for swallowing (Szczesniak 1987). Foods eaten at temperatures higher than or lower than body temperature also undergo physical changes as thermal equilibration occurs. The fatty component of meat products, for example, can solidify and give a change to the mouthfeel characteristics. More importantly, chilled and frozen foods undergo changes leading to melting of any ice component and extensive structural breakdown; this can happen rapidly even in the absence of any chewing action.

It is clear that no realistically priced mechanical instrument can operate in such a way as to simulate this wide range of conditions. At best an instrument can only simulate a first bite, and even then only with an oversimplified mechanical action. Moreover, few instruments can reach the high force application rate experienced in the mouth; most tests in the industry are operated at 3 cm/min. Since most solid foods are viscoelastic in nature, the rate of force application is an important factor in measuring physical properties but is rarely taken into account.

An elegant alternative approach to solving the problems surrounding sensory assessments using people and objective measurements using instruments would be to make objective measurements on people during eating.

3.1 Monitoring Human Subjects

Early studies (Drake 1963) on food-crushing sounds showed that sounds from crisp foods differed from those of non-crisp foods, primarily in terms of amplitude. Frequency and duration played a less important role. A subsequent paper (Vickers and Bourne 1976) presented studies of the acoustical properties of tape-recorded biting sounds of wet and dry crisp foods. Amplitude–time plots indicated that both sound amplitude and the number of sounds produced in a given bite distance discriminated between different levels of crispness. A series of papers on crispness appeared subsequently (reviewed in Vickers 1988). In one of this series (Vickers 1985) it was shown that pitch was a useful parameter in distinguishing between crispness and crunchiness; crisp sounds tended to be higher in pitch than crunchy sounds.

Parts of this research programme examined the hypothesis that perceived crispness is a purely auditory sensation, but the work showed clearly that crispness has both auditory and non-auditory components. In a later study (Vickers 1987) multiple linear regression was used to select combinations of mechanical and acoustical measurements that would best correlate with sensory crispness. For potato crisps, the following relationship was found:

$$\text{crispness} = -15.6 + 5.35NP + 133MHP - 6.21\,peak$$

where NP is the number of sounds during one bite, MHP is the mean height of peaks (both taken from oscilloscope displays of bite sounds) and *peak* is the peak of the force–deformation curve from a Kramer shear cell. For crisp breakfast cereals, the best relationship found was as follows:

$$\text{crispness} = 538 + 539\,(\log MHP) - 222\,(peak)$$

In spite of the considerable improvement in our understanding of crispness that has resulted from these extensive studies, a practical method of measuring crispness has not yet appeared. Such an approach is, of course, limited both to foods that emit sounds during chewing and also to a small number of textural attributes.

Electromyography (EMG) has been used for some time for studying muscular activity (Boyar and Kilcast 1986), but little prior work has been carried out to use the method for determining the activity of muscles used during chewing and relating the activity to food texture. In the following section, the recent developments in texture measurement by EMG will be covered in detail. Other new approaches in food texture assessment, such as sonic input techniques, will be referred to in Section 5.

4 Texture Measurement by Electromyography

EMG involves the measurement of electrical signals generated in muscles that are active during mastication. For certain muscles that lie close to the surface of the skin, for example the masseter muscle, which is active during the chewing of solid foods, this activity can be related to a specific muscle.

Other oral activity, for example tongue movement, is controlled by groups of muscles that are deeper lying. Monitoring of signals from this latter type of musculature ideally requires implanted electrodes, whereas signals from the masseter muscle can be readily recorded using surface electrodes.

Early attempts to use EMG in the study of food texture were limited by difficulty in interpreting the complex data patterns produced. In the absence of suitable computerized acquisition and analysis equipment, visual inspection of the raw data was carried out. For example, motor pauses (or silent periods) were more frequent with hard foods than with soft foods (Boyar and Kilcast 1986). The development of more sophisticated EMG equipment and computer systems, however, has permitted much deeper analysis of EMG data and their relationship to food texture. An added potential advantage of this technique is studying changes in food texture in the mouth throughout the whole chewing cycle. The remainder of this chapter mainly covers experiments carried out at the Leatherhead Food Research Association.

4.1 Instrumentation and Measurement Procedure

Electromyographic patterns were recorded using a Grass Polygraph (Model 7D). The system has a regulated power supply, consisting of two DC driver amplifiers. One of these is connected to a pre-amplifier and displays the raw signal. The other is connected to a pre-amplifier and an integrator module and displays the integrated data. Both driver amplifiers include a 50 Hz filter rejecting interference from AC sources. All recordings were made with the amplifiers' frequency bands as wide as possible (10 Hz to 40 kHz). This prevents the chart from undulating with the movement of the jaw and ensures a flat frequency response up to 200 Hz, this being the maximum response rate of the pens. The time base was set at 0.2 s at the recommendation of the manufacturer.

A schematic diagram of the equipment is shown in Figure 11.7. Although analysis of raw data using Fourier techniques is possible, computational demands are complex, and consequently analysis of integrated data only is described in this chapter.

Surface electrodes were used to detect the electrical signals from the masseter muscle. The skin on the cheek and earlobe was cleansed with 95 per cent ethanol to remove traces of dust and perspiration, which may interfere with the signal. A position was located on the cheek at the maximum point of inflection of the masseter muscle, and two electrodes were located above and below this point in line with the muscle and approximately 0.5 cm apart. A third electrode was placed on the earlobe, a point of no muscular activity, which acted as an earth. Electrode cream, a conductive paste, was applied to the electrode surface in all cases. Arrangement of subject and EMG is shown in Figure 11.8.

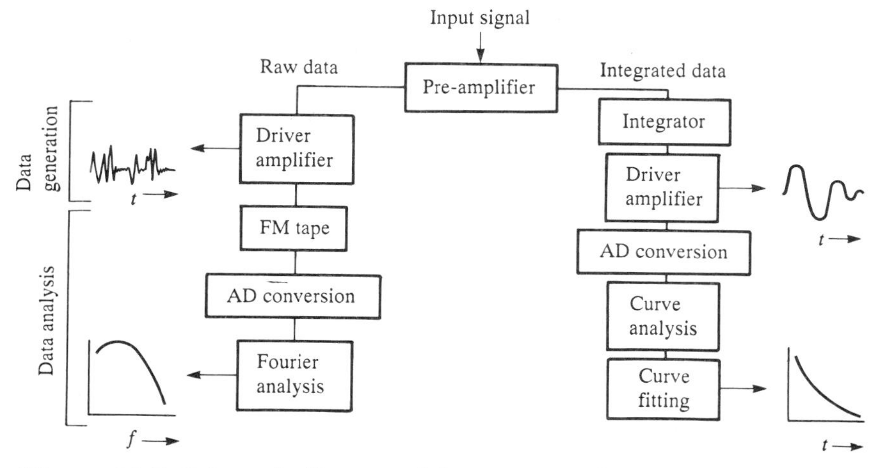

Figure 11.7 Schematic diagram of electromyograph

Following connection of the terminal ends of the electrodes to the polygraph, the subject was presented with the sample to chew. The data acquisition system was capable of recording information from up to three muscles simultaneously; however, the experiments described in this paper were carried out using a single masseter muscle, and consequently the subjects were asked to eat on the side of that muscle only. All samples were assessed in triplicate and all samples were of the same size and geometry.

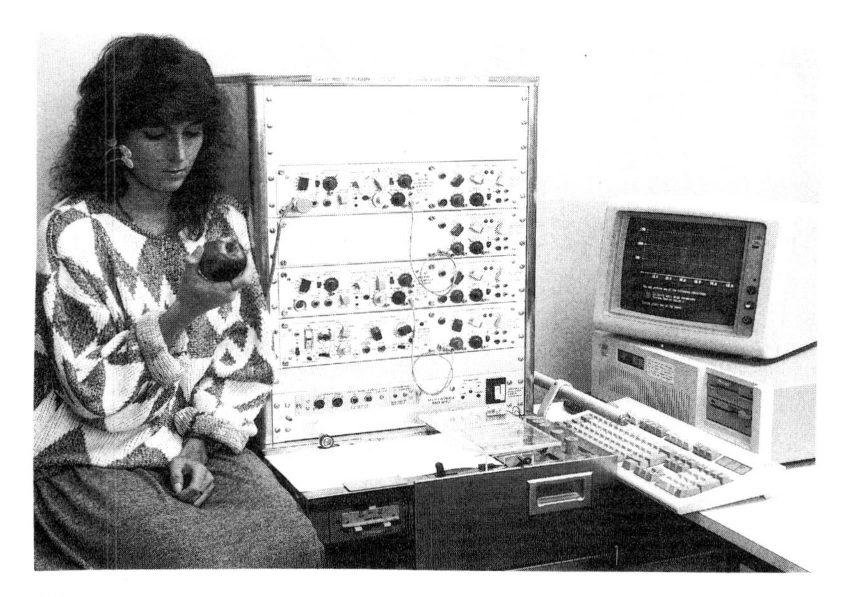

Figure 11.8 General arrangement of subject and EMG

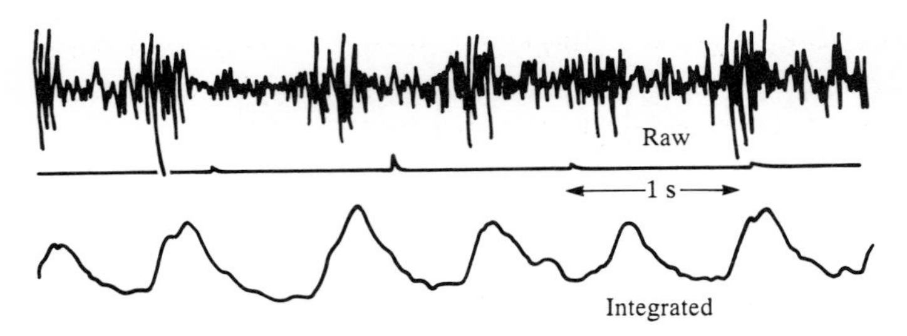

Figure 11.9 *EMG output signals*

Data were recorded from the time the sample was put into the mouth to the time of swallowing.

The raw EMG signal and its integrated form were recorded on a *Y–t* recorder, together with a time signal (Figure 11.9). The integrated wave was digitized and stored on an IBM PC microcomputer using the commercial data acquisition package ASYST (Keithley Instruments, Reading), modified for this purpose. A digitization rate of 50 points per second was used, and from the data a number of curve parameters were calculated: peak height *PH*, pre-maximum gradient *PRG*, post-maximum gradient *POG* and area under peak *A* (Figure 11.10).

Gradients and area were calculated from minima rather than baseline, and all parameters were adjusted in relation to a pre-recorded baseline. Any of the trace parameters can be plotted against time, and the package permits the fitting of a range of curves to the data. With the exception of the post-maximum gradient, however, the change of most EMG parameters with

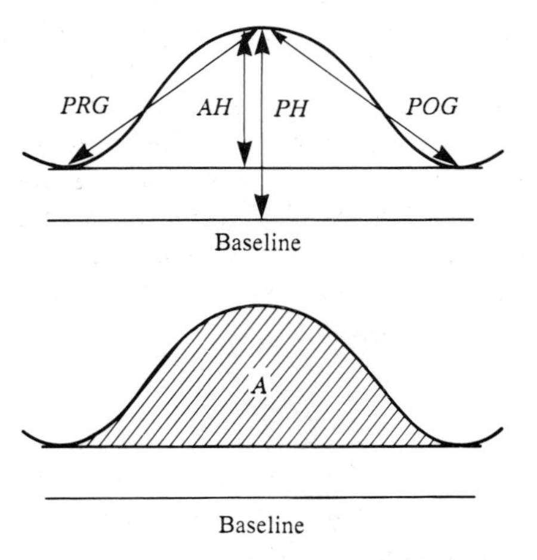

Figure 11.10 *Measured MG parameters on the integrated signal: PH* peak height; *AH* adjusted height; *PRG* pre-maximum gradient; *POG* post-maximum gradient; *A* area

time can be described either by a falling exponential curve or by a rising straight line followed by a falling exponential curve. The following 'broken stick' equation was consequently fitted to peak height data:

$$y = \begin{cases} Ae^{-at} + Q(t-u) & \text{if } t < u \\ Ae^{-at} & \text{if } t \geqslant u \end{cases}$$

4.2 Reproducibility

4.2.1 Between Subjects

Thirty subjects (approximately one-third male, two-thirds female) were asked to chew samples of fruit pastilles whilst linked to the polygraph. Each subject chewed a total of three replicate samples, and integrated peak height data were averaged and plotted against time. Results showed that 26 subjects gave data that could be described as a broken stick relationship (group 1), although subjects within this group exhibited a wide range of amplitudes and chew times. The other four subjects fell into two groups: those who gave data best represented by a single exponential decay curve (group 2) and those who produced increasing forces during chewing (group 3). Illustrative examples of the three curve types are shown in Figure 11.11.

Experience of the technique has shown that the choice of subjects for EMG work is not critical provided they are chosen from group 1. The behaviour characteristic of group 2 may be a result of aggressive biting behaviour; the upward phase exhibited by the other group is thought to be a consequence either of positioning the food before biting proper, or of a tentative biting of foods that are unfamiliar or that are recognized as hard or chewy. Research has concentrated predominantly on the breakdown phase, but valuable information on oral manipulation may be contained in the upward phase. The group 3 behaviour, which implies increased biting forces as food is broken down, appears to be anomalous. This may reflect different biting behaviour, or possibly a breakdown in the peak height–biting force relationship, and needs further investigation.

4.2.2 Between Occasions

Five subjects were asked to eat 4 g samples of toffee (three replicates) on three separate occasions. Integrated peak height was measured for each subject for each day and a broken stick equation was fitted.

An example of the results obtained is shown in Figure 11.12. The results apply to one subject only; similar results were, however, obtained for other subjects. Although differences were apparent in the initial phases of

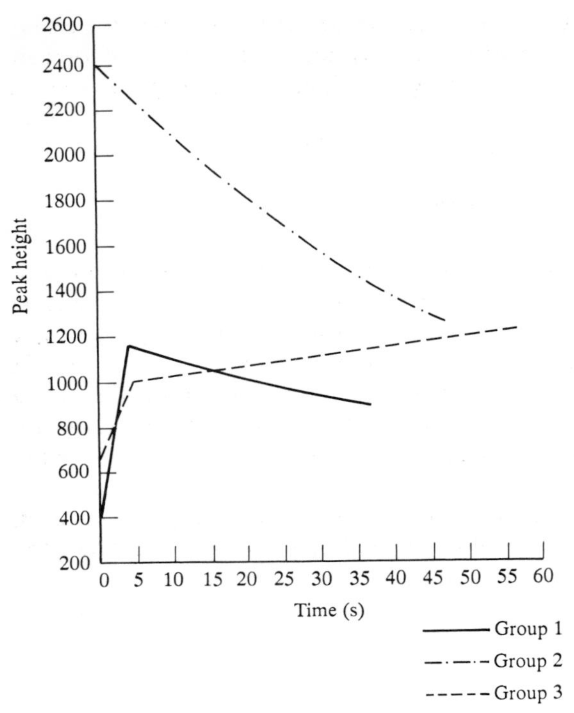

Figure 11.11 *Plots of integrated peak height against time, showing behaviour of 26 subjects*

chewing, the breakdown rates of the samples over the three days remained the same. A slight shift is seen in the amplitude of the plot on day 3; this was probably due to slightly inaccurate electrode placement. Chew time on day 1 was slightly shorter than on the following two days.

From these curves it would seem that from day to day the results do not vary considerably. Methods by which the amplitude of peaks can be maintained are under consideration, for instance the use of a standard material on which a subject would chew prior to any study.

4.3 Data Interpretation

4.3.1 Relationship to Biting Force

In an experiment designed to investigate the physical significance of integrated peak height, a strain gauge force transducer was constructed and calibrated under compression using a Stevens Compression Response Analyser. A subject was linked to the polygraph and asked to bite on the tip of the strain gauge using the molars. The voltage produced from the strain gauge was recorded, together with the integrated peak height. A plot of biting force against peak height is shown in Figure 11.13. A good straight line relationship is seen over most of the range tested ($r = 0.96$), indicating that the peak height gives a good measure of biting force in the mouth.

4.3.2 Relationship to Sensory Data

Any non-subjective method of food texture assessment can be of practical value to the food industry only if the results of the data measured can be interpreted in subjective terms. Although the use of integrated peak height as a measure of biting force is of great value, it is clear that a considerable amount of additional textural information is contained within the EMG traces. A preliminary study was carried out in an attempt to relate the texture profile of three types of commercial confectionery gums to measured EMG parameters.

Profile analysis was carried out on three commercial fruit pastilles using the QDA method. A panel of twelve people derived textural terms for the

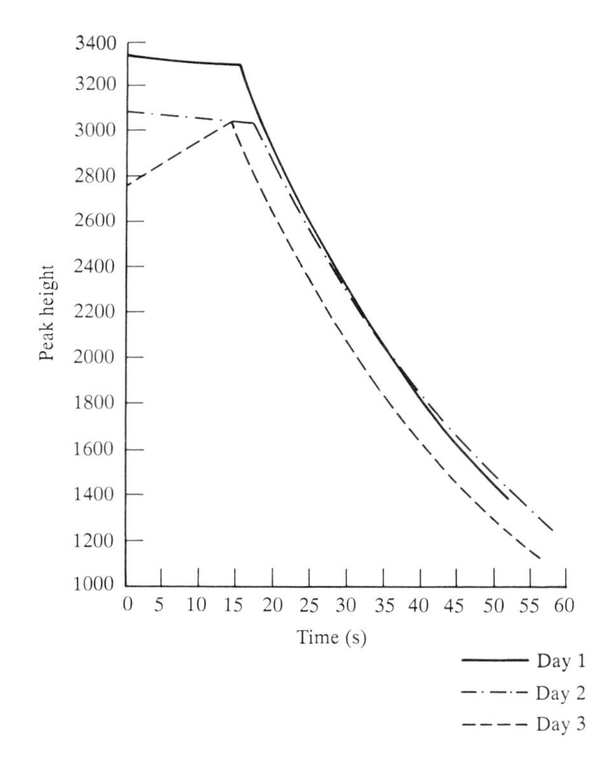

Figure 11.12 *Plots of integrated peak height against time, showing reproducibility when measured on different occasions*

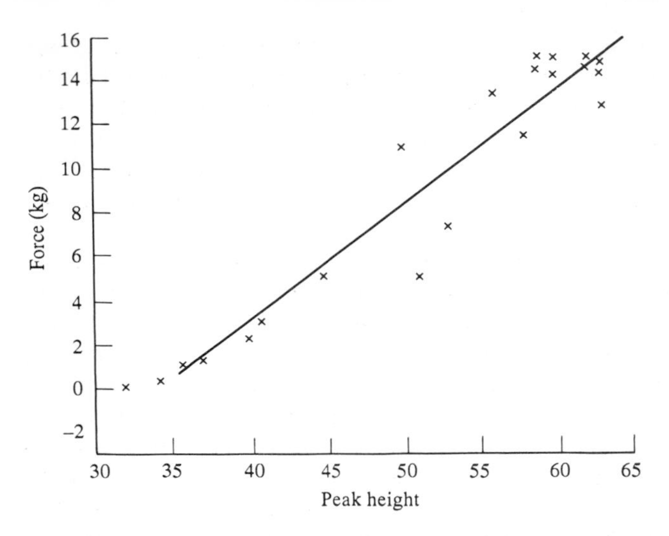

Figure 11.13 *Relationship between integrated peak height and measured bite force*

pastilles (jelly, tough, rubbery, chewy, moist and sticky) and scored their intensities in five replicate sessions. Eleven of the twelve panellists then chewed the pastilles in three replicate sessions whilst wired to the EMG. Initial attempts were made to find simple correlations between profile terms and EMG parameters. It was found, however, that each individual profile term showed high correlations with several EMG terms and that these correlations could not be explained logically. Similarly, attempts to relate each profile term to linear combinations of EMG parameters using multiple linear regression required, typically, more than five EMG parameters to give a satisfactory fit. Such ambiguous relationships are of little practical value to the food industry, and consequently the EMG data were examined further using a multivariate statistical technique, canonical variates analysis (CVA).

Multivariate methods seek to examine patterns in data points by finding linear combinations of the original experimental variables that account for most of the variations in the data. CVA operates in a similar way, but defines groupings in the data by maximizing the ratio of the variation between groups to the variation within groups. The first canonical variate is the direction that maximizes this ratio; the second is the orthogonal direction that maximizes the remaining ratio; and so on.

Figure 11.14 shows the CVA plot from analysis of EMG parameters for the three commercial fruit pastilles. Each point represents one subject, and the dashed lines represent the 95 per cent confidence limits for each grouping. The plot separates the three pastilles into three non-overlapping groups, with most subjects falling within the 95 per cent confidence limits.

Since CVA is designed to maximize the separation between groups, a check was carried out on the risk that the groupings might be an artefact of the method. Nine identical samples of wine gums were chewed, and the EMG data were treated as three different samples chewed in triplicate. The resultant CVA plots are shown in Figure 11.15. The overlap of the 95 per cent confidence circles shows that no distinct groupings were found in the data.

In a second set of experiments designed to examine further the use of the CVA method, profile and EMG analyses were carried out on a set of confectionery materials formulated and manufactured to give a controlled range of structures and textures. The set comprised two wine gum formulations (50/50 gelatin/starch and 100 per cent gelatin, respectively), a gelatin chew and a whipped chew. The QDA profile generated a total of fifteen textural terms. EMG data for peak height are shown in Figure 11.16, and CVA plots on all EMG parameters are shown in Figure 11.17. The CVA plots show the wg50 sample to be well separated from the other three products, all of which overlap. The reason for this separation may be associated with the unique breakdown characteristics seen in Figure 11.16, in which the wg50 sample exhibits rapid breakdown with no initial upward phase.

CVA plots on the profile data are shown in Figure 11.18. This analysis separates wg50 from the gelatin and whipped chews, but not from the wg100 sample. One reason for the difference between the CVA plots on the EMG and the profile data may be that the profile data contain textural information on attributes, such as grainy and floury, that may not influence the EMG data acquired from the masseter muscle alone.

Although considerable effort remains to be made in understanding the

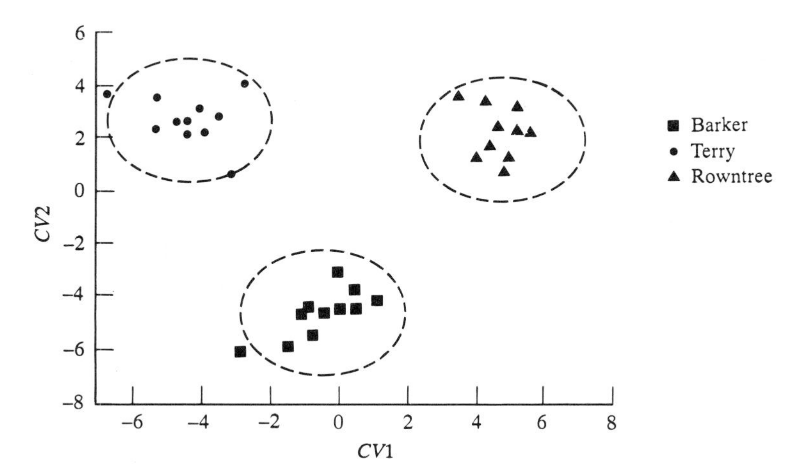

Figure 11.14 *CVA plots on EMG data measured on commercial pastilles*

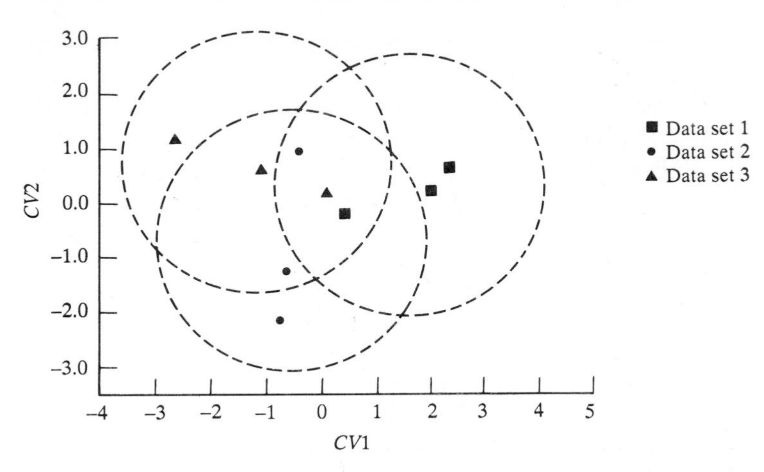

Figure 11.15 *CVA plots on EMG data from nine replicates of the same type of wine gum*

potential and the limitations of EMG, the use of CVA enables foods of different textural characteristics to be readily visualized in a low dimensionality texture space. This gives the product developer a powerful tool in attempting to match the textural characteristics of foods and in identifying novel textures by finding areas in CVA texture space not occupied by existing products.

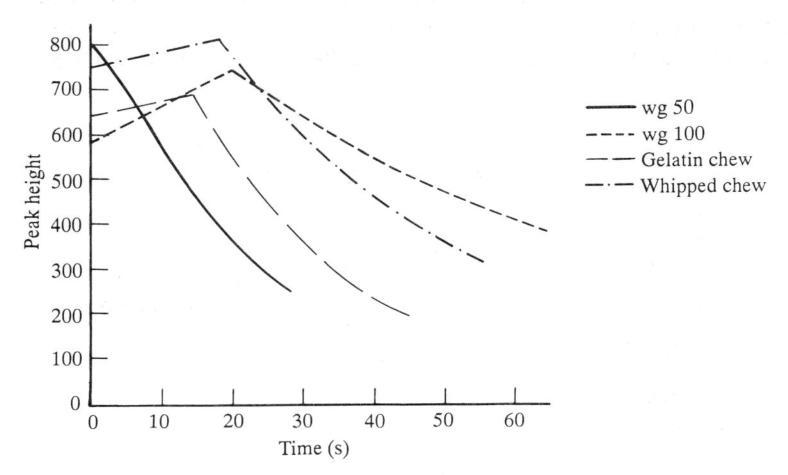

Figure 11.16 *Peak height data from four confectionery products with different structural and textural characteristics*

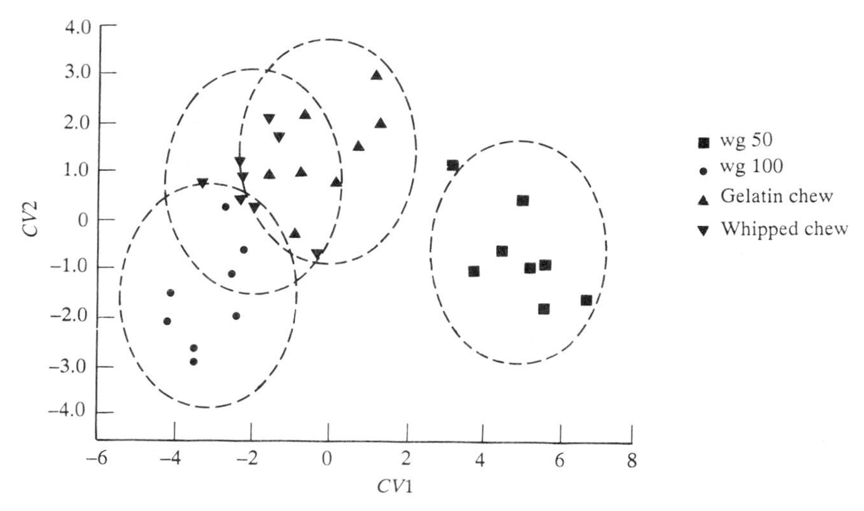

Figure 11.17 CVA plots on EMG data from four confectionery products

5 Future Developments in Texture Measurement

It is unlikely that the food industry will see any revolutionary developments in texture measurement in the near future. There will be a continuing demand for new empirical instruments for specific purposes; one of the authors has recently been asked to develop a simple hand-held instrument for measuring the texture of ratatouille. The industry will also increasingly

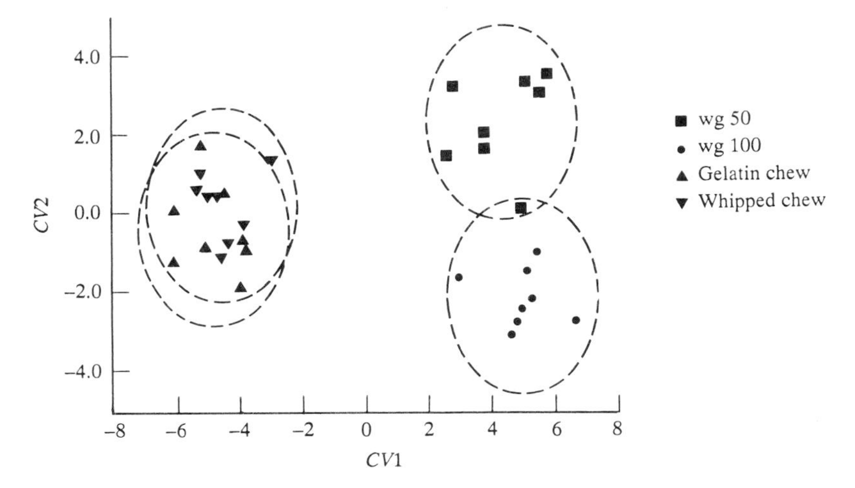

Figure 11.18 CVA plots on profile data from four confectionery products

need to rely on precision, general purpose machines, particularly with increasing product diversification. Although substantial improvements have been made to such instruments, particularly in terms of computer control and analysis, they still remain too complex (and expensive) for routine quality control use in many companies. Simplification and cost reduction are, however, not generally compatible with the required versatility. One important requirement for further development in such instrumentation will be the capability to operate accurately at the high rates of force application prevailing in the mouth. A summary of the characteristics of the main classes of texture measurement is shown in Table 11.1.

The industry is becoming increasingly aware that detailed texture assessment requires sensory methods. Relatively few companies, however, have the resources needed to operate trained profile panels. More companies may be able to operate free choice profiling procedures, with their minimal training requirements, but simplifications of the computer software is needed for analysing the profile data. Another sensory method that may have an important role is time–intensity (T–I) testing. T–I methods involve assessors recording the intensity of specific sensory attributes as a function of time (Lee and Pangborn 1986). Such methods have the advantage of following textural change during chewing, and have been used to investigate melting foods such as ice cream. A method has been developed at the Leatherhead Food Research Association for recording such data directly on to computer using a digitizer pad.

EMG also records, indirectly, textural change in the mouth. Most work to date has been associated with solid foods. Preliminary studies at Leatherhead have shown that information on viscous and semi-solid foods can be gathered from the group of muscles under the chin controlling tongue movement, although data acquisition is more difficult. This would also open up the possibility of monitoring the textural change of foods undergoing temperature-related phase changes, such as the melting of chocolate or frozen desserts, by monitoring both masseter and tongue muscle activity. At present EMG should be regarded as an important research tool, rather than as a quality control technique. Considerable simplification of equipment is feasible, but the need to rely on human subjects would limit its use as a routine quality control method. It has potential, however, as a tool in food product development and for spot checks.

With the exception of measurement of viscosity, there has been little progress in the development of on-line methods. One area in which future developments may be made is in the use of acoustic measurements. Research has been carried out into the use of sonic input methods for measuring the texture of biscuits (Povey and Harden 1981), cheese curd (Hatfield 1982) and soft gels (Ring and Stainsby 1985). The last mentioned is based on timing the passage of shear waves through soft gels, which form a difficult class of material for instrumental texture measurement. A

Table 11.1 *Summary of the main classes of texture assessment*

Assessment class	Food types	Initial costs	Operating costs	Operating environment	Development status	Consumer relevance
Sensory[a]	All	High	High	Laboratory	Mature/ continuing	High
Empirical[b]	All	Low	Low	Production/ laboratory	Mature/ continuing	Low[f]
Imitative[c]	Most solids	Moderate	Low	Laboratory	Obsolescent	Low/moderate[f]
Fundamental[d]	All	High	Low	Laboratory	Mature	Low[f]
Sound emission	Brittle solids	Moderate/high	Moderate	Laboratory[e]	Continuing	Moderate[f]
EMG	Solids, possibly viscous foods	Moderate	Moderate	Laboratory[e]	Continuing	Moderate[f]

[a]e.g. trained profile panels.
[b]e.g. hand-held penetrometers.
[c]e.g. General Foods Texturometer.
[d]e.g. Instron under tensile conditions.
[e]Further research may identify potential uses in production.
[f]Consumer relevance depends on the degree to which test data can be correlated with subjective textural data.

research programme is being carried out at the Leatherhead Food Research Association on the use of such methods.

6 Conclusions

As is appropriate for a food characteristic that is perceived by the human senses, texture assessment is increasingly being carried out using human subjects, alongside the assessment of other organoleptic characteristics such as appearance, odour and taste. Inevitably, however, the food industry has sought more rapid and less expensive methods of texture assessment based on instrumental measurement of the physical properties of food. Such methods can be judged to be reliable only if the measured instrumental properties can be shown to relate to relevant sensory attributes. This consideration applies to the simple tests based on empirical methods through to measurement of fundamental parameters using sophisticated instrumentation. Unfortunately, texture instrumentation is frequently used indiscriminately and without sufficient regard to the end user – the consumer. There is no reason why the wide range of instrumentation now available could not be used more effectively if more thought was given to the nature of the measurement process involved, and its relevance to the processes that occur in the mouth. It is probable that misuse of instrumental measurements is partly a result of a relatively poor understanding of the physiology of texture perception.

Recent developments involving measurement of sound emission and muscular activity on human subjects have demonstrated an alternative route that may be seen as bridging subjective and instrumental methods. In addition, such methods also highlight the importance of changes in food texture during chewing. Further exploitation of physiological principles and methods will lead to other innovations, but progress will require close cooperation between food scientists, physiologists and researchers from other disciplines.

References

Arnold, G. M. and Williams, A. A. (1986) The use of generalised Procrustes techniques in sensory analysis. In *Statistical Procedures in Food Research*, J. R. Piggott (ed), Elsevier, London, 233–253

Bourne, M. C. (1982) *Food Texture and Viscosity*. Academic, New York

Boyar, M. M. and Kilcast, D. (1986) Food texture and dental science. *Journal of Texture Studies*, **17**, 221–252

Brandt, M. A., Skinner, E. Z. and Coleman, T. A. (1963) Texture profile method. *Journal of Food Science*, **28**, 404–409

Brennan, J. G. (1988) Texture perception and measurement. In *Sensory Analysis of Foods*, 2nd ed, J. R. Piggott (ed), Elsevier, London, 69–102

Brennan, J. G., Jowitt, R, and Williams, A. (1975) An analysis of the action of the General Foods Texturometer. *Journal of Texture Studies*, **6**, 83–100

Civille, G. V. and Szczesniak, A. G. (1973) Guidelines to training a texture profile panel. *Journal of Texture Studies*, **4**, 204–223

Drake, B. K. (1963) Food crushing sounds: an introductory study. *Journal of Food Science*, **28**, 233–241

Friedman, H. H., Whitney, H. and Szczesniak, A. S. (1963) The Texturometer – a new instrument for objective texture measurement. *Journal of Food Science*, **28**, 390–403

Hatfield, D. S. (1982) A new instrument to measure cheese curd rigidity. *Journal of Physics E, Scientific Instrumentation*, **15**, 108–113

Jowitt, R. (1974) Terminology of food texture. *Journal of Texture Studies*, **5**, 351–358

Lee, W. K. III and Pangborn, R. M. (1986) Time–intensity: the temporal aspects of sensory perception. *Food Technology*, **40**, (11), 71–78, 82

Munoz, A. M. (1986) Development and application of texture reference scales. *Journal of Sensory Studies*, **1**, 55–83

Peleg, M. (1980) A note on the sensitivity of fingers, tongue and jaws as mechanical testing instruments. *Journal of Texture Studies*, **10**, 245–251

Povey, M. J. W. and Harden, C. A. (1981) An application of the ultrasonic pulse echo technique to the measurement of crispness in biscuits. *Journal of Food Technology*, **16**, 167–175

Proctor, B. E., Davison, S., Malecki, G. J. and Welch, M. (1955) A recording strain gauge tenderometer for foods. I: Instrument evaluation and initial tests. *Food Technology*, **9**, 471–477

Ring, S. G. and Stainsby, G. (1985) A simple method for determining the shear modulus of food dispersions and gels. *Journal of the Science of Food and Agriculture*, **36**, 607–613

Scott-Blair, G. W. (1958) Rheology in food research. *Advances in Food Research*, **8**, 1–56

Sherman, P. (1970) The correlation of rheological and sensory assessment of consistency. In *Industrial Rheology*, Academic, London, 371–391

Stone, H., Sidel, J., Oliver, S., Woolsey, A. and Singleton, R. C. (1974) Sensory evaluation by quantitative descriptive analysis. *Food Technology*, **28**, (11), 24, 26, 28, 29, 32, 34

Szczesniak, A. S. (1963a) Objective measurement of food texture. *Journal of Food Science*, **28**, 410–420

Szczesniak, A. S. (1963b) Classification of textural characteristics. *Journal of Food Science*, **28**, 385–389

Szczesniak, A. S. (1987) Relationship of texture to food acceptance and nutrition. In *Food Acceptance and Nutrition*, Academic, London, 157–172

Szczesniak, A. S., Brandt, M. A. and Friedman, H. H. (1963) Development of standard rating scales for mechanical parameters of texture and correlation between the objective and sensory methods for texture evaluation. *Journal of Food Science*, **28**, 397–403

Szczesniak, A. S. and Kahn, E. L. (1971) Consumer awareness and attitudes to food texture. I: Adults. *Journal of Texture Studies*, **2**, 280–295

Tornberg, E., Fjelkner-Modig, S., Ruderus, H., Glantz, P., Randow, K. and Stafford, D. (1985) Clinically recorded masticatory patterns as related to the sensory evaluation of meat and meat products. *Journal of Food Science*, **50**, 1059–1066

Vickers, Z. M. (1985) The relationships of pitch, loudness and eating technique to judgements of the crispness and crunchiness of food sounds. *Journal of Texture Studies*, **16**, 85–95

Vickers, Z. M. (1987) Sensory acoustical and force-deformation measurement of potato chip crispness. *Journal of Food Science*, **52**, 138–140

Vickers, Z. M. (1988) Evaluation of crispness. In *Food Structure – its Creation and Evaluation*, J. M. V. Blanshard and J. R. Mitchell (eds), Butterworths, Oxford

Vickers, Z. M. and Bourne, M. C. (1976) A psychoacoustical theory of crispness. *Journal of Food Science*, **41**, 1158–1164

Volodkevich, N. N. (1938) Apparatus for measurement of chewing resistance or tenderness of foodstuffs. *Food Research*, **16**, 73–82

12 *Water Activity and its Measurement in Food*

Wolfgang Rödel

Contents

1 Definition

Food should be stable and must be safe. These requirements mean that the products must not endanger the health of the consumer with micro-organisms or their toxins, or deteriorate owing to enzymic or microbial activity, at any stage from production through storage and retail to consumption. Factors determining microbial deterioration may be differentiated as intrinsic factors, process factors and extrinsic factors. Intrinsic factors include water activity a_w, pH value and redox potential E_h, and extrinsic factors cover temperature and humidity as well as atmospheric influences and partial pressures of gases during food storage. The techniques in food technology that affect shelf life by altering the conditions for microbial growth in the product are described as process factors. The control points critical for production can be deduced by analysing the hygienic risks of a food. Then measurements of the critical variables can be taken as part of process control, compared with standard levels and corrected where necessary. This concept of process control is known as hazard analysis critical control point (HACCP: Kaufmann and Schaffner 1974; Bonberg and David 1977; Bryan 1980).

For food, there are several factors that have a bearing upon any assessment of microbiological stability, and thus upon the shelf life and safety of a product. Water activity a_w is a particularly important parameter for risk analysis as defined by the HACCP concept, as are the pH value, the F_0 value and the redox potential (see Chapter 1, Section 1.2). These intrinsic factors of a food can be measured more or less accurately. Of the physical parameters, the pH value and the water activity of food may be reliably determined; equipment suitable for measuring the a_w level has been developed in recent years. As a consequence, the concept of water activity with all its significance has become ever more widely established in research and especially in industrial applications.

Water is essential for the growth and metabolic activity of micro-organisms. But not all of the water present in food is in fact available for the biological activity of micro-organisms or for other chemical and enzyme reactions. The concept of 'water activity' (Scott 1957) has generally been accepted as a parameter for the concentration conditions in the aqueous part of food. The water activity is defined as the ratio

$$a_w = p/p_0$$

where p represents the actual partial pressure of water vapour and p_0 the maximum possible water vapour pressure of pure water (saturation pressure) at the same temperature. The a_w level is therefore dimensionless; pure water has a level of 1.0, and a completely water-free substance has a level of 0.0. The relationship between the equilibrium relative humidity *ERH* in a food and the water activity is

$$a_w \times 100 = ERH$$

The a_w level is expressed as a fraction of 1, the equilibrium relative humidity as a percentage.

2 Significance of Water Activity

2.1 Effect of Water Activity on Food Quality

For foods with a high level of water activity, the shelf life is limited mainly by microbiological activity. Products with a_w levels below about 0.70 may well be stable microbiologically and consequently have a longer shelf life, but now the slower, enzyme-related breakdown processes come to the fore. It is mainly chemical reactions that determine the quality and stability of these foods. Figure 12.1 clarifies the mechanisms of food deterioration as a

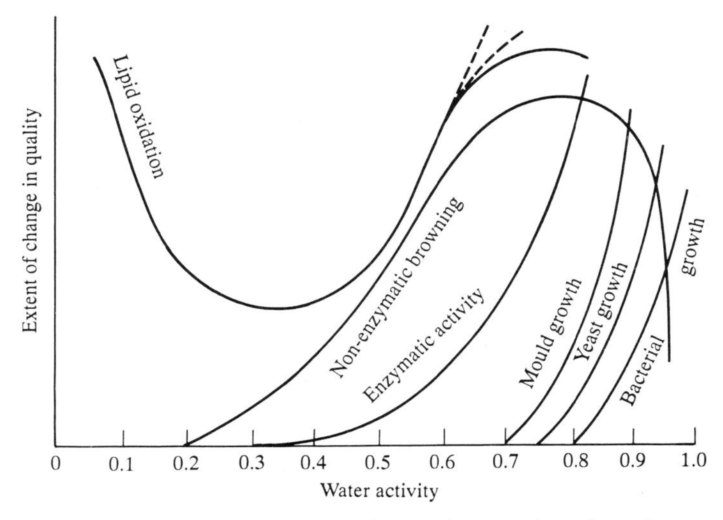

Figure 12.1 *Extent of change in quality as a function of water activity (from Heiss and Eichner 1971, Labuza et al. 1972b). The figure represents bacteria, yeasts and moulds of average tolerance. Individual strains can have exceptional a_w tolerance (see Table 12.1)*

function of water activity (Heiss and Eichner 1971; Labuza et al. 1972b). As shown in the figure, the shelf life of products with very low water activity is limited primarily by a marked fat oxidation (Maloney et al. 1966), whereas non-enzymic browning (Maillard reaction) is dominant, with a pronounced maximum in the range of intermediate water activities. Labuza et al. (1972b) also observed a further increase in fat oxidation in certain cases within this intermediate range. In foods with even higher a_w levels, the rate of reaction of enzyme-catalysed oxidation and hydrolysis also increases (Hunter et al. 1951; Acker 1962; Acker and Huber 1970), as there is now enough water available to transport the substrate to the enzyme. For water activities over 0.70, changes in the food are mainly caused by the growth of micro-organisms (bacteria, yeasts and moulds).

2.2 Effect of Water Activity on Food Stability

2.2.1 Water Activity Tolerance of Bacteria, Yeasts and Moulds

Micro-organisms, like people, contain over 70 per cent water. A very important function of water is maintaining osmotic pressure within the cell of the micro-organism and transporting nutrients. This transport mechanism works principally because the necessary osmotic forces required for osmosis between the inside of the cell and its surroundings are present. In this environment, the endogenous and exogenous enzymes produced by the micro-organisms can play their role in the microbial metabolism. By means of exogenous enzymes, larger molecules, which may not pass through the cell membrane of the micro-organism, may be split up into smaller fragments, which can then diffuse inside the cell through the osmotic barriers, aided by active transport mechanisms. Once here, the fragments are then either further oxidized directly by endogenous enzymes or prepared for oxidation in several stages.

If this ordered, highly complicated cooperation between different enzyme systems in the living cell is disturbed, for example by a reduction in the water activity, the reproduction, metabolic activity, resistance and survival of the micro-organisms in the food are affected.

As shown in Figure 12.2, many traditional food preservation processes, such as salting, sugaring, drying and freezing, alter the concentration of the particles dissolved in the water of the product and thus its a_w level (Rödel et al. 1979). The transport of nutrients into the cell interior of the micro-organism is affected by the reduction in water activity, since the osmotic pressure in the cell or its water activity can only be changed and adapted to environmental conditions within a limited individual range. The result is retarded growth of the micro-organism, or its death, thus producing a stabilizing or preserving effect on the food.

Micro-organisms occurring in food are frequently responsible for spoil-

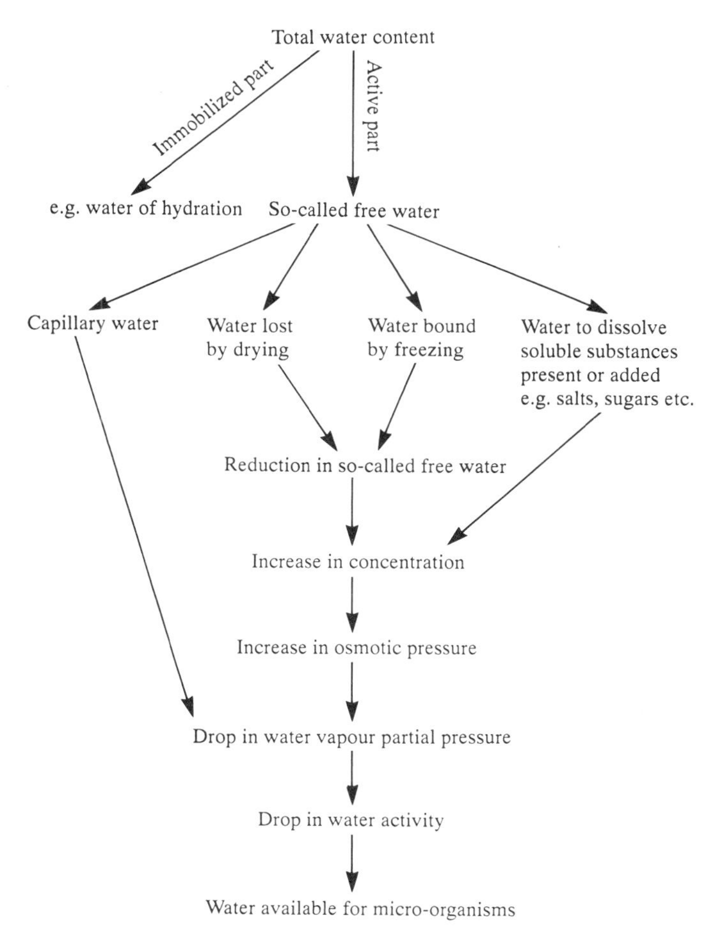

Figure 12.2 *Comparison between the water available to micro-organisms and the total water content of foods*

age, and under certain conditions also for food-induced infections or food poisoning. They may, however, be desirable, for example to preserve and add flavour to meat products (raw sausage and raw ham) or to dairy products by fermentation. All these desirable and undesirable microbial activities take place only if the water activity of the product permits multiplication of the appropriate micro-organisms. Table 12.1 gives the minimum a_w levels for the growth of various species of bacteria, yeasts and moulds. This table was compiled by Leistner et al. (1981) from data by various authors.

As can be seen from the table, bacteria in general require higher water activity in the substrate than yeasts, and yeasts higher levels than moulds. The micro-organisms under discussion are no longer capable of reproduc-

Table 12.1 *Minimum water activity a_w for multiplication of micro-organisms associated with foods (Leistner et al. 1981)*

a_w	Bacteria	Yeasts	Moulds
0.98	*Clostridium*[b], *Pseudomonas*[a]	—	—
0.97	*Clostridium*[c], *Pseudomonas*[a]	—	—
0.96	*Flavobacterium, Klebsiella, Lactobacillus, Proteus*[a], *Pseudomonas*[a], *Shigella*	—	—
0.95	*Alcaligenes, Bacillus, Citrobacter, Clostridium*[d], *Enterobacter, Escherichia, Propionibacterium, Proteus, Pseudomonas, Salmonella, Serratia, Vibrio*	—	—
0.94	*Bacillus*[a], *Clostridium*[e], *Lactobacillus, Microbacterium, Pediococcus, Vibrio, Streptococcus*[a]	—	*Stachybotrys*
0.93	*Bacillus*[f], *Micrococcus*[a], *Lactobacillus*[a], *Streptococcus*	—	*Botrytis, Mucor, Rhizopus*
0.92	—	*Pichia, Rhodotorula, Saccharomyces*[a]	—
0.91	*Corynebacterium, Streptococcus*	—	—
0.90	*Bacillus*[g], *Lactobacillus*[a], *Micrococcus, Staphylococcus*[h], *Vibrio*[a]	*Hansenula, Saccharomyces*	—
0.88	—	*Candida, Debaryomyces, Hanseniaspora*	*Cladsosporium*
0.87	—	*Debaryomyces*[a]	—
0.86	*Micrococcus*[a], *Staphylococcus*[i], *Vibrio*[j]	—	—
0.84	—	—	*Alternaria, Aspergillus*[a], *Paecilomyces*
0.83	*Staphylococcus*	*Debaryomyces*[a]	*Penicillium*[a]
0.81	—	*Saccharomyces*[a]	*Penicillium*
0.79	—	—	*Penicillium*[a]
0.78	—	—	*Aspergillus, Emericella*
0.75	*Halobacterium, Halococcus*	—	*Aspergillus*[a], *Wallemia*

Table 12.1—continued

a_w	Bacteria	Yeasts	Moulds
0.70	—	—	*Aspergillus*[a], *Chrysosporium*
0.62	—	*Saccharomyces*[a]	*Eurotium*[a]
0.61	—	—	*Monascus*

[a]Some isolates. [b]*Clostridium botulinum* type C. [c]*C. botulinum* type E, and some isolates of *C. perfringens*. [d]*C. botulinum* type A and B, and *C. perfringens*. [e]Some isolates of *C. botulinum* type B. [f]Some isolates of *Bacillus stearothermophilus*. [g]*B. subtilis* under certain conditions. [h]*Staphylococcus aureus* anaerobic. [i]*S. aureus* aerobic. [j]Some isolates of *Vibro costicolus*.

Sources: Stille 1948; Snow 1949; Burcik 1950; Bullock and Tallentire 1952; Christian and Scott 1953; Scott 1953; 1957; Williams and Purnell 1953; Christian 1955a; Wodzinski and Frazier 1960; 1961; Christian and Waltho 1962; 1964; Lanigan 1963; Riemann 1963; Blanche Koelensmid and van Rhee, 1964; Gough and Alford 1965; Hobbs 1965; Matz 1965; Brownlie 1966; Limsong and Frazier 1966; Segner et al. 1966; 1971; Baird-Parker and Freame 1967; Ohye and Christian 1967; Ohye et al. 1967; Kushner 1968; McLean eet al. 1968; Pitt and Christian 1968; Pivnick and Thatcher 1968; Emodi and Lechowich 1969; Kang et al. 1969; Mossel 1969; Bem and Leistner 1970; Strong et al. 1970; Troller 1971; 1972; Jakobsen et al. 1972; Rödel et al. 1973; Tomcov et al. 1974; Beuchat 1974; Pitt 1975; Leistner and Rödel 1975; 1976a; 1976b; Jakobsen and Murrell 1977; Troller and Christian 1978; Christian 1981; Rüegg and Blanc 1981.

tion below these a_w levels. The test results of the cited authors do not always agree on the a_w level limits for individual strains, partly because of the different experimental conditions. Therefore, the values in Table 12.1 must be seen as something of a compromise. As Table 12.1 shows, reproduction of most of the Gram-negative rods is inhibited in foods with an a_w level lower than 0.95, and this is also the case for most bacilli and clostridia and for germination of their spores. Neither can *Shigella, Salmonella, Escherichia coli* or most *Vibriona* multiply, so the most common causes of spoilage by microbial activity are eliminated, together with food-related infections and food poisoning. *Staphylococcus aureus,* also a food-poisoning organism, can tolerate a_w levels as low as 0.86, but under conditions of reduced oxygen this type of cell is inhibited at a level of 0.91.

If water activity in the substrate is adjusted not with NaCl or sugar but with glycerol, then different micro-organisms, such as *Clostridium botulinum* types A, B and E (Baird-Parker and Freame 1967), *Clostridium perfringens* (Kang et al. 1969), *Bacillus cereus* (Jakobsen et al. 1972; Jakobsen and Murrell 1977), *Salmonella oranienburg* (Christian 1955b;

Marshall et al. 1971; Rödel and Lücke 1983) and *Vibrio parahaemolyticus* (Beuchat 1974), grow if water activity is lower. This is worth mentioning because glycerol is frequently used in place of NaCl or sugar to reduce a_w in products of intermediate moisture content.

The tolerance of individual micro-organisms to water activity is in general lower if other factors in the foodstuff such as temperature, pH value, redox potential, oxygen and carbon dioxide concentration deviate from the optimum, or if the product has been treated with preservatives. This 'hurdle effect' (Leistner and Rödel 1976a; Leistner 1977; 1978) is of fundamental importance for food preservation because it may be used to prevent food-related infections, food poisoning and deterioration due to microbes, and the fermentation of food can be controlled more easily.

The water activity of food influences the toxin-forming ability of micro-organisms as well as their growth. According to Leistner et al. (1981) the limits for producing toxins by *Clostridium botulinum*, *Staphylococcus aureus* and toxinogenic moulds are as presented in Table 12.2. *Clostridium botulinum* types A, B, and E tolerate a_w levels of 0.95, 0.94 and 0.97; type C tolerates 0.98. The limit for the formation of enterotoxin C by *Staphylococcus aureus* is 0.94, whilst for enterotoxin B it is as low as 0.90. The staphylococci that cause most food poisoning are represented by type A, which only loses its toxin-forming ability at water activities below 0.87.

2.2.2 Water Activity Tolerance of Trichinae and Bovine Bladderworms

The sensitivity of trichinae to reduced a_w levels should also be mentioned. In model tests on sausage artificially infected with trichinae, Lötzsch and Rödel (1974) found that there were no further invasive trichinae when, by curing and drying the sausages, a_w levels fell below 0.93. In further tests Lötzsch and Leistner (1977) proposed a_w limits of 0.90 for sausage and 0.87 for ham as a protection against the possibility of trichinae in meat products.

A human parasite found worldwide is the beef tapeworm *Taenia saginata*. It is estimated that 40 million people are carriers of this intestinal parasite. In the (former) Federal Republic of Germany, on average 1 per cent of slaughtered cattle are infected with *Cysticercus bovis* or bladderworm, the larval stage of the beef tapeworm (Krauss and Weber 1986). This cysticercus is the sexless immature stage of the tapeworm, and lives in the muscles of cattle, its intermediate host. About 1 per cent of the population of the (former) Federal Republic of Germany are estimated to be tapeworm carriers (Grossklaus 1977). If any bladderworms are overlooked during the legally prescribed meat inspection then, despite all the care taken, if this meat is consumed raw – for example, in the form of steak tartare, rare steak or pink roast beef – the bladderworms grow into tapeworms in the human intestine. With this in mind, Schmidt and Rödel (1987) carried out tests to find out the a_w levels needed to destroy

Table 12.2 *Minimum water activity for toxin production by micro-organisms*

Micro-organism	*Minimum a_w*	*Source*
Clostridium botulinum		
Type C	0.98	a
Type E	0.97	b
Type A	0.95	b
Type B	0.94	b
Staphylococcus aureus		
Enterotoxin C	0.94	c
Enterotoxin B	0.90	d,e,f,g
Enterotoxin A	0.87	h
Mycotoxins		
Penitrem A	0.94	i,j
Citrinin	0.90	i,j
PR-toxin	0.90	i
Patulin	0.88	i
Cyclopiazonic acid	0.87	i,j
Roquefortine	0.87	i,j
Citreoviridin	0.86	j
Ochratoxin A	0.85	k,m
Griseofulvin	0.85	j
Verrucosidin	0.84	j
Aflatoxin	0.83	l
Penicillic acid	0.80	j,k

[a]Segner et al. 1971.
[b]Ohye and Christian 1967.
[c]Genigeorgis et al. 1971.
[d]Genigeorgis and Sadler 1966.
[e]McLean et al. 1968.
[f]Genigeorgis et al. 1969.
[g]Troller 1971.
[h]Lotter and Leistner 1978.
[i]Beuchat 1983.
[j]Lötzsch and Trapper 1979.
[k]Bacon et al. 1973.
[l]Northolt et al. 1977.
[m]Harwig and Chen 1974.

Table 12.3 *Expected inactivation time of bovine bladderworm (with and without bladder) by different NaCl concentrations (Schmidt and Rödel 1987)*

NaCl (%)	*With bladder (hours)*	*Without bladder (hours)*
1.72 (a_w 0.990)	96	72
3.19 (a_w 0.981)	15	10
4.90 (a_w 0.972)	3	1

Cysticercus bovis. The results are given in Table 12.3. The high NaCl sensitivity of the beef bladderworm, observed in the table, also permits statements on the possible risk of invasion through raw sausage. In NaCl concentrations of about 3 per cent bladderworms will not survive after 24 hours, and even in NaCl concentrations of 2.5 per cent they will most probably die after 2 days at the most. Human ingestion of viable bladderworms through eating sausage is almost totally avoided by a curing period of at least 7 days, which is allowed even for quick-cured raw sausage products. For other meat products such as, for example, raw ham with the usual common salt content, it can also be concluded that any potential bladderworm will die after 2 days at the most and there will be no danger to the consumer.

2.3 Legal Requirements

The importance attached to water activity as one of the few parameters easily measurable in food, for assessing chemical and microbiological stability, is underlined in many countries by legal specifications. These regulations contain limits for water activity alone as well as in combination with the pH value. These two parameters are then used to designate product stability. In Canada, meat products may be stored at room temperature if the a_w level is less than or equal to 0.90 and the pH value less than or equal to 5.4 (*Meat Hygiene Manual*, Meat Hygiene Division, Agriculture Canada, 1987). In Japan, dried or salted, smoked and dried meat products must have a_w levels less than 0.86 to be stored above 10°C, or lower than 0.94 to be stored below 10°C (*Food Sanitation Law*, Ministry of Health and Welfare). Limits for food in the USA, which only refer to the water activity, stipulate an a_w of less than 0.85, with no pH requirement (Johnston and Lin 1987; FDA 1979; 1985). These few examples show clearly how important it is to control the water activity of foodstuffs; and, particularly where sausage and ham are being produced for export into the above-mentioned countries, it is absolutely essential to control the water activity of these products. Compliance with the required limits is often strictly and severely enforced by the importing countries.

3 Water Activity Levels in Food and their Control

The water activity levels of fruits, vegetables, milk products and meat, measured by various authors in separate studies, have been published in the comprehensive work of Chirife and Ferro Fontan (1982). Alzamora and Chirife (1983) published water activity levels of different types of canned food such as fruits, vegetables and meat products. Detailed information on a_w levels in German meat products has been published by Rödel (1975).

3.1 Water Activity Levels in Food of Animal Origin

The water activity of meat and meat products is at the top of the a_w scale because of their high water content. Fresh meat has the highest water activity level, but this declines to a greater or lesser extent during processing into meat products. In this process, the a_w level is characterized particularly by the content of common salt in the aqueous phase of the product. The water activity of salted and dried meat products, such as raw sausage and ham, is therefore correspondingly lower. The levels for different meat products can be found in Table 12.4 (Leistner et al. 1981).

Fresh meat comes top of the list, with an a_w of 0.99 in the lean part. This level is not affected by the type of animal, muscle group, or water-holding capacity. Meat products have a lower water activity than fresh meat and therefore in general a longer shelf life. Measures such as, for example, adding common salt, extracting water or adding fat have the greatest effect in reducing the a_w level of meat products. The addition of fat indirectly influences the a_w level of meat products, since fat contains very little water in comparison with lean meat, as shown in Table 12.5. Meat products that are very rich in fat therefore contain relatively little water, so that the same amount of salt added to these products produces a sharper decrease in the a_w level than in products with a greater lean portion (Rödel et al. 1980). Meats with a relatively high a_w level, and therefore short shelf life, are frankfurter-type products; this is attributed to a large and variable excess of water, according to recipe, which is a processing requirement for this product group. Meat products produced without heating, and so consumed raw, must have lower water activities to guarantee the required stability towards microbial spoilage of the product, and to ensure safety by avoiding any threat to the health of the consumer. Allowance is made for this demand in the traditional production processes for sausage and ham and similar products by expensive drying processes, for example. The water activity of raw ham and bacon comes within the wide range of 0.88 to 0.96. This variation is principally due to the difference in the degree of drying. There is less variation with dried beef. The a_w range for sausages is also quite wide. Hungarian and Italian salamis have the lowest a_w levels, and the shelf life for these products is limited only by chemical changes, such as rancidity.

Bone (1973) and Karmas and Chen (1975) demonstrated in their work that low-molecular-weight soluble compounds in food, in contrast to high-molecular-weight compounds, have a considerable effect on the water activity. The number of dissolved particles is critical for the a_w. It is not only in meat products but also in cheese that proteins of high molecular weight are found alongside compounds of low molecular weight. Some of these, according to Rüegg and Blanc (1977), develop in cheese during the maturing process; others are added during production, for example sodium chloride. Table 12.6 gives the a_w levels of different cheeses. Detailed assay data on different types of European cheese were published by Rüegg and Blanc (1977; 1981) and by Marcos et al. (1981).

3.2 Water Activity Levels in Food of Vegetable Origin

Most bakery products have water activities that do not allow bacteria or yeasts to grow. Moulds, on the other hand, may still develop. Individual

Table 12.4 *Water activity range of fresh meat and some representative meat products (Leistner et al. 1981)*

Product	*Minimum*	*Maximum*	*Average*
Fresh meat	0.98	0.99	0.99
Bologna sausage	0.87[a]	0.98	0.97
Liver sausage	0.95	0.97	0.96
Blood sausage	0.86[b]	0.97	0.96
Raw ham	0.80[c]	0.96	0.92
Dried beef[d]	0.80	0.94	0.90
Fermented sausage	0.65[e]	0.96[f]	0.91

[a]Tiroler.
[b]Speckwurst.
[c]Country cured ham.
[d]Bündener Fleisch.
[e]Hard sausage.
[f]Fresh Mettwurst.

Table 12.5 *Water activity of porcine and bovine fat (Rödel et al. 1980)*

Species and location	*Treatment*	a_w	H_2O (%)	*NaCl* (%)
Pork, back fat	Fresh	0.991	9.6	0.1
	Chilled	0.982	6.5	0.1
	Salted and smoked	0.724	2.4	1.2
Beef, tallow	Fresh	0.993	13.6	0.2
	Chilled	0.984	4.2	0.2

Table 12.6 *Water activity of various European cheeses (Rüegg and Blanc 1981)*

Type	a_w *(25°C)*	*Standard deviation*
Appenzeller	0.962	0.011
Brie	0.980	0.006
Camembert	0.982	0.008
Cheddar	0.950	0.010
Cottage cheese	0.988	0.006
Edam	0.960	0.008
Emmentaler[a]	0.972	0.007
Fontal	0.962	0.010
Gorgonzola	0.970	0.017
Gouda	0.950	0.009
Gruyère[a]	0.948	0.012
Limburger	0.974	0.015
Münster	0.977	0.011
St Paulin	0.968	0.007
Parmesan	0.917	0.012
Quarg	0.990	0.005
Sbrinz[a]	0.940	0.011
Tilsiter	0.962	0.014
Processed cheese	0.975	0.010

[a]Values for Emmentaler, Gruyère and Sbrinz were measured after ripening periods of 4–5, 6–7 and 10–11 months, respectively. The other values were determined using commercially available samples.

Table 12.7 *Water activity range and water content of some breads and pastries (Flückiger and Cleven 1978)*

Product	a_w	H_2O (%)
White bread	0.92	40
Plundergebäck[a]	0.86	26
Cake	0.83	26
High-ratio cake	076	26
Gingerbread	0.63	16
Rusk	0.38	6
Waffles	0.30	5

[a]Danish pastry.

types of mould again require different minimum a_w levels for growth in bakery products. The kinds of *Penicillium* forming a velour-type cushion may still develop well on bread and brioches, for example, whereas they are no longer able to multiply on cake because of the lower water activity. However, cake might still be affected by types of *Aspergillus*, which impart the appearance of a spider's web. It is possible to influence the water activity of bakery products principally by means of sugar (sucrose), invert sugar, fructose, glucose, sorbitol and salt, as well as by reducing the overall water content (Flückiger and Cleven 1978). Table 12.7 presents a general survey of the a_w levels of different bakery products.

Chirife and Ferro Fontan (1982) give an overview of the water activity levels of produce in a paper in which the a_w levels of over 80 fruit and vegetable products are listed.

3.3 Control of the Water Activity Level

The modification of the water activity in foods is not only limited by flavour considerations, but also restricted more or less rigidly by the legal regulations in individual countries. However, numerous publications have described technologies for new types of foods termed 'intermediate-moisture foods' (Heiss and Eichner 1971; Labuza et al. 1972a; 1972b; Bone 1973; Ross 1975; Davies et al. 1976; Simatos and Multon 1985). These products, in which mostly glycerol is used to retain moisture, have achieved great significance in a similar form in animal feeds. The moisture retainers or humectants, such as glycerol or propylene glycol, used in these products combine with or replace part of the water in the product and therefore reduce the a_w level without the product losing its tenderness.

These products can be stored for a long period even without refrigeration if a fungal growth inhibitor is applied. For traditional meat products, the water activity can be reduced mainly by salting, adding fat and drying. For Asian products the water activity is also in many cases reduced additionally by large quantities of sugar.

According to Flückiger and Cleven (1978) there are two ways of reducing the water activity of bakery products to below the critical limit of 0.75 for these products (lower limit for mould growth). During bakery production the water content of the product can be greatly reduced either through the choice of recipe and the process, or by the addition of sugar or sugar substitutes. In Great Britain and the USA the latter option is used for so-called high-ratio cakes. When special flours (cake flour) are used, sugar may be added at levels up to 160 per cent in proportion to the flour. The a_w level of these very moist cakes lies in the range between 0.70 and 0.76. Unopened, these products are protected from mould growth even after prolonged storage. The fact that cakes containing raisins or candied fruit are hardly affected by mould is likewise due to a reduction in the water activity brought about by the soluble sugars present in the fruits.

3.4 Example: Regulating Raw Sausage Ripening by Controlling the Water Activity Level

It is not only from a microbiological point of view that water activity in food is of interest. This parameter may also be used to optimize products and to save energy in the processing steps involved in the fermentation of raw sausages such as salami. Conventional sausage ripening is currently based mainly on empirical principles. Thus, established processing conditions are regarded as optimal if there are no significant losses in the form of faulty products, for example with overdried edge zones or with tears. The ripening conditions (that is, temperature, humidity, air flow rate and time) are still altered mainly on the basis of sensory impressions, for example elasticity and dampness of the skin, external colour etc. As a safeguard, a spot check is sometimes made on weight loss, pH value or water activity. Only in conditioned curing rooms in larger plants are continuous records of temperature and humidity standard practice. The character of modern curing plants has, however, changed with the increased use of microprocessors. In these new-generation microprocessor-controlled air-conditioned curing chambers, particularly those used in the production of raw slicing sausage, it is necessary to replace the older 'time control' of the curing chamber with process control, involving continuous assessment of selected fermentation parameters to allow continuous feedback to the curing process. It is then possible to react to desired – or particularly to undesired – changes and processes in the sausage during the long fermentation process by regulating the curing plant. The measurements necessary for this are supplied to the microprocessor by sensors (Rödel and Stiebing 1987; Stiebing and Rödel 1989).

4 Measuring the Water Activity Level

4.1 Background

The following methods for measuring water activity in foods, with the exception of the freezing point technique, operate by determining the equilibrium moisture content. In this context, equilibrium means that equality has been reached between the water activity of the food and the relative humidity of the air enclosed in a measuring chamber impermeable to water vapour. The following conditions are therefore essential for accurate practical measurement of a_w. The measuring chamber must be sealed to prevent the effects of humidity in the external air on the eqilibrium humidity in the chamber, and to prevent water vapour losses from the chamber. The water content of the samples should be practically identical before and after achieving equilibrium in the measuring chamber, which presupposes that the volume of air enclosed with the food in the measuring chamber is small. The rate of equilibration (determining the measuring period) is also increased for small volumes of air.

4.2 Water Activity as a Function of Temperature

The water activity of foods decreases as the temperature drops (Ross 1975; Van den Berg 1975; Fennema 1978). Research carried out by Krispien and Rödel (1976) and Rödel and Krispien (1977) on meat and meat products has shown that if these products are cooled from 25°C down to the chilled and frozen range, there is a reduction in the a_w level. At temperatures above the onset of freezing the decrease in a_w level is insignificant ($\sim$0.000 15 per K), whereas below this point the decrease is considerable ($\sim$0.008 per K). Below the freezing point of meat and meat products, the a_w level equals the a_w level of ice at the particular freezing temperature (Table 12.8). The a_w level of foods at freezing temperatures can be read from this table. Table 12.9 illustrates the processes at decreasing temperatures for meat and meat products. All frozen foods, including frozen water, have the same a_w level at the same temperature.

The temperature above freezing point at which the a_w is measured therefore makes no difference, as the effect of temperature on the vapour pressure ratios is very slight. It is only necessary to ensure that calibration and sample measurement take place at the same temperatures. A measurement temperature of 25°C has proved practical, as there is a great deal of information in the literature on calibration references at this temperature. Particular attention must be paid to the constancy of temperature while the a_w is being measured. Any difference in temperature between measuring

Table 12.8 *Water activity of meat at freezing temperatures (calculated data from Moran 1936; Storey and Stainsby 1970; Fennema and Berny 1974)*

Temp (°C)	a_w
−1	0.990
−2	0.981
−3	0.971
−4	0.962
−5	0.953
−6	0.943
−7	0.934
−8	0.925
−9	0.916
−10	0.907
−11	0.899
−12	0.889
−13	0.881
−14	0.873
−15	0.864
−16	0.856
−17	0.847
−18	0.839
−19	0.831
−20	0.823
−21	0.815
−22	0.807
−23	0.799
−24	0.792
−25	0.784
−26	0.776
−27	0.769
−28	0.761
−29	0.754
−30	0.746

Table 12.9 *Water activity of meat and meat products at freezing temperatures (Rödel and Krispien 1977)*

	a_w of product examples			
Temperature (°C)	Fresh Meat	Bologna sausage	Liver sausage	Fermented sausage
25	0.993	0.980	0.970	0.870
5	essentially unchanged	essentially unchanged	essentially unchanged	essentially unchanged
0	↓	↓	↓	↓
−1	↓	↓	↓	↓
−2	0.981	↓	↓	↓
−3	0.971	0.971	↓	↓
−4	0.962	0.962	0.962	↓
−5	0.953	0.953	0.953	↓
−10	0.907	0.907	0.907	↓
−15	0.864	0.864	0.864	0.864
−20	0.823	0.823	0.823	0.823

chamber, sensor and food may cause gross errors in the measurement. The higher the a_w level of the sample, the greater the error.

4.3 Influence of Equilibration Periods and Sample Properties

It is not only the humidity equilibrium that can be adversely affected by variations in temperature during measurement; most electronic a_w sensors also have a typical temperature response. Thus all water activity measurements must be taken at ambient temperatures that are as constant as possible (maximum ± 0.2K fluctuation), which requires the use of temperature-controlled cabinets or Peltier-cooled boxes.

The accuracy and in particular the reproducibility of many methods are adversely affected by inadequate measuring periods. If the a_w level of the sample is determined by the process of equilibration, the measuring period for foods is generally about 2 to 4 hours. For better control during measurement, the equilibration should be checked with a recorder. If, however, the a_w level of a sample is established by determining the freezing point, then about 8 to 30 minutes will be necessary, depending on the level of water activity.

In the measurement of water activity by certain techniques, reactions may occur at the sensor due to various chemical compounds introduced with the sample, and these reactions can then compromise the results (Rödel et al. 1979; Pollio et al. 1986); glycerol, propylene glycol and similar compounds are principally responsible for such reactions.

Sample inhomogeneity can also influence the result, particularly for air-dried sausages or hams. For such products, the average water activity is of little significance. To determine the storage stability, the sample portion with the highest a_w (namely the sample core) needs to be tested. The a_w profile can help to study the drying technology employed.

4.4 Instrument Calibration

Both saturated and unsaturated solutions of various salts are suitable as standards for testing or calibrating a measurement technique. For calibration, Stoloff (1978) recommends saturated salt solutions in the form of salt slurries (Table 12.10). Tables of a_w levels (percentage relative humidity) of saturated salt solutions at different temperatures can be found in Greenspan (1977), Resnik et al. (1984), Kitic et al. (1986b) and Pollio et al. (1987). Unsaturated NaCl solutions varying in concentration are particularly suitable for calibrating instruments in the a_w range from 0.75 to 0.99, critical for the microbial stability of foods. These solutions are easily made up and are relatively unaffected by temperature over a wide range (Chirife and Resnik 1984). The a_w levels of NaCl solutions of differing molality and differing percentage levels (according to Robinson and Stokes 1965; Krispien and Rödel 1976) are brought together in Table 12.11. Saguy and Drew (1987) report on the adequate statistical evaluation of calibration data determined by different techniques for a_w measurements.

Table 12.10 *Water activity of salt slurries at 25°C (Stoloff 1978)*

Salt	a_w	*Salt*	a_w
$MgCl_2$	0.328	KBr	0.809
K_2CO_3	0.432	$(NH_4)_2SO_4$	0.810
$MgNO_3$	0.529	KCl	0.843
NaBr	0.576	$Sr(NO_3)_2$	0.851
$CoCl_2$	0.649	$BaCl_2$	0.902
$SrCl_2$	0.709	KNO_3	0.936
$NaNO_3$	0.743	K_2SO_4	0.973
NaCl	0.753		

Table 12.11 *Water activity of NaCl solutions with various molalities at 25°C (from Robinson and Stokes 1965)*

Molality	*NaCl* (% w/w)	a_w
0.1	0.58	0.996
0.2	1.15	0.993
0.3	1.72	0.990
0.4	2.28	0.986
0.5	2.84	0.983
0.6	3.39	0.980
0.7	3.93	0.976
0.8	4.47	0.973
0.9	5.00	0.970
1.0	5.52	0.966
1.2	6.55	0.960
1.4	7.56	0.953
1.6	8.55	0.946
1.8	9.52	0.938
2.0	10.46	0.931
2.2	11.39	0.924
2.4	12.30	0.916
2.6	13.19	0.908
2.8	14.06	0.901
3.0	14.92	0.893
3.2	15.75	0.885
3.4	16.58	0.876
3.6	17.38	0.868
3.8	18.17	0.860
4.0	18.95	0.851
5.0	22.62	0.807
6.0	25.97	0.760

5 Measurement Technique

5.1 Manometric Method

Numerous methods of measuring water activity have been treated in detail in review articles (Troller and Christian 1978; Prior 1979; Rödel et al. 1979; Troller 1983a; Wolf 1984; Weisser et al. 1985). A special measuring technique, the direct manometric measurement of vapour pressure, has been described by Legault et al. (1948), Taylor (1961), Sood and Heldman (1974), Lewicki et al. (1978), Troller (1983b), Nunes et al. (1985), Benado and Rizvi (1987) and Saguy and Drew (1987). For this method, the comminuted sample is evacuated in a desiccator for several minutes, and after 1 hour the water vapour pressure in equilibrium with the sample is measured by means of an oil or capacitance manometer. This method requires very accurate temperature control. Sample volatiles other than water, if present, will influence the measurement.

5.2 Gravimetric Method

In the isopiestic method, one determines the water activity of foods from the sorption isotherms of suitable materials (Landrock and Proctor 1951; Smith 1965; Gur-Arieh et al. 1965). In the so-called Fett-Vos method, the water activity of food is determined by means of dried reference materials (proteins, microcrystalline cellulose) (Fett 1973; Vos and Labuza 1974; Vansteenkiste and van Hoof 1982). The dried reference material is equilibrated with the sample in an evacuated desiccator, and any weight alteration in the reference substance is then recorded. The water activity in the sample is calculated from the change in weight and the known sorption isotherms of the reference substances. In a comparable method, Steele (1987) used polyols as references, determining the change in water content refractometrically rather than gravimetrically. These methods are relatively easy to perform. If equilibration is carried out in a static atmosphere they are not, however, suitable for measuring perishable foods because of the long adjustment period of more than 24 hours. But if the work is carried out in a dynamic, conditioned air stream, this period is substantially reduced (Multon et al. 1980).

Lang et al. (1981), McCune et al. (1981), Lenart and Flink (1983) and Palacha and Flink (1987) describe a proximity equilibration cell (PEC) method, which is not very expensive and makes use of the change in weight of filter paper to determine the a_w level of the sample.

5.3 Psychrometric Method

Prior et al. (1977) and Wiebe et al. (1981) outline a psychrometric technique for foods (Wescor MJ-55 microvoltmeter and C-52 test chamber, or Wescor HR-33T dew-point hygrometer and C-52 test chamber, from Wescor Inc., 459 South Main Street, Longan, UT 84321, USA) for the a_w ranges of microbiological interest, from 0.935 to 1.0 or 0.99 to 0.60. The equipment comprises a microvoltmeter and a test chamber with a thermal element. A sample is equilibrated in the chamber for 2 hours and the thermal element (Peltier cooling) is cooled, causing water vapour from the air in the chamber to condense on it. The rate at which water vapour condenses on the thermal element is a function of the a_w level of the sample, and proportional to the psychrometer reading. The a_w level of the sample is determined with the aid of a standard curve obtained from salt solutions of known a_w level. The Wescor psychrometer gives a reproducibility (coefficient of variation; CV) of between 0.18 and 0.35 per cent. For the a_w range above 0.90, a standard deviation of 0.0017 to 0.0035 a_w is given by Prior et al. (1977). An a_w SC-10 psychrometer with an MT-3 nanovoltmeter from Decagon Devices Inc. (PO Box 835, Pullman, WA 99163, USA) works on a similar principle.

5.4 Hygrometric Methods

5.4.1 Salt Method

The a_w level may be determined relatively cheaply using apparatus with the salt/filter-paper method of Kvaale and Dalhoff (1963), as modified by Northolt and Heuvelman (1982) and Hilsheimer and Hauschild (1985). The method is based on the fact that dried salt adhering to filter paper is dissolved if the humidity of the surrounding air has reached a point equal to the saturation humidity of the salt. If salts of different saturation humidities are used, the a_w level of a sample may be estimated.

5.4.2 Thread Hygrometers

This method is based on the hygroscopicity of the polyamide thread, which reacts to an increase in moisture by elongating noticeably. The instrument shown in Figure 12.3 for measuring a_w (a_w level meter 5803, made by Lufft, Gutenbergstrasse 20, D-7012 Fellback-Schmiden, Germany, and Abbeon-Cup, Abbeon Cal. Inc., 123-1A Gray Avenue, Santa Barbara, CA 93101, USA) consists of a sample-scale pan and an attachment, which is connected from the pan to a measuring unit by means of a lever system. The polyamide thread is inside the attachment. The change in thread length is converted to a scale on an indicator by means of an axle and lever mechanism.

Measuring water activity with this instrument (which takes about 3–4 hours) requires constant temperature. This method has been extensively tested and described by Rödel and Leistner (1971), Rödel et al. (1975), Labuza et al. (1977), Bousquet-Ricard et al. (1980), Jakobsen (1983), Gerschenson et al. (1984) and Stroup et al. (1987). The accuracy of the instrument is ± 0.01 a_w and the reproducibility is also 0.01 a_w (Rödel et al. 1975; Bousquet-Ricard et al. 1980; Jakobsen 1983) at a constant temperature of 25°C, with a sufficient equilibration period and using an adsorptive procedure. Foods that contain glycerine or similar volatile organic compounds cannot be measured using this equipment.

5.4.3 Electric Hygrometers

There is a relatively large number of different electric hygrometers on the market for measuring the water activity of foods. Since they cannot all be mentioned here, details will be given only of those employed in our own measurements on foods. The electric or electronic hygrometers fall into the category of capacitive hygrometers, conductivity hygrometers and dew-point hygrometers according to the principle on which they function.

Capacitive hygrometers In capacitive hygrometers the capacitance of a polymer capacitor in a measuring chamber changes as a function of humidity. The instrument shown in Figure 12.4 (Hygromess Labo type 47, made by Steinecker Elektronik, Daimlerstrasse 9, D-6052 Mühlheim/ Main, Germany) works on this principle to determine a_w. The equipment consists of a display unit connected to a measuring chamber. This measuring station may be extended to include up to seven chambers monitored by means of a multiplexer (type 4720). It may be used to take reliable a_w measurements for foods in the range 0 to 1.00, with the aid of microprocessor-controlled linearization of the calibration curve linked to temperature compensation of both sensor and electronics, and a provision for calibration storage. The instrument has digital data output (serial interface RS 232-V24), enabling further computerized data processing.

The measuring principle is based on a capacitor of very low mass (rapid

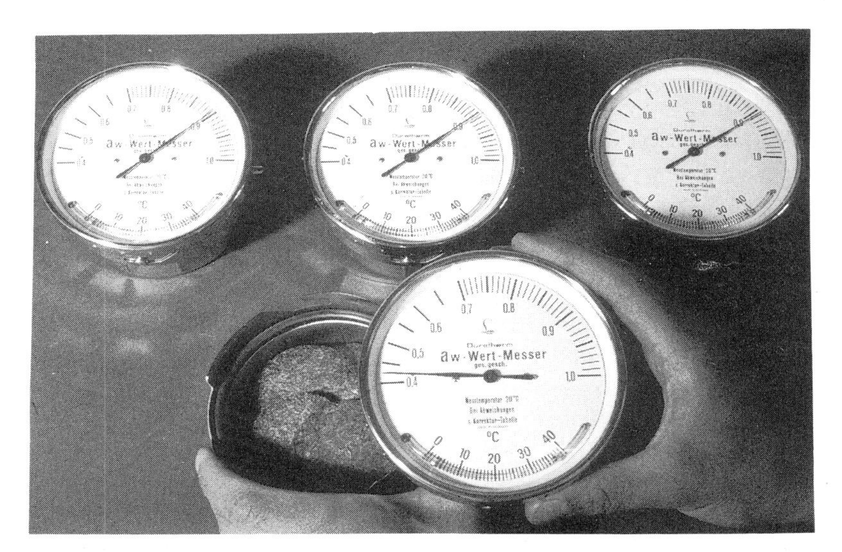

Figure 12.3 *Thread hygrometer Lufft type 5803*

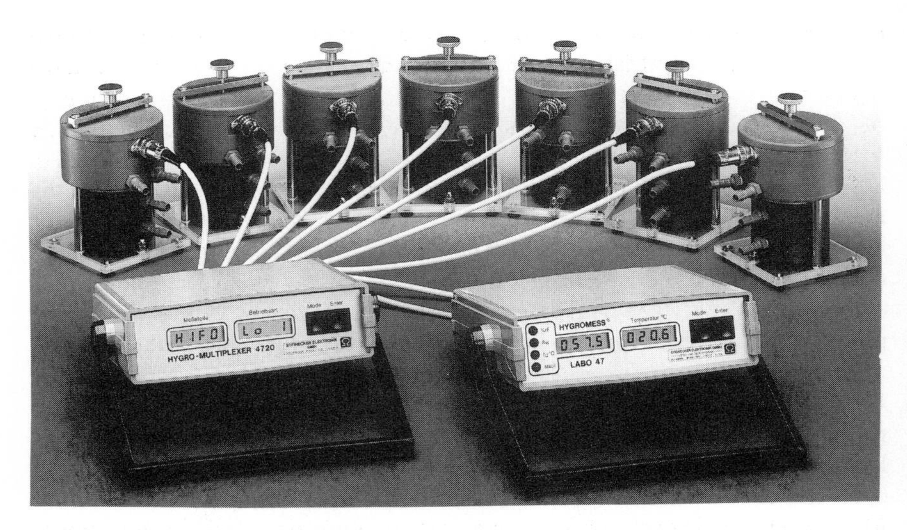

Figure 12.4 *Capacitive humidity meter Steinecker Hygromess Labo type 47 (front right) and a multiplexer 4720 with seven probes for water activity measurement*

temperature adjustment) acting as a humidity sensor. The dielectric is a hygroscopic polymer, which has a thin permeable sputtered layer of metal on both sides. Any alteration in relative humidity triggers a very rapid proportional change in capacitance. The adjustment period for equilibrium humidity in the measuring chamber is optimized by means of an integral fan, the period being about 2.5 hours for foods. The measuring head with the capacitance-measuring cell is pressed by a compression system against the surface of the sample dish so that it is water-vapour tight. Potentiometer adjustments are not necessary because of software-controlled equipment calibration. Calibration points for a_w from 0.00 to 0.95 are programmed into the digital electronics of the instrument. The accuracy of the system over this range of measurements is better than $\pm 0.015\ a_w$, and there is good reproducibility in the 0.95 a_w range under investigation (standard deviation 0.001, CV=0.11 per cent for $n=6$) (Rödel 1989, unpublished data).

The a_w measuring equipment Humicap HMI 32 with the DK 159 sorption probe (made by Vaisala OY, PL 26 SF-00421, Helsinki 42, Finland) also works on the principle of capacitance measurement. The system consists of the digital humidity/temperature display attached to the special sorption probe. A miniature fan is fitted to this probe to reduce the equilibration time. The head, with its capacitance-measuring cell and complete probe electronics, is simply laid on to the surface of a sample dish. A measuring period of 30 minutes for the range of 0.752 to 0.974 at 25°C is suggested by Favetto et al. (1983). In our own tests with unsaturated NaCl solutions in a measuring range from 0.80 to 0.95 (ten measurements at each of 0.80, 0.85, 0.90, 0.95) using this technique there was a mean standard deviation of 0.002 a_w (Rödel 1989, unpublished data). The cell can change its characteristics if it comes into contact with various volatile organic substances, and this can be ascertained from the manufacturer. Favetto et al. (1984) named certain alcohols in this connection (ethyl alcohol, methyl alcohol) as well as 1-2 propylene glycol and 2-3 butylene glycol. However, glycerol had no interference effect on the sensor.

Conductivity hygrometers With conductivity hygrometers, the measurement of electric impedance of a liquid hygroscopic substance is used for direct reading of the relative humidity or water activity in foods. The liquid hygroscopic materials include salt solutions or mixtures of various salt solutions, depending on the make. These sensors are thus also called electrolytic cells. The principle on which these cells are based relies (with only minor variations between individual makes) on a very precisely defined tiny quantity of the hygroscopic material sandwiched between a pair of electrodes mounted on a support plate. This substance tends towards equilibrium with the ambient humidity. The electrolyte produces a defined water vapour pressure at its surface, depending on its temperature and water content. If there are differences between this water vapour pressure and that in a test chamber, there is a water vapour exchange until the two pressures are identical. The water content of the electrolyte thus changes, depending on the temperature and water vapour pressure in the monitored surroundings, that is as a function of the relative humidity of the air or, in equilibrium, of the a_w level of a food. The impedance of the system, measured by an electronic bridge with a high-frequency signal, is proportional to the water content. The effects of temperature on the hygroscopic material are largely corrected by electronic compensation, which is built directly into the cell. The electrolytic measuring cells are calibrated against salt solutions with known a_w levels and provide a signal that is directly dependent on the measured a_w level.

When this type of a_w measurement system is used, in addition to the upper temperature limit, the following conditions should generally be observed. Direct contact of the test cells with water or salt solutions should be avoided under all circumstances. This means that the a_w level of calibration liquids and of foods may only be measured in the air space above their surfaces. Measuring chambers for samples are designed accordingly for different systems. The electrolytic cells must not be subjected to any heavy shock condensation such as may occur with larger swings from low to high temperatures. Heavy mechanical impacts when measuring above a_w 0.80 must be avoided.

Electrolytic measuring cells, in common with capacitance cells, may react to volatile inorganic and organic substances by changing their characteristics, and these changes can be difficult to interpret. Low concentrations ($\sim$100 ppm) are normally tolerated, with the operating range for temperature and humidity also having an influence. To protect the measuring cell, some manufacturers offer special filters (active carbon filters) which are particularly recommended for use when measuring the a_w of petfoods, as these animal foods are frequently produced with a propylene glycol additive (Pollio et al. 1986). If filters are not used, high concentrations of organic vapours may dissolve in the hygroscopic material and its characteristics may change. Some chemicals have only a temporary effect on the cell (that is the measuring cell regenerates when they evaporate from the electrolyte), but others have an irreversible effect on the electrolyte, destroying the measuring cell. Oil and fat volatiles are also harmful to the cell as these materials can condense in the sensor and prevent it from functioning.

Figure 12.5 illustrates the Hygroskop DT electrolytic a_w measuring system with the WA-40 measuring station (made by Rotronic AG, Badenerstrasse 435, CH-8040 Zürich, Switzerland, and Kaymount Instrument Corp., Huntington Station, NY, USA). This system is particularly suitable for measuring water activity of foods over an a_w range of 0 to 1.0. Because the chamber is of solid metal construction, temperature differences within it are eliminated and rapid changes in temperature are compensated. The chamber is well sealed (leakage rate lower than 0.005 a_w per 24 h), permitting exact measurements of foods even with very long equilibration periods. The test cell is calibrated against reference salt solutions under the same conditions as the actual a_w measurement.

The samples to be measured are placed in small polystyrene dishes in the bottom half of the measuring station. The top part of the chamber contains the measuring head and is locked from above with a lever compression system to seal the measuring station against a neoprene O-ring. The lever system allows rapid opening and closing of the chamber.

The WA-40 measuring station, together with the Hygroskop DT, gives

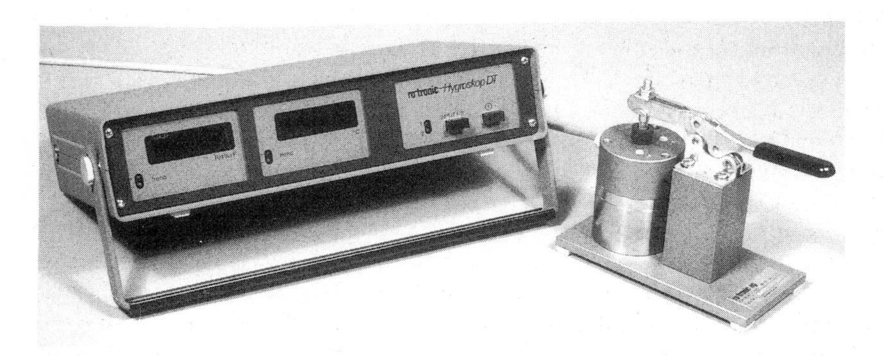

Figure 12.5 *Conductivity humidity meter Rotronic Hygroskop DT with special probe WA-40*

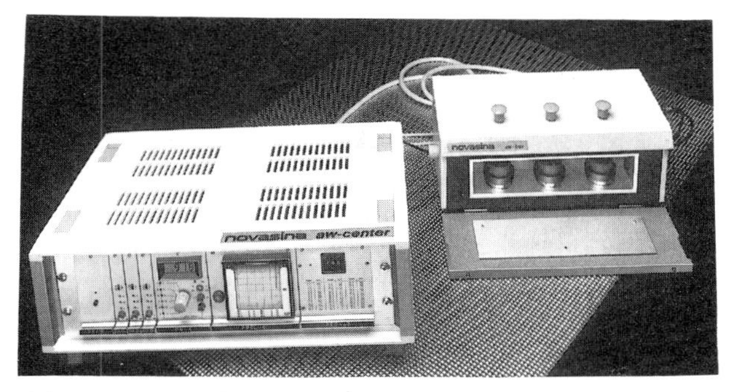

Figure 12.6 *Conductivity humidity Novasina a_w-centre, with on the right the temperature-controlled chamber for three samples*

very accurate a_w measurements of foods because of the high stability and very good linearity of the instrument. However, one requirement is that the temperature must be kept very constant, which is only possible in precision heating/cooling cabinets. Temperature and a_w levels are indicated simultaneously in digital form on the instrument; two samples may be measured at the same time. Dual outputs for voltage and current allow continuous signal recording.

This measurement system was investigated in detail under practical conditions by Rödel et al. (1979), Vansteenkiste and van Hoof (1982), Stamp et al. (1984), Saguy and Drew (1987) and Stroup et al. (1987). With regard to the influence of volatiles, Yamada et al. (1984) did not note any effect of ethanol on the measuring cell. In comparing experiments, Rödel et al. (1988) were able to record good reproducibility of the resulting measurements (eleven measurements each for NaCl solutions with a_w levels 0.90 and 0.95). The standard deviations were 0.0004 a_w (CV 0.04 per cent) or 0.0007 a_w (CV 0.07 per cent). As a further development of this a_w measuring system, signals will in future be processed digitally, which will simplify the calibration of the instrument with salt solutions, as well as offering other advantages such as computer interfacing.

The electrolytic a_w measuring system made by Novasina AG (Thurgauerstrasse 50, Ch-8050 Zurich, Switzerland) and Beckman Instrument Inc. (Cedar Grove, NJ, USA) works on a very similar principle. Figure 12.6 shows the Novasina a_w centre. Water activity of food in the range 0.10 to 1.00 may be measured thermostatically in this a_w centre. This measuring system comprises the a_w centre and the thermostatic device with three sensors. The thermostatic device is a well-insulated metal box in which an air-cooled Peltier system thermostatically controls the interior, and thus the sensors and the samples, with great precision. This measuring system is independent of the ambient temperature at the measurement location because of the thermostat, which is electronically controlled by the a_w centre. The a_w levels and temperatures of three food samples can be measured simultaneously with the a_w centre, with data being automatically registered by an integral recorder for control purposes (for example for checking equilibration). All measured variables can be switched sequentially to the digital display of the instrument.

This measuring system may be supplied with an AD converter with serial interface (RS 232C). It is then possible to acquire data from the a_w centre and correct the measurements immediately for temperature variations by means of a computer. With automatic correction, a_w measurements may be taken continuously during the production of food (for example raw sausage) and these signals may be used for on-line control of the air-conditioned curing rooms (Rödel and Stiebing 1987). The mean standard deviation for this measuring system in an a_w range from 0.80 to 0.95 (ten measurements each from NaCl solutions with a_w levels 0.80, 0.85, 0.90, 0.95) was 0.0008 a_w (Rödel 1989, unpublished data); at a_w levels 0.90 and 0.95, standard deviations of 0.0005 (CV 0.05 per cent) or 0.0006 (CV 0.06 per cent) were observed (Rödel et al. 1988).

A similar system, without an AD converter and with only one measuring head, is known as the Novasina Humidat-TH2 Thermoconstanter. With this a_w instrument, thermostatic control is again provided by a Peltier system. As tests on this system have shown, a_w levels of foods may be determined with high reproducibility over a wide range by automatic thermostatic control of test chamber and measuring cell. The standard deviation varied with the level of a_w when salt solutions were measured, and was between 0.0004 and 0.0008 (Kitic et al. 1986b) in the a_w range from 0.52 to 0.97. The standard deviation in all the measurements increased when products containing organic volatiles were measured, despite the use of filters (Pollio et al. 1986).

Another conductivity meter for measuring the a_w of foods is illustrated in Figure 12.7. This compact instrument, called the a_w centre WA 10-100, is manufactured by AP-Elektronik (Joseph-Maria-Lutz-Anger 34, D-8000 Munich 83, Germany). Up to three different samples in hermetically sealed chambers can be tempered to a tolerance of 0.05K by the measuring head, which is temperature regulated by Peltier cooling (range 10–40°C). This system permits very accurate a_w measurements over the range 0.09 to 1.00 a_w. By turning the bayonet socket ring on the cable-free measuring head, the system is opened for changing the sample, and by turning the measuring head by 120°, the measuring cell is set up for the next pre-tempered sample. The measured water activity and temperature of the food sample are indicated simultaneously in digital form. The accuracy over the whole range is ±0.02 a_w, the reproducibility ±0.003.

Dew-point hygrometer Protimeter Ltd (Meter House, Fieldhouse Lane, Marlow, Bucks, SL7 1LX, UK) offers a dew-point hygrometer for determining the water activity of various foods. The range of measurement is from 0°C to 40°C dew point and ambient temperature (option of extending to -20°C), so that a_w levels from 0.15 to 1.00 may be determined (Stamp et al. 1984). This optoelectronic dew-point hygrometer has a 50 mm high sensor (diameter 40 mm) with an electrically cooled gold mirror (Peltier cooling). The gold mirror determines optically the dew point of the air in the chamber, which is in humidity equilibrium with the sample. Three buttons are pressed to indicate the levels of dew-point temperature, chamber temperature or water activity. This instrument is very reliable owing to the optoelectronic measurement principle with a gold mirror.

5.5 Thermometric Technique

The freezing point of a food is closely linked in a physical/chemical sense with the water activity of the product, this being shown diagrammatically for meat products in Figure 12.8. All those processes in a product that reduce the water activity also lower the freezing point of the food.

The point at which foods begin to freeze can be measured and from this the a_w level at 25°C can be calculated. The a_w Kryometer AWK-10, made by Nagy Meßsysteme GmbH (Siedlerstrasse 34, D-7046 Gäufelden 1, Germany) works on this principle. With this instrument (Figure 12.9) water activity, particularly for meat products, can be determined thermometrically. The measuring system includes an electronic indicator module with a microprocessor, a cylindrical sample chamber and a cooling bath at a temperature of approximately −50°C. The measuring period for this cryoscopic a_w test on meat products depends on the a_w level. The higher

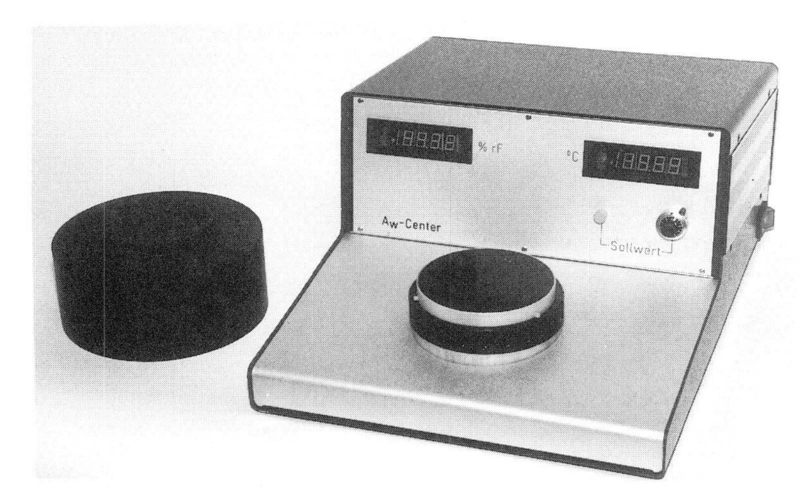

Figure 12.7 *Conductivity humidity meter AP-Elecktronik a_w centre type WA 10–100*

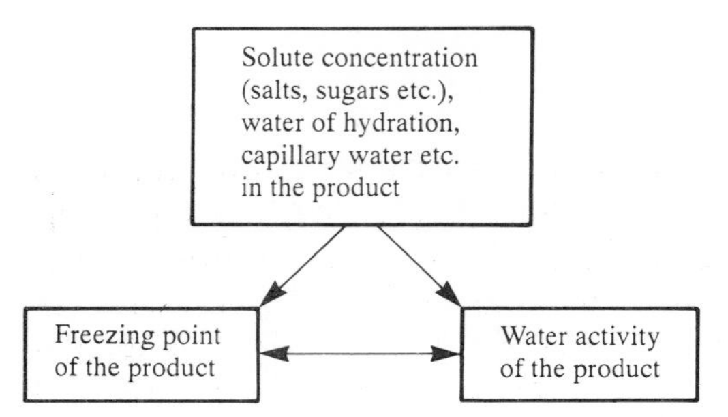

Figure 12.8 *Relation between freezing point and water activity of meat products*

the water activity of the sample, the shorter the measuring period. It is between about 8 and 30 minutes with an a_w range from 0.999 to 0.80.

Because the freezing point of a sample is identified without operator intervention, the equipment automatically finishes the measuring procedure after calculating and displaying the water activity level, and it is then ready to commence a new measurement. Sample-specific effects, for example from humectants such as glycerol or similar materials, do not pose any problem with thermometric a_w measurement, and the method has good reproducibility. When salt solutions with a_w levels of 0.90 and 0.95 were used, the standard deviation on eleven measurements was 0.0002 (CV 0.03 per cent) and 0.0001 (CV 0.01 per cent), respectively (Rödel et al. 1988).

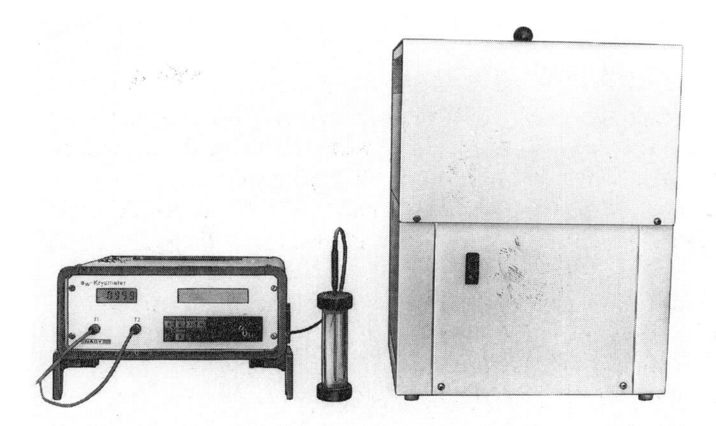

Figure 12.9 *Thermometric device Nagy a_wKryometer type AWK-10 (on the left) with cylindrical probe and cooling bath*

6 Conclusions

In this chapter, an attempt has been made to present a general survey of water activity in food to the student of food sciences, the engineer in the food industry, and also to manufacturers of measuring equipment. The significance of water activity for the quality and stability of food has been outlined, and a description and a discussion of the current possibilities provided by modern instrumentation for determining the water activity of foods have been given. From the many methods of a_w measurement cited in the literature, those that are suitable for practical applications and which have already been proven in science and industry have been chosen for detailed discussion. There is no intention of discounting other methods that are not mentioned in this survey, some of which may well be better for an intended application.

The cost of the individual a_w measurement methods varies greatly. The gravimetric methods in particular are less expensive, but cannot all be used for perishable foods because of the relatively long measuring period involved. The acquisition of electronic instruments to determine the water activity of foods means higher costs, but they generally produce good reproducibility. These instruments are therefore suitable for scientific and industrial applications.

References

Acker, L. (1962) Enzymic reactions in food of low moisture content. In *Advances in Food Research*, vol. 11, C. O. Chichester, E. M. Mrak and G. F. Stewart (eds), Academic Press, New York, 263–330

Acker, L. and Huber, L. (1971) Über das Verhalten der Glucoseoxidase in wasserarmem Milieu. *Lebensmittel-Wissenschaft und -Technologie*, **3**, 33–36

Alzamora, S. M. and Chirife, J. (1983) The water activity of canned foods. *Journal of Food Science*, **48**, 1385–1387

Bacon, C. W., Sweeney, J. G., Robbins, J. D. and Burdick, D. (1973) Production of Penicillic acid and Ochratoxin A on poultry feed by *Aspergillus ochraceus:* temperature and moisture requirements. *Applied Microbiology*, **26**, 155–160

Baird-Parker, A. C. and Freame, B. (1967) Combined effect of water activity, pH and temperature on the growth of *Clostridium botulinum* from spore and vegetative cell inocula. *Journal of Applied Bacteriology*, **30**, 420–429

Bem, Z. and Leistner, L. (1970) Die Wasseraktivitätstoleranz der bei Pökelfleischwaren vorkommenden Hefen. *Fleischwirtschaft*, **50**, 492–493

Benado, A. L. and Rizvi, S. S. H. (1987) Water activity calculation by direct measurement of vapor pressure. *Journal of Food Science*, **52**, 429–432

Beuchat, L. R. (1974) Combined effects of water activity, solute and temperature on the growth of *Vibrio parahaemolyticus*. *Applied Microbiology*, **27**, 1075–1080

Beuchat, L. R. (1983) Influence of water activity on growth, metabolic activities and survival of yeasts and molds. *Journal of Food Protection*, **46**, 135–141

Blanche Koelensmid, W. A. A. and van Rhee, R. (1964) Salmonella in meat products. *Annales de l'Institute Pasteur de Lille*, **15**, 85–97

Bonberg, B. J. and David, B. D. (1977) HACCP models for quality control of entrée production in foodservice systems. *Journal of Food Protection*, **40**, 632–638

Bone, D. (1973) Water activity in intermediate moisture foods. *Food Technology*, **27**, 71–76

Bousquet-Ricard, M., Quayle, G., Pham, T. and Cheftel, J. C. (1980) Étude comparative critique de trois méthodes de mesure de l'activité de l'eau des aliments à humidité intermédiaire. *Lebensmittel-Wissenschaft und -Technologie*, **13**, 169–176

Brownlie, L. E. (1966) Effect of some environmental factors on psychrophilic microbacteria. *Journal of Applied Bacteriology*, **29**, 447–454

Bryan, F. L. (1980) Foodborne disease in the United States associated with meat and poultry. *Journal of Food Protection*, **43**, 140–150

Bullock, K. and Tallentire, A. (1952) Bacterial survival in systems of low moisture content. Part 4: The effects of increasing moisture content on heat resistance, viability and growth of spores of *B. subtilis*. *Journal of Pharmacy and Pharmacology*, **4**, 917–931

Burcik, E. (1950) Über die Beziehung zwischen Hydratur und Wachstum bei Bakerien und Hefen. *Archiv für Mikrobiologie*, **15**, 203–235

Chirife, J. and Ferro Fontan, C. (1982) Water activity of fresh foods. *Journal of Food Science*, **47**, 661–663

Chirife, J. and Resnik, S. L. (1984) Unsaturated solutions of sodium chloride as reference sources of water activity at various temperatures. *Journal of Food Science*, **49**, 1486–1488

Christian, J. H. B. (1955a) The influence of nutrition on the water relations of *Salmonella oranienburg*. *Australian Journal of Biological Sciences*, **8**, 75–82

Christian, J. H. B. (1955b) The water relations of growth and respiration of *Salmonella oranienburg* at 30°C. *Australian Journal of Biological Sciences*, **8**, 490–497

Christian, J. H. B. (1981) Specific solute effects on microbial water relations. In *Water Activity: Influences on Food Quality*, L. B. Rockland and G. F. Stewart (eds), Academic, New York, 825-854

Christian, J. H. B. and Scott, W. J. (1953) Water relations of salmonellae at 30°C. *Australian Journal of Biological Sciences*, **6**, 565-573

Christian, J. H. B. and Waltho, J. A. (1962) The water relations of staphylococci and micrococci. *Journal of Applied Bacteriology*, **25**, 369-377

Christian, J. H. B. and Waltho, J. A. (1964) The composition of *Staphylococcus aureus* in relation to the water activity of the growth medium. *Journal of General Microbiology*, **35**, 205-213

Davies, R., Birch, G. G. and Parker, K. J. (1976) *Intermediate Moisture Foods*, Applied Science, London

Emodi, A. S. and Lechowich, R. V. (1969) Low temperature growth of type *Clostridium botulinum* spores. 1: Effects of sodium chloride, sodium nitrite and pH. *Journal of Food Science*, **34**, 78-81

Favetto, G., Resnik, S. L. and Chirife, J. (1984) Sensor contamination with organic volatiles during water activity measurements with an electric hygrometer. *Journal of Food Science*, **49**, 514, 515, 546

Favetto, G., Resnik, S., Chirife, J. and Ferro Fontan, C. (1983) Statistical evaluation of water activity measurements obtained with the Vaisala humicap humidity meter. *Journal of Food Science*, **48**, 534-538

Fennema, O. R. (1978) Enzyme kinetics at low temperature and reduced water activity. In *Dried Biological Systems*, J. H. Crowe and J. S. Clegg (eds), Academic Press, New York, 297-322

Fennema, O. R. and Berny, L. A. (1974) Equilibrium vapor pressure and water activity of food at subfreezing temperatures. *Proceedings of the IV International Congress of Food Science and Technology* (Madrid), **2**, 27-35

Fett, H. M. (1973) Water activity determination in foods in the range 0.80 to 0.99. *Journal of Food Science*, **38**, 1097-1098

Flückiger, W. and Cleven, F. (1978) Wasseraktivität. Ihre Bedeutung für die Haltbarkeit von Gebäcken. *Backtechnik*, **26**, 13-15

Food and Drug Administration (1979) Current Good Manufacturing Practices. *Federal Register 44 (53): 16209, 44 (53): 16230, 44 (112): 33238*. Washington DC, US Government Printing Office

Food and Drug Administration (1985) *Title 21 Code of Federal Regulations Parts 58, 108, 113 and 114*. Washington DC, US Government Printing Office

Genigeorgis, C. and Sadler, W. W. (1966) Effect of sodium chloride and pH on enterotoxin B production. *Journal of Bacteriology*, **92**, 1383-1387

Genigeorgis, C., Riemann, H. and Sadler, W. W. (1969) Production of enterotoxin B in cured meats. *Journal of Food Science*, **34**, 62-68

Genigeorgis, C., Foda, M. S., Mantis, A. and Sadler, W. W. (1971) Effect of sodium chloride and pH on enterotoxin C production. *Applied Microbiology*, **21**, 862-866

Gerschenson, L., Favetto, G. and Chirife, J. (1984) Influence of organic volatiles during water activity measurement with a fiber-dimensional hygrometer. *Lebensmittel-Wissenschaft und -Technologie*, **17**, 342–344

Gough, B. J. and Alford, J. A. (1965) Effect of curing agents on the growth and survival of food-poisoning strains of *Clostridium perfringens*. *Journal of Food Science*, **30**, 1025–1028

Greenspan, L. (1977) Humidity fixed points of binary saturated aqueous solutions. *Journal of Research of the National Bureau of Standards A, Physics and Chemistry*, **81A** (1), 89–96

Grossklaus, D. (1977) Lebensmittelhygienische Aspekte der Zoonosenbekämpfung. *Fleischwirtschaft*, **57**, 1649, 1652

Gur-Arieh, C., Nelson, A. I., Steinberg, M. P. and Wei, L. S. (1965) A method for rapid determinations of moisture adsorption isotherms of solid particles. *Journal of Food Science*, **30**, 105–110

Harwig, J. and Chen, Y. K. (1974) Some conditions favoring production of ochratoxin A and citrinin by *Penicillium viridicatum* in wheat and barley. *Canadian Journal of Plant Science*, **54**, 17–22

Heiss, R. and Eichner, K. (1971) Die Haltbarkeit von Lebensmitteln mit niedrigen und mittleren Wassergehalten. *Chemie Mikrobiologie Technologie der Lebensmittel*, **1**, 33–40

Hilsheimer, R. and Hauschild, A. H. W. (1985) A modified method for ascertaining water activities within defined limits. *Journal of Food Protection*, **48**, 325–326

Hobbs, B. C. (1965) *Clostridium welchii* as a food poisoning organism. *Journal of Applied Bacteriology*, **28**, 74–82

Hunter, I. R., Houston, D. F. and Kester, E. B. (1951) Development of free fatty acids during storage of brown (husked) rice. *Cereal Chemistry*, **28**, 232–239

Jakobsen, M. (1983) Filament hygrometer for water activity measurement: interlaboratory evaluation. *Journal of the Association of the Offical Analytical Chemists*, **66**, 1106–1111

Jakobsen, M., Filtenborg, O. and Bramsnaes, F. (1972) Germination and outgrowth of the bacterial spore in the presence of different solutes. *Lebensmittel-Wissenschaft und -Technologie*, **5**, 159–162

Jakobsen, M. and Murrell, W. G. (1977) The effect of water activity and a_w-controlling solute on sporulation of *Bacillus cereus* T. *Journal of Applied Bacteriology*, **43**, 239–245

Johnston, M. R. and Lin, R. C. (1987) FDA views on the importance of a_w in good manufacturing practice. In *Water Activity: Theory and Application to Food*, L. B. Rockland and L. R. Beuchat (eds), Marcel Dekker, New York, 287–294

Kang, Ch. K., Woodburn, M., Pagenkopf, A. and Cheney, R. (1969) Growth, sporulation, and germination of *Clostridium perfringens* in media of controlled water activity. *Applied Microbiology*, **18**, 798–805

Karmas, E. and Chen, C. C. (1975) Relationship between water activity and water binding in high and intermediate moisture foods. *Journal of Food Science*, **40**, 800-801

Kaufmann, F. L. and Schaffner, R. M. (1974) Hazard analysis, critical control points and good manufacturing practices regulations (sanitation) in food plant inspections. *Proceedings of the IVth International Congress on Food Science and Technology*, 402-407

Kitic, D., Favetto, G. J., Chirife, J. and Resnik, S. L. (1986a) Measurement of water activity in the intermediate moisture range with the Novasina Thermoconstanter humidity meter. *Lebensmittel-Wissenschaft und -Technologie*, **19**, 297-301

Kitic, D., Pereira Jardim, D. C., Favetto, G. J., Resnik, S. L. and Chirife, J. (1986b) Theoretical prediction of the water activity of standard saturated salt solutions at various temperatures. *Journal of Food Science*, **51**, 1037-1041

Krauss, H. and Weber, A. (1986) *Zoonosen*. Deutscher Ärzte, Köln

Krispien, K. and Rödel, W. (1976) Bedeutung der Temperatur für den a_w-Wert von Fleisch und Fleischerzeugnissen. *Fleischwirtschaft*, **56**, 709-714

Kushner, D. J. (1968) Halophilic bacteria. In *Advances in Applied Microbiology*, W. W. Umbreit and D. Periman (eds), Academic Press, New York, 73-99

Kvaale, O. and Dalhoff, E. (1963) Determination of the equilibrium relative humidity of foods. *Food Technology*, **17**, 151-153

Labuza, T. P., Cassil, S. and Sinskey, A. J. (1972a) Stability of intermediate moisture foods. 2: Microbiology. *Journal of Food Science*, **37**, 160-162

Labuza, T. P., McNally, L., Gallagher, D., Hawkes, J. and Hurtado, F. (1972b) Stability of intermediate moisture foods. 1: Lipid oxidation. *Journal of Food Science*, **37**, 154-159

Labuza, T. P., Kreisman, L. N., Heinz, C. A. and Lewicki, P. P. (1977) Evaluation of the Abbeon cup analyzer compared to the VPM and Fett-Vos methods for water activity measurement. *Journal of Food Processing and Preservation*, **1**, 32-41

Landrock, A. H. and Proctor, B. E. (1951) A new graphical interpolation method for obtaining humidity equilibria data, with special reference to its role in food packaging studies. *Food Technology*, **5**, 332-337

Lang, K. W., McCune, T. D. and Steinberg, M. P. (1981) A proximity equilibration cell for rapid determination of sorption isotherms. *Journal of Food Science*, **46**, 936-938

Lanigan, G. W. (1963) Silage bacteriology. I: Water activity and temperature relationships of silage strains of *Lactobacillus plantarum, Lactobacillus brevis*, and *Pediococcus cerevisiae*. *Australian Journal of Biological Sciences*, **16**, 606-615

Legault, R. R., Makower, B. and Talburt, W. F. (1948) Apparatus for measurement of vapor pressure. *Analytical Chemistry*, **20**, 428-430

Leistner, L. (1977) Microbiology of ready-to-serve foods. In *How Ready Are Ready-to-Serve Foods?*, K. Paulus (ed), Karger, Basel, 260–272

Leistner, L. (1978) Hurdle effect and energy saving. In *Food Quality and Nutrition*, W. K. Downey (ed), Applied Science, London, pp. 553–557

Leistner, L. and Rödel, W. (1975) The significance of water activity for micro-organisms in meats. In *Water Relations of Foods*, R. B. Duckworth (ed), Academic, London, 309–323

Leistner, L. and Rödel, W. (1976a) Inhibition of micro-organisms in food by water activity. In *Inhibition and Inactivation of Vegetative Microbes*, F. A. Skinner and W. B. Hugo (eds), Academic, London, 219–237

Leistner, L. and Rödel, W. (1976b) The stability of intermediate moisture foods with respect to micro-organisms. In *Intermediate Moisture Foods*, R. Davies, G. G. Birch and K. J. Parker (eds), Applied Science, London, 120–137

Leistner, L., Rödel, W. and Krispien, K. (1981) Microbiology of meat products in high- and intermediate-moisture ranges. In *Water Activity: Influences on Food Quality*, L. B. Rockland and G. F. Stewart (eds), Academic, New York, 855–916

Lenart, A. and Flink, J. M. (1983) An improved proximity equilibration cell method for measuring water activity of foods. *Lebensmittel-Wissenschaft und -Technologie*, **16**, 84–89

Lewicki, P. P., Busk, G. C., Peterson, P. L. and Labuza, T. P. (1978) Determination of factors controlling accurate measurement of a_w by the vapor pressure manometric technique. *Journal of Food Science*, **43**, 244–246

Limsong, S. and Frazier, W. C. (1966) Adaptation of *Pseudomonas fluorescens* to low levels of water activity produced by different solutes. *Applied Microbiology*, **14**, 899–901

Lotter, L. P. and Leistner, L. (1978) Minimal water activity for enterotoxin A production and growth of *Staphylococcus aureus*. *Applied and Environmental Microbiology*, **36**, 377–380

Lötzsch, R. and Leistner, L. (1977) Überleben von *Trichinella spiralis* in Rohwurst und Rohschinken in Abhängigkeit von der Wasseraktivität (a_w-Wert). *Proceedings of the XXIIIrd European Congress of Meat Research Workers*, Moscow, **I 6**

Lötzsch, R. and Rödel, W. (1974) Untersuchungen über die Lebensfähigkeit von *Trichinella spiralis* in Rohwürsten in Abhängigkeit von der Wasseraktivität. *Fleischwirtschaft*, **54**, 1203–1208

Lötzsch, R. and Trapper, D. (1979) Minimale Wasseraktivitäten (a_w-Werte) für die Bildung von zehn Mykotoxinen bei *Penicillium* spp. Poster at the 4th International IUPAC Symposium on Mycotoxins and Phycotoxins, Lausanne

Maloney, J. F., Labuza, T. P., Wallace, D. H. and Karel, M. (1966) Autoxidation of methyl linoleate in freeze-dried model systems. I: Effect of water on the auto-catalyzed oxidation. *Journal of Food Science*, **31**, 878–884

Marcos, A., Alcalá, M., León, F., Fernández- Salguero, J. and Esteban, M. A. (1981) Water activity and chemical composition of cheese. *Journal of Dairy Science*, **64**, 622–626

Marshall, B. J., Ohye, D. F. and Christian, J. H. B. (1971) Tolerance of bacteria to high concentrations of NaCl and glycerol in the growth medium. *Applied Microbiology*, **21**, 363–364

Matz, S. A. (1965) *Water in Foods*. AVI, Westport, CT

McCune, T. D., Lang, K. W. and Steinberg, M. P. (1981) Water activity determination with the proximity equilibration cell. *Journal of Food Science*, **46**, 1978–1979

McLean, R. A., Lilly, H. D. and Alford, J. A. (1968) Effects of meat-curing salts and temperature on production of staphylococcal enterotoxin B. *Journal of Bacteriology*, **95**, 1207–1211

Moran, T. (1936) The state of water in tissues. *Report of the Food Investigation Board for the Year 1935*, HMSO, London, 20–24

Mossel, D. A. A. (1969) Nahrungsmittel als Umwelt für Mikroorganismen, die Lebensmittel gesundheitsschädlich machen. *Alimenta*, **8**, 8–16

Multon, J. L., Savet, B. and Bizot, H. (1980) A fast method for measuring the activity of water in foods. *Lebensmittel-Wissenschaft und -Technologie*, **13**, 271–273

Northolt, M. D. and Heuvelman, C. J. (1982) The salt crystal liquefaction test: a simple method for testing the water activity of foods. *Journal of Food Protection*, **45**, 537–540, 546

Northolt, M. D., van Egmond, H. P. and Paulsch, W. E. (1977) Differences between *Aspergillus flavus* strains in growth and aflatoxin B1 production in relation to water activity and temperature. *Journal of Food Protection*, **40**, 778–781

Nunes, R. V., Urbincain, M. J. and Rotstein, E. (1985) Improving accuracy and precision of water activity measurements with a water vapor pressure manometer. *Journal of Food Science*, **50**, 148–149

Ohye, D. F. and Christian, J. H. B. (1967) Combined effects of temperature, pH and water activity on growth and toxin production by *Cl. botulinum* types A, B, and E. *Proceedings of the 5th International Symposium of Food Microbiology*, Moscow, 1966, 217–223

Ohye, D. F. and Christian, J. H. B. (1967) Influence of temperature on the water relations of growth of *Cl. botulinum* type E. *Proceedings of the 5th International Symposium of Food Microbiology*, Moscow, 1966, 136–143

Palacha, Z. and Flink, J. M. (1987) Revised PEC method for measuring water activity. *Journal of Food Science and Technology*, **22**, 485–490

Pitt, J. I. (1975) Xerophilic fungi and the spoilage of foods of plant origin. In *Water Relations of Foods*, R. B. Duckworth (ed), Academic, London, 273–307

Pitt, J. I. and Christian, J. H. B. (1968) Water relations of xerophilic fungi isolated from prunes. *Applied Microbiology*, **16**, 1853–1858

Pivnick, H. and Thatcher, F. S. (1968) Microbial problems in food safety with particular reference to *Clostridium botulinum*. In *The Safety of Foods*, AVI, Westport, CT, 121–140

Pollio, M. L., Kitic, D., Favetto, G. J. and Chirife, J. (1986) Effectiveness of available filters for an electric hygrometer for measurement of water activity in the food industry. *Journal of Food Science*, **51**, 1358–1359

Pollio, M. L., Kitic, D., Favetto, G. J. and Chirife, J. (1987) Prediction and measurement of the water activity of selected saturated salt solutions at 5°C and 10°C. *Journal of Food Science*, **52**, 1118–1119

Prior, B. A. (1979) Measurement of water activity in foods: a review. *Journal of Food Protection*, **42**, 668–674

Prior, B. A., Casaleggio, C. and van Vuuren, H. J. J. (1977) Psychrometric determination of water activity in the high a_w range. *Journal of Food Protection*, **40**, 537–539

Resnik, S. L., Favetto, G. J., Chirife, J. and Ferro Fontan, C. (1984) A world survey of water activity of selected saturated salt solutions used as standards at 25°C. *Journal of Food Science*, **49**, 510–513

Riemann, H. (1963) Safe heat processing of canned cured meats with regard to bacterial spores. *Food Technology*, **17**, 39–49

Robinson, R. A. and Stokes, R. H. (1965) *Electrolyte Solutions*. Butterworth, London

Rödel, W. (1975) Einstufung von Fleischerzeugnissen in leicht verderbliche, verderbliche und lagerfähige Produkte aufgrund des pH-Wertes und a_w-Wertes. Dissertation (thesis), Freie Universität Berlin, West Germany

Rödel, W. and Krispien, K. (1977) Der Einfluß von Kühl- und Gefriertemperaturen auf die Wasseraktivität (a_w-Wert) von Fleisch und Fleischerzeugnissen. *Fleischwirtschaft*, **57**, 1863–1867

Rödel, W. and Leistner, L. (1971) Ein einfacher a_w-Wert-Messer für die Praxis. *Fleischwirtschaft*, **51**, 1800–1802

Rödel, W. and Lücke, F. -K. (1983) Vermehrung von *Staphylococcus aureus* und *Salmonella* spp. bei niedriger Wasseraktivität. *Mitteilungsblatt der Bundesanstalt für Fleischforschung, Kulmbach*, **82**, 5707–5714

Rödel, W. and Stiebing, A. (1987) Kontinuierliche Messung des Reifungsverlaufs von Rohwurst. *Fleischwirtschaft*, **67**, 1202–1211

Rödel, W., Herzog, H. and Leistner, L. (1973) Wasseraktivitäts-Toleranz von lebensmittelhygienisch wichtigen Keimarten der Gattung *Vibrio*. *Fleischwirtschaft*, **53**, 1301–1303

Rödel, W., Krispien, K. and Leistner, L. (1979) Messung der Wasseraktivität (a_w-Wert) von Fleisch und Fleischerzeugnissen. *Fleischwirtschaft*, **59**, 831–836

Rödel, W., Krispien, K. and Leistner, L. (1980) Die Wasseraktivität von Fetten tierischer Herkunft. *Fleischwirtschaft*, **60**, 642, 644–648, 650

Rödel, W., Ponert, H. and Leistner, L. (1975) Verbesserter a_w-Wert-Messer zur Bestimmung der Wasseraktivität (a_w-Wert) von Fleisch und Fleischwaren. *Fleischwirtschaft*, **55**, 557–558

Rödel, W., Scheuer, R. and Wagner, H. (1988) Neues Verfahren zur Bestimmung der Wasseraktivität bei Fleischerzeugnissen. *Mitteilungsblatt der Bundesanstalt für Fleischforschung, Kulmbach*, **100**, 7979–7986

Ross, K. D. (1975) Estimation of water activity in intermediate moisture foods. *Food Technology*, **2**, 26–34

Rüegg, M. and Blanc, B. (1977) Beziehungen zwischen Wasseraktivität, Wasser-Sorptionsvermögen und Zusammensetzung von Käse. *Milchwissenschaft*, **32**, 193–201

Rüegg, M. and Blanc, B. (1981) Influence of water activity on the manufacture and aging of cheese. In *Water activity: Influences on Food Quality*, L. B. Rockland and G. F. Stewart (eds), Academic, New York, 791–823

Saguy, I. and Drew, B. (1987) Statistical calibration of instruments using water activity determination as an example. *Journal of Food Science*, **52**, 767–771

Schmidt, U. and Rödel, W. (1987) Einfluß der Wasseraktivität (a_w-Wert) auf das Überleben der Rinderfinne (*Cysticercus bovis*). *Mitteilungsblatt der Bundesanstalt für Fleischforschung, Kulmbach*, **97**, 7550–7556

Scott, W. J. (1953) Water relations of *Staphylococcus aureus* at 30°C. *Australian Journal of Biological Sciences*, **6**, 549–564

Scott, W. J. (1957) Water relations of food spoilage microorganisms. In *Advances in Food Research*, vol VII, E. M. Mrak and G. F. Stewart (eds), Academic Press, New York, 83–127

Segner, W. P., Schmidt, C. F. and Boltz, J. K. (1966) Effect of sodium chloride and pH on the outgrowth of spores of type E *Clostridium botulinum* at optimal and suboptimal temperatures. *Applied Microbiology*, **14**, 49–54

Segner, W. P., Schmidt, C. F. and Boltz, J. K. (1971) Minimal growth temperature, sodium chloride tolerance, pH sensitivity, and toxin production of marine and terrestrial strains of *Clostridium botulinum* type C. *Applied Microbiology*, **22**, 1025–1029

Simatos, D. and Multon, J. L. (1985) *Properties of Water in Foods in Relation to Quality and Stability*. Martinus Nijhoff, Dordrecht

Smith, P. R. (1965) A new apparatus for the study of moisture sorption by starches and other foodstuffs in humidified atmospheres. In *Humidity and Moisture*, A. Wexler and W. A. Wildhack (eds), Reinhold, New York, 487–494

Snow, D. (1949) The germination of mould spores at controlled humidities. *Annals of Applied Biology*, **36**, 1–13

Sood, V. C. and Heldman, D. R. (1974) Analysis of a vapor pressure manometer for measurement of water activity in nonfat dry milk. *Journal of Food Science*, **39**, 1011–1013

Stamp, J. A., Linscott, S., Lomauro, C. and Labuza, T. P. (1984) Measurement of water activity of salt solutions and foods by several electronic methods as compared to direct vapor pressure measurement. *Journal of Food Science*, **49**, 1139–1142

Steele, R. J. (1987) Use of polyols to measure equilibrium relative humidity. *International Journal of Food Science and Technology*, **22**, 377–384

Stiebing, A. and Rödel, W. (1989) Kontinuierliches Messen der Oberflächen-Wasseraktivität von Rohwurst. *Mitteilungsblatt der Bundesanstalt für Fleischforschung, Kulmbach*, **104**, 221–227

Stille, B. (1948) Grenzwerte der relativen Feuchtigkeit und des Wassergehaltes getrockneter Lebensmittel für den mikrobiellen Befall. *Zeitschrift für Lebensmitteluntersuchung und -Forschung*, **88**, 9–12

Stoloff, L. (1978) Processed vegetable products. Calibration of water activity measurement instruments and devices: collaborative study. *Journal of the Association of the Official Analytic Chemists*, **61**, 1166–1178

Storey, R. M. and Stainsby, G. (1970) The equilibrium water vapour pressure of frozen cod. *Journal of Food Technology*, **5**, 157–163

Strong, D. H., Foster, E. F. and Duncan, C. L. (1970) Influence of water activity on the growth of *Clostridium perfringens*. *Applied Microbiology*, **19**, 980–987

Stroup, W. H., Peeler, J. T. and Smith, K. (1987) Evaluation of precision estimates for fiber-dimensional and electrical hygrometers for water activity determinations. *Journal of the Association of the Official Analytic Chemists*, **70**, 955–957

Taylor, A. A. (1961) Determination of moisture equilibria in dehydrated foods. *Food Technology*, **15**, 536–540

Tomcov, D., Bem, Z. and Leistner, L. (1974) Minimalne a_w vrednosti odabranih bakterijskih vrsta. *rim*, **6**, 3–9

Troller, J. A. (1971) Effect of water activity on enterotoxin B production and growth of *Staphylococcus aureus*. *Applied Microbiology*, **21**, 435–439

Troller, J. A. (1972) Effect of water activity on enterotoxin A production and growth of *Staphylococcus aureus*. *Applied Microbiology*, **24**, 440–443

Troller, J. A. (1983a) Methods to measure water activity. *Journal of Food Protection*, **46**, 129–134

Troller, J. A. (1983b) Water activity measurements with a capacitance manometer. *Journal of Food Science*, **48**, 739–741

Troller, J. A. and Christian, J. H. B. (1978) *Water Activity and Food*, Academic Press, New York

Van der Berg, I. C. (1975) Thermodynamical aspects of water activity in intermediate moisture foods. Course on Intermediate Moisture Foods Cycle CPCIA Europe, Seminaire E.5

Vansteenkiste, J. P. and van Hoof, J. (1982) Measurement of water activity of dry sausages: a comparison of the isopiestic method and the electric hygrometer. *Archiv für Lebensmittelhygiene*, **33**, 116–118

Vos, P. T. and Labuza, T. P. (1974) Technique for measurement of water activity in the high a_w range. *Journal of Agricultural and Food Chemistry*, **22**, 326–327

Weisser, H., Roth, Th. and Harz, H. -P. (1985) Spezielle Methoden zum Bestimmen der Wasseraktivität. *Zeitschrift für Lebensmittel-Technologie und -Verfahrenstechnik*, **3**, 170, 172, 176, 178–179

Wiebe, H. H., Kidambi, R. N., Richardson, G. H. and Ernstrom, C. A. (1981) A rapid psychrometric procedure for water activity measurement of foods in the intermediate moisture range. *Journal of Food Protection*, **44**, 892–895

Williams, O. B. and Purnell, H. G. (1953) Spore germination, growth, and spore formation by *Clostridium botulinum* in relation to the water content of the substrate. *Food Research*, **18**, 35–39

Wodzinski, R. J. and Frazier, W. C. (1960) Moisture requirements of bacteria. I: Influence of temperature and pH on requirements of *Pseudomonas fluorescens*. *Journal of Bacteriology*, **79**, 572–578

Wodzinski, R. J. and Frazier, W. C. (1961) Moisture requirements of bacteria. III: Influence of temperature, pH, and malate and thiamine concentration on requirements of *lactobacillus viridescens*. *Journal of Bacteriology*, **81**, 359–365

Wolf, W. (1984) Ermittlung des a_w-Wertes in Lebensmittelproben. *Getreide Mehl und Brot*, **38**, 116–119

Yamada, J., Tanaka, A., Shinmura, Y. and Aibara, K. (1984) Influence of ethanol on determination of water activity. *Journal of Food Hygienic Society of Japan*. **25**, 118–124

13 *Instrumental Methods in the Chemical Quality Control Laboratory*

P. T. Slack

Contents

5.6 System Based on Oven Drying: Computrac Max 50 Moisture Analyser 440
5.7 Instrument for Moisture Determination by Oven Drying and Ash by Combustion: Leco MAC-400 442
5.8 Instrument for Moisture by Microwave Drying and Fat by Solvent Extraction: CEM Meat Analysis System 443
5.9 Instrument for Fat Determination by Solvent Extraction: Foss-Let System 445

6 Dedicated Instruments for Nitrogen Determinations: for the Calculation of Protein and Meat Content 445
6.1 Instrument for Nitrogen Determination by Acid Digestion: 446
6.1.1 Kjel-Foss Automatic 446
6.1.2 Kjeltec Auto/Labtec System 448
6.1.3 Büchi Kjeldahl System 449
6.2 Instrument for Nitrogen Determination by Combustion: Leco FP-228 452

7 Further Development of Instrumentation for Food Analytical Laboratories 454

References 454

1 Introduction

In the food industry, there is pressure to keep costs down by making every stage of the production process as efficient as possible. Quality control is no exception, and instrument manufacturers have responded by developing a wide range of analytical instruments that are claimed to cut costs by making analysis less labour-intensive and by providing results quickly, so avoiding costly delays.

'It is better to be roughly right on time than precisely right too late' goes the saying. Certain instruments fulfil this criterion in that they give approximate results within minutes. However, current and proposed food legislation makes it increasingly important for quality control analysts to have robust, reliable instruments that quickly produce results of both reasonable accuracy and reproducibility.

The widespread application of microprocessor technology in control systems is the major reason for the vast range of new instruments now available. Many instruments contain built-in minicomputers, which can be interfaced with personal computers, mainframe computers or purpose-built data management systems and are pre-programmed with standard equations. The problem is that the rate of progress in the application of microprocessor technology has outstripped development at the sharp analytical end. The analytical performance of instruments is not always as good as that claimed by the manufacturers.

Instrument performance can vary considerably between makes and models even when they are based on the same analytical method. This makes it dangerous to generalize about the choice of instrumentation. Consequently, the instruments described in this chapter are confined to those evaluated over the last few years at the Leatherhead Food Research Association (LFRA), mostly by the author and his co-workers. While readers should not suppose these to be the only instruments on the market capable of doing the job, prudent potential purchasers do not order expensive instruments until they have either seen an independent evaluation or, indeed, carried one out themselves. This is probably the most important piece of advice that can be given to anyone proposing to buy. As is indicated in this chapter, there are many good instruments available that have performed well in such evaluations and are now finding widespread use in food laboratories. Before these are discussed in detail, some general comments are made about certain instrument types finding general use in food laboratories. Such versatile instruments include those for high performance liquid chromatography (HPLC), gas liquid chromatography (GLC), spectrophotometry and ion chromatography. The trend towards dedication of these instruments is discussed, as is the emergence of truly dedicated instruments such as nitrogen analysers. Where evaluations have been carried out by the author or his colleagues, such instruments are discussed in detail.

2 Versatile Instruments

The versatile category includes some of the most widely used instrumentation in food analytical laboratories, namely spectrophotometers, HPLC and GLC equipment. Detailed discussion of examples in the versatile category is outside the scope of this chapter since this category is too broad. A good general reference on food analysis is Egan et al. (1981). Various books have chapters devoted to the underlying principles of instrumental analysis: Charalambous (1984), King (1984), Macleod (1973), Pomeranz and Meloan (1978), Stewart and Whitaker (1984), Willard et al. (1974). If a single publication is sought, Gruenwedel and Whitaker (1987) is probably the most up-to-date and comprehensive treatise available; Volumes 2 and 4 contain chapters covering the most commonly used instrumental methods. The earlier references will give the reader a flavour of how instrumental methods have developed over the last 20–30 years. For the purpose of this chapter, the author therefore confines comments to overall trends and observations rather than detail.

The performance of the versatile instruments is extremely dependent on the method and the operator. A vast range of instruments is available, all with their own characteristics, and new models are constantly arriving on the scene.

Spectrophotometers are often used with chemically or enzymically based colorimetric methods. Although the design of an instrument can have a bearing on its usefulness, overall performance is more likely to be determined by the specificity of the method used and the operator's competence in performing the extraction and assay.

This also applies to atomic absorbance, which is a widely used spectrometric technique for the determination of metals in foodstuffs, often at ppb (μg/kg) levels. Good technique and careful operation of the instrument are essential for accurate and repeatable results.

Other more advanced spectrophotometric techniques (Fourier transform infrared (FTIR) and near infrared (NIR)) are currently under active investigation for their potential as analytical instruments, for a wide range of macrocomponents. This section concentrates more on those instruments that have found, or will most imminently find, routine use in food laboratories. (Instruments for on-line NIR applications are discussed in Chapter 5.)

2.1 Gas and Liquid Chromatography

GLC is now the most advanced column chromatographic technique, with a wide range of application in the food analytical laboratory. Again, the technique is only as good as the operator is in performing the extraction and, if necessary, derivatization steps. Correct optimization of instrumental parameters for any given method is essential. They include, for example: gas flow rates; types of injector; type of detector fitted; choice of oven; programming of injector and detector settings; and choice, age and history of column. Conventional packed column instruments have now largely been replaced by capillary column instruments.

Capillary columns avoid the hit and miss element involved in preparing one's own packed columns, although packed columns cope better with crude extracts. Pre-prepared packed columns can be fragile in transit and make it difficult to obtain good resolution. Capillary columns are more robust and capable of higher resolution, although samples need a more thorough clean-up for good results. The more advanced research techniques clean up extracts using disposable solid-phase extraction cartridges. Alternatively, packed GLC or megabore capillary columns or even HPLC columns can be installed on-line for pre-column clean-up. One such instrument is the AC-MUSIC system made by Chrompack (megabore column clean-up, capillary analytical column). The other techniques are also beginning to emerge as commercial instruments, for example HPLC and high resolution gas chromatography (HRGC).

Perhaps the greatest advantage of GLC is the ease with which it is coupled to mass spectrometry (MS) and other detectors. HPLC is particularly widely used in analytical laboratories to measure thermally unstable or non-volatile components, and components that cannot be readily derivatized to produce sufficiently volatile or thermally stable derivatives for GLC. As with GLC, the technique of HPLC is only as good as the operator is in making the extract and setting up the instrument. The history and age of the column are even more important in this technique since columns tend to be chemically modified by the extracts injected into them, particularly if these are not properly cleaned up. Consequently, the best advice that can be given to the reader is to approach some of the companies offering advanced chromatography instrumentation, for example Chrompack, Hewlett-Packard, Fisons and Perkin-Elmer. These often offer excellent courses and symposia ranging from basic techniques to the most advanced research developments. They could also help the reader with the best choice of instrument for a specific application.

Specific application of liquid and gas chromatography are described in Slack (1987) and in Charalambous and Inglett (1983).

2.2 Limitations of Gas and Liquid Chromatography

For HPLC, compounds or their derivatives normally need to be ultraviolet (UV) absorbing. There is a real need for a wider range of specific detectors to be developed for HPLC. The technique, which has been in use for a long time now, is being hampered in its development towards more dedicated instrumentation packages by the lack of specificity and sensitivity of the detectors available. Dedicated instruments would be less operator dependent.

Attempts to move HPLC into a similar technological league to GLC have been frustrated by this limited number of detectors that can be used, and by their lack of sensitivity (i.e. visible light and UV detectors, refractive index detectors). Improved detectors will be necessary before capillary and microbore HPLC columns that have already been developed find broad application. Detection systems under development include flame atomic absorption spectrometry, electronic atomic absorption spectrometry and inductively coupled plasma atomic emission spectrometry. The arrival of thermospray devices makes HPLC-MS possible, and this technique is now beginning to find some application. Much work is under way to extend the range of analyses amenable to HPLC through development of on-line derivatization technologies. Detectors have been slow to be applied, which underlines the specific difficulties of coupling these new detection systems to HPLC.

Although widely applied in the food industry, and considered as versatile and powerful in skilled hands, the technique of gas chromatography has probably reached the limit of its development. GLC requires compounds to be volatile, or easily made so by derivatization. Compounds also need to be thermally stable. This alone limits the scope for further development.

2.3 Supercritical Fluid Chromatography

Perhaps the most exciting development in chromatography is the arrival of supercritical fluid chromatography (SFC). This technique enables flame ionization detection (FID) to be used in liquid chromatography. This sensitive detection method had previously been restricted to GLC instrumentation. Easier coupling to mass spectrometers, mass selective and other detection systems is also now being realized. SFC is a relatively new technique with few written methods, or indeed instruments, available so far.

The idea of SFC was conceived from theoretical work surrounding the development of industrial liquid carbon dioxide extraction processes. Supercritical fluids are, in practice, made by heating an appropriate pressure-liquefied gas above its critical temperature. For carbon dioxide, the critical temperature is 31.3°C and the fluid must have been pressurized to at least 72.9 atmospheres to become supercritical. The resulting fluids have the solvating power of a liquid and the flow characteristics of a gas. These two properties make them ideal for chromatography since these fluids can be pumped through chromatography columns with great velocity. Fast separations are therefore possible.

Supercritical carbon dioxide fluids have polarities in the range of non-polar organic solvents. Precise polarity varies somewhat with density (that is, pressure). Thermostatic control of the entire system is necessary since temperature also has a considerable effect on retention times.

Solvating power also increases with pressure (density); 300–400 atmospheres are common. One advantage is the relatively low temperature range over which the technique can operate compared with GLC (commonly 50–100°C). Almost any capillary GLC column coating or HPLC stationary phase (in packed megabore capillary columns) can be used for SFC, although retention mechanisms are unclear. Specially prepared SFC versions of these columns should, however, be used. The technique is ideal for thermally labile, high-molecular-weight, non-polar compounds. It can also be applied to more polar compounds if these are derivatized to make them less polar. Alternatively, organic modifiers (e.g. methanol) are often added to reduce interaction between silanol groups of silica packing materials and polar compounds. This enables the chromatograph operator to take advantage of the selectivity of separation possible with packed SFC columns. The inability to use FID when organic modifiers are present is regarded by some, however, as a serious drawback.

The rapidly growing number of applications has included the separation of caffeine in coffee, of opiates in drug preparations and of alkaloids, for example ergotamine, in foodstuffs. The technique's main strengths are the variety of detectors that can be used and the ease with which it can be interfaced with a mass spectrometer. Perhaps the fastest growing side of the technology now is supercritical fluid extraction (SFE). This has great potential for the development of rapid analytical methodology, particularly when executed on-line. SFE couples easily to SFC and GLC, and all of the exciting possibilities for method development are thoroughly described in a recent book edited by Lee and Markides (1990). SFE/SFC and SFE/GLC instruments are now beginning to be introduced into food laboratories, and the reader might well expect to see this trend continue with a number of more sophisticated instruments arriving on the scene and finding routine use in years to come. However, for the moment, SFE/SFC remains very much the experts' technique, with no dedicated instrumentation yet available.

2.4 Capillary Electrophoresis

Early indications suggest that capillary electrophoresis (CE), which operates principally at the other end of the polarity scale to SFC, is another research technique that will find application in food laboratories. It is theoretically possible to apply this emerging technique to separations of any group of charged polar molecules, for example halide ions, organic acids and proteins. Indeed, CE is expected to become a powerful complementary technique to SFC in that, between them, they will enable the analyst to develop the most straightforward means of performing low-temperature separations of most thermally labile, non-volatile compounds right across the polarity spectrum.

CE systems are basically composed of a fused silica capillary filled with an electrolyte immersed in a reservoir (also containing the same electrolyte). At one end can be placed either optical detectors (using the capillary as a flow cell) or, less commonly, electrochemical or conductivity detectors arranged with an appropriate electrode within the capillary. When a sample is introduced and a high voltage applied, two effects are observed. Firstly, an induced negative charge on the capillary wall causes cations, including protons (which are naturally hydrated), firstly to associate loosely with this wall and then to move along it towards the cathode. This effect induces bulk liquid flow towards the cathode and is termed electro-osmosis. Conversely, anions are attracted towards the anode against this liquid flow. The ability of any one anion to compete against the flow will depend upon its electrophoretic mobility, a value that obviously varies from anion to anion. In similar fashion, the ability of any one cationic solute to move faster than the flow is similarly governed. These two effects therefore produce a considerable separating potential.

Instruments designed for versatility are already on the market (e.g. made for example by Waters Associates, Beckmann and Spectra-Physics). A variety of capillary types and conditions are available which allow method optimization for different classes of analytes. Indeed, one or two of these manufacturers are already producing application notes for foodstuffs. It is surely only a matter of time before instruments become available that are dedicated, that is pre-optimized for one particular determination such as organic acids in foods.

3 The Trend towards Dedication

As outlined above, the power of gas and liquid chromatographic procedures stems from a combination of the skill of the operator and the precision of the instrument. If either is at fault, poor analytical values are obtained. Although there will always be a place for this partnership in the more complex laboratory-based analytical procedures, economic forces demand that more routine quality control procedures be automated as far as possible. Instruments are emerging that are designed to reduce and ultimately eliminate the 'thinking' from analytical techniques, so enabling less skilled staff to be used in their operation. This is particularly important in view of the trend in the food industry towards removing the more routine quality assurance procedures from the analytical laboratory and placing them under the control of the production manager. If instruments are to be sited near the production line, they must be robust, compact, easy to operate, reliable and accurate, and give reproducible results. Instruments should avoid glass components in their design, or at least isolate them from production staff and, most importantly, the product. They should do similarly with corrosive and 'wet' chemicals. It is now appropriate to describe in detail instruments that illustrate these trends. Examples are chosen from the fields of gas chromatography and liquid chromatography. The development of the YSI Industrial Analysers is used as a further illustration.

3.1 GLC System for the Analysis of Pesticides: Nordion Analysis System

A good example of the trends towards dedication and automation is shown by this GLC system, which is designed for the analysis of pesticides.

The Nordion Analysis System (Figure 13.1) consists of a number of units. These are the Micromat high-resolution gas chromatograph, a printer/plotter and a personal computer.

The gas chromatograph has three main parts, namely the pneumatic unit, the oven unit and the electronics unit. The pneumatic system is used for the supply of gases to the injector, from where it carries the vaporized sample to the columns and then to the detectors. It includes on/off switches, pressure gauges and pressure regulators for all gases as well as needle valves for the adjustment of gas flow to the detectors. Adjustment valves for the split flow and septum purge are located in the oven lid unit, together with the split shut-off valve (operated from the keyboard).

The oven is the central unit of the gas chromatograph. Access is obtained by lifting the detachable oven lid unit, allowing columns to be installed and changed conveniently. The oven temperature is electronically controlled. The injector and both detectors are installed in the oven lid unit.

The microprocessor in the electronics unit controls the implementation of the temperature program, stores operational parameters, integrates detector signals and transfers data to the computer. The keyboard/display unit allows the operator to program chromatographic conditions, parameters etc. It indicates the status of the gas chromatograph (for example, waiting, running, ready or inactive) and asks for instructions by displaying various messages.

In routine use, extracts of food samples would normally be mixed with an internal standard solution and injected into the high-resolution gas chromatograph, fitted with two capillary columns of different polarity

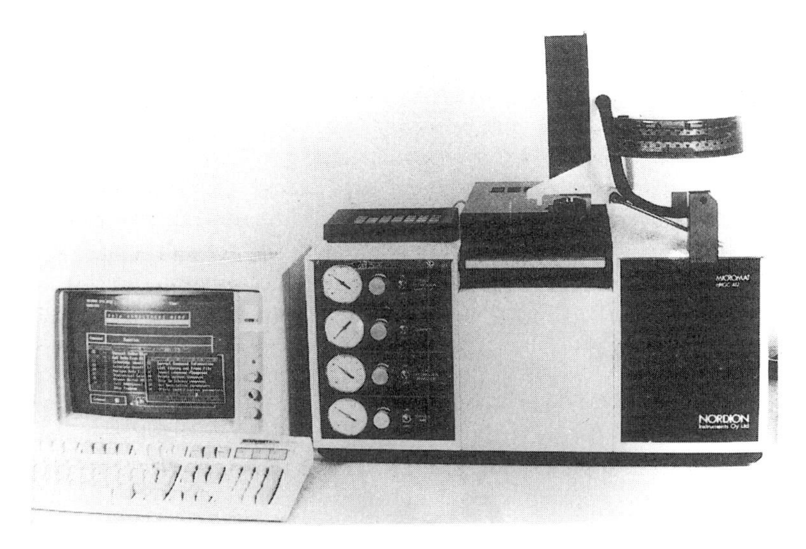

Figure 13.1 *The Nordion Analysis System for the assay of pesticides (courtesy of Dionex (UK) Ltd)*

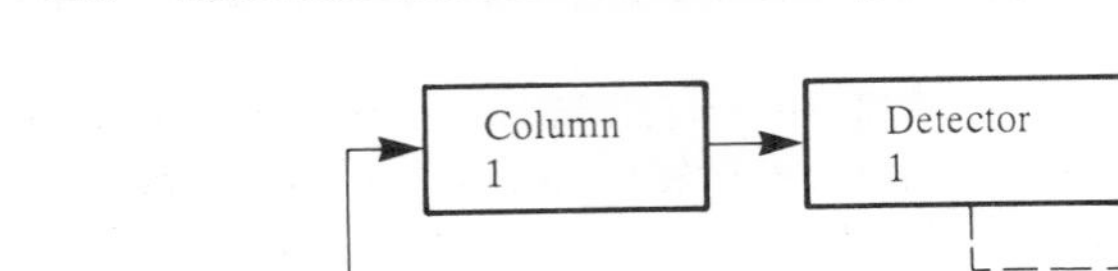

Figure 13.2 *The Nordion Analysis System. The system chromatographs any given sample using two GC columns of differing polarity. To be identified as present and to be quantified, a peak must be present in the correct place for each column. It is highly unlikely that any interfering compound will chromatograph in the defined position on columns of such differing polarity*

(Figure 13.2). For the purpose of the evaluation described here, mixed standards of pestcides were added to extracts of various foodstuffs prior to the further addition of the internal standard solution.

The sample solution vaporizes in the injector and is swept on to the columns by carrier gas. Compounds are separated on the two columns and carried to two separate detectors. For the analysis of organophosphorus pesticides, two alkali thermal ionization detectors (ATDs) were used, whereas for organochlorine pesticides two electron capture detectors (ECDs) were employed.

For all recommended detector types, signals proportional to the amount of the compounds of interest are obtained for each of the two columns. These are represented as channels 1 and 2 on the chromatograms. A built-in dual-channel integrator provides retention times, peak areas and area percentages for each peak, together with a code indicating how each peak has been integrated. These results are output to a printer/plotter, which also provides a chromatogram, where the magnitude of the signal is plotted against time. The raw data are then transferred to the computer and the software searches the chromatogram for the series of internal standards. Retention indices for every peak in the chromatogram are calculated and the internal library is searched for any compounds with matching indices. The concentrations of identified compounds are then calculated and final results printed out.

The analysis system is quite simple to use and can be operated, after some initial training, by relatively inexperienced staff. When the gas chromatograph is used in combination with the computer and associated software, the speed and accuracy of identification and quantification are improved by eliminating the time required to interpret complex chromatograms. Although the software was found to be easy to use and many 'help' messages were available if required, it is considered that staff with experience in pesticide analysis would be required to check results and, if necessary, interpret data. This would be especially important in those cases where the two columns gave conflicting values and the chromatograms needed to be examined and interpreted by an analyst. Mixed pesticide standards are supplied by the manufacturer and other compounds can be added to the library quite easily provided that standards are available.

The equipment fits on to a laboratory bench and the only additional space required is for any gas cylinders. If hydrogen is to be used, the room needs adequate ventilation. Columns and detectors are easily accessible and quickly interchangeable. Columns can be changed outside the oven itself, which was found to be very beneficial.

Any combination of two columns and two detectors can be used, which means that many other multi-residue determinations are possible. Although these were not investigated during the evaluation, application notes were available for determinations such as fatty acid methyl esters, free fatty acids in water, alcohols in water, polychlorinated biphenyls, aromatic hydrocarbons and flavour compounds. An autosampler was also available if required, but again this was not tested during the course of the evaluation.

The instrument still needs the operator's skill in preparing the extract, setting up the instrument and interpreting the results. However, the unit has been primarily designed with pesticides in mind, which means that it is doing some of the analyst's thinking for him.

In the evaluation, the system was used to identify and quantify a wide range of organochlorine and organophosphorus pesticides. In combination with the computer and associated software, identification and quantification can be obtained accurately and rapidly. In general, the results obtained using both electron capture (for organochlorine pesticides) and alkali thermionic detection (for organophosphorus pesticides) were found to have acceptable mean analytical values and repeatability.

3.2 HPLC System for the Assay of Ions and Sugars: Dionex 2000i Series

A branch of HPLC showing the trend towards dedication most noticeably is that of high performance ion chromatography (HPIC), considered by many to be a field in its own right. Ion chromatographs were originally conceived as modular instruments, that is with individual modules for pumps, columns and one or more types of detectors. Consequently, the evaluations carried out were with instruments made up of modules relevant for the analytical determination in question. Configurations for a range of ions and for sugars were evaluated in 1985. Some examples of configurations are given in Figure 13.3. Design improvements have since been made by the manufacturer, Dionex.

3.2.1 Configuration for the Detection of Anions

The chromatograph contains no metallic parts in contact with the eluents.

An aliquot of sample containing the anions of interest is injected into the analytical column (separator column), which is packed with a pellicular anion exchange resin, that is one bearing cationic sites. Anions in solution compete for these sites to varying extents, the most effective competitors moving most slowly along the column and vice versa. Consequently, a separation is effected.

Various methods of detection can now be used. Ions with an absorbence in the ultraviolet or visible light regions of the spectrum can be detected with a UV-VIS detector. A post-column reactor can be placed between the analytical column and this detector to enable light-absorbing derivatives to be made where appropriate. Other detectors can be used instead of, or placed in series with, the UV-VIS detector provided that the post-column reactor is removed. An electrochemical detector is used for bromide since this ion can be oxidized.

The Dionex system (Figure 13.4) is often referred to as a dual-column ion chromatograph because a second column, called an ion-fibre suppressor, is used to make conductimetric detection possible. This column, which is described below, either is placed immediately after the analytical column or follows in series after one of the other detectors mentioned above. The conductimetric detector, which detects a wide range of ions through their elecrical charge, can now be connected to the end of this chain before the eluent stream goes to waste.

An eluent stream passes through a cation-permeable fibre, outside of which flows a counter-current of dilute sulphuric acid. Metallic cations are exchanged for protons. These then associate with OH^- (which would have contributed strongly to eluent conduction) producing neutral water mol-

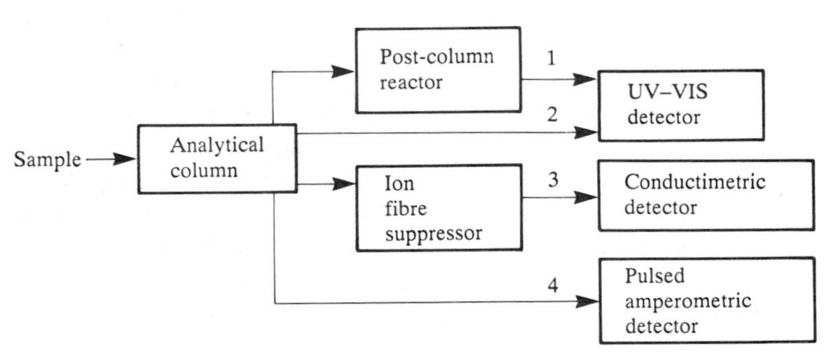

Figure 13.3 *Possible configurations of the Dionex Ion Chromatograph. Examples: (1) phosphates and polyphosphates in cured meats (2) aspartame in soft drinks (3) chloride and other anions in foodstuffs (4) sugars in foodstuffs. Different analytical columns are required for these applications*

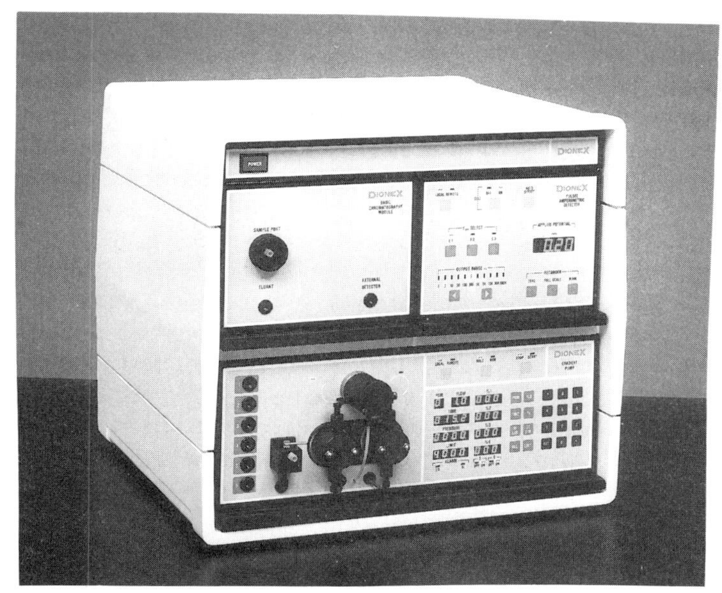

Figure 13.4 *The Dionex 2000i Series Ion Chromatographs for the assay of ions and sugars: the configuration for the assay of sugars is shown here*

ecules (H_2O). Additionally, strongly conducting acids (for example HCl) replace salts in the sample (chlorides, bromides). Consequently, background conductivity due to the eluent is suppressed, while amplification of the signal from target ions occurs simultaneously. Ion chromatography was found to be a useful technique for measuring nitrite and nitrate in bacon and for bromide (resulting from methyl bromide treatment) in peanuts. This was assessed by comparing values with those of a reference method in the case of nitrate and nitrite. An in-house method involving derivatization followed by coupled gas chromatography and mass spectrometry was used to check bromide levels.

The modular approach allows the instrument to be configured for different determinations. For example, for bromide alone the ion-suppression need not be used since this is only necessary for conductimetric detection. Alternatively, this can be added on after the electrochemical detector as an additional means of detection. It is in any case essential for non-oxidizable ions such as chloride and sulphate.

3.2.2 Configuration for the Detection of Sugars

Another example of the flexibility of ion chromatography is in the modular configuration for sugars. The principle behind this determination is as follows.

Sugars in aqueous solution are injected into a stream of 0.16 mol/l odium hydroxide solution. Under these conditions (pH > 12) hydroxyl groups on the sugars dissociate, which results in the sugars becoming negatively charged. The sugars are then separated on an anion exchange column. The concentration of the sodium hydroxide eluent determines the retention time of sugars in two ways. Firstly, high concentrations reduce retention times because hydroxyl ions compete with sugars for sites on the anion exchange column. Secondly, it influences the ratio of charge to mass of the ionized sugars since dissociation constants (pK_a values) of hydroxyl groups vary from one sugar to another. The result of these two effects is seen with, for example, glucose and lactose, where increasing the sodium hydroxide concentration from 0.05 mol/l to 0.2 mol/l decreases retention times but improves their resolution. This is an unusual observation in liquid chromatography.

Detection is effected with an electrochemical detector referred to as a pulsed amperometric detector. This operates by sequential application of differing potentials to an electrochemical cell, each sequence making up one cycle. Three cycles are applied per second.

Carbohydrates are detected by oxidation of primary hydroxyl groups in their dissociated form at the surface of gold electrodes, where a positive measuring potential is applied. Undissociated hydroxyl groups cannot be oxidized under the conditions used (see below). An amperometric signal results from oxidation. A second but higher positive potential is applied, removing reaction products that would otherwise quickly poison the surface of the electrode if allowed to accumulate. Since some oxidation of the gold also occurs at these potentials, a third and negative potential is then applied to regenerate the gold on the surface (Figure 13.5). Each cycle generates one brief electrical signal during the period when the measuring potential is applied. Such pulses, the strength of which relates to the concentration of the sugar in solution, are electronically integrated within the detector.

In order to evaluate the Dionex Ion Chromatograph for the analysis of mixtures of sugars, it was necessary to choose an established method to check the results from the instrument. Apart from other chromatographic methods, the Boehringer enzymic test combination is almost the only practical and relatively inexpensive way to determine mixtures of sugars. Consequently, it is now widely used for this purpose in many quality control laboratories in the food industry, to good effect.

The problem with the enzymic methods is that they are manual procedures, with a considerable number of pitfalls for inexperienced operators. Meticulous technique is a prerequisite of accurate and reproducible results. This is particularly true for the analysis of mixtures of sugars. Often, more than one kit is necessary. In the case of the sucrose/glucose/fructose test combination, values for individual sugars are obtained through subtraction of a series of absorbance measurements. In order to cope with widely differing concentrations of individual sugars, the written method required some modification. Such modifications add to the difficulties mentioned above but do enable all three sugars to be analysed in one run.

Training of staff to this high standard can be expensive and, for this reason, HPLC with refractive index detection is sometimes used. This technique is unsuitable for many foodstuffs where individual sugars are

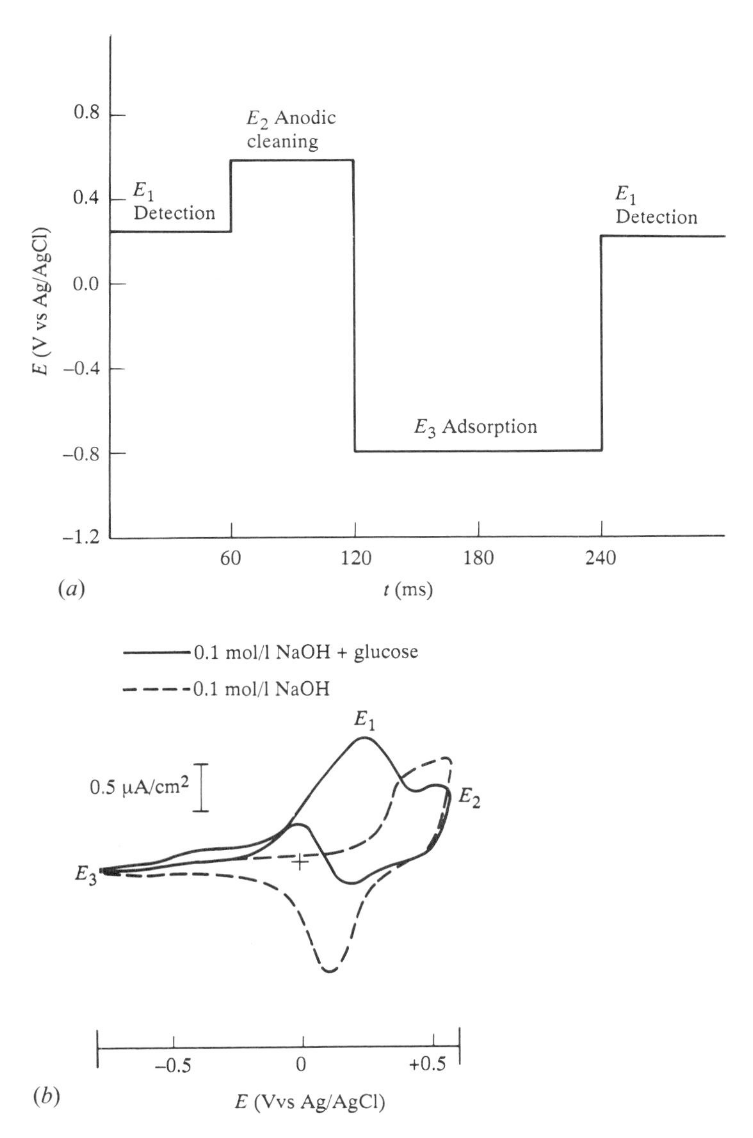

Figure 13.5 *Glucose determination by pulsed amperometric detection: (a) applied potential sequence (b) cyclic voltammetry of glucose on a gold electrode*

present in small amounts (less than 1 per cent m/m) because of the lack of specificity of the detector. GLC can be used to good effect, especially when sugars are present at high levels in the foodstuffs and extracts can be diluted prior to derivatization. Derivatization to trimethylsilyl oximes is time-consuming and would make the analysis comparatively expensive. This problem is exacerbated where sugar contents are low. It is then necessary to clean up extracts considerably, which may mean complete precipitation of polymeric constituents from the extracts, followed by preparative thin-layer chromatography of the sugars. In summary then, the routine determination of sugars in foodstuffs by GLC can be very expensive. Finally, rapid classical methods for sugars, i.e. titrimetric methods, are not suitable since they will not differentiate between individual sugars and other reducing constituents.

The Dionex Ion Chromatograph overcomes all of these problems by combining a novel chromatographic principle with a sensitive and specific method of detection – the pulsed amperometric detector. The chromatograph can resolve and detect both reducing and non-reducing sugars on one trace. One great advance applies to the case of sucrose determinations, where the sugar is measured directly and not determined by difference measurements after hydrolysis. This is probably the main reason that there is such a marked improvement in the reproducibility of sucrose determinations when the ion chromatograph is used. Most importantly, the instrument is simple to operate, requiring only brief staff training.

The manufacturers claim that the ion chromatograph will also resolve and determine oligosaccharides up to ten glucose units, polyhydric alcohols and sugar alcohols such as xylitol, sorbitol and mannitol. Dionex also produces a range of application notes showing potential uses of the more conventional ion chromatography of anions and cations monitored through conductimetric detection.

The above comments relate principally to the 2000i Series. However, the arrival of the 4000i Series (also a modular system) has brought with it improved ion suppression and the ability to perform gradient ion chromatography. The resolution of a greater number of components in any one analytical class now seems possible.

Returning to the main theme of the trends in instrument design, the modular approach allows a versatile instrument to become dedicated. Dionex will sell a system configured just to determine sugars, for example. However, the theme is probably best illustrated by the Wescan Sulphite Analyser recently evaluated at the LFRA.

3.3 Ion Chromatograph for Sulphite: Wescan Sulphite Analyser

The equipment consists of three separate units. The primary units are the pump and the analyser system itself. A chart recorder or integrator may then be attached to record chromatograms and, in the latter case, calculate peak areas and results.

Sulphite is extracted from samples by homogenization with a buffer solution consisting of 20 mmol/l disodium hydrogen orthophosphate and 10 mmol/l D-mannitol (pH 9). Free sulphur dioxide is not present at this pH value. When sodium metabisulphite, the most commonly used source of sulphur dioxide, is dissolved in water it hydrolyses to bisulphite and an equilibrium is set up between sulphite (SO_3^{2-}), bisulphite (HSO_3^-) and sulphurous acid (H_2SO_3):

$$2HSO_3 \rightarrow H_2SO_3 + SO_3^{2-}$$

At basic pH values sulphite is the predominant species, and above pH 8 dissociation of reversibly bound sulphites is complete. The mannitol minimizes the oxidation of sulphites once they are extracted into solution.

Extracts of samples are then injected on to the HPLC. The solution is carried on to a column filled with a cation exchange resin, which separates the sulphite from other oxidizable species. Finally, the sulphite is carried to a platinum electrode, where it is oxidized and detected. The change in current, generated by the oxidation, across the electrodes in the electrochemical detector cell is proportional to the concentration of sulphite. When evaluated against the Tanner method, the instrument produced similar analytical values and gave acceptably repeatable results.

3.4 Enzyme Electrode Analysers for Sugars and Alcohols: YSI Model 27 and Model 2000 Analysers

These instruments have an interesting history. They were originally developed as the Model 23A Clinical Glucose Analyser, and this led to the Model 27 Industrial Analyser.

Development of a series of kits enabled the instrument to be converted from one dedicated analytical instrument to another, whilst retaining versatility. The glucose kit renders the instrument a glucose analyser; the alcohol kit does the same with alcohol. Other determinations are also possible with the appropriate kit. A standard amperometric gas probe is supplied with the instrument; selectivity is achieved by using the enzyme membranes in the kit.

The system configured as an alcohol analyser is chosen by way of illustration. Model 27 is shown in Figure 13.6a. The following additional items are, however, required for the determination of alcohol. The YSI 2391 temperature adaptor is a small plug-in cartridge, which is placed in the power supply line to the temperature block. This brings the sample chamber to 27°C, which is the required temperature for alcohol oxidase activity. The 5 μl Syringepet (YSI 2393) is a specifically designed automatic

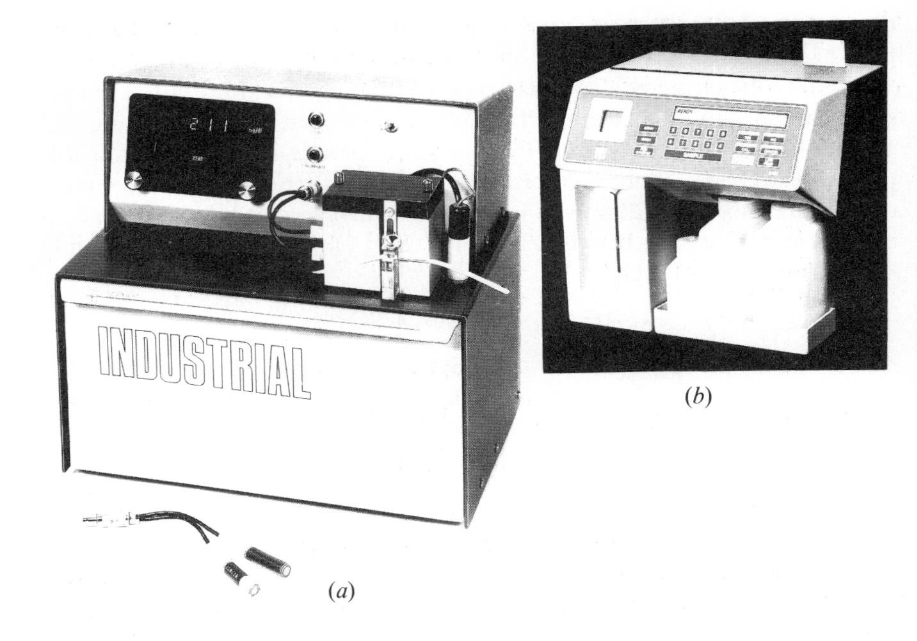

Figure 13.6 *YSI (a) Model 27 and (b) Model 2000 Analysers (enzyme electrode method): Model 27 was evaluated. A spare peroxide probe is shown with model 27, and the instrument itself is mounted over a large accessories drawer.*

pipette for injection into the sample chamber. Finally, the YSI 2390 alcohol kit contains:

- 4 vials buffer, each reconstituted to 450 ml (YSI 2387)
- 4 alcohol oxidase membranes (YSI 2386), stored frozen
- 4 vials (3 ml each) 200 mg per 100 ml ethanol standard (YSI 2388)
- 4 vials (3 ml each) 320 mg per 100 ml ethanol standard (YSI 2389)
- 100 microbeaker sample cups
- 1 bottle (120 ml) potassium ferrocyanide (YSI 2363) used to test the integrity of the membrane; injecting an aliquot should not lead to a reading unless the membrane is damaged
- 1 booklet Teflon membranes

Importantly, there are no glass components in this kit, which makes it relatively safe for use near a production line.

Molecules of ethanol or methanol diffuse through a 5 μm layer of

Nucleopore polycarbonate membrane material, via 0.015 μm diameter pores. They then encounter a 1 μm layer of immobilized alcohol oxidase, where the following reaction occurs:

$$C_2H_5OH + O_2 \xrightarrow[\text{oxidase}]{\text{alcohol}} H_2O_2 + CH_3CHO$$

Oxygen is present in excess and the reaction is limited by the ethanol concentration. Other alcohols such as methanol also react. Molecules of hydrogen peroxide diffuse towards an anode, where they are oxidized electrochemically as follows:

$$H_2O_2 \longrightarrow 2H^+ + O_2 + 2e^-$$

The circuit is completed by a silver cathode thus:

$$4H^+ + O_2 \longrightarrow 2H_2O - 4e^-$$

The resulting amperometric signals are amplified and processed by the instrument. The results are displayed on an LED readout.

Experience in the LFRA Member Services Laboratory shows that when the instrument is set up and in standby mode, Model 27 can be calibrated and six samples determined in about 1½ hours. (In contrast, distillation is a much longer process, requiring bulky apparatus and the continual attention of a technician. Six samples take one day to complete.) Membranes are judged to have failed when the instrument can no longer be calibrated.

Model 27, fitted with the appropriate Analytical Kit, was found to be a rapid, accurate and reproducible method for the determination of alcohol in beverages and lactose in milk. It should be noted, however, that the application notes supplied by the manufacturers relate to the foodstuffs for which they are written only. The instrument can be used with other foodstuffs but only if careful checks of accuracy and repeatabilites against a suitable reference method are made.

4 Dedicated Instruments

The biggest factor influencing the development of any instrument is whether it will sell profitably. For this reason, dedicated instruments for food applications have tended to be restricted to the most commonly used analytical tasks. These are moisture, ash, fat and protein assays (proximate analysis). They are the only analyses that are performed with high sample throughput by most food manufacturing companies. One word of warning is now necessary. All of the instruments detailed in Sections 5 and 6 are intended primarily for routine quality control purposes since they provide fast throughput of samples.

A considerable number of official methods are available for proximate analysis, and these should be used for determinations connected with food labelling. The use of certain instruments is acceptable for this purpose, however (for example the CEM Meat Analysis System for meat and meat products, the Kjel-Foss nitrogen analyser for protein, and the Foss-Let fat extraction system) since these are AOAC approved for certain foodstuffs. Outside this, instruments should only be used where they have been evaluated against an official method by an independent organization. Official methods usually relate to specified foodstuffs and take account of their particular analytical problems. Adequate performance by an instrument with one foodstuff does not guarantee acceptable results with another. It is therefore now appropriate to consider in detail some of the state-of-the-art instrumentation available.

Where instruments perform more than one of those analyses, they are discussed together since confusion could result from trying to separate functions.

5 Dedicated Instruments for the Determination of Moisture, Ash and Fat

There are two fundamental methods for the determination of moisture in foodstuffs – directly by chemical reaction as in the Karl Fischer procedure, and indirectly through loss of weight on drying. There are, in turn, three ways of applying heat in order to dry a sample: infrared irradiation; thermal conduction and convection (as in oven drying); and microwave energy. Ash is really a measure of the mineral content of the sample. Since, in foodstuffs, minerals are dispersed amongst and sometimes complexed with organic material, combustion is the only way an assay of such material can be achieved. Finally, there are three ways in which fat can be measured. Whilst all involve extraction of the sample with a non-polar solvent, they then proceed by weighing the insoluble residue, by drying and weighing extracted fat directly, or by measuring the specific gravity of the fat containing solvent.

Following a discussion of the underlying principles, five state-of-the-art instruments are described as representatives of these methods, based on evaluations.

5.1 Moisture by Drying Methods

A comprehensive discussion on laboratory moisture measurement can be found in Egan et al. (1981). This topic is again broad enough for a book devoted to it alone. Consequently, one or two general principles only are given here.

Oven drying procedures are generally the best and most common methods, which is why most official methods use them. However, unsaturated oils, for example, can gain weight from oxidation and so a gentle oven drying regime (preferably under nitrogen) is advisable. Vacuum oven drying allows lower temperatures to be used for delicate samples. Infrared dryers are definitely not suitable here, although they do perform better with flour and baked products. Charring can be a particular problem with infrared heat applied to meat samples and other foodstuffs high in fat, where there is also a risk of loss of sample through splattering. Lumpy samples dry unevenly and so dispersion of the sample is important prior to drying. Foodstuffs containing volatile components can be a problem in that 'moisture' may not equate to 'water', for example in the analysis of coffee or certain edible oils. In this case, and for very low moisture contents, the analyst needs to decide in advance what needs to be measured, then choose the appropriate method (see also Section 5.2). Milk powder can give erroneous results owing to the water of crystallization associated with its major component, lactose. This problem is overcome by wetting the sample with water, then redrying, since such water is only associated with one crystalline form of lactose. Analysts will find that official methods are usually written for given foodstuffs and use the best procedures and optimized conditions.

Careful application of microwave energy seems to work well for meat in automated instrumentation (CEM system), although the Leco system, which has precise control of oven heating, also gives good results. Generally, each type of sample has its own particular problems of sample preparation and optimum drying regime. The LFRA *Analytical Methods Manual* (Slack 1987) indicates the variation in methodology available.

5.2 Moisture by Karl Fischer Titration

The Karl Fischer procedure can be carried out manually but this is tedious. The classic Karl Fischer reagent consists of pyridine, iodine, sulphur dioxide and methanol. The chemical reaction is represented by the following equations:

$$3C_5H_5N + I_2 + SO_2 + H_2O \rightarrow C_5H_5N^+SO_3^- + 2C_5H_5N^+HI^- \quad (1)$$

$$C_5H_5N^+SO_3^- + CH_3OH \rightarrow C_5H_5NHSO_3OCH_3 \quad (2)$$

Since pyridine is toxic, many laboratory suppliers now sell alternative reagents. The end-point is determined electrometrically. If a small voltage is applied across two platinum electrodes in the reaction mixture, a current will flow while iodide ions are present. (The iodide removes protons and depolarizes the cathode.) When there is no free iodide around the cathode, the current decreases to almost zero and the end-point is reached.

Karl Fischer titration is the method of choice for samples with low moisture levels, around 1–2 per cent m/m. Here, the technique is sensitive and more accurate than oven drying. It is particularly useful for measuring moisture in confectionery, and measures all the water of crystallization in lactose, for example. Such samples can be easily dissolved or dispersed in methanol based solvents, whereas they are difficult to oven dry. Unsaturated oils and fats can also be assayed for moisture since there is no risk of losses from volatile fatty acids, or indeed weight increases due to oxidation and then polymerization. The Karl Fischer procedure is not, however, universally applicable since certain chemicals (for example, aldehydes in flavour preparations) can interfere with the result.

For the Karl Fischer method, automated versions (as in Section 5.5) are much quicker than oven drying but the technique is sometimes too sensitive for foodstuffs with a high moisture content, for example meat products. Although better for powders and, say, oil samples, the technique is widely used for moisture in syrups, sugars and sweets.

5.3 Ash by Combustion Methods

Ashing involves firstly drying the sample under a gentle regime, then combusting (usually in a furnace) to remove all organic material and hence liberate mineral oxides, hydroxides, carbonates, etc. The sample residue is then weighed. For ashing by combustion, there is an enormous number of official methods tailored for each given foodstuff. These are written to solve problems such as sample inhomogeneity, tendency to splatter, and the question of whether acid or alkali inclusion is necessary to prevent sublimation of oxides, carbonates and hydroxide components of any particular ash sample. An automated instrument is described in Section 5.7.

5.4 Fat by Solvent Extraction

Fat content can be measured in a number of ways. The fat can be extracted into solvent, the solvent removed and the fatty residue weighed. Alternatively, the specific gravity of the solvent/fat solution can be measured since this is related to the fat concentration. Thirdly, the loss of weight of a dried sample can be measured following extraction of the fat with solvent and redrying the solid residue. This option is featured in the CEM system described in Section 5.8. There is not, to the author's knowledge, any instrument that automates the first option but the specific gravity option forms the basis of a well established instrument in food laboratories, namely the Foss-Let (see Section 5.9).

Apart from this, certain extraction aids are marketed that cannot be described as instruments. Again, a vast number of official methods are available. The Foss-Let and CEM instruments are effective with free fat, that is that readily extracted with solvents. Hydrolytic methods need to be used where fat is complexed, or indeed chemically combined, for example, in cereals such as wheat or maize and in starchy foodstuffs. There is also a risk that these instruments will give erroneously high results with foodstuffs containing high levels of unsaturated fats, especially fish oils. Oxygen uptake is most likely during drying steps. Again, the analyst must be sure of the application before investing in such instruments.

5.5 System Based on the Karl Fischer Procedure: Baird and Tatlock Turbotitrator

This instrument (Figure 13.7) is a compact, bench-top unit. A Perspex cabinet isolates the assembly from the user to contain toxic vapours. Closing the door at the front initiates blending and titration sequences. Titrations, performed by the titration unit, are controlled and monitored by an autometering unit.

The instrument console has a series of controls for programming the motor, autotitration and autometering units. Values from the electrochemical sensor and the metering unit feed back to the console via microprocessors.

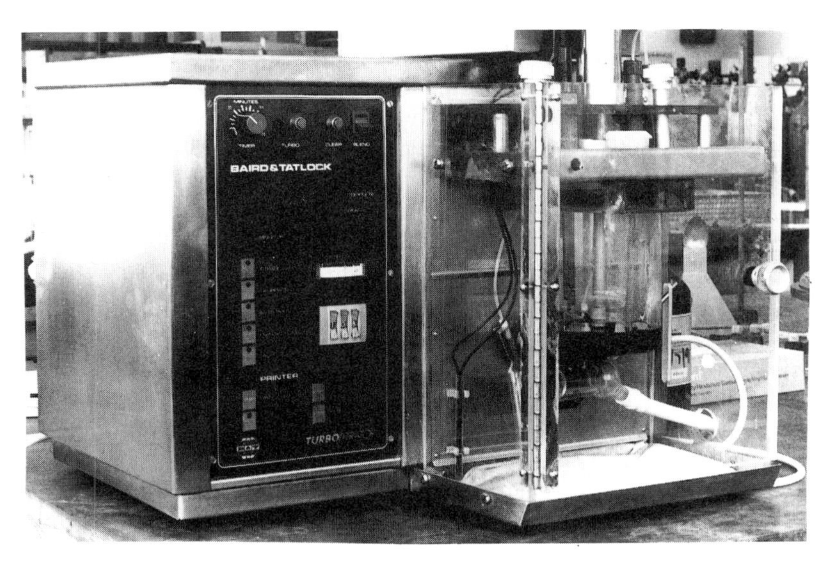

Figure 13.7 *The Baird and Tatlock Turbotitrator for the determination of water content, after many years of use in the Member Services laboratory*

Activity centres around a glass vessel (the blender), which contains a motor-driven blade capable of operation at two speeds. A known weight of solid or liquid is introduced into the vessel, which contains dry methanol. The total contents are homogenized at high speed for a given time. The instrument then reverts to a slower speed and automatically titrates Karl Fischer reagent in the same vessel. This is rapidly consumed to begin with but, as the end-point is approached, the rate of titration exceeds that of consumption. The water-dependent release of iodide ceases, as does the flow of current through the solution. This causes titration to stop momentarily, then stop and start as the end-point is reached. This process is automatically repeated until the current does not rise over a 12 second period.

The Turbotitrator AF3H takes about five minutes per sample to measure the moisture content of solid or liquid samples automatically. It was evaluated for a small range of foodstuffs, namely canned meat, palm acid oil, jelly sweets, salad cream and mayonnaise.

Apart from the experiments given in this report, the Turbotitrator AF3H has been used at the Leatherhead Food Research Association for the routine determination of moisture by the Karl Fischer principle. Determination using this instrument is much quicker than that using conventional procedures for moisture measurement or when using earlier instruments based on the Karl Fischer principle. One sample every five minutes has been found to be the average throughput in routine use (from weighing to result). The instrument can be operated for long periods without attention, and little training of staff was necessary prior to operation.

The instrument is a compact, robust unit, which could be operated by semi-skilled staff after training. The Turbotitrator was found to produce results with acceptable reproducibilities on duplicates. Single readings are, at most, within 5 per cent of the mean value for samples with a moisture content of 10 per cent or above. Difficult samples with very low moisture contents (for example, palm acid oil) may produce greater variation over the mean value. At these levels of moisture, this is probably not analytically significant. Standard deviations show that greater precision can be achieved by analysing several replicates.

5.6 System Based on Oven Drying: Computrac Max 50 Moisture Analyser

A sample is placed on to an electronic weighing system referred to as a force restoration balance. This consists of a circular balance pan mounted above a metal shaft surrounded by a magnetic coil and resting on spring-loaded bellows. Application of force compresses the bellows, causing an electronic signal to be passed to the computer. The computer then controls the application of a magnetic force until the shaft is returned to its original

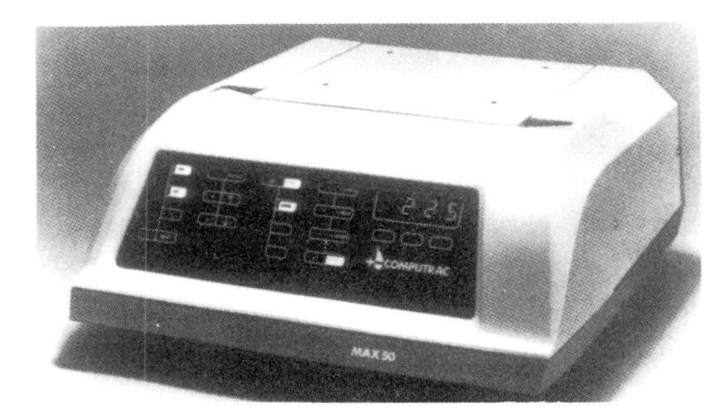

Figure 13.8 *Computrac Max 50 Moisture Analyser (oven-drying method)*

position. The amount of electrical energy necessary to resist the downward force caused by the sample on the pan is measured by the computer and expressed as a weight. This type of balance is claimed to be a very sensitive system for measuring changes in weight on drying.

Application of radiated heat from an electric element causes the sample temperature to be quickly raised to a pre-set level. The computer monitors this temperature by means of a heat sensor placed immediately above the sample, which controls the current in the element. Loss of weight is continually monitored and the computer periodically predicts final moisture content through extrapolation of a loss-on-drying curve. Initially, weight loss is rapid, causing measured and predicted values to differ considerably. As the rate of weight loss decreases, these two values converge until a point is reached when actual and predicted values are close. Once the variation in end-point prediction falls below a defined value (0.015 per cent for low moistures near 0 per cent, 0.12 per cent for high moistures near 100 per cent), the computer makes a final prediction, which is displayed on the LED as percentage moisture content. The sample is not necessarily dried to constant weight.

The Computrac Max 50 (Figure 13.8) is the most sophisticated of a range of instruments originally developed for moisture determinations on chemical and pharmaceutical products. Operation of this robust unit is straightforward, producing moisture values extremely quickly compared with conventional drying procedures. The Max 50 eliminates the need for repetitive calculations. The instrument is, however, unsuitable for oils.

The analytical value obtained is sometimes dependent on the amount and distribution of sample across the metal trays. This is especially so for bacon, where even distribution is essential for a correct result. The recommended amount of sample (7–10 g) is obviously too great in these cases. Smaller amounts (4–5 g) are clearly easier to spread, thus avoiding lumps, which will tend to trap moisture. Lumps probably reduce the rate of drying to a level that could lead to premature termination of the drying programme.

Conventional oven drying methods avoid this problem by dispersing the sample with sand. The technique is not applicable to the Max 50 since it does not have any facility to compensate for the added weight of sand. Consequently, apparent moisture values would have to be manually corrected, thus removing one of the major advantages of the instrument.

Generally speaking, the Max 50 produces more variable results than the reference methods. This may be because the instrument is more susceptible to slight differences in the distribution of the sample from one run to the next. There would not appear to be any relationship between drying times and the variability in analytical values.

The Computrac Max 50 could well find use in a routine quality control laboratory, where continuous assessment of a few products is necessary. The speed at which values are obtained is a clear advantage of the system. However, it is recommended that results from the instrument should be checked for each food product against a reference method. This should be done on samples of the product covering a range of moisture contents to ensure that the drying temperature chosen will give comparable results. If this is not so, some adjustment to the pre-set temperature might be necessary. It is certainly inadvisable to rely solely on the temperature chosen from a moisture versus temperature drying curve to give comparable results; this procedure should be used merely as a guide. The instrument is unlikely to be suitable for a laboratory that handles a wide range of different food products.

The Computrac Max 50 gave reasonable mean analytical values for moisture in flour, oilseeds and bacon provided that care was taken to calibrate it against a reference method. This would need to be done prior to its use as a routine quality control instrument, where its main advantage is speed.

5.7 Instrument for Moisture Determination by Oven Drying and Ash by Combustion: Leco MAC-400

The Leco MAC-400 is a compact bench-top instrument, which consists of three units. These are the furnace unit, the electronics unit and the control console.

Up to nineteen empty sample crucibles and one reference crucible are placed on a turntable positioned within a furnace. The furnace top then closes, the turntable rotates, and crucibles are automatically weighed, one at a time, on a balance unit situated underneath the furnace. Samples are then placed into each sample crucible and the weighing process is repeated.

The furnace is purged with nitrogen and its internal temperature is then raised to the pre-programmed level for determination of moisture (approximately 100°C). The reference crucible is also weighed and this weight is used to correct for any drift of the balance due to the change in temperature. Moisture analysis is terminated when the change in successive sample weights is within pre-set limits and the moisture results are then computed and printed out.

Ashing commences with the temperature of the furnace being raised to another pre-programmed level. The furnace atmosphere is changed from nitrogen to oxygen and the sample is combusted. Crucibles and contents are continuously weighed as before, and when successive weights are again within pre-set limits the analysis is terminated and the ash results computed. Moisture and ash results for each sample are then printed out.

The Leco MAC-400 is a robust, bench-top instrument that requires minimal training of staff. The system requires a 20 ampere supply for single-furnace operation and a 40 ampere supply for dual-furnace work. The instrument is simple to operate and the manual provides step-by-step instructions that are easy to follow. Moisture and ash results on a variety of foodstuffs tested showed a good agreement with reference methods in most cases. Low values were obtained for moisture from the MAC-400 (in sausages and cheese) and probably reflect incomplete dispersion of the sample with the associated formation of a water-impermeable crust. Reference methods use sand to disperse the sample, thereby ensuring better moisture loss. Consequently, mean values for the MAC-400 could possibly be brought closer to the reference method values by the use of sand, although this would not permit the determination of ash on the same run. In all cases except one (ash in cheese), where a number of replicates are advised, repeatability was acceptable, meaning that duplicate determinations would normally be adequate.

Times taken for the moisture and ash determinations on the samples tested varied from about 2 hours (milk powder and ham) to about 6 hours (sausages and cheese). This compares favourably with the reference methods, which required a minimum of 3–4 hours for the moisture determination, with repeated weighings along the time course of the drying period, followed by an overnight ashing period.

The Leco MAC-400 Proximate Analyser determines moisture and ash in up to nineteen samples simultaneously. The instrument produces mean analytical values for ham, sausages, cheese and dried milk which, in most cases, compare satisfactorily with values from established reference procedures and which are acceptably reproducible. The instrument is much quicker (2–6 hours from loading to result) than the established methods (usually requiring overnight drying or ashing).

5.8 Instrument for Moisture by Microwave Drying and Fat by Solvent Extraction: CEM Meat Analysis System

The system is assembled as two units. The moisture/solids analyser is a compact bench-top unit (oven). It consists of a balance unit mounted within a microwave drying system (oven), which also incorporates a microprocessor digital computer. The other, floor-standing unit consists of an automatic extraction system containing an extractor and a solvent recovery system.

The sample is spread thinly on a pad and placed on an electronic balance in the oven. The initial weight is stored by the computer and the sample is then exposed to microwave irradiation, causing the sample to heat up and the water to be driven off. The computer continually monitors and displays the sample weight, enabling the operator to judge when constant weight is reached (usually 4–5 minutes). These data are stored and the sample is then transferred to the floor-standing fat extraction unit, where it is automatically homogenized with dichloromethane. This residue is separated from the fat/solvent mixture by filtration and initially dried under compressed air. The residue and pad are returned manually to the oven, where a short (approximately 2 minutes) exposure to more microwave energy removes any remaining solvent. The computer calculates the percentage fat from the further loss in weight and prints out the moisture and fat contents (in per cent). The computer can also be programmed to print out the protein content (in per cent) calculated by difference if known values for the ash and total carbohydrate contents (in per cent) are entered.

The CEM Meat Analysis System determines moisture and fat in meat and meat products in about 9 minutes for a single sample. Three types of meat were analysed and the results were compared with those obtained by standard methods. Mean values for moisture were found to agree with the reference results, whilst mean values for fat were found to be consistently 0.3–0.4 per cent lower by the CEM instrument, and it is recommended that a correction should be applied.

The CEM Meat Analysis System performs the determination of moisture and fat in meat products more quickly than established reference methods. Values of fat will need to be increased by 0.3–0.4 per cent to make them comparable with the British Standard method. However, a computerized system does help to eliminate errors that could occur as a result of having to perform repetitive calculations.

The instrument is a robust unit that requires both bench and floor space. Relatively inexperienced staff could be trained to use it in a short period.

More importantly, the solvent recovery system allows economic operation in which filter pads and the occasional addition of solvent are the only consumables. One further advantage comes from the enclosure of the extraction process, which means that it should be possible to site the instrument in the vicinity of a production line.

The instrument is considered suitable for a quality control laboratory provided that fat results are first compared with those obtained by a reference method to determine the degree of bias for any given product. The manufacturers claim that other types of foodstuffs can be analysed using the instrument, but this was not investigated.

5.9 Instrument for Fat Determination by Solvent Extraction: Foss-Let System

One of the best established instruments in the food analytical laboratory is the Foss-Let system for total fat determination.

The Foss-Let consists of a number of units. Firstly, there is a balance for weighing the sample and the oil for a 50 per cent calibration liquid. Secondly, there is a dispenser for dispensing the exact amount (120 ml) of perchlorethylene. The dispenser is provided with an expansion device to compensate for the temperature-induced expansion of the liquid. In addition, there is a cooler for cooling the extraction weights during continuous operation.

Grinding and extraction of the sample are performed in a homogenizer housed in a lockable chamber with a movable weight. This extraction chamber is fastened to a holder in the homogenizer and is vibrated automatically. A measuring unit measures the specific gravity of the fat/perchlorethylene by means of a specially designed swimmer. This swimmer is loaded with a small magnet. By regulation of an external electromagnetic field, the buoyancy is adapted to the specific gravity of the solution.

The movement of the swimmer is converted to yield a display showing the oil content directly in per cent. Since the specific gravity of the liquid changes with temperature, the measuring chamber is mounted in a thermostatically controlled unit adjusted to 37°C. If required, the temperature can be set at 32°C or 42°C by mean of a switch.

This instrument is widely acceptable following its AOAC evaluation and approval.

6 Dedicated Instruments for Nitrogen Determinations: for the Calculation of Protein and Meat Content

The conventional Kjeldahl analysis is a manual procedure and can be divided into three phases:

Digestion The sample is boiled in concentrated H_2SO_4 containing a metallic catalyst. This destroys the sample and releases the nitrogen bound as ammonium sulphate.

Distillation After cooling and dilution with water, the acid is neutralized with NaOH. The now free ammonia is distilled into dilute acid in a receiver flask.

Titration The ammonia, now bound by acid, is back titrated with alkali to determine the amount of acid consumed. This will in turn be proportional to the amount of ammonia released.

In general, 16 per cent of the protein content of food and feeding stuff is nitrogen. Thus the protein content of a sample can be determined by multiplying the nitrogen content by a constant (often 6.25). This value is valid for most samples, even though the protein compositions may differ. They also form the basis of lean meat content calculations after multiplication with constant values, which vary from one meat species (for example beef) to another (for example chicken). A portion of the fat is then added for the calculation of total meat content.

There are some excellent instruments available for automated or semi-automated nitrogen determinations. Instruments for nitrogen determination represent an area of active development. Most are based upon the Kjeldahl method, a procedure that was hitherto laboriously and hazardously carried out in food laboratories for many years by the manual procedure. One slight problem with Kjeldahl-based instruments is that they use different catalysts. Copper is recommended for the Kjeltec system and mercury for the Büchi and Kjel-Foss Systems. The choice of catalyst can influence analytical values slightly. One instrument is also available that liberates free nitrogen from the sample then measures it in a thermal conductivity cell.

6.1 Instrument for Nitrogen Determination by Acid Digestion

6.1.1 Kjel-Foss Automatic

The best established instrument in food laboratories is the Kjel-Foss Automatic (Foss Electric) (Figure 13.9). The Kjel-Foss system represents a slightly modified and automated version of the manual procedure. The instrument is designed to perform rapid and automatic protein measurements based on the well-known Kjeldahl method. Samples in the range 500–1000 mg are used in order to contain sampling errors. The operation follows the conventional pattern for Kjeldahl measurements, with slight modification, in order to speed up the operation. An interval of only 12–15 minutes from sample addition until the final result is available.

Six special Kjeldahl flasks are placed around a bucket. The system turns 60° clockwise every 3 minutes in a carousel arrangement. A sample is inserted in the flask available in position 1. Three Kjel Tabs (containing a metallic catalyst), a selected amount of H_2SO_4 and a fixed amount of H_20_2 are also injected into the flask in this position, prior to insertion of the sample, and immediately before the system begins to turn. As the flask approaches position 2, the gas burner at this position is lit. The sample is now digested for 3 minutes and the waste products are sucked through the side tube into the waste bucket.

At the end of this period, a new sample is inserted into the flask now present in position 1, and so on. As the system turns 60° once more, the digestion of this new sample is begun in position 2, while the original sample is moved to position 3, where digestion of this sample is completed. The next system movement brings this sample to the cooling position 4, where a high-speed fan cools the flask. At the end of this period, the lid is opened and the sample is diluted with 140 ml water. The sample is then moved to position 5, where NaOH is injected into the flask.

There now begins the most important modification from the manual procedure. Steam distillation with 100 per cent recovery of the ammonia is commenced; the ammonia is condensed in a water-cooled condenser and dropped through a strainer into the titration glass. Indicator liquid is already dispensed into the glass and titration of the ammonia is now performed by means of diluted sulphuric acid delivered from a motor-controlled syringe. The motor is controlled by a photocell under the glass, the photocell detecting a change in colour and applying an amount of acid proportional to the amount of ammonia present. The motor spindle is also connected to a potentiometer, the output of which will be proportional to the nitrogen content of the sample. This value is calculated and displayed on a three-digit readout. The readout is also available in binary-coded decimal (BCD) forms as input to a printer.

In position 6 the flask is emptied by means of air pressure. Owing to the NaOH introduced in position 5, the remaining acid is neutralized and the waste product is near neutral. $Na_2S_2O_3$ is added to the NaOH in order to

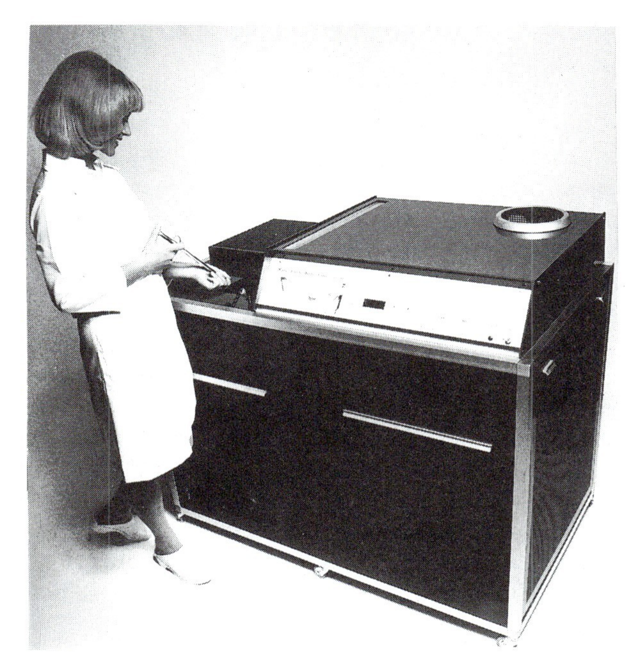

Figure 13.9 *The Kjel-Foss Automatic system for nitrogen (and protein) determination by acid digestion*

precipitate the mercury used as a catalyst. The mercury sulphide is insoluble in water and could, with some danger of pollution, be pumped out into the waste. It is, however, accumulated in a special container connected to a waste outlet tube. After being emptied in position 6, the flask moves to position 1, ready to receive a new sample.

Since the Kjel-Foss instrument is such a well-established instrument in food laboratories, it was not formally evaluated at the LFRA.

6.1.2 Kjeltec Auto/Labtec System

The Kjeltec Auto 1030 Analyser (Perstorp Analytical) is a microprocessor-controlled bench-top unit capable of performing automatic distillation and titration of digested samples. The instrument consists of several subunits, most of which are hidden from the analyst. The two most important are distillation and titration units. A Labtec microcomputer system, an interfaced electronic balance, an external printer and a 1005 Heating Unit for digestion of the samples were also evaluated as a package.

The distillation unit contains a distillation head and a condenser. Before distillation takes place, sodium hydroxide solution is automatically dispensed into a digestion tube from a reservoir. The titration unit consists of the titration vessel and a burette cylinder, which dispenses dilute acid into the vessel until a neutral end-point is reached. The end-point is determined by a photocell positioned beneath the vessel, which measures the amount of light passing through the solution from a lamp above. Neutral 'receiver solution' is pumped from a reservoir into the titration vessel just before distillation commences. The steam generation unit consists of the steam generator, containing two electrodes, which pass a current between them to heat the water, and an expansion vessel, which allows a constant, steady flow of water to pass through the system. Also contained within the body of the instrument are the microprocessor units.

The interfaced analytical balance, supplied by Tecator, is a top-loading balance containing an LED display unit, an internal calibration system and an RS 232 interface.

Samples for analysis are first registered on the Labtec microcomputer system, where they can be assigned various identifying codes and names as well as limits within which a result is expected to lie. The samples are then weighed on a balance directly interfaced with the computer, thereby allowing the weight to be stored automatically for further use. After digestion of the sample with sulphuric acid and hydrogen peroxide in a heating block, the digestion tube is placed in the Kjeltec Auto 1030 Analyser, where the samples are automatically steam distilled and titrated with acid until the neutral end-point is reached as described above. The computer then acquires and utilizes the titration volume and original sample weight, together with a number of other pre-programmed factors such as the molarity of the acid, the blank titre and the nitrogen factor (constant value) for that type of product, to calculate and print out the percentage protein in the sample.

Six types of foodstuff were analysed for protein and the results compared with those obtained on a Kjel-Foss Automatic, an instrument that is routinely used at the LFRA (see Section 6.1). The evaluation assesses the combined operation of the Kjeltec Auto with the ability of the Labtec system to pick up raw data from the Auto and the balance, and then to calculate and display a final result. Results of acceptable accuracy and reproducibility were obtained across the range of foodstuffs tested.

The evaluation has shown that a Labtec system, interfaced with a Kjeltec Auto and an analytical balance, can produce acceptable values for protein in a variety of foodstuffs when compared with a well-established reference instrument. It eliminates time-consuming calculation and errors involved in multiple calculations because of its data-handling and reporting capabilities.

Comparison of results obtained using the Kjeltec Auto 1030 and the Kjel-Foss Automatic show that the two instruments give similar results for most of the foodstuffs.

The Labtec system, when used in conjunction with the Kjeltec Auto and an analytical balance, does not speed up analysis times, but it was found to eliminate the time that would normally be required to carry out and check calculations as well as to eliminate errors that can occur in multiple calculations. Savings are likely to be greater with more complex sets of calculations arising from a variety of different determinations.

The Kjeltec manual was found to be easy to follow and could be understood and used by relatively inexperienced staff after a short training time. The instrument allows an incoming sample to be registered and data for that sample to be stored during analysis. Results can then be automatically calculated and a printout of all results for that sample can be obtained. The automatic transfer of data between the balance, Kjeltec Auto and Labtec units was not found to have any effects on the results obtained for protein on the samples tested.

6.1.3 Büchi Kjeldahl System

The Büchi Kjeldahl System (Büchi Instruments) consists of several main units (Figure 13.10). These are the control unit B-343, the distillation unit B-322, the digestion unit B-430, the fume scrubber B-412 and the cooling fan B-410. In addition, a Metrohm 665 Dosimat (automatic titrator) and an Epson LX-80 printer were used in the evaluation.

The control unit is the central part of the system. It uses an interactive dialogue, which enables the operator to enter sample weights, calibrate the pH electrode and select the distillation parameters such as the volume of distilled water and alkali to be added to the flask. The time between starting

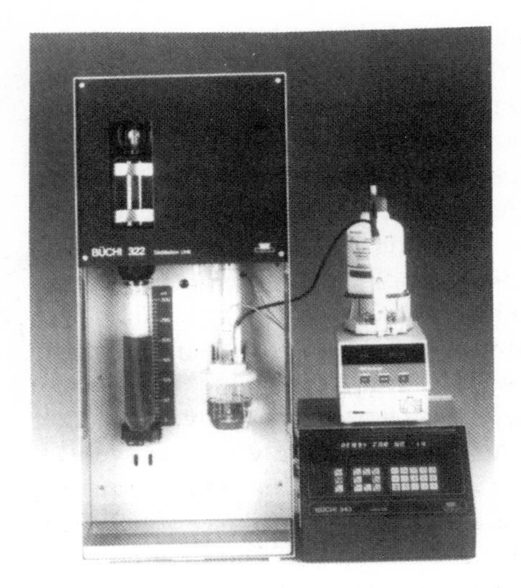

Figure 13.10 *The Büchi Kjeldahl System for nitrogen and protein determination by acid digestion*

the distillation and titration and the molarity of the acid used for titration are also entered in the same way.

The distillation unit has a built-in steam generator and a condenser leading to the receiving vessel. Residues can automatically be aspirated from the distillation flask to waste. The metering of liquids and the distillation time are preselected on the control unit and controlled automatically.

The digestion unit has eight positions with individual heating cups. It allows a stepless control of the heating power and reaches a maximum temperature of approximately 650°C in ten minutes. The inclusion of a suction tube, a manifold with gaskets to fit into the top of the digestion tubes and the scrubber unit means that the system can be used in the laboratory without a fume cupboard. Gases and vapours produced during the digestion are aspirated by an air pump through a Woulfe bottle, which separates out liquid portions. The gases and vapours are then washed and neutralized in an aqueous sodium carbonate solution. Any particles and residual gases are then trapped downstream by an active carbon filter. Waste air is passed into the atmosphere.

The 665 Dosimat is an automatic titration unit. In the evaluation described in this report it was fitted with a 20 ml burette for dispensing 0.1 mol/l hydrochloric acid. It has three push buttons on the front for filling the burette from the reservoir, resetting the volume display to zero and dispensing the acid. The rate at which the acid is dispensed can also be varied using the knob situated at the front of the unit.

Samples are weighed on to nitrogen-free paper and placed into digestion tubes. Catalyst tablets, glass beads and a volume of sulphuric acid are then added and the tubes connected to a manifold unit and suction tube. The tubes and a manifold unit are then placed into a pre-heated digestion block and left for a set time. While the digestion is taking place the parameters for the distillation and titration, together with the sample weights, are entered into the control unit.

After the digestion is complete, the tubes are removed from the heating block and cooled using a fan. When cool, they are placed sequentially into the distillation unit following a prompt from the control unit. When the start button is pressed, automatic steam distillation and titration to a previously set pH end-point are carried out. The result for that sample is automatically calculated and printed out and the sample residue is aspirated to waste. The instrument is then ready to analyse the next sample.

One chemical, with a known nitrogen content, and four types of foodstuff were analysed. Results of acceptable accuracy and reproducibility were obtained across the range of foodstuffs tested. The results were compared with those obtained using reference methods. The evaluation also assessed the speed, efficiency and ease of use of the instrument.

The Büchi Kjeldahl System is simple to use and could be operated by relatively inexperienced staff after a minimal period of training. However, it was found that the manuals were quite difficult to follow at first and this could lead to problems if any faults were to occur. The interactive dialogue on the control unit compensated for this by giving step-by-step instructions for setting up a method, which were found to be easy to follow. All of the apparatus is robust and fits easily on to a laboratory bench, ensuring that corrosive chemicals and digests do not have to be transported far.

The scrubber unit obviates the use of a fume cupboard, which would be of use to smaller laboratories. During digestion, care has to be taken that the cotton wool plug, used to seal one end of the manifold system, is not too loose or too tight. If it is too loose it can be forced out, leading to the release of acidic fumes into the laboratory atmosphere. Conversely, irregular boiling can occur in the digestion tubes when the plug is too tight, leading to the loss of nitrogen from some samples. The digestion unit contained eight places for tubes, and a four-place unit is also available. A laboratory with many determinations to perform might utilize its time most efficiently by acquiring further digestion units. Even more useful would be a digestion unit that contained, say, sixteen or twenty places for tubes.

The distillation unit improved the time required for distillation and titration of samples by a minimum factor of 2. The facility of simultaneous titration is also an obvious advantage. Further time saving could be achieved by coupling the control unit to a balance so that sample weights could be directly input. This facility is available with the system but was not investigated during the evaluation.

The calculation routine available with the instrument is also very helpful. Results can be printed out as percentage nitrogen or percentage protein contents (with the inclusion of the appropriate factor) as desired. The automatic calculation can also save time involved in manual calculation and checking of results. It also avoids any human errors that can occur in multiple calculations

6.2 Instrument for Nitrogen Determination by Combustion: Leco FP-228

This instrument, orginally developed by the Leco Corporation for the coal industry, utilizes the thermal conductivity principle of measurement.

The FP-228 (Figure 13.11) is a bench-top unit which contains electronics, furnace, thermal conductivity cell, displays and printer in a single cabinet. Three gas cylinders are also required nearby, namely oxygen for combustion, helium for carrier gas, and an inert gas or air to operate the pneumatics. The major difference between this and Kjeldahl-based systems is the liberation of organically combined nitrogen as free nitrogen rather than ammonia. It follows that the system needs to remove other combustion gases prior to detection of N_2. The heart of the system is

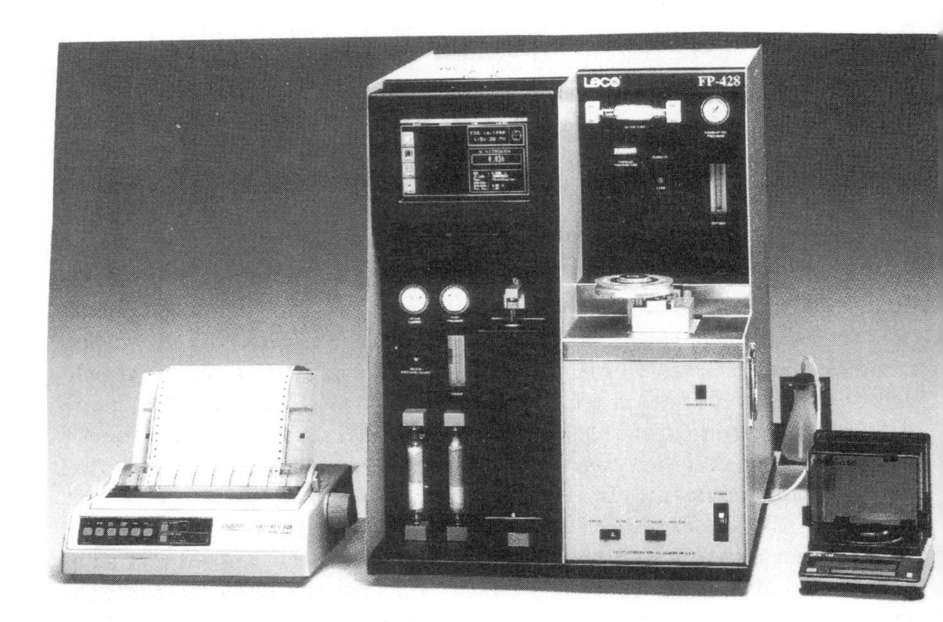

Figure 13.11 *The Leco FP-428 instrument for nitrogen and protein determination by combustion. The FP-428 is an upgraded version of the FP-228 evaluated here*

therefore a series of dry catalysts (referred to as the 'reagent train') mounted at the centre of the instrument. The effect of these on the combustion gas mix is described later.

The sample (nominally 50–300 mg) is weighed into a tin capsule. The sample weight and identity are electronically entered, then the capsule is placed in the loading head. Pressing the 'analyse' key causes the loading head to be sealed and purged of any atmospheric gases. The sample is then automatically dropped into a crucible in the furnace and combusted in pure oxygen at 1000°C. Products of combustion (mainly H_2O, CO_2, O_2, NO_2, NO and N_2) are collected in a 'ballast volume'. This ballast volume, described by a free-floating piston, allows the gas products to homogenize. When all the gases are collected, the piston forces out a 10 ml aliquot (the remainder of gases go to exhaust), which is carried by helium through a series of solid chemical phases. Firstly, the gases pass through the reagent train, beginning with a filter tube of iron chips, which removes any dust particles. Gases then pass through a hot copper catalyst in an oven, where all oxides of nitrogen are reduced to nitrogen, before finally entering a tube containing Ascarite (sodium hydroxide coated on asbestos) to remove carbon dioxide and Anhydrone (magnesium perchlorate) to remove water. Only helium and nitrogen remain. These gases pass on to the thermal conductivity cell, which has a reference and sample filament to generate heat. Nitrogen is detected as a change in the thermal conductivity of the helium stream. The reference cell, which is not connected to the sample line, provides a constant background reading of the helium.

The Leco FP-228 is an innovation as far as nitrogen measurement in foodstuffs is concerned. It totally eliminates the corrosive wet chemicals so essential for Kjeldahl methods. The instrument performed with accuracy and repeatability in the author's evaluations with flour, beef and sausages. There is no reason to suppose that other foodstuffs would not also give reasonable results, but tests should be carried out before routine application to other foodstuffs. One slight problem was with the small sample size (150–300 mg), which meant that very careful homogenization was necessary for repeatable results. The manufacturers have apparently converted the instrument to work with larger sample sizes, and hence this problem may have been overcome. The best way to calibrate the instruments is by using a standard flour of known nitrogen content, measured using an established official method. The FP-228 would be ideal for setting up close to a production line.

The Leco FP-228 performs the determination of nitrogen in foodstuffs in about 3 minutes provided that suitable care is taken over sample preparation. The instrument is a robust bench-top unit, which can be used by relatively inexperienced staff with a minimum of training, and involves no hazardous chemicals. It is judged to be acceptable for use as a rapid quality control procedure on those foodstuffs that were tested.

7 Further Development of Instrumentation for Food Analytical Laboratories

This chapter has attempted to give the reader the type of information about the instruments available to the food analyst that can only be obtained from a formal evaluation. It has also described systems in some detail. Many of the instruments described here may well have been updated to improve their performance at the time this chapter is read. The trends observed are clear and continuing. Firstly, there is a movement towards greater dedication to a particular task. Effort here will be targeted towards production of a greater range of easily used instruments for industrial quality control laboratories. However, there will also be a significant niche for specialist instruments for the food industry's high technology trouble-shooting laboratories. Just as the technique of GC-MS has already become established, so may NIR, FTIR, supercritical fluid extraction and chromatography and perhaps capillary electrophoresis. Many of these require the use of more sophisticated computer hardware and software. Indeed, training associated with the installation of such new pieces of equipment is becoming focused upon an understanding of software control and data handling as the first pressing objective.

As the use of wet chemical methods diminishes, and instrumental methods increase, the analyst is becoming more remote from the task in hand and may acquire a false sense of security from the increasing array of buttons, LEDs, flashing lights and consoles. It must be remembered that every determination of a component in foodstuffs requires some manual sample preparation, whether this is merely subsampling and homogenization or is followed by extraction and the preparation of standards. There are frequently no instruments available to assist with these steps, and no amount of technological sophistication in the instrumental stages of an analytical method will compensate for poor manual execution of these early stages. Instruments are merely tools that can be either used or abused. In well-trained hands they are revolutionizing, and will continue to revolutionize, the food analytical laboratory.

References

Charalambous, G. (ed.) (1984) *Analysis of Foods and Beverages: Modern Techniques*. Academic, London

Charalambous, G. and Inglett, F. (eds) (1983) *Instrumental Analysis of Foods: Recent Progress, vols 1 and 2*. Academic, London

Gruenwedel, D. W. and Whitaker J. R. (eds) (1987) *Food Analysis:*

Principles and Techniques. Vol. 1: Physical Characterization. Vol. 2: Physicochemical Techniques. Vol. 3: Biological Techniques. Vol. 4 Separation Techniques. Marcel Dekker, New York, Basel

King, R. D. (ed.) (1984) *Developments in Food Analysis Techniques*, Vol. 1. Elsevier, London

Kirk, R. S. and Sawyer, R. (1991) *Pearson's Chemical Analysis of Foods*, 9th edn. Longman

Lee, M. L. and Markides, K. E. (eds) (1990) *Analytical Supercritical Fluid Chromatography and Extraction.* Chromatography Conferences, Department of Chemistry, Brigham Young University, Provo, UT

Macleod, A. J. (1973) *Instrumental Methods of Food Analysis.* Elek, London

Pomeranz, Y. and Meloan, C. E. (1978) *Food Analysis: Theory and Practice*, rev. edn. AVI, Westport, CT

Slack, P. T. (ed.) (1987) *Analytical Methods Manual*, 2nd edn. Leatherhead Food Research Association, Leatherhead, UK (available to non-members)

Stewart, K. K. and Whitaker, J. R. (eds) (1984) *Modern Methods of Food Analysis*, IFT Basis Symposium Series, Proceedings of the Symposium on Modern Methods of Food Analysis. AVI, Westport, CT

Willard, H. H., Merritt, L. L. and Dean, J. A. (1974) *Instrumental Methods of Analysis*, 5th edn. Van Nostrand, New York

14 *Impedance Techniques for Microbial Assay*

D. M. Gibson and A. C. Jason

Contents

Nomenclature

b	number of bits
c	mass concentration of medium, g ml^{-1}
f	frequency, Hz
k	initial specific conductivity, S m^{-1}
l	effective separation of electrodes, m
n	population density, cfu ml^{-1}
n_o	inoculum density, cfu ml^{-1}
n_s	maximum possible population density, cfu ml^{-1}
n_D	arbitrary detection level of population density, cfu ml^{-1}
$\dot{n}$	growth rate, cfu h^{-1}
$\dot{n}/n$	specific growth rate, h^{-1}
$(\dot{n}/n)_o$	initial specific growth rate, h^{-1}
r	correlation coefficient
t	time, h
t_g	generation time, h
t_g'	$1/(\dot{n}/n)_o$, h
t_D	time after inoculation at which population density grows to a value n_D, detection time, h
t_L	lag time, h
A	electrode area, m^2
A_D	conversion gain
C_D, R_D	series capacitance and series resistance due to alignment of polar dipoles in double charge layer in fluid, F, Ω
C_{ox}, R_{ox}	series capacitance and series resistance resulting from the presence of an oxide layer at the electrode surfaces, F, Ω
C_{se}	lumped electrode capacitances C_D and C_{ox} (for R_D, $R_{ox} \ll R_s$),F
G_o	initial conductance of medium in test cell, S
G_s	subsequent conductance of medium in test cell, S
K	increase in specific conductance associated with production of single bacterial cell, per unit volume, S m^{-1} (cfu ml^{-1})$^{-1}$
N	number of colonies counted on agar plate
R_{fs}	full-scale range of resistance, Ω
R_s	true resistance of medium in test cell, Ω
S	substrate mass, g
S_o, S_s	initial and final stationary mass of substrate, g
ΔS	mass of substrate metabolised by one bacterial cell, g
V, i	instantaneous voltage, current, V, A
$\|V\|$,$\|i\|$	peak voltage, current, V, A
$\|V_m\|$,$\|i_m\|$	mean voltage, current, modulus, V, A
X_s	reactance of C_s, Ω
X_{se}	reactance of C_{se}, Ω

Y	admittance, S
Z	impedance of R_s and C_s in series, Ω
ϕ	phase angle
γ_D	number of generations after inoculation at which growth is detected

1 Introduction

Almost a century ago Stewart (1899) found that the electrical conductance of a medium in which bacteria were growing increased significantly as growth proceeded. He showed that the change in conductance was proportional to the concentration of the electrolytes produced, suggested the nature of some of these and deduced that the conductance change was thus proportional to the concentration of active bacteria. He speculated that 'sufficiently great and constant differences might be observed between different kinds of bacteria when grown in the same or different liquid media to enable us to use the method as a supplement to our present means of diagnosing between nearly related forms.'

It is interesting to note the design of the equipment he used and the experimental conditions he thought desirable. He measured the electrical resistance of a cell containing the medium by means of a Wheatstone bridge under alternating current conditions using a 'telephone' for zero detection. It was necessary to prevent evaporation of the medium and to maintain it at constant temperature during growth so as to avoid artefactual changes in resistance. Pure bacterial cultures were grown in a peptone medium. Stewart though it desirable to use 'tubes furnished with platinum electrodes fused through the glass, or perhaps passing through the plug' to avoid contamination.

He was aware of the essential features that have made modern impedance methods so successful today but was prevented from developing the method further because of the limitations imposed by the technology of his time. Despite sporadic interest (see Easter and Gibson 1989) the subject lay virtually dormant until 1973 when Ur and Brown (1975) and Cady (1975) again drew attention to the potential application of electrical impedance measurement and designed apparatus for sensitively displaying its change during bacterial growth. They demonstrated the distinctive effects of different organisms grown in a given broth and the influence of different growth media on the shape of the impedance curves. The possibilities for the rapid automation of microbiological tests were clearly indicated by these authors.

We propose to describe the current situation of rapid microbiological methods and then to give an outline of the underlying principles of impedance measurement for cultures, to discuss factors influencing sensitivity and to describe various growth analysers currently available commercially. We shall also delineate the ways in which the method has been used to obtain microbial assays and so provide quantitative indices for food quality control. (See also Chapter 15 for practical advice on the applications of impedance microbiology.)

2 Rapid Microbiological Methods: an Overview

There are two objectives in food microbiology; firstly to determine the total load or numbers of microbes in a sample, and secondly to determine the presence or absence of a particular microbial species. The total load is determined as the *total viable count* (TVC) or the number of *colony forming units* (cfu), and relates to quality and stability of the food. Enumeration of specific groups is associated with safety or freedom from the risk of food poisoning.

The time taken to complete conventional microbiological assays has changed little this century as it depends on the time taken for microbes to grow and multiply so as to give a visible colony in the surfaces of a nutrient solid medium or obvious turbidity in a liquid medium. Much effort has accordingly been devoted to shortening assay times and to replacing the visible end points with alternative measurements. The procedures used in the conventional assay involve the following stages. First a sample, normally 10 to 25 g, is homogenized in a buffer or medium to release the microbes from any solid material and to dilute out any natural antimicrobial compounds. Then the extract is further diluted in serial decades in buffer and inoculated in small volumes either on to the surface of a medium, or into a warm liquid medium which solidifies on cooling, or into a broth. After incubation for say 1–7 days, depending on the particular assay, visible colonies which grow on a solid medium are counted on the most suitably diluted sample and the turbidities are noted. Further tests are often done especially to confirm the identity of pathogens or organisms indicative of their presence. Thus much of the routine of food microbiology is laborious, technically demanding and slow to yield results. Some of the steps have been mechanized, for example using diluter/spreaders and colony counters, but the overall assay times have altered little, even with the advent of improved and more selective media.

The birth of rapid methods dates from 1973 when the first symposium on the subject was held (Héden and Illeni 1975a; 1975b). Many new approaches were described, since when some have become available commercially, while others have been discarded. Recently, a few new techniques have appeared. Apart from mechanization of various manual steps in assays, the main techniques for which there are instruments available are those based on electrical methods (conductance or impedance measurement, electrochemical assay of the culture broth), on chemical methods such as the determination of bacterial adenosine triphosphate (ATP), on colorimetric measurement of dye reduction, on filter techniques (direct epifluorescent filter (DEFT), hydrophobic grid membrane), and on agglutination methods (immunological, limulus amoebocyte lysate (LAL)). These and others have been reviewed in Adams and Hope (1989), Huis in't Veld et al. (1988) and Stannard et al. (1990). The characteristics of some of the techniques as perceived by users are shown in Table 14.1. There are no instant results from any of these assays; all involve some extract preparation and then assay time.

Of all the methods, the electrical assays have been most widely accepted by the food industry (Huis in't Veld et al. 1988). As described later, the net metabolic activity of the culture is represented by such assays and this correlates well with the microbial load in the material. Specific media can be used for particular organisms. Results can be obtained in 2–4 h with non-sterile foods. The instruments are simple to use; an extract is made, added to the measurement cell and connected to the instrument, which then monitors the assay and reports the result. Up to 500 samples can be examined simultaneously on current machines. (Figure 14.1). Thus extensive sampling can be done with few staff, so justifying the capital outlay. Indeed, the payback time for equipment has been less than a year in some instances. The output data are accepted by many companies in place of conventional data.

The ATP test depends on the reaction between ATP and the enzyme luciferase, producing light which is measured photometrically with a claimed sensitivity down to 10^{-16} mol l^{-1}. It is a rapid test taking only minutes to complete. The main problem with most foods is that they, like the microbes present, contain ATP, and that the ATP content of microbial cells is variable depending on their physiological state. Manufacturers have produced reagents to selectively destroy somatic ATP, mainly by using detergents to lyse the cells and ATPase to destroy them. The assays are useful in materials which have little intrinsic ATP such as cooked meat products, carbonated beverages and fruit juices. The proceedings of a symposium on the subject have been published (Stanley et al. 1990).

The DEFT test was originally devised for the liquid milk industry (Pettipher 1983) to provide a rapid test for raw milk payment schemes. It depends on the uptake by cells of acridine orange. Viable cells fluoresce orange under ultraviolet light owing to their ribonucleic acid (RNA) content, whereas dead or non-growing cells fluoresce green owing to interactions of the dye with deoxyribonucleic acid (DNA). Samples are concentrated by filtration, stained and viewed microscopically. Image analysis equipment can be used to automate counting. The test has been applied mainly in the dairy industry on products which have not undergone a heat treatment which could give misleading results, and other foods which can be easily filtered. Procedures have been devised for preparing

Table 14.1 *Characteristics of some rapid methods in relation to industrial needs (modified from Jarvis and Easter 1987)*

	Method			
Criteria	*Conductance/impedance*	*ATP*	*DEFT*	*LAL*
Simplicity[a]	+	+/−	+	+
Capital cost	High	Moderate/high	Low/high	Very low
Running costs	Low	High	Low	Moderate
Stable reagents[b]	NR	+/−	+	+/−
Labour costs	Low	High	High	Moderate
Data compatibility[c]	Indirect	Indirect	Direct	Indirect
Computer interface	Yes	Possible	Possible	None
Computer networkable	Possible	Possible	Possible	None
Throughput (no. h^{-1})	>120	10–30	10–15	10
Speed (elapsed time)	>4 h	>5 min	30 min	1–2 h
Sensitivity (cfu g^{-1})	$>10^4$	10^5	10^3–10^4	10^4
Specificity	Moderate/high	Very low	Low	Low

[a] + simple; − complex.
[b] + stable; − unstable; NR not relevant.
[c] With standard plate count.
DEFT: direct epifluorescent filter technique.
ATP: adenosine triphosphate.
LAL: limulus amoebocyte lysate.

Figure 14.1 *Bactometer microbial monitoring system (Vitek Systems Ltd)*

suitable filtrates from solid foods, but correlations with conventional assays have often been too low. The hydrophobic grid membrane technique also relies on filtration of a sample or an extract such that the microbes are deposited on a special filter membrane on which a grid of hydrophobic lines is imposed. The filters are incubated until each microbe has multiplied to occupy a square, facilitating simple automated counting techniques. There are procedures for handling various foods, incorporating filter aids, using enzymes to destroy spurious particles etc. Selective media can be used and the grids stained with dyes or immunochemicals after incubation.

The LAL test depends on the ability of lipopolysaccharide (LPS) to activate a clotting enzyme in the blood of the horseshoe crab and so produce a gel. Any LPS, from living or dead cells, is active. The test has not been widely applied in the food industry, although owing to its apparent simplicity it has been evaluated in many laboratories for a variety of products. It may have a role in indicating the earlier microbial load of heat treated products such as UHT milk, or checking parenteral feeds for pyrogens (fever-inducing bacterial polysaccharides). It has not yet been automated.

The potential market for rapid automated microbiological analyses is large but the users have been reluctant to purchase instruments, so that development costs have not been recovered and many companies pioneering the field have had financial difficulties. Thus, from the wealth of ideas and prototypes described in the past decade, only a few have reached the market place. However, because microbiological testing is so labour intensive, and the current demand for increased food safety requires even more testing, the need for rapid methods becomes ever more pressing. Further research and development are clearly desirable.

Impedance techniques have, however, already found successful practical applications in the food industry, and commercial instruments are available to the industrial quality control laboratory.

2.1 Impedance as an Indicator of Microbial Load

With impedance techniques, an increase in the conductance, capacitance or admittance (having both conductance and capacitance components) can be measured as microbial growth proceeds in the medium. The conductance increase is due to charged metabolites being produced by the microbes from uncharged substrate, and these will contribute to the capacitance increase. An increase in the measured capacitance can also be observed when biomass attaches to the electrodes. Such an increase in capacitance would occur later in the growth cycle and could be due to cells already inactivated or dead. The conductance increase, on the other hand, always indicates the presence of active microbial cells and is observed earlier in the growth cycle. Change of pH of the medium caused by metabolism without growth can change the conductance and capacitance.

Conductance monitoring is therefore the method preferred by the authors. Capacitance measurement is, on the other hand, often used for the assay of yeasts and moulds where ionic metabolites are not produced with standard growth media. With such capacitance methods, care must be taken to avoid errors due to artefacts related to the condition of the electrodes. A replacement with conductance measurements employing specially formulated media is often advantageous if time for media engineering is available.

It should be noted that the frequency of measurement will affect the choice between conductance or capacitance methods. This chapter is based largely on experience gained in the development of and studies with the Malthus Analyser, which operates at a frequency of 10 kHz. Other commercial instruments operate at frequencies between 2 and 10 kHz. The Bactometer, for example, operates at 2 kHz, and for this frequency the capacitance change with microbial numbers is more pronounced.

The choice of capacitance or impedance measurement instead of conductance monitoring will be discussed in more detail in Section 4. The description of impedance techniques in this chapter will refer to conductance measurement unless noted otherwise. However, in order to determine the conductance of the growth medium accurately, it is necessary to examine the impedance of the system consisting of the fluid and the electrodes. In certain cases, and with appropriate instrument design, it is also possible to measure admittance as an approximation to conductance. These considerations will now be discussed in detail.

3 Principles of Electrical Conductance Methods

3.1 Impedance and its Component Variables

If we consider the current that flows between two electrodes immersed in a fluid when an alternating potential is applied, there are three components that limit its flow. These are: (1) the true resistance R_s of the fluid; (2) a capacitance C_{ox} in series with a resistance R_{ox} resulting from the presence of an oxide layer at the surface of each electrode; and (3) a capacitance C_D in series with a resistance R_d generated close to each electrode as a consequence of the alignment of polar dipoles in the fluid in a double charge layer (Figure 14.2a). For convenience the electrode capacitances are lumped together as a single series capacitance C_{se}. The resistance values R_{ox} and R_D are generally small in relation to R_s and may be neglected in some less critical applications. In more demanding circumstances, such as the investigation of early growth characteristics, the need for the absence of an oxide layer and minimum electrode polarization indicates the use of platinum electrodes. The combined effect of components (1), (2) and (3) as 'seen' by a measuring system is therefore an *impedance* consisting of a single capacitance C_{se} in series with a single resistance R_s (Figure 14.2b).

The instantaneous current i that flows through a pure resistance R_s is proportional to, and in phase with, the applied sinusoidal voltage V such that

$$\begin{aligned} i &= V/R_s \\ &= (|V|) \cos 2\pi ft/R_s \end{aligned} \qquad (1)$$

where $|V|$ is the amplitude (peak value) of the voltage, f is its frequency and t is time.

The mean current $|i_m|$ averaged over a complete cycle is

$$|i_m| = |V_m|/R_s \qquad (2)$$

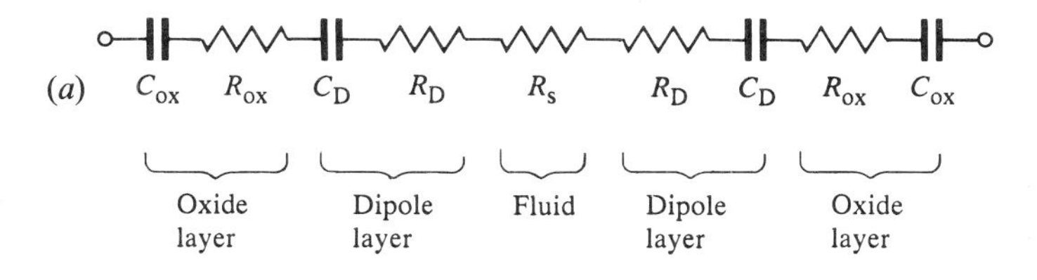

***Figure 14.2** Impedance measured across two metal electrodes immersed in a fluid: (a) impedance components for the fluid (R_s), the metal oxide (C_{ox}, R_{ox}) and the dipole double charge layer (C_D, R_d) (b) equivalent circuit for R_{ox}, $R_D \ll R_s$*

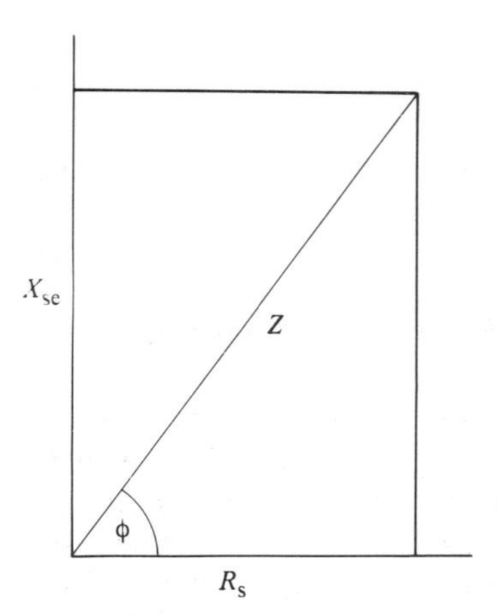

Figure 14.3 *Vector representation of impedance* Z *for* R_s *and* C_{se} *placed in series:* $X_{se} = (2\pi fC_{se})^{-1}$ *and* ϕ *is the phase angle*

where $|V_m|$ is the mean voltage modulus averaged over a complete cycle.

For a pure capacitance C_{se} it is the charging current that determines i, and because the maximum current flows when $V=0$ and no current flows when $V=|V|$, i flows in advance of V by an angle $\pi/2$. Hence

$$i = C_{se}(\mathrm{d}V/\mathrm{d}t)$$

$$= C_{se}\frac{\mathrm{d}}{\mathrm{d}t}(|V| \cos 2\pi ft)$$

$$= 2\pi fC_{se}|V| \cos (2\pi ft + \pi/2)$$

Therefore

$$i_m = 2\pi fC_{se}|V_m| \quad (3)$$

$$= |V_m|/X_{se}$$

where $X_{se} = (2\pi fC_{se})^{-1}$ is termed the *reactance* of the capacitance C_{se} and has the dimensions of resistance.

The effect of placing R_s and C_{se} in series is to reduce the angle by which the current leads the voltage measured across the terminals to a value such that $0 < \phi < \pi/2$. ϕ is termed the *phase angle*. The total impedance is a vector quantity Z which is the resultant of R_s and $-X_{se}$. Note that the negative value of X_{se} indicates that the reactance vector leads the resistance vector by an angle $\pi/2$ and that it has no component on R_s.

Referring to Figure 14.3, we see that

$$R_s = |Z| \cos \phi \quad (4)$$

$$-X_{se} = |Z| \sin \phi \quad (5)$$

so that

$$Z = (R_s^2 + X_{se}^2)^{1/2}$$
$$= [R_s^2 + (2\pi f C_{se})^{-2}]^{1/2} \quad (6)$$

3.2 Cell Design and Geometry and Composition of Electrodes

The shape of the vessel and the size and disposition of the electrodes are not critical in the design of impedance cells unless high resolution is required. The materials of which they are constructed should not contaminate the growth medium, should not affect the growth of microbes, should withstand sterilization and, for reusable cells, should be easily cleaned. Electrodes a few square millimetres in area, separated by a distance of 3 to 50 millimetres, will result in easily measurable impedances for common culture media. In general, it is desirable to position both electrodes in a vertical plane above the base of the cell to prevent detritus from coating them during the period of measurement. Alternatively the electrodes may be coated in agar to eliminate direct deposition, or the viscosity of the medium may be increased with dextran (Curtis et al. 1985). Nevertheless, these requirements may be relaxed if the time of detection of some high level threshold of growth, say 5×10^6 cfu ml^{-1}, is the desired measurement; then electrodes protruding through the base or situated in the plane of the base form an acceptable arrangement. It is essential that electrical connections to the electrodes consist of low resistance self-cleaning contacts since variations of more than 0.01 Ω may be easily detected.

The electrode material must of course be non-toxic to microbes and must be corrosion-resistant. A cheap option is stainless steel but, if a linear representation of bacterial number is necessary, the requirements are more demanding. To obtain such linearity it is preferable to measure R_s directly. Attention must be paid to minimizing electrode polarization and to avoiding oxide formation on the electrodes, since these produce variable and unpredictable impedance effects. Schwann (1963) has investigated the measurement of biological impedances in considerable depth, and has concluded that a platinum surface coated with platinum black offers the lowest possible interfacial impedance but that a plain platinum surface is acceptable for the measurement of all but the lowest impedances. All other materials are inadequate in relation to platinum and their deficiencies become more pronounced as the measurement frequency is reduced.

In order to give a more quantitative indication of the importance of the above considerations we have calculated, from the measurements of Richards et al. (1978), the total impedances Z measured between platinum electrodes at 10 kHz of a number of growth media (Table 14.2). We show also the corresponding resistance R_s and capacitance C_{se} values. So as to

Table 14.2 *Impedance measurements of various growth media at 10 kHz*

				Measurement with platinum electrodes				Calculated values with highly polarized electrodes[b]	
Medium	*pH*	c^a	T (°C)	R_s (Ω)	C_{se} (μF)	X_{se} (Ω)	Z (Ω)	X_{se} (Ω)	Z (Ω)
Oxoid PPLO broth	7.2	0	0	724.6	2.41	6.6	724.6	33.0	725.4
	7.2	0	37	307.9	1.87	8.5	308.0	42.6	310.8
Oxoid CM1 broth	4.0	0.5	0	522.7	1.05	15.2	522.9	75.8	528.2
	7.2	0.5	37	278.6	1.08	14.7	279.0	73.7	281.8
	9.0	0.5	0	357.0	1.53	10.4	357.2	52.0	360.8
	9.0	0.5	37	251.6	1.82	8.8	251.8	43.7	255.4
Oxoid CM1 broth with added NaCl	4.0	15.0	0	47.4	1.61	9.9	48.4	49.4	68.5
	4.0	15.0	37	28.7	1.88	8.5	29.9	42.7	51.2
	7.2	15.0	37	27.2	2.12	7.5	28.2	37.5	46.4
	7.2	15.0	53	22.5	2.14	7.4	23.7	37.2	43.5
	9.0	15.0	0	47.3	2.12	7.5	47.9	37.5	63.5
	9.0	15.0	37	27.8	2.18	7.3	28.7	36.5	45.9
	9.0	15.0	53	22.9	2.19	7.3	24.0	36.3	43.0
NaCl solution	6.0	0.5	0	938.1	1.44	11.1	938.2	55.3	939.7
	6.0	7.5	0	80.7	1.74	9.1	81.2	45.7	92.8
	6.0	15.0	0	48.0	1.94	8.2	48.7	41.0	43.1
Mineral salts medium plus glucose[c]	7.2	—	37	200.3	0.87	18.3	201.1	91.5	220.2

[a]g NaCl per 100 ml H_2O.
[b]See text.

Table 14.3 *Influence of measuring frequency on series impedance (CM1 broth, pH 7.2, 37°C, 0.5g NaCl per 100g H_2O)*

Frequency (kHz)	Z (Ω)	R_s (Ω)	X_s (Ω)
10	279.0	278.6	14.7
5	280.1	278.6	29.4
2	288.1	278.6	73.5
1	315.0	278.6	147.0
0.5	405.0	278.6	294.0

illustrate the influence of large electrode polarization effects, such as might arise from the use of stainless steel electrodes, we have calculated the consequences of a fivefold reduction in the value of C_{se}. In Table 14.3 we show the effect of reducing the measuring frequency under otherwise favourable conditions.

It is evident that at 10 kHz the capacitance C_{se} varies over a small range only (from 0.87 μF to 2.41 μF for platinum electrodes), while the true resistance of the fluid R_s varies from 22.5 Ω to 938.1 Ω under all conditions. Also the impedance Z approximates quite closely to R_s for large values of R_s but, in the case of highly polarized electrodes, diverges seriously for the lower values. The use of electrodes composed of metals other than platinum, or the reduction of measuring frequencies below 10 kHz, leads to errors in assuming Z and R_s are equivalent. Unpredictable fluctuations in C_{se} will also give rise to a certain lack of discrimination if impedance is measured instead of resistance. Thus, unless the requirements are not exacting, it is preferable to measure R_s when the medium is highly conducting.

3.3 Relationship between Test Cell Conductance and Bacterial Growth

After a medium is inoculated with the sample, there is an initial period, the lag phase, in which cells metabolize but the bacterial numbers remain unchanged and, as a rule, there appears to be no detectable change in the conductance of the growth medium. After cell division commences, charged metabolites pass into solution and the conductance increases (Figure 14.4). During its existence as an individual, a single bacterium gives rise to a fixed number of ions which does not vary in each succeeding generation. The introduction of ions enhances the conductance G_s of the medium, and the enhancement δG_s is proportional to the growth in population density $n-n_0$ and the mobility of the ions. Thus δG_s may be regarded as a conductance shunting the initial conductance of the medium G_0, as illustrated in Figure 14.5, its value being

$$\delta G_s = 1/R_s - 1/R_0 \tag{7}$$

where R_0 is the initial value of the resistance R_s of the medium. The unit in which G_s is measured is the siemens (S). An example showing the linearity of the relationship to bacterial numbers is presented in Figure 14.6.

If we consider a cell containing a medium of initial specific conductivity k, in which electrodes of area A and effective separation l are immersed, then

$$G_s = kA/l \tag{8}$$

The production of a single bacterial cell in unit volume of medium is associated with an increase of specific conductivity K, so that

$$\begin{aligned} \delta G_s &= (KA/l)(n-n_0) \\ &= (KG_s/k)(n-n_0) \end{aligned} \tag{9}$$

For Figure 14.6, the constant of proportionality KG_s/k is approximately

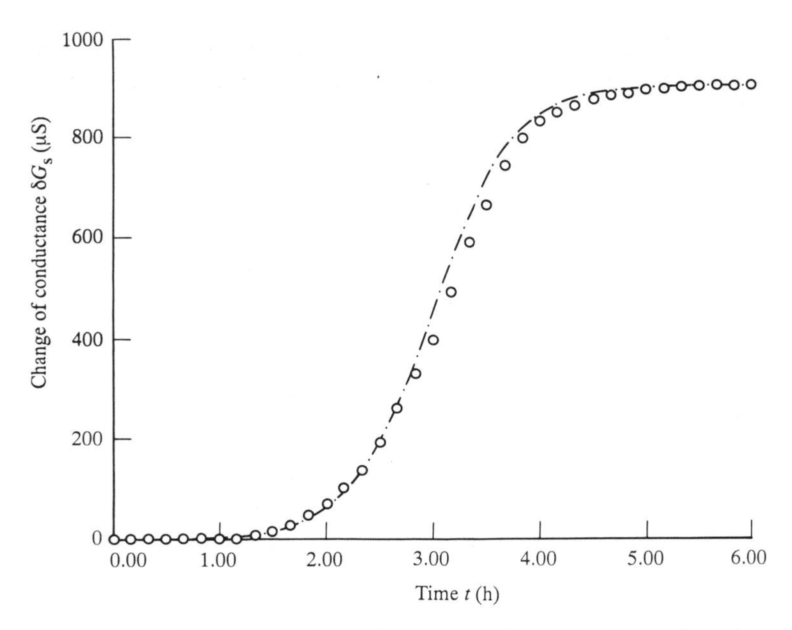

Figure 14.4 *Change of conductance δG_s with time after inoculation of nutrient broth (Oxoid CM1) with* E. coli *(8×10^5 cfu ml^{-1}) The broken line represents equation 16 in which* KG_s/k *is* 1pS*(cfu ml^{-1})$^{-1}$*

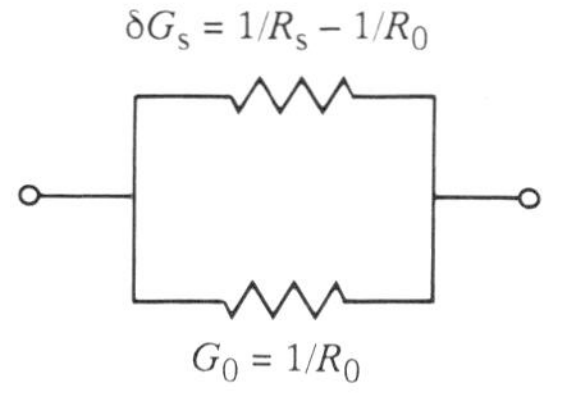

Figure 14.5 *Shunting effect of conductance increase due to bacterial growth on initial broth conductance*

1 pS (cfu ml^{-1})$^{-1}$, so that the value of K may be obtained by substituting the value of k for the CM1 broth (88 S cm^{-1}) and G_s measured at 10 kHz (derived from Table 14.3, $R_s = 1/G_s = 279\,\Omega$):

$$K = 2.45 \times 10^{-8}\ \mathrm{S\ cm^{-1}\ (cfu\ ml^{-1})^{-1}}$$

The magnitude of K generally lies within ± 20 per cent of this value for a wide range of bacteria in this medium and is of the same order for bacteria in other media (Gibson and Hobbs 1987).

3.4 Bacterial Number Resolution of Analysers

All analysers used in the food industry employ digital techniques to measure either impedance Z or its inverse, admittance Y, or, when true

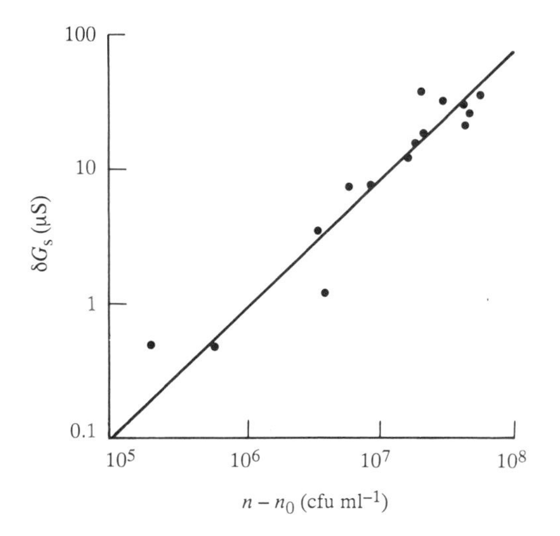

Figure 14.6 *Change of conductance δG_s with increase in cell population density $n - n_0$ for* E. coli *in CM1 broth at 37°C*

linearity may be required, either resistance R or conductance G. Their general design is described below.

The numerical resolution of an analyser is limited by the resolution ΔV of the analogue-to-digital (A/D) voltage converter employed to display whichever variable is being measured. If, say, R_s is the measured variable and R_{fs} is the full-scale range corresponding to the full-scale voltage range, then the smallest value of resistance that can be resolved is

$$\Delta R_s = R_{fs}/2^b$$

where the voltmeter has a range of b bits. The corresponding resolution of conductance is therefore

$$\begin{aligned}\Delta G_s &= -\Delta R_s/R_s^2 \\ &= -R_{fs}/R_s^2 2^b \qquad (10)\end{aligned}$$

Substituting this value into equation (9) and denoting Δn as the limiting numerical resolution corresponding to ΔG_s, we have

$$\Delta n = kR_{fs}/KR_s 2^b \qquad (11)$$

Equation (11) shows that the resolution is determined by the ratios k/K and R_s/R_{fs} and by b. The dominant factor controlling k/K is the concentration of the salts in the medium.

The resolutions for various salt concentrations, bit values and ratios R_s/R_{fs} are shown in Table 14.4. We see immediately that in order to achieve the highest possible resolution three conditions must be met:

- low specific conductivity of growth medium
- value of ratio R_s/R_{fs} close to unity
- high resolution of analogue-to-digital converter.

3.5 Temperature Control

The temperature coefficient of conductance of media is very high. For an isotonic medium it is $0.016K^{-1}$, for example, so that the conductance of a cell of say 3000 μS would change by 0.48 μS for each 0.01 K change of temperature. Clearly it is essential to maintain cell temperature constant to within a few millidegrees of a desired value. This is readily achieved by immersing cells in a thermostat controlled water bath (Figure 14.7) or by placing them in deep holes bored in an aluminium or copper block similarly controlled. One manufacturer employs an air incubator (Figure 14.1) and a system of balanced pairs of cells, one of which (as described below) is a sterile control to minimize temperature effects, but the practice is only

Table 14.4 *Numerical resolution of conductance analyser*

NaCl molarity	k/K[a] *(cfu ml^{-1})*	*Bit no.* b	*Resolution (10^6 cfu ml^{-1}) for* $R_s/R_{fs}=$ *0.1*	*0.3*	*0.6*	*0.9*
0.146M	3.59×10^9	14	2.19	0.73	0.37	0.24
		16	0.55	0.18	0.09	0.06
		18	0.14	0.05	0.02	0.01
		20	0.03	0.01	0.01	0.004
0.060M	1.53×10^9	14	0.94	0.31	0.16	0.10
		16	0.23	0.08	0.04	0.03
		18	0.06	0.02	0.01	0.006
		20	0.01	0.005	0.002	0.002
0.028M	7.18×10^8	14	0.44	0.15	0.07	0.01
		16	0.11	0.04	0.02	0.01
		18	0.03	0.01	0.005	0.003
		20	0.007	0.002	0.001	0.001

[a]Inverse of normalized conductivity increase.

satisfactory for low resolution determinations. Both the water bath and the aluminium block can control cell temperatures within ± 1mK and fluctuations and drift in conductance measurement can be lower than 0.1 μS h^{-1}.

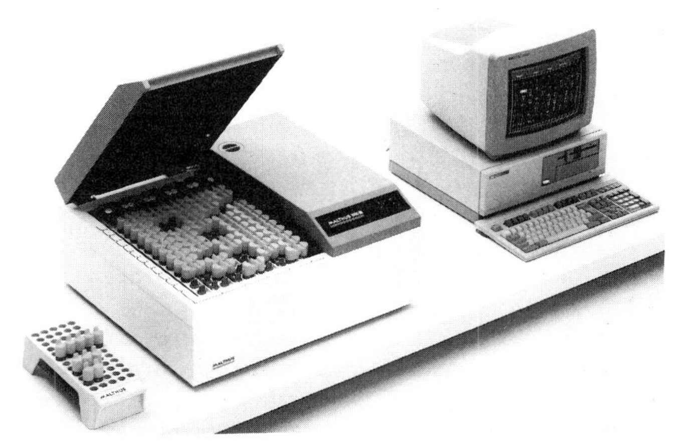

Figure 14.7 *Malthus Microbiological Growth Analyser (Malthus Instruments Ltd)*

4 Capacitance versus Conductance Measurement

4.1 Selection of Display Variable Early in the Growth Cycle

There has been much controversy regarding which is the correct electrical variable to be displayed. The key data are found in the papers by Richards et al. (1978) and Firstenberg-Eden and Zindulis (1984), with a more recent contribution by Pethig and Kell (1987). Cady (1975) originally measured the modulus of Z, $|Z|$, using gold electrodes, which reduce the large and uncertain contribution of the oxide layer to C_{se} attendant on the use of stainless steel electrodes. Richards et al. (1978) measured series conductance G_s at 10 kHz, in the manner previously described. They stated: 'Experimentally it was found that C was subject to fluctuations which did not correlate well with any measured variable, such as temperature. These more or less *random* fractional changes in C were of the same order of magnitude as the systematic fractional changes in G which could be ascribed to bacterial growth; and it became clear that measuring C was most unlikely to be a reliable means of observing bacterial growth.' (See, however, Sections 4.2 and 2.1 on the changes later in the growth cycle and on the influence of frequency.)

In the work reported, later repeated and extended, Jason and Ogden (personal communication) measured the changes in G_s and C_{se} of microbial cultures using a high quality transformer ratio bridge. It is clear from the results reproduced in Figure 14.8a that G_s initially remains constant before change due to growth occurs. But even when microbial numbers are changing, fluctuations in C_{se} measured at 2 kHz remain within noise limits (Figure 14.8b), referred to by Richards et al. (1978) as the 'more or less random fractional changes'. In other words there is no systematic change in C_{se} when changes in G_s are already observed. This was evident for both bacteria and yeasts. This is not to imply that C_{se} does not change later but, for early detection of growth, C_{se} is a less sensitive indicator than G_s at 2 kHz and its influence on variations of Z introduce a proportional uncertainty. At 10 kHz, on the other hand, an increase in both G_s and C_{se} early in the growth cycle is observed (Figure 14.8c and d).

4.2 Mechanisms of Changes Observed Later in the Growth Cycle

Firstenberg-Eden and Zindulis (1984) examined the electrochemical changes in conventional peptone media caused by the growth of *E. coli* and *Saccharomyces cerevisiae* using stainless steel electrodes in a 100 ml bottle. They used a transformer-type admittance bridge to measure equivalent parallel conductance and capacitance, which were then converted to the equivalent series resistance and capacitance. It is clear from the figures they

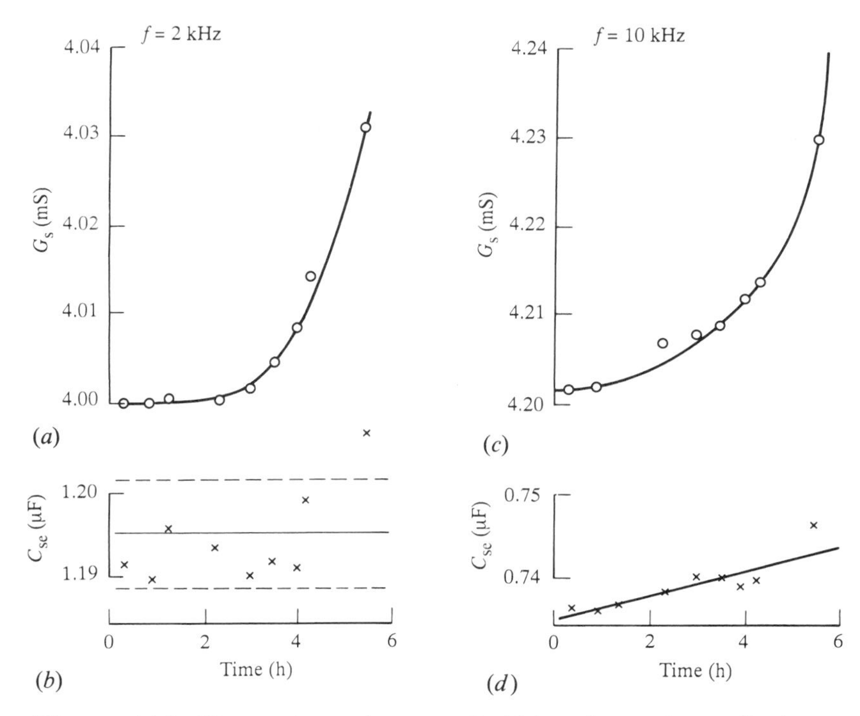

Figure 14.8 *Changes in conductance* G_s (○) *and capacitance* C_{se} *(×) early in the growth of* E. coli *in nutrient broth: (a), (b) at 2 kHz (c), (d) at 10 kHz (A. C. Jason and I. D. Ogden, personal communication)*

present that both conductance and capacitance change significantly only after microbial numbers exceed 10^7 cfu ml^{-1}. Table 14.5 shows that while the relative changes in G_s were often quite small compared with those in C_{se}, the actual conductance change calculated from their data was above 40 μS throughout. Since the accuracy and sensitivity of the bridge measurement is such that $G_s \geqslant 10$ μS is easily detected in all cases, measurement of the conductance change is well within the resolution of all the commercial instruments. Of course if the baseline data for C_{se} are fluctuating, then a large percentage change in C_{se} is needed to determine when growth has initiated.

Kell (1987) indicates that cells no longer viable contribute to the magnitude of the capacitance C_{se}, so that dead or nearly dead cells contribute to C_{se}. They attach themselves to the electrodes when densities exceed 10^7 cfu ml^{-1}. This accounts for the constantly rising magnitude of C_{se} when viable numbers are constant or falling. This is confirmed by Firstenberg-Eden and Zindulis (1984). Careful examination of their figures, for example Figure 14.9 reproduced here, shows that the cessation of the growth of *E. coli* at the end of the growth cycle is accompanied by a

Table 14.5 *Changes in polarization capacitance C_{se} and solution conductance G_s due to microbial growth (Modified from Firstenberg-Eden and Zindulis 1984 and reproduced with permission from the authors and publisher: the actual conductance change calculated from their data has been added)*

Organism	*Medium*	*Medium conductance (mS)*	Change in C_{se} (%)	*Change in G_s (%)*	*Actual change in G_s (μS)*
E. coli	PCB	1.3	56	35	460
E. coli	PCB+0.4% glucose	1.7	56	35	595
E. coli	BHI	8.0	70	7	560
E. coli	TSB	6.9	90	10	690
Pseudomonas aeruginosa	PCB	1.4	55	40	560
Pseudomonas aeruginosa	BHI	7.9	23	10	790
Streptococcus pyogenes	TSB	6.7	42	16	107
Staphylococcus epidermidis	TSB	6.8	52	13	884
Saccharomyces cerevisiae	TSB	7.0	103	4	280
Saccharomyces cerevisiae	YCB	4.5	30	2	90
Candida utilis	TSB	6.7	20	1	67
Kloeckera apiculata	YCB	4.7	60	1	47

PCB: plate count broth.
BHI: brain heart infusion
TSB: tryptic soy broth.
YCB: yeast carbon base.

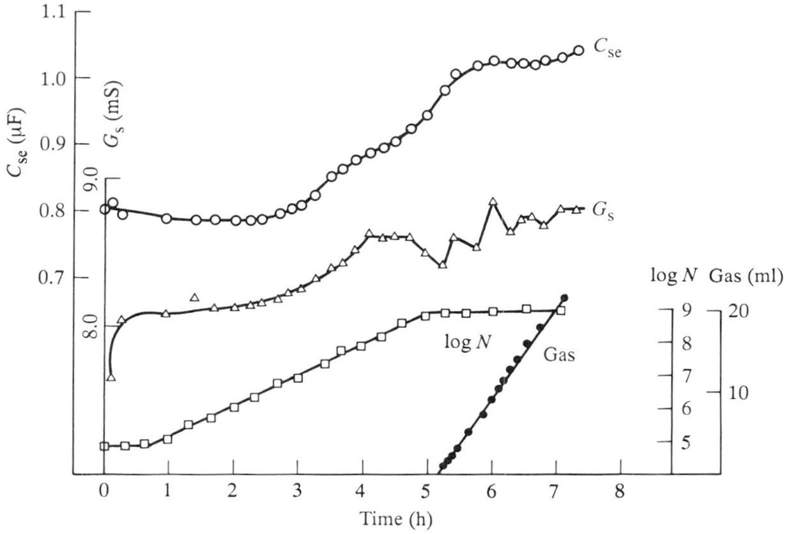

***Figure 14.9** The effect of the growth of* E. coli *in brain heart infusion on capacitance* C_{se}*, conductance* G_s*, number* N *of bacteria and gas evolution (from Firstenberg-Eden and Zindulis 1984, by courtesy of the authors and publisher)*

continued rise in C_{se}. With yeast, C_{se} changes before numbers change, an unsatisfactory situation probably due to precipitating cells. However, these results were obtained with laboratory apparatus and not with the commercial instrument, and so may not be typical.

In a routine quality control the magnitude of the electrical signal should correlate with change in microbial number at all times. For example, food spoilage or deterioration leading to its rejection by consumers is based on the activities of microbes rather than their biomass, as many foods such as cheese or yoghurt contain large numbers of non-metabolizing non-growing cells. There are no reports of false decisions based on changes in conductance, but Jarvis (1985) showed that wrong decisions were made when a change in capacitance obtained with fruit materials was found. The product was in fact sterile and the change was ascribed to the 'unauthorized use of an acidulant'. If added acids can give continuous drift of the measured variable due to corrosion of the electrodes, then a more robust display variable is desirable, or totally inert electrodes. Capacitance measurement has had an important role to play in the detection of some yeasts and moulds, but the development of new media formulations enhancing conductance signals (such as described by Connolly et al. 1988; Adams et al. 1989) is generally a safer route forward; simultaneous measurement of C_{se} and G_s may be safer than C_{se} alone in certain applications.

5 Instrument Design

5.1 Method of Measurement

There are several ways in which the impedance of media may be measured. The most direct method employs a four-terminal bridge measurement (Kent and Jason 1975) in which the influence of interfacial electrode impedances are eliminated and the impedance of the medium is measured directly. However, the switching arrangement for connecting many cells in sequence (multiplexing) is overcomplicated, so that the bridge method has only been attempted for a single-channel instrument (Lovitt et al. 1986). Fortunately, at audio frequencies (20 Hz to 20 kHz) the impedance of culture media is entirely resistive, so that it is possible to devise an arrangement which enables each of the component variables to be measured directly as desired while at the same time permitting the operation of a simple multiplexing arrangement.

Essentially it is necessary to measure the resistance R_s, its equivalent conductance G_s or (in some situations discussed in Section 2.1) the capacitance C_{se}. The resultant impedance Z or admittance Y ($Y=1/Z$) is often measured as well as, or instead of, R_s or G_s. The basic circuit for making such measurements is presented in Figure 14.10. Here, an oscillator generates a voltage $V_1\cos 2\pi ft$ which is applied across a high resistance R_1 in series with the unknown impedance Z. The current flowing in Z produces a voltage $V_1\cos(2\pi ft+\phi)$ at the input of a high impedance operational amplifier (impedance Z_{in}) which forms part of a detector arrangement. The amplified output is chopped in phase with the oscillator voltage $V_1\cos 2\pi ft$ by means of a squarer, which operates on a phase-sensitive demodulator that enables the positive in-phase signal to appear at the input of the DC amplifier. The squarer also switches the output of the

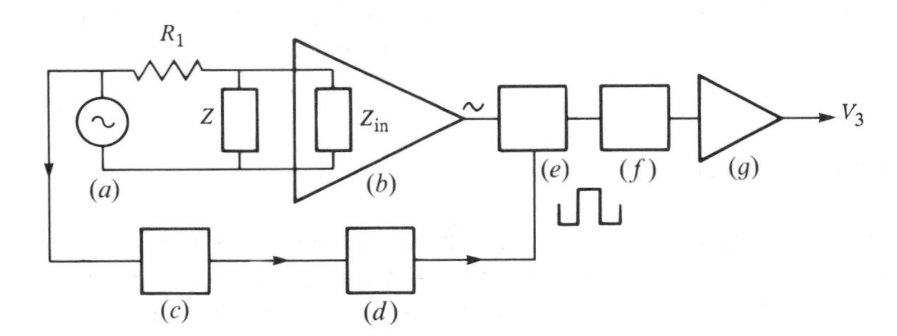

Figure 14.10 *Circuit arrangement for the measurement of the unknown impedance Z: (a) oscillator (b) high impedance operational amplifier (c) phase adjuster (d) squarer (e) phase-sensitive demodulator (f) low pass filter (g) DC amplifier;* R_1 shunt resistance

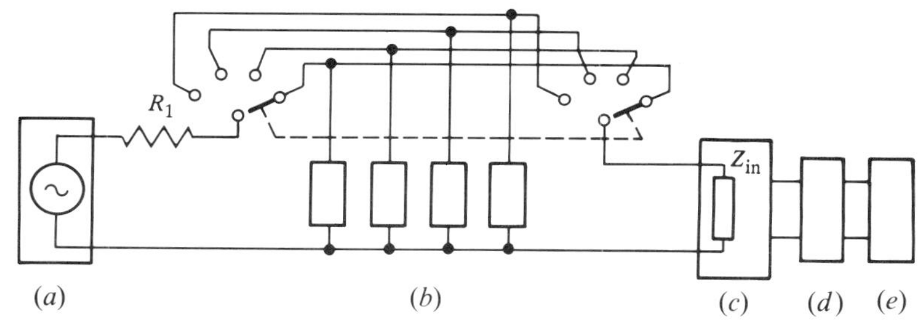

Figure 14.11 *Arrangement for multiplexed cell impedance measurement (solid state switches are chosen): (a) oscillator (b) cells (c) detector (d) A/D converter (e) microcomputer;* R_1 *shunt resistance*

AC amplifier in opposite phase to the oscillator voltage (push-pull operation) and enables a similar signal to be inverted in the phase-sensitive demodulator to give a full-wave rectified signal

$$V_3 = (A_d V_1/R_1)|Z|\cos\phi \qquad (12)$$

at the output of the DC amplifier, where A_D is the conversion gain of the detector. The purpose of the low pass filter is to exclude unwanted signals from all sources of frequency greater than f.

Combining equations (4) and (12) we have

$$V_3 = (A_D V_1/R_1)R_s$$

It is a simple matter to convert successive readings of V_3 obtained during bacterial growth either to a resistance change $\delta R_s = R_s - R_0$ or to a conductance change $\delta G_s = 1/R_s - 1/R_0$ by means of an A/D recorder and a computer.

Impedance change δZ or admittance change δY are similarly obtained from readings of V_3 by disconnecting the phase-sensitive demodulator from the circuit and connecting the output of the high impedance operational amplifier directly to the input of the DC amplifier.

5.2 Multiplexed Cell Switching

Automated data logging of several cell impedances requires each cell to be connected in turn to a measurement system. Mechanical switching can be employed but this brings with it a risk of variable contact resistance and a limited component lifetime. It is therefore preferable to utilize a system of solid state switches in the arrangement shown in Figure 14.11, in which measurement is largely independent of switch resistance (typically 100 Ω 'on' and greater than 10 MΩ 'off'). A dual-ganged switch, driven by a suitably timed clock pulse, first connects a stable oscillator to each of the cells in turn via a high value resistor R_1 (typically 100 kΩ) and then feeds the signal voltage produced into a high impedance detector. This, in turn, feeds an analogue-to-digital converter, the output of which is accumulated in a buffer for subsequent processing by a computer. A cell resistance of 1000 Ω produces an output of 10 V DC into the A/D converter. If, for example, this is a 16 bit converter, the resolution of 153 μV corresponds to 0.0152 μS. The zero drift of the electronic circuits is less than 10 μV and the noise is about 10 μV, so that this resolution can be handled with confidence.

The rate of data acquisition is slow in terms of the capability of microcomputers. In fact, the limitation to the speed of switching is imposed by the settling time of measurement, that is the time required after switching for the measuring current to approach its steady value. This is related to the time constant of the cells and circuitry, that is the product of bulk resistance and capacitance. For 100 μV resolution the settling time is of the order of 1 second, so that 240 cells can be scanned every 4 minutes. It is common practice to scan ten times an hour. In most equipment the cell data are initially stored in a buffer and then rapidly read into memory in batches.

5.3 Commercial Instrumentation

It is evident that an easily operated method of measuring impedance or its derivatives is required for quality control in the food industry.

Electrical detection of the presence of microbes in media or of their growth rate has become well established technology, especially in the food industry where rapid measurement is essential for perishable products. This approach was reinforced at the First International Symposium on Rapid Methods and Automation in Microbiology and Immunology in 1973 (Héden and Illeni 1975a; 1975b), where equipment based on the measurement of the electrical impedance of microbial cultures was described and has since been recognized as having made the most significant commercial advance. Although the method is rapid relative to conventional microbiological analysis it is still slow, taking some hours to obtain a positive result, but the usefulness of the data has been proven and is ahead of that from many other techniques. As will be shown below, there is still considerable scope for much speedier performance. Most users adopt the ‘black box’ approach: introduce sample into conventional medium, then wait for result. At this level the understanding of the science involved is low, but this does not detract from its usefulness or its accuracy in routine applications, once established.

At the 1973 Symposium, the machines providing automated impedance measurement were the Bactometer (Cady 1975) and the Bactobridge (Ur and Brown 1975). Later Richards et al. (1978) described an instrument which became known as the Malthus Growth analyser (because the plots of conductance change over time were regarded as similar in appearance to the population dynamics curves published by the philosopher Thomas Robert Malthus in 1798). The present authors, part of the latter team, have continued to be involved in its subsequent development.

The Bactometer (Vitek Systems Ltd, Henley-on-Thames, UK) and the Malthus (Malthus Instruments Ltd, Crawley, UK) instruments are currently in production. They are highly automated multi-channel instruments containing dedicated microcomputers for processing the measurement of the impedance of up to 500 cells. They sit easily on a bench-top or on their own desk-size tables. They offer sophisticated output display of numerical and graphic results and diagnostic information which may be replicated as hard copy by a printer/plotter. A similar instrument marketed by Don Whitley Scientific (Shipley, Yorkshire, UK) is known as the RABIT (Figure 14.12). The Bactometer and the RABIT have a resolution of 10^6 cfu ml^{-1}, the Malthus 10^5 cfu ml^{-1}. Both the Malthus and the RABIT analysers allow the introduction and removal of single samples. The importance of close temperature control has been taken into account in the design of the three instruments. The Malthus incorporates a water bath incubator (see Figure 14.7) controlled to ± 2 mK, the volume of water being large enough to minimize any perturbations caused by the addition of cold or hot samples. The RABIT similarly controls the temperature of individual cells (Figure 14.13) held in metal blocks. The Bactometer uses an air incubator (see Figure 14.1) with pairs of cells; one of each pair is inoculated and the other serves as a reference. There is only a limited advantage in using such a balanced cell arrangement as the temperature coefficient is not identical from cell to cell. (Incidentally it is

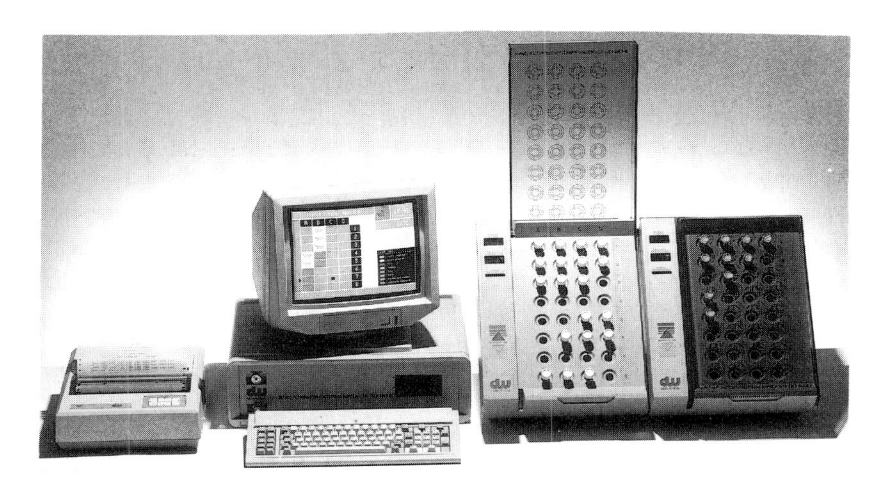

Figure 14.12 *RABIT: rapid automated bacterial impedance technique (Don Whitley Scientific Ltd)*

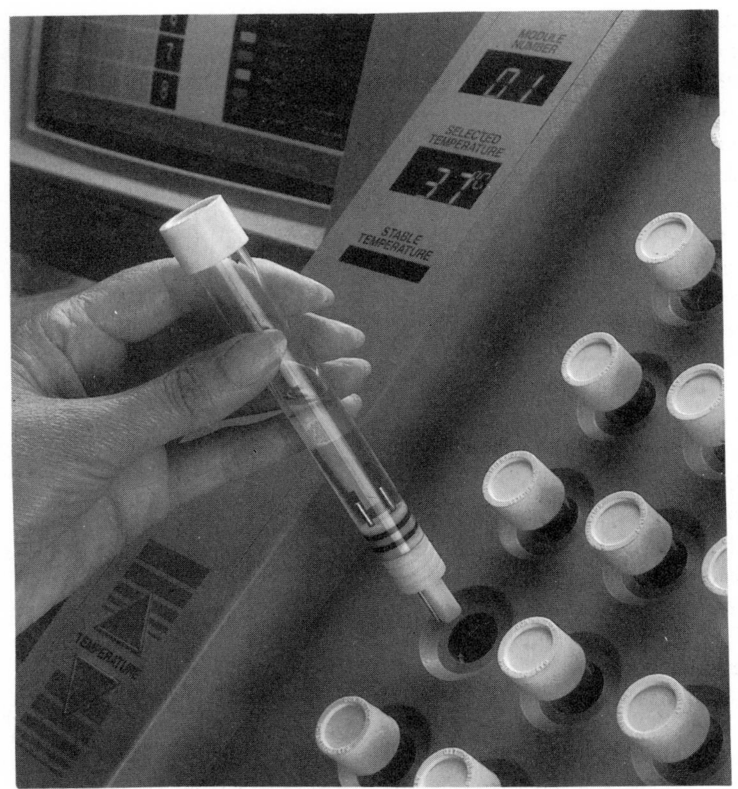

Figure 14.13 *Test cell for the RABIT*

interesting to note that Stewart (1899) employed this balanced cell method; his procedure was to thermostat the cells crudely by holding them under running tap water.)

The instruments each have their own type of electrode design and the cells are not interchangeable. The electrodes used in the Malthus Growth Analyser consist of a small area exposed at the tip of each of two thin platinum strips deposited on a ceramic substrate and insulated (except at the tip) by glass frit. The top of the ceramic substrate is moulded into a plastic cap which is screwed on to standard glass tubes: together they form a conductance cell. The Bactometer electrodes are formed of two stainless steel strips in the base of each of sixteen wells in a plastic tray; each well and its electrode constitute a cell, and one electrode of each pair of cells is interconnected so as to constitute a sample cell and a sterile reference cell. The RABIT electrodes consist of two stainless steel studs protruding through a plastic plug inserted in the base of a cylindrical glass tube. Suitable edge connectors couple the electrodes to the measuring circuitry in each system.

The current commercial situation concerning the Bactobridge is unclear. Other electrical growth analysers have been described by Ackland et al. (1984), Wilkins et al. (1978) and D. B. Kell's group (Harris et al. 1987). The first apparently operates on the self-generation of a population-dependent EMF generated by electrodes of dissimilar metals. It appears to be very insensitive to bacterial growth, responding only to a change in population density of more than 10^7 cfu ml^{-1}. The Wilkins et al. machine is based on electrochemical detection of growth; many Enterobacteriaceae produce hydrogen gas during growth, and this changes the redox potential of a sensing electrode. This method, too, is somewhat insensitive, having a lower detection limit of 10^8 cfu ml^{-1}. Since both instruments are not based on impedance measurement, they will not be discussed further here. The analyser designed by Kell (marketed as the Bugmeter, or more recently as the Biomass Monitor) makes a single-channel measurement of the true capacitance of the growth medium using a four-terminal measuring system. It is designed to operate in the presence of high microbial loads, as in fermenters, at about 10^8 cfu ml^{-1} and higher.

The Bactometer displays the change of impedance ratio $Z_r/(Z_r+Z_s)$, where Z_r is the impedance of the reference cell and Z_s is the impedance of the sample or experimental cell, and has the option of displaying the conductance and capacitance normalized similarly (Eden and Eden 1984). The Malthus Analyser measures series resistance and displays conductance change. The importance of measuring series conductance in high resolution growth determination has been emphasized earlier, and it has been pointed out that this must be coupled with the use of platinum electrodes and close temperature control. Only the Malthus offers this combination, though it must be emphasized that this is of advantage only when very early detection of growth is necessary.

6 The Evaluation of Conductance Data

6.1 Bacterial Growth in Batch Culture

Consider a population of initial density n_0 which commences to grow in a medium of mass concentration c at lag time t_L after inoculation. Each bacterial cell metabolizes an equal mass ΔS of the substrate during growth, and growth eventually ceases owing to the accumulation of a limiting concentration of toxic metabolites when the total number of cells present is n_s. The number of cells that have grown in this period is

$$n_s - n_0 = (S_0 - S_s)/\Delta S$$

where S_0 and S_s are the initial and stationary values respectively.

The initial specific growth rate is $(\dot{n}/n)_0$, where $\dot{n}=dn/dt$ is constant for a given strain growing in a given medium at a given temperature (Jason 1983). It has the dimension of $(\text{time})^{-1}$, so that we write

$$t'_g = 1/(\dot{n}/n)_0 \qquad (13)$$

where t'_g is characteristic of the reproduction and growth of a bacterium in the absence of any self-generated inhibitory substance, n is the population density at time t and $\dot{n}$ signifies the growth rate dn/dt.

The specific growth rate declines from its initial value to zero when $\dot{n}=n_s$. Furthermore, $(n_s-n_0)\propto(S_0-S_s)$, so we may assume that

$$\dot{n}/n = (\dot{n}/n)_0[1-(n-n_0)/(n_s-n_0)]$$

Thus

$$(\dot{n}/n)/(\dot{n}/n)_0 = (n_s-n)/(n_s-n_0) \qquad (14)$$

Combining equations (13) and (14) we have

$$\dot{n} = (n/t'_g)[(n_s-n)/(n_s-n_0)] \qquad (15)$$

On integration we obtain the relationship between population growth and time t after inoculation. Integration is over the limits $[n_0 \rightarrow n]$ and $[t_L \rightarrow t]$ and we approximate $n_s \pm n_0 \simeq n_s$ and write in $\dot{n}=dn/dt$.

$$n = n_0 n_s / \{n_0 + n_s \exp[(t_L - t)/t'_g]\} \qquad (16)$$

The sigmoidal curve associated with monophasic growth (shown in Figure 14.4) is closely followed by this equation, where the constant of proportionality (equation (9)) is $1\,\text{pS}(\text{cfu ml}^{-1})^{-1}$. Equation (16) thus scaled is represented by the broken line in the figure.

Equation (16) reduces to the familiar exponential form for the early stages of growth.

$$n = n_0 \exp[(t-t_L)/t'_g] \qquad (17)$$

when the approximation $n \ll n_s$ is used in integrating equation (15). The generation time t_g is defined as the time during which the population doubles during the initial stages of growth, that is when the population is well below the saturation level $(n \ll n_s)$. It therefore follows from equation (17) that

$$t_g = t'_g \ln 2 \qquad (18)$$

On logarithms, see Appendix B14 on page 711.

6.2 Determination of Inocula

Growth in bacterial density follows the relationship of equation (16), but often suddenly changes a few generations after the peak growth rate has been reached. The culture then continues to grow at a slower rate which is also described by equation (16) but with a larger value of t_g and with other parameters changed. There can be several such growth phases, but for present purposes we shall consider only exponential growth in the first phase.

Rearranging equation (17) we have

$$t - t_L = t'_g \ln(n/n_0)$$

Substituting for t'_g from equation (18) and writing $\ln(n/n_0) = 2.303 \lg(n/n_0)$, we have (see Appendix B14 on page 711)

$$\lg n_0 = \lg n - 0.301(t - t_L)/t_g \quad (19)$$

If the detection time t_D is the time at which a batch culture grows to an arbitrary fixed population density n_D, then the relationship between $\lg n_0$ and t_D is linear:

$$\lg n_0 = \lg n_D - 0.301(t_D - t_L)/t_g \quad (20)$$

We therefore have a means of determining an inoculum of a given organism from a previously obtained calibration simply by finding the time at which the conductance change ΔG_D, corresponding to n_D, occurs and by reading off the value of n_0.

The generation time of the organism is

$$t_g = -0.301/m$$

where m is the slope of the line.

It is customary to select a value of ΔG_D corresponding to about 10^7 cfu ml^{-1}, but its magnitude is unimportant as long as t_D can be determined accurately from the growth curve to within ± 0.1 h.

The calibration for pure cultures is best obtained from the determination of a series of values of t_D corresponding to values of n_0 obtained from successive dilution of a parent inoculum of high concentration. In this way, the scatter inherent in the determination of successive individual values of n_0 by plate counts is avoided. The statistical error implicit in obtaining a plate count of N colonies grown from a sample volume of a true inoculum n_0 is $\pm N^{1/2}$ or $\pm 100N^{1/2}/N$ per cent, an error that remains unchanged whatever the dilution of the parent inoculum. From a large value of N (say >100), the error in determining the daughter inocula is insignificant in relation to the parent value, being attributable only to the small error of dilution. A calibration curve derived in this way has no detectable scatter (Figure 14.14). In most situations, however, mixed cultures are present in foods and the data are more scattered (Figure 14.15). This has the effect of introducing an uncertainty in estimating the inoculum concentration from the value of a specific detection time.

The degree of uncertainty is calculated from the standard error of estimate ($SE_{\ln n_0}$) of $\lg n_0$ on t_D. Its value is given approximately by the expression

$$SE_{\ln n_0} = \sigma_{\lg n_0}(1 - r^2)^{1/2}$$

where $\sigma_{\lg n_0}$ is the standard deviation of all lg n_0 values and r is the correlation coefficient. Thus 95 per cent of all values lie within plus or minus two standard errors of estimate of the values given by the regression equation, and practically all values lie within plus or minus three standard errors of estimate. (See also Chapter 15, Section 3.1 on acceptable, suspect and reject categories.)

6.3 Choice of Growth Media

In many assays for growth of microbes in undefined media containing protein extracts, beef extract, yeast hydrolysates etc., the conductance change cannot be ascribed to any particular ions, compounds or reactions. In some cases the main source is known and has been confirmed by chemical analysis of fresh and spent media, or by obtaining a quantitative response to altering the concentration of the suspected chemical. For example, Easter et al. (1982) showed that the reduction of trimethylamine oxide, an uncharged molecule, to trimethylamine, a very basic molecule, resulted in a quantitative conductance change (Figure 14.16); the compound has been used in assays for fish quality and for assays for salmonellas for food safety. Owens and Wacher-Viveros (1986) suggested that organic buffers could be employed, and this approach has been particularly useful in media formulations for yeast assays. Connolly et al. (1988) used ammonium tartrate and Adams et al. (1989) succinate buffer to good effect.

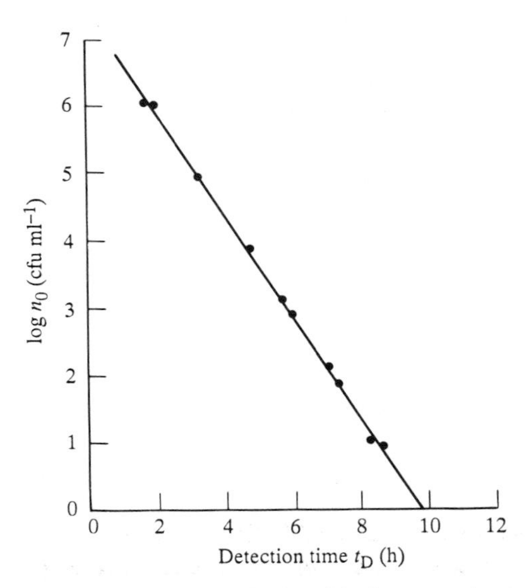

Figure 14.14 *Relationship between inoculum concentration n_0 and detection time t_D for* E. coli *grown in Oxoid nutrient broth at 37°C*

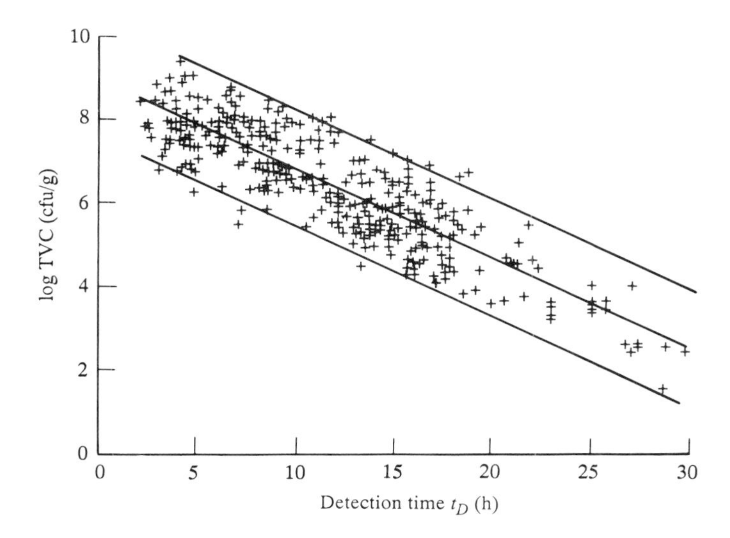

Figure 14.15 *Relationship between total viable count (TVC) of various fish products (cod, haddock, plaice, trout and scampi) and detection times* (t_D) *assayed in brain heart infusion at 20°C: the 95 per cent confidence limits are indicated (from Gibson and Ogden 1987, by courtesy of the authors and publisher)*

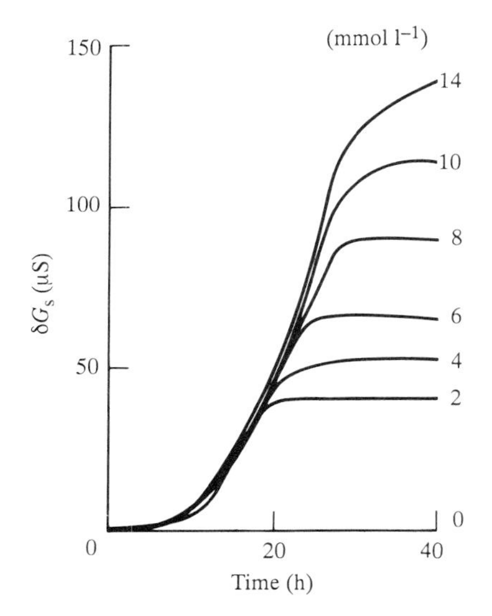

Figure 14.16 *The effect of trimethylamine oxide (TMAO) concentration on conductance change* δG_s: Shewanella putrefaciens *NCMB 400 was grown in broth containing 0–14 mmol* l^{-1} *TMAO (from Easter et al. 1982, by courtesy of the authors and publisher)*

The ionization of inorganic ions may change during the course of the assays owing to changing pH and may affect the conductance (Owens et al. 1985; Gibson 1990).

Most studies on the source of the conductance change have been done with pure cultures in defined media. In the analysis of foods and beverages, there is a significant nutritional contribution to the medium from the sample being tested, and this can influence the composition of the medium and the metabolism of the organisms present. Indeed, to obtain data on the stability or spoilage of products, Gibson (1990) has argued that the medium should contain the constituents of or reflect the composition of the material under test.

For special applications, a measurement of evolved carbon dioxide by a variant of the conductance method has been suggested. During growth, all micro-organisms produce CO_2. Numerous ways to measure this have been proposed (Dixon and Kell 1989) including some conductimetric methods (Hill and Evans 1986; Bruckenstein and Symanski 1986). Linear quantitative responses were found over a range of concentrations of up to 100 mM. Owens et al. (1989) modified Malthus and RABIT electrodes for CO_2 measurement. They placed the electrodes in a solution of KOH or NaOH which was kept separate from the growth media. Carbon dioxide produced by growth was trapped in the alkali, causing a change in conductance. The response with *E. coli* was slower than by direct conductance assay, but Owens et al. propose its use for assays of microbes which give a poor response by direct assay. There could of course be interference by CO_2 produced by tissue respiration.

6.4 Correlation with Conventional Microbiological Data

Conductance data are being used in place of conventional microbiological data for three purposes, to assess the quality and safety of products, and also to predict their shelf life. Hitherto, quality has usually been derived from the standard plate count (TVC or cfu ml^{-1}) and, while it is one of the most used assays in the food industry, it is of limited value in quality control or assurance, with results available days after sampling. Sharpe (1980) has said that 'no other analytical microbiological procedure in the scientific world can correlate with it.' Thus it is not possible for any novel technique, regardless of its scientific merits, to give data fully comparable with the conventional assay; indeed, it is not desirable that it does so, for the plate count has many drawbacks. It is a measure of the ability of microbes to form colonies or biomass on the test medium. Theoretically each microbe gives rise to a colony, but it is well known that colonies can arise from clumps of perhaps hundreds of cells. In a liquid assay, each viable cell in the inoculum has an individual contribution and, even if it is not replicating, its metabolism end products may contribute to the conductance change. Thus the magnitude of correlation coefficients between conventional and conductance or impedance data is not necessarily a valid criterion for the acceptability of the latter. Calibration curves relating plate counts and other conventional methods to, say, detection times have been published by many authors. Usually they cover a wide range of microbiological quality, far greater than is found for particular products in commerce. Samples are modified to give a range of qualities, for example by holding for longer than usual at higher than normal temperatures. While such procedures were acceptable in the early days of automated microbiology, they are not needed now. Accept/reject decisions can be based on the actual detection time without recourse to the calibration curve and conversion to microbial numbers, by comparison with a target detection time varying with product and desired quality. Of high importance is the reproducibility of the assay, allowing for the uneven distribution of microbes in many materials and products. In conventional assays involving homogenization or extraction, dilution and plating, the amount of starting material actually assayed is in the milligram range. In electrical assays, only a primary dilution is needed (to eliminate antimicrobial effects of the constituents of the sample), and in an assay of 10 ml volume, gram quantities are present. It would be reasonable to assume that it is better to analyse larger rather than smaller quantities of heterogeneous materials.

Table 14.6 lists some products and product categories which are known to be analysed by electrical methods in commerce.

The safety of foods relates to the absence of specific bacteria or groups of organisms. Assays have been developed for the organisms listed in Table 14.7. The electrical assays have been particularly useful in this respect. Many conventional assays are slow, laborious and even contentious; for example, the effects of diffferent selective agents for salmonellas are disputed. Conductance assays have been accepted for use in industry to give rapid clearance (or otherwise) of raw materials and products. Indeed, it has been said that the only economic justification in the food field for such expensive equipment is product safety.

It has rarely been possible to take conventional media and use them in electrical assays. An example is the assay for *Salmonella* spp. in foods. Easter and Gibson (1985) tested most of the commonly used media for salmonellas (which incidentally do not give equivalent results). They found that many, owing to their high ionic strength, were outside the measuring range of the conductance instruments, and others failed to give a distinct conductance change. In their earlier work on fish spoilage, Easter et al. (1982) had observed that the reduction of trimethylamine oxide (TMAO) gives a large conductance change, and it was evident from the literature that the ability to reduce TMAO was common to salmonellas. They added TMAO to the most promising media, selenite-cystine broths, and produced, with other modifications, a medium in which most salmonellas are

Table 14.6 *Some applications of conductance and impedance methods for food quality*

Dairy products:	
Raw milk	O'Connor (1979); Gnan and Luedecke (1982); Firstenberg-Eden and Tricario (1983)
Pasteurized milk	Visser and de Groote (1984); Kamei et al. (1988)
Dried milk powders	McMurdo and Whyard (1984)
Fermented milk	Okigbo and Richardson (1985)
Cream	Griffiths and Phillips (1984)
Cheese	Waes and Bossuyt (1984)
Fish and fish products	Gibson et al. (1984); Ogden (1986); van Spreckens and Stekelenburg (1986); Jørgensen et al. (1988)
Brewing	Evans (1985)
Wine	Henschke and Thomas (1988)
Niacin	Einarrson and Snygg (1986)
Meat	Firstenberg-Eden (1983); Bulte and Reuter (1984)
Vegetables	Hardy et al. (1977)
Confectionery	Pugh et al. (1988)
Fruit Juice	Jarvis (1985)
Shelf life	Gibson (1985) Gibson and Ogden (1987) Jørgensen et al. (1988)

detected, along with some *Citrobacter freundii*. Accordingly, alternative media have been formulated that are based on the ability of salmonellas to decarboxylate lysine (Arnott et al. 1988; Ogden 1988). All these media have been tested extensively under industrial conditions and it can be concluded that their performance is very good – better than many media used in conventional assays, and only limited in reliability by the statistics of sampling. The original Easter-Gibson formulation has been tested in a ring trial run on behalf of the International Dairy Federation on 2500 samples (Prentice et al. 1990). Smith et al. (1989) have tested it, along with Ogden's formulation, on animal feeds. On the basis of these results, standard methods have been agreed and are being implemented. With the upsurge in the requirements for testing for salmonella (and other pathogens), laboratories can only cope with the workload by using automated methods such as described here.

7 Future Possibilities

A criticism frequently levelled at the application of impedance techniques in microbial assay is the long period before detection is possible, even under the most favourable conditions. The detection time, of course, depends on the inoculum concentration and on the arbitrary level of conductance change for detection (Figure 14.16). In this example, a detection level is registered 2 hours after inoculation with 10^6 cfu ml^{-1} and

Table 14.7 *Published conductance and impedance methods for organisms of public health significance*

Salmonella spp.	Easter and Gibson (1985); Gibson D. M. (1987); Ogden (1988); Arnott et al. (1988); Pugh et al. (1988); Bullock and Frodshaw (1989)
E. coli/coliforms	Silverman and Munoz (1979); Gibson et al. (1984); Martin and Selby (1980)
Enterobacteriaceae	Petitt (1983)
Staph. aureus	Prentice and Neaves (1987)
Clostridium botulinum	Gibson A. M. (1987)
Histamine producers	Klausen and Huss (1987)
Faecal streptococci (enterococci)	Neaves et al. (1988)
Listeria spp.	Phillips and Griffiths (1989)
Mycotoxigenic moulds	Adak et al. (1987)
Yeasts	Connolly et al. (1988); Schaertel et al. (1987)
Bacteriophage	Waes and Bossuyt (1984)

Table 14.8 *Threshold number of generations γ_D obtainable on 16 bit* AD *converter for various* n_0 *($R_s/R_{fs}=0.9$)*

n_0 (cfu ml^{-1})	10^4	10^3	10^2	10^1	10^0	10^{-1}
γ_D (no.)	0	3.3	6.6	10.0	13.2	16.6

8 hours after inoculation with 10^1 cfu ml^{-1}. It is clearly desirable to improve this performance for more speedy results.

In the absence of baseline drift, growth is detected when the change in conductance exceeds ΔG_s (as defined in equation (10)). This occurs after the elapse of γ_D generations of growth of any given organism, and it may be shown that

$$\gamma_D = 3.32\,(\log\Delta n - \log n_0) \qquad (21)$$

where Δn corresponds to ΔG_s and n_0 is the inoculum concentration. Reference to equation (10) shows that when $R_s = R_{fs}$ a 16 bit A/D voltage converter resolves conductance to within 0.015 μS, corresponding to a numerical resolution of 1.5×10^4 cfu ml^{-1}. Electronic drift and noise are equivalent to less than one-tenth of this resolution, so that without resorting to elaborate measures it is possible to anticipate a resolution of about 10^3 cfu ml^{-1}. Thus, taking a value of $\Delta n = 10^3$ cfu ml^{-1} and assuming that a detection level of ten times this resolution denotes the detection time, we can predict the number of generations that must elapse before growth is detected for a range of inocula (Table 14.8). Therefore it seems reasonable to expect almost immediate detection of the growth after the lag phase when the inoculum is $>10^4$ cfu ml^{-1}. If the inoculum is 10^3 cfu ml^{-1} and the generation time is 0.3 h, then detection would occur one hour later. Thus, by using a cell of resistance R_s close to the maximum value R_{fs} and a 20 bit A/D converter, a very early detection (following the lag phase) would be possible for determining quite low bacterial loads. Such a possibility represents only a small design effort on the part of manufacturers and can confidently be expected in the future.

Cell design, too, is likely to be improved. Electrodes, for example, could be screened from deposition of detritus from the food sample which will lead to a more stable baseline. Thus earlier detection will be possible with samples that are at present difficult to monitor.

It is also advantageous to develop media which give rise to the shortest possible generation times, since detection times are correspondingly reduced.

In many countries the expected shelf life, during which the food should be of good quality, has to be marked on packages. Generally a considerable error margin is allowed for in the absence of data, and this can lead to losses of perfectly edible food. For foods in which microbes are the main cause of deterioration or spoilage, rapid assays can yield data relating to the predicted shelf life of the products. Gibson (1985), Gibson and Ogden (1987) and Jørgensen et al. (1988) have used conductance detection times for this purpose for fish. They based their method on the time-temperature relationships of Ratkowsky et al. (1983) for microbial spoilage at chill temperatures. The method is promising and, with some refinement, could lead to an electrical assay indicating not only the quality at the time of testing but also the expected shelf life at any selected storage temperature. It should be supplemented with careful monitoring of the storage temperature or the use of time-temperature integration labels.

References

Ackland, M. R., Manvell, P. M. and Bean, P. R. (1984) A rapid electrical method to detect microbial growth automatically. *Biotechnology Letters*, **6**, 137-142

Adak, G. K., Corry, J. E. L. and Moss, M. O. (1987) Use of impedimetry to detect tricothecene mycotoxins. 1: Screen for susceptible microorganisms. *International Journal of Food Microbiology*, **5**, 1-13

Adams, M. R., Bryan, J. J. and Thurston, P. J. (1989) A medium designed for monitoring pitching yeast contamination in beer using a conductimetric technique. *Letters in Applied Microbiology*, **8**, 55-58

Adams, M. R. and Hope, C. F. A. (1989) Rapid methods in food microbiology. In *Progress in Industrial Microbiology*, vol. 26, Elsevier, Amsterdam

Arnott, M. L., Gutteridge, C. S., Pugh, S. J. and Griffiths, J. L. (1988) Detection of salmonellas in confectionery products by conductance. *Journal of Applied Bacteriology*, **64**, 409-420

Bruckenstein, S. and Symanski, J. S. (1986) Continuous conductimetric sensor for carbon dioxide. *Analytical Chemistry*, **58**, 1766-1770

Bullock, R. D. and Frodshaw, D. (1989) Rapid impedance detection of salmonellas in confectionery using modified LINCR broth. *Journal of Applied Bacteriology*, **66**, 385-391

Bulte, M. and Reuter, G. (1984) Impedance measurement as a rapid method for the determination of the microbial contamination of meat surfaces, testing two different instruments. *International Journal of Food Microbiology*, **1**, 113-125

Cady, P. (1975) Rapid automated bacterial identification by impedance measurements. In *New Approaches to the Identification of Microorganisms*, eds C. G. Héden and T. Illeni, Wiley, London, 74-99

Clowes, R. P. and Hayes W. (1968) *Experiments in Microbial Genetics*. Blackwell Scientific, Oxford

Connolly, P., Lewis, S. J. and Corry, J. E. L. (1988) A medium for the detection of yeasts using a conductimetric method. *International Journal of Food Microbiology*, **7**, 31–40

Curtis, G. D. W., Thomas, C. D. and Johnston, H. H. (1985) A note on the use of dextran in blood cultures monitored by conductance methods. *Journal of Applied Bacteriology*, **58**, 571–575

Dixon, N. M. and Kell, D. B. (1989) The control and measurement of CO_2 during fermentations. *Journal of Microbiological Methods*, **10**, 155–176

Easter, M. C. and Gibson, D. M. (1985) Rapid and automated detection of salmonella by electrical measurements. *Journal of Hygiene (Cambridge)*, **94**, 245–262

Easter, M. C. and Gibson, D. M. (1989) Detection of microorganisms by electrical measurements. *Progress in Industrial Microbiology*, **26**, 57–100

Easter, M. C., Gibson, D. M. and Ward, F. B. (1982) A conductance method for the assay and study of bacterial trimethylamine oxide reduction. *Journal of Applied Bacteriology*, **52**, 357–365

Eden, R. and Eden, G. (1984) *Impedance Microbiology*. Research Studies Press, Letchworth, UK

Einarrson, H. and Snygg, B. G. (1986) Niacin assay by monitoring changes in electrical conductance caused by microbial growth. *Journal of Applied Bacteriology*, **60**, 15–19

Evans, H. A. V. (1985) A note on the use of conductivity in brewery microbiology control. *Food Microbiology*, **2**, 19–22

Firstenberg-Eden, R. (1983) Rapid estimation of the number of microorganisms in raw meat by impedance measurement. *Food Technology*, **37**, 64–70

Firstenberg-Eden, R. and Tricario, M. K. (1983) Impedimetric determination of total mesophilic and psychrophilic counts in raw milk. *Journal of Food Science*, **48**, 1750–1754

Firstenberg-Eden, R. and Zindulis, J. (1984) Electrochemical changes in media due to microbial growth. *Journal of Microbiological Methods*, **2**, 103–115

Gibson, A. M. (1987) Use of conductance measurement to detect growth of *Clostridium botulinum* in a selective medium. *Letters in Applied Microbiology*, **5**, 19–21

Gibson, D. M. (1985) Predicting the shelflife of packaged fish from conductance measurements. *Journal of Applied Bacteriology*, **58**, 465–470

Gibson, D. M. (1987) Some modifications to the media for rapid automated detection of salmonellas. *Journal of Applied Bacteriology*, **63**, 299–304

Gibson, D. M. (1990) Optimisation of automated conductance assays. In *Rapid Microbiological Methods for Foods, Beverages and Pharmaceuticals*, Society for Applied Bacteriology technical series 25, eds C. J. Stannard,

S. B. Petitt and F. A. Skinner, Blackwell Scientific, Oxford, 87–99

Gibson, D. M. and Hobbs, G. (1987) Some recent developments in microbiological methods for assessing seafood quality. In *Seafood Quality Determination*, eds D. E. Kramer and J. Liston, Elsevier, Amsterdam, 283–298

Gibson, D. M. and Ogden, I. D. (1987) Estimating the shelf life of packaged fish. In *Seafood Quality Determination*, eds D. E. Kramer and J. Liston, Elsevier, Amsterdam, 437–445

Gibson, D. M., Ogden, I. D. and Hobbs, G. (1984) Estimation of the bacteriological quality of fish by automated conductance measurements. *International Journal of Food Microbiology*, **1**, 127–134

Gnan, S. and Luedecke, L. O. (1982) Impedance measurements in raw milk as an alternative to the standard plate count. *Journal of Food Protection*, **25**, 4–7

Griffiths, M. W. and Phillips, J. D. (1984) Detection of post-pasteurization contamination of cream by impedimetric methods. *Journal of Applied Bacteriology*, **57**, 107–114

Hardy, D., Kraeger, S. J., Dufour, S. W. and Cady, P. (1977) Rapid detection of microbial contamination in frozen vegetables by automated impedance measurements. *Applied and Environmental Microbiology*, **34**, 14–17

Harris, C. M., Todd, R. W., Bungard, S. J., Lovitt, R. W., Morris, J. G. and Kell, D. B. (1987) The dielectric permittivity of microbial suspensions at radio frequencies. A novel method for the real-time estimation of microbial biomass. *Enzyme and Microbial Technology*, **9**, 181–186

Héden, C. G. and Illeni, T. (1975a) *New Approaches to the Identification of Microorganisms*. Wiley, New York

Héden, C. G. and Illeni, T. (1975b) *Automation in Microbiology and Immunology*. Wiley, New York

Henschke, P. A. and Thomas, D. S. (1988) Detection of wine-spoiling yeasts by electronic methods. *Journal of Applied Bacteriology*, **64**, 123–133

Hill, M. O. and Evans, D. F. (1986) Conductimetric measurement of respiration rates, with observations on the physics and chemistry of absorption and conductivity change. *Pedobiologia*, **29**, 247–250

Huis in't Veld, J., Hartog, B. and Hofstra, H. (1988) Changing perspectives in food microbiology: implementation of rapid microbiological analyses in modern food processing. *Food Reviews International*, **4**, 271–329

Jarvis, B. (1985) A philosophical approach to rapid methods for industrial food control. In *Rapid Methods and Automation in Microbiology and Immunology*, ed. K. -O. Habermehl, Springer, Berlin, 593–602

Jarvis, B. and Easter, M. C. (1987) Rapid methods in the assessment of microbiological quality; experiences and needs. *Journal of Applied Bacteriology Symposium Supplement*, **63**, 115S–126S

Jason, A. C. (1983) A deterministic model for monophasic growth of batch cultures of bacteria. *Antoine van Leeuwenhock*, **49**, 513–536

Jørgensen, B. R., Gibson, D. M. and Huss, H. H. (1988) microbiological quality of shelf life prediction of chilled fish. *International Journal of Food Microbiology*, **6**, 295–307

Kamei, T., Sato, J. Kodama, Y., Omata, Y. and Noda, K. (1988) Application of the conductance method to detection of post-pasteurization contamination of pasteurized milk. *Nippon Shokulin Kogyo Gakkaishi*, **35**, 226–234

Kell, D. B. (1987) Forces, fluxes and the control of microbial growth and metabolism. *Journal of General Microbiology*, **133**, 1651–1665

Kent, M. and Jason, A. C. (1975) Dielectric properties of foods in relation to interactions between water and the substrate. In *Water Relations of Food*, ed. R. B. Duckworth, Academic, London, 221–231

Klausen, N. K. and Huss, H. H. (1987) A rapid method for detection of histamine producing bacteria. *International Journal of Food Microbiology*, **5**, 137–146

Lovitt, R. W., Walter, R. P., Morris, J. G. and Kell, D. B. (1986) Conductimetric assessment of the biomass content in suspensions of immobilised (gel-entrapped) microorganisms. *Applied Microbiology and Biotechnology*, **23**, 168–173

McMurdo, I. H. and Whyard, S. (1984) Suitability of rapid microbiological methods for the hygienic management of spray dried plant. *Journal of the Society for Dairy Technology*, **34**, 4–9

Malthus, T. R. (1798) *An Essay on the Principle of Population*. Reprinted 1973, Dent, London

Martin, S. B. and Selby, M. J. (1980) Evaluation of a rapid method for the quantitative examination of coliforms in meat by impedimetric procedures. *Applied and Environmental Microbiology*, **39**, 518–524

Neaves, P., Waddell, M. J. and Prentice, G. A. (1988) A medium for the detection of Lancefield Group D cocci in skimmed milk powder by measurement of conductance changes. *Journal of Applied Bacteriology*, **65**, 437–448

O'Connor, F. (1979) An impedance method for the determination of bacteriological quality of raw milk. *Irish Journal of Food Science and Technology*, **3**, 93–100

Ogden, I. D. (1986) Use of conductance methods to predict bacterial counts in fish. *Journal of Applied Bacteriology*, **61**, 263–268

Ogden, I. D. (1988) A conductance medium to distinguish between *Salmonella* and *Citrobacter* spp. *International Journal of Food Microbiology*, **7**, 287–297

Okigbo, O. N. and Richardson, G. H. (1985) Detection of penicillin and streptomycin in milk by impedance microbiology. *Journal of Food Protection*, **48**, 979–981

Owens, J. D., Miskin, D. R., Wacher-Viveros, M. C. and Benge, L. C. A.

Sources of conductance changes during bacterial reduction of trimethylamine oxide to trimethylammonium in phosphate buffer. *Journal of General Microbiology*, **131**, 1357–1361

Owens, J. D., Thomas, D. S., Thompson, P. S . and Timmerman, J. W. (1989) Indirect conductimetry: a novel approach to the conductimetric enumeration of microbial populations. *Letters in Applied Microbiology*, **9**, 245–249

Owens, J. D. and Wacher-Viveros, M. C. (1986) Selection of pH buffers for use in conductimetric microbiological assays. *Journal of Applied Bacteriology*, **60**, 395–400

Pethig, R. and Kell, D. B. (1987) The passive electrical properties of biological systems: their significance in physiology, biophysics and biotechnology. *Physics in Medicine and Biology*, **32**, 933–970

Petitt, S. (1983) Detection of 'coliforms' and Enterobacteriaceae. *Journal of Applied Bacteriology*, **55**, vii

Pettipher, G. L. (1983) *The Direct Epifluorescent Filter Technique for the Rapid Enumeration of Microorganisms*. Research Studies Press, Letchworth, UK

Phillips, J. D. and Griffiths, M. W. (1989) An electrical method for detecting *Listeria* spp. *Letters in Applied Microbiology*, **9**, 129–132

Prentice, G. A. and Neaves, P. (1987) Detection of *Staphylococcus aureus* in skimmed milk powder using the Malthus 128H Microbiological Growth Analyser. In Fifth International Symposium on Rapid Methods and Automation in Microbiology and Immunology, Florence, Italy, abstract p128

Prentice, G. A., Neaves, P., Jervis, D. I. and Easter, M. C. (1992) An interlaboratory evaluation of an electrical method for detection of salmonellae in milk powders. *Journal of Applied Bacteriology*, in press

Pugh, S. J., Griffiths, J. L., Arnott, M. L. and Gutteridge, C. S. (1988) A complete protocol using conductance for rapid detection of salmonellas in confectionery materials. *Letters in Applied Microbiology*, **7**, 23–27

Ratkowsky, D. A., Lowry, R. K., McMeekin, T. A., Stokes, A. N. and Chandler, R. E. (1983) Model for bacterial culture growth rate throughout the entire biokinetic temperature range. *Journal of Bacteriology*, **154**, 1222–1226

Richards, J. C. S., Jason, A. C., Hobbs, G., Gibson, D. M. and Christie, R. H. (1978) Electronic measurement of bacterial growth. *Journal of Physics E: Scientific Instruments*, **11**, 560–568

Schaertel, B. J., Tsang, N. and Firstenberg-Eden, R. (1987) Impedimetric detection of yeast and mold. *Food Microbiology*, **4**, 155–163

Schwann, H. P. (1963) Determination of biological impedances. *Physical Techniques in Biological Research*. Academic, London

Sharpe, A. N. (1980) *Food Microbiology: a Framework for the Future*. Charles P. Thomas, Springfield, MA

Silverman, M. P. and Munoz, E. F. (1979) Automated electrical impedance technique for rapid enumeration of faecal coliforms in effluents from sewage treatment plants. *Applied and Environmental Microbiology*, **37**, 521–526

Smith, P. J., Boardman, A. and Shutt, P. C. (1989) Detection of salmonella in animal feeds by electrical conductance. *Journal of Applied Bacteriology*, **67**, 575–588

Stanley, P. E., Smither, R. and McCarthy, B. J. (1990) *ATP Luminescence: Rapid Methods in Microbiology*. Society for Applied Bacteriology technical series 26, Blackwell Scientific, Oxford

Stannard, C. J., Petitt, S. B. and Skinner, F. A. (eds) (1990) *Rapid Microbiological Methods for Foods, Beverages and Pharmaceuticals*. Society for Applied Bacteriology technical series 25, Blackwell Scientific, Oxford

Stewart, G. N. (1899) The changes produced by the growth of bacteria in the molecular concentration and electrical conductivity of culture media. *Journal of Experimental Medicine*, **4**, 235–247

Ur, A. and Brown, D. F. J. (1975) Monitoring bacterial activity by impedance measurements. In *New Approaches to the Identification of Microorganisms*, eds C. G. Héden and T. Illeni, Wiley, London, 61–71

Van Spreckens, K. J. A. and Stekelenburg, F. K. (1986) Rapid estimation of the bacteriological quality of fresh fish by impedance measurements. *Applied Microbiology and Biotechnology*, **24**, 95–96

Visser, I. J. R. and de Groote, J. (1984) The Malthus microbiological growth analyser as an aid in the detection of post-pasteurization contamination of pasteurized milk. *Netherlands Milk Dairy Journal*, **38**, 151–156

Waes, G. H. and Bossuyt, R. G. (1984) Impedance measurements to detect bacteriophage problems in Cheddar cheesemaking. *Journal of Food Protection*, **47**, 349–351

Wilkins, J. R., Young, R. and Boykin, E. (1978) Multichannel electrochemical microbial detection unit. *Applied and Environmental Microbiology*, **35**, 214–215

Addendum to References

The proceedings of the 1990 symposium on rapid methods in microbiology have now been published (*Rapid Methods in Microbiology and Immunology*, ed. A. Vaheri, R. C. Tilton and A. Balows, Springer-Verlag, Berlin, 1991). They contain recent reviews of automated methods in food microbiology including papers on conductance methods ('Conductance and Impedance Methods for the Detection of Foodborne Pathogens, by M. C. Easter and A. L. Kyriakides). The conductance method for detecting salmonellas in foods has been accepted as an official method ('Automated Conductance Method for the Detection of *Salmonella*', in *Foods: Collaborative Study*, D. M. Gibson, P. Coombes, and D. W. Pimbley, *Journal of AOAC International*, **75**, 293–302, 1992).

15 *Impedance Microbiology in Food Quality Control*

Mike L. Arnott

Contents

1 Introduction 499

2 Development of Protocols 501

3 Application to the Detection of Micro-organisms in Foods 503
 3.1 Correlation Procedures 503
 3.2 Total Counts 509
 3.3 Selective Procedures 510
 3.4 Shelf Life Prediction 514

4 Choice of Instrumentation 515

5 Future Developments 516

References 517

1 Introduction

The quality control (QC) microbiologist needs to monitor the whole of the food process and provide guidance and direction to ensure a microbiologically sound finished product. In order to do this satisfactorily the microbiologist will need to carry out certain checks, at key points in the process, on certain raw ingredients and semi-processed materials and inevitably on finished products. It used to be said that a microbiologist could achieve more with a sharp pair of eyes and a thermometer on the factory floor than by all the complex tests he could run in the laboratory. Factory floors seem to have disappeared in favour of production sites and in the modern plant with computer controlled processing equipment this approach may be more difficult, but it does of course depend where the thermometer is inserted!

As a general principle it is important that microbiologists should be familiar with all stages of production; how else can they make a sensible interpretation of the results they obtain? Traditionally, microbiologists have taken environmental and plant swabs, ingredients, process and finished product samples, and subjected them to various tests which involve growing the micro-organisms to detectable levels in culture media, either to produce visible colonies on the surface of solid media or to produce a colour change.

These procedures are very labour intensive and, because of the necessary incubation periods, take anything from two to five days or more for a result. In terms of today's high volume production this represents a lot of product and a huge stock holding facility if finished products are held pending QC release or raw materials stored for clearance before use. More rapid and less labour intensive methods are obviously desirable but, whatever methods are adopted, they will probably be required to correlate in some way with the entrenched plate count methods on which most product specifications are based.

There are a number of techniques in current use including measurement of microbial adenosine triphosphate (ATP), direct microscope count methods, membrane filtration, and electrical impedance measurement. Of these methods, electrical measurement is not the most rapid but it is quicker than conventional methodology. It also offers versatility, automation and computer compatibility, and measures the total metabolic activity of the micro-organisms rather than the somewhat spurious colony forming unit, which may represent a single organism or twenty! It is not surprising that, with these admirable attributes, electrical measurement methods have been adopted in the food industry to replace traditional plating methods.

The principles and theory of impedance or electrical measurement microbiology have been described in Chapter 14, but a brief summary is appropriate here. As micro-organisms grow in a medium they break down the constituents, which in turn changes the electrical resistance of the medium. When the microbial population reaches a level of approximately 10^6 cells/ml an exponential change in the electrical signal with time takes place. The time taken to reach this exponential stage is known as the detection time, and is inversely proportional to the logarithm of the initial numbers of micro-organisms.

Impedance is the resistance to flow of an alternating current through a conducting material, in our case a microbial growth medium. It is in fact a vectorial combination of a conductive element and a capacitive element. In practice it is usually conductance or capacitance that is measured rather than the overall impedance. The signal to be monitored will be determined by the type of micro-organisms to be detected, the culture medium and the particular instrument being used.

There has been a certain amount of debate over which electrical variable – conductance, capacitance or impedance – gives the most accurate measurement of microbial numbers. In fact it is probably possible to develop media which would give an adequate detection with any chosen variable, be it conductance, capacitance or impedance. Pragmatically, if a developed protocol consistently gives the desired correlation with the replaced traditional method but in a shorter time, it probably does not matter.

2 Development of Protocols

As the electrical technique measures metabolic activity rather than visible biomass, as in a colony, the preparation procedures and growth conditions for specific organisms may be different from those used for conventional cultural procedures.

It is essential that good quality detection curves are produced, otherwise it will be difficult to identify a clear detection time either by eye or by computer algorithm. A good quality curve should have a stable baseline with little drift of the curve upwards followed by a period of sharp acceleration. It is at the point of acceleration that the detection time is registered, so it is important that the curve exhibits a clean takeoff (Figure 15.1). Curve characteristics and parameters have been described in detail by Firstenberg-Eden and Eden (1984).

Probably the most important factor in the production of good quality curves is the choice of medium. Micro-organisms break down the constituents of growth medium to end products which will be different according to the metabolism of the organisms and the nutrients present in the

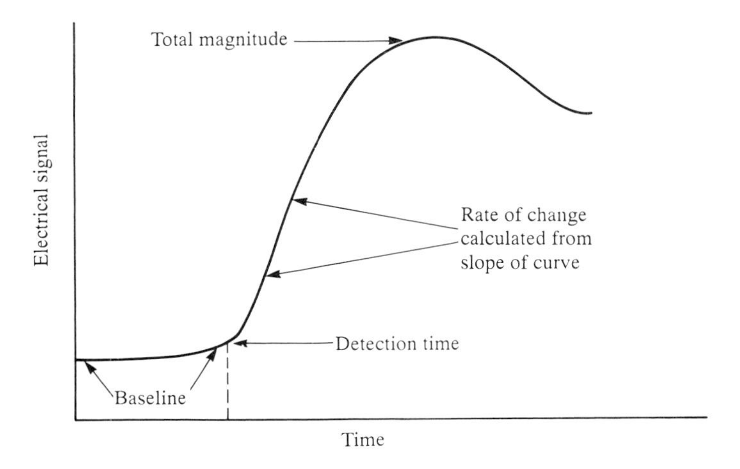

Figure 15.1 *Typical detection curve*

medium. It is desirable that the end products give a strong change in the electrical signal. In practice this may be achieved either by a substrate being broken down into more conductive end products or, less often, by a conductive substrate being metabolized to less conductive products. In either case it is the degree and reproducibility of the change that is important.

Carbon and nitrogen sources, ions and buffers all play an important role in conventional media, but it is even more important to achieve the correct balance in impedance media. It is obvious that media can be tuned to improve their performance and by the judicious choice of selective agents can be made to favour the growth of one type of organism over another or to give a better signal for a specific group of organisms. Some commercially available media may give good results with electrical techniques but not all, so it is necessary to check their performance before using them or to develop media from basic ingredients, to give the desired electrical response.

When developing a medium using basic ingredients, consideration should also be given to the physiological and biochemical characteristics of the important micro-organisms in the products to be tested, so that a suitable substrate can be provided for the production of high signal metabolites. A good example of this principle was the use of trimethylamine oxide (TMAO) by Easter et al. (1982). TMAO is a neutral molecule which is reduced to the highly charged basic trimethylammonium ($TMAH^+$) moiety, and results in a large change in the charge carrying capacity of the medium and hence in a strong electrical response curve.

Another important factor to consider is the generation time of the micro-organisms under consideration. If the objective of the test is to detect a specific group of micro-organisms then the conditions can be tailored to produce the optimum growth conditions and hence the shortest generation time and shortest detection time for that group of micro-organisms. (In this case it would be a positive benefit if other groups of micro-organisms either did not grow at all or grew at a much slower rate.) However, if it is necessary to detect mixed populations of micro-organisms, as in an assay for total microbial count, then conditions should be selected which bring the generation times for all the micro-organisms as close to one another as possible. This can be most readily achieved by altering the incubation temperature of the assay. In theory, the predominant microflora of the products under test should be isolated and generation times at different temperatures determined. The temperature at which these generation times are most similar could then be selected for the assay. In practice this can be accomplished by correlating plate count results with electrical assay results at a range of temperatures and selecting the temperature which gives the best correlation.

As mentioned earlier, the choice of the electrical variable measured can also affect the quality of the curves produced. It is often better to measure conductance in low conductivity media and capacitance in more conductive media. It is usually possible to produce a reasonable curve with either of these admittance components by manipulating the composition of the media. However, some instruments such as the Bactometer (Vitek Systems Ltd) can measure all three variables, impedance, conductance or capacitance, and this can save a considerable amount of time when establishing a new test protocol.

3 Application to the Detection of Micro-organisms in Foods

3.1 Correlation Procedures

In traditional microbiological testing most results are expressed in terms of the numbers of micro-organisms present in the sample as either organisms or colony forming units (cfu) per gram or millilitre or square centimetre of sample. Consequently microbiological specifications have been based on the concept of products containing below or above a defined acceptable level of organisms. The determination of these levels has been based on the ability of micro-organisms to multiply on the surface of suitable solid growth media to produce visible colonies which can then be counted. The fact that a colony can be formed from a single cell or a group of cells is often ignored, as is the inherent variability of plate counts. Although the validity of plate counts is somewhat dubious, it is still true that any innovative method has to overcome the hurdle of relating the measurement unit of the new method to that of the traditional plate count. As electrical methods measure metabolic activity as a function of the number of micro-organisms in the sample, they should in fact more accurately reflect the total microbial load than a plate count. However, it is possible to correlate the two methods, although the shortcomings of the plate count procedures should be borne in mind when comparing the results obtained.

A calibration curve can be produced by testing samples using plate counts and electrical measurements in parallel and plotting the logarithm of the colony forming units against the electrical detection times. The number of samples used to produce a calibration curve will be a compromise between the optimum number of data points required for accurate generation of the calibration curve and the time and effort available to carry out parallel testing. Obviously, the more samples used for the production of the calibration curve, the greater the accuracy and the confidence that can be placed in the calibration. It is generally thought that 60–100 samples are necessary for the production of a reliable calibration curve with a significant correlation coefficient.

Perhaps a more difficult requirement is that the samples used for calibration should cover a four to five log cycle range of micro-organisms. Furthermore, Firstenberg-Eden and Eden (1984) recommended that at least 20 per cent of the samples used for the generation of the calibration curve should have levels of micro-organisms above the specified acceptable limit, and that at least 10 per cent of samples should have levels at least one log above the specified limit. When calibration curves are produced from samples with a narrow range of one to two log cycles, the regression line is very sensitive to the addition of extra data at certain points on the curve.

The correlation coefficient can change considerably with the addition of data from one or two extra samples, and caution is necessary when extrapolating beyond the actual data points. It can be a considerable practical problem to obtain sufficient samples which cover the desired range of microbial levels. Low count samples can be fairly easily prepared by diluting high count samples with low count or sterile samples. High count samples can be prepared either by abusing the product in some way or by inoculating the samples with micro-organisms isolated from the product. It is important to mimic both the balance of the natural flora and the normal physiological state of the organisms, which implies a knowledge of the normal microflora of the product and effect of processing. In either case it would be required that the data points from these artificial samples should give a reasonable fit on the regression line with the data previously obtained with naturally contaminated samples. If this does not happen then a different approach will have to be tried until a good fit on the regression line is obtained with the artificially prepared samples.

Calibration curves can be prepared for all types of food products, making impedance microbiology a genuine alternative to conventional microbiological testing. For a food company manufacturing a wide variety of products, the production of calibration curves for every product might present a formidable task. Fortunately in many cases, by adopting a sterility test approach or a simpler sample distribution procedure, it will not be necessary to carry out a full calibration before using electrical methods.

Even when a full calibration is considered desirable it is often possible to use a single calibration curve for more than one type of product providing the data points fit the same regression line. An example of this cumulative approach when producing calibration curves for total counts in flavoured beverages is shown in Figure 15.2. Separate calibration curves were produced for chocolate and malt beverages, and the data from these product types were found to be sufficiently similar to combine into a single regression line.

Any changes in recipe or new products could be accommodated by evaluating samples using conventional plate counts in parallel with the impedance method and checking the results against the established regression line. If the results from the new recipe or new product can be fitted

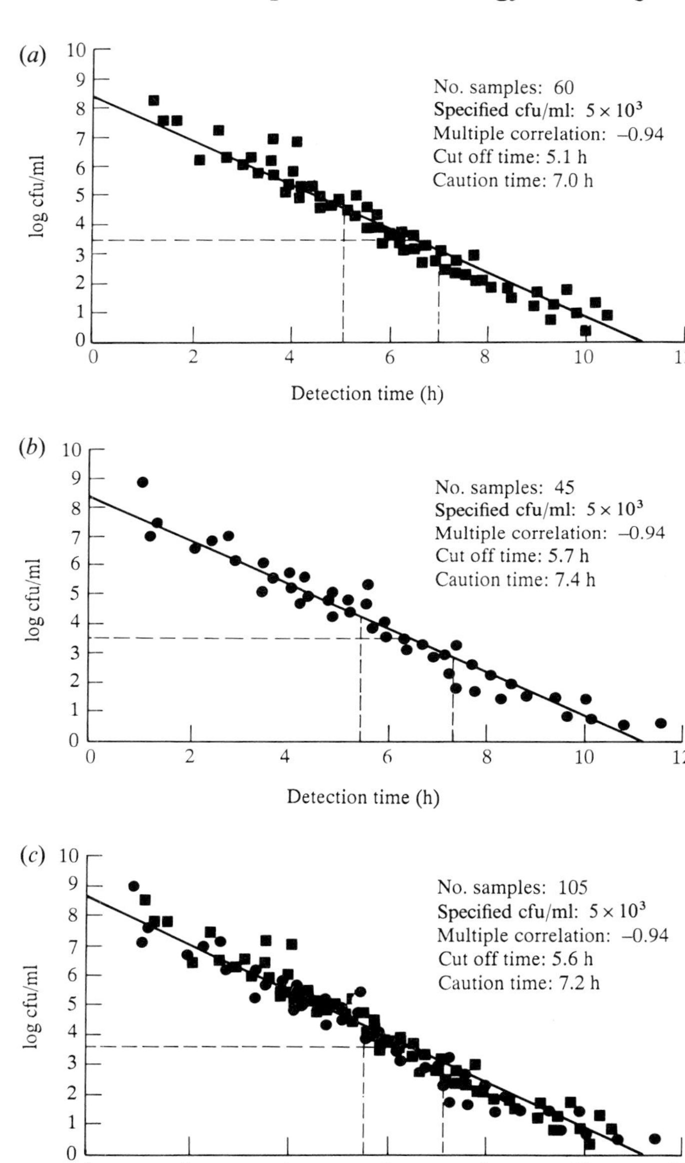

***Figure 15.2** Combination of calibration curves for total counts in flavoured beverages: (a) chocolate beverage (b) malt beverage (c) combination for flavoured beverages*

without changing the regression line then it would not be necessary to carry out a full calibration for the new product variant. It is still crucial to optimize the impedance test protocol by developing a medium that will give a good quality curve with the selected variable, to ensure accurate detection times, and by using a temperature that will minimize any difference in generation times of the microflora of the test products.

Often it is not necessary for the QC microbiologist to know the precise numbers of organisms present in samples providing they meet established specifications. The calibration curve can be used to give detection times which would segregate samples into acceptable, suspect and reject categories by taking one standard deviation either side of the detection time for the specified limit. For example, in Figure 15.2 the specification for the chocolate product is that it should not contain more than 5000 cfu/g. Using the individual calibration curve for the chocolate beverage, this gives a cutoff time of 5.1 hours (any samples detecting before this time would be rejected) and a caution time of 7.0 hours (any samples detecting later than this would clearly be acceptable). Samples which give detection times between the cutoff and caution times would normally be regarded as suspect and resampled or quarantined pending further investigation.

It would be possible to interpret results in the same manner as the ICMSF two-class specifications (ICMSF 1974). In these plans certain parameters are defined:

- n the number of samples to be taken per unit of production (batch or shift)
- c the maximum number of defective samples allowable
- m the defective level, which would usually indicate a problem with some aspect of production but not a health risk
- M the reject level, which would indicate either a gross fault in the process or the possibility of a health risk to the consumer.

Typically for a test such as a total plate count, there would be the recommendation to take five samples per unit of production ($n=5$) with up to three of these samples being allowed to exceed the defective level ($c=3$) providing none exceeded the reject level.

The values designated for m and M would obviously vary according to the product under test. In the case of calibration curve data the cutoff time could be equated to M, the reject level, and the caution time to m, the defective level. If existing two-class specifications based on plate counts were already in use when the impedance methodology was introduced, there is no reason why the specification levels being used should not be retained. After all, these specifications will have been established for sound practical or commercial reasons. The specification levels can be converted into detection times by reading off from the calibration curve, and these detection times used for QC purposes instead of the generated cutoff and caution times.

An alternative to the full calibration procedure is to plot the distribution of detection times for samples with above specification counts and samples with below specification counts in the form of a histogram (Figure 15.3). This method is useful for samples with low count specifications because it allows use to be made of the data for samples giving no detection times which could not be used in a normal calibration.

The impedance and standard plate counts are run in parallel on the same samples as with calibration, and a record is made of the number of samples with counts above and below specification and the period in which they give detection. These results are plotted as the percentage distribution of samples with above specification levels and the percentage distribution of samples with below specification levels combined on the same histogram against detection times, as shown in Figure 15.3.

A time can be selected from the histogram after which no above specification samples were detected. This time would be equivalent to the caution time in the calibration curve, and it could reasonably be expected that samples with detection times in excess of this time would be within specification for that product. In the case of the example in Figure 15.3 the caution time would be 7 hours, and for the samples plotted this would mean no false negatives and 86 per cent of below specification samples correctly classified.

A cutoff or rejection time could then be selected from the histogram,

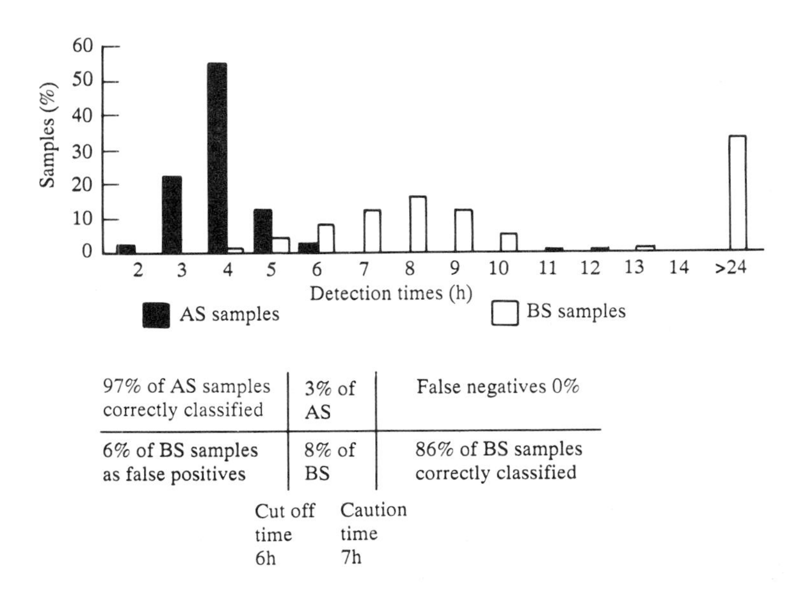

Figure 15.3 *Combined distribution plot of above specification (AS) and below specification (BS) samples (courtesy Vitek Systems Ltd)*

balancing the risk of accepting a sample with a count above specification against the possibility of rejecting false positive samples having counts within specification. Selecting a cutoff time of 6 hours in the example would mean that 97 per cent of the above specification samples would be correctly classified and rejected, but 6 per cent of the below specification samples would be misclassified false positives. In the example, 3 per cent of above specification samples and 8 per cent of below specification samples fell in the period between 6 and 7 hours; these samples were all close to the specification level. If this is the case with a large number of samples, the distribution system becomes difficult to apply and it may be necessary to carry out a full calibration.

If a cutoff time of 5 hours was selected, this would reduce the percentage of below specification samples rejected (i.e. false positives) to 1 per cent. However, it would increase the percentage of above specification samples in the caution zone between 5 and 7 hours to 16 per cent and the percentage of the below specification samples in this area to 13 per cent. Therefore a larger number of samples may fall into the caution zone. The decision whether to go for a 5 hour or 6 hour cutoff would probably depend on the procedure for handling samples which fell in the caution area.

If an ICMSF two-class specification plan was in operation, it might not make much difference which cutoff time was chosen depending on the value selected for c (the maximum number of defective, but not reject, samples allowed). If a costly resampling and retesting regime was applied to all the samples which fell in the caution zone, it might be better to go for the cutoff time which provides an immediate decision for most of the samples.

A third procedure can be adopted when the product is intended to be sterile or functionally sterile. In sterile products, all viable organisms should have been eliminated, so the detection of any organisms would trigger rejection of the product. It is necessary to develop media that would allow the growth and detection of any organisms that might contaminate the product and then to determine the maximum period needed to obtain detection time.

This will obviously be governed by the time taken for low levels of the slowest growing contaminant to trigger detection. Subsequently in routine testing any samples which fail to give a detection within this maximum period can be considered to be sterile. Because this period may be a fairly long one, it may be necessary to use pre-incubation to save valuable sample space in the impedance instrument, or to use a concentration step to increase the number of contaminants presented to the instrument and hence reduce the impedance detection times.

In the case of functional sterility, the organisms have not been completely eliminated from the products but have been reduced to such a level that there is no health risk to the consumer or no chance of spoilage within an established shelf life period. An impedance test can give early warning of contamination providing that there is a large difference in numbers of micro-organisms between acceptable samples and contaminated or spoiled samples. Again the method relies on the development of a suitable test protocol that will give detection of contaminated samples within a short period while not giving detection times for the low levels of the normal microflora.

3.2 Total Counts

Probably the most widely used application of electrical measurement methods is to generate detection times which correlate to traditional plate counts and can be used in the same way to give some measure of the total number of micro-organisms present in a sample of food. Protocols have been developed for a great variety of food products – beverages, cereals, chilled foods, confectionery, dairy products, fish, frozen foods, meat, milk, pasta, spices, herbs and vegetables.

The first stage in any method of enumeration is sample preparation. A known weight or aliquot of sample is taken and typically, in a standard plate count method, dilutions of the sample are made in a diluent such as 0.1 per cent peptone or saline. These dilutions would then be pipetted into petri dishes, and molten medium tempered to 45 °C added, mixed and allowed to set; or the dilution would be surface plated directly on to solid medium in petri dishes. These petri dishes (hundreds of them!) would then be incubated at a selected temperature until visible colonies appeared which could be counted. Often for electrical measurements it is possible to make a single dilution of the sample directly in the growth medium and to inoculate the empty test cell.

Typically a 10^{-1} dilution of the food product is made in the test medium, which is often a commercially available medium such as tryptone soya broth or brain heart infusion, with perhaps minor modifications to improve curve quality. In some cases, a combination of liquid inoculum on to a solid medium is used. This technique has been used to minimize base drift with particulate or fatty samples which can cause interference with the electrical signal in liquid media. If the initial stabilization period in the instrument is excessive there is more chance of a baseline drift, and therefore less chance of a clean acceleration which would trigger a detection by the computer algorithm.

The adoption of a pre-incubation stage before inoculating the electrical test cells may be beneficial, although this does introduce another handling procedure which is undesirable. However, pre-incubation obviously makes sense when very low initial numbers of organisms and hence long detection times are expected. In this case, valuable test space in the measurement instrument is optimized by carrying out part of the incubation of samples prior to introduction to the instrument.

Another point to consider is the effect of the food product on the characteristics of the test medium. The medium may have been developed using pure cultures of the most relevant micro-organisms. The addition of the food sample may introduce new substrates which the micro-organisms may metabolize in a different way, or the food material itself may directly influence the electrical response signal. Obviously the less dilute the sample homogenate, the greater any such effect is likely to be. Therefore in the development of a medium the likely food samples and the level of addition of such foods should be considered along with the composition of the actual medium.

The temperature of incubation is a critical parameter, and it should not be assumed that the temperature used for the corresponding plate count will be suitable. As discussed earlier, the relative generation times of the normal microflora will need to be considered. It is often advisable to temper the medium to the incubation temperature before the addition of the product sample, as this can help to reduce the stabilization period.

In addition to the incubation temperature, the degree of aeration of the medium may be important. Some micro-organisms require aerobic conditions whilst others are strictly anaerobic or micro-aerophilic. With large volume test cells of 10 ml or more this may not be a problem, but with test cells of 1 ml or 2 ml capacity it may be necessary to overlay with liquid paraffin or agar, after inoculation, to produce anaerobic conditions. Vitek Systems Ltd have developed an impedance medium for anaerobic organisms which uses an oxygen reducing membrane fraction. They claim that this will produce conditions suitable for obligate anaerobes without the necessity of placing the instrument's incubator in an anaerobic cabinet.

Either the sample dilution, mixed or homogenized, is inoculated directly into the electrical measurement cell, or further dilutions are made before inoculation. It is necessary to prepare further dilutions when a high level of micro-organisms is known or suspected to be present in the sample to prevent an immediate detection, which may be missed by the instrument if the initial level exceeds the impedance detection threshold (usually around 10^6 cfu/ml). Dilutions must also be prepared with products such as cocoa, spices and herbs which are known to contain antimicrobial substances, so that these will be diluted sufficiently to render them ineffective. However, this will also dilute the level of organisms introduced to the instrument, so there may be a tradeoff in terms of prolonged detection times and sensitivity.

3.3 Selective Procedures

In conventional microbiology, certain groups of micro-organisms are used as indicator organisms. For instance, coliforms and Enterobacteriaceae are regarded as indicators of hygiene during the manufacture of food products.

Their presence is taken sometimes as evidence of direct potentially hazardous contamination and sometimes as indicators of post-process contamination, as they are not particularly heat resistant organisms. Faecal streptococci have been used in a similar way, and these groups of organisms are routinely tested for in QC laboratories by using selective media. In addition to these indicator groups of micro-organisms, many raw materials and products are tested directly for the presence of food poisoning organisms such as *Salmonella*, *Staphylococcus aureus* and *Listeria* species.

Protocols have also been developed for electrical measurement of many groups of organisms by using selective media which favour the growth of the selected organisms or inhibit the growth of competitors. Table 15.1 gives an indication of the range of organisms and variety of food products for which these protocols have been developed.

With indicator organisms it is often necessary to obtain a count in order to assess the extent of the problem, be it contamination or a processing fault. In this case the procedures used for the development of a protocol are essentially the same as for total count determinations, except that the media are obviously tuned to favour the growth of the selected organisms. Calibration curves or distribution plots will be established for the specific organisms in specific food products and used to determine detection time criteria which can be applied to specification levels.

In cases where a count is not required, as in the detection of *Salmonella*, it is necessary to develop media which will either only allow the growth of the selected organisms or alter the characteristics of the detection curve in some way to enable discrimination. For example, Figure 15.4 shows the growth of six organisms in lysine decarboxylase broth using the Bactometer instrument. Two of the cultures, *Proteus mirabilis* and *Citrobacter freundii*, have not produced recognizable detection curves, but *Escherichia coli* and *Klebsiella pneumoniae* produced detection curves similar to the two salmonellas. Obviously this medium would need to be modified before it could be used for the detection of salmonellas.

A possible modification might be the addition of sodium biselenite, which is used as a selective agent in conventional media for the detection of salmonellas. Figure 15.5 shows the effect of adding differing concentrations of sodium biselenite to lysine decarboxylase broth on the detection curves produced by *Escherichia coli*. It can be seen that increasing concentrations of sodium biselenite reduced the quality of the curve, and at 0.06 and 0.08 per cent sodium biselenite no discernible curve was produced.

In this manner various additions can be made to a basal medium so that only the selected organisms give a detection time or produce a detection curve which meets established criteria such as rate and magnitude. An interesting technique described by Pugh et al. (1988) was the addition of bacteriophage specific to salmonellas, Felix 01, to produce a delay in detection time in dulcitol broth (Figure 15.6).

Because the curve criteria can be evaluated at the selective enrichment stage there is a considerable time saving over conventional methods for the detection of salmonellas, where it is necessary to grow visible colonies on solid selective media after selective enrichment. It is unlikely that any

Table 15.1 *Protocols for specific organisms in various food products*

Coliforms	
Cheese	Khayat et al. (1988)
Powdered dairy products	Fryer and Forde (1989)
Meat	Martins and Selby (1980)
Dairy products	Firstenberg-Eden et al. (1984)
Sewage treatment	Munoz and Silverman (1979)
Waste waters	Strauss et al. (1984)
Enterobacteriaceae	
Dairy products	Cousins and Marlatt (1989)
Staphylococci	
Cheese	Khayat et al. (1988)
Skimmed milk powder	Prentice and Neaves (1987)
Meat	Kahn and Firstenberg-Eden (1985)
Streptococci	
Skimmed milk powder	Neaves et al. (1988)
Lactic acid bacteria	
Fruit juices	Schaertel and Firstenberg-Eden (1985)
Beer	Evans (1985)
Pseudomonas	
Chilled foods	Banks et al. (1989)
Fish	Ogden (1986)
Salmonella	
General	Easter and Gibson (1985)
Confectionery	Arnott et al. (1988)
Milk powders	Prentice et al. (1988)
Yeasts/moulds	
Orange juice	Zindulis (1984)
Fruit yoghurt	Fleischer et al. (1984)
Cheese	Williams and Wood (1986)
Wine	Henschke and Thomas (1988)
Listeria	
Dairy products	Phillips and Griffiths (1989)

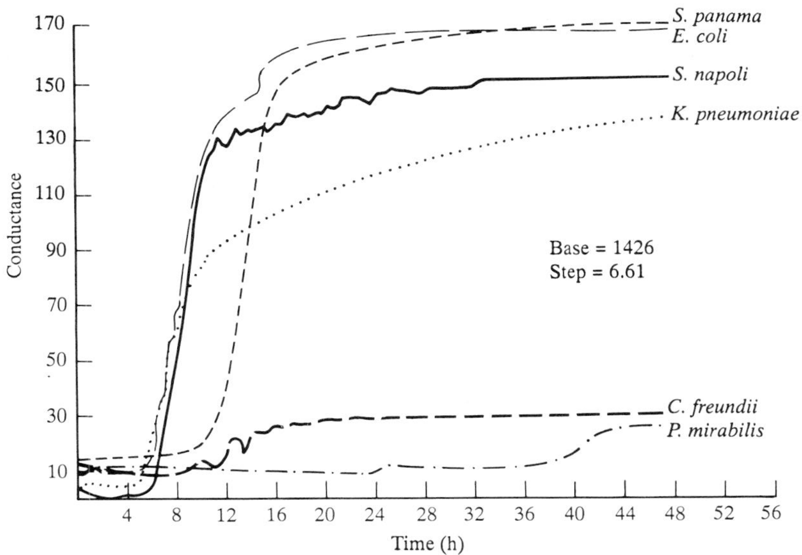

Figure 15.4 *Detection curves of pure cultures in lysine decarboxylase broth*

single medium would have the necessary selectivity and specificity but, by measuring the electrical changes associated with microbial growth in a number of media at the same time, a characteristic pattern can be obtained and used to identify organisms.

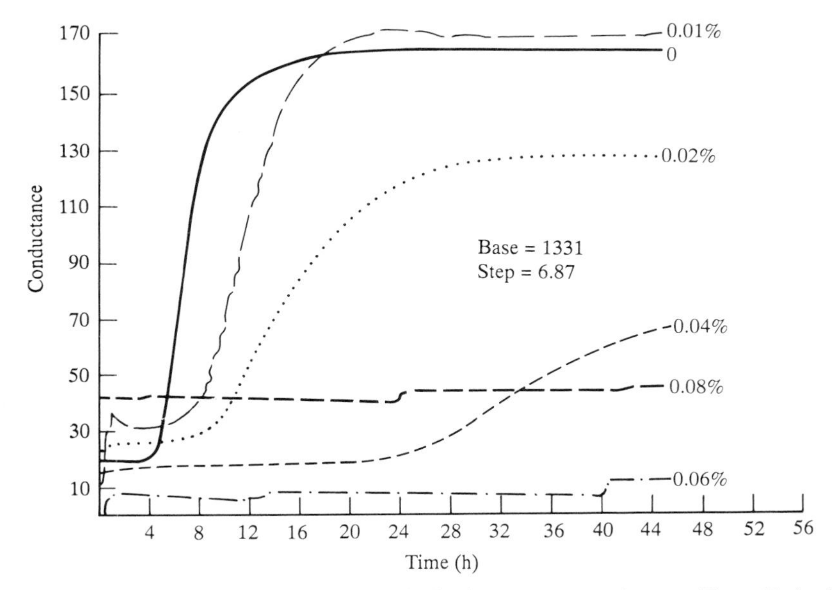

Figure 15.5 *Effect of sodium biselenite concentration on* E. coli *in lysine decarboxylase*

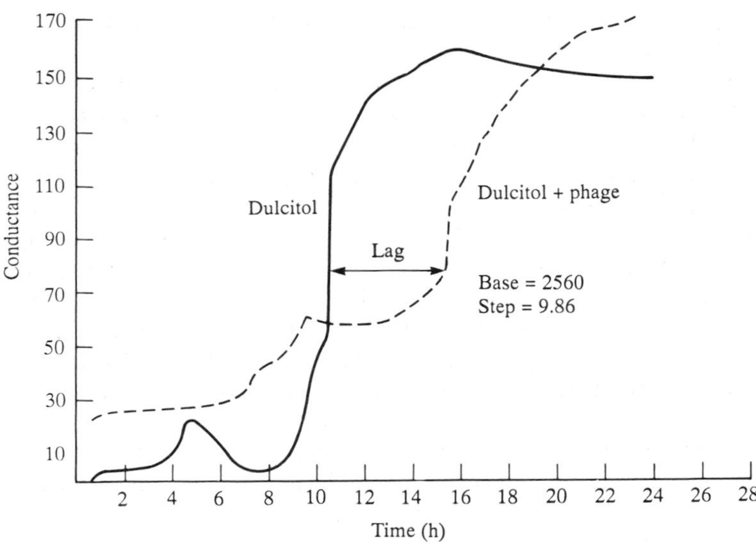

***Figure 15.6** Effect of 01 bacteriophage on* Salmonella thompson *in dulcitol broth*

Protocols for the detection of specific organisms often include a non-selective pre-enrichment stage in order to allow injured organisms a period of resuscitation before introducing them to selective media. There are two points to consider here. First, the introduction of the pre-enrichment inoculum to the selective medium must not upset the substrate balance of the selective medium, either by the carry-over of food particles or by components of the pre-enrichment medium itself. If this is the case then it may be necessary to carry out a dilution of the pre-enrichment before inoculation of the selective medium so as not to impair the production of good quality detection curves. The second point to consider is whether greater use can be made of the pre-enrichment medium as a conditioning medium, for instance to induce enzyme systems in organisms by incorporating substrates in the pre-enrichment medium which the organisms will subsequently encounter in the selective medium. This may give shorter detection times by reducing the stabilization period of the selective media.

3.4 Shelf Life Prediction

Traditionally, prediction of shelf life has been based on microbial counts on finished product and practical experience of storage conditions. The microbiologist tries to assess the likelihood of the growth of spoilage organisms and the likely fluctuations in the storage conditions. With the trend in the food industry to higher volume production, wider distribution and increasing use of chill distribution, the prediction of product shelf life has become increasingly important.

Hankin and Stephens (1972) amongst others have reported the dubious value of initial bacterial counts as predictors of the keeping quality of perishable foods. Because impedance microbiology methods measure metabolic activity as a function of the total organisms present in the sample, they are likely to be much more accurate predictors of shelf life than plate counts. Bishop et al. (1984) investigated potential predictors of shelf life for pasteurized whole milk samples, and found that an impedance method gave a correlation with the shelf life with a coefficient of 0.88, compared with the best plate count method which had a correlation coefficient of −0.67. Gibson (1985) developed a system for predicting the shelf life of chilled packaged fish. He demonstrated that detection times using conductance measurement correlated well with other spoilage parameters including chemical and sensory data. Protocols would have to be developed which gave the shortest detection times for potential spoilage organisms in specific products. Often in shelf life studies the initial numbers of organisms are very low and therefore they may take some time to grow to detection threshold levels; there may be some benefit in adopting a pre-incubation stage before transferring the sample cultures to the impedance instrument. Other tests could be run in the instrument while the pre-incubation of the shelf life samples was taking place in conventional incubators.

4 Choice of Instrumentation

There are three commercially available instruments for the detection of micro-organisms by electrical measurement: the Bactometer (Vitek Systems Ltd), the Malthus (Malthus Instruments Ltd) and the RABIT (Don Whitley Scientific Ltd). All three instruments measure the change, caused by the metabolism of micro-organisms, in the electrical impedance of a growth medium under a small applied voltage. They are all computer controlled modular systems which can be expanded as required for up to 512 test cells in the case of the Bactometer and the RABIT and up to 256 test cells in the Malthus. There have been considerable improvements in the computer systems and software, and all offer continuous monitoring of sample with current status information on video display units with a variety of colour graphics packages. The test data can be displayed as colour coded detection times or as detection curves, and the analysis and comparison of data, including regression analysis, can be carried out within the system. The computers can be interfaced to other computers so that data can be displayed in production areas as well as in the laboratory. This enables production management to be immediately aware of test results and provides faster clearance of acceptable samples.

The three systems do have certain differences, and one major difference is the type of test cell used. The Bactometer employs a sixteen-well plastic disposable module with stainless steel electrodes, and each well has a capacity of 2 ml. They are supplied sterile, used once and then discarded; thus there are no problems associated with washing and resterilization of the test cells, but there is an ongoing disposables cost. The Malthus offers reusable glass test tubes and bottles of 1 ml, 2 ml, 10 ml or 100 ml capacity, with platinum electrodes on a ceramic strip. The manufacturers claim that the sample cells, when handled correctly, are reusable hundreds of times, but they do of course require cleaning and sterilizing between uses. The RABIT also uses reusable glass cells of 2–10 ml capacity with stainless steel electrodes embedded in a plug which forms one end seal of the test cell.

The Bactometer uses air incubators with temperatures controlled to ± 0.1 °C; the Malthus uses water baths to ± 0.002 °C, and the RABIT uses alloy blocks controlled to ± 0.005 °C. Bactometer users have the facility to select impedance, conductance or capacitance as the electrical variable to be measured. The Malthus is specifically designed to measure conductance, and the RABIT measures admittance.

All three systems offer the considerable advantages over conventional methodology of rapidity, automation, accuracy, versatility and cost effectiveness. Impedance microbiology can provide an overall reduction in elapsed test time of at least 50 per cent. The ability to rapidly identify unacceptable production can significantly minimize downtime on production plant and allow fast corrective action. Finished product stock holding can be reduced considerably, with positive clearance giving greater confidence as well as cost savings. More samples can be tested without the expense of increasing staff, thus providing a better control of product quality.

5 Future Developments

Electrical methods have gained a wider acceptance in QC laboratories than any of the other alternative rapid methods. A major factor in this success is the versatility of the technique. Almost every test that is carried out in QC laboratories by conventional methods can be replaced by an electrical measurement method, and in many cases at a lower cost per test.

In the future it should be possible to develop protocols for any chosen organism. At the moment the large sample capacity and automation of the instruments lend them to screening tests, with confirmation for specific organisms being carried out subsequently by some other test such as enzyme linked immunosorbent assays. If the electrical detection sensitivity of the systems could be improved then tests based on selective media impedance monitoring could be directly linked to provide identification and confirmation of specific organisms simultaneously.

At present some use is made of impedance microbiology for production control, particularly in the dairy industry for checking cheese process culture activity. It is likely that, in the future, measurements will be made on-line through the use of sensors, such as biosensors, chemical sensors or immunosensors. These may take the form of miniaturized self-contained instruments on the production site or measurement cells which transmit the information back to central computers. It would be possible to monitor the progress of any food fermentation process and identify the presence of undesirable contaminants immediately. It should also be possible to detect the presence of indicator organisms and perhaps even food poisoning organisms by direct measurement systems.

It could be argued that some of the measurement systems envisaged, such as miniaturized biosensors, are not directly comparable with the electrical measurement systems described here. However, it is true that impedance measurement microbiology has given tremendous impetus to the development and implementation of rapid methods, which provide a valid alternative to traditional methods in food quality control laboratories and will continue to be of value in the QC laboratory for some time to come.

References

Arnott, M. L., Gutteridge, C. S., Pugh, S. J. and Griffiths, J. L. (1988) Detection of salmonellas in confectionery products by conductance. *Journal of Applied Bacteriology*, **64**, 409–420

Banks, J. G., Rossiter, L. M. and Clark, A. E. (1989) Selective detection of Pseudomonas in foods by a conductance technique. In *Rapid Methods and Automation in Microbiology and Immunology*, eds A. Ballows, R. C. Tilton and A. Turano, Brixia Academic, Brescia, Italy, 725–727

Bishop, J. R., White, C. H. and Firstenberg-Eden, R. (1984) A rapid impedimetric method for determining the potential shelf life of pasteurized whole milk. *Journal of Food Protection*, **47**, 471–475

Cousins, D. L. and Marlatt, F. (1989) The use of conductance microbiology to monitor Enterobacteriaceae. *Dairy, Food and Environmental Sanitation*, **9**, (10), 599

Easter, M. C. and Gibson, D. M. (1985) Rapid and automated detection of salmonella by electrical measurements. *Journal of Hygiene (Cambridge)*, **94**, 245–262

Easter, M. C., Gibson, D. M. and Ward, F. B. (1982) A conductance method for the assay and study of trimethylamine oxide reduction. *Journal of Applied Bacteriology,* **52,** 357–365

Evans, H. A. V. (1985) A note on the use of conductimetry in brewery microbiological control. *Food Microbiology,* **2,** 19–22

Firstenberg-Eden, R. and Eden, G. (1984) *Impedance Microbiology.* Research Studies Press, Letchworth, UK

Firstenberg-Eden, R., Van Sise, M. L., Zindulis, J. and Kahn, P. (1984) Impedimetric estimation of coliforms in dairy products. *Journal of Food Science,* **49,** 1449–1452

Fleischer, M., Shapton, N. and Cooper, P. J. (1984) Estimation of yeast numbers in fruit mix for yoghurt: comparison of impedance and its components using the Bactometer 32 and M120B. *Journal of the Society of Dairy Technology,* **37,** (2), 63–65

Fryer, S. M. and Forde, K. (1989) Impedimetric procedures for estimating total bacterial numbers and coliforms in powdered dairy products. *Journal of the Society of Dairy Technology,* **42,** (3), 88–89

Gibson, D. M. (1985) Predicting the shelf life of packaged fish from conductance measurements. *Journal of Applied Bacteriology,* **58,** 465–470

Hankin, L. and Stephens, G. R. (1972) What tests usefully predict keeping quality of perishable foods? *Journal of Milk and Food Technology,* **35,** 574–576

Henschke, P. A. and Thomas, D. S. (1988) Detection of wine spoiling yeasts by electronic methods. *Journal of Applied Bacteriology,* **64,** 123–133

ICMSF (International Commission on Microbiological Specifications for Foods) (1974) *Micro-organisms in Foods. Vol. 2: Sampling for Microbiological Analysis: Principles and Specific Applications,* 1st edn. University of Toronto Press, Toronto

Kahn, P. and Firstenberg-Eden, R. (1985) An impedimetric method for the estimation of *Staphylococcus aureus* concentrations in raw ground beef. Paper presented at Institute of Food Technologists Annual Meeting

Khayat, F. A., Bruhn, J. C. and Richardson, G. H. (1988) A survey of coliforms and *Staphylococcus aureus* in cheese using impedimetric and plate count methods. *Journal of Food Protection,* **51,** (1), 53–55

Martins, S. B. and Selby, M. J. (1980) Evaluation of a rapid method for the quantitative estimation of coliforms in meat by impedimetric procedures. *Applied and Environmental Microbiology,* **39,** (3), 518–524

Munoz, E. F. and Silverman, M. P. (1979) Rapid, single-step most probable number method for enumerating fecal coliforms in effluents from sewage treatment plants. *Applied and Environmental Microbiology,* **37,** (3), 527–530

Neaves, P., Waddell, M. J. and Prentice, G. A. (1988) A medium for the detection of Lancefield Group D cocci in skimmed milk powder by measurement of conductivity changes. *Journal of Applied Bacteriology,* **65,** 437–448

Ogden, I. D. (1986) Use of conductance methods to predict bacterial counts in fish. *Journal of Applied Bacteriology*, **61**, 263–268

Phillips, J. D. and Griffiths, M. W. (1989) An electrical method for detecting *Listeria* spp. *Letters in Applied Microbiology*, **9**, 129–132

Prentice, G. A. and Neaves, P. (1987) Detection of *Staphylococcus aureus* in skimmed milk powder using the Malthus 128H microbiological growth analyser. In *Rapid Methods and Automation in Microbiology and Immunology*, Programme of the Fifth International Symposium on Rapid Methods and Automation in Microbiology and Immunology, Florence 4–6 November 1987, Tipolitografia Queriniana, Brescia, Italy

Prentice, G. A., Neaves, P., Jervis, D. I. and Easter, M. C. (1988) An interlaboratory evaluation of an electrometric method for detection of salmonellas in milk powder. In *Rapid Methods for Foods, Beverages and Pharmaceuticals*, Society for Applied Bacteriology technical series 25, Academic, London

Schaertel, B. J. and Firstenberg-Eden, R. (1985) A medium for the growth and impedimetric measurement of lactic acid bacteria in fruit juices. Paper presented at Institute of Food Technologists Annual Meeting

Strauss, W. M., Malaney, G. W. and Tanner, R. D. (1984) The impedance method for monitoring total coliforms in waste waters. Part 1: Background and methodology. *Folia Microbiology*, **29**, 162–169

Williams, A. P. and Wood, J. M. (1986) Impedimetric estimation of moulds. In *Methods for the Mycological Examination of Food*, eds A. D. King, J. I. Pitt, C. R. Beuchat and J. E. L. Corry, Proceedings of the NATO Advanced Research Workshop, Boston, July 1984, Plenum, Chapter 7, 230–238

Zindulis, J. (1984) A medium for the impedimetric detection of yeasts in foods. *Food Microbiology*, **1**, (3), 159–167

Part III

New Sensors for Applications in the Food Industry

16 *The Marker Concept: Frying Oil Monitor and Meat Freshness Sensor*

Erika Kress-Rogers

Contents

1 Introduction to the Concept and Overview on Applications

Improved methods for the assessment of food conditions such as freshness, or conversely of changes due to microbial spoilage, oxidative rancidity or heat-induced degradation, have been high on the priority list of food manufacturers for many years. Food products must be tested and their shelf life predicted correctly to ensure that they will reach the consumer in good condition. Food ingredients need to be assessed prior to processing, so that the resulting product will be enjoyable and wholesome for the duration of the specified shelf life. This applies not only to industrial food processing, but also to the larger institutional catering establishments or chains of smaller catering outlets.

Conventional methods for the determination of microbial load require incubation, and results often become available when the food is already well into its shelf life or perhaps even consumed. Rapid methods in microbiology have been developed which can now provide answers within a day, or in some cases within an hour (see Chapters 14 and 15), but these are still not always rapid enough for all desirable applications, and they are usually confined to the quality control laboratory. If instead a chemical or physical property that is linked to the progress of deterioration can be identified as an indicator, one can then examine the possibility of developing a sensor for this indicator or marker property that will give a result in minutes or seconds.

An indicator long established for the assessment of fish freshness is a low trimethylamine content, but this has been measured with conventional laboratory assays (Bullard and Collins 1980). A biosensor development for this application is complicated by the presence of dimethylamine (Askar 1976) which is not a specific indicator of deterioration. A fish freshness indicator system proposed more recently for a biosensor measurement is based on the decomposition of ATP into the purines inosine and hypoxanthine (Watanabe et al. 1983), and this has since been further developed (see for example Suzuki et al. 1989) particularly for fish to be eaten raw.

In dark-fleshed fish such as tuna or herring, bacterial decomposition of histidine (which occurs particularly in warm ambient environments) produces histamine. This is a heat-resistant indicator of past bacterial activity and also constitutes a toxin in scombroid poisoning. (Some data on concentration ranges and toxicity thresholds and further references on amines and purines in fish can be found in Kress-Rogers 1985a, Sections AIII.8 and AIII.7.)

In the deterioration of meat, as for fish, amines are produced by microbial decomposition of amino acids (Wortberg and Woller 1982), but when these have reached a level that can readily be detected, a predictive test is no longer applicable. Other indicators for fish freshness have been the subject of much research and development in Japan (see for example Karube and Tamiya 1987).

Meat has, in general, a higher glucose level than fish, and glucose rather than amino acids are decomposed at the pre-spoilage stage of microbial growth. This glucose decomposition is the basis of the probe development for the assessment of chilled meat freshness described in Section 3.

In the assessment of frying oil, some simple, rapid tests are available in addition to the more sophisticated and time-consuming laboratory methods. These simple tests are, however, strongly dependent on oil type, fried food type and frying conditions. Also, they require a sample to be taken and cooled first, and some of the simple tests require reagents.

A sensor for frying oil quality that can be used *in situ* in the hot oil, that can give quasi-instant answers or can be configured for continuous in-line operation, and that requires no reagents, has now been developed. This measures the viscosity increase of the oil that accompanies the oxidation and polymerization on frying-induced deterioration. The research and development resulting in the first generation of prototypes are described and the evaluation results are discussed in Section 2 of this chapter. Further development of the probe towards the commercial prototype is now in progress and is briefly reported at the end of Section 2 and on page 714.

The marker approach for the development of novel instruments for the rapid assessment of complex food conditions has great potential, particularly in the determination of food freshness. It can offer unique solutions to the problems of determining the food status with respect to microbial or oxidative spoilage. However, a substantial initial research effort is needed to identify valid indicators and probe designs suitable for their measurement, and then to examine the applicability to different variants of the food type under consideration.

This effort is worthwhile in offering the possibility of acquiring information that was previously inaccessible, rather than just a more convenient way of acquiring and processing data that have been accessible to routine measurement for many decades. The latter approach has led to the appearance of many 'new' instruments on the market in recent years that provide a user-friendly and convenient way of performing a conventional measurement. The marker approach, on the other hand, offers the possibility of assessing food conditions such as microbial or oxidative deterioration with a rapid measurement in place of a lengthy or complex procedure.

This chapter describes primarily the nature and scope of the development of the concept, the feasibility assessment, the establishment of specifications and the subsequent adaptation of sensors intended initially for other industries or the medical sector; the evaluation for the novel application is then discussed. For information on the general principles underlying the probe design, the reader will be referred to other chapters or articles. Beyond the two examples described here, many more applications of the marker approach have been suggested, investigated and developed to different stages. Table 16.1 lists some applications of the marker approach.

A study on the feasibility of a non-destructive measurement of oxidative rancidity has been carried out. A preliminary GC-MS study of the volatile compounds present in the headspace of lard, butter oil and potato flakes

Table 16.1 *Some applications of the marker approach to the assessment of food conditions* (adapted from Kress-Rogers 1985a)

Indicator	*Condition to be assessed*	*Other significance of the indicator*
Measurement with a mechanical resonance probe		
Viscosity	Frying oil quality *in situ* in hot oil	Rheology
Measurements with biosensors in an aqueous phase		
Glucose profile	Meat freshness (pre-spoilage stage)	Sugar contamination of effluent
	Browning potential (Maillard reaction)	Progress of 'sugar inversion' (sucrose→glucose)
Ethanol	Yeast contamination	Progress of intended fermentation
Penicillin	Mastitis treatment	Inhibitor of dairy fermentation
Amines:		
Trimethylamine	Loss of freshness (fish)	Toxin, allergen
Histamine, tyramine	Past microbial activity (meat, fish, cheese)	
Cadaverine, putrescine	Microbial spoilage (advanced)	
Purines: ATP, inosine, hypoxanthine	Fish freshness (for very fresh fish)	
Measurements with sensors for gases and volatiles (in headspace or through membrane)		
Ethanol	Yeast contamination	Progress of intended fermentation
Aldehydes: pentanal, hexanal	Oxidative rancidity	
Hydrogen, a range of volatiles	Microbial spoilage of CAP and vacuum-packed meat	
Ethylene	Fruit ripening	

While many of the entries in this list are similar to those in Table III of the earlier survey on novel sensors and their applications (Kress-Rogers 1985a), one has been replaced and several have been added. Viscosity is now listed in place of free fatty acids as an indicator of frying oil quality in accordance with the results of the study described in Section 2. Pentanal and hexanal have been identified as indicators of oxidative rancidity for oils, fats and products containing fat (see Section 1). Hydrogen, measured non-intrusively by placing a sensor on to the covering film of modified atmosphere packaged (MAP) meat, has been observed to evolve as spoilage progressed, although this study is at a very preliminary stage (see Section 1). Glucose as an indicator of meat freshness has remained in the list, because the glucose profile as a function of depth has been shown to be a good marker of meat freshness (see Section 3).

during storage suggested that compounds of the aldehyde group could be useful indicators of oxidative rancidity. If the relevant compounds could be detected with a biosensor in contact with headspace volatiles, then a continuous monitoring of food products during storage would be possible (Kress-Rogers and D'Costa 1986).

A more detailed GC-MS study was then carried out at the Leatherhead Food Research Association (LFRA) and identified hexanal and pentanal as the most appropriate indicators (Dagnall et al. 1989). Routes to the development of a biosensor for these aldehydes were investigated in an experimental study by E. D'Costa and A. P. F. Turner at Cranfield Biotechnology Centre. A prototype development for the relevant aldehydes would have exceeded the scope of that project owing to the difficulties presented by the low solubility of hexanal and pentanal in aqueous solution and the unavailability of a suitable commercial enzyme preparation. Since these analytes were not of interest in clinical applications, they had not been the subject of any prior biosensor development work on which an adaptation could have been based. However, further examination of routes based on the biosensor designs currently available could bring this work to fruition.

This illustrates the potential of technology transfer, that is the adaptation of a sensor that has already been developed for another sector. The development of a sensor from basic principles for dedicated food industry applications alone is often too expensive. It may pay, therefore, to take a lateral approach and look for a matching pair of a useful indicator and a current sensor in another sector, rather than follow the straight and narrow path of identifying *the* optimum indicator and then setting out to develop a sensor for this. The suitability of different biosensor designs for adaptation to food industry applications, and alternative approaches to full biosensor optimization, are discussed in Chapter 17.

A highly sensitive microelectronic gas sensor (a CHEMFET; see Chapter 17 for principles) for hydrogen and a small range of other gases has been employed in a further feasibility study for an instrument based on the marker approach. Hydrogen (together with carbon dioxide) is produced by the decomposition of glucose (via lactate) by the microbial flora on meat and fish packed under anaerobic conditions as they are encountered in vacuum packaging or in controlled atmosphere packaging (CAP). Molecular hydrogen evolved by oxygen-limited cultures of *Escherichia coli* had earlier been observed with a CHEMFET gas sensor by Cleland, Hoernsten, Elwing, Enfors and Lundstroem (1984) and by Hoernsten, Danielsson, Elwing and Lundstroem (1986). (See also Danielsson and Winquist 1989.)

In a preliminary study, a CHEMFET provided by Thorn EMI was used in the original configuration intended for other applications. The sensor was placed on to the covering film of CAP meat and fish without damaging the pack in any way. This study showed that hydrogen permeating through the high gas barrier film could indeed be observed in the advanced stages of deterioration and increased with storage. The signal was weaker for fish steaks than for pork steaks under the same conditions, as would be expected from the lower glucose content of fish. A high response was observed for whole gutted mullet, consistent with a higher Enterobacter count in the gut region (Kress-Rogers 1987).

The sensing head was then adapted to provide a higher response amplitude under the relevant conditions. In a brief subsequent study on commercially packed CAP meat, this was shown to yield a response observable on the sell-by date and increasing thereafter. Much work still remains to be done to establish whether the hydrogen production attributed to Enterobacter in the flora will be encountered in all relevant samples. Some literature data on hydrogen production by bacteria were located, but these had been taken with conventional instruments of poor sensitivity compared with the CHEMFET. Further experimental studies are therefore required to resolve this question. In the case of a positive outcome, a non-destructive rapid method of assessing vacuum and CAP meat (and possibly other foods packed under such conditions) could result.

2 A Novel *In situ* Monitor for Frying Oil

2.1 The Need for a New Probe

Deep-fat frying is widely used in catering establishments, in institutional catering in hospitals for example, and also in the industrial production of snack foods such as potato crisps (chips to American readers) and of foods such as prepared frozen potato chips (French fries in the USA) or fish cakes for both domestic and catering use. Fried products in catering include chips (French fries), fish or meat products coated in batter and/or breading, doughnuts and fritters (Weiss 1983; Berger 1984).

With an increase in health awareness, there is a trend towards the use of polyunsaturated oils in preference to the traditional frying fats. These oils are less stable than the traditional frying media and need to be assessed more frequently. In the frying-induced deterioration of oils and fats, the component triglycerides are hydrolyzed, liberating free fatty acids. Unsaturated fatty acid chains, either free or in the form of triglycerides, are oxidized and polymerized. A number of undesirable changes occur in the advanced deterioration of frying oil (Gillatt et al. 1991):

(a) The free fatty acid (FFA) content increases, leading to off-flavours.
(b) Hydroperoxides are formed and then decompose with the formation of volatiles including aldehydes, alcohols and ketones. Initially, the formation of the aldehydes is desirable in imparting a pleasant fried

food flavour. In fact, food fried in very fresh oil will be considered as bland by many consumers. However, when the oil is used for too long, then off-flavours are formed, and these are unpleasant both for the frying operator and for the consumer.

(c) Non-volatile breakdown products are formed, mainly polymerized and oxidized fatty material (POM). Cyclic compounds are found in deteriorated oils and these are nutritionally undesirable.

(d) The smoke point decreases, rendering the frying process unpleasant for the operator.

(e) The viscosity increases, leading to a higher oil uptake into the food.

In order to produce a wholesome, enjoyable food product, it is therefore essential to maintain the quality of the frying oil or fat. Often, the operators will dispose of the oil when it is still in good condition, because they have no access to laboratory facilities for oil assessment. Simple tests such as that for FFA content or for colour indication values can give results varying by several hundred per cent for different oils of the same POM content, even when the fried food type and the frying conditions are identical. This is borne out by results from the trial described in Sections 2.5 and 2.6, shown here for illustration (Figure 16.1).

Using the same limit value for FFA content, a medium such as partially hydrogenated vegetable oil containing a silicone additive would have been discarded at a lower POM content than palm olein under the frying conditions in our trial (which were akin to those in a catering establishment rather than in a continuous operation). Sunflower oil would have been badly deteriorated when reaching one of the commonly used limit values for FFA content.

The development of off-flavours in our trial (described below) was correlated with POM content rather than FFA content or colour values. There will, however, be cases where a high FFA value in itself is a reason to discard an oil, for example when the smoke point is decreased to the frying temperature. This is typically 170 to 190 °C, but can extend over the range 130 to 250 °C for certain foods and frying equipment types (Weiss 1983; Berger 1984).

Oil is an expensive component in the fried food, and early disposal can increase costs considerably. Later disposal on the other hand will adversely affect the quality of the fried food product. In catering establishments, there is therefore a requirement for a simple, rapid method that has a much lesser dependence on the frying oil type, fried food type and frying conditions than current simple tests.

Conditions in Batch and Continuous Frying

For products that will be stored for several months in the freezer (by the distributor, consumer or catering establishment) prior to use, the reduction

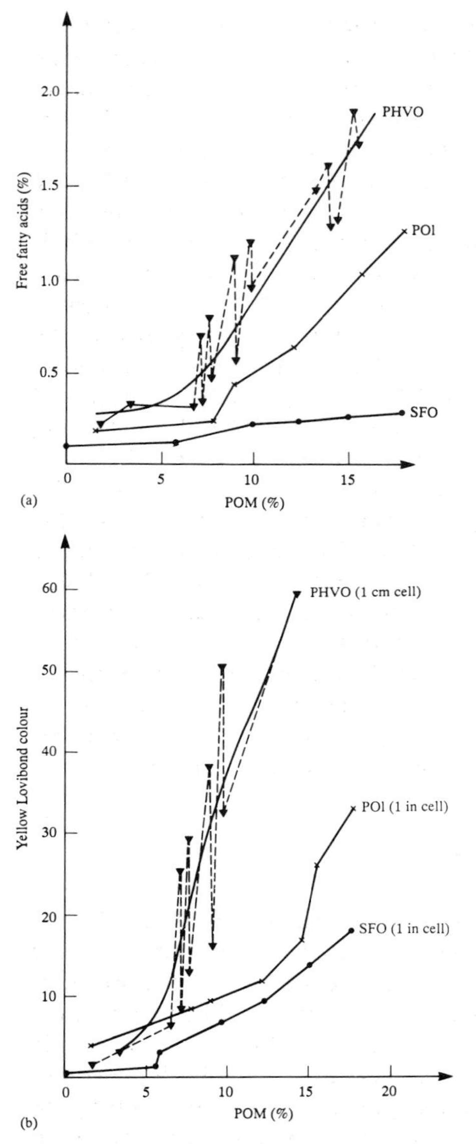

***Figure 16.1** Simple indicators as a function of polymerized and oxidized material (POM) content for palm olein (POl), sunflower seed oil (SFO), partially hydrogenated vegetable oil (PHVO; longer-term study, values for fryers 1 and 2 averaged). For potato chips fried in the oils, off-flavours were noted when 15 ± 0.5 per cent POM had been reached; for less than ~4 per cent POM, chips fried in PHVO were considered as bland. (a) Free fatty acid content (b) Yellow Lovibond colour: 1 in cells for POl and SFO, but 1 cm cell for PHVO (Kress-Rogers et al. 1990a; 1990b)*

in shelf life of a food prepared with oil of inadequate quality will cause complaints and wastage. With some high fat frozen foods, most complaints are related to fat rancidity (Rossell, 1992). An increased fat oxidation in foods can be observed at temperatures below the freezing point of water due to the freeze concentration of reactants (Ranken, 1989). For industrial fryers, an in-line monitor of frying quality would facilitate the maintenance of a consistent oil quality. The trial described below was, however, designed primarily with the examination of catering applications in mind.

For an extension to continuous industrial fryers, further trials under differing frying conditions would be carried out. These would include continuous frying for the period of a longer heating cycle, topping up the oil at regular intervals, and the choice of a selection of oils typically used in industrial frying operations. The partially hydrogenated vegetable oil (PHVO) used in this trial contained a silicone additive. This allowed frying for a period of over five times that of the other two oils in the trial. Towards the end of this long frying life, at a POM content of 10–15 per cent, the production of fried potato chips of good flavour was still possible with this oil. However, it would have been unsuitable for the industrial frying of products with a long shelf life on account of the elevated FFA level. However, the silicone additive is less likely to be used in industrial fryers in any case, since its protective effect is due to the formation of a surface barrier to oxidation that is useful only in discontinuous batch frying in catering establishments (Berger 1984). The other two oils were below the limit value of 1 per cent FFA which is widely used. All three oils were well below the limit values given for certain applications (Rossell 1986).

2.2 Development of the Probe Concept

Both simple tests and sophisticated laboratory methods (Table 16.2) were examined for the feasibility of developing them into a probe meeting the specifications above (Kress-Rogers 1988). Of the properties determined in the chemical laboratory, hydroperoxide content was considered for determination with a biosensor that could have been developed. However, this was then ruled out because the peroxide content not only decreases in advanced deterioration after an initial rise, but can also change during the sampling and measurement period. (Hydroperoxides decompose at frying temperatures, making peroxide values an unreliable criterion of frying oil quality during use; Rossell, 1989; Robards, Kerr and Patsalides, 1988.) For other sophisticated chemical assays, the route to the development of a chemical sensor was less obvious.

Of the simple tests, FFA content and colour were ruled out on account of their known strong dependence on oil type, food type and/or frying conditions. Odour and taste, smoke point and foam persistence were considered impractical for sensor development. The dielectric constant of a cooled oil sample is measured in one commercial instrument, but there are indications that this may also be dependent on variables other than the deterioration of the oil, especially the water content, an important variable during the frying operation.

Viscosity was chosen for a detailed study, since we were aware of a study on samples collected from catering establishments representing a wide range of frying media, showing that the samples could be divided into acceptable and unacceptable frying media (using the content of polar compounds as reference method) by setting a cutoff value for the viscosity measured at 50 °C (Table 16.3 and Figure 16.2a). Earlier data on assorted fats used in bakery operations had also shown a correlation between the oil viscosity (of cooled samples) with an accepted indicator of deterioration, namely the non-urea-adducting fatty acid content (see Figure 16.2b). We were also aware of an *in situ* probe that GEC Marconi Research Centre had developed for hot boiler oil (Figure 16.3a, b). This had the advantages of a rugged, simple construction and *in situ* operation in hot oil.

2.3 Study to Assess Feasibility and Define Specifications

Data from earlier LFRA studies and from the literature were scanned to find viscosity data for fresh frying oils and fats and data for used frying

Table 16.2 *Indicators of frying oil quality* (Kress-Rogers 1988; Kress-Rogers et al. 1990a; 1990b)

Indicators determined by sophisticated assays		*Indicators determined by simple tests*
Polymeric triglycerides	Pol	Odour and taste
Polymerized and oxidized material	POM	Smoke point (SP)
Petroleum-ether-insoluble oxidized fatty acids	PIOFA	Foam persistence
Polymeric fatty acids	PFA	Free fatty acids (FFA) content
Non-urea-adducting fatty acids	NAF	Colour
Acid value	AV	Viscosity
Polar components	PC	Dielectric constant
Carbonyl value	CV	
New chemical compounds	NCC	
Iodine value	IV	
Anisidine value	AnV	
Peroxide value	PV	

Table 16.3 *Comparison of viscosity and polar components content for mixed used frying media from restaurants* (Kress-Rogers 1988; Kress-Rogers et al. 1990a; 1990b; data from Figure 16.2(a))

Oil assessment	*No. of samples*	*Indicator A: polar components, limit 27%*	*Indicator B: viscosity at 50 °C, limit 47 mPl*
Acceptable	24	Below limit	Below limit
Unacceptable	20	Above limit	Above limit
Doubtful	1	Up to 8% above limit	Up to 10% below limit
Doubtful	5	Up to 7% below limit	Up to 11% above limit

media combined with data on an indicator for their degradation. The search produced such data of viscosity and degradation indicator for seventeen specific frying oils or fats and also for assorted non-specified hydrogenated fats and for mixed frying media samples from restaurants. The data had been taken by twenty different laboratories, and these had used different oil temperatures for the viscosity measurement, and chosen a different indicator for the freshness or degradation.

Not all had given the limit value for their indicator that would determine the discard point for the oil. Limit values were therefore taken from other literature sources or legal requirements where applicable. The limit values used here to assess the data are given in Table 16.4. It was now possible to select the data for the two groups of interest, namely oils and fats that were (1) fresh and (2) at the limit of usability as defined by an indicator measurement. Over 100 samples fell into these two groups. These are shown in Table 16.5, together with some further data which illustrate that for severe degradation beyond the limit of usability, the viscosity increases further. Such a monotonic dependence of a prospective indicator variable is, of course, important to avoid misinterpretation.

Oil temperatures were from 19 to 100 °C. To compare the viscosity values, it was necessary to plot them in accordance with the expected temperature dependence of oil viscosity. This was taken to be the relationship followed by most Newtonian liquids, namely

$$\log \eta = b/T + c$$

where b and c are constants. The two groups, fresh frying media and frying media at the limit of usability, fell into two bands (Figure 16.4), and two lines (line 1 and line 2 on the figure) with a tentative extrapolation to higher

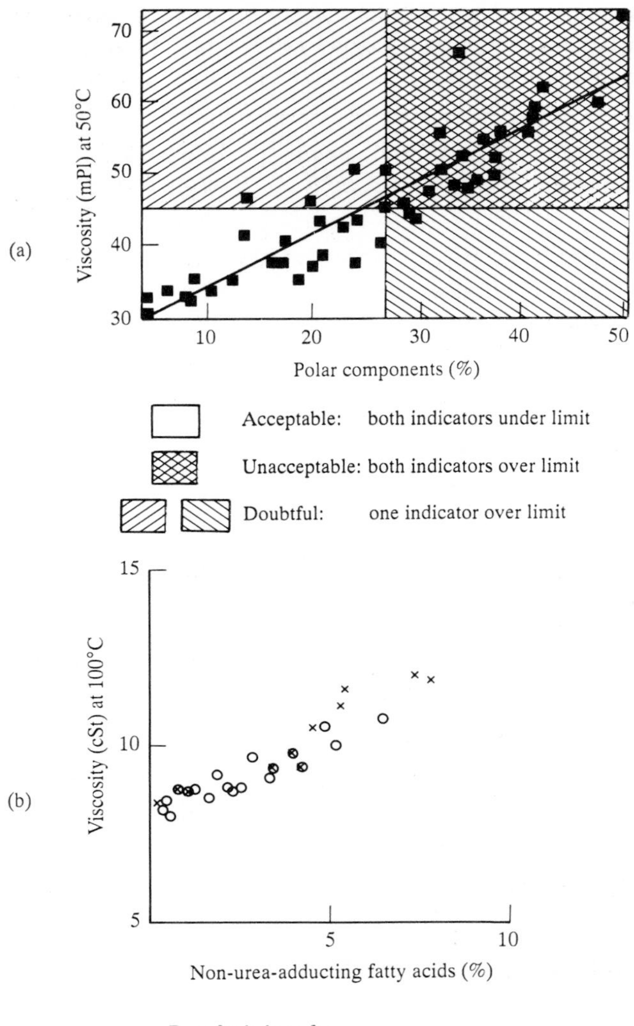

***Figure 16.2** (a) Viscosity and polar components used to indicate acceptability of mixed used frying oils from restaurants. (b) Viscosity as a function of the non-urea-adducting fatty acids content as an indicator of the deterioration of assorted frying fats from bakeries (Kress-Rogers 1988; Kress-Rogers et al. 1990a; data from Battaglia and Mitiska 1983 and from Rock and Roth 1966)*

temperatures were drawn through them. The extrapolation is certainly inaccurate owing to the diversity of the samples and the indicator variables used by the twenty laboratories. It is likely that there was also a varying accuracy of the methods used for the viscosity measurements, as suggested by the examination of dependence on oil type (see below). Also, in some cases, the oil temperature during the viscosity measurement was not unambiguously clear. Nevertheless, the extrapolation sufficed to provide specifications for a prototype sensor that could be used for a study at frying temperatures, namely having a range of approximately 1–10 mPl and a resolution of 0.5 mPl or better (mPl is the unit symbol for millipoiseuille: see notes to Table 16.5 and Appendix B to the book. (mPl = mPa s = cP)).

From the data extracted as above, it was also possible to give an

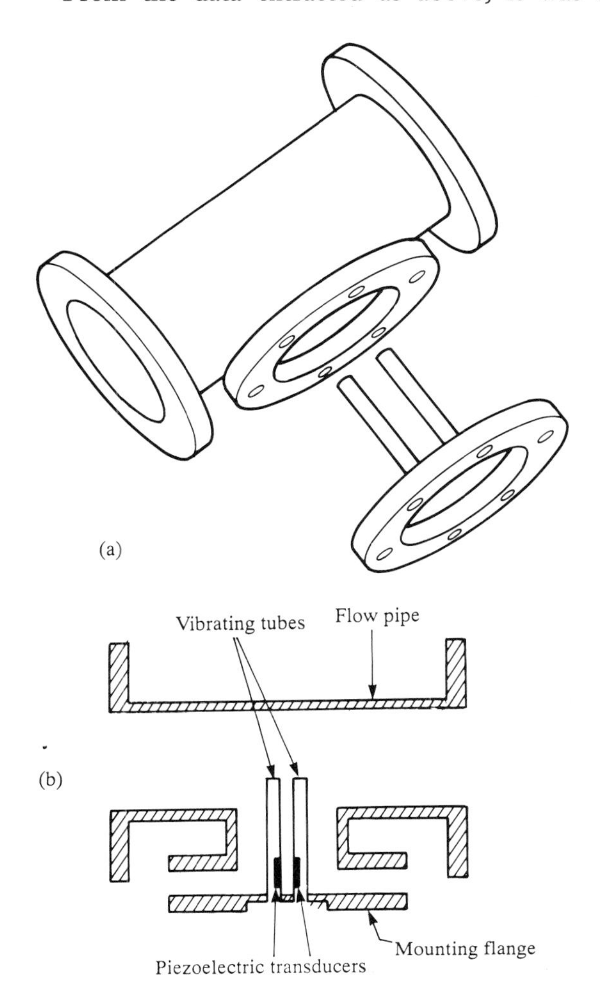

Figure 16.3 *(a) Experimental viscosity sensor using a low-frequency vibrating head inside a liquid flow pipe. (b) Schematic diagram of GEC viscosity/density probe for flange-in mounting in pipes (courtesy GEC Marconi Research Centre)*

approximate estimate of the dependence of the viscosity on oil type compared with the dependence on the oil degradation. The variability between measurements for the same oil type by different laboratories was comparable to the difference between the average value for each oil type (see Table 16.6). On Figure 16.4 the average values for the fresh oils varied from being 15 per cent below line 1 for sunflower oil to being 16 per cent above the line for rapeseed oil. This is well below the separation between lines 1 and 2 which amounts to approximately 80 per cent.

These data, although not suitable as a basis for an accurate prediction of the calibration of viscosity as a freshness indicator, nevertheless suggest a more favourable relationship than for free fatty acid determination as a rapid test. Moreover, viscosity has the added advantage of a potential in-line measurement in the hot oil. It was therefore concluded that the results of this study (Kress-Rogers 1987; 1988) were encouraging and that the construction of a specially adapted viscosity sensor for an experimental study at frying temperatures was now appropriate.

To the specifications on the range and resolution of the viscosity measurement, and on the operating temperatures, further specifications for the probe were added in order to provide food compatibility (Table 16.7). The probe surfaces in contact with the oil had to consist of food-grade material that would not leach any compounds or catalyse any reactions such as oxidation in the oil at the elevated temperatures used in frying. There should be no fragile components that might enter food in the case of breakage. The material should be easily cleanable with the standard procedures used in the frying operation. A food-grade stainless steel was the material of choice.

Table 16.4 *Reference values used to indicate the discard point* (Kress-Rogers et al. 1990a; 1990b)

Indicator	*Limit* %	*Source**
Polymeric triglycerides	10–15	O, BGW
Petroleum-ether-insoluble oxidized fatty acids	1	BGW
Free fatty acids	1	B
Non-urea-adducting fatty acids	11	RR
Acid value	4.5	O
Polar components	27	BM

*For sources see Table 16.5.

Table 16.5 *Viscosity of fresh and used frying media* (Kress-Rogers et al. 1990a)

		*Viscosity measurement**			*Freshness assessment*		
Oil or fat type	*Source*	T (°C)	*1000/*T *(1/K)*	η *(mPl)*	*State*†	*Indicator values*‡	
Cottonseed oil	KS	100	2.68	6.3	Fresh	Pol. = 1.5%	FFA = 0.10%
				11.4	Limit	13.8%	1.01%
				26.2	Degrad.	20.6%	1.13%
Hydrogenated lards	RR	100	2.68	7.2	Fresh	NAF = 0.2%	
				10.3	Used	7.4%	
Assorted hydrogenated fats	RR	100	2.68	7.0–7.5	Fresh	NAF = 0.2–0.5%	
				9.2–10.8	Used	6.5–7.4%	
				10.6		11.2%	
Soya-bean oil	Me	99	2.69	6.4	Fresh		
Soya-bean oil	W	99	2.69	7.0	Fresh		
Sunflower oil	Me	99	2.69	6.4	Fresh		
Cottonseed oil	W	99	2.69	8.5	Fresh		
Cottonseed oil	Me	99	2.69	7.8	Fresh		
Groundnut oil	Me	99	2.69	8.7	Fresh		
Olive oil	Me	99	2.69	8.0	Fresh		
Lard	W	99	2.69	9.5	Fresh		
Soya-bean oil	W	91	2.75	8.0	Fresh		
Hydrogenated cottonseed oil	M	90	2.75	12	Fresh		

Table 16.5 *continued*

Oil or fat type	Source	Viscosity measurement* T (°C)	1000/T (1/K)	η (mPl)	Freshness assessment State†	Indicator values‡
Sunflower-seed oil	MR	70	2.92	12	Fresh	FFA=0.15%
Hydrogenated sunflower-seed oil	MR	70	2.92	13	Fresh	FFA=0.16%
Olive oil	C	70	2.92	12	Fresh	
Soya-bean oil	W	51	3.09	20	Fresh	
Groundnut oil	M	50	3.10	24	Fresh	IV=91
Hardened groundnut oil	M	50	3.10	29	Fresh	IV=72
Mixed oils from restaurants	BM	50	3.10	30–33	Fresh	PC=0%
				40–50	Limit	27%
				55–60	Degrad.	45%
Palm oil	T	45	3.14	33	Fresh	
Palm olein	T	45	3.14	33	Fresh	
Soya-bean oil	T	45	3.14	27	Fresh	
Maize oil	Me	40	3.19	30	Fresh	
Lard	LZBa	40	3.19	35	Fresh	PC=2%
				87	Limit	27%
Soya oil	Me	38	3.22	26	Fresh	
Soya-bean oil	W	38	3.22	29	Fresh	
Sunflower oil	Me	38	3.22	30	Fresh	
Sunflower oil	DPGR	38§	3.22	29	Fresh	PC=4%
				66	Degrad.	37%
				735	Sev.	

Cottonseed oil	Me	38	3.22	34	Fresh			
Cottonseed oil	HM	38	3.22	50	Fresh	AV = 0		
				70	Limit	4.5		
Groundnut oil	Me	38	3.22	44	Fresh			
Groundnut oil	HM	38	3.22	47	Fresh	AV = 0		
				70	Limit	4.5		
Olive oil	Me	38	3.22	42	Fresh			
Olive oil	DPGR	38§	3.22	35	Fresh	PC = 3%		
				67	Limit	29%		
				105	Degrad.	51%		
Rapeseed oil¶	Me	38	3.22	41–48	Fresh			
Lard	W	37	3.23	44	Fresh			
Cottonseed oil	W	37	3.23	36	Fresh			
Maize	Me	30	3.30	45	Fresh			
Trilinolein	PC	30	3.30	33	Fresh	FFA = 0.04%		
				181	Degrad.	2.6%		NAFE = 26.3%
Soya-bean oil	W	29	3.31	41	Fresh			
Palm oil liquid fraction	BDK	25	3.36	49	Fresh	Pol. = 0.4%		PIOFA = 0%
				103	Limit	10.6%		3.02%
Safflower oil	SS	25	3.36	52	Fresh	FFA = 0.09%		SP = 250 °C
Cottonseed oil	KS	25	3.36	67	Fresh	Pol. = 1.5%		FFA = 0.10%
				130	Limit	13.8%		1.01%
				190	Degrad.	20.6%		1.13%
Refined groundnut oil	SS	25	3.36	63	Fresh	FFA = 0.1%	IV = 93	SP = 242 °C
				211	Used	0.5%	78	215 °C
				575	Sev. deg.	0.8%	55	180 °C
						and OFA = 15%, NAF = 28%		

Table 16.5 *continued*

Oil or fat type	*Source*	*Viscosity measurement**			*Freshness assessment*		
		T (°C)	*1000/T (1/K)*	*η (mPl)*	*State†*	*Indicator values‡*	
Groundnut oil	BDK	25	3.36	50	Fresh	Pol. = 0.2%	PIOFA = 0.03%
				102	Limit	12.6%	> 1%
Groundnut oil	LZBb	25	3.36	60	Fresh	IV = 98	
				149	Used	92	
Soya-bean oil	LMH	20	3.41	57–62	Fresh		
Soya-bean oil	T	20	3.41	64	Fresh		
Sunflower oil	LMH	20	3.41	65–67	Fresh		
Maize oil	Me	20	3.41	67	Fresh		
Cottonseed oil	W	20	3.41	71	Fresh		
Cottonseed oil	M	20	3.41	70	Fresh		
Cottonseed oil	KS	20	3.41	70	Fresh		
Olive oil	LMH	20	3.41	76–81	Fresh		
Olive oil	AM	20	3.41	81	Fresh		
				112	Heated	NAF = 7.6%	
(cod)				135	Used	8.8%	
(battered cod)				139	Limit	10.3%	
(sprat)				148	Limit	11.8%	
Olive oil	M	20	3.41	84	Fresh		
Palm olein	T	20	3.41	85	Fresh		
Soya-bean oil	W	19	3.42	69	Fresh		

Notes

*mPl (millipoiseuille) = mPa s = mN s/m^2 = cP (centipoise).

†Fresh — before frying, or reference values indicate a fresh oil
Limit — reference values indicate that the oil has reached or is close to the point where it should be discarded or topped up
Degrad. — degraded, or unacceptable for further use
Sev. deg. — severely degraded, or indicator value 100 per cent or more over limit
Used — used for frying

Table 16.4 gives the reference values used to indicate the limit.

‡Pol. — polymer content
FFA — free fatty acids
PIOFA — petroleum-ether-insoluble fatty acids
OFA — oxidized fatty acids
NAF — non-urea-adducting fatty acids
NAFE — NAF esters
PC — polar components
AV — acid value
SP — smoke point
IV — iodine value

§ Temperature estimated from density of fresh oil.
¶ Probably low erucic acid rape (LEAR) variety.

Sources

AM — Alim and Morton 1974. Temperature of viscosity measure not found; assumed room temperature = 20 °C
B — Bailey 1985
BDK — Bracco et al. 1981. Limit viscosity: for oil after frying time between times when OFA = 1 and when Pol. = 15%
BGW — Billek et al. 1978
BM — Battaglia and Mitiska 1983

Table 16.5 *Sources continued*

C	CRC 1976–77
DPGR	Dobarganes et al. 1985
HM	Hussain and Morton 1974, using density values from Table 16.8. Samples with higher values for acid value and viscosity have also been observed, but have been omitted here because of the unusually high density values for these samples
KS	Khalil and Steiner 1979. For 100 °C, values as given. For 25 °C, relative increase of viscosity with frying from KS, scale by comparison of fresh oil values from HM, W and C
LMH	Luque et al. 1982
LZBa	Lorusso et al. 1984
LZBb	Lorusso et al. 1982, using density values given by authors
Me	Meara 1978
M	Muller 1973
MR	Morrison and Robertson 1978, using density value 0.88 g/ml at 70 °C to calculate mPl from cSt (centistokes)
O	Olieman 1983
PC	Paulose and Chang 1973, using density value 0.9 g/ml at 30 °C to calculate mPl from cSt
RR	Rock and Roth 1966, using density value 0.86 g/ml at 100 °C to calculate mPl from cSt
SS	Sulthana and Sen 1979
T	Timms 1985
W	Weiss 1983, p. 12

Data on the densities of common frying oils (Table 16.8) were added to these specifications. Further data (in both tabular and graph form) from the feasibility study can be found in Kress-Rogers (1988) and in Kress-Rogers et al. (1990a). The latter is a detailed report on the research that provided the concept and specifications for the new probe for frying oil quality, and also on the evaluation of the resulting prototype. A briefer account has been given in Kress-Rogers et al. (1990b), and a discussion centering on the oil properties and reference methods can be found in Gillatt et al. (1991).

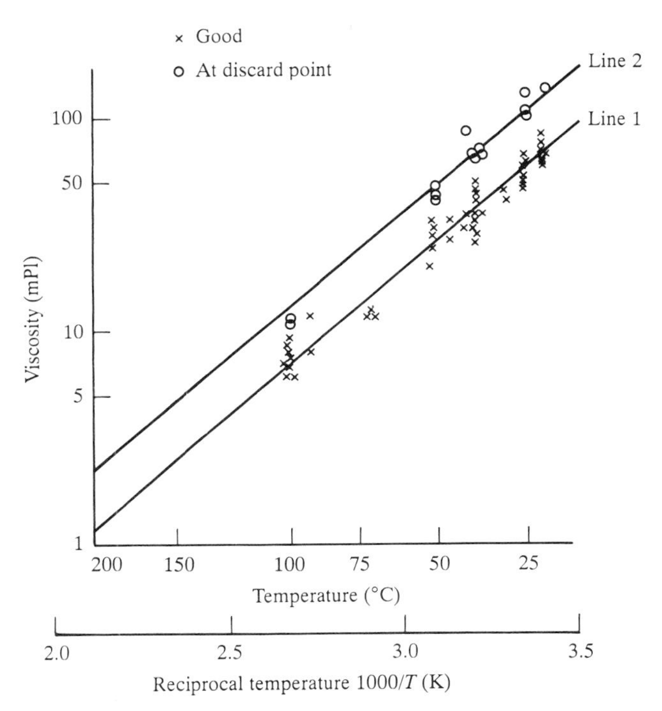

Figure 16.4 *Extrapolation of viscosity data to frying temperatures for a wide range of frying oils and shortenings. The oil quality was assessed with the help of indicator assays combined with limit values given elsewhere (see text for details). Data from Table 16.5: 100 samples analysed by 20 laboratories. Lines 1 and 2 have been drawn in tentatively to obtain an approximate estimate of the viscosity of fresh and used frying oils at elevated temperatures. In the absence of further data, two parallel lines were drawn (after Kress-Rogers 1988; also in Kress-Rogers et al. 1990a; 1990b; Kress-Rogers 1987, with some changes) (mPl = mPa s = cP)*

Table 16.6 *Viscosity dependence on oil type and degradation, listed as factors of deviation of viscosity values in Table 16.5 from line 1 (fresh oil) and line 2 (oil at the limit of usability) in Figure 16.4. The deviation factors averaged for each oil (last two columns) should be compared with the factor of 1.8 which divides lines 1 and 2 to gain an impression of the variation between oils as compared with the change with degradation (Kress-Rogers et al. 1990a)*

Oil	Source*	T(°C)	Deviation factors r1(fresh) η/line1	Deviation factors r2(limit) η/line2	Averages for oil type r1	Averages for oil type r2
Soya-bean oil	Me	99	0.89			
	W	99	0.97			
	W	91	0.92			
	W	51	0.80			
	T	45	0.82			
	Me	38	0.68			
	W	38	0.75			
	W	29	0.80			
	LMH	20	0.81–0.89			
	T	20	0.91			
	W	19	0.96		0.85	
Sunflower oil	Me	99	0.89			
	MR	70	0.80			
	Me	38	0.78			
	DPGR	38§	0.75			
	LMH	20	0.93–0.96		0.85	
Safflower oil	SS	25	0.87		0.87	
Maize oil	Me	40	0.86			
	Me	30	0.92			
	Me	20	0.97		0.92	
Cottonseed oil	KS	100	0.90	0.88		
	W	99	1.18			
	Me	99	1.08			
	Me	38	0.88			
	HM	38	1.30	1.01		
	W	37	0.90			
	KS	25	1.12	1.20		
	W	20	1.01			
	KS	20	1.00			
	M	20	1.00		1.04	1.03
Groundnut oil	Me	99	1.21			
	M	50	0.92			
	Me	38	1.14			
	HM	38	1.22	1.01		
	BDK	25	0.83	0.94		
	LZBb	25	1.00			
(refined)	SS	25	1.05		1.05	0.98
Olive oil	Me	99	1.11			
	C	70	0.80			

Table 16.6 *continued*

	Me	38	1.09			
	DPGR	38§	0.91	0.97		
	LMH	20	1.09–1.16			
	AM	20	1.16	1.11–1.18		
	M	20	1.20		1.07	1.09
Palm oil	T	45	1.10		1.10	
Rapeseed oil‡	Me	38	1.06–1.25		1.16	
Palm olein	T	45	1.10			
	T	20	1.21		1.16	

*For sources see Table 16.5.
‡Probably low erucic acid rape (LEAR) variety.
§Temperature estimated from density of fresh oil.

2.4 Characteristics of the New Probe

The new probe was adapted from a viscometer developed for the monitoring of boiler oils by GEC Marconi Research Centre (Chelmsford, UK).

Table 16.7 *Specifications for the probe*

Probe characteristics
Range of viscosities: 1–10 mPl
Resolution of viscosities: 0.5 mPl or better
Operation at an oil temperature of up to 205°C (for the preproduction model)
Operation at a sample density of approximately 0.8 g/ml

Additional sensors
Integral sensor for sample temperature (and temperature compensation in the final prototype)

Probe configuration and materials
Immersibility in hot oil
Absence of any traces of copper on the probe surface
Food-grade stainless steel surfaces in contact with the sample
Smooth surfaces in contact with the sample
Tolerant of cleaning in detergents and hot caustic solutions
Robust, non-fragile
Long handle; sensing head small enough for oil depth in fryer

Table 16.8 *Density of common frying oils* (Kress-Rogers et al. 1990a)

Oil	*Density** *(g/ml at 20 °C)*	*Approx. density** *(g/ml at 200 °C)*	*Source†*
Soya-bean	0.908	0.804	LMH, W
Sunflower	0.906	0.802	LMH
Cottonseed	0.904	0.800	Me, W
Groundnut	0.904	0.800	LZBb
Hydrogenated soya	0.902	0.798	W
Olive	0.899	0.795	LMH
Palm	0.898	0.794	T
Hydrogenated cotton	0.897	0.793	W

*Approximate slope: 0.065 g/ml per 100 °C. Extrapolated where data for the temperature of interest were not given by the referenced source.
†For sources see Table 16.5.

The latter probe (shown in Figure 16.3a, b) is designed for a higher range of viscosities up to 2000 mPl. This probe was adapted by GEC Marconi to allow measurements over the required viscosity range, as derived above, and reconfigured (Figure 16.5a, b) to allow immersion in hot oil in a batch fryer. (Configuration for continuous industrial fryers is also possible.)

The probe is of a simple rugged construction, consisting of two short vibrating stainless steel tubes. The piezoelectric transducers drive the pair of steel tubes to resonate in the manner of a tuning fork, and pick up the response characteristics. By measuring the amplitude of the tube vibration over a band of frequencies around the resonance frequency, the resonance curve (Figure 16.6) is mapped out. This is characterized by the position of the frequency f_0 where the amplitude is at a maximum and by the bandwidth BW. The instrument electronics determine the points of the resonance curve where the amplitude is 3 dB below the maximum and allow a measurement of the associated frequencies f_1 and f_2. The bandwidth is then determined as the frequency difference:

$$BW = f_2 - f_1$$

This bandwidth is indicative of the damping of the vibration by friction with the viscous liquid. It is related to the viscosity η and the density ρ of the liquid by the equation

$$\eta = k_1 (BW - k_2)^2 / \rho$$

where k_1 and k_2 are constants for a given probe design. This relationship is

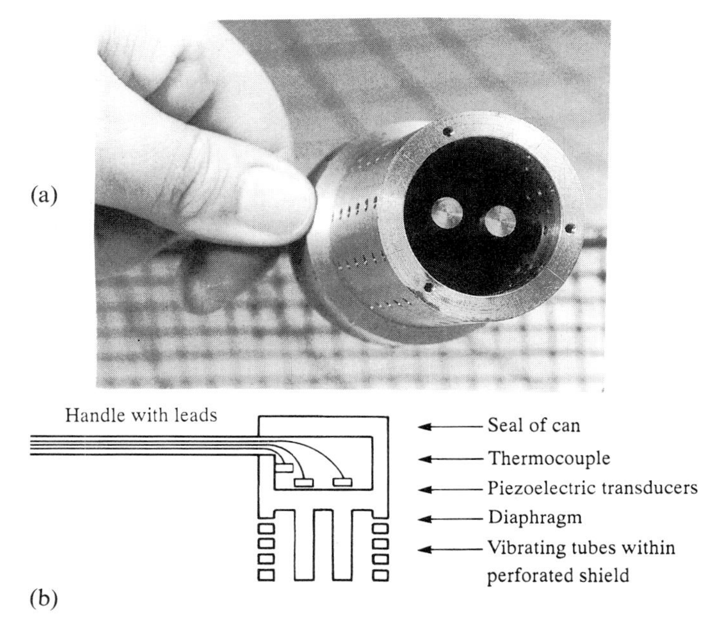

Figure 16.5 *(a) Prototype of the new probe: covering plate removed to show vibrating rods. The probe has a long handle containing the probe leads; this handle is concealed in the close-up photograph. (b) Schematic drawing of cross-section through the probe (Kress-Rogers et al. 1990a; 1990b). Refer to Acknowledgements*

illustrated by the characteristics of prototype 0 of the new probe for frying oil (Figure 16.7).

It is possible to determine both the viscosity and the density of a liquid

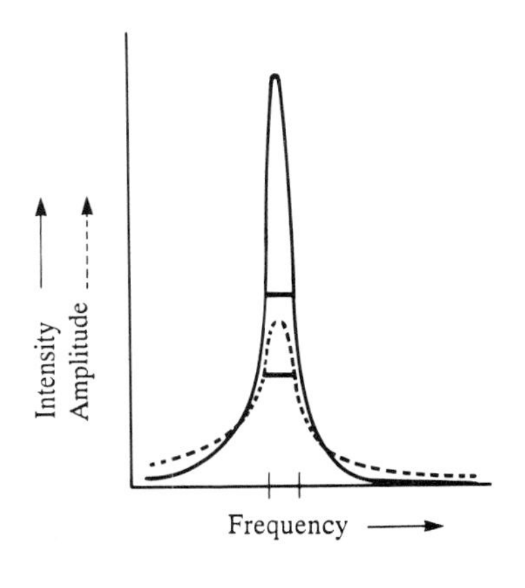

Figure 16.6 *Resonance curves showing bandwidth*

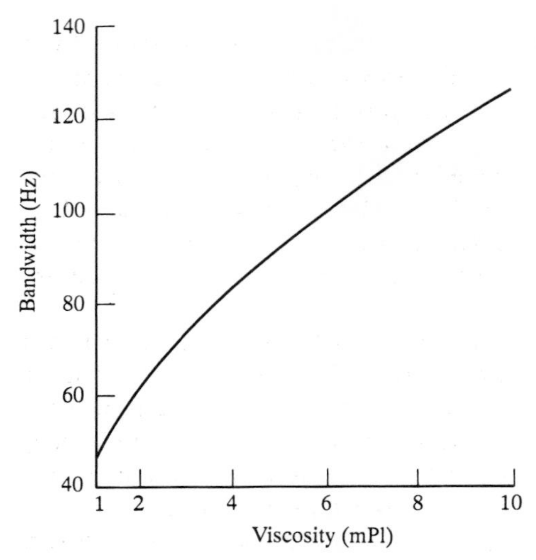

Figure 16.7 *Calibration for prototype 0 of the novel frying oil quality probe, $\rho = 0.82 g/ml$; see text for equation and parameters (Kress-Rogers et al. 1990a; 1990b)*

with resonance probes (Langdon 1985, 1987). However, the literature study had indicated that the density of frying oil was likely to change only by approximately 1 per cent during degradation from the fresh state to the discard point, whereas the viscosity nearly doubled during this time (Kress-Rogers et al. 1990a). Consequently, a probe output indicative of the oil density was not included in the study. Nevertheless, the literature data had been sparse and one source reported stronger changes, so that a measurement of the density at the beginning and the end of the frying trial was undertaken. This showed an increase of 0.7 per cent from the fresh condition to the discard point (at POM = 15 per cent) of partially hydrogenated vegetable oil.

One of the advantages of measuring viscosity with a mechanical resonance probe is that the signal is in the form of a frequency output. This means good signal to noise characteristics and inexpensive interfacing to a digital signal processing system. The new probe, adapted for frying oil, conforms with the recommendations for food industry equipment, exposing only food-grade stainless steel surfaces to the hot frying oil. (The boiler probe contained copper welds that would have catalysed oxidation and thus promoted rancidity.)

The probe can be cleaned in the same manner as the frying equipment, using either detergents or hot caustic solutions. The piezocrystals are mounted inside the probe head so that only smooth easily cleanable steel surfaces are in contact with the oil. Some advantages of the new probe for frying oil quality are listed in Table 16.9.

For the trial described below, the piezocrystals were accessible by a bolted gasket-sealed flange to permit adjustments. After initial test runs in a batch fryer, the probe was modified slightly, so that the characteristics for the probe used in the trial (prototype 1) described below differed slightly from those shown in Figure 16.7 for the initial configuration (prototype 0). A recalibration of bandwidth as a function of viscosity was not undertaken, as the trial was intended to show the dependence of probe output on frying oil quality as determined by a chemical reference method, and absolute viscosity values were not needed.

The method for bonding the piezocrystals to the inner probe surface allowed operation up to 180 °C in prototype 1. (Prototype 2, constructed more recently, is briefly described in Section 2.7.) A thermocouple mounted internally allows a monitoring of the probe temperature which, after equilibration, is indicative of the oil temperature, thus allowing a compensation of the measured viscosity for temperature changes.

Test runs with prototype 1 (Figure 16.8) showed that the run-to-run reproducibility was satisfactory and that the dependence of viscosity on oil temperature followed the same pattern at temperatures between 100 and 180 °C as for temperatures below 100 °C.

2.5 Protocol for the Trial with Prototype 1

In order to obtain reproducible results, an experimental procedure had to be followed that took account of the limitations of the preliminary design of prototype 1. As the probe was not fully immersible without risk of hot oil leakage into the probe, it had to be clamped in a position that would leave the seal just above the oil surface. This meant that part of the probe was exposed to ambient air temperature, with implications for temperature equilibration between the probe and the oil. To minimize this, the air circulation above the fryer was reduced by covering it. Further, measure-

Table 16.9 *Advantages of the new probe*

Less dependent on oil type than current simple tests
In situ operation in hot oil
Instant or continuous readout possible
Simple rugged construction
No serviceable moving parts
No reagents or disposable components
Tolerant of standard cleaning procedures for fryers
Only crevice-free food-grade stainless steel surfaces exposed to the oil
Frequency output: good signal noise, inexpensive interfacing to digital systems

ments were undertaken when the fryer thermostat was switched off so that the oil followed an even, slow cooling pattern.

The positioning of the probe to keep the seal out of the oil also necessitated a vertical orientation of the vibrating tubes. This orientation is prone to the trapping of gas bubbles at the upper end of the tubes where they join the probe can. Also, stirring was not possible so that the dislodging of gas bubbles and the acceleration of the exchange of oil between the oil within the probe shield and the bulk of the oil was not possible.

For each measurement, therefore, the probe output was measured for a range of oil temperatures. This allowed an examination of the temperature dependence and also a reduction of random error by using a value derived from fitting a curve to the probe output as a function of temperature rather than an individual point. It also provided an indication of the resolution with respect to degradation as a function of temperature for the bandwidth measurement (Figure 16.9). This procedure was lengthy and restricted the number of oils that could be studied with prototype 1. Three oils representing very different chemical characteristics within the frying oil range were selected for the trial (Table 16.10).

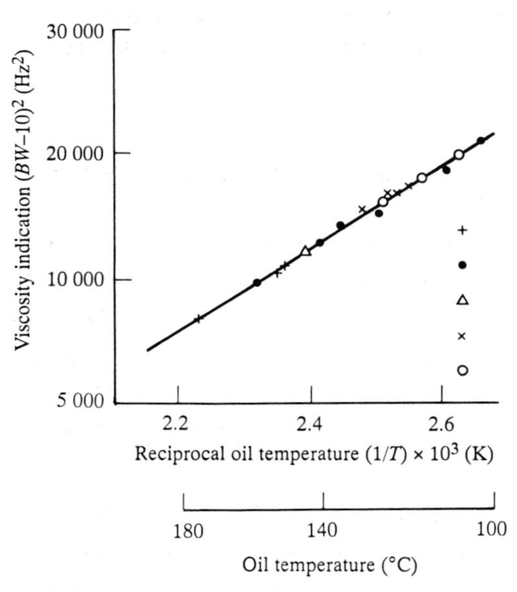

Figure 16.8 *Viscosity indication as a function of temperature for the test runs in used groundnut oil. Variable* (BW – *10*)^2 *is directly proportional to viscosity, neglecting a small deviation due to the density change with temperature. Test runs 1–5 marked by different symbols (Kress-Rogers et al. 1990a; 1990b)*

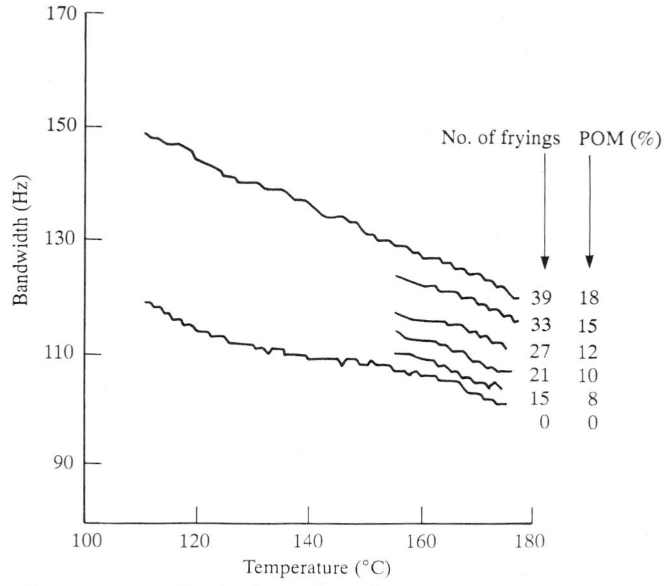

Figure 16.9 *Typical set of probe output versus temperature curves taken for an oil to verify correct positioning of the probe and gain improved accuracy for bandwidth value at 170 °C by interpolation (Kress-Rogers et al. 1990a)*

Table 16.10 *Characteristics of the oils used in the trial* (Kress-Rogers et al. 1990a; 1990b)

	Sunflower-seed oil	*Palm olein*	*PHVO**
Free fatty acids (as oleic)	0.11%	0.19%	0.10%
Polyunsaturated fatty acids	67.0%	10.5%	15%
Monounsaturated fatty acids	21.1%	44.2%	70%
Saturated fatty acids	11.9%	45.3%	15%
Iodine value (calculated from FAC)†	134.4	55.4	86

*The partially hydrogenated vegetable oil is a blend of processed hydrogenated vegetable oils containing polydimethyl siloxane. This additive inhibits oxidation so that the oil is more stable and, for the same frying use, shows a better quality as assessed by polymeric acids content and foaming tests (Berger 1984). In our trials, the frying life of this oil was longer by a factor of ~5.5 compared with the other two oils (see, for example, Figure 16.10). See also the notes on this oil in Section 2.1.
†FAC: fatty acid composition.

As the reference method, the determination of polymerized and oxidized matter (POM) by GLC (as non-eluted material on the GLC column) was chosen as the best indicator of frying oil quality independent of oil type or food fried. GLC analysis also provided a further indication of oil quality, the FFA composition.

Two simple tests used frequently on the shop floor at present were included for comparison. These are the FFA content and the colour (red and yellow 'content' measured by the Lovibond tintometer). The considerations in choosing the reference method (POM) and the other tests undertaken for the oil samples and the assay procedures used are described in Kress-Rogers et al. (1990a; 1990b) and in Gillatt et al. (1991).

Potatoes were chosen as the food fried, as both potato chips and potato crisps (known as French fries and chips, respectively, in the USA) occupy a sizeable market volume in many countries worldwide. In each frying, 200 g of potatoes were fried until golden brown. The trial was carried out until an oil had reached a POM content in excess of 15 per cent. Measurements with the new viscosity probe were undertaken regularly during the trial and samples taken for the reference methods. The taste of the chipped potatoes fried in the oil during the trial was assessed by an informal taste panel of six.

2.6 Results of the Trial with the New Probe

The trial confirmed that the new probe was capable of monitoring changes in the viscosity of the oils as they were degraded during the frying trials. Moreover, the change in probe output with degradation from the fresh state to the discard point (as determined by POM) was very similar for the three (chemically very different) oil types included in the trial. By contrast, the changes in the FFA content and colour tests differed considerably between the three oil types for the same degree of degradation as determined by POM (see Table 16.11 and Figures 16.1a, b, 16.10, 16.11a). The taste panel detected off-flavours in the fried food when the oil had reached a POM content of between 15 and 16 per cent, that is the upper of the two limit values (of 10 and 15 per cent) in use (Table 16.4). A POM content of 15 per cent was reached after approximately 30, 33 and 180 fryings for palm olein, sunflower oil and PHVO, respectively.

It was possible to define a cutoff value for the probe output, which could be regarded as the discard point. For this cutoff value of bandwidth $BW = 111$ Hz, all three oils had a POM content between the two limit values of 10 and 15 per cent (Figure 16.11b) that are used as a guide for the end of usability. (A lower discard point will be chosen when the fried product is to be stored in the freezer for several months.) For either FFA or colour test values, it was impossible to define such a common cutoff value for the three oil types.

Table 16.11 *Values for oils with 10 per cent polymerized and oxidized material (POM) (Kress-Rogers et al. 1990a; 1990b)*

	Palm olein	*Sunflower-seed oil*	*PHVO*
Absolute values			
No. of fryings	15	21	120
Red Lovibond colour	1.5	1.2	5.5
Yellow Lovibond colour	10	7	36
FFA(%)	0.6	0.2	1.0
Bandwidth (Hz)	106	109	110
Increases			
ΔBandwidth (Hz)	6	6	5
ΔPOM (%)	8.5	10	7
ΔYellow Lovibond colour	6	6	34

The precise calibration of the probe output against POM for the three oil types could not be achieved in this trial owing to the limitations of the preliminary prototype 1. These restricted the number of oil samples studied and limited the accuracy of the measurement of the bandwidth, which is indicative of the oil viscosity.

A parallel trial with two identical oils in two adjacent fryers showed that the FFA and colour test values indicated a noticeable difference between the two fryers, whereas the probe output was the same for both fryers. The POM content values, although appearing to have a high random error in this study, showed no systematic difference between the two fryers.

2.7 Conclusions and Outlook

The results of the trial indicate that the viscosity of frying oil, measured at a temperature of 170 °C, is a good indicator of the frying-induced deterioration of the oil. Moreover, it shows much less dependence on oil type and frying conditions than FFA content or colour tests. The measurement with the new probe for frying oil quality, once optimized for commercial production, is likely to be even simpler than the FFA or colour test, and can already be carried out *in situ* without having to take a sample. No reagents or other disposable components are needed. The probe is robust enough for use in any frying environment. A change in oil type will affect the new probe much less than the current shop floor tests.

The calibration is to be established in further trials with probe prototypes developed further. Prototype 2 has been constructed and tested

(Figure 1 on page 714). The temperature range has been extended to allow measurements at 200 °C. Most importantly, the probe can is now welded to allow full immersion and movement in the hot oil. This has simplified the measurement procedure considerably, and enhanced the reproducibility and resolution. The development programme is now well into phase VI of Figure 16.12. Scientists from many different backgrounds have contributed (Figure 16.13) to its progress towards the first probe for

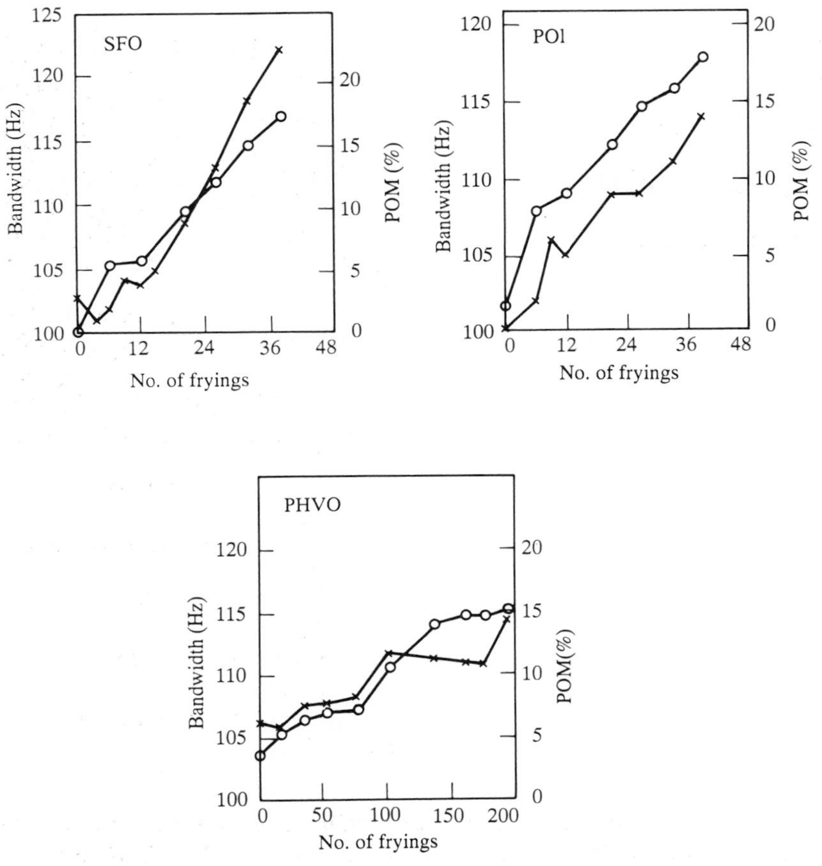

Figure 16.10 *Results of* in situ *monitoring of viscosity and reference assay (polymerized and oxidized material (POM) content) values during frying-induced deterioration of sunflower oil (SFO), palm olein (POl) and partially hydrogenated vegetable oil (PHVO; longer-term study, values for fryers 1 and 2 averaged): × viscosity as bandwidth output; O POM content by GLC (Kress-Rogers et al. 1990a; 1990b). Shown here are data for prototype 1, see page 714 for the more recent results with prototype 2.*

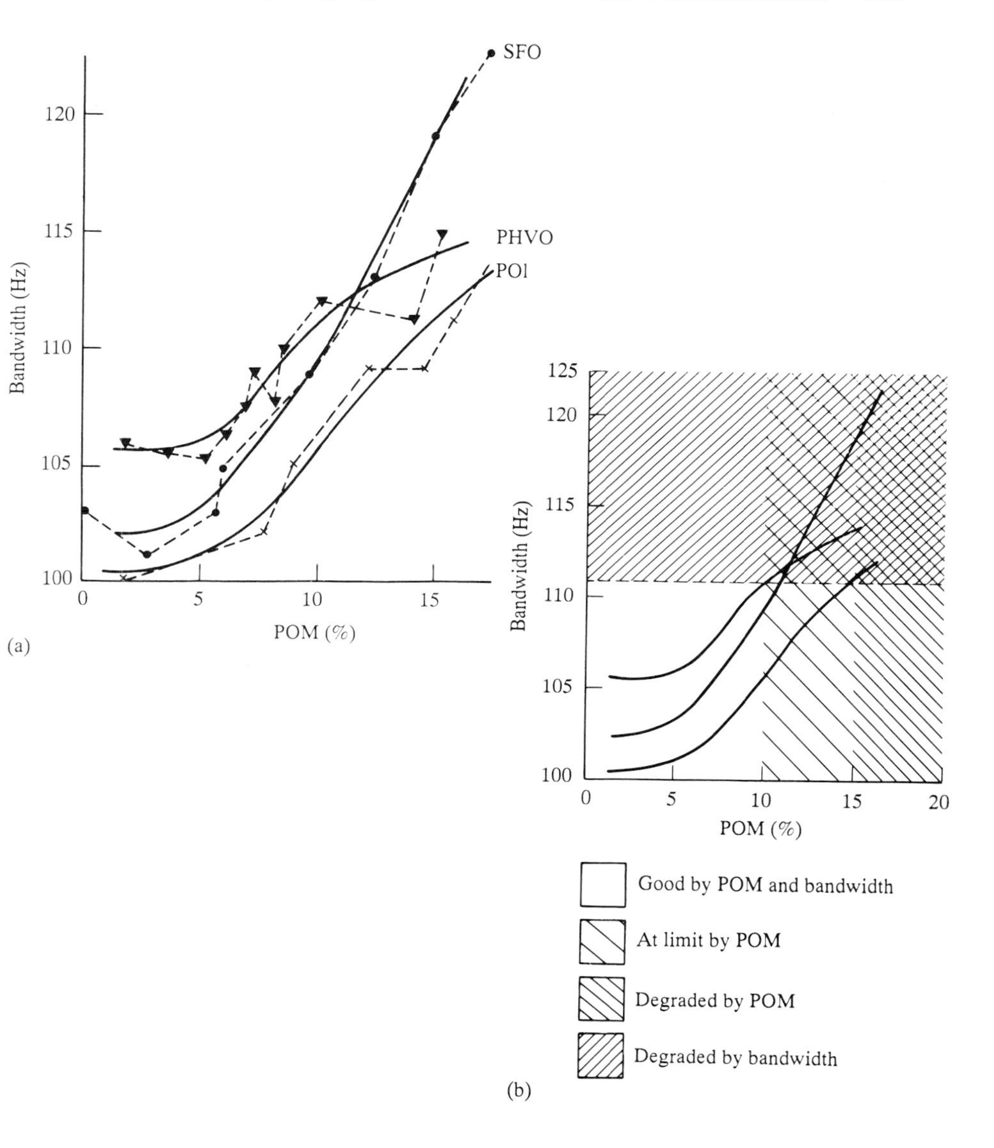

Figure 16.11 *(a) Comparison of the calibration curves taken* in situ *with the prototype viscometer (for abbreviations see Figure 16.10). Solid lines are drawn to help guide the eye through the overlapping curves. At 15 ± 0.5 per cent POM, off-flavours were noted for potato chips fried in the three oils. Below ~4 per cent POM, chips fried in PHVO were considered as bland. (b) Classification of oils by POM content and prototype sensor output (curves from (a)). The dashed line indicates the cutoff value of 111 Hz for which the three calibration curves fall between the two limiting values of 10 and 15 per cent POM (Kress-Rogers et al. 1990a; 1990b) (for prototype 1)*

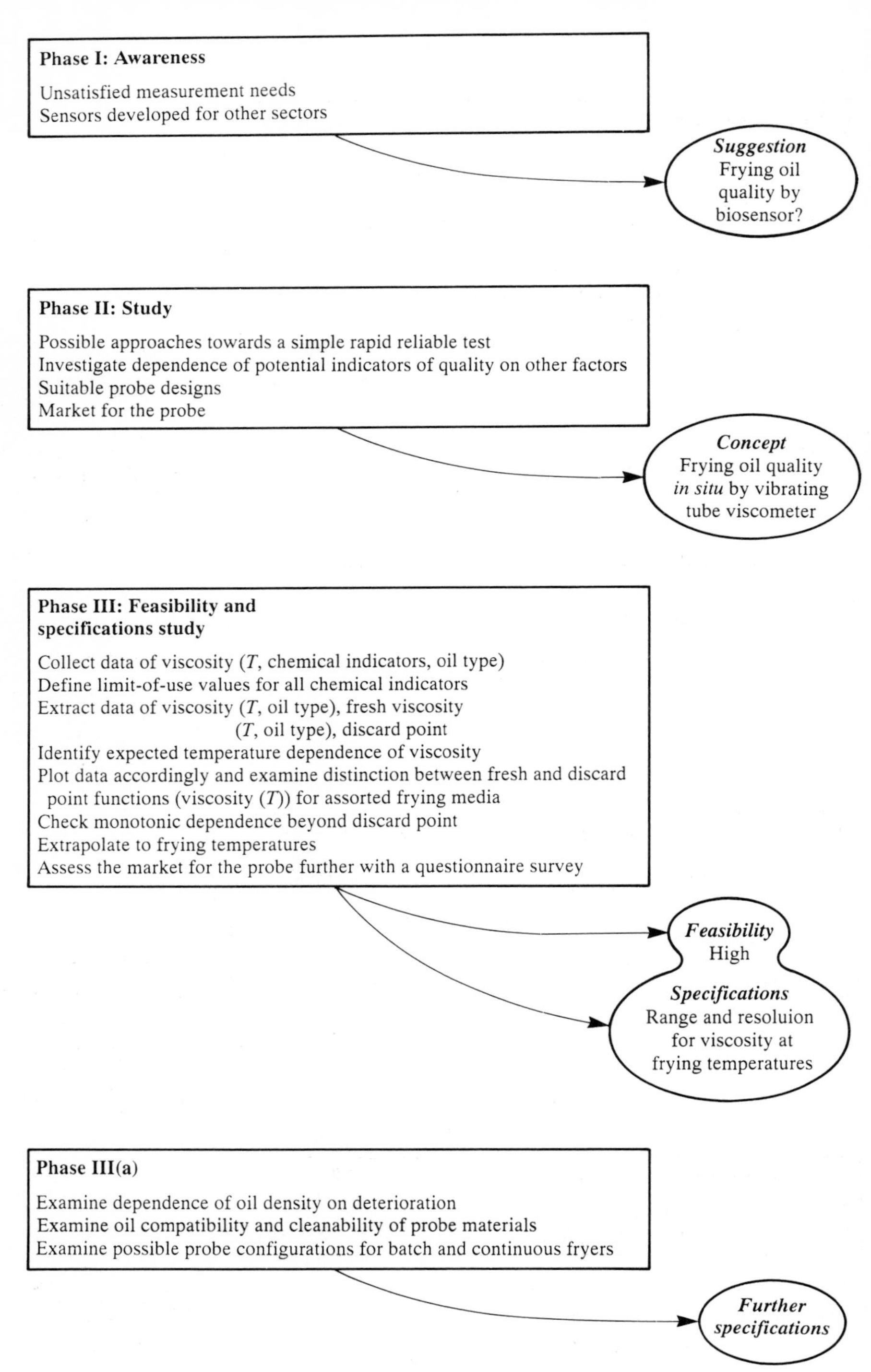

Figure 16.12 *Development of the frying oil monitor. Status in 1992: phase VI*

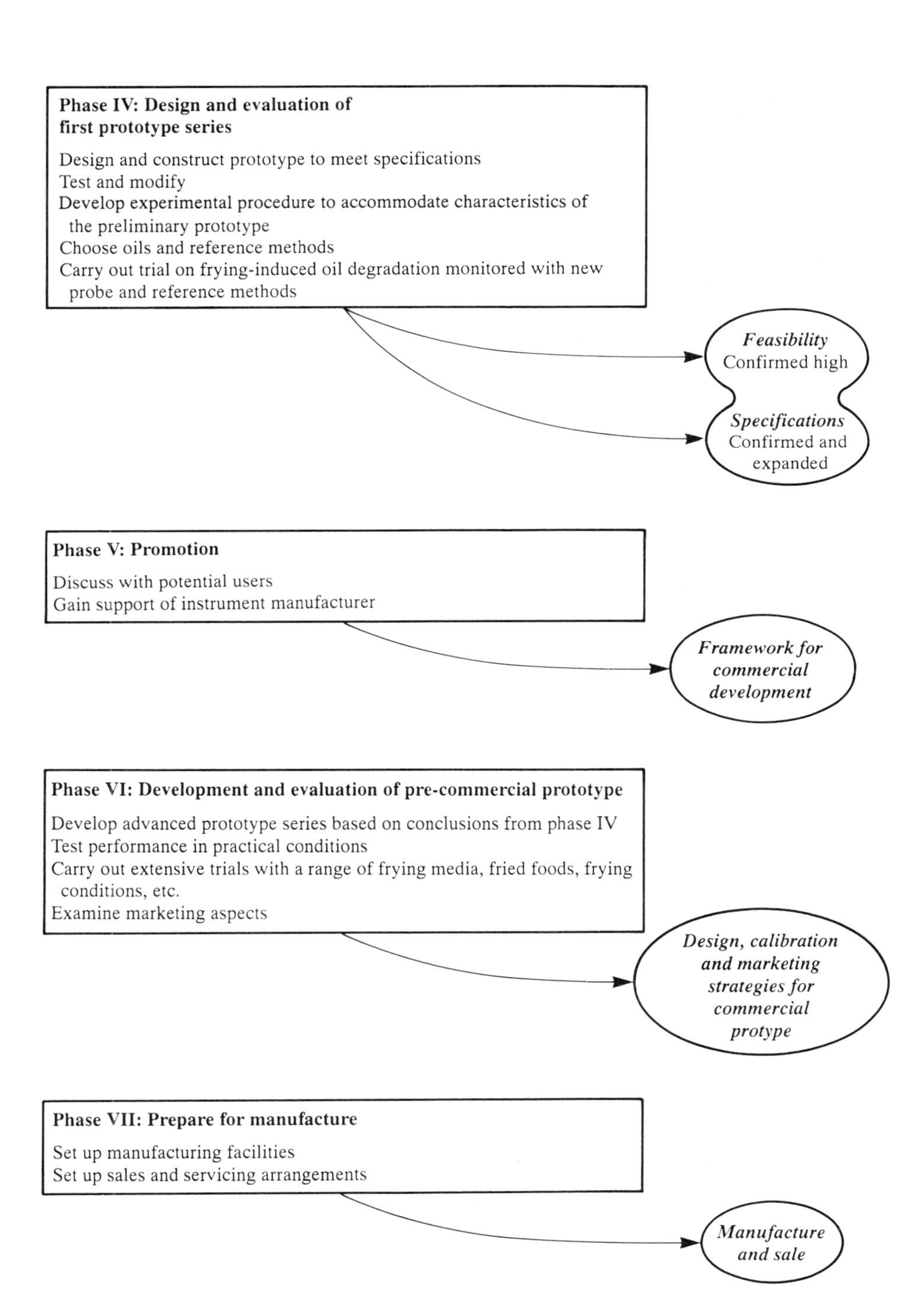

Figure 16.12 *Continued*

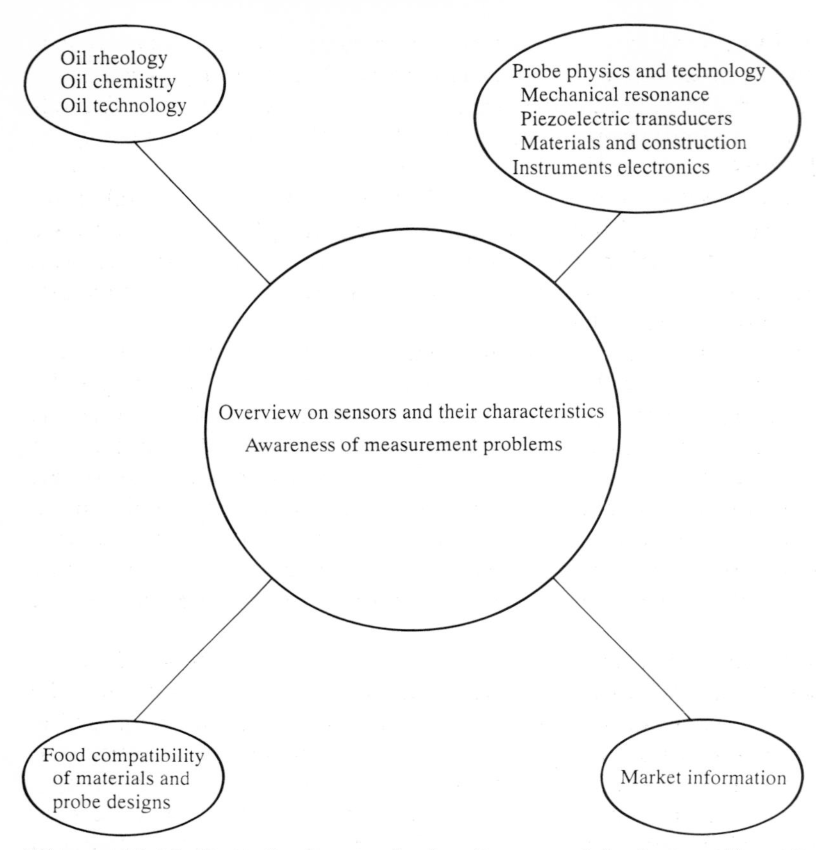

Figure 16.13 *Contributions to the development of the frying oil monitor (see acknowledgements section for information on the contributors)*

the quality of frying oil that can be used *in situ* in the hot oil. The preproduction model is shown in Figure 2 on page 714.

3 A Novel Knife-Type Probe for Meat Freshness

3.1 Development of the Probe Concept

Conventional methods for the assessment of microbial load on food (plate counts, that is incubation until visible colonies have formed) take several days. Such methods are not applicable to the distinction between meat that can be expected to have a remaining shelf life of two days, or meat that can be expected to be fit for another four days for retail. Nor are they of use in the individual grading of meat joints or carcasses by specifications that establish the suitability as ingredients for a particular meat product.

Rapid methods have been developed in recent years, notably the impedance techniques described in Chapters 14 and 15, but these still require several hours from sampling to result, and are intended for laboratory rather than shop floor implementation. The capital expenditure needed will prevent localized use. Other rapid methods such as ATP by luminescence (see Chapter 14, Section 1) can provide results within an hour, but again the instrument price and sample preparation procedures will preclude application on the shop floor.

We were looking for a method that would lend itself to the development of a compact user-friendly instrument that could be used, for example, by staff accepting a delivery of raw meat and having to decide whether the meat was of acceptable quality for the intended purpose, be it packing for retail or processing into a ready meal. The instrument should be usable both in the production environment of an industrial food processing plant and on the premises of a large catering establishment. The operation should be simple, there should be no need for sample preparation, and a reading should be displayed within minutes of inserting the probe into the meat.

To achieve such a truly rapid test without sample preparation, the approach taken was to look for an indicator compound that could be measured with a sensor to be developed (Table 16.12). As we were aiming for an instrument that would predict shelf life for meat that showed no obvious signs of spoilage, it was decided to examine the chemical changes in the meat at the pre-spoilage stage. At this time, glucose is still present at the surface of the meat, and is used preferentially by the bacterial flora of chilled raw meat. When the supply of glucose becomes the limiting factor, the bacterial flora switches to the decomposition of nitrogenous compounds and the production of amines. These amines are perceived as off-odours and are taken to signify microbial spoilage of the meat (Shelef 1977; Gill and Newton 1977; 1978; 1980; Newton and Gill 1978).

As the flora thrives and multiplies, the glucose concentration at the surface decreases, and glucose from the bulk of the meat diffuses towards the surface. Thus a gradient in glucose concentration is formed, with a high level of glucose in the bulk of the meat, gradually decreasing towards the surface. The glucose profile (that is the glucose concentration as a function of depth into the meat) is characterized by the slope and the value at the surface. Earlier work for lamb that was sterilized and then inoculated with controlled microbial flora had shown a correlation between the increase in the microbial load and the formation of the glucose profile (Gill 1976; see Figure 16.14).

We were also aware of a biosensor (Figure 16.15) developed for clinical applications at the Cranfield Biotechnology Centre (in collaboration with other universities at the early stages; see Turner 1983, Cass et al. 1984) that could measure glucose in blood in the presence or absence of oxygen. A company had taken on the commerical development of a clinical instrument based on the sensor. This was to be simple enough to be used by the patients themselves with a minimum of training. The analytical range was not too dissimilar from that likely to be encountered in the juices of meat. The sensor had also been constructed in miniature form so that arrays containing several adjacent biosensors could be prepared.

It was therefore concluded that it was appropriate to attempt the development of a meat freshness probe based on the determination of the

Table 16.12 *Microbial assessment*

Method	*Time needed**	*Comments*†
Plate count	2–3 days	Low capital expenditure Laboratory method Sample preparation needed QC applications only (spot checks on batches already moved on)
Impedimetry	4–12 hours	Medium capital expenditure Laboratory method Sample preparation needed Some QA applications possible with holding times
Bacterial ATP	1 hour	Medium capital expenditure Generally a laboratory method, and sample preparation needed For beverages without endogenous ATP: at-line possible, faster (30 minutes)
Marker chemical by biosensor	1 minute	Low capital expenditure Field applications At-line applications In-line possible Direct measurement possible Specific to a well-defined group of foods for each marker

*From sampling to result, including sample preparation, incubation, assay.
†Capital expenditure: low, medium, high indicate cost of instrument of the order of £1000, £10 000, £100 000 respectively (cost after development costs have been recovered in the initial period of marketing in the case of the biosensor).

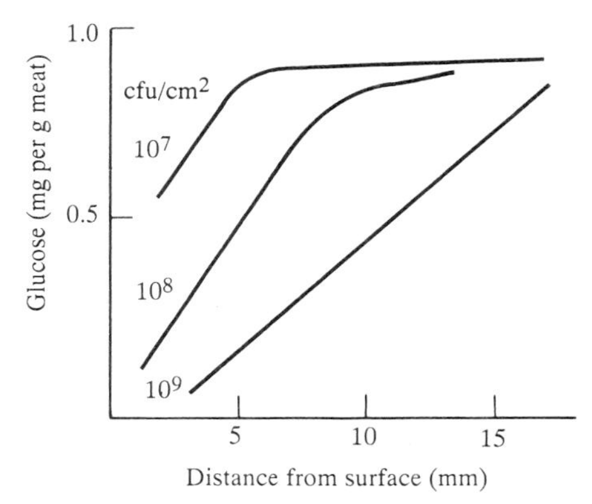

Figure 16.14 *Change of glucose depth profile with increasing bacterial counts for inoculated lamb (after Gill 1976; shown in Kress-Rogers 1985b)*

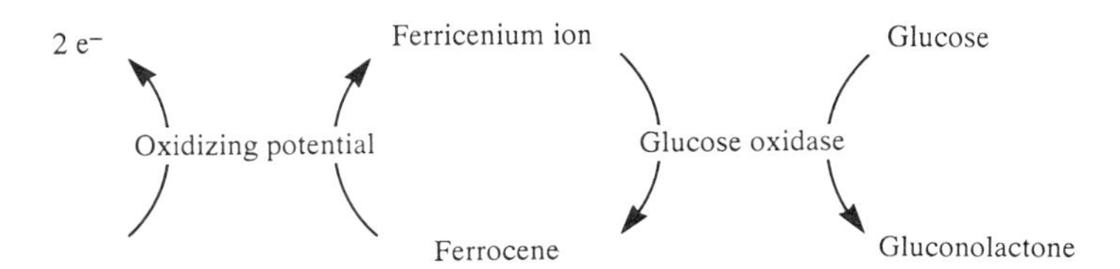

Figure 16.15 *Principle of the Cranfield glucose sensor*

glucose profile by measuring the glucose concentration at several depths into the meat with an array of biosensors mounted on a knife-type probe.

3.2 Study to Assess Feasibility and Define Specifications

It was now necessary to establish whether the observations for inoculated lamb applied to lamb, beef and pork with their native flora. This had to be achieved with an experimental study based on laboratory assays (equipment Sigma 510) for the determination of the glucose content of fine slices of the meat, each a millimetre or less in thickness, taken from the surface down into the bulk of the meat. The aim of the study was to provide both an assessment of the feasibility of the proposed development and also the specifications for the biosensor adaptation. In parallel with this study, work at Cranfield was to characterize the chosen biosensor type with respect to the performance at the pH values and temperatures relevant to the chilled meat application.

Experiments at the LFRA for lamb, beef and pork joints with their native bacterial flora exhibited a decrease in the glucose concentration at the surface relative to that deeper in the meat, as observed earlier by Gill

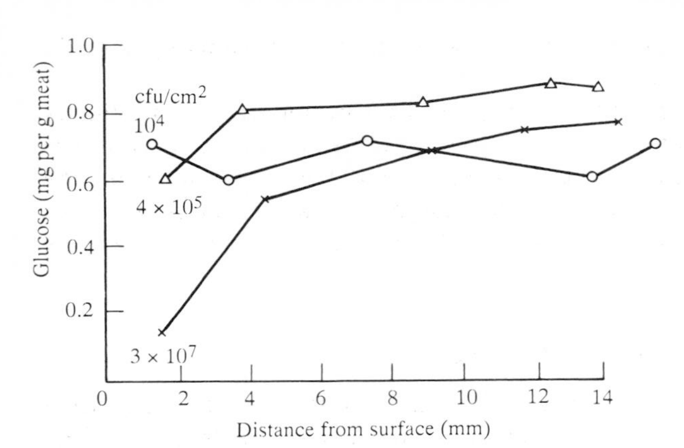

Figure 16.16 *Development of glucose profile in lamb joint with increasing bacterial surface load (shown in Kress-Rogers and D'Costa 1986)*

(1976) for the model system (inoculated lamb). However, in the meat with the native flora, as received from the slaughterhouse at 24 to 48 hours *post mortem*, an increase in glucose concentration throughout the depth of the meat was observed within the first days of chill storage. This was attributed to the production of glucose by hydrolysis of glycogen owing to the continuing activity of endogenous enzymes present in the muscle (Dolrymple and Hamm 1975).

As a consequence of the increase in the overall glucose content, the absolute value of the glucose concentration at the surface rose in the early storage days. Subsequently, as the bulk glucose concentration stabilized and the microbial load at the meat surface showed a pronounced increase, the surface level decreased and a concentration gradient towards the bulk of the meat joint developed (Figure 16.16).

The results demonstrated the formation of a glucose gradient resulting from the activity of the native chilled meat flora at the pre-spoilage stage (under aerobic conditions). The formation of a glucose gradient was first observed at a microbial surface load of 10^5 to 10^6 cfu/cm^2 when the microbial population had entered the rapid multiplication stage of growth ending the initial lag phase (cfu: colony forming units).

The results also provided specifications for the biosensor array to be developed, particularly on the glucose concentration range and resolution needed to suit different meat types and on the required spacing of the individual sensors forming the array for depth profiling. The initial glucose concentration in pork, beef and lamb varied with meat type, consistent with data by Gill (1976) and Lawrie (1985). It varied also between samples of the same meat type, covering a range of 0.2 to 3.6 mg glucose per gram of meat for pork loin and 0.3 to 0.7 mg glucose per gram of meat for lamb loin

(1 mg glucose per gram of meat corresponds to approximately 7.4 mM glucose in the meat juices).

The variation of the pH value along the depth profile was also studied as a function of storage time and compared with the pH dependence of the biosensor to see whether the change in pH with depth would mask the glucose profile as determined with the biosensor array. The change in pH with depth was found to be too small to affect the measured glucose profile significantly.

3.3 Characteristics of the New Probe

Based on the specifications obtained from the study above, the adaptation of the clinical glucose sensor was now undertaken at the Cranfield Biotechnology Centre. An array formed by four individual glucose sensors was developed (Figure 16.17) based on the amperometric mediated enzyme electrode (AMEE) developed earlier at Cranfield for clinical applications (see Figure 16.15; the principles and references are given in Chapter 17). Four layers of porous carbon material were sandwiched alternately with an inert material and then cut across the layers with a diamond saw such that a high enzyme and mediator loading on the interfaces could be obtained. This provided the required sensitivity near the meat surface where glucose

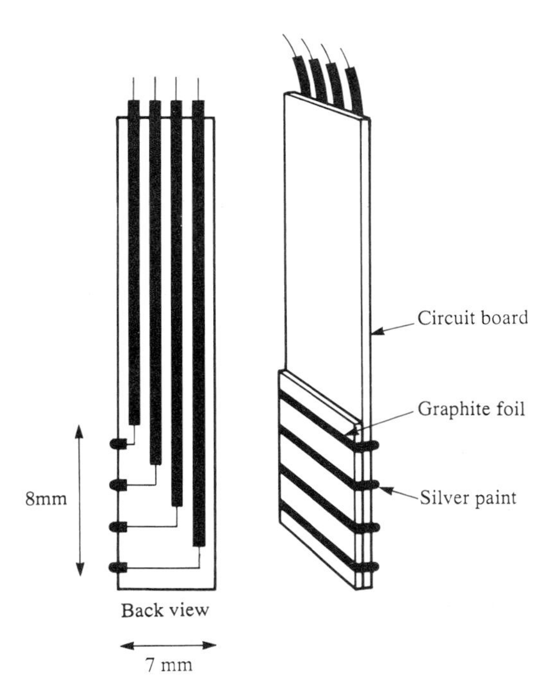

Figure 16.17 *Laminate electrode. Refer to Acknowledgements*

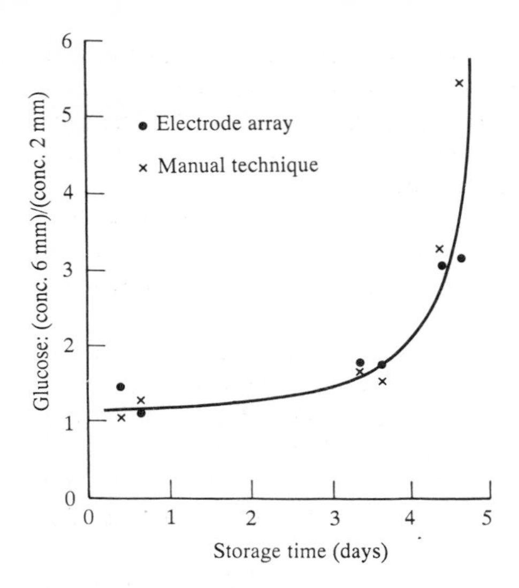

Figure 16.18 *Glucose concentration gradient in pork leg steak (Kress-Rogers et al. 1987; 1988; see also Kress-Rogers 1987). Refer to Acknowledgements*

levels are low. Procedures for the immobilization of reagents, for the initial conditioning in solution and for storage were developed at Cranfield, in order to provide the required characteristics. As a common reference electrode for the glucose sensor array, a silver wire carrying a silver chloride film was used.

After the successful development of a prototype for laboratory use, a preliminary trial was undertaken to test the performance in meat and establish the relationship of the biosensor readings to assays carried out with a standard enzymic assay kit. The Sigma kit employing the same enzyme as the biosensor (glucose oxidase) was used. There was good agreement between the two methods for the gradient as determined from the ratio of glucose concentration at 6 mm depth and at 2 mm depth (Figure 16.18).

An interface to a personal computer and software for monitoring the sensor arrays during calibration and measurement completed the system, which was then transferred to the LFRA for the main trial.

3.4 Trial with the New Probe

The deterioration of pork loins during chill storage at 3 °C ± 1 °C was followed over a period of two weeks, comparing the biosensor array readings with results from the Sigma assay. The pork muscles (longissimus dorsi) were received fresh (within 24 hours *post mortem*), they had a microbial load of approximately 10^4 cfu/cm^2.

Each day in the second week of the trial and less frequently in the first week, a slice was removed from each pork loin for the measurements. This allowed glucose measurements to be taken at room temperature for the slice (which was later disposed of) whereas the remainder of the pork loin was left in a continuously controlled environment.

The glucose profile was determined directly by insertion of the biosensor array into the meat after making an incision with a knife (Figure 16.19). The measurement was carried out at room temperature immediately after calibration.

The Sigma assay was carried out on slices cut parallel to the meat surface. For the Sigma assay, a plug of meat was cut from the meat adjacent to the position where the sensor array had been inserted. This was cut into four slices tallying with the depths of the sensor positions in the array. Each slice measured approximately 7 mm × 7 mm × 2 mm. A larger top slice with a surface area of 20 mm × 20 mm was used for the microbial assay. An organoleptic assessment was also carried out.

Consistent with the earlier measurements (Section 3.2), the glucose concentration throughout the meat increased during the first days of storage. This rise was also observed for the measuring point closest to the surface (at 2 mm depth). After seven days, the glucose concentration for the deeper measuring points had stabilized, and after eight days, a slight decrease in the glucose level measured nearest to the surface was first observed (Figure 16.20). The microbial flora had reached the phase of maximum growth rate at this time (Figure 16.21).

Figure 16.19 *Trial with glucose sensor arrays in pork loin slices*

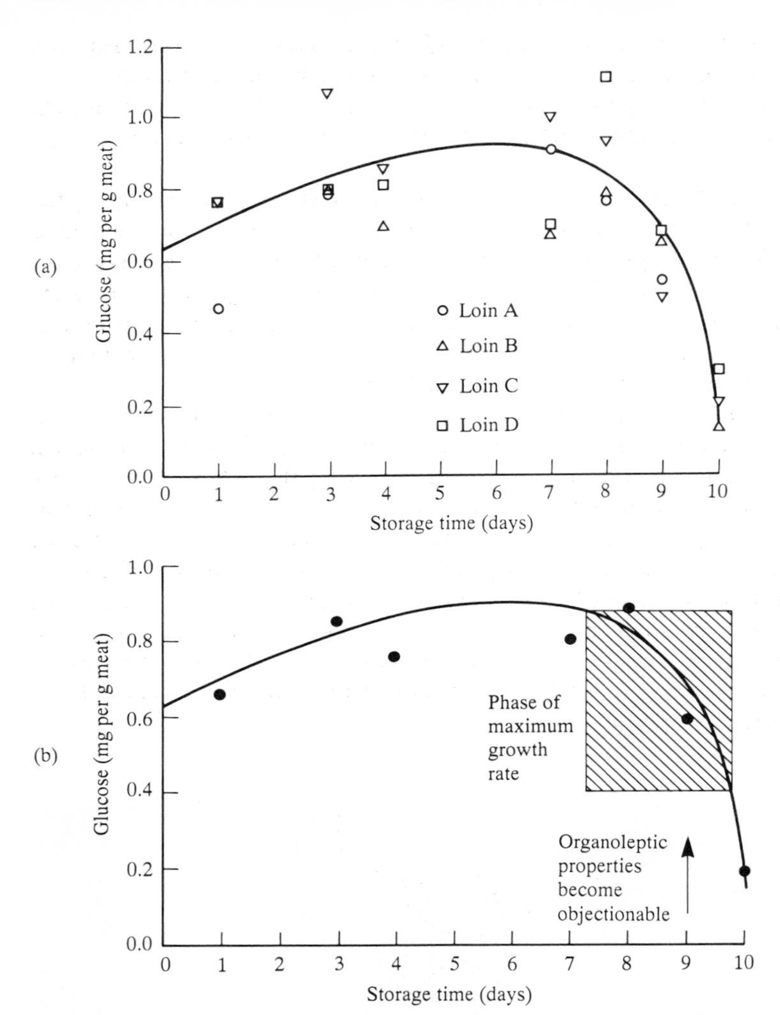

***Figure 16.20** Change in glucose concentration at 2 mm depth with storage time for four pork loins: (a) loins A–D shown separately (b) average of loins A–D. Fitted average curve on both graphs. Refer to Acknowledgements*

The Sigma assay showed the initial formation of the gradient more clearly. However, at the later stages the glucose concentration in the small surface slices was too small for a reproducible assay with the enzymatic assay kit, whereas the biosensor was still capable of giving readings (Figure 16.22).

At this stage, an accurate measurement of the glucose gradient was not possible with the first laboratory prototype. One of the problems encountered was the collection of drip juice from the meat in the cut made to allow insertion of the sensor array. Although the measurement took only approximately 5 minutes, the data from the biosensor at the deeper points (at 6 and 8 mm depth) in the array were not consistent with those from the Sigma assay. The gradient between 2 and 4 mm depth only was determined instead.

In future prototypes this problem can be addressed by constructing the biosensor array on a disposable knife blade, so that no prior incision is needed. Also, a later prototype would be expected to exhibit a reproducible time dependence in reaching the equilibrium current after contacting the sample. This would allow a shorter measurement period of, for example, half a minute, without the need to reach the equilibrium current stage (as is the case with current commercial clinical probes).

The glucose gradient, averaged for the four loins, showed the first increase as the microbial flora approached the phase of maximum growth rate. In the subsequent days of the trial, during this phase, the glucose concentration gradient increased and soared as the saturation load was approached (Figures 16.23, 16.24). A glucose depletion by a factor of three over the 2 mm of the profiled depth (from 2 to 4 mm distance from the surface) was observed when the load was a little over 10^8 cfu/cm^2. The glucose concentration at the point closest to the surface had fallen to a fifth of the maximum value (that had occurred at the end of the first week) at this time. The organoleptic properties had now changed to exhibit unpleasant odours and colour changes.

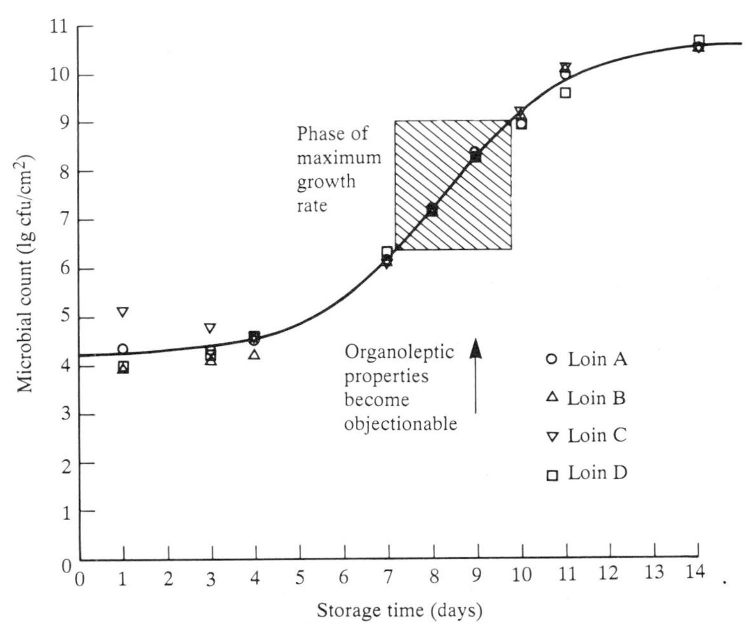

Figure 16.21 *Microbial growth values and fitted urve for four pork loins*

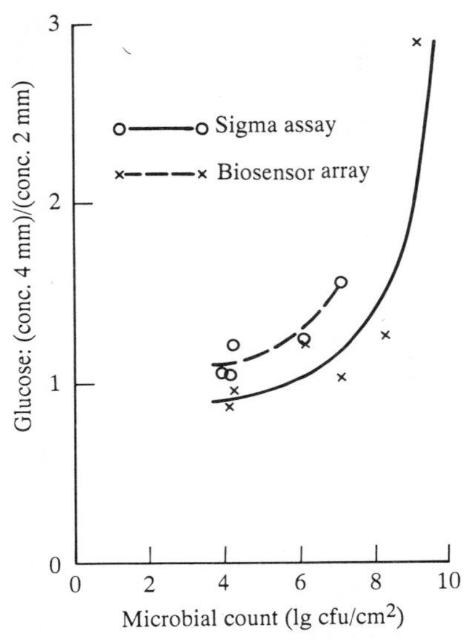

***Figure 16.22** Measurement of glucose concentration gradient for pork loin B by Sigma assay and with the new biosensor array (laboratory prototype, to be further optimized)*

Note that the lines in Figures 16.20–16.24 are tentative. No attempt was made to formulate a model for the form of the glucose depth profile as a function of microbial growth since the number of parameters to be fitted would have been too large compared with the number of data points. This is because the formation of the glucose concentration gradient is dependent on the metabolic activity of the microbial flora and on the diffusion rate from the bulk to the surface of the meat. The diffusion rate is itself a function of the gradient and of other factors, and assumptions on the change of the glucose decomposition rate and of the diffusion coefficients during storage would have to be made. The data in Figures 16.20–16.24 have also been used in other figures published in Kress-Rogers et al. (1988).

To obtain an estimate of the time at which the formation of a glucose concentration gradient could first be observed with an optimized prototype, data obtained by averaging the results for the four pork loins in the trial were examined. The prediction from these data is that a gradient would be measurable at 10^6 cfu/cm^2, that is two days before clear organoleptic changes occurred. It must be stressed that this estimate will be subject to confirmation with a more extensive trial that will have to be carried out with a more advanced prototype. Such an advanced prototype should be manufactured on a pilot-scale facility similar to a commercial operation, and would not only provide a deeper depth profile and a better

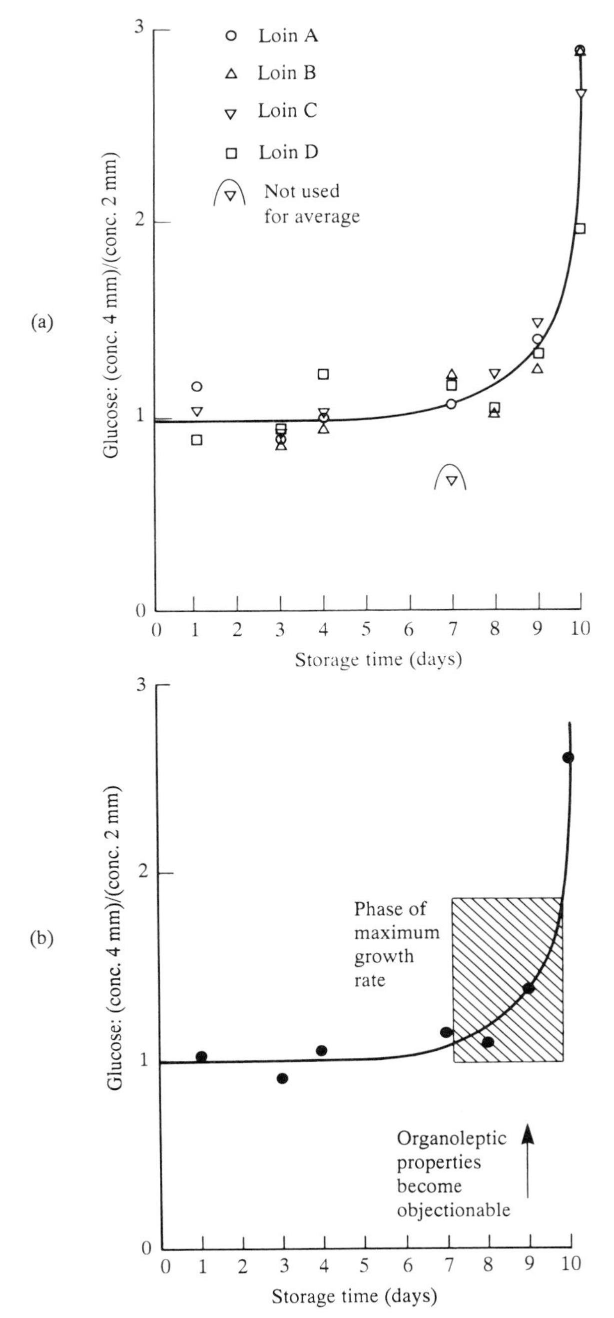

Figure 16.23 *Increase in glucose concentration gradient 4 mm/2 mm with storage time for four pork loins: (a) loins A–D shown separately (b) average of loins A–D. Fitted average curve on both graphs. Refer to Acknowledgements*

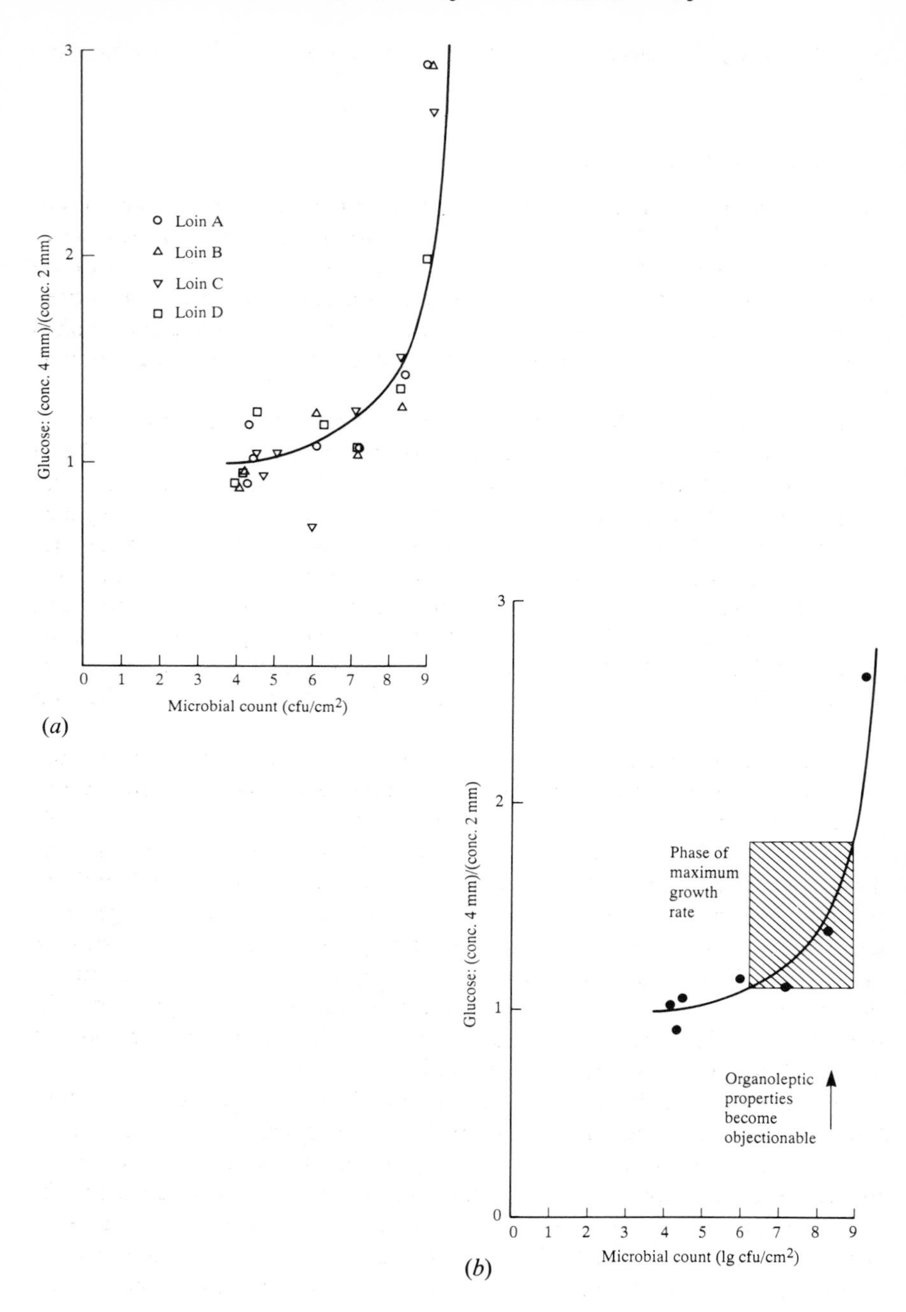

Figure 16.24 *Increase in the glucose concentration gradient 4 mm/2 mm with microbial growth for four pork loins: (a) loins A–D shown separately (b) average of loins A–D. Fitted average curve on both graphs*

reproducibility, but also render feasible the acquisition of more extensive data.

3.5 Conclusions and Outlook

The trial supports the validity of the initial concept of a correlation between the microbial load of chilled meat with a native chilled meat flora, and the glucose depth profile. Further, it confirms that this can be measured within minutes with a glucose sensor array based on the chosen biosensor type. It is now necessary to carry out an extended trial with a more advanced prototype to be developed in accordance with the conclusions from the initial trial.

Such a trial is needed to provide more detailed information on the influence of the type of meat (species) and of the cuts (muscle type) and also of the slaughterhouse environment. It is expected that, if using a universal calibration of glucose gradient or surface glucose level, the probe response would give a different assessment from the microbial assay for meat that was low in endogenous glycogen at the time of slaughter. This would be a bonus, however, in that the probe would predict a short shelf life and lead to the rejection of such low-quality meat. Further guidance could be obtained with a novel pH probe developed specifically for food applications on the basis of a microelectronic (ISFET) device (see Chapter 17, Section 2.3.7).

For guidance on the possible performance of a commercially developed prototype, it is useful to examine the advances achieved with the clinical probe based on the same principle. The clinical probe has become available commercially after a substantial development effort which has resulted in disposable probes that are sufficiently reproducible to be used without prior calibration for determining the glucose level in the blood of patients with diabetes. The probes are in the form of paper strips that plug into the pen-shaped instrument (see Chapter 17, Section 3.2 for details). On the probe strips, a carbon film is deposited that carries the enzyme and the mediator. A silver/silver-chloride film acts as the reference electrode. A drop of blood is placed on to the sensing area of the paper strip.

In analogy to this clinical instrument, but with an array of four biosensors constructed on a disposable knife blade, and with a rather more robust construction of both the knife-type sensing heads and the instrument, the configuration of the meat freshness probe is now envisaged for later prototypes (Figure 16.25). Also, instead of a digital display of glucose concentrations, a simple traffic light system would provide the answer: fresh (green), nearing the end of shelf life (amber) or shelf life expired (red).

Clinical biosensors of the AMEE type have now benefited from many years of intensive development, and from considerable experience in the commercial mass production of the probes and in the performance of the probes and instruments in the hands of a large number of patients. The resulting advances in technology could now be transferred to the food industry application described, where the sample shares some of the chemical properties of the clinical sample. Surveys carried out in 1984 and 1988 (at the outset of the development programme for the meat freshness probe, and after completion of the trial with the first prototype) have shown and later confirmed a considerable interest from the meat industry. This interest, together with technical advances, could promote the transition to phase VI of the instrument development shown in Figure 16.26.

The meat freshness probe could be teamed up with the new ISFET solid-state pH probe designed specifically for meat applications (see Chapter 17, Section 2.3.7). Applications of the glucose profile probe could extend beyond the determination of meat freshness; the sensor could also be employed in the study of glucose diffusion rates in meat and their relationship to meat quality or a prior freeze-thaw history.

The food application has a number of features which are more demanding than the clinical measurement on a droplet of blood placed on to the sensor strip. One of these is the insertion into meat of a knife-type probe. It may be necessary to include a retractable sheath in the design to prevent contact between juices in the vicinity of the surface and the sensors positioned on the lower part of the knife probe. Another demanding specification is the operation at chill temperature, and this could be addressed by optimization of the biosensor kinetics or by including a heater film in the knife probe. With a sheath retracted on insertion to the correct position and reapplied subsequently, it may be possible to obtain a reading after removing the probe from the meat, if the probe surfaces are porous. Before these engineering problems are addressed, however, it is necessary to carry out an extensive trial with meat representing a satisfactory range of

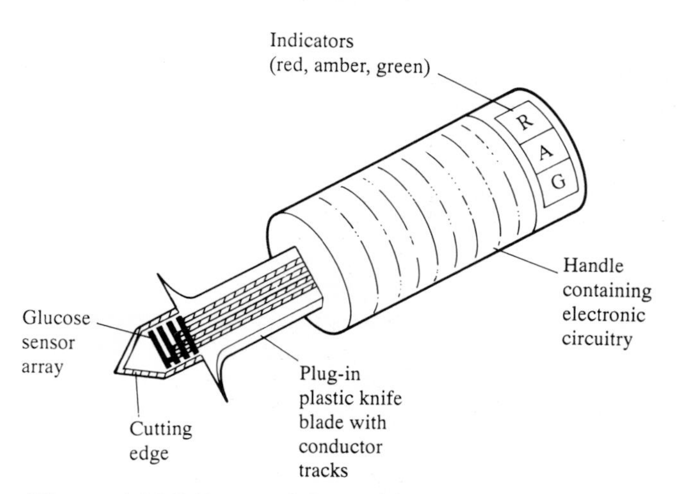

Figure 16.25 *Proposed form of instrument for the assessment of chilled meat by the glucose depletion profile (Kress-Rogers 1987)*

species, muscles and compositions of the microbial flora and the variations in initial glycogen content resulting from pre-slaughter conditions and breed. This trial could be carried out with an advanced but not fully optimized prototype.

Contributions from diverse scientific, technical and information areas have made the development of the current prototype possible (Figure 16.27) and these will again be needed in a further development towards a commercial probe.

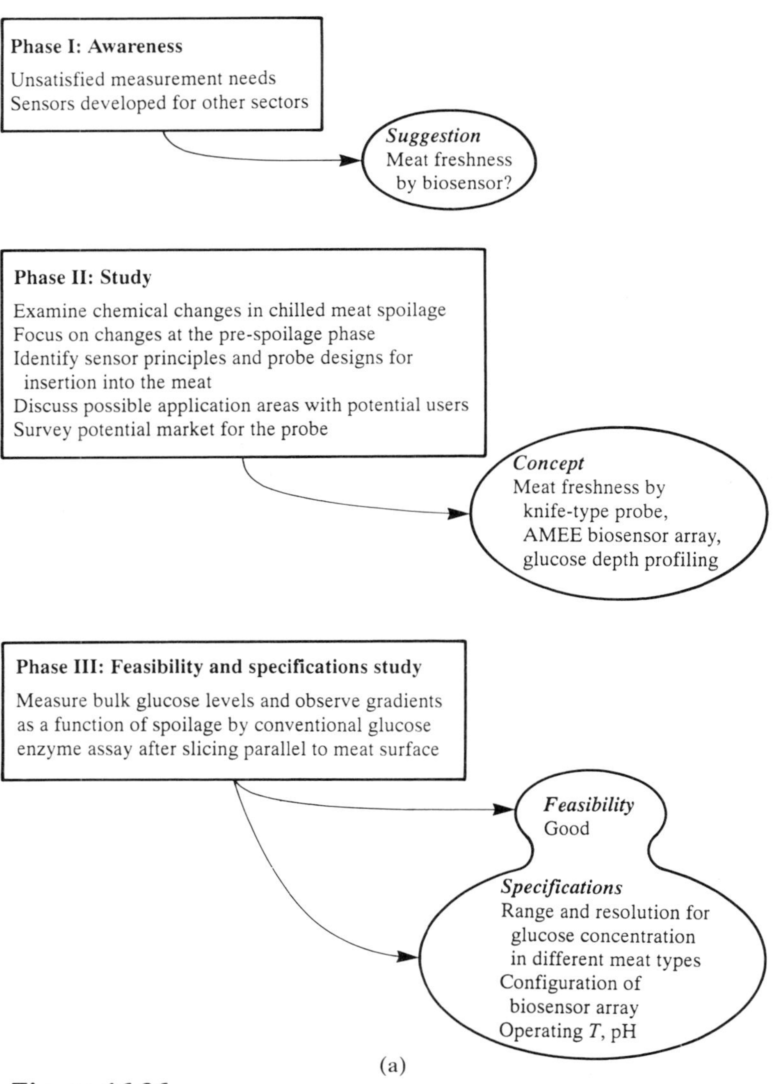

Figure 16.26

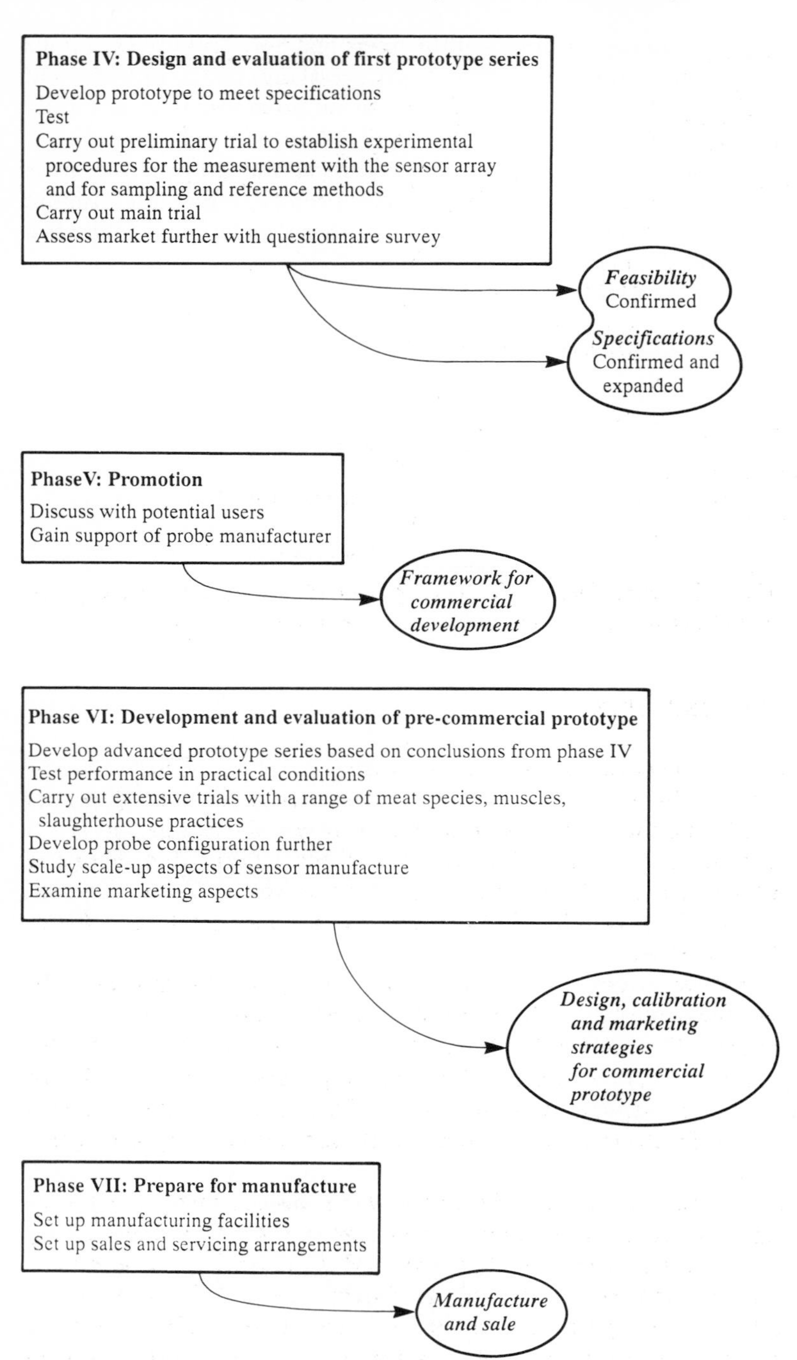

***Figure* 16.26** *Development of the meat freshness probe. Status in 1992: phase V*

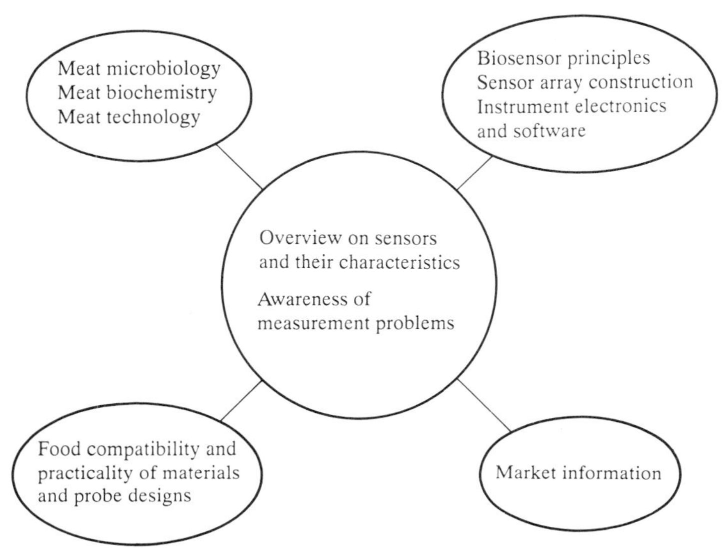

Figure 16.27 *Contributions to the development of the meat freshness probe (see acknowledgements section for information on the contributors)*

Acknowledgements

Project M51 on the development of novel sensors for the food industry was based on study M45 on the new generation of sensors and their potential applications in the food industry (reported in Kress-Rogers 1985a). Both were carried out on behalf of the Ministry of Agriculture, Fisheries and Food and have received further support from the Research Coordination Committee (RCC, now STPC: Science and Technology Policy Committee) of the LFRA.

Project M45 was suggested by R. T. Roberts (LFRA). During subsequent feasibility studies with chemical sensors, the author has benefited from his advice. The author should like to thank J. N. Fry (LFRA) for his encouragement and advice during the decisive phases of project M51. In both projects, the continued support from A. W. Holmes (LFRA) has been invaluable. Feedback from LFRA member companies has helped to guide the projects.

The development of the meat freshness probe was the first project to be undertaken in the sensors development programme (M51, coordinated by the author). Discussions with R. T. Roberts, P. A. Gibbs, J. M. Wood and A. W. Holmes at the LFRA and with I. J. Higgins and A. P. F. Turner at the Cranfield Biotechnology Centre and with representatives of LFRA member companies have been important in the early stages of this project (phase II in Figure 16.26). The study on feasibility and specifications (phase III) was carried out by J. M. Wood and N. Church (LFRA); microbial assays were supervised by P. A. Gibbs and C. J. Stannard (LFRA).

Biosensor arrays for the determination of the glucose sensor profile in meat were developed (phase IV, part a: design in Figure 16.26) by E. J. D'Costa and A. P. F. Turner (Cranfield Biotechnology Centre). The trial with the biosensor arrays (phase IV, part b: evaluation) was carried out by J. E. Sollars (LFRA, working with the author on project M51). Biosensors were prepared by E. J. D'Costa, who also offered valuable help and advice during the trial. Microbial assays were supervised by R. M. Blood and advice on microbial aspects was offered by P. A. Gibbs (both LFRA).

Of the sensor developments resulting from project M51, the most advanced to date is that of the frying oil monitor which has entered the commercial development phase recently (phase VI in Figure 16.12). Discussions with J. B. Rossell (LFRA; oils and fats aspects) and with P. E. Stephens and R. M. Langdon (GEC Marconi Research Centre; mechanical resonance sensors) have been essential in the early stages of this project. The author has also been able to rely on the advice of J. B. Rossell during the feasibility and specifications study (phase III).

For the design and construction (phase IV, part a: design in Figure 16.12) of the mechanical resonance viscosity probe specially adapted to meet the specifications from the feasibility study, the author is indebted to P. E. Stephens and R. M. Langdon (GEC Marconi Research Centre). In the evaluation of the probe, the author wishes to express her thanks to P. N. Gillatt (LFRA) for his efficient collaboration, and J. B. Rossell for his advice. The pre-commercial development of the probe is now being led by N. O'Brien (GEC Sensor Ltd), together with P. E. Stephens and R. M. Langdon at the GEC Marconi Research Centre and with the author and J. B. Rossell at the LFRA.

For the determination of oxidative rancidity in oils, fats and products containing oils and fats, a feasibility and specifications study based on GC-MS monitoring of the headspace was carried out by W. J. Reid and P. M. Dagnell (LFRA). An experimental study on routes to a biosensor development for the aldehydes identified as possible markers in the GC-MS study was carried out by E. J. D'Costa and A. P. F. Turner (Cranfield).

The study on the non-intrusive measurement of hydrogen as an indicator of microbial spoilage of fish and meat packed under modified atmospheres was made possible by a microelectronic sensor (a CHEMFET) provided by D. Pedley and C. Baker of Thorn EMI (who have also offered similar support in the study of microelectronic sensors (ISFETs) for pH and ions). Measurements were carried out by J. E. Sollars and C. Hobbs. The author should like to thank R. T. Roberts and P. A. Gibbs for their advice during this study.

On the last day of 1990 Dr. John Wood, who was then managing the Food Technology section at the LFRA, sadly died unexpectedly. He is still missed by the author and her colleagues and by colleagues in the food industry for his friendly and competent advice.

Several of the other contributors to the sensors work, both at the LFRA and at collaborating research laboratories, have since moved to positions in the food industry.

References

Alim, H. and Morton, I. D. (1974) Oxidation in foodstuffs fried in edible oils. *Proceedings of the IVth Congress on Food Science and Technology*, vol. 1, 345–356

Askar, A. (1976) Amine und Nitrosamine. Fortschritte in der Lebensmittelwissenschaft 4, Technische Universität Berlin, Institut für Lebensmittelwissenschaft

Bailey, A. E. (1985) *Bailey's Industrial Oil and Fat Products*, vol. 3, ed. T. H. Applewhite. Wiley, New York

Battaglia, R. and Mitiska, J. (1983) Zur Beurteilung von gebrauchten Frittierfetten *Mitteilungen aus dem Gebiet der Lebensmitteluntersuchung und Hygiene*, **74**, 156–159

Berger, K. G. (1984) The practice of frying. PORIM Technology 9, Palm Oil Research Institute of Malaysia

Billek, G., Guhr, G. and Waibel, J. (1978) Quality assessment of used frying fats: a comparison of four methods. *Journal of the American Oil Chemistry Society*, **55**, 728–733

Bracco, U., Dieffenbacher, A. and Kolarovic, L. (1981) Frying performance of palm oil liquid fractions. *Journal of the American Oil Chemistry Society*, **58**, 6–12

Bullard, F. A. and Collins, J. (1980) An improved method to analyse trimethylamine in fish and the interference of ammonia and dimethylamine. *Fishery Bulletin, National Oceanic and Atmospheric Administration*, **78**, 465–473

Cass, A. E. C., Davis, G., Francis, G. D., Hill, H. A. O., Aston, W. J., Higgins, I. J., Plotkin, E. V., Scott, L. D. L. and Turner, A. P. F. (1984) Ferrocene-mediated enzyme electrode for amperometric determination of glucose. *Analytical Chemistry*, **56**, 667–671

Cleland, N. Hoernsten, E. G., Elwing H., Enfors, S. and Lundstroem, I. (1984) Measurement of hydrogen evolution by oxygen-limited *Escherichia coli* by means of a hydrogen sensitive PD-MOS sensor. *Applied Microbiology and Biotechnology*, **20**, 268–270

CRC (1976–77) *Handbook of Chemistry and Physics*, 57th edn, ed. R. C. Weast. CRC Press, Cleveland

Dagnall, P. M., Holgate, J. H. and Reid, W. J. (1989) Use of GC-MS to investigate organic compounds produced with the onset of rancidity in fats. Leatherhead Food Research Association research report 642

Danielsson, B. and Winquist, F. (1989) Biosensors based on semiconductor gas sensors, In: eds. A. P. F. Turner, I. Karube and G. Wilson.

Biosensors: Fundamentals and Applications, Oxford University Press, 531–548

Dobarganes, M. C., Perez-Camino, M. C., Gonzalez-Quijano, R. G. and Repetto, M. (1985) Acietes calentados. Estudios de toxicidad cronica. I. Evaluacion quimica de las muestras. *Grasas Y Acietes*, **36,** no. 1, 30–34

Dolrymple, R. H. and Hamm, R. (1975) Post-mortem glycolysis in pre-rigor ground bovine and rabbit muscle. *Journal of Food Science*, **40,** 850–853

Gill, C. O. (1976) Substrate limitation of bacterial growth at meat surfaces. *Journal of Applied Bacteriology*, **41,** 401–410

Gill, C. O. and Newton, K. G. (1977) The development of aerobic spoilage flora on meat stored at chill temperatures. *Journal of Applied Bacteriology*, **43,** 189–195

Gill, C. O. and Newton, K. G. (1978) The ecology of bacterial spoilage of fresh meat at chill temperatures. *Meat Science*, **2,** 207–217

Gill, C. O. and Newton, K. G. (1980) Development of bacterial spoilage at adipose tissue surfaces of fresh meat. *Applied and Environmental Microbiology*, **39,** 1076–1077

Gillatt, P. N., Kress-Rogers, E. and Rossell, J. B. (1991) A novel sensor for the measurement of frying oil quality. *Lipid Technology*, **3,** 78–82

Hoernsten, E. G., Danielsson, B., Elwing, H. and Lundstroem, I. (1986) Sensorized on-line determinations of molecular hydrogen in *Escherichia coli* fermentations. *Applied Microbiology and Biotechnology*, **24,** 117–121

Hussain, S. S. and Morton, I. D. (1974) Absorption by food of frying oil. *Proceedings of the IVth Congress on Food Science and Technology*, vol. 1, 322–334

Karube, I. and Tamiya, E. (1987) Biosensors for the food industry. *Food Biotechnology*, **1,** 147–165

Khalil, S. A. and Steiner, E. H. (1979) Chemical and physical changes in heated cottonseed oil during simulated frying conditions and their correlation with foaming properties. Leatherhead Food Research Association research report 313

Kress-Rogers, E. (1985a) Technology transfer. II: The new generation of sensors. Leatherhead Food Research Association scientific and technical survey 150

Kress-Rogers, E. (1985b) Sensors in quality control in the food industry. Two examples: microbes and moisture. *Proceedings of the Conference on Sensors and their Applications*, Southampton, 10–12 September, Institute of Physics, 44–45

Kress-Rogers, E. (1987) Sensors and their applications to meat technology. *Proceedings of the International Symposium on Trends in Modern Meat Technology 2*, eds B. Krol, P. S. Van Roon and J. H. Houben, 23–25 November, Den Holder; Pudoc Wageningen, 33–39

Kress-Rogers, E. (1988) Viscosity sensors for assessment of oil quality during use. In *Frying,* Leatherhead Food Research Association symposium proceedings 35, 54–60

Kress-Rogers, E. and D'Costa, E. J. (1986) Biosensors for the food industry. *Analytical Proceedings,* **23,** 149–151

Kress-Rogers, E., D'Costa, E. J., Wood, J. M. and Roberts, R. T. (1987) Glucose sensor for meat freshness. Poster and demonstration at SAB Demonstration Meeting on Rapid Methods, 30 September, Bath

Kress-Rogers, E., Gillatt, P. N. and Rossell, J. B. (1990a) Development of a novel sensor for the *in situ* assessment of frying oil quality. Leatherhead Food Research Association (confidential to LFRA members; less detailed information is available in Kress-Rogers 1988; Kress-Rogers et al. 1990b; Gillatt et al. 1991)

Kress-Rogers, E., Gillatt, P. N. and Rossell, J. B. (1990b) Development and evaluation of a novel sensor for the *in situ* assessment of frying oil quality. *Food Control,* **1,** 163–178

Kress-Rogers, E., Sollars, J. E., D'Costa, E. J., Wood, J. M. and Turner, A. P. F. (1988) Meat freshness assessment using a biosensor array. Proceedings of the 34th International Congress of Meat Science and Technology, 29 August to 2 September, Brisbane, 508–510

Langdon, R. M. (1985) Resonator sensors: a review. *Journal of Physics E: Scientific Instruments,* **18,** 103–115

Langdon, R. M. (1987) Resonators – a review, In: *Current Advances in Sensors,* ed. B. E. Jones, Adam Hilger, 19–32

Lawrie, R. A. (1985) *Meat Science,* 4th edn. Pergamon, Oxford

Lorusso, S., Zelinotti, T. and Betto, P. (1982) Caratteristiche chimico-fisiche di interesse bromatologico di oli sottoposti a riscaldamento: olio di arachide. *La Revista Italiana Delle Sostanze Grasse,* **LIX,** 141–148

Lorusso, S., Zelinotti, T. and Betto, P. (1984) Caratterizzazione degli oli sottoposti al rescaldamento: strutto. *La Revista Italiana Delle Sostanze Grasse,* **LXI,** 667–673

Luque, V. F., Martin, J. C. and Herrera, C. G. (1982) Variaciones de la viscosidad y de la densidad con la temperatura en aceites de oliva, girasol y soya espagnoles. *Grasas Y Acietes,* **33,** no. 6, 334–339

Meara, M. L. (1978) Physical properties of oils and fats. Leatherhead Food Research Association scientific and technical survey 10

Morrison, W. H. and Robertson, J. A. (1978) Hydrogenated sunflowerseed oil: oxidative stability and polymer formation on heating. *Journal of the American Oil Chemistry Society,* **55,** 451–453

Muller, H. G. (1973) *An Introduction to Food Rheology.* Heinemann, London

Newton, K. G. and Gill, C. O. (1978) The development of the anaerobic spoilage flora of meat stored at chill temperatures. *Journal of Applied Bacteriology,* **44,** 91–95

Olieman, N. W. (1983) Inspection of frying oils and fats. In *Fat Science,* Proceedings of the 16th ISF Congress, Budapest, 1983, 757–767

Paulose, M. M. and Chang, S. C. (1973) Chemical reactions involved in deep fat frying of foods. VI: Characterisation of nonvolatile decomposition products of trilinolein. *Journal of the American Oil Chemistry Society,* **50,** 147–154

Ranken, M. K. (1989) Rancidity in meats. In *Rancidity in Foods,* eds J. C. Allen and R. J. Hamilton, 2nd edn, Elsevier Applied Science, 225–236, (see p. 231)

Robards, K., Kerr, A. F. and Patsalides, E. (1988) Rancidity and its measurement in edible oils and snack foods. *Analyst,* **113,** 213–224

Rock, S. P. and Roth, H. (1966) Properties of frying fat. I: The relationship of viscosity to the concentration of non-urea adducting fatty acids. *Journal of the American Oil Chemistry Society,* **43,** 116–118

Rossell, J. B. (1986) Factors affecting the quality of frying oil. Leatherhead Food Research Association technical note 47

Rossell, J. B. (1989) Measurement of rancidity. In *Rancidity in Foods,* eds J. C. Allen and R. J. Hamilton, 2nd edn, Elsevier Applied Science, 23–52

Shelef, L. A. (1977) Effect of glucose on the bacterial spoilage of beef. *Journal of Food Science,* **42,** 1172–1175

Sulthana, S. N. and Sen, D. P. (1979) Studies on deep fat frying: changes during heating of oil. *Journal of Food Science and Technology,* **16,** 208–213

Suzuki, M., Suzuki, H., Karube, I. and Schmid, R. D. (1989) Disposable micro hypoxanthine sensors for freshness estimation. In *Biosensors Applications in Medicine, Environmental Protection and Process Control,* eds R. D. Schmid and F. Scheller, GBF Monographs vol. 13, VCH, Weinheim, 107–111

Timms, R. E. (1985) Physical properties of oils and mixtures of oils. *Journal of the American Oil Chemistry Society,* **62,** no. 2, 241–249

Turner, A. P. F. (1983) Applications of direct electron transfer bioelectrochemistry in sensors and fuel cells. *Biotech 83,* Online Publications, Northwood, UK

Watanabe, E., Ando, K., Karube, I., Matsuoka, H. and Suzuki, S. (1983) Determination of hypoxanthine in fish meat with an enzyme sensor. *Journal of Food Science,* **48,** 496–500

Weiss, T. J. (1983) Food oils and their uses. Ellis Horwood, Chichester

Wortberg, B. and Woller, R. (1982) Zur Qualität und Frische von Fleisch und Fleischwaren im Hinblick auf ihren Gehalt an biogenen Aminen. *Fleischwirtschaft,* **62,** 1457–1463

17 Chemosensors, Biosensors and Immunosensors

Erika Kress-Rogers

Contents

1	Introduction	582
1.1	Chemical Sensor Types and their Building Blocks	582
1.2	Recent Developments	587
1.3	Sensor Specifications for Food Applications	588
2	Chemically Sensitive Semiconductor Devices: Solid-State Sensors for pH, Acidity, Ions, Gases and Volatiles	589
2.1	Introduction to Novel pH Sensors	589
2.2	Alternative Approaches to pH Measurement	591
2.2.1	Solid Contact to a Glass Membrane	591
2.2.2	Solid Contacts to Other Ion-Sensitive Membranes	594
2.2.3	Electrodes Based on Electronically Conducting Metal Oxides	595
2.2.4	Microelectronic Devices	595
2.2.5	Fibre-Optic Chemical Sensors	598
2.3	Microelectronic Chemical Sensors Based on the FET: Principles, Characteristics and Applications	599
2.3.1	Introduction to the ISFET and the CHEMFET	599
2.3.2	Nomenclature: MOS and MeOx, IC and Smart Sensor	600
2.3.3	How Do the ISFET and the CHEMFET work?	601
2.3.4	Ion-Sensitive Membranes for the ISFET Family	603
2.3.5	The REFET Reference Device	605
2.3.6	Development for Industrial Applications	606
2.3.7	A Commercial ISFET/REFET System for pH Measurement in the Food Industry	608
2.3.8	Trials of pH ISFET Probes in Meat Processing	608
2.3.9	Acidity Sensors	608
2.4	Gas Sensing Semiconductor Devices	611
2.4.1	Introduction	611

1 Introduction

1.1 Chemical Sensor Types and their Building Blocks

Biosensors were beginning to feature in newspapers in the early 1980s and are now the subject of dedicated conferences. Commercial instruments based on advanced biosensor designs are already on the market in the clinical sector. In the food industry, notably in Japan, simpler designs are used at present since the diversity of this market does not, in general, favour a large investment on a single sensor specification.

Biosensors are defined here as sensors that rely on the interaction of a biocatalyst, usually an isolated and purified enzyme, with the analyte: they represent just one sector within the area of novel chemical sensors, (Figure 17.1). (Many other definitions of the term 'biosensor' (Section 3.1) are implied elsewhere; often they are much wider and thereby less distinctive.) Biosensors have been developed using a wide range of base devices monitoring the enzyme-catalysed reaction either by the gases or small ions produced or consumed in the reaction, or by the heat evolved, or by coupling to other reactions involving redox couples or luminescent compounds, for example.

Both traditional and novel gas or ion probes or heat-sensitive devices have been used as base devices for biosensors. Novel gas and ion probes are, however, also of interest to the food industry in their own right. Without the biological characteristics but also without the added complication and development cost due to the inclusion of the biocatalyst, these devices could help solve problems such as those associated with the in-line use of conventional ion-selective electrodes in the food industry.

Chemical sensors are based on the interaction of the analyte with a chemically sensitive layer immobilized on a base device (Table 17.1). A wide range of interactions have been employed in the design of chemical sensors, including reactions catalysed by an immobilized inorganic catalyst or enzyme, binding reactions with an immobilized antibody or lectin, or reactions with an immobilized dye or fluorescent compound. Specific adsorption or absorption based on size or polarity, for example, is also used. The reaction or absorption process results in a change in the electrical, optical or acoustic properties of the sensing surface, and these control the output of the base device.

A wide range of microelectronic devices have been chemically sensitized by the incorporation of ion-sensitive membrane layers or inorganic catalyst films for the assay of small ions and gases. For example, the change in surface charges formed at the interface of a chemically sensitive insulator with a sample containing the analyte may be used to change the characteristics of a microelectronic device such as the ISFET (Section 2.3). This can be configured to provide a voltage proportional to the logarithm of the analyte concentration.

The tip of an optical fibre can be sensitized by immobilizing a reagent more commonly used in conventional colorimetric or fluorescent assays. If a reversible reaction is chosen, continuous measurement of the analyte concentration is possible.

Piezoelectric crystal balances have been coated with films that absorb or adsorb certain chemical species with a greater or lesser degree of specificity.

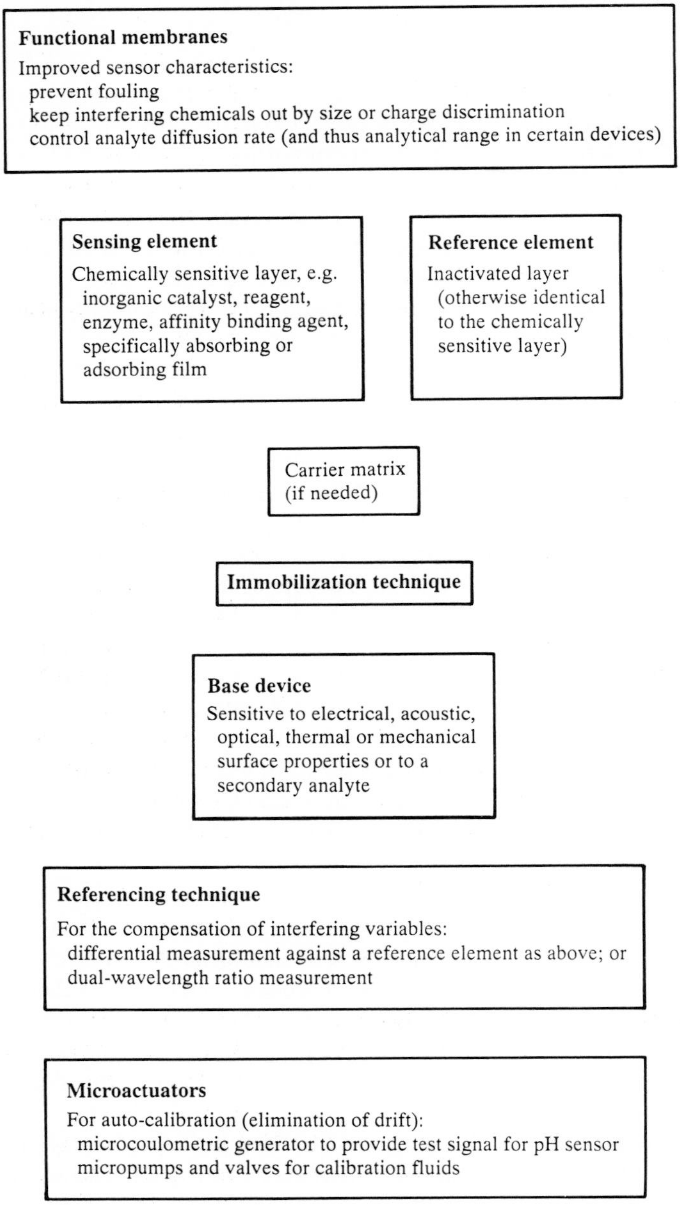

Functional membranes

Improved sensor characteristics:
- prevent fouling
- keep interfering chemicals out by size or charge discrimination
- control analyte diffusion rate (and thus analytical range in certain devices)

Sensing element

Chemically sensitive layer, e.g. inorganic catalyst, reagent, enzyme, affinity binding agent, specifically absorbing or adsorbing film

Reference element

Inactivated layer (otherwise identical to the chemically sensitive layer)

Carrier matrix (if needed)

Immobilization technique

Base device

Sensitive to electrical, acoustic, optical, thermal or mechanical surface properties or to a secondary analyte

Referencing technique

For the compensation of interfering variables:
- differential measurement against a reference element as above; or
- dual-wavelength ratio measurement

Microactuators

For auto-calibration (elimination of drift):
- microcoulometric generator to provide test signal for pH sensor
- micropumps and valves for calibration fluids

Figure 17.1 *Building blocks for chemical sensors: some can be omitted and others added depending on sensor type and application*

Table 17.1 *Chemically sensitive layers*

Sensitizing layer	*Device examples*	*Analyte examples*
Inorganic catalyst (for example Pd)	CHEMFET	Hydrogen, ammonia, methane, ethylene
Metal oxide semiconductors SnO_2, ZnO_2	Resistor	Oxidizing and reducing gases (methane, alcohols, aldehydes ...)
Metal oxides (MeOx) (electronically conducting oxide) PtO_2, IrO_2, SbO_2	Metal-MeOx electrode	pH (oxidizing and reducing agents as interference)
Insulating oxides SiO_2, Ta_2O_5, Al_2O_3, further oxides	ISFET	pH, sodium, calcium, fluoride
Crown compounds, e.g. valinomycin	ISFET, ISE	Many ions, e.g. potassium
LB films	ISFET	A wide range of analytes after further R&D
Chemically sensitive dye or fluorescent compound	FOP	pH (a wide range of other analytes after further R&D)
Enzyme	AMEE, AIDMEE, ENFET, ENFOP, ENTHERM, ENISE	Specific saccharides, alcohols, amines, amino acids, aldehydes, organic acids
Whole cell incorporating enzymes and cofactors	CISE	Nitrate, BOD, mutagenicity, herbicides
Antibody or lectin	SPR, TIR, SAW	Specific bacteria, vitamins, pesticides, moulds, meat species
DNA probes		Specific microbes (including virus), speciation, sourcing

Table 17.1 *continued*

As a special case, the measurement of acidity (continuous titration curve, in-flow) requires the integration of ISFET or metal-MeOx electrode with microcoulometric generating electrode (sensor-actuator integration)

List of Abbreviations for Table 17.1

Further details on the devices can be found in the sections indicated.

CHEMFET	Chemically sensitive field effect transistor (Section 2.3.4)
MeOx	Metal oxide semiconductor (a) ion-sensitive (Section 2.2.3) (b) gas-sensitive (Section 2.4.2) (see also Section 2.3.2)
ISFET	Ion-selective field effect transistor (Section 2.3)
ISE	Ion-selective electrode (conventional probe)
FOP	Fibre-optic probe (Sections 4.4 and 2.2.5)
AMEE	Amperometric mediated enzyme electrode (Section 3.2)
AIDMEE	Amperometric indirect dual membrane enzyme electrode (Section 3.3)
ENFET	Enzyme-sensitized field effect transistor (Section 3.4)
ENFOP	Enzyme-sensitized fibre-optic probe (Section 4.4)
ENTHERM	Enzyme-sensitized heat-sensitive device (Section 3.5)
ENISE	Enzyme-sensitized ion-selective electrode
CISE	Ion-selective electrode carrying whole microbial or tissue cells instead of isolated enzymes and co-factors
SPR	Surface plasmon resonance device (sensitive to small changes in optical surface properties) (Section 4.2)
TIR	Total internal reflection device (sensitive to small changes in optical surface properties) (Section 4.3)
SAW	Surface acoustic wave device (sensitive to small changes in acoustic surface properties) (Section 4.6)

The resulting increase in the mass of the device is measured. Such films have also been applied to devices which detect a change in the acoustic or optical surface properties as binding occurs. When the specific binding agent is an antibody to the analyte, or another affinity binding agent, as employed in immunoassay kits, the device is referred to as an immunosensor.

Chemical sensors that do not include a biocatalyst or affinity binding agent, will be labelled here as chemosensors to distinguish them from biosensors and immunosensors as defined in this chapter.

1.2 Recent Developments

All the devices described above already existed when novel sensor developments were surveyed in the early eighties to see whether they could be adopted for applications in the food industry. (For a detailed description, see Kress-Rogers, 1985a in Section 2.6.) However, since then development and engineering have progressed to provide clinical biosensors sufficiently advanced for use by the patients themselves, for example. Optical immunosensors are now approaching a performance standard such that manufacture for the clinical market can be expected shortly. Also, innovative instrument designs for the assessment of complex chemical or microbial food conditions have been proven at the prototype stage (an overview and two examples are given in Chapter 16).

Microelectronic and fibre-optic sensors for pH, ions and gases have experienced a patchy commercial development. Designs for applications in surgery have been on the market for several years. These are produced in small numbers at a high cost per sensing head and instrument and cannot be transferred to industrial applications without further development. The latter has been hampered primarily by difficulties in defining the markets for such devices so as to justify the appreciable costs associated with the development and subsequently the setting up of a production line. However, industrial pH probes based on transistor devices are now gaining a firmer foothold in the commercial field, and fibre-optic pH probe designs have become more refined recently. Considering the hazards and maintenance problems associated with the use of traditional glass membrane pH electrodes in the food industry, this is a very welcome development.

A development which will benefit a wide range of sensors is the recent progress in the design of special polymer membranes that exclude interfering chemical species, control analyte diffusion and have sites for the covalent bonding of reagents.

As more effort is now directed at developing chemical sensors for industrial applications, it has become more evident that the optimization to commercial standard of just one sensor for one specific in-line application can be very expensive. This has led to an interest in techniques such as neural network analysis that make the best of the sensors available. There is, however, the risk of necessitating renewed expensive setting-up procedures when food recipes or suppliers of ingredients, for example, change. There is, therefore, room for a limited range of fully optimized sensors for food applications where a wide market exists with similar application details. Such optimized sensors can also be applied in situations such as regular check-ups in store rooms or delivery vehicles where neural networks are less applicable.

1.3 Sensor Specifications for Food Applications

In addition to the requirements in many other industries, instruments for the food industry need to satisfy high standards in hygienic design. An in-line sensor should, for example, be flush fitting so that bacteria cannot be trapped in any crevices. It should also withstand cleaning-in-place (CIP) with hot caustic solutions. The sensor must not represent a hazard of potential food contamination with either harmful reagents or sharp fragments. The latter condition is not met by the conventional glass membrane electrode, for example. In principle, a tight mesh could be mounted over the electrode, but this would then restrict the access both for the product flow and for the cleaning solution. Further, a reference electrode containing a glass frit needs laborious maintenance attention so as to ensure a stable electrode calibration. Suitably designed microelectronic or fibre-optic sensors for pH and ion concentrations could help here.

As an alternative to a CIP-resistant, long-life probe it is in some applications possible to use disposable probes if these can be obtained inexpensively. This is an option particularly for biosensors where hot caustic solutions would lead to inactivation of the enzyme. Another alternative is robotic sampling from the line (see Section 5.3).

In contrast to clinical applications, chemical sensors for food applications have to operate over a wide temperature range (chill/ambient/hot). Chill applications and temperature variations represent a considerable difficulty in the engineering of probes governed by enzyme kinetics.

For many applications, it is possible to choose biosensor design principles that are instead governed by diffusion kinetics and have a lesser temperature dependence (Section 3.3). Chemical sensors without a biological element can be applied in hot products. The upper limit of the temperature range here is often defined by adhesives and functional membranes (for size exclusion or protection against fouling, for example) used in the construction of the device rather than by the properties of the chemically sensitive layer. For applications where a resilient chemical sensor cannot be developed, a non-contact technique such as near-infrared analysis (Chapter 5) may be considered as an alternative. This may involve the determination of a different chemical variable, for example a precursor variable measured upstream (see Chapter 1).

Principles, characteristics and practical considerations in the application of chemical sensors in the food industry will now be described. Solid-state sensors for pH, acidity, ions, gases and volatiles will be discussed in Section 2; amperometric, potentiometric and thermometric biosensors in Section 3. Immunosensors based on chemically sensitive optical and acoustic devices are treated in Section 4 together with gas sensors based on the same base devices. Finally, techniques used in the installation and signal interpretation of sensors are discussed in Section 5.

2 Chemically Sensitive Semiconductor Devices: Solid-State Sensors for pH, Acidity, Ions, Gases and Volatiles

2.1 Introduction to Novel pH Sensors

In the discussions on the development of novel sensors for the food industry, biosensors for sugars and amines or immunosensors for vitamins and microbes feature prominently. Yet the comparably simple measurement of the pH value is still carried out off-line in many food processing operations because the conventional pH electrode is not suitable. The consequent infrequency of and delay in obtaining pH values for the food process stream leads to a delayed correction for out-of-specification values or necessitates holding times. New pH sensor designs suitable for in-line applications in the food industry are therefore of considerable interest. The pH value not only governs chemical reactions during food processing, but is also important as one of the factors controlling microbial growth, and particularly the growth of pathogenic bacteria. The pH can also indicate the progress of fermentation processes.

Potentiometric pH glass electrodes have benefited from over 80 years of development and are generally reliable sensors. However, they have a number of disadvantages, particularly for clinical and food applications (see Table 17.2). One of their disadvantages resides in the glass membrane, which can present a potential hazard in a food processing area. Also, the performance in food is often adversely affected by fouling of the calomel or silver/silver chloride reference electrode usually employed in combination with the glass membrane electrode. A number of alternative pH sensors are being developed with the aim of overcoming these problems.

A microelectronic device, the ion-selective FET (ISFET), together with a solid-state reference electrode (REFET), is now available in a configuration specifically designed for food applications. This device will be described in detail in this chapter after a comparison of the approaches under investigation. Among the advantages of the all-solid-state ISFET/REFET system are robustness, fast response and the possibility of specific multi-ion sensing with a compact device. It can either be configured for the measurement of pH microenvironments in foods, or else in a more chunky form for insertion into tough foods. Moreover, a microtitrator that can be employed for in-line acidity measurements is under development, based on the ISFET together with a microactuator. Acidity is important in contributing to the flavour of a food, and analysis of the titration curve provides information on the nature and concentration of specific acids.

The chemical and microelectronic aspects of solid-state pH sensors are discussed in detail here in view of the importance of the pH value and the unfamiliarity of the new technology, now commercially available for in-line

Table 17.2 *Problems with conventional pH probes*

Property of the pH electrode or calomel reference electrode	*Implication for application in food processing area or QC laboratory*
Fragility of the liquid-filled glass membrane	Foreign body hazard
Filling solutions	Inconvenience Leakage
Drift in liquids with biofouling properties	Problems in maintaining a reliable operation Laborious cleaning and recalibration
Minimum size (3 mm)	Not suitable for the study of pH gradients or pH profiles Not suitable for detection of high-pH pockets (assessment of microbial hazard) Not easily incorporated in multi-ion sensors Interference compensation not easily arranged
High impedance	Heavy shielded cables and pH meters needed
Limited operation temperature (-10 to $+110$ °C for Ag/AgCl ref.el. and sat. KCl bridge) (0 to $+60$ °C for Hg/Hg_2Cl_2 ref.el. and sat. KCl bridge)	Not suitable for elevated temperatures (in solutions at excess pressures) Not easily sterilized
Slow response (5 s)	Delayed indication of change in process flow (where in-line use is possible) Not suitable for flow injection analysis (FIA)
Long time to hydration equilibrium	Needs to be stored wet

food applications. If a brief introduction to solid-state pH sensors is preferred, this can be found in Kress-Rogers (1991). ISFETs under development for the measurement of ions such as sodium, calcium or potassium are introduced in Section 3.3.4; biosensors based on the ISFET are described in Section 3.

The pH ISFET represents a significant development in the history of instruments for the measurement of pH which has been dominated by the glass membrane electrode for several decades (Figure 17.2). It is therefore appropriate to look back to the years when the glass electrode was becoming established in the 1940s after several decades of development. The standard method at that time was the measurement of the potential difference between two hydrogen or calomel electrodes positioned in the sample solution and a reference solution, respectively. Strohecker wrote in 1943 (translated, and with small inserts taking account of preceding paragraphs): 'Nowadays, the glass electrode, a thin-walled glass vessel with an inner and outer electrical contact, is often used to good effect. A potential difference depending on the pH of the inner and outer solutions is formed at the thin glass wall. The method requires very sensitive instrumentation (preferably for an electrostatic measurement); also it is not as reliable as the measurement with hydrogen electrodes.'

The users of the early glass electrodes had to consult tables of logarithms to relate the H^+ concentration, or more accurately, the H^+ activity, to the pH value and to identify the correct glass electrode to be used for the pH range in question. Fifty years on, modern glass electrodes with built-in reference electrodes and pH meters are, of course, much simpler to use, more versatile, more reliable and more compact. They will doubtless remain standard laboratory instruments for many years to come. However, in many industrial and clinical applications, solid-state devices will offer the better or the only alternative. More details are given in Section 2.3 and after an introduction in 2.2.4 and 2.2.5.

2.2 Alternative Approaches to pH Measurement

2.2.1 Solid Contact to a Glass Membrane

Some of the remaining problems with the glass electrode could be reduced if a stable solid contact could be made instead of the internal reference buffer between the inner glass membrane surface and the Ag/AgCl electrode immersed in this buffer. There have been a number of attempts to achieve such a solid contact; these have not had convincing success, in that the repeatable large-scale production of stable electrodes has not been achieved. Fjeldly and Nagy (1985) have been more successful than many others in approaching this difficult problem. They used a reaction with silver fluoride while heating in a gas flame to establish a gradual junction forming the inner contact of a glass membrane.

The problem consists in creating a stable solid contact providing a reversible transition from the ionic to the electronic part of the sensor. A new approach has been developed by Kreuer (1990) based on the reversible transference of Li^+ (to the inner membrane surface, hence not in direct

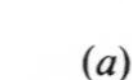

***Figure 17.2** Developments in instrumentation for pH measurement. (a)* Der Chemist: *The chemist of a bygone era, controlling the process by stoking the fire guided by feedback from his senses: visual, audio, tactile (from the 1968 calendar of Scholven-Chemie AG, Gelsenkirchen). (b) The food chemist of the 1940s occasionally measures pH with a simplified apparatus after Roeder, avoiding the cost of sensitive electrostatic instrumentation. Base is added until the capillary electrometer indicates elimination of the potential difference between sample and reference solutions contacted by solid chinhydron electrodes with platinum contacts (after Strohecker 1943). (c) The QC laboratory chemist measures pH with a modern pH glass electrode with integral reference electrode. This is dependent on the maintenance of the filling solutions, storage in an appropriate liquid medium and avoidance of fouling the liquid junction with proteins or fats. The fragile glass membrane needs to be protected from impact. The photo shows a Philips pH electrode with integral reference and provided with a tight-fitting cap carrying the liquid storage medium. (d) A novel pH probe is now available for the laboratory. This is based on an ISFET pH sensor and can be stored dry. Due to the absence of filling solutions maintenance is simplified, and the all-solid construction provides robustness and eliminates glass fragment hazards. The probe shown is produced by UNIFET (USA). (An earlier Thorn EMI (GB) ISFET device and an ISFET probe for industrial applications by Sensoptic (NL) are shown in Figures 17.6(a) and (b).)*

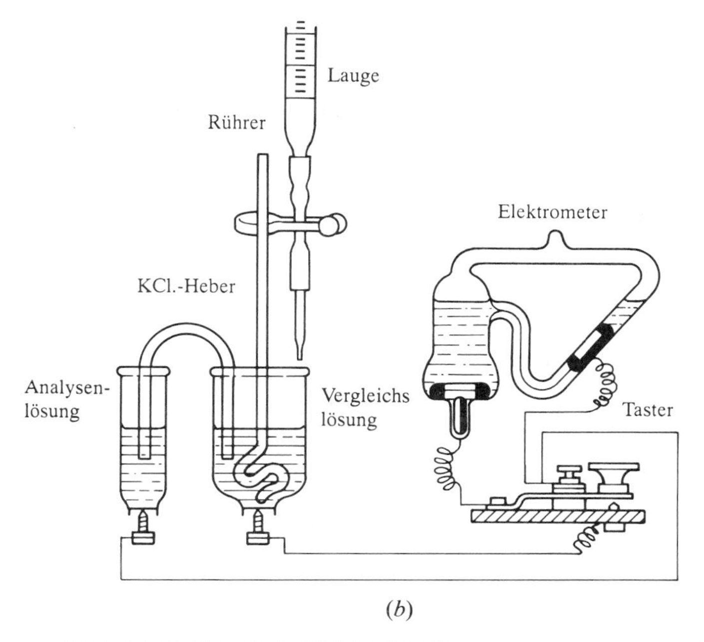

(*b*)

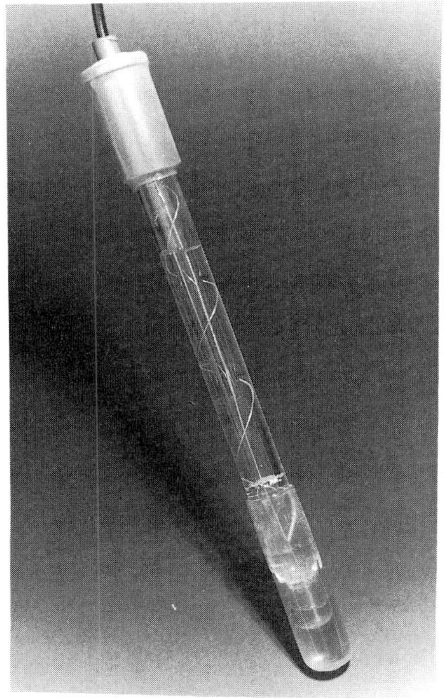

(*c*)

Figure 17.2 *Continued*

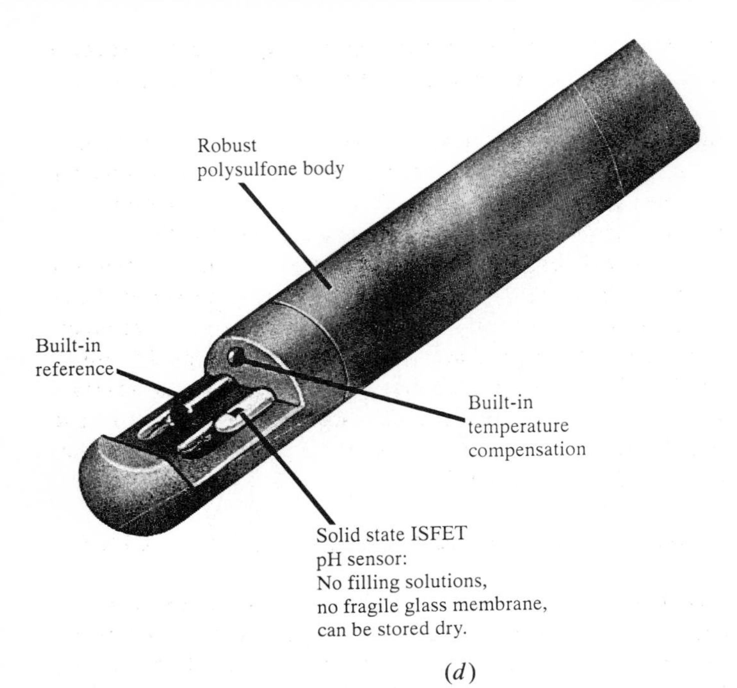

Figure 17.2 *Continued*

contact with the sample). He prepared contacts based on the ternary system Li-Ag-I. (Li ions play a role in the ion exchange processes in glass membrane pH sensors. They are transferred between the hydrated gel layer and the bulk of the glass membrane.)

The new pH electrode is stable at temperatures up to 150 °C, that is 30 K higher than the conventional electrodes; the lower temperature limit, on the other hand, is less favourable at 20 °C. The minimum size of the electrode is reduced to 0.5 mm, and one would expect this device to be less fragile than the liquid-filled glass bulb. The new electrode type is proposed for development for applications where temperatures are elevated during operation or where sterilization is needed, for example in fermenters.

2.2.2 Solid Contacts to Other Ion-Sensitive Membranes

In addition to the glass membrane, a number of ion-sensitive membranes exist for the assay of larger ions such as sodium, potassium, calcium or nitrate/nitrite. These membranes may have the form of sintered pellets or polymers carrying embedded electroactive compounds, for example. They combine ionic conductivity with a specific ion sensitivity at the interfaces. For all these membranes, it is necessary to establish a stable contact between the ionic conduction world and the electronic conduction world of the metal electrode (Figure 17.3). The most natural way of achieving this is by positioning the membrane between the sample liquid and a second liquid that provides a stable potential at the interface of an immersed Ag/AgCl electrode.

Efforts to replace this liquid contact between ionic and electronic conduction by a solid junction have been largely unsuccessful, not only for the glass membrane (see above) but also for other ion-selective membranes. Coated wire electrodes, for example, are simple to produce (by dipping the metal into a membrane polymer mixed with plasticizer and carrying a specific ion exchanger), but stability has often been poor. Exceptions have been attributed to the (often accidental) formation of an intermediate layer, such as an oxide, between the membrane and the metal (Bergveld 1988). This may explain the difficulties in achieving the repeatable large-scale production of stable devices. These considerations also apply to the extended-gate FET (Van der Spiegel et al. 1983), where the coated metal electrode is integrated with a MOSFET device.

2.2.3 Electrodes Based on Electronically Conducting Metal Oxides

It is known that oxidized metals behave as stable electrodes, and this has led to the development of the metal/metal-oxide ion-sensitive electrodes (De Rooij and Bergveld 1980). These consist of films of electronically conducting oxides such as PtO_2, IrO_2 or SbO_3 produced either on their native metal or on another metal. The pH sensitivity of these oxides has been attributed to hydroxyl groups at the surface as for the more familiar oxide of the conventional glass membrane. Most of the work in this area has centred on the Ir/IrO_2 system. The metal oxides discussed here, however, are also sensitive to reducing or oxidizing agents, and this is a serious drawback in many applications (Bergveld 1988). Disagreement still exists on the exact mechanism of the iridium oxide redox system, and efforts continue to stabilize the offset and sensitivity (voltage change per pH decade change) of the device and to overcome the interference due to the redox sensitivity (Olthuis et al. 1990; Tarlov et al. 1990).

Favourable characteristics of the sensor are stability in agressive environments at high temperatures and pressures, low impedance and a short response time. Also, film deposition alows miniaturization and integration with other sensors. The Ir/IrO_2 system has also been applied in the construction of a microtitrator (see acidity sensors in Section 2.3) both in the actuator and in the pH sensor parts of the device (Olthuis, et al. 1991). Before the advantages of the Ir/IrO_2 probe can be fully utilized, the stability and selectivity will need to be optimized based on a full characterization of the underlying mechanisms.

2.2.4 Microelectronic Devices

A number of microelectronic semiconductor devices can be chemically

Sample
Ion-conducting membrane with specific ion-sensitivity at the interfaces
Intermediate layer: internal electrolyte
Galvanic contact
Ag/AgCl electrode
(a)

Sample
Electronically conducting metal oxide with redox-sensitive exposed surface
Ohmic contact
Metal electrode
(b)

Sample
Ion-sensitive film or membrane
Insulating film
Capacitive contact
Channel
Substrate
(c)

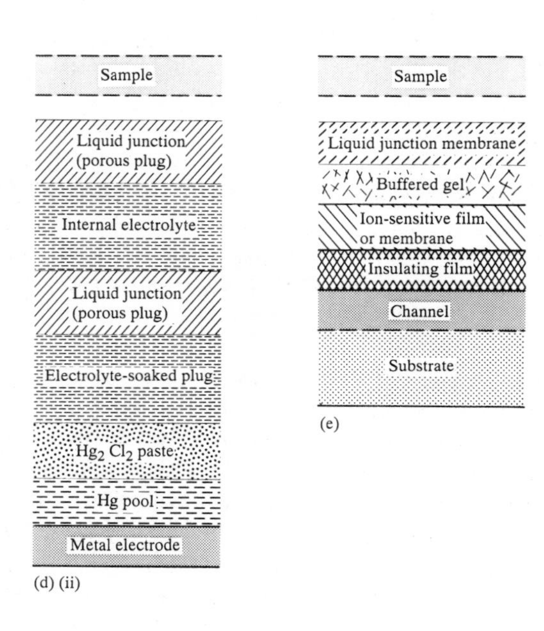

Figure 17.3 *Membrane configurations. To fit in with the usual presentation of the microelectronic devices with planar structures formed on a substrate surface, the ion-selective electrodes appear upside-down. (a) Ion-selective membrane in galvanic contact with metal electrode. Examples for membrane types: pH-sensitive glass membrane; sintered pellet ion-selective membranes. The connection of pH-sensitive glass membranes to a silver|silver-chloride electrode via a gradual solid junction has been made. In general, it has been difficult to achieve a stable potential for solid junctions between the ionically conducting membrane and the electronically conducting metal electrode. This also applies to coated wire electrodes, where the formation of an intermediate oxide layer between the ion-selective membrane and the metal wire can improve stability. Balance of electrochemical potentials:* $\mu(H^+ \ sol) = \mu(H^+ \ glass)$. *(b) pH-sensitive metal oxide in ohmic contact with metal electrode. Examples*

for membrane materials: PtO_2, IrO_2, Sb_2O_3. Reaction with hydrogen ions: $MO_x + 2\,\delta\,H^+ + 2\,\delta\,e^- \rightleftharpoons M\,O_{x-\delta} + \delta\,H_2$. (c) Ion-selective membrane in capacitive contact with conducting layer (as in the ISFET). Examples for membrane types: thin pH-sensitive inorganic films; thin ion-selective organic membranes. The channel is a sheet of high electronic conductance formed in metal-insulator-semiconductor (MIS) device at appropriate gate bias values. This is known as the field effect. For a p-type substrate, the channel exhibits n-type conductance in the usual inversion mode. A common MIS device is the field-effect transistor (FET), a pre-amplifier element. Measurement principle: the surface charges at the membrane–sample interface induce a charge increment in the channel region. This charge can readily be measured when the MIS structure is part of an FET device. (d) Conventional reference devices. (i) Ag|AgCl reference electrode (not shown). As diagram (a), but the ion-selective membrane is replaced with a liquid junction formed by a porous plug, glass frit or other 'leaky' structure. (ii) Calomel reference electrode (shown). (e) ISFET-based reference devices. (i) REFET, configuration 1 (not shown). As diagram (c), but the ion-sensitive membrane is replaced with a similar membrane of weaker ion sensitivity (under development). (ii) REFET, configuration 2 (shown). As diagram (c), but the ion-sensitive membrane is covered with a buffered gel and a membrane fulfilling the role of a liquid junction

sensitized for highly sensitive and specific measurements, but most of the work in this area has concentrated on the field effect transistor (FET). This device has been turned into the ion-selective FET (ISFET) by incorporating an ion-sensitive insulating film. Insulating oxides such as SiO_2, Ta_2O_5 or Al_2O_3 are known (from colloid chemistry) to be ion-sensitive. This property was earlier assessed by potentiometric acid or base titration of a suspension of the oxide. In this experiment, the surface charge on the oxide as a function of pH could be determined. One of the models applied to this system is the site dissociation model, which describes the equilibrium between the surface OH groups and the H^+ ions in the vicinity of the surface. Surface potentials are also formed in specific ionophores such as the crown ethers, which can be synthesized for the selective detection of ions (Bergveld 1988).

While the traditional method of colloid chemistry for the assessment of surface charge density is time-consuming, microelectronics can offer a device that indicates electronic changes induced in the surface layer of a semiconductor as a function of surface charge density on an adjacent insulator film. At the same time, the selection of *p*- or *n*-doped contacts allows a choice of measuring either the properties of the substrate bulk or the properties of the surface layer of minority charge carriers (known as the channel) induced by the charges in the insulator. This allows a very sensitive measurement of the insulator charge density. At the same time, the rather high impedance of the insulator is converted to a very low channel impedance. This impedance conversion is, in fact, the usual bread-and-butter job of the FET device as an element of electronic circuits. Hence the ISFET was developed, and this is described in more detail in Section 2.3.

2.2.5 Fibre-Optic Chemical Sensors

An alternative approach to the solution of the problem associated with the use of conventional pH electrodes in the food industry is the construction of fibre-optic pH probes (Figure 17.4). Here, the fibre tip carries a reagent that will change either colour or fluorescence properties with a change in pH. Good-quality, low-drift probes of this type are already used in surgery, but need to be adapted for the pH range and environment encountered in the food industry. Some of the many experimental probe designs drift owing to factors such as leaching of adsorbed dye and photobleaching of dye or by drying out of gels physically entrapping the dye, as well as because of optoelectronic drifts. It has been demonstrated by many other recent experimental probes and by the commercial clinical probes that these problems can be overcome by techniques such as covalent bonding on special polymer films that can be grown on the fibre tip, and further by suitable referencing arrangements. Most reagents will cover only a limited pH range of typically two decades, but this can be addressed by combining reagents within one tip or by including a bundle of fibre tips each carrying a different reagent within the same cable. This may not be required for many process control operations, since a small range of pH values will normally be encountered at an in-line measuring point. The chemical response characteristics of the fibre-optic pH probe will differ from those of either the conventional glass membrane electrode or the microelectrode ISFET probe. Instead, they will be more akin to those of colorimetric assays.

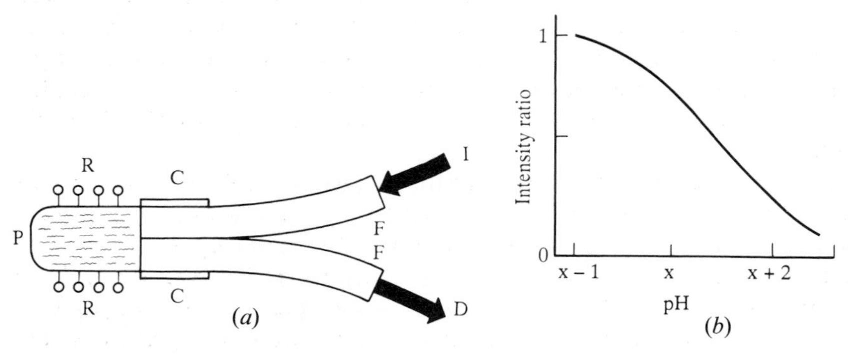

Figure 17.4 *(a) Principle of the fibre-optic chemical sensor: F optical fibres; C cladding; I input radiation; D detected radiation; R reagent immobilized on polymer; P polymer grown on fibre tip. (a) Schematic diagram. (b) Typical response of single fibre-optic pH probe over two pH decades*

The advantages of fibre-optic probes include immunity to electromagnetic noise as it is encountered in manufacturing plants. They can be operated in microwave ovens if required (fibre-optic thermometers are already used in industrial microwave cookers). Also, fibre-optic probes can be constructed as long flexible cables that can readily be suspended into the centre of a tank containing a food liquid, or into a liquid at the bottom of a conduit for processing waste liquid, for example. Fibre-optic probes for a wide range of sensors for chemical and physical properties can be constructed (including immunosensors) so that there is the option of a multivariable measurement system, initially perhaps for ions and dissolved gases. (Fibre-optic biosensors and immunosensors are briefly described in Section 4.4.)

2.3 Microelectronic Chemical Sensors Based on the FET: Principles, Characteristics and Applications

2.3.1 Introduction to the ISFET and the CHEMFET

The ISFET offers an elegant solution to the old problem of connecting the ionic conduction world to the electronic conduction world. Instead of attempting to produce a stable conductive contact between an ionically conducting membrane and an electronically conducting metal, the ISFET uses a capacitive contact to monitor an ion-sensitive insulator. The ISFET provides a direct, sensitive and convenient method of measuring the surface potential at the insulator/electrolyte interface (Bergveld 1988).

The FET is a common device in integrated circuits, where it acts as a pre-amplifier by converting a high-impedance signal to a low-impedance signal, which can then be transmitted or processed conveniently with good signal to noise characteristics. The standard FET device has earlier been incorporated into ion-selective electrodes to pre-amplify the signal at source so as to achieve a low-noise signal and avoid the need for heavy shielded cables and pH meters. The logical next step was the modification of the FET to act as a sensor in its own right, as first suggested by Bergveld (1970).

Arrays of several ISFETs, each measuring a different ion, can be constructed on a very small device and monitored with a single very compact instrument. This is useful not only for multi-ion assays but also for cross-sensitivity compensation. The small size is also attractive in itself not only to surgeons, but also for the study of the microenvironment of foods. An ISFET probe can be configured with a microtip that allows the identification of small regions of pH values that contribute to conditions favourable for the growth of pathogens in an otherwise safe food. Another advantage of the ISFET is the fast response (particularly for the ISFET with inorganic membranes), which is superior to that governed by the bulk diffusion processes in a conventional thick glass membrane. This fast response is a particular advantage in flow injection analysis (FIA) systems (Bergveld 1991).

In principle, it is possible to produce the ISFET at a price comparable with that of electronic components such as the close relative, the FET pre-amplifier, but this requires a very large annual production volume. As with other devices produced in planar silicon technology, the cost of setting up and running a line producing 100 devices per year is not much smaller than that for a line producing 100 000 per year, with obvious implications for the price of the individual device.

Related to the ISFET is the CHEMFET (also known as the GasFET). This family of chemical sensors provides a highly sensitive and specific response. Target analytes have been mainly gases such as hydrogen, ammonia, methane or ethylene (each with a different CHEMFET variant), but the range is now being expanded to other gases and volatiles. Gas sensors based on the ISFET have also been constructed, using Severinghouse-type designs for CO_2 measurement, for example.

Other microelectronic semiconductor devices developed into chemical sensors include the gate-controlled diode which has been turned into an ion-controlled diode in analogy to the ion-selective FET (ISFET), which is described in more detail later. The Schottky diode, in its simplest form a metal film in contact with a semiconductor, has been adapted as a gas sensor by using a catalytic metal as in the CHEMFET (see later). In its more usual form, the Schottky diode is a rectifier; it is the successor of the crystal and metal pin used in old radio sets.

2.3.2 Nomenclature: MOS and MeOx, IC and Smart Sensor

The ISFET is based on the metal-oxide-semiconductor FET (MOSFET) structure. The classical MOS devices are made by growing an insulating film of silicon oxide on a semiconducting silicon substrate. Over the oxide a thin metal film is deposited, thus forming a capacitor. Etching allows access from the top surface (through the oxide and metal films) so that contacts to the substrate can be made next to the contact with the metal film to form a planar device. This planar silicon technology allows the fabrication of many different circuit elements on a single silicon chip and is the basis of integrated circuits (ICs). When one or more silicon sensors are integrated on-chip with circuitry for signal conditioning (such as linearization of the sensor calibration, compensation for temperature changes, or analogue/digital conversion), the sensor is called a smart sensor. These can be inexpensive if produced in sufficient numbers to offset the tooling-up costs.

The metal-oxide-semiconductor (MOS) structure is sometimes confused with the metal oxide (MeOx) semiconductor, but has very little in common with the latter, which is a single material consisting of a metal oxide that is semiconducting. The most well-known materials of this type are tin oxide (SnO_2) and zinc oxide, which, at elevated temperatures, exhibit a conductivity change in the presence of oxidizing and reducing gases.

Currently, such MeOx semiconductors are usually manufactured in the form of a thick film printed on to a ceramic substrate, which also incorporates a heater and is often referred to as the Taguchi or Figaro sensor. This device acts as a gas sensor of a broad specificity, which can be tuned to a certain degree by temperature selection and chemical treatments. (Arrays of MeOx gas sensors with broad overlapping specificities are the subject of many studies on volatile assays by pattern recognition methods with the aim of constructing so-called odour meters; see Section 2.4.) In general, the MeOx semiconductor gas sensors are less specific and less sensitive than the CHEMFETs (or GasFETs). Another group of metal oxides (and here principally iridium oxide) has already been discussed above as one of the solid-state pH sensor types.

2.3.3. How Do the ISFET and the CHEMFET Work?

The applied gate–substrate voltage, above the threshold value, attracts minority *n*-type carriers from the *p*-type substrate to the area just under the insulating film that separates the gate metal from the substrate (Figure 17.5a). These minority carriers form a sheet of *n*-type conductance, known as the 'channel', which provides electrical continuity between the *n*-type source and drain contacts (provided that the device is switched on, that is the gate voltage is above the threshold value). The gate–substrate voltage sensitively controls the conductance in the channel. For gate voltages above the threshold region, the device thereby converts a high gate–substrate impedance into a low source–drain impedance, so that the FET acts as a pre-amplifer.

The unprotected device is very sensitive to impurities entering the thin (typically 200 nm) gate insulator layer or interacting with its surface. Such impurities shift the device characteristics (channel conductivity versus gate voltage) as shown in Figure 17.5b. Whilst this was the Achilles' heel of the early FET devices (prior to passivation with silicon nitride), it has now been used to good effect in the design of the chemically sensitive ISFET or CHEMFET. These are two families of highly sensitive chemical sensors; levels in the parts per billion range can be specifically detected by some CHEMFET gas sensors, for example. In the latter devices, the usual gate metal of the FET (such as aluminium) is replaced by a catalytic metal such as palladium or platinum or by a catalytic metal alloy.

In the ISFET, on the other hand, the gate metal film has been removed altogether, and an ion-sensitive film has been deposited over the gate insulator (Figure 17.5c). (In the early ISFET devices, the exposed usual gate insulator film acted as the pH-sensitive layer.) Ions diffusing into the gate area will polarize the interface of the gate with the fluid and thus

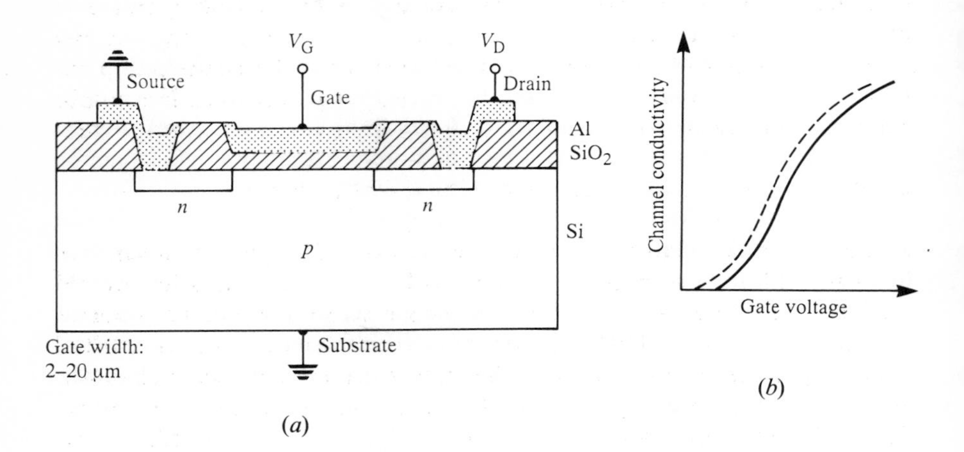

V_G
V_D
Source
Gate
Drain
Al
SiO_2
n
n
Si
p
Gate width:
2–20 μm
Substrate
(a)
Channel conductivity
Gate voltage
(b)

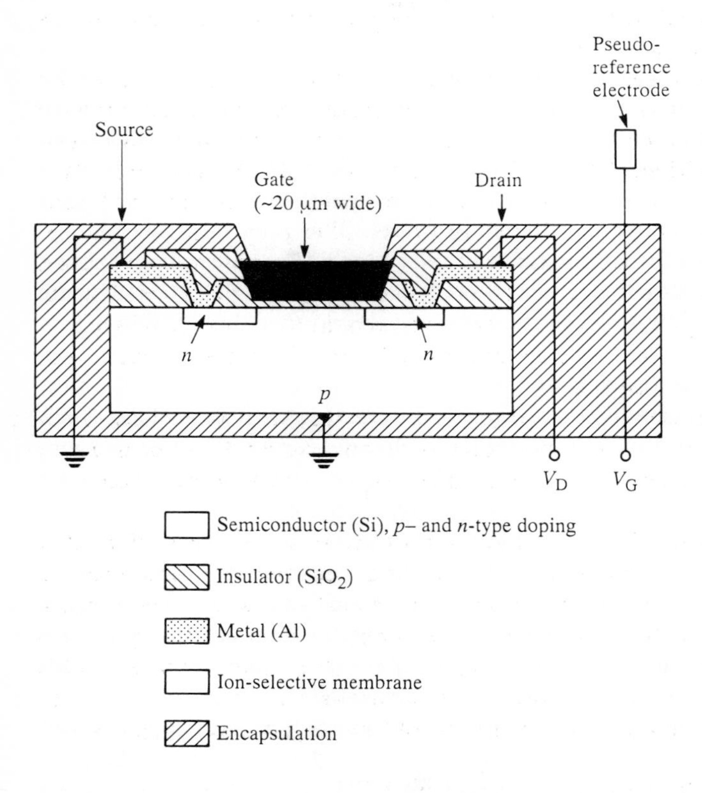

Pseudo-reference electrode
Source
Gate
(~20 μm wide)
Drain
n
n
p
V_D
V_G
Semiconductor (Si), p– and n-type doping
Insulator (SiO_2)
Metal (Al)
Ion-selective membrane
Encapsulation

(c)

control the device characteristics. A gate voltage is now applied to a reference electrode. In principle, the device would then be operated at constant gate voltage and the current measured. In practice, the channel current (or source–drain current) is kept constant at the isothermal point with a feedback loop and the required gate voltage provides the output signal. In this mode, the operation is largely independent of temperature.

2.3.4 Ion-Sensitive Membranes for the ISFET Family

A number of insulating oxides and some other inorganic ion-sensitive insulating films have been tested on ISFET devices, primarily for pH sensing, and for the assay of some ions such as sodium, calcium or fluoride (Clechet 1991; Klein 1991). In order to produce a wider range of ISFET devices, polymer membranes carrying specific ionophores such as crown ethers have also been applied. With the deposition of inorganic films, IC-compatible technologies can be used. For the polymer films, new deposition techniques have been proposed in recent years to overcome problems associated with dip-coated devices, such as poor membrane adhesion after prolonged operation.

At present, the response times reported for ISFETs carrying polymer membranes are much longer than for the inorganic films; the latter can have response times (5–5 per cent) of 5 ms, for example. Time effects have been investigated by Klein (1989). The lifetime of ISFETs carrying PVC (polyvinylchloride) membranes similar to those in conventional ion-selective electrodes (ISEs) is limited by poor adhesion of the PVC membrane to the silicon nitride surface of the ISFET gate. Some membranes for ISFETs also require the incorporation of a hydrogel layer for the exclusion of CO_2. New membrane types are now being developed to optimize the performance of a wider range of ISFET devices, beyond those based on inorganic membranes.

Improved membrane adhesion to silicon nitride was reported by Geun Sig Cha et al. (1991) for polyurethane- or silicone-rubber-based membranes, and adhesion was further enhanced by the incorporation of a silanizing agent.

Van den Berg et al. (1991) have reported on-wafer fabrication of a photopolymerized membrane in the construction of a calcium-sensitive ISFET. The ion-sensitive membrane is chemically bound to the sensing surface. An intermediate polyHEMA membrane is applied (also by chemical bonding) to ensure a thermodynamically well-defined interface. This new technique is expected to provide a calcium-sensitive ISFET with an operational lifetime comparable with that of conventional counterparts, but

Figure 17.5 *(a) Schematic diagram of the FET and the CHEMFET. (b) Device characteristics in the presence and absence of ions in the gate area. (c) Schematic diagram of the ISFET*

with the possibility of multi-ion sensor fabrication and other advantages already realized for pH sensing with ISFETs (low impedance, small sensor allowing convenient probe shape adapted to application, etc.; see Table 17.3).

A new approach has been taken by Vogel et al. (1990). They succeeded in depositing a new type of Langmuir-Blodgett (LB) film, which lacks the amphiphilic nature that had been thought of as necessary for such films. The new LB film type exhibits long-term stability (in contrast to the usual amphiphilic LB films). Further work (Vogel et al. 1991) has demonstrated the possibility of tailoring the pH sensitivity of the new LB films so that an

Table 17.3 *Advantages of the pH ISFET/REFET system*

Characteristic of the system	*Implications*
All solid state	No inconvenience with filling solutions No leakage Robust
Non-fragile	No foreign body hazard Suitable for insertion into tough materials Convenient handling
No liquid junction (if optimum REFET is used)	Easy maintenance
Small sensing element (0.2 mm with encapsulation)	Convenient probe shape can be chosen depending on application Microenvironment can be characterized (detection of high-pH pockets) Compact multi-sensor arrays can be produced
No hydration time	Can be stored dry
Temperature-independent over a wide range	No adjustment for temperature changes needed
Fast response	Suitable for flow injection analysis (FIA) No delay in indication of change (good for in-line monitoring)
Low impedance	Good signal to noise with light cables and compact instruments
IC integration possible	Smart sensors can be produced

ISFET-REFET system could be developed in the first instance. The potential of the technique resides, however, in the wide range of applications that have been suggested for LB films (Roberts 1990) using techniques such as the incorporation of doping molecules into the films. The practical applications of such films had earlier been limited by instability. The new structure has its own designation and abbreviation: electrolyte/LB-film/oxide/semiconductor (ELBOS).

2.3.5 The REFET Reference Device

Initially, a conventional reference electrode was employed in conjunction with the ISFET, either in standard size and immersed at a convenient point in contact with the sample, or in miniature form mounted with the ISFET device(s). After the development of pH-sensitive ISFETs in the 1970s, their performance was limited primarily by the shortcomings of the reference electrode, which gave rise to drift. Also, the operating period of miniature Ag/AgCl electrodes was limited. This has not prevented the application in surgery, where intermittent operation with regular calibration with the help of microvalves was employed (Sibbald et al. 1984). It was, however, recognized that a reference ISFET needed to be developed for optimum ISFET performance, particularly for industrial applications.

An early reference ISFET was designed by the Janata group (Comte and Janata 1978; Janata and Huber 1979). This was based on a pH-sensitive ISFET carrying buffered agarose and a miniature capillary liquid junction. Subsequently, various ion-blocking membranes were tested in the development of reference ISFETs (Matsuo and Nakajima 1984; Nakajima et al. 1982). A non-blocking layer on the REFET has been used by Bergveld et al. (1989). A promising approach is the construction of reference devices consisting of pH ISFETs with a weakened pH response (Wilhelm et al. 1991; Clechet 1991). These REFETs can readily provide a thermodynamically well-defined interface to the sample and represent liquid-free, all-solid-state IC-compatible desensitized reference ISFET devices. The characteristics (other than pH response) of such REFETs are also very similar to those of the actual pH ISFETs, and this is an ideal condition for sensor/reference pairs. A number of such REFET devices are under development. (The REFET of a current commercial pH ISFET system consists of another pH-sensitive ISFET, which carries a buffer gel layer over the ion-sensitive membrane. The buffer layer in turn is separated from the sample by a membrane which acts as a liquid junction.)

Whatever REFET design is used, the output signal is gained by differential amplification of the ISFET and REFET signals. The metal can housing of the devices now acts as a necessary but no longer performance-determining pseudo-reference electrode. Such a solid-state-based differential ISFET pH probe is free from leakage problems and requires neither reagents nor recovery periods for reagent recycling (Table 17.3).

2.3.6 Development for Industrial Applications

The potential advantages of the ISFET for food applications were examined in the early 1980s. In our discussions with the food industry we have encountered great interest, particularly for the pH-sensitive ISFETs. (The interest in fibre-optic probes has also been high for certain applications, and these probes are briefly discussed in Section 2.2.5.) The pH is, after all, important as one of the factors controlling both food processing operations and product stability (see Section 1), and conventional pH probes have many disadvantages for applications in food processing areas (see Table 17.2).

The ISFET, together with a solid-state reference electrode, offers the potential of solving these problems. Moreover, it can provide the advantages of fast response and specific multi-ion sensing with a compact device. The potential benefits for applications in the food industry were noted earlier (Kress-Rogers 1985a; 1985b). Yet the commercial development of the ISFET and CHEMFET was hesitant in the 1980s, but it is now coming to fruition. The active interest from the meat industry in the potential application of novel sensors in this sector (Kress-Rogers 1987) has helped to provide the motivation for the commercial production of a probe specifically designed for meat. A further impetus was provided by the recognition of the possibility of developing an in-line acidity sensor for food liquids based on a microtitrator device employing the pH ISFET. A dipstick microtitrator device was subsequently developed. (See Section 2.3.9 on Acidity.)

A number of ISFET applications for continuous blood analysis during surgery have been reported, and instruments for this purpose have also been commercially available in Japan. However, these instruments were not suited to industrial applications and were also expensive owing to their construction in small numbers for a specialist market. Only instruments intended for industry will be discussed here.

In the mid 1980s, Thorn EMI in the UK released a range of pre-commercial prototype instruments based on CHEMFET and ISFET devices. (An application for the ammonia-detecting CHEMFET was found in fire-detection systems; Anon. 1985.) The ISFETs were mounted in a flow cell. A small pump was included in the instrument to provide a continuous flow of sample fluid through the cell. A dual device for pH and sodium determination was available; devices for potassium, calcium and nitrate were tested in a clinical environment. A conventional, miniaturized reference electrode was mounted in the flow system (Figure 17.6(a)). Later, Thorn developed a cartridge containing an ISFET array and associated electronics intended for the determination and display of the concentrations of four ions, a reservoir of calibration fluid, a flow system with valves and a pump, all within the size of an audio cassette. The complete cassette was intended for disposal after 50 samples had been analysed.

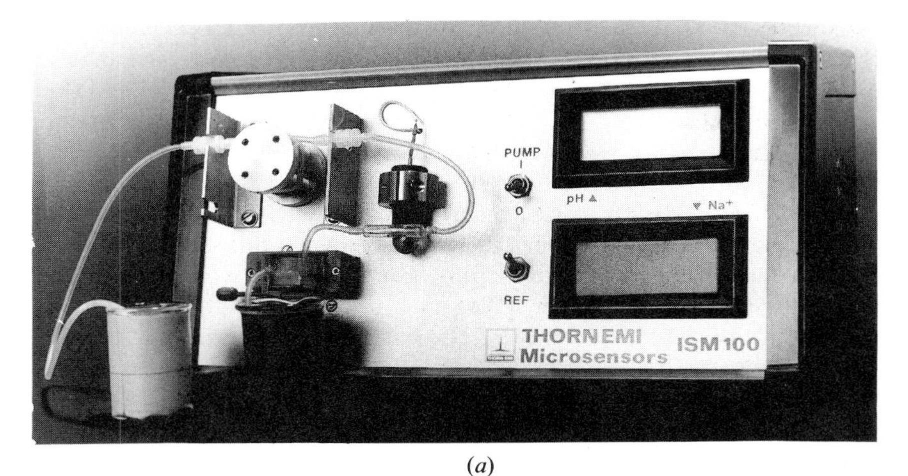

(a)

(b)

Figure 17.6 *(a) Early ISFET test kit developed by Thorn EMI. The front panel of the instrument carries a flow cell fitted with a dual ISFET device for the measurement of both pH and sodium, a miniaturized conventional reference electrode mounted in the sample flow stream, a miniature pump and displays for the ISFET outputs. (b) Commercial ISFET/REFET probe designed for applications in the food industry. The device is mounted in a robust stainless steel stab probe that can be inserted into meat carcasses and meat products, for example (courtesy of Sensoptic BV). A range of redesigned ISFET probes is to be launched by Sentron BV in 1993.*

A commercial instrument designed for laboratory use by the company CHEMFET (USA) was on the market in 1990 (but is no longer available). This still had a miniaturized conventional reference electrode. It had the advantage of dry storage, but was suitable for only a limited continuous operation time of about half an hour. After this, a rest period and a recalibration run were required. The instrument is now back on the market with improved design features, manufactured by UNIFET (USA) as shown in Figure 17.2d. A lighter ISFET probe with a fine tip is produced by Microsens (Neufchatel).

2.3.7 A Commercial ISFET/REFET System for pH Measurement in the Food Industry

Now, a pH ISFET with a REFET is commercially available (from the Sensoptic company situated in The Netherlands, where the Bergveld group earlier pioneered the ISFET development). Thus a potential source of drift through clogging up and loss of reagent associated earlier with the conventional reference electrode is eliminated. Differential amplification of the signals from the ISFET and the REFET also provides a substantial improvement in the sensor performance by eliminating a number of other sources of drift and cross-sensitivity. The supplier reports a stability to 0.05 pH over 24 hours. The ISFET and REFET devices are kept at the isothermal point of the channel current (see above) for temperature stability, and operation over the temperature range 0–70 °C is specified for the probe (Figure 17.2d). The upper limit is defined by the properties of the glue (non-toxic) used for connections.

The probe is supplied with a robust pointed stainless steel tip, and successful test runs in meat and dairy processing plants and in sauce manufacture (including highly viscous sauces rich in fat or dispersed solids, salts and acids) have been reported by the supplier. The sturdiness of the probe can be judged from the specifications, which include a minimum of 1 000 000 measurements in meat for an individual sensor.

Maintenance is by cleaning with a toothbrush and soap solution or alcohol. Corrosive materials as used in cleaning-in-place (CIP) will shorten the probe life significantly, which can otherwise extend over many months. The sensitivity of the pH ISFET to the concentration of larger ions is similar to that of conventional glass membrane electrodes. This can be eliminated by calibration in buffer solutions of matched specific ion concentrations.

2.3.8 Trials of pH ISFET Probes in Meat Processing

Applications in slaughterhouses include the identification of dark firm dry (DFD: high end pH, short shelf life) beef carcasses and the control of the effectiveness of electrostimulation treatment of beef carcasses. For pork, the identification of pale soft exudative (PSE; fast drop of pH) meat and the prediction of yields for cooked ham preparation as well as the prediction of salt penetration in raw ham curing can be achieved. The rate of change and the end value of the pH during the first hours *post mortem* are characteristic for normal meat and for meat with the quality defects DFD or PSE. It is essential that the pH measurement on the carcase is carried out at the right time of the *post mortem* (PM) period. Approximately 45 minutes PM, the difference in pH between PSE and normal meat by pH is greatest; approximately 4 hours PM, there is a clear distinction between normal meat and both PSE and DFD meat (Eikelenboom 1990a).

Two new pH probe designs for meat applications that have come on to the market recently have been highlighted by Eikelenboom (1990b) and compared by Den Reijer (1990). The first (manufactured by Neukum in Germany, imported by Gullimex to The Netherlands) is a conventional pH probe but is armed with a metal shaft and mounted together with the associated pH meter instrument in a pistol form. The second is the ISFET/REFET system by Sensoptic (see above), with its glass-free steel tip configuration. Both sensors were found to be robust in meat applications, whereas four other types of portable pH meter in the test were considered to have non-robust sensors.

The ISFET/REFET system was found to have the following advantages over the armed conventional probe: it requires no temperature compensation adjustment because it is not influenced by temperature over the specified range; it can be inserted through artificial sausage skins as well as into plain meat. The response of the ISFET/REFET system for meat was fast as one would expect, but for slicing sausage a slower response was observed (Den Reijer 1990). Maintenance was easier for the ISFET/REFET than for the conventional pH probes (Den Reijer 1990; Eikelenboom 1990b).

The supplier of the ISFET/REFET probe reports that by measuring the pH of meat carcasses at the most relevant time of the PM period, correlation with the colour after chilling is obtained. A probe designed for the measurement of the colour of translucent materials such as meat (the CTM probe) can also be supplied with this system.

2.3.9 Acidity Sensors

Whereas pH is important as one of the factors influencing microbial growth, and also controls chemical reactions in certain food processing operations, acidity is of interest primarily for the flavour of a food. Based on the ISFET pH sensor, it is possible to construct a coulometric microtitrator by combining the ISFET with an integrated gold electrode as the actuator that generates the titrant (Figure 17.7). This device can provide rapid titration owing to the close proximity of the actuator to the pH sensor. A whole array of such microtitrators can be constructed on a silicon chip machined to form a channel. As the sample flow passes the individual titrator device pairs of the array, it is being consecutively titrated; each ISFET in the array measures the pH at a different point of the titration curve (Van der Schoot and Bergveld 1985). The microtitrator arrangement was initially intended as a means of providing a test signal for an *in situ* (*in vivo*) pH ISFET, so as to provide automatic calibration control (Van der Schoot and Bergveld 1988).

The microtitrator array does, however, also offer the possibility of a truly continuous acidity measurement, with a real-time titration curve being displayed as the liquid flows through the device array. For optimum performance, the dimensions of the flow channel are such that only non-viscous liquids without larger particles can be analysed. For single titrator devices, this limitation need not apply. With the single device, of course, there is no truly continuous titration curve update, but a cyclic measurement can provide updates at intervals of the order of seconds.

The potential of the microtitrator for in-line acidity measurement in the food industry has been discussed (Kress-Rogers 1986; 1987) and a dip-stick configuration has subsequently been constructed by Dr. Bart Van der Schoot. In this form, the device is now suitable for further develop-

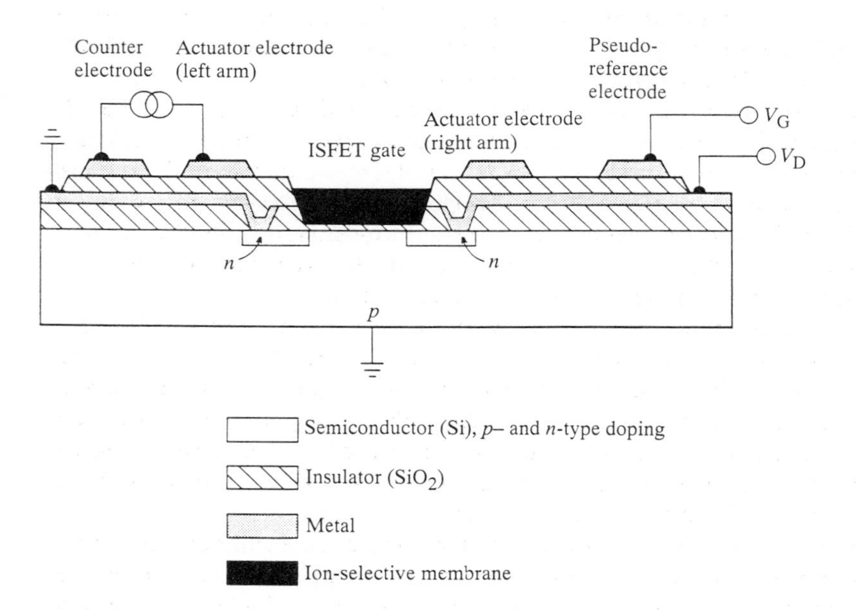

Figure 17.7 *Acidity sensor principle, schematic diagram. As developed by Van der Schoot and Bergveld.*

ment as an in-line acidity sensor for both light and viscous food liquids. Further work on the dipstick acidity sensor has been described by Olthuis et al. (1989). With this single device, the titration time is 0.1 to 10 seconds for acid or base titrations from 0.5×10^{-3} to 10^{-4} mol/l. For low-viscosity samples such as certain beverages, it may be possible to adapt the titrator array for continuous in-line real-time acidity determination. The commercial development of microtitrators could become viable as more and more food processing operations are converted from batch to continuous mode.

In the meantime, different actuator electrodes for the titrator are being studied as alternatives to the initial gold electrode. Platinum was tried but shown to be sensitive to chloride ions; tungsten trioxide was found to be free of this interference (Van Kerkhof et al. 1991). As a further possibility, an all-iridium-oxide actuator-sensor system is also being investigated (Olthuis et al. 1991).

2.4 Gas Sensing Semiconductor Devices

2.4.1 Introduction

As single elements, the gas sensors described here are generally more suited to industrial and domestic safety applications (Table 17.4). There are possible exceptions; the highly sensitive and specific CHEMFET hydrogen sensor could aid in the monitoring of microbial food spoilage under anaerobic conditions (see Chapter 16) and the ethylene-sensitive CHEMFET could help in monitoring of fruit ripening.

It is possible to apply MeOx sensors to the headspace assay of ethanol, although problems can be experienced owing to the current non-specificity and drift of these devices. Many of the volatiles to which the MeOx sensors respond are of interest in the monitoring of foods, but owing to the broad specificity it is often better to consider a biosensor with the narrow specificity that the enzyme can provide for the determination of, for example, specific aldehydes, alcohols or acetic acid.

There is, however, considerable potential in combining several broadly specific gas sensors into an array of sensor elements. The sensors in the array are all of the same type, for example MeOx gas sensors, conducting polymer gas sensors or acoustic devices with coatings that adsorb with a broad specificity. The sensors within the array are however either prepared differently, or operated at different temperatures, so that they have different, broadly overlapping specificities.

The response pattern of such a sensor array can be characteristic for a particular volatile, and analogies to the response mechanisms of receptors in the human nose have been drawn. In the response of these receptors in the animal nose, the size, shape, charge and dipole moment of the molecules of volatile compounds are important. As yet, no gas sensing material with the same response characteristics has been found. However, the principle of analysing response patterns generated by an array of sensing elements with broadly overlapping specificities for the recognition of complex volatile mixtures is being copied by sensor designers.

The term 'odour sensors' is often used for such gas sensor arrays, although current commercial instruments are still far removed from the 'electronic nose' (Persaud and Dodd 1982). Nevertheless, successes have been cited, for example, for the application of a 50-element gas sensor array of conducting polymers in the assessment of coffee roasting. Encouraging preliminary results have been reported for trials on the application in the search for truffles, using the sensor array (and associated pattern recognition system) in place of the traditional truffle pig. (With reference to private communications with K. C. Persaud, UMIST, Manchester, UK; for a description of the sensor design, see Persaud and Pelosi, 1992.)

Table 17.4 *Some gases and volatiles in food applications*

O_2, N_2, CO_2	Constituents of the modified atmosphere in modified atmosphere packaging (MAP)
CO_2	Indicator of fermentation, carbonation
SO_2	Can be an indicator of sulphite addition, but generally sample preparation is needed to obtain total SO_2
H_2, H_2S, NH_3	Can indicate microbial spoilage
Ethylene	Indicates fruit ripening
H_2O	Can indicate the progress of baking operations

Complex mixtures of gases and volatiles

End point indication or aroma of final product: in cooking, baking, roasting operations

Indication of oxidative or microbial spoilage processes: during storage or distribution, assessment of ingredients

Relevant sensor types

For the applications listed above, the semiconductor gas sensors or sensitized FET devices described in this section can be considered. For the following volatiles, biosensors would be suitable to provide a specific determination:

Alcohols	Headspace analysis of volatile content
Aldehydes	Pentanal and hexanal indicate oxidative rancidity
Amines	Indicators of microbial spoilage

Further volatile sensors are described in Section 4. Near infrared spectroscopy is another alternative for many volatiles.

2.4.2 Gas Sensor Characteristics

The highly specific and sensitive microelectronic CHEMFET gas sensor has already been described in Section 2.3. There is also a group of semiconducting materials which can be applied as gas sensors in a very simple configuration, without the necessity of turning them into microelectronic devices. These are the metal oxide semiconductors and the organic semiconductors which respond to certain gases and volatiles with a conductivity change.

Commercial MeOx gas sensors are usually configured as simple thick-film resistors on a heated ceramic substrate. The most well-known example is the Taguchi gas sensor by Figaro Engineering of Japan, responding to a range of oxidizing and reducing gases. A commercial odour meter based on MeOx sensors is also on the market; in its current form it is used primarily in the monitoring of processes such as resin curing, although application in the monitoring of coffee roasting has also been reported.

In current practical applications for gas detection, the metal oxide (MeOx) semiconductor resistors are widely used. (The abbreviation MOX is also used, but is not adopted here to avoid confusion with the MOS device; see Section 2.3.) At elevated temperatures, reducing gases interact with the surface of these semiconductors and donate electrons or release electrons bound by adsorbed oxygen. This results in a drastic increase in surface conductivity, while for an oxidizing gas the conductivity will decrease.

These MeOx gas sensors have been used particularly in safety and automotive applications, and are sometimes referred to as Figaro or Taguchi sensors. To a certain extent, their selectivity can be controlled by their operating temperature and by the film thickness and catalytic additives used in their construction, but it is broad compared with that of the much more specific CHEMFET gas sensors, and the MeOx sensors are usually less sensitive. Typical materials for MeOx gas sensors are tin oxide and zinc oxide. The analytes include combustible and hazardous gases such as methane or carbon monoxide, and hydrocarbons such as ethanol. Ambient humidity influences the sensor response. The catalytic reactions and electronic processes at the MeOx gas sensor surface have been examined by Kohl (1989a; 1989b) with a view to improving the specificity.

The mechanisms in the gas sensitivity of MeOx devices are complex. Charge transfer reactions occur at the MeOx surface in contact with the gas, and also at interfaces with the metal contacts, at grain boundaries and in the bulk oxide. Depending on the techniques used for the deposition of the metal contacts on the MeOx material, the contact can be ohmic or have Schottky character, that is the resistance will be voltage dependent (Schierbaum et al. 1991).

Conducting polymers used in gas sensing have been constructed primarily in the form of sensor arrays for odour sensing (see the following section).

They include substituted polypyrroles and polyanilines (Bartlett et al. 1989). They are generally of broad specificity, and thus suited to combinations of sensor arrays with pattern analysis. The electronic structure of these materials is still the subject of research. Owing to the chain-like structure of the material, charge transport is characterized by quasi-one-dimensional behaviour. Dopant molecules can diffuse into the structure between the chains. It has been suggested that volatile polar compounds adsorbed on to the surface of conducting polymers may act as reversible dopants (see Persaud and Pelosi 1992; Persaud 1991).

Organic semiconductor gas sensors based on metal phthalocyanines, on the other hand, have a narrow specificity and high sensitivity. They are usually in the form of thin films and have been incorporated in microelectronic devices such as the FET. Lead phthalocyanine (PbPc) is an organic *p*-type semiconductor. When adsorbing oxidizing gases, its conductivity changes. It is sensitive to oxygen, chlorine and nitrogen dioxide (NO_2), but insensitive to carbon dioxide and to many hydrocarbons (Heilmann et al. 1991; Hamann et al. 1991; Bott and Jones 1984). Chlorinated hydrocarbons have been determined after decomposition on a heated platinum coil, thus releasing chlorine (Unwin and Walsh 1989).

Metal phthalocyanine sensors of differing characteristics can be obtained by using different metals (for example, lead and copper in PbPc and CuPc, less commonly ZnPc, NiPc, CoPc or the metal-free H_2Pc) and different operating temperatures (Cranny et al. 1991; Sadaoka et al. 1991). As in the case of the MeOx sensors, this has been applied in the construction of arrays for the analysis of gas mixtures (see later).

In Table 17.5 the main characteristics of the gas sensors described in this section are briefly summarized. Response times and recovery times are further important characteristics. They depend on the thickness and mode of preparation of the film and on the operating temperature as well as on the nature of the gas sensor material. The required operating temperature is of practical significance in influencing the power consumption and compactness of a device. For the MeOx sensors and the CHEMFETs based on catalytic metals, elevated temperatures are needed, but these can be achieved with a heater film applied to the device for the CHEMFET and also for MeOx sensors in film form. Conducting polymers and phthalocyanines can operate at room temperature.

Thin films of both MeOx semiconductors and organic semiconductors have been incorporated in microelectronic devices such as the FET rather than simply used as resistor devices. These more complex devices can be integrated with other microelectronic sensors or signal processing elements (the latter turns them into smart sensors), can offer better signal to noise characteristics in signal transmission and in signal processing, and are more easily interfaced with computing systems, for example. However, for the microelectronic devices, a large number of devices per year must be produced if the price is to be low.

Table 17.5 *Gas sensor types based on semiconductor devices. In addition to the determination of the analytes listed, the sensors have also been configured into arrays for the recognition of complex mixtures of gases and volatiles*

CHEMFET and related microelectronic devices	
Principle	Catalytic metal or alloy as gate metal in an FET semiconductor MOS device or as the metal in a Schottky diode (LB film of organic semiconductor for Cl_2, NO_2)
Analytes	Small range: hydrogen, hydrogen sulphide, ammonia, methane, ethylene, etc.
Specificity	Very narrow
Sensitivity	Very good
MeOx gas sensor	
Principle	Semiconducting metal oxide as sintered pellet or as thick film or thin film resistor
Analytes	Combustible or hazardous gases such as hydrogen, methane, carbon monoxide, oxygen, etc.; hydrocarbons such as alcohols, aldehydes, etc.; many other volatiles
Specificity	Very broad
Sensitivity	Good
Interference	Humidity
Conducting polymer gas sensor	
Principle	Conducting organic film (polypyrrole or polyaniline) as thick-film or thin-film resistor
Analytes	Polar volatile compounds including alcohols, amines, ethyl acetate, pyridine, etc.
Specificity	Very broad
Sensitivity	Good
MePb gas sensor	
Principle	Metal phthalocyanine thin film (or less often thick film) resistor; deposition techniques include Langmuir–Blodgett and molecular beam epitaxy
Analytes	NO_2, Cl_2, O_2
Specificity	Narrow
Sensitivity	Very good
Interference	Humidity, SO_2 and CO reduce the response

After pyrolytic decomposition, chlorinated hydrocarbons can be detected.

Langmuir-Blodgett (LB) techniques allowing the deposition of highly ordered films (one monomolecular layer at a time) have been applied in the preparation of films of metal phthalocyanines and also of porphyrins. Such LB films can be incorporated into the gate of FET devices for the construction of CHEMFETs. It used to be thought that the LB technique was restricted to amphiphilic films. A new, more stable LB structure has been developed recently and could allow the high potential versatility of LB films to be applied in practical devices (see Section 2.3).

2.4.3 Gas Sensor Arrays for the Measurement of Odours

In recent years, substantial research and development have been carried out to tailor the selectivity and sensitivity of the MeOx devices by special treatments during their manufacture. Inspired by the concept of the electronic nose (see earlier), they have also been combined into arrays of gas sensors that differ in their characteristics so that broad overlapping specificities result. The difference in the characteristics of the array elements can be achieved by operating them at different temperatures, for example.

These arrays can provide an assessment of volatile mixtures. The set of outputs from the individual devices in the array forms a pattern characteristic for a particular volatile (although this can be difficult to analyse for mixed volatiles). This pattern can be compared with a fingerprint library or analysed with pattern recognition techniques, or neural networks can be trained to interpret it (Section 5.2). This approach has also been applied to gas sensors based on organic semiconductors such as the phthalocyanine films.

Current commercial 'odour meters' are based on MeOx arrays operated at different temperatures. Beyond the reported applications in the curing of resins, for example, or in the monitoring of coffee roasting, the assessment of aroma has been suggested. A cautious approach would be advisable in the latter application, since the response pattern of the sensor array will not necessarily attach the same importance to particular compounds in the spectrum of gases and volatiles as the human nose. However, current advances in device design and pattern recognition techniques (Section 5.2) will make aroma assessment more feasible in the future (Shurmer 1990).

Some of the MeOx array designs are based on conventional Taguchi devices; others employ more advanced fabrication techniques. Gardner et al. (1991) have developed an integrated thin-film tin oxide sensor array. Studies on the discrimination between alcohols, beverages and tobaccos with more conventional sintered pellet tin oxide sensors have been carried out by the same group (Gardner 1991; Shurmer, et al. 1990; see also Schild 1990, p. 131).

Arrays with up to 20 elements (and more recently 50 elements) have been constructed with gas sensors based on the conducting polymer polypyrrole

(Persaud et al. 1990). In order to improve the reproducibility of such arrays and to reduce the future manufacturing costs, a new technique for the achievement of differing response characteristics between the array elements has been developed by Shurmer et al. (1991). They have applied molecular sieves over the polymer films, using Langmuir-Blodgett (LB) techniques. Monomolecular LB films have also been used by Mueller (1991) to coat polymer films so as to modify the sensitivity of array elements for better selectivity in combination with pattern recognition. The effect on the sensor response was found to be more complex than that of a passive molecular sieve.

CHEMFET gas sensors (see Section 2.3) have also been configured as arrays for the evaluation of multiple gas mixtures (hydrogen, ammonia, ethylene, ethanol) with the help of pattern recognition techniques (Sundgren et al. 1990).

With current commercial instruments (and those close to the commercial stage at present) for odour assessment, an individual trial and calibration for each application is advisable, and this would have to be checked and updated regularly to allow for any changes in the product (other than the intended analyte volatile) or the environment. Until proven for the intended application, there is no guarantee that an array made up of broadly specific sensors will attach particular importance to a volatile species that the human nose will pick out against a variable background as a pleasant aroma or unpleasant taint.

2.5 Conclusions

Reliable though they generally are, pH glass membrane electrodes have a number of disadvantages for the food industry. In particular, the fragile glass membrane is a potential hazard when applied in-line, and maintenance is laborious. A microelectronic pH probe (namely an ISFET device), designed specifically for food applications, has recently appeared on the market and is already being used in meat processing and dairy plants. The ISFET is free of fragile components and is now available in a robust stainless steel insertion probe configuration. Another new commercial ISFET probe is suitable for at-line measurements. Configurations for measurements of the pH microenvironment within a food are also possible, although no specifically adapted commercial probe for this purpose is available at the moment.

The chemical and microelectronic aspects of the ISFET and their implications for practical applications have been discussed here in detail in view of the importance of the pH value for the microbial stability of foods and also in the control of food processing operations. A number of other solid-state probes for pH, developed or emerging as alternatives for pH monitoring applications in surgery, in fermentation and in food processing, have been briefly reviewed. Further ISFET devices are under development now for the monitoring of ions such as calcium or potassium. A novel acidity probe based on a coulometric microtitrator is also at the research stage. Acidity is of importance for the flavour of foods.

Biosensors based on the ISFET device have been developed, but their application area is primarily the medical field. Together with biosensor designs more suited to food applications at present, they are described in the following sections.

Among the applications for novel sensors for gases and volatiles, measurements of indicators of spoilage and of cooking processes are the subject of current research and development. An area of active research is that of sensor arrays for the analysis of complex volatile mixtures with the aim of assessing the aroma of foods. Pattern recognition techniques including neural network applications feature strongly in this development area.

References for Sections 2.1 to 2.3

Anon. (1985) Thorn develops microsensors. *Processing Control & Instrumentation,* October, p. 10.

Bergveld, P. (1970) Development of an ion-sensitive solid-state device for neurophysiological measurements. *Trans. Biomedical Engineering,* **BME-14,** 70–71

Bergveld, P. (1988) Development and application of chemical sensors in liquids. In *Sensors and Sensory Systems for Advanced Robots,* NATO ASI Series F43, ed. P. Dario, Springer, Berlin, 397–414

Bergveld, P. (1991) Future applications of ISFETs. *Sensors and Actuators,* **B4,** 125–133

Bergveld, P., Van den Berg, A., Van der Wal, P. D. Skowronska-Ptasinska, M., Sudhoelter, E. J. R. and Reinhoud, D. N. (1989) How electrical and chemical requirements for REFETs may coincide. *Sensors and Actuators,* **18,** 309–327

Clechet, P. (1991) Membranes for chemical sensors. *Sensors and Actuators,* **B4,** 53–63

Comte, P. A. and Janata, J. (1978) A field effect transistor as a solid-state reference electrode. *Analytica Chimica Acta,* **101,** 247–252

De Rooij, N. F. and Bergveld, P. (1980) The iridium/anodic iridium oxide film (Ir/AIROF) electrode as a pH-sensor. *Proceedings of the International Conference on Monitoring of Blood Gases, Blood Ion Concentrations and Respiratory Gas Exchange,* Nijmegen, 156–165

Den Reijer, M. (1990) Goede pH-meters verkriegbaar. *Vleesdistributie en vleestechnologie,* **25,** 34–39

Eikelenboom, G. (1990a) Toepassingen van de pH-meting bij vers vlees. *Vleesdistributie en vleestechnologie*, **25**, 19–23

Eikelenboom, G. (1990b) Het meten van de pH van vlees. *Vleesdistributie en vleestechnologie*, **25**, 24–33

Fjeldly, T. A. and Nagy, K. (1985) Glass electrodes with solid-state membrane contacts and their application in differential potentiometric sensors. *Sensors and Actuators*, **8**, 261–269

Geun Sig Cha, Dong Liu, Meyerhoff, M. E., Cantor, H. C., Midgley, A. R., Goldberg, H. D. and Brown, R. B. (1991) Electrochemical performance, biocompatibility and adhesion of new polymer matrices for solid-state ion sensors. *Analytical Chemistry*, **63**, 1666–1672

Janata, J. and Huber, R. J. (1979) Ion-sensitive field-effect transistors. *Ion-Selective Electrode Reviews*, **1**, 31–79

Klein, M. (1989) Time effects of ion-sensitive field-effect transistors. *Sensors and Actuators*, **17**, 203–208

Klein, M. (1991) Calcium-sensitive field-effect transistor with inorganic layer. *Sensors and Actuators*, **B4**, 141–144

Kress-Rogers, E. (1985a) Technology transfer. II: The new generation of sensors. Leatherhead Food Research Association scientific and technical survey 150

Kress-Rogers, E. (1985b) Seeking sensor improvements. *Food Processing*, September, 37–39

Kress-Rogers, E. (1986) Sensors for measurement of food properties and for quality control. COST 91bis subgroup workshop, 25 November, Chipping Campden (summary in Berichte der Bundesforschungsanstalt für Ernährung BFE-R-87-01, *Sensors and Measurement of Product Properties: Instrumentation and Process Control*, ed. K. O. Paulus, February 1987, 11–14

Kress-Rogers, E. (1987) Sensors and their applications to meat technology. *Proceedings of the International Symposium on Trends in Modern Meat Technology 2*, eds B. Krol, P. S. van Roon and J. H. Houben, November, Den Holder; Pudoc Wageningen, 33–39

Kress-Rogers, E. (1991) Solid-state pH sensors for food applications. *Trends in Food Science & Technology*, **2**, 320–324. Please note typesetting errors: Drift should read 0.05 pH units over 24 hours (on p. 323); pH biosensors should read pH sensors (in table 1). See also *TFST*, **3**, 91.

Kreuer, K.-D. (1990) Solid potentiometric pH electrode. *Sensors and Actuators*, **B1**, 286–292

Matsuo, T. and Nakajima, H. (1984) Characteristics of reference electrodes using a polymer gate ISFET. *Sensors and Actuators*, **5**, 293–305

Mueller, R. (1991) High electronic selectivity obtainable with nonselective chemosensors, *Sensors and Actuators*, **B4**, 35–39

Nakajima, H., Esashi, M. and Matsuo, T. (1982) The cation concentration response of polymer gate ISFET. *Journal of the Electrochemical Society*, **129**, 141–143

Olthuis, W., Robben, M. A. M., Bergveld, P., Bos, M. and Van der Linden, W. E. (1990) pH sensor properties of electrochemically grown iridium oxide. *Sensors and Actuators,* **B2,** 247–256

Olthuis, W., Van der Schoot, B. H., Chavez, F. and Bergveld, P. (1989) A dipstick sensor for coulometric acid-base titrations. *Sensors and Actuators,* **17,** 279–283

Olthuis, W., Van Kerkhof, J. C., Bergveld, P., Bos, M. and Van der Linden, W. E. (1991) Preparation of iridium oxide and its application in sensor-actuator systems. *Sensors and Actuators,* **B4,** 151–156

Roberts, G. G. (1990) Potential applications of Langmuir-Blodgett films as chemical sensors. *Thin Solid Films,* **160,** 413–429

Schasfoort, R. B. M., Streekstra, G. J., Bergveld, P., Kooyman, R. P. H. and Greve, J. (1989) Influence of an immunological precipitate on DC and AC behaviour of an ISFET. *Sensors and Actuators,* **18,** 119–129

Sibbald, A., Whalley, P. D. and Covington, A. K. (1984) A miniature flow-through cell with a four-function CHEMFET integrated circuit for simultaneous measurements of potassium, hydrogen, calcium and sodium ions. *Analytica Chemica Acta,* **159,** 47–62

Strohecker, R. (1943) *Methoden der Lebensmittelchemie (Methods in Food Chemistry)*. De Gruyter, Berlin

Tarlov, M. J., Semancik, S. and Kreider, K. G. (1990) Mechanistic and response studies of iridium oxide pH sensors. *Sensors and Actuators,* **B1,** 293–297

Van den Berg, A., Grisel, A. and Verney-Norberg, E. (1991) An ISFET-based calcium sensor using a photopolymerized polysiloxane membrane. *Sensors and Actuators,* **B4,** 235–238

Van der Schoot, B. H. and Bergveld, P. (1985) An ISFET-based microlitre titrator: integration of a chemical sensor-actuator system. *Sensors and Actuators,* **8,** 11–22

Van der Schoot, B. H. and Bergveld, P. (1988) Coulometric sensors, the application of a sensor-actuator system for long-term stability in chemical sensing. *Sensors and Actuators,* **13,** 251–262

Van der Spiegel, J., Lauks, I., Chan, P. and Babic, D. (1983) The extended gate chemically sensitive field-effect transistor as multi-species microprobe. *Sensors and Actuators,* **4,** 291–298

Van Kerkhof, J. C., Olthuis, W. Bergveld, P. and Bos, M. (1991) Tungsten trioxide (WO_3) as an actuator electrode material for ISFET-based coulometric sensor-actuator systems. *Sensors and Actuators,* **B3,** 129–138

Vogel, A., Hoffmann, B., Sauer, Th. and Wegner, G. (1990) Langmuir-Blodgett films of phthalocyaninato-polysiloxane polymers as a novel type of CHEMFET membrane. *Sensors and Actuators,* **B1,** 408–411

Vogel, A., Hoffmann, B., Schwiegk, S. and Wegner, G. (1991) Novel Langmuir-Blodgett membranes for silicon-based ion sensors. *Sensors and Actuators,* **B4,** 65–71

Wilhelm, D., Voigt, H., Treichel, W., Ferretti, R. and Prasad, S. (1991) pH sensor based on differential measurements on one pH-FET chip. *Sensors and Actuators,* **B4,** 145-149

References for Section 2.4

Bartlett, P. N., Archer, P. B. M. and Ling-Chung, S. K. (1989) Conducting polymer gas sensors. 1: Fabrication and characterization. *Sensors and Actuators,* **19,** 125-140

Bott, B. and Jones, T. A. (1984) A highly sensitive NO_2 sensor based on electrical conductivity changes in phthalocyanine films. *Sensors and Actuators,* **5,** 42-53

Cranny, A. W. J., Atkinson, J. K., Burr, P. M. and Mack, D. (1991) A comparison of thick- and thin-film gas-sensitive organic semiconductor compounds. *Sensors and Actuators,* **B4,** 169-174

Gardner, J. W. (1991) Detection of vapours and odours from a multisensor array using pattern recognition. I: Principal component and cluster analysis. *Sensors and Actuators,* **B4,** 109-115

Gardner, J. W., Shurmer, H. V. and Corcoran, P. (1991) Integrated tin oxide odour sensors. *Sensors and Actuators,* **B4,** 117-121

Hamann, C., Mrwa, A., Mueller, M., Goepel, W. and Rager, M. (1991) Lead phthalocyanine thin films for NO_2 sensors. *Sensors and Actuators,* **B4,** 73-78

Heilmann, A., Mueller, M., Hamann, C., Lantto, V. and Torvela, H. (1991) Gas sensitivity measurements on NO_2 sensors based on lead phthalocyanine thin films. *Sensors and Actuators,* **B4,** 511-513

Kohl, D., (1989a) Catalytic reactions and electronic processes relevant in gas sensing: an extended abstract. *Sensors and Actuators,* **17,** 309-311

Kohl, D. (1989b) Surface processes in the detection of reducing gases with SnO_2-based devices. *Sensors and Actuators,* **18,** 71-113

Persaud, K. C. and Pelosi, P. (1992, in press) Sensor arrays using conducting polymers for an artificial nose. In *Electronic Noses,* NATO ASI Series, eds P. G. Gardner, G. H. Dodd, J. Bartlett, Kluwer Press (Proc. of NATO Workshop on Electronic Noses, Reykjavik, Iceland, August 1991)

Persaud, K. C. (1991) Odour detection using sensor arrays. *Analytical Proceedings,* **28,** 339-341

Persaud, K. C. and Dodd, G. H. (1982) Analysis of discrimination mechanisms in the mammalian olfactory system using a model nose. *Nature,* **299,** 352-255

Sadaoka, Y., Matsuguchi, M., Sakai, Y., Mori, Y. and Goepel, W. (1991) Effect of crystal form on the conductance in oxidative gases of metal-free and some metal phthalocyanines. *Sensors and Actuators,* **B4,** 495-498

Schierbaum, K. D., Kirner, U. K., Geiger, J. F. and Goepel, W. (1991) Schottky-barrier and conductivity gas sensors based upon Pd/SnO_2 and Pt/TiO_2. *Sensors and Actuators,* **B4,** 87–94
Schild, D. (1990) *Chemosensory Information Processing*. Springer, Berlin
Shurmer, H. V. (1990) Basic limitations for an electronic nose. *Sensors and Actuators,* **B1,** 48–53
Shurmer, H. V., Gardner, J. W. and Corcoran, P. (1990) Intelligent vapour discrimination using a composite 12-element sensor array. *Sensors and Actuators,* **B1,** 256–260
Shurmer, H. V., Corcoran, P. and Gardner, J. W. (1991) Integrated arrays of gas sensors using conducting polymers with molecular sieves. *Sensors and Actuators,* **B4,** 29–33
Sundgren, H., Lundstroem, I., Winquist, F., Lukkari, I., Carlsson, R. and Wold, S. (1990) Evaluation of a multiple gas mixture with a simple MOSFET gas sensor array and pattern recognition. *Sensors and Actuators,* **B2,** 115–123
Unwin, J. and Walsh, P. T. (1989) An exposure monitor for chlorinated hydrocarbons based on conductometry using lead phthalocyanine films. *Sensors and Actuators,* **18,** 45–57

3 Amperometric, Potentiometric and Thermometric Biosensors

3.1 Introduction: Biosensor Types

There is a spectrum of different definitions for the term 'biosensor'. In a wider sense (but by no means the widest of the definitions proposed), it is understood as a 'self-contained analytical device that responds selectively and reversibly to the concentration or activity of chemical species in biological samples'. In a narrower sense, it is 'an analytical device that incorporates a biologically active material in intimate contact with an appropriate transduction element for the purpose of detecting (reversibly and selectively) the concentration or activity of chemical species in any type of sample' (both definitions given by Arnold and Meyerhoff 1988).

For practical purposes in selecting a sensor for food applications, the second of these definitions is more useful, since characteristics such as the operational life and the tolerance to the thermal and chemical environment of the sensor will be influenced quite strongly by the presence of a biological sensor component such as an enzyme or antibody layer (or indeed by whole biological cells or tissues; see Rechnitz and Ho 1990; Karube 1990). This second definition is therefore adopted here, narrowed further to include only sensors based on the interaction of biocatalysts (enzymes either in isolated form or contained in whole cells or tissues).

Sensors based on the interaction of antibodies with the analyte (or microbe) or other immobilized layers with an affinity for the analyte will be classified separately as immunosensors to take account of their distinct set of design and performance characteristics. Similar in design to the immunosensors are sensors based on selective sorption of the analyte (see Section 4).

It is worth noting that both definitions exclude indicator strips that display, for example, an irreversible colour change that is then read either by eye or with the help of a separate instrument. This does not detract from the usefulness of such indicator strips for spot-check sampling mainly in the clinical area, but possibly in the future also in industrial spot checks (see for example Diebold et al. 1991).

The first biosensor was developed by Updike and Hicks (1967). They mounted a membrane carrying glucose oxidase on a Clark-type oxygen probe. In the vicinity of the enzyme membrane, glucose and oxygen in the sample react to form gluconic acid and hydrogen peroxide. The oxygen consumption is monitored with the gas probe to indicate the glucose concentration of the sample. Glucose has remained a favourite analyte for biosensor designers, partly because of its clinical significance, partly because it is a stable, long-lived enzyme. Many more analytes have been the subject of biosensor research and development (Table 17.6).

Many indirect biosensor designs have been developed, based on amperometric gas probes for oxygen, peroxide or ammonia, or on potentiometric ion-selective probes for pH or ammonium, for example. On these, one or more enzymes (and cofactors if needed) are immobilized to catalyse a reaction involving the analyte. Usually the analyte is the substrate of the enzyme, but the analyte can also be an inhibitor of the enzyme, for example. The enzyme-catalysed reaction produces or consumes the secondary analyte to which the base device responds. The resulting biosensors are often referred to as amperometric or potentiometric depending on the nature of the probes for the secondary analyte.

Microelectronic as well as traditional ion-selective electrodes and gas probes in the usual or in miniaturized form can be used as the base devices. Immobilization of enzymes on microelectronic pH sensors (ISFET) is described in Section 3.4 below and the resulting ENFET is classed among the potentiometric biosensors. (The actual raw output signal would be an electrical current if it was not for the feedback circuit; see Section 2.3. Nevertheless, the signal is related to the potential drop at the interface between the ion-sensitive membrane and the sample.) The traditional Clark gas probe can be constructed in planar technology and will then have a size similar to the ISFET. These micro-Clark gas probes are the basis for miniaturized amperometric indirect biosensors (Section 3.3).

The respective advantages of amperometric and potentiometric biosensors used to be the subject of a heated debate, often decided in favour of the followers of the amperometric school. A 'noisy DC signal' used to be held against the potentiometric devices, and was contrasted with the more

Table 17.6 *Analytes and properties for which biosensors can be developed in the food industry.*
*Symbol * indicates potential immunosensor or DNA probe development. The list is not exhaustive (see also Table 16.1; Kress-Rogers 1985a; Pearson's Food Analysis* 1991)

Carbohydrates
glucose, sucrose, fructose, galactose, maltose, lactose, raffinose, starch

Alcohols, phenols, lipids
methanol, ethanol
cholesterol, tri/di/monoglyceride, glycerol, sorbitol

Aldehydes
acetaldehyde, pentanal, hexanal

Organic acids
lactic, malic, oxalic, acetic, ascorbic, citric, pyruvic, glutamic

Amino acids
total amino acids
specific amino acids, particularly essential amino acids: phenylalanine, glutamine, lysine, histidine, tyrosine, cysteine, methionine, asparagine, alanine, arginine, tryptophane

Imino acids
hydroxyproline

Purines
inosine, hypoxanthine

Amines
putrescine, cadaverine, tyramine, histamine, trimethylamine

Nucleosides, nucleotides, glycosides
adenosine, ATP, ADP, amygdalin

Miscellaneous
creatinine
misc. veterinary antibiotics (penicillin, etc.)
misc. agricultural residues (pesticides, fertilizers)*
misc. enzymes (amylases, proteases)
heavy metals
minerals
vitamins*
preservatives (nitrate/nitrite, sulphite)
antioxidants
emulsifiers
sweeteners (aspartame)

mycotoxins*
sea food toxins

Chirality

Mutagenicity

BOD, COD

Microbes (wanted and unwanted species)
total count and/or identification*
bacteria, moulds, yeasts, virus
bacteria: Salmonella, Listeria, Staphylococcus, etc.; lactic acid bacteria

Predicted shelf life
progress of microbial or rancid spoilage processes

Predicted degree of Maillard browning

(For the significance of many of these analytes and properties see Chapter 16, Kress-Rogers (1985a), and *Pearson's Composition and Analysis of Foods,* eds Kirk and Sawyer (1991)

favourable AC current signal of the amperometric device. As far as the ENFET is concerned, this argument does not hold, given that the intrinsically pre-amplified output of FET devices has favourable signal to noise characteristics. Nor is a conventional reference electrode necessary now that efficient REFET devices are available. Moreover, the ISFET can be complemented with a microactuator electrode that keeps the local pH constant. The actuator signal needed to maintain constant pH is the output signal of this device. In contrast to most other biosensor types, the pH dependence of the enzyme is then no longer a problem, and the device is insensitive to the buffer strength of the sample. The ENFET has held its own in the medical biosensors league. For industrial applications, it is too early to judge its relative position.

In addition to the amperometric, potentiometric and microelectronic base devices for biosensors, optical sensors can also be employed. The enzyme-catalysed reaction can be linked to reactions that produce luminescence or fluorescence and can then be used in conjunction with fibre-optic technology to construct fibre-optic biosensors (Section 4.4).

As an alternative to such secondary detection approaches, the enzyme-catalysed reaction can be monitored with a heat-sensitive device (Section 3.5). Another approach is to measure the current generated by electron transfer to an appropriately biased electrode carrying a mediator compound immobilized together with the enzyme. The mediator is a redox couple that will, for example, replace oxygen in the decomposition of glucose in the presence of glucose oxidase. The latter approach results in the amperometric mediated enzyme electrode (Section 3.2) and will be abbreviated here as AMEE.

The AMEE used to be known as the amperometric direct biosensor, but this is no longer used to avoid confusion with biosensors constructed by immobilizing enzymes on conducting organic salt electrodes that perform the functions of an electron mediator as well as conducting the measured current (see Kulys and Svirmickas 1980; Albery and Bartlett 1984; Bartlett 1989; Sim 1990). At one time, biosensors were classified as first generation (indirect), second generation (mediated) and third generation (based on organic conducting electrodes). This is used less now as the indirect biosensors have become more sophisticated and can actually offer performance advantages in certain applications over the mediated biosensors. For the organic conducting electrode biosensors, it is too early to judge their practical performance characteristics.

The choice among these varied designs will certainly influence the characteristics of the biosensor, and each base sensor will introduce its own cross-sensitivities unless compensated with a reference device. This, incidentally, applies also to the AMEE which is sensitive to certain antioxidants, for example. Reactions in biosensors can be controlled by the enzyme kinetics or by a limitation of analyte diffusion through a membrane. This choice of the mode will determine sensitivity to factors such as viscosity, pH or temperature. The method of immobilization will influence the shelf and operational life and the storage conditions. To a large extent the nature of the enzyme(s) and cofactors will, however, define the specificity and limit the temperature range and the chemical environment that the biosensor will tolerate.

The development of biosensors in recent years has expanded rapidly and it would be impossible to cover this adequately within one section or even one chapter. The reader may consult the references in this chapter and the current literature. In the remaining part of Section 3 and in Section 4, food-relevant aspects of biosensor design will be discussed, and immunosensors will be introduced (see also General Reading, pages 668–9).

3.2 Biosensors Based on the Amperometric Mediated Enzyme Electrode (AMEE)

In the first biosensors of the indirect amperometric type, a membrane entrapping the enzyme was loosely attached on a conventional Clark-type gas probe for oxygen or peroxide to monitor gas evolution or consumption in the presence of the analyte. Many laboratory prototypes of glucose biosensors were developed, helped by the high stability of glucose oxidase and motivated by the considerable market for blood glucose monitoring. These inspired the vision of the artificial pancreas (Albisser and Leibel

1984; Yanchinski 1983) for diabetes sufferers. This is to be implanted *in vivo* to dispense insulin as needed, controlled by a built-in continuously operating glucose sensor. For this purpose, it was necessary to reduce the dependence of the initial biosensor designs on the oxygen pressure in the sample. Also, the base device needed to be miniaturized.

Both objectives were to be achieved with 'direct' biosensors (later known as mediated), which were based on electron transfer to a plain carbon electrode via a mediator compound that could be immobilized together with the enzyme on to the electrode material. Immobilization now involved dipping into reagents rather than cumbersome retaining membranes held on with O-rings. Miniaturization was thus possible, restricted mainly by the enzyme loading necessary for the required signal to noise ratio at the relevant temperature. Miniature Ag/AgCl wire electrodes served as reference electrodes. Oxygen dependence was also much reduced and, using dimethylferrocene as the mediator compound, the electrode bias could be kept at about 0.1 V, thus eliminating the interference from most electroactive compounds in blood. Ascorbic acid was still an interference, but ascorbic acid levels in blood *in vivo* do not reach levels that would cause concern (Turner 1983; Cass et al. 1984).

The amperometric mediated enzyme electrode (abbreviated below as AMEE) is the basis of the ExacTech system for blood glucose determination. This consists of a pen- or credit card-sized instrument into which disposable single-use electrode strips are inserted (Figure 17.8). It is used by diabetic patients to measure their own blood glucose level using a droplet of fresh capillary blood. The manufacturing process for this disposable AMEE produces biosensor strips of consistent characteristics, so that a calibration of the individual AMEE strip is not required. Upon starting a new batch, the patient calibrates the instrument with a batch-specific calibrator strip (a dual resistor strip) supplied with the biosensor pack. The reading of blood glucose can be obtained 30 seconds after starting the instrument function, without awaiting the establishment of an equilibrium current. This is possible due to a reproducible time dependence of the current equilibration pattern exhibited by these AMEE when they are contacted by the sample liquid and the potential is applied.

Based on the AMEE principle, a range of biosensors utilizing oxidases have been developed, among them sensors for alcohols. These are most sensitive to methanol, less sensitive to ethanol and with a decreasing response to higher alcohols. (For a sensor responding to ethanol, but not to methanol, see Section 3.4.) Among the saccharides, galactose as well as glucose is an analyte (Yokoyama et al. 1989). For a multi-saccharide sensor, including for example sucrose among the analytes, other types of device will compete including the AIDMEE (Section 3.3) and the ENTHERM (Section 3.5).

Based on a biosensor array using the same principle for the individual glucose sensors in the array, a novel meat freshness probe has been

(*a*)

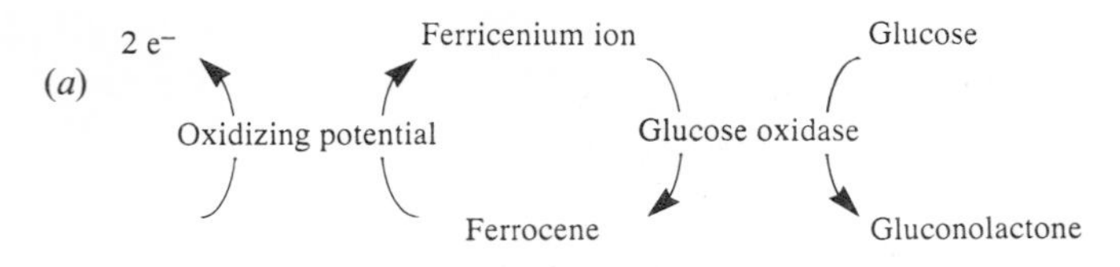

(*b*)

Figure 17.8 *(a) Principle of the amperometric mediated enzyme electrode (AMEE). (b) A commercial instrument based on the AMEE principle: the ExacTech system for blood glucose measurement by diabetic patients. The photograph shows the pen-type instrument and the disposable biosensor strips which are inserted into the instrument.*

developed by the Leatherhead Food Research Association together with the Cranfield Biotechnology Centre. This is based on the consumption of glucose by the microbial flora on the surface of chilled meat (under aerobic conditions) at the pre-spoilage stage. The resulting glucose depth profile has been measured initially with an enzyme assay kit after slicing in parallel to the meat surface, and subsequently with a specially adapted glucose biosensor array. Both trials have shown a correlation between the glucose concentration gradient and the microbial load. A simple-to-use robust meat freshness probe based on a knife-type design is envisaged after further probe optimization which would allow more extensive trials to be carried out for a wide range of meat types (Figure 17.9: details are given in Chapter 16).

This instrument development has made use of the marker principle, that is the identification of a property that can be measured rapidly and indicates a complex food condition that would otherwise require lengthy

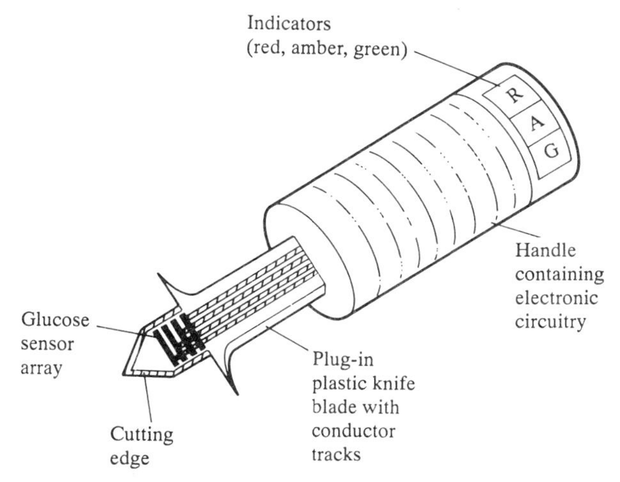

Figure 17.9 *Proposed form of instrument for the assessment of chilled meat by the glucose depletion profile. The development of the prototype for this knife-type meat freshness probe is described in Chapter 16.*

and expensive measurements. The technology transfer of the biosensor design for medical applications to the application in meat freshness assessment has been facilitated by the similarities in the analytical range and chemical background of the meat juice and the clinical sample.

Food samples such as fruit or vegetables, on the other hand, are high in sugars, thus requiring a considerable extension of the analytical range, and also high in ascorbic acid, which interferes with the AMEE output. Further, their pH is often both variable and unfavourably low for enzyme activity. Although it is possible to adapt the AMEE for operation in such conditions, the solutions can be expensive and not viable for a probe specification that will serve a small market only. When a probe for a new analyte is to be developed, there is the added complexity and cost of developing a new enzyme-mediator-electrode system.

Several alternative approaches can be undertaken instead of AMEE optimization for a particular food application. It is possible to develop a simple and rapid sample preparation procedure that allows the application of the ExacTech system for blood glucose determination to the assay of glucose in fruit and vegetable juices. Figures 17.10a–c show the response of the system when challenged in non-blood samples. Figure 10d shows the calibration of the system for fruit juices diluted with a suitable buffered solution. For vegetable juices, the sample treatment consisted instead in the elimination of ascorbate. Such preparation procedures could be presented in the form of a small kit also containing the ExacTech system and special calibrator strips. This approach could be applied to other clinical analysers and is related to the robotics approach (Section 5.3). A further alternative is

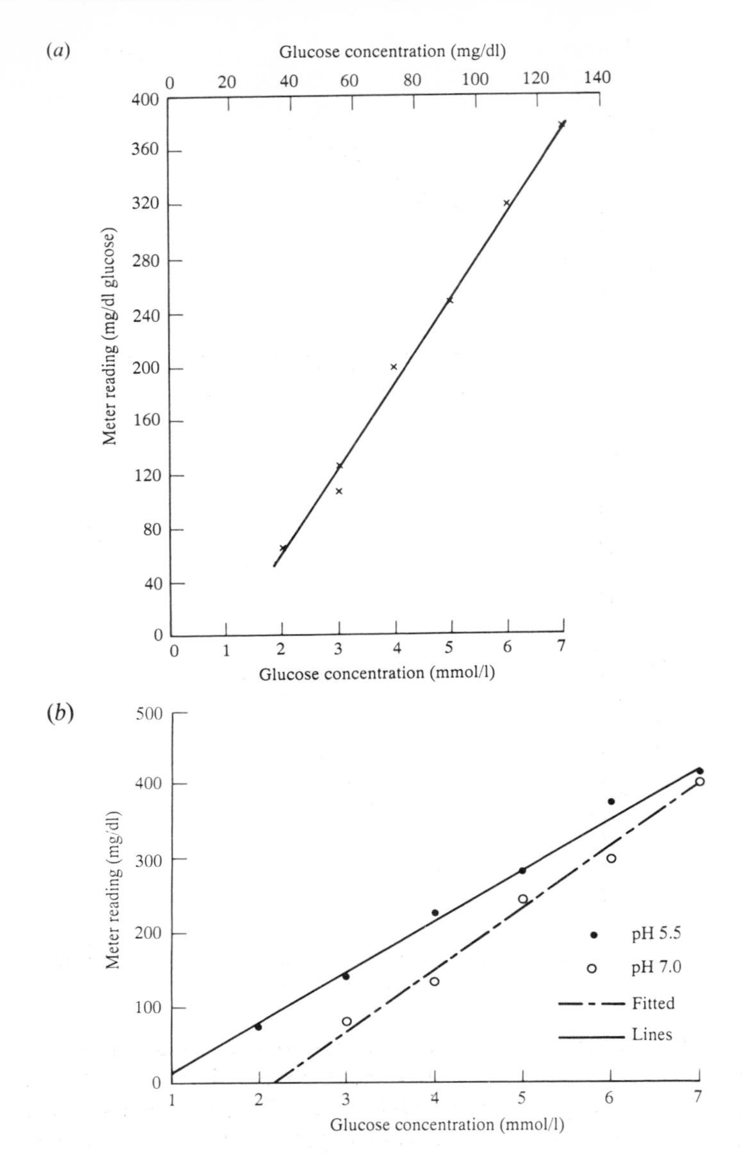

Figure 17.10 *Characteristics of the ExacTech system for blood glucose measurement when challenged in non-blood samples. Each point in the graphs corresponds to an individual disposable biosensor strip used without prior calibration.*

Source for Figures 17.10(a)–(d): Results reported at LFRA Panel Meetings 1988/1989. (Experiments at the LFRA; J. E. Sollars, E. Kress-Rogers, S. L. Lassetter). (a) Calibration in 20 mM phosphate buffer (pH 7). Comparison of the glucose concentration in the buffer (upper x-axis) and the

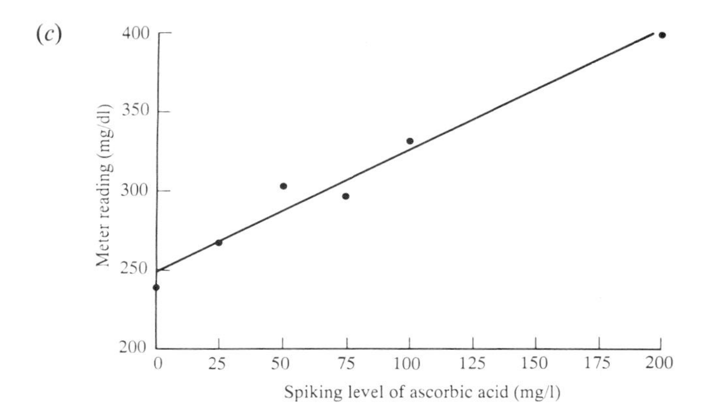

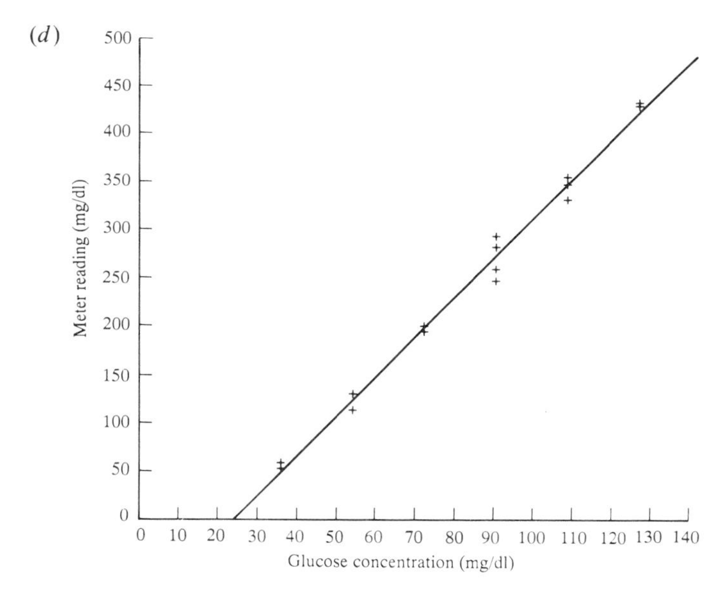

reading of the instrument, as calibrated for glucose in blood, shows the dependence on the sample type. (b) Measurements taken at the pH values typical for fresh blood (pH 7) and for meat after the early post-mortem phase (pH 5.5, lower trace) show the dependence on pH. (c) Response to spiking with ascorbic acid for a buffered 5 mM glucose solution. The spiking levels correspond to ascorbic acid concentrations that may be encountered in the juices of a potato tuber. Ascorbate removal for such a sample prior to glucose measurement with the sensor type tested here is therefore essential and can be accomplished with a simple rapid procedure. (d) Calibration for assorted fruit juices after dilution (1:30) with a buffered solution. The reference measurements (x-axis) were taken with Dionex (HPLC) and Yellowsprings (enzyme electrode based analyser) instruments. (See Chapter 13 for a description of these instruments).

the development of another biosensor type which is more inexpensively adapted to a particular specification (see, for example, Section 3.3).

For continuous or multi-use operation, AMEE usually need to be conditioned after applying the enzyme and mediator (the latter is often by dipping or printing). The most notable commercial success of the AMEE has, however, been in the form of a single-use no-calibration enzyme electrode as part of a user-friendly system. Continuous operational stability is not required in this configuration. A good shelf life and strip-to-strip reproducibility of both equilibrium current and equilibrium pattern are the essential features here, and these have been shown to be achievable.

Ferrocene-based mediators are generally preferred due to their apparent lack of toxicity, but other mediators have been developed to widen the range of AMEE analytes and also to overcome the limitations imposed by the patent on ferrocene as an AMEE mediator. AMEE devices based on the mediator tetrathiafulvalene (TTF) have been developed for lactate, for example. They are to be further developed as disposable blood analyser devices (Palleschi 1989, Palleschi and Turner, 1990). The development of the mediator TTF has been described earlier by Turner, Hendry and Cardosi (1987). In addition to the ferrocene derivatives and TTF, electron mediators studied for use in AMEE's include quinone derivatives, phenazonium ions, hexacyanoferrate, various organic dyes and polypyrrole.

The ferrocene-based AMEE biosensor (Cass, Davis, Francis, Hill, Aston, Higgins, Plotkin, Scott and Turner 1984) for glucose concentration has been developed into an *in-situ* monitor for the glucose content of bacterial culture broth in fermentation vessels (Brooks, Ashby, Turner, Calder and Clarke, 1987). This fermenter probe configuration, but with a modified immobilization procedure, was later tested in molasses samples injected into a buffered electrolyte (Bradley, Kidd, Anderson, Dear, Ashby and Turner, 1989). A high correlation with GLC-AA assays (gas-liquid chromatography of acetic anhydride derivatives) was observed, with the advantage of a far more rapid assay. A ferrocene-based AMEE was also tested successfully in banana extracts (D'Costa, Dillon, Hodgson and Quantick, 1988).

An extended linear range up to 20 g/l with the fermenter probe was achieved by a continuous internal buffer flow in conjunction with a variation in the membrane pore size (Bradley and Schmid 1991). This was at the cost of an increase in response time, but enhanced stability was reported for this probe (attributed to the constant ionic strength and pH environment provided by the buffered electrolyte). By injecting glucose into the internal flow solution, *in-situ* calibration became possible. For operation in molasses samples, frequent updating of the calibration (approximately hourly) is carried out. For clinical *in vivo* trials and many potential industrial applications, continuous operation with longer intervals between calibrations is desirable. The factors contributing to the difficulties in achieving long-term stability without calibration updates include the leaching of minute amounts of mediator from the device. To some extent, this can be overcome by conditioning prior to use and frequent re-calibration in use.

So far, AMEE have been applied less often in continuous *in vivo* trials than the indirect amperometric biosensors. The problems still to be solved before the AMEE design can be regarded as suitable for *in vivo* applications are discussed by Pickup, Shaw and Claremont (1987, 1989) and by Shaw, Claremont and Pickup (1991) and Alcock, Karayannis and Turner (1991). Many of the aspects considered there are also relevant for on-line applications in food processing. Turner (1989) has suggested that a polymeric form of the ferrocene mediator should be developed for *in vivo* applications. An approach aimed at the improvement of AMEE stability in continuous operation is being followed by Tamiya, Karube, Hattori, Suzuki and Yokoyama (1989). They immobilize the enzyme and dimethylferrocene mediator onto a polypyrrole-modified electrode.

3.3 Biosensors Based on the Amperometric Indirect Dual-Membrane Enzyme Electrode (AIDMEE)

In the food industry, a wide range of specifications exists for probes wanted for chemical analysis. Both the physical and chemical properties of the samples and their environment and the desired analyte and range are diverse in this industry. For each probe specification the market is, in general, small compared with that for a blood glucose probe, for example.

A biosensor type suitable for adaptation to such applications is the amperometric indirect dual-membrane enzyme electrode (abbreviated here as AIDMEE). In common with the early biosensors, this device is based on an amperometric gas probe. The new device design makes use of advances in membrane technology, however. The enzyme layer is situated between an inner and an outer membrane (Figure 17.11a). Each of the two membranes is tailored to a set of device characteristics. One of the membranes can also be engineered to provide binding sites for enzyme immobilization (see Vadgama 1986; Mullen et al. 1986; Churchouse et al. 1986; Battersby and Vadgama 1987; MacDonnell and Vadgama 1989; Vadgama 1990; Tang and Vadgama 1990; Tang et al. 1990).

The base electrode (detecting H_2O_2) is protected by an inner membrane from interfering species such as ascorbate by size exclusion. This does not entail a significant increase in response time, in contrast to the membranes that would have to be applied on AMEE devices for this purpose between the sample and the enzyme layer.

On the other side of the enzyme layer is the outer membrane. This is tailored to control diffusion of the analyte from the sample to the enzyme layer such that the analytical range can be extended significantly to provide direct measurements on undiluted food samples for sugars, for example

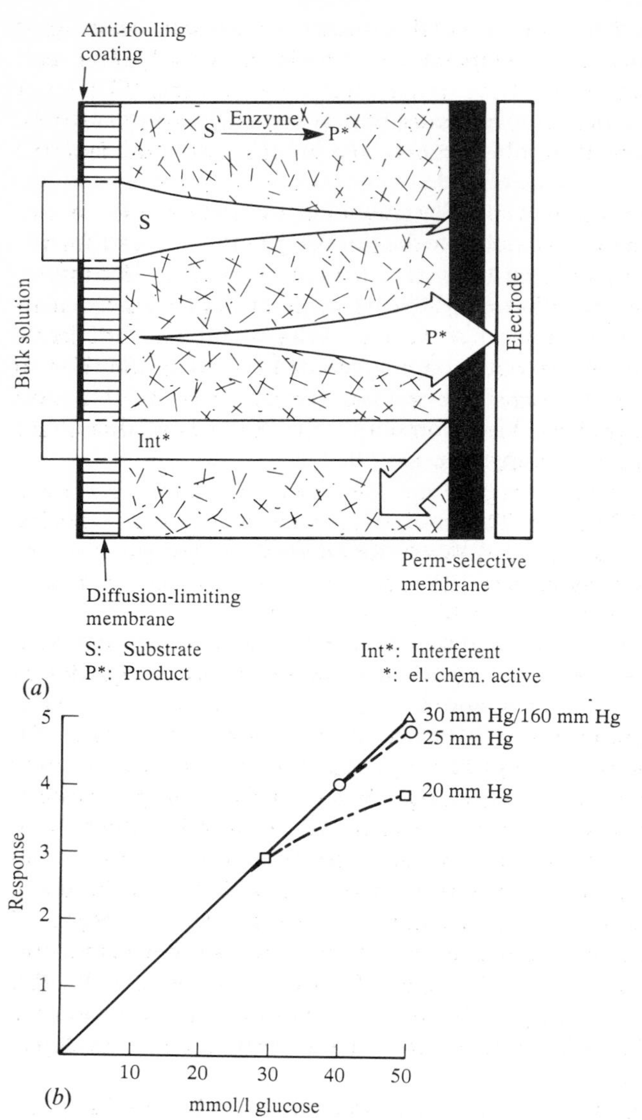

Figure 17.11 *The amperometric indirect dual membrane enzyme electrode (AIDMEE). (a) Schematic design. After MacDonnell and Vadgama (1989). (b) Extended linear range and reduced oxygen dependence achieved by tailoring of the outer membrane (diffusion-limiting polycarbonate membrane). A linear range up to 500 mmol/l was reported in Mullen, Keedy, Churchouse and Vadgama (1986). (Reprinted from Vadgama (1986) by permission from the author and the journal* Measurement, *published for the International Measurement Confederation by The Institute of Measurement and Control, London.)*

(Figure 17.11b). Another advantage of this control is the small loss in signal amplitude with a reduction in temperatures. This is particularly important for foods as these are often kept under chill temperatures. The outer membrane can also be designed to reduce biofouling by the sample. Again, this is relevant for food samples which often contain proteins, fats and starch particles.

A mediator is not used in this device; this can reduce the cost of developing probes for a new analyte substantially, since the need for the development of a new enzyme-mediator-electrode system is eliminated. The latter development can be quite expensive, as it would require a non-toxic mediator that is insoluble in the sample and that will efficiently mediate electron transfer between the enzyme and the electrode. For an AMEE, both reagents (enzyme and mediator) need to be immobilized together without a significant loss of activity. The choice of mediators for AMEE can also be restricted by patents.

Against that is the disadvantage of oxygen dependence in the operation of the basic AIDMEE device. This problem can be overcome by tailoring the outer membrane. Oxygen independence down to oxygen pressures of 20 mm Hg has been demonstrated already. For lower oxygen pressures, aeration can be provided by a suitable probe design.

The AIDMEE can readily be constructed in needle form and has been used in many *in vivo* trials. Although the conditions in food industry applications would differ from the clinical implantation trial, the latter is nevertheless an indication of the general suitability of a device for continuous operation in a biological sample. Drift due to reagent leaching or due to biofouling, for example, are equally relevant in both applications.

Indirect amperometric biosensors can also be constructed on the basis of Clark-type oxygen electrodes (instead of peroxide probes). Hendry et al. (1990) point to the potential problems associated with the evolution of peroxide which accompanies the consumption of oxygen in the presence of the analyte. Nevertheless, glucose sensors of this type have been comparatively successful in *in vivo* trials (Gough 1989; Ege 1989; Mascini 1989b). Based on the Clark-type oxygen electrode, a set of biosensors for inosine and hypoxanthine for fish freshness assessment has also been developed (Watanabe et al. 1986).

In the *in vivo* trials so far the membranes of the devices have not necessarily been optimized, so that improvements in performance can be expected for the future. Christie et al. (1989) point to the extension of linear range, avoidance of oxygen limitation, enhanced selectivity and reduced biofouling for AIDMEE needle-type electrodes (for glucose and ethanol in the reported case) with suitably fabricated membranes.

Devices of the AIDMEE type have also been constructed using miniature Clark-type peroxide probes produced in planar silicon technology as the base device (Gernet et al. 1989; Koudelka et al. 1989). The overall dimensions of this planar biosensor were 0.8 mm × 3 mm. An outer polyurethane membrane was dip coated to reduce pO_2 dependence. The device had a linear range up to 40 mM glucose (equivalent to approximately 0.7 per cent m/m in aqueous solution). The device has been implanted subcutaneously and tested *in vivo* in a clinical trial (Koudelka et al. 1989). (The base device itself, that is the micro-Clark probe, was earlier described by Engels and Kuypers 1983.)

A glutamate sensor based on an amperometric microglucose electrode and a glucose sensor based on a micro-peroxide electrode have been reported by Karube et al. (1987). An improved microglucose electrode was the basis of a glucose sensor by Suzuki et al. (1988).

Biosensors have also been constructed on the basis of CHEMFET microelectronic gas probes for hydrogen and ammonia (Danielsson and Winquist 1989; Mattiasson and Danielsson 1989). Biosensors based on the ammonia-sensitive CHEMFET detected urea and creatinine with a sensitivity of down to 10^{-7} mol/l. A Teflon membrane, separating the gas sensor from the solution in which the biochemical reaction took place, allowed the work with crude sample solutions and protected the sensor from fouling. Based on the ammonia-sensitive CHEMFET device, L-asparagine, L-glutamate, L-aspartate and adenosine can also be detected. Biosensors based on the hydrogen-sensitive CHEMFET were suggested for alcohols and aldehydes. The unmodified H_2-sensitive CHEMFET (without an enzyme layer) was used to monitor the metabolic activity of *E. coli* and *Clostridium acetobut.* and to assess the susceptibility of Enterobacteriaceae to ampicillin. (See also Chapter 16 on the potential application of the hydrogen-sensitive CHEMFET for the monitoring of microbial spoilage processes in food under anaerobic conditions.)

Whether a commercial AIDMEE probe will be engineered to be sufficiently reproducible for use without calibration remains to be seen. This will depend on both technical and economic developments in the area of membrane technology. As a reproducible base device, the micro-Clark cell or the CHEMFET would be suitable. The stability of advanced AIDMEE sensors is helped by the integral outer membrane, the covalent immobilization of the enzyme and the absence of a mediator.

A commercial example of an indirect amperometric biosensor is the YSI Analyser (Yellowsprings Co., USA). This is a benchtop instrument for industrial QC laboratories based on the peroxide electrode and exchangeable enzyme membranes. It was initially configured mainly for saccharides; membranes are also available for the assay of starches and alcohols. More recently, the assay of aspartame with this instrument has been suggested.

3.4 Biosensors Based on the FET: the ENFET

The ENFET has already been introduced in Section 3.1 and compared with the amperometric biosensors. The advantages of intrinsic pre-amplification and of combination with the REFET reference device have been indicated in comparing the ENFET with biosensors based on conventional potentiometric ion-sensitive electrodes. Some examples of ENFET devices will now be given. Based on the pH ISFET, a number of biosensors have been constructed by immobilizing enzymes on the pH-sensitive gate. These so-called ENFETs (enzyme-sensitized FETs) or BioFETs have been developed for the assay of penicillin, glucose, urea and other compounds (Caras and Janata 1980; Van der Schoot and Bergveld 1987–88; Karube and Sode 1989; Hintsche et al. 1990; Brand et al. 1991; Karube et al. 1990). A review of the techniques used in the construction of ENFETs has been given by Kimura and Kuriyama (1990).

An ISFET-based system for the determination of ATP has been studied (Gotoh et al. 1986). A multi-enzyme layer in the form of a cell membrane of acetic acid bacteria on an ISFET has been reported for ethanol, intended for blood analysis (Tamiya et al. 1988). Interestingly, this sensor responds to ethanol but not to methanol, in contrast to sensors based on isolated alcohol oxidase. This property could be of interest in the analysis of alcoholic beverages. Work on an ENFET for hypoxanthine, intended for fish freshness assessment, has been reported by Gotch et al. (1988) using xanthine oxidase immobilized on an ISFET. An ENFET for inosine, also relevant in the assessment of fish freshness, has been reported by the same group (Gotoh et al. 1988).

An interesting variant has been developed by Van der Schoot and Bergveld (1987). They have incorporated a microactuator electrode surrounding the gate area of a pH ISFET. A feedback circuit keeps the pH in the immobilized enzyme layer constant, and the signal needed for the actuator represents the output. (This operation is reminiscent of the constant current operation of the plain pH ISFET to eliminate temperature dependence; see earlier.) This design has the advantage of eliminating the dependence on the buffer capacity of the sample.

In clinical applications or in flow injection analysis systems, the ENFET will compete with other biosensor designs such as the amperometric mediated enzyme electrode (Section 3.2), the amperometric indirect dual-membrane enzyme electrode (Section 3.3) or the enzyme thermistor (Section 3.5). Each of these biosensor types has its own set of characteristics which define the most appropriate applications for the design. The ENFET or BioFET can be integrated with other FET devices into multiple FET sensor arrays for the assay of ions, gases and organic compounds. The wide range of specifications for food applications of biosensors, together with the increase in the cost of FET devices with decreasing production volume for a particular specification, would limit the commercial viability of ENFETs for food applications.

The construction of an immunosensor based on the FET, that is an FET device carrying immobilized antibodies and responding to the corresponding antigen, has been suggested. The search for a viable design for this IMFET is still on (Schasfoort et al. 1989).

3.5 The Enzyme Thermistor and Related Devices (ENTHERM)

The exothermic nature of many biological reactions, and particularly the generality of enthalpy changes in enzymatic reactions, have been used in conventional calorimetry. The advantages of this detection technique compared with colorimetric, fluorescent or luminescent techniques include independence of the optical properties of the sample, and the obviation of labels. An attractive feature of the calorimetric assay is also the replacement of multi-enzyme assays with a single-enzyme assay as, for example, in the measurement of sucrose. The assay technique is particularly valuable when a large range of different bioanalytes is to be assayed with a single instrument. The calorimetric method is more versatile than luminescent or amperometric methods, for example, since there is no need for matching luminescent labels, electron mediators or secondary gas or ion sensors.

The construction of sensors based on a calorimetric principle has been pioneered by Mosbach and Danielsson (1974). These calorimetric biosensors combine the advantages of the conventional calorimetric assay with simpler operation. The enzyme can be directly immobilized on a temperature sensor such as the thermistor (ENTHERM) or, for more efficient detection, on a small column located in the sample flow leading to the thermistor. A split-flow device with an enzyme column and a reference column helps to reduce the effect of heat from other sources. The thermal biosensor principle has been applied to a wide variety of enzyme-substrate systems (Danielsson 1990).

Possible fields of application for thermal biosensors include the determination of metabolites in physiological fluids or process mixtures; the assessment of toxicity in environmental monitoring; the activity of enzymes in biotechnology; the metabolism of microbial cells in suspension; and the determination of proteins and hormones. The column carries immobilized enzymes or whole cells or is empty in these instruments. For the determination of proteins and hormones, for example, a thermometric enzyme-linked immunoassay (TELISA) has been demonstrated. (See reviews by Mosbach and Danielsson 1981; Danielsson 1990.)

For high sensitivity, unfortunately, the enzyme thermistor required a temperature-controlled environment to eliminate interference from external temperature variations, even when a reference device (a thermistor without the enzyme) was used. The typical water bath arrangement prevented a compact instrument design when high sensitivity was needed.

This residual influence from external temperature variations has been attributed to the difficulty of finding perfectly matched thermistor pairs for the enzyme thermistor and the reference device. By using a heat-sensitive device that has been manufactured in utilizing microcircuit processes, it is possible to overcome this problem. One such device is the thermopile, consisting of an array of multiple microthermocouples in series. Nieveld (1983) has described a monolithic silicon thermopile comprising 152 couples fabricated on a single chip sized 4 mm × 2.5 mm and providing an output of 76 mV/°C (see also Kress-Rogers 1985a, Sections BIII and DII.5).

The successful application of the thermopile as the base device for a calorimetric sensor with high sensitivity at room temperature has been reported by Muehlbauer et al. (1990a) (see also Muehlbauer et al. 1989; 1990b). An attractive application of advanced biothermal sensors for the food industry would be a multi-saccharide sensor for the simultaneous assay of glucose, sucrose, fructose, maltose and galactose.

The possibility of developing an implantable glucose sensor based on the thermopile design has been suggested by Guilbeau et al. (1987). A portable flow injection system for biothermal analysis is under development by Danielsson (1990). An integrated circuit design has been described by Muramatsu et al. (1987).

References for Sections 3.1 to 3.3

Albery, W. J. and Bartlett, P. N. (1984) An organic conductor electrode for the oxidation of NADH. *Journal of the Chemical Society,* no. 4, 234–236

Albisser, A. M. and Leibel, B. S. (1984) Closed-loop control of diabetes. In *Theory, Design and Biomedical Applications of Solid State Chemical Sensors,* eds P. W. Cheung, D. G. Fleming, M. R. Neuman and W. H. Ko, CRC Press

Alcock S. J., Karayannis, M. and Turner, A. P. F. (1991) The design and development of new chemical sensors for *in vivo* monitoring, *Biosensors and Bioelectronics,* **6,** 647–652

Arnold, M. A. and Meyerhoff, M. E. (1988) Recent advances in the development and analytical applications of biosensing probes. *CRC Critical Reviews in Analytical Chemistry,* **20,** 149–196

Bartlett, P. N. (1989) Conducting organic salt electrodes. In *Biosensors: a Practical Approach,* ed. A. E. G. Cass, IRL Press, 47–96

Battersby, C. M. and Vadgama, P. (1987) A lactate needle enzyme electrode for whole blood measurement. *Diabetes, Nutrition and Metabolism,* **1,** 87–92

Bradley, J., Kidd, A. J., Anderson, P. A., Dear, A. M., Ashby, R. E. and Turner, A. P. F. (1989) Rapid determination of the glucose content of molasses using a biosensor. Analyst, **114,** 375–379

Bradley, J. and Schmid, R. D. (1991) Optimization of a biosensor for *in situ* fermentation monitoring of glucose concentration. *Biosensors and Bioelectronics,* **6,** 669–674

Brooks, S. L., Ashby, R. E., Turner, A. P. F., Calder, M. R. and Clarke, D. J. (1987) Development of an on-line glucose sensor for fermentation monitoring. *Biosensors,* **3,** 45–56

Cass, A. E. G., David, G., Francis, G. D., Hill, H. A. O., Aston, W. J., Higgins, I. J., Plotkin, E. V., Scott, L. D. L. and Turner, A. P. F. (1984) Ferrocene-mediated enzyme electrode for amperometric determination of glucose. *Analytical Chemistry,* **56,** 671–677

Christie, I., Koochaki, Z., Rosenberg, M., Tang, L. X., Treloar, P. and Vadgama, P. (1989) Electromechanical sensors for analytical biochemical measurements. In Mascini 1989a, section 'Concerted action'

Churchouse, S. J., Battersby, C. M., Mullen, W. H. and Vadgama, P. (1986) Needle enzyme electrodes for biological studies. *Biosensors,* **2,** 325–342

Danielsson, B. and Winquist, F. (1989) Biosensors based on semiconductor gas sensors. *Biosensors: Fundamentals and Applications,* eds A. P. F. Turner, I. Karube and G. Wilson, Oxford University Press, 532–548

D'Costa, E., Dillon, M., Hodgson, F. J. A. and Quantick, P. C. (1988) Rapid prediction of banana fruit maturation using a glucose biosensor. *Analyst,* **113,** 225–228

Diebold, E., Rapkin, M. and Usmani, A. (1991) Chemistry on a stick. *Chemtech,* August, 462–467

Ege, H. (1989) Construction and testing of a needle glucose electrode with a membrane formed from a polyurethane dispersion. In Mascini 1989a, section 'Experience of implanted sensors'

Engels, J. M. L. and Kuypers, M. H. (1983) Medical applications of silicon sensors. *Journal of Physics E: Scientific Instruments,* **16,** 987–994

Gernet, S., Koudelka, M. and De Rooij, N. F. (1989) Fabrication and characterization of a planar electrochemical cell and its application as a glucose sensor. *Sensors and Actuators,* **18,** 59–70

Gough, D. A. (1989) Experience with an implanted electrochemical glucose sensor. In Mascini 1989a, section 'General topics'

Hendry, S. P., Higgins, I. J. and Bannister, J. V. (1990) Amperometric biosensors. *Journal of Biotechnology,* **15,** 229–238

Karube, I. (1990) Microbial sensor. *Journal of Biotechnology,* **15,** 225–266

Karube, I., Sode, K. and Tamiya, E. (1990) Microbiosensors. *Journal of Biotechnology,* **15,** 267–282

Karube, I., Tamiya, E., Murakami, T., Gotoh, M. and Kagawa, Y. (1987) Microsensor based on silicon fabrication technology. *Annals of the New York Academy of Sciences* (Enzyme Engineering 8), no. 501, 256–264

Kirk, R. S. and Sawyer, R. (1991) *Pearson's Composition and Analysis of Foods*, 9th ed. Longman Scientific, Harlow, UK

Koudelka, M., Gernet, S. and De Rooij, N. F. (1989) Planar amperometric enzyme-based glucose microelectrode. *Sensors and Actuators,* **18,** 157–165

Koudelka, M., Rohner-Jeanrenaud, F., Terrettaz, J., Bobbioni-Harsch, E.,

Jeanrenaud, B. and De Rooij, N. F. (1989) *In vivo* calibration of the subcutaneously implanted glucose sensor. In Mascini 1989a, section 'Concerted action'

Kulys, J. J. and Svirmickas, G. J. S. (1980) Reagentless lactate sensor based on cytochrome b_2. *Analytica Chimica Acta,* **117,** 115-120

MacDonnell, M. B. and Vadgama, P. (1989) Membranes: separation principles and sensing. *Selective Electrode Reviews,* **11,** 17-67

Mascini, M. (ed.) (1989a) *Strategies for* In Vivo *Sensing.* Abstracts of the 2nd Workshop of Biomedical Engineering Action of the European Community on Chemical Sensors for *In Vivo* Monitoring, Firenze, 12-15 November 1989

Mascini, M. (1989b) Biosensors development and applications. In Mascini 1989a, section 'Concerted action'

Mattiasson, B. and Danielsson, B. (1989) Biosensor research in Lund with relevance for development towards *in vivo* monitoring. In Mascini 1989a, section 'Concerted action'

Mullen, W. H., Keedy, F. H., Churchouse, S. J. and Vadgama, P. (1986) Glucose enzyme electrode with extended linearity, application to undiluted blood measurements. *Analytica Chimica Acta,* **183,** 59-66

Palleschi, G. (1989) Biosensor applications in blood analysis. In Mascini 1989a, section 'Concerted action'

Palleschi, G. and Turner, A. P. F. (1990) Amperometric tetrathiafulvalene-mediated lactate electrode using lactate oxidase adsorbed on carbon foil, *Analytica Chimica Acta,* **234,** 459-463.

Pickup, J. C., Shaw, G. W. and Claremont, D. J. (1987) Implantable glucose sensors: choosing the appropriate sensing strategy. *Biosensors,* **3,** 335-346

Pickup, J. C., Shaw, G. W. and Claremont, D. J. (1989) Potentially implantable amperometric glucose sensors with mediated electron transfer: improving the operating stability. *Biosensors,* **4,** 109-119

Rechnitz, G. A. and Ho, M. Y. (1990) Biosensors based on cell and tissue material. *Journal of Biotechnology,* **15,** 201-218

Shaw, G. W., Claremont, D. J. and Pickup, J. C. (1991) *In vitro* testing of a simply constructed, highly stable glucose sensor suitable for implantation in diabetic patients. *Biosensors and Bioelectronics,* **6,** 401-406

Sim, K. W. (1990) Development of a sensor for ethanol, *Biosensors and Bioelectronics,* **5,** 311-325.

Suzuki, H., Tamiya, E. and Karube, I. (1988) Fabrication of an oxygen electrode using semiconductor technology. *Analytical Chemistry,* **60,** 1078-1080

Tamiya, E., Karube, I., Hattori, S., Suzuki, M. and Yokoyama, K. (1989) Micro glucose sensors using electron mediators immobilized on a polypyrrole-modified electrode. *Sensors and Actuators,* **18,** 297-307

Tang, L. X. and Vadgama, P. (1990) Optimization of enzyme electrodes. *Medical and Biological Engineering and Computing,* **28,** B18-B24

Tang, L. X., Koochaki, Z. B. and Vadgama, P. (1991) Use of a composite liquid membrane for enzyme electrode construction. *Analytica Chimica Acta,* **232,** 357–365

Turner, A. P. F. (1983) Applications of direct electron transfer bioelectrochemistry in sensors and fuel cells. *Biotech 83,* Online Publications, Northwood, UK

Turner, A. P. F. (1989) Cranfield Biotechnology Centre. In Mascini 1989a, section 'Concerted action'

Turner, A. P. F., Hendry, S. P. and Cardosi, M. F. (1987) Tetrathiafulvalene: a new mediator for amperometric biosensors. In *The World Biotech. Report on Biosensors, Instrumentation and Processing,* I, Online Publications, Pinner, UK, 125–137

Updike, S. J. and Hicks, G. P. (1967) The enzyme electrode. *Nature,* **214,** 986

Vadgama, P. (1986) Enzyme electrodes and their potential for medical exploitation. *Measurement,* **4,** 154–159

Vadgama, P. (1990) Biosensors: adaptation for practical use. *Sensors and Actuators,* **B1,** 1–7

Watanabe, E., Endo, H., Hayashi, T. and Toyama, K. (1986) Simultaneous determination of hypoxanthine and inosine with an enzyme sensor. *Biosensors,* **2,** 235–244

Yanchinski, S. (1983) Step towards automatic insulin plant. *New Scientist,* **98,** 628

Yokoyama, K., Sode, K., Tamiya, E. and Karube, I. (1989) Integrated microbiosensor for determination of glucose and galactose. *Analytica Chimica Acta,* **218,** 137–142

References for Section 3.4

Brand, U., Reinhardt, B., Ruether, F., Scheper, T. and Schuegerl, K. (1991) Bio-field-effect transistors for process control in biotechnology. *Sensors and Actuators,* **B4,** 315–318

Caras, S. and Janata, J. (1980) Field effect transistor sensitive to penicillin. *Analytical Chemistry,* **52,** 1935–1937

Gotoh, M., Tamiya, E., Karube, I. and Kagawa, Y. (1986) A microsensor for adenosine-5′-triphosphate (using a) pH-sensitive field effect transistor. *Analytica Chimica Acta,* **187,** 287–291

Gotoh, M., Seki, A., Suzuki, M., Tamiya, E. and Karube, I. (1988) Hypoxanthine sensor based on an amorphous silicon ISFET. *Analytica Chimica Acta,* **215,** 301–305

Gotoh, M., Tamiya, E., Seki, A., Shimizu, I. and Karube, I. (1988) Inosine sensor based on an amorphous silicon ISFET. *Analytical Letters,* **21,** 1783–1800

Hintsche, R., Dransfield, I., Scheller, F., Pham, M. T., Hoffmann, W., Hueller, J. and Mortiz, W. (1990) Integrated differential enzyme sensor using hydrogen and fluoride ion sensitive multigate FETs. *Biosensors and Bioelectronics*, **5**, 327–334

Karube, I. and Sode, K. (1989) Biosensors for fermentation process control. In *Bioprocess Engineering: the First Generation*, ed. T. K. Ghose, Ellis Horwood, Chichester, 207–219

Kimura, J. and Kuriyama, T. (1990) FET biosensors. *Journal of Biotechnology*, **15**, 239–254

Kirk, R. S. and Sawyer, R. (1991) *Pearson's* Composition and Analysis of Foods, 9th edn, Longman, Harlow, UK

Tamiya, E., Karube, I., Kitagawa, Y., Ameyama, M. and Nakajima, K. (1988) Alcohol FET-sensor based on a complex cell membrane enzyme system. *Analytica Chimica Acta*, **207**, 77–84

Van der Schoot, B. H. and Bergveld, P. (1987) The pH-static enzyme sensor. *Analytical Chimica Acta*, **199**, 157–160

Van der Schoot, B. H. and Bergveld, P. (1987–88) ISFET-based enzyme sensors. *Biosensors*, **3**, 161–186

References for Section 3.5

Danielsson, B. (1990) Calorimetric biosensors. *Journal of Biotechnology*, **15**, 187–200

Guilbeau, E. J., Towe, B. C. and Muehlbauer, M. J. (1987) A potentially implantable thermoelectric sensor for measurement of glucose. *Transactions of the American Society for Artificial Internal Organs*, **XXXIII**, 329–335

Mosbach, K. and Danielsson, B. (1974) An enzyme thermistor. *Biochimica Biophysica Acta*, **364**, 140–145

Mosbach, K. and Danielsson, B. (1981) Thermal bioanalysers in flow streams: enzyme thermistor devices. *Analytical Chemistry*, **53**, 83A–94A

Muehlbauer, M. J., Guilbeau, E. J. and Towe, B. C. (1990a) Applications and stability of a thermoelectric enzyme sensor. *Sensors and Actuators*, **B2**, 223–232

Muehlbauer, M. J., Guilbeau, E. J., Towe, B. C. and Brandon, T. A. (1990b) Thermoelectric enzyme sensor for measuring blood glucose. *Biosensors and Bioelectronics*, **5**, 1–12

Muehlbauer, M. J., Guilbeau, E. J. and Towe, B. C. (1989) Model for a thermoelectric enzyme glucose sensor. *Analytical Chemistry*, **61**, 77–83

Muramatsu, H., Dicks, J. M. and Karube, I. (1987) Integrated-circuit biocalorimetric sensor for glucose. *Analytical Chimica Acta*, **197**, 347–352

Nieveld, G. K. (1983) Thermopiles fabricated using silicon planar technology. *Sensors and Actuators*, **3**, 179

4 Chemically Sensitive Optical and Acoustic Devices

4.1 Introduction

The sensors described in Sections 1 and 3 relied on a chemical reaction between the analyte and a reagent immobilized on the sensor or a material that formed the sensor surface. This chemical reaction caused (directly or via a secondary reaction) a change in the electronic or ionic conduction properties of the sensing device that would give rise to a signal indicating the concentration of the analyte. Alternatively, the heat of reaction was measured by the base transducer.

Most of the sensors in this section rely instead on changes in the surface characteristics of the sensor that occur when the analyte is bound to the sensitizing layer on the device by sorption or by the formation of a complex with an affinity binding agent. Sorption can take place at the external, or for a porous material also the internal surface of a material (adsorption), or involve migration into the material itself (absorption).

This sorption or affinity binding causes changes in the dielectric, acoustic or mechanical properties of the sensitized surface. The dielectric properties affected include those in the optical refractive index, and minute changes of this can be measured with special optical devices (Sections 4.2, 4.3). Changes in the mechanical properties include changes in the layer thickness, mass, density or elasticity. These can be sensitively measured by their influence on the propagation of acoustic surface waves (Section 4.6), or the adsorbed mass can be monitored as change in the resonance frequency of an oscillating piezocrystal (Section 4.5). The latter is known as the crystal microbalance, a highly sensitive mass monitor.

An exception are those fibre-optic probes (FOP) where the optical fibre acts as a carrier for an optical signal created by a colourimetric, fluorescent or luminescent reaction of the analyte with a reagent immobilized on the fibre tip. FOP are nevertheless included here (in Section 4.4), because it is also possible to form optical devices such as the SPR and TIR devices described in Sections 4.2 and 4.3 on an optical fibre.

Many of the sensors described here measure volatiles with the help of an adsorbing or absorbing film with a broad specificity responding to properties such as the polarity, solubility and/or size of the analyte. The sensitivity of these sensors is often exceptionally good due to the choice of a base device responding to minute changes in mass or refractive index. For volatiles, concentrations at the ppb level can often be detected. The properties of chemically selective sorbent coatings for sensors are described in the review by Grate and Abraham (1991). Analytes include volatile anaesthetics, organic solvents, organophosphorous compounds and many others. Many of these coatings are humidity-sensitive and this needs to be compensated for.

The same sensor configurations have also been used to carry antibodies or other affinity binding reagents for the specific determination of many volatiles as well as analytes carried in fluids, particularly vitamins, pesticides or specific microbes. In analogy to the immuno assay kits, these sensors are known as immunosensors. A more accurate, but also more unwieldy description is receptor-based chemical sensors.

They gain their selectivity from the natural affinity of certain proteins or protein fragments towards specific target species (complementary ligands). The formation of the complex is governed by size and shape, as well as charge-charge interactions, hydrogen bonding and Van der Waals interactions (Arnold and Meyerhoff, 1988). The receptor types include antibodies to specific bacteria or to smaller antigens such as hormones if these have first been attached to a high molecular weight antigen. They include also endogenous binding proteins and lectins that will form complexes with ligands such as drugs, small ions and gases or with polysaccharides and smaller sugar molecules, respectively. Some receptors need to be incorporated into natural or synthetic biomembranes including Langmuir–Blodgett films (see Sections 2.3.4. and 2.4.2). These receptors respond, for example, to drugs and to amino acids.

It is possible to choose reversible binding agents for a continuous operation at the cost of other performance criteria such as sensitivity and detection limit. It is also possible to recycle bound antibodies with a stronger affinity with the help of chemical treatments, but often disposable devices are preferred to avoid the labour costs incurred by cleaning and recalibration or to avoid a lesser performance of the assay.

Some designs use isolated devices which are filled or smeared with the sample and left to incubate for a few minutes before they are presented to an instrument for the reading. Strictly speaking, these cannot be regarded as sensors but would be classified as indicators. Based on the same basic device principles, it is, however, also possible to construct instruments that give a continuous reading and that can follow the process of binding as it occurs. This can mean an almost instantaneous indication of analyte concentration without prior incubation provided that the analyte concentration is not too low. The useful analytical range for the rate-of-change mode of operation has been estimated as being in the nanomolar to the micromolar range (Eddowes 1988). A recent fibre-optic immunosensor design uses a continuous supply of antibody to the sensing surface to achieve a continuous output combined with the advantages of a high affinity binding agent (Section 4.4).

Immunosensors based on heat-sensitive devices (TELISA) have been briefly introduced in Section 3.5. Immunosensors based on bioelectrochemical detection techniques are not considered in this chapter. (A description of these is given in Kress-Rogers and Turner, 1986 and in Green, 1987.)

For the immunosensors described here, it is important to ensure that non-specific binding of proteins to the sensing surface is avoided and this can be achieved by coatings and by using reference devices carrying non-specific antibodies.

In general, the immunosensor devices reduce the incubation time and labour required for an assay compared with an equivalent immuno assay kit. The washing step that separates bound from unbound antibody in the kit, for instance, is not needed when the device responds only to the properties in the immediate vicinity of the sensing surface and is insensitive to any changes in the properties of the bulk of the sample. For example, the separation step is also eliminated, when the bound material is at the sensing surface where its fluorescent properties are measured, whereas the unbound material is in the bulk of the sample which does not influence the signal.

Where an immunosensor is available, considerable savings compared with the immuno assay kit can therefore be achieved in the labour cost for an assay, the time needed for a result can be reduced, the determination can be carried out in locations remote from laboratories with experienced staff, and in some cases a continuous measurement can be performed. The commercial development of immunosensors (by companies such as Pharmacia, Sweden; Ares-Serono, Switzerland or Amersham, UK) is intended primarily for medical analyses. Immunosensors are also of great interest to the environmental monitoring and defence sectors. A number of applications exist in the food industry where a specially adapted immunosensor development would be viable.

This would be the case for pathogens such as Salmonella and Listeria, where a high number of assays are carried out regularly so that considerable benefits would accrue from a lower labour cost per assay. Moreover, a reduction in the assay time would allow more products to be screened before releasing a batch for distribution. It would be particularly valuable, if not only the actual assay time but also the pre-incubation time could be reduced. This could be achieved if the sensitivity could be increased compared to the present assay kits. For many food samples, the confirmation of absence of Salmonella can take 5 days while a series of preparation, pre-incubation, and assay steps is carried out with current methods.

Other areas of high interest for the food industry are the determination of moulds (where the Howard Test for mould in tomato products, for example, is exceptionally tedious), or of vitamins and toxins that require lengthy and therefore expensive analysis procedures at present. Further possible areas are the speciation of meat (is it actually beef?) or the identification of proteins from vegetable sources (is soya protein mixed with beef mince?).

In the field of volatile sensors based on sorption of the analyte to the sensing surface, the assessment of aroma and the monitoring of cooking processes are of particular interest.

4.2 The Surface Plasmon Resonance (SPR) Device

The SPR device detects minute changes in the refractive index of the sensing surface and its immediate vicinity. Laboratory prototypes of SPR devices may consist of a prism on a glass slide carrying a thin metal layer. For more compact devices, a diffraction grating structure can be formed instead of using a prism. The metal layer on the glass substrate carries a sensitizing layer, and this is in contact with the sample which can be either a gas or a liquid. In this device, a collective excitement (the surface plasmons) of electrons in the metal film occurs and leads to a total absorption of light at a particular angle of incidence (Raether 1977). This angle is dependent on the refractive indices on either side of the metal film. The refractive index of the sensitizing layer and a thin layer immediately adjacent to it can be measured as a shift in the angle of total absorption of light (Figure 17.12). Continuous measurement during the incubation process is possible.

The application of chemically sensitive layers on SPR devices was demonstrated for the determination of halothane gas (using a sorption layer) and for the determination of anti-IgG and anti-HSA in the early 1980s and subsequent work has included diffraction grating devices (Nylander, Liedberg and Lind 1982; Liedberg, Nylander and Lundstrom 1983; Flanagan and Pantell 1984; Pettigrew 1984; Cullen, Brown and Lowe 1988). An SPR study aimed at estimating the sensitivity in the rate-of-change mode for rapid analysis has been carried out by Daniels, Deacon, Eddowes and Pedley (1988). A method for enhancing the sensitivity of SPR immunosensors has been reported by Evans and Charles (see Mitchell 1990). Here, antibodies labelled with refractive index probes are deposited on to a layer of analyte conjugate on the SPR sensing surface; labelled antibodies are released from the device when the analyte binds to the antibodies. LB films have been incorporated into SPR devices for the determination of nitrogen dioxide (Zhu, Petty and Harris 1990).

4.3 The Total Internal Reflectance (TIR) Device

The TIR device consists of a light guide carrying a sensitizing layer. When light coming from an optically denser medium is incident on an optically rarer medium (for example, on the interface of glass or heavy plastic to air or water), then it will be totally reflected if the incidence angle is above the critical angle. Under total reflection conditions, an evanescent wave exists that penetrates only a fraction of a wavelength into the optically rare

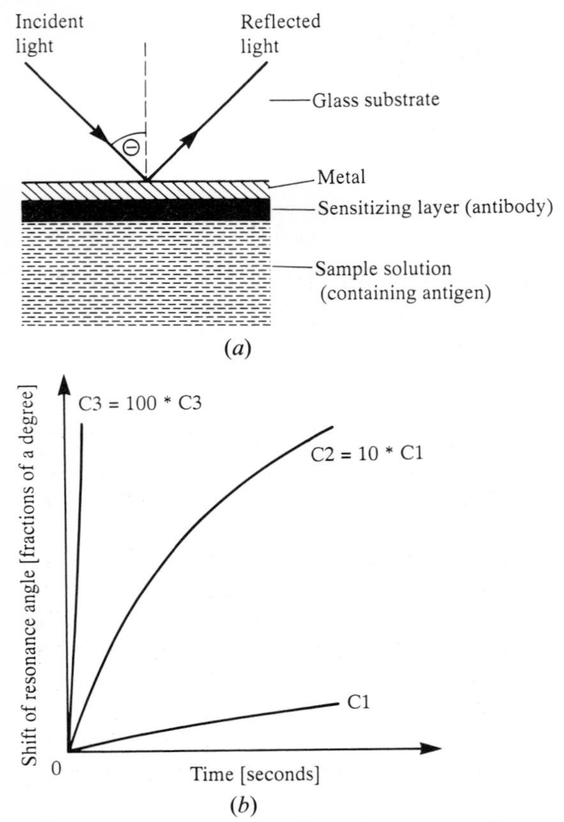

Figure 17.12 *The Surface Plasmon Resonance (SPR) device for the measurement of minute changes in the refractive index. (a) The device can be configured as a highly sensitive gas sensor or as an immunosensor by the immobilization of sensitizing layers on the device. (From Kress-Rogers and Turner 1986.) (b) In the immunosensor configuration, it is possible to monitor the antibody-antigen binding process directly without a prior incubation period unless the antigen concentration is very low (as for the concentration c_1). The sensitivity for kinetic monitoring can be enhanced with labelling techniques (see text). (Adapted from Liedberg, Nylander and Lundstroem 1983.)*

medium. The sensitizing layer of the TIR device and the sample layer in its immediate vicinity are in the field of this evanescent wave. The device is also known as evanescent wave immunosensor.

Changes in the refractive index or absorptivity in this surface layer can be observed as a reduction in the light intensity transmitted through the guide (Figure 17.13). This technique is known as ATR or attenuated total reflectance. Fluorescent techniques can be used to advantage since the fluorescent evanescent wave originating from fluorescent complexes at the

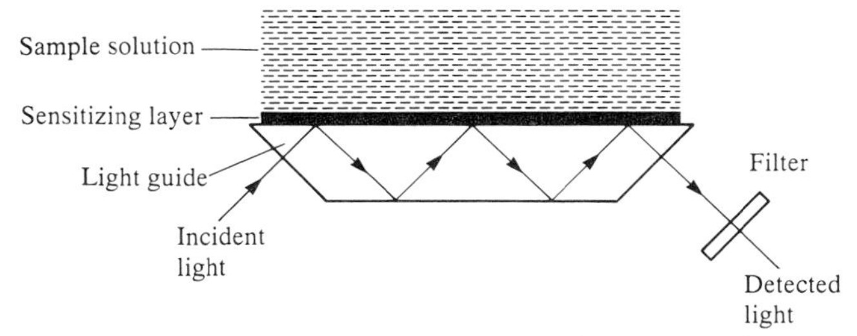

Figure 17.13 *The Total Internal Reflection Device (TIR) makes use of the effect of analyte binding to an immobilized sensing layer, on the evanescent wave. (From Kress-Rogers and Turner 1986.)*

surface is coupled back into the guide, yielding a high fluorescence intensity at the angle of total internal reflection. This technique is known as TIRF or total internal reflection fluorescence. (See Place, Sutherland and Dahne 1985; Dessy 1989; Parriaux 1991.)

Continuous measurement during incubation and separation of free labels in the sample solution is unnecessary. Disposable devices with automatic definition of sample volume by capillary tubes have been constructed. With suitable geometries of the capillary tubes (similar to the cells used in LCD displays), sensitive detection can be achieved with short incubation times (Shanks 1986; Badley, Drake, Shanks, Smith and Stephenson 1987). Remote immunosensing using TIRF optics inside a single optic fibre has been demonstrated (Andrade 1985; Wang, Christensen, Brynda, Andrade, Ives and Lin 1989).

Rather than measuring the intensity of light transmitted through the guide, one can also measure the phase using an interferometric technique. This allows very sensitive detection of changes in the refractive index profile in the evanescent wave field. Immunosensors on this basis have been demonstrated by Heideman, Koodyman and Greve (1991).

An immunosensor system intended for the simultaneous detection of different analytes or for detection at different sensitivities has been demonstrated. This consists of an array of TIRF devices combined with a CCD imaging device for spatially resolved detection (Hlady, Lin and Andrade 1990).

4.4 Fibre-Optic Probe (FOP) Devices

A light beam travelling through an optical fibre can be modulated in optical intensity, phase, polarization, wavelength and spectral distribution. Intrinsic fibre-optic sensors rely on an effect of the variable to be measured on the fibre and its transmission characteristics. An intrinsic sensor for strain, for example, can rely on the effects of fibre bending. (This particular fibre-optic sensor type can be sensitive to interference from vibrations, but many other fibre-optic sensor types are mechanically robust and can be handled without special care.) Extrinsic fibre-optic probes rely on sensitizing layers or devices applied to the fibre tip to modify the optical input transmitted by the fibre which subsequently relays the output signal. A well-known example is the fibre-optic temperature sensor based on fluorescent rare earth compounds applied to the tip. This probe is used in microwave ovens in the food industry. (See Gratten 1991, on FOP temperature sensors.)

Many chemically sensitive fibre-optic probes (FOP) are of a similar construction. A reagent which changes its colour or fluorescent characteristics in the presence of the analyte is immobilized on the fibre-tip. Interference from changes in ambient light or temperature and drift due to photobleaching of a dye are addressed by techniques such as using an AC-light signal, a reference wavelength measurement, in the case of fluorescence also by observing a time-constant rather than an intensity. In addition, it is important to optimize the immobilization technique for the reagent, and preferably to use a covalent bonding technique to avoid losses of reagent in use. (See Figure 17.4.)

Using such techniques, it is possible to develop stable FOPs for pH, for example, and some of these are commercially available and successfully used in surgery. In laboratory prototypes designed primarily for the exploration of new reagents or of particular FOP probe characteristics, on the other hand, severe drift is sometimes present because the optimization of the immobilization procedure was not a priority. While the clinical pH probes serve to demonstrate the feasibility of developing stable chemical sensors based on FOP, they are not suitable for direct transfer to the food industry because the clinical probes serve a very narrow pH range only and are not sufficiently robust nor designed for inexpensive mass production.

An adaptation for food applications could involve a bundle of fibres carrying different reagents to cover a wider pH range, for example. It would require either an inexpensive disposable tip or be tolerant to rigorous cleaning procedures. The surface would need to be designed to minimize biofouling by using a suitable physical structure and/or a chemical surface treatment. Much progress has been made recently in the areas of immobilization techniques and functional membranes for sensors, and this could well promote the development of food industry-compatible FOP probes, in the first instance for pH. As material for the optical fibre, plastic rather than glass may be chosen to reduce foreign body hazards, alternatively, heavy shielding of the fibre is possible exposing only the chemically sensitive tip which can be polymer-based. (See Section 2.1 on problems with current pH glass electrodes in the food industry.) An interesting approach has been taken by Wolfbeis and Offenbacher (1986) in constructing a dual sensor for monitoring ionic strength and physiological pH values. In the two sensing elements, they immobilized a pH indicator in two micro-environments that maximize and minimize, respectively, the influence of ionic strength on the pH reading of the FOP. For a review on pH FOP see, for example, Leiner and Wolfbeis (1991).

A range of further analytes can be measured with FOP of similar construction as for pH, for example dissolved gases such as oxygen, carbon dioxide. Such FOP are also known as optrodes (in analogy to electrodes) or sometimes as optodes. FOP for ammonia or dissolved metal ions including aluminium, magnesium, zinc or cadmium can be produced with a similar design. (Lübbers and Opitz 1983; Seitz, 1984.) Approaches for the construction of a FOP for the detection of paralytic shellfish poison have been explored by Guevremont, Quigley and Schweitzer (1991). A recent review of FOP has been given by Arnold (1990) and a two-volume book dedicated to FOP has appeared recently (Wolfbeis, 1991a).

Many colorimetric and fluorescent assays can, in principle be converted into FOP designs, provided that the reagent acts reversibly with a short response time and that it can be immobilized reliably without losing its activity. Advantages of the FOP are robustness (reduced foreign body hazard), tolerance to high microwave intensities (in ovens), tolerance to electrical mains noise, safe operation in environments where any electrical connections are undesirable or unreliable.

Enzymes can be co-immobilized with a dye sensitive to pH, oxygen or ammonia for the construction of biosensors (see Section 3 for analytes). FOP glucose biosensors based on an oxygen optrode and an FOP biosensor for ethanol, for example, have been reported by the group of Wolfbeis (Wolfbeis and Posch 1988; Moreno-Bondi, Wolfbeis, Leiner and Schaffar 1990; Wolfbeis 1991b).

Bioluminescence-based FOP using immobilized firefly luciferase have been constructed for the measurement of ATP (adenosine triphosphate)

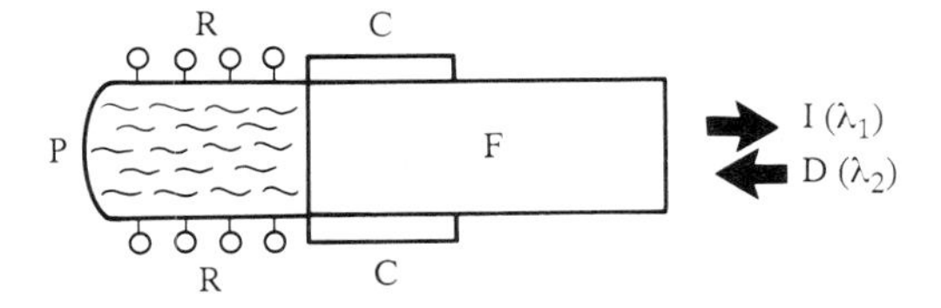

Fibre-optic chemical sensor, alternative configuration (see also Figure 17.4(a)). F: optical fibre. C: cladding. I, D: input and detected radiation at different wavelengths λ_1 and λ_2. P: polymer grown on fibre tip. R: reagent immobilised on polymer.

and NADH (nicotinamide adenine dinucleotide) and for glutamate. FOP based on the fluorescence of NADH have been produced for the detection of lactate and pyruvate (Arnold 1990).

A more complex design can be used for the measurement of glucose, for example by competitive binding. The glucose binding agent concanavalin A (ConA) is immobilized on the walls of the fibre outside the illuminated volume. The competitive binding agent, dextran labelled with fluorescein, is displaced from the ConA by the analyte glucose, causing it to diffuse into the illuminated solution volume. This leads to an increase in measured fluorescence related to the concentration of glucose in the sample. This principle can also be applied to other receptor-based assays. The possibility of constructing SPR and TIR devices directly on an optical fibre has already been indicated above.

Immunosensors based on FOP are described in Vo-Dinh, Griffin and Stepaniak (1991). An interesting approach is the use of a continuous supply of antibody to the sensing surface to achieve a continuous output without having to use a low affinity binding agent (Walt, Barnard and Luo 1991). They use a controlled-release polymer developed earlier for long lasting FOP pH sensors (Luo and Walt 1989).

4.5 Piezocrystal Balance Devices

This device (also known as quartz microbalance or piezoelectric oscillator) is based on the dependence of the resonance frequency of a vibrating quartz crystal (or other piezoelectric crystal) on its mass. Such a crystal can be coated with a sensitizing film to measure the concentration of volatiles. The mass increase on adsorption of the analyte is observed as frequency shift. The estimated detection limit is 1 picogram. Chemical sensors on this basis have early been constructed to detect explosives, and sensitivities in the ppb range were achieved. Odorant volatiles have been detected with lipid-coated crystals. (King 1964; Hlavay and Guilbault 1977; Tomita, Ho and Guilbault 1979; Guilbault 1980; Alder and McCallum 1983; Guilbault and Jordan 1988; Mierzwinski and Witkiewitz 1989; Muramatsu, Tamiya and Karube 1989a; McCallum 1989).

The crystal will also oscillate when immersed in a liquid, and has been used to determine analytes such as trace iodide or nanogram quantities of silver or mercury. The signal for the immersed crystal will, however, depend on properties such as the density, viscosity and conductivity of the liquid, and changes in these need to be compensated for by use of a reference crystal (with a non-specific coating), for example. Immunosensors based on the piezocrystal balance have also been constructed. (Nomura and Iijima 1981; Nomura and Mimatsu 1982; Muramatsu, Tamiya and Karube 1989b).

Alternatively, rather than compensating for changes in density or viscosity as an interference, these can be monitored as indicators for a reaction induced by a reagent with the analyte. This has been applied to the determination of endotoxin by gelation of Limulus Amebocyte Lysate (Muramatsu, Tamiya, Suzuki and Karube 1988).

High concentrations of bacterial or yeast cells, as they occur in fermenters, can be monitored by a measurement of the ultrasonic velocity of the culture broth. This velocity is dependent on the density of the broth, and a measurable change in velocity will be observed at high biomass concentrations. Piezoelectric crystals will be the basis of the transmitter and receiver transducers, but this measurement type is quite distinct from the highly sensitive quartz microbalance technique. (Compare Chapters 8 and 9; see also Clarke, Blake-Coleman and Calder 1987.)

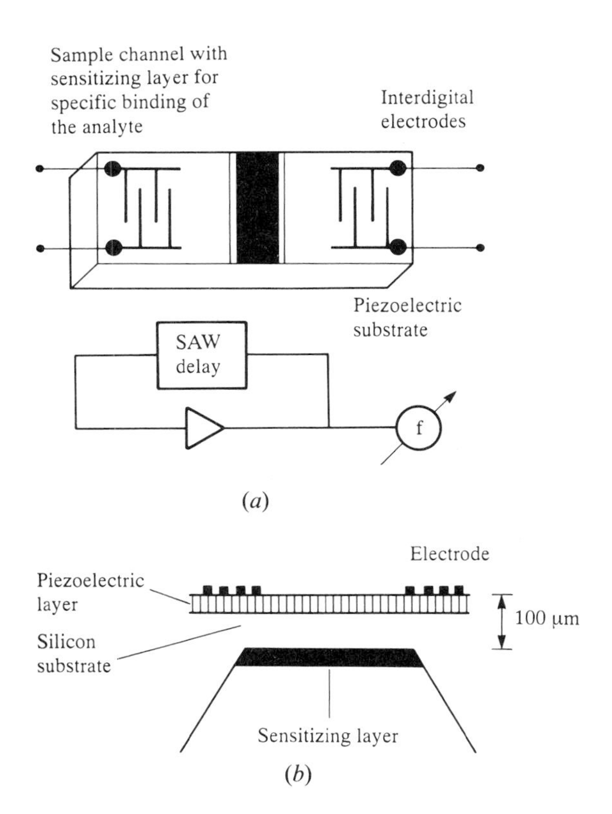

Figure 17.14 *The Surface Acoustic Wave (SAW) device. (a) Schematic design of a chemical sensor based on a SAW delay line device. The basic circuit for monitoring the sensor response is also shown. (b) Schematic design of SAW membrane device. (Both from Kress-Rogers and Turner 1986.)*

4.6 Surface Acoustic Wave (SAW) Devices

The SAW device consists of a piezoelectric crystal, such as quartz or lithium niobate, carrying thin-film interdigital electrodes. Radio frequency excitation of the electrode pair creates a synchronous mechanical surface wave. This is propagated on the surface of the piezoelectric substrate and received by another electrode pair (if the device is a SAW delay line) or by the same pair after reflection (if the device is a SAW resonator device). The basic SAW device is typically encountered as a component in the VHF circuit of a television. SAW sensors for temperature and pressure are commercially available now, and much research has been carried out into the application of SAW devices as chemical sensors. (Wohltjen and Dessy 1979a, b, c; Chuang, White and Bernstein 1982; D'Amico, Palma and Verona 1982; Bryant, Poirier, Riley, Lee and Vetelino 1983; Wohltjen 1984; Barendsz, Vis, Nieuwenhuizen, Nieuwkoop, Vellekoop, Ghijsen and Venema 1985; Ballantine and Wohltjen 1989; Nieuwenhuizen and Venema 1989; D'Amico and Verona 1989). Two types of SAW devices configured as chemical sensors are shown in Figures 17.14(a) and (b). The membrane configuration has the advantage of separating the electrical contacts from the sensing surface which is in contact with the sample.

The propagation of the acoustic surface wave is highly sensitive to small changes in the density, elasticity and electrical conductivity of the surface of the SAW device. These properties will in turn be affected by the sorption or affinity binding of an analyte to a sensitizing layer applied to the device. Most readily, the construction of sensors for volatiles can be achieved with SAW sensors. (Liquids in contact with the sensing surface attenuate the signal considerably.) The application of the SAW device as an immunosensor operating in liquids was pioneered by Roederer and Bastiaans (1983). The sensitivity was limited to the detection of 10 micrograms at that time. (Some confusion was caused by a review in 1985 which quoted nanograms, a more typical figure for the desirable detection limit.) Progress has since been achieved in the development of SAW sensors for analytes in liquids, but the optical devices (SPR and TIR) are still dominant for immunosensing applications in liquids.

In analogy to the 'odour meters' based on arrays of MeOx semiconductor gas sensors with broad overlapping specificities, the broad specificity of films that bind volatiles by sorption can be employed in the development of analysers for complex gas mixtures and ultimately for the assessment of aroma or the monitoring of cooking operations. Pattern recognition techniques can be combined with SAW sensor arrays (Ballantine, Rose, Grate and Wohltjen 1986). Langmuir–Blodgett (LB) films (as described for electronic devices in Sections 2.3.4, 2.4.2; see also Roberts, Petty, Baker, Fowler and Thomas 1985) can also be applied to SAW devices. Karube's group has incorporated odorant recepting LB films into SAW devices (Chang, Ebert, Tamiya and Karube; 1991). (Compare Section 2.4.)

References for Section 4

Alder, J. F. and McCallum, J. J. (1983) Piezoelectric crystals for mass and chemical measurements. *Analyst* (London), **108,** 1169–1189

Andrade, J. D., VanWagenen, R. A., Gregonis, D. E., Newby, K. and Lin, J. N. (1985) Remote fibre-optics biosensors based on evanescent-excited fluoro-immunoassay: concept and progress. *IEEE Transactions on Electronic Devices ED-22,* 1175–1180

Arnold, M. A. (1990) Fibre-optic biosensors. *Journal of Biotechnology,* **15,** 219–228

Arnold, M. A. and Meyerhoff, M. E. (1988) Recent advances in the development and analytical applications of biosensing probes. *CRC Critical Reviews in Analytical Chemistry,* **20,** 149–196

Arnold, M. A. and Wangsa, J. (1991) Transducer-based and intrinsic biosensors. Chapter 16, 193–216 in Wolfbeis (1991a)

Badley, R. A., Drake, R. A. L., Shanks, I. A., Smith, A. M. and Stephenson, P. R. (1987) Optical biosensor for immunoassays: the fluorescence capillary-fill device. *Philosophical Transactions Royal Society London B316,* 143–160

Ballantine, D. S. Jr. and Wohltjen, H. (1989) Surface acoustic wave devices for chemical analysis. *Analytical Chemistry,* **61,** 704A–715A

Ballantine, D. S. Jr., Rose, S. L., Grate, J. W. and Wohltjen, H. (1986) Correlation of surface acoustic wave device coating responses with solubility properties and chemical structure using pattern recognition. *Analytical Chemistry,* **58,** 3058–3066

Barendsz, A. W., Vis, J. C., Nieuwenhuizen, M. S., Nieukoop, M. J., Ghijsen, W. J. and Venema, A. (1985) A SAW chemosensor for NO_2 gas concentration measurement. IEEE Ultrasonics Symposium, San Francisco 1985, 586–590

Bryant, A., Poirier, M., Riley, G., Lee, D. L. and Vetelino, J. F. (1983) Gas detection using surface acoustic wave delay lines. *Sensors and Actuators,* **4,** 105–111

Chang, S. M., Ebert, B., Tamiya, E. and Karube, I. (in press) Development of chemical vapour sensor using SAW resonator oscillator incorporating odorant recepting LB films. *Biosensors and Bioelectronics*

Chuang, C. T., White, R. M. and Bernstein, J. J. (1982) A thin-membrane surface acoustic wave vapour-sensing device. *IEEE Electronic Device Letters 3,* 145–148

Clarke, D. J., Blake-Coleman, B. C. and Calder, M. R. (1987) Principles and potential of piezoelectric transducers and acoustical techniques. 551–571. In: *Biosensors, Fundamentals and Applications,* ed. A. P. F. Turner, I. Karube and G. S. Wilson, Oxford University Press, New York

Cullen, D. C., Brown, R. G. W. and Lowe, C. R. (1987/88) Detection of immuno-complex formation via surface plasmon resonance on gold-coated diffraction gratings. *Biosensors,* **3,** 211–226

Daniels, P. B., Deacon, J. K. Eddowes, M. J. and Pedley, D. G. (1988) Surface plasmon resonance applied to immunosensing. *Sensors and Actuators,* **15,** 11–18

D'Amico, A. and Verona, E. (1989) SAW (Surface Acoustic Wave) sensors. *Sensors and Actuators,* **17,** 55–66

D'Amico, A., Palma, A. and Verona, E. (1989) Palladium surface acoustic wave interaction for hydrogen detection. *Applied Physics Letters 55–66;* see also *Sensors and Actuators,* **3,** 31–35

Dessy, R. E. (1989) Waveguides as chemical sensors. *Analytical Chemistry,* **61,** 1079a–1094a

Eddowes, M. J. (1987/88) Direct immunochemical sensing: basic chemical principles and fundamental limitations. *Biosensors,* **3,** 1–15

Flanagan, M. T. and Pettigrew, R. H. (1984) Surface plasmon resonance and immunosensors. *Electronics Letters,* **20,** 968–970

Grate, J. W. and Abraham, M. H. (1991) Solubility interactions and the design of chemically selective sorbent coatings for chemical sensors and arrays. *Sensors and Actuators,* **B3,** 85–111

Gratten, K. T. V. (1991) Fibre-optic techniques for temperature sensing. Chapter 15, 151–192 in Wolfbeis (1991a)

Green, M. (1987) Electromechanical immunoassays. *Philosophical Transactions Royal Society London B316,* 135–142

Guevremont, R., Quigley, M. N. and Schweitzer, M. (1991) Strategies in the design of an effective fibre-optic sensor for the detection of paralytic shellfish poison. *Analytical Proceedings,* **28,** 241–244

Guilbault, G. G. (1980) Uses of the piezoelectric crystal detector in analytical chemistry. *Ion-selective Electrode Review,* **2,** 3–16

Guilbault, G. G. and Jordan, J. M. (1988) Analytical uses of piezoelectric crystals: a review. *CRC Critical Reviews in Analytical Chemistry,* **19,** 1–28

Heidman, R. G., Kooyman, R. P. H. and Greve, J. (1991) Development of an optical waveguide interferometric immunosensor. *Sensors and Actuators,* **B4,** 297–299

Hlady, V., Lin, J. N. and Andrade, J. D. (1990) Spatially resolved detection of antibody-antigen reaction on solid/liquid interface using total internal reflection excited antigen fluorescence and charge-coupled device detection. *Biosensors and Bioelectronics,* **5,** 291–301

Hlavay, J. and Guilbault, G. G. (1977) Applications of the piezoelectric crystal detector in analytical chemistry. *Analytical Chemistry,* **49,** 1890–1894

King, W. H. (1964) Piezoelectric sorption detector. *Analytical Chemistry,* **36,** 1735–1739

Kress-Rogers, E. and Turner, A. P. F. (1986) Immunosensors based on acoustic, optical and bioelectrochemical devices and techniques. Leatherhead Food RA Technical Notes No. 49. See also p. 227–253 in *Immunoassays for Veterinary and Food Analysis*, ed. B. A. Morris, M. N. Clifford and R. Jackman (1988), Elsevier Applied Science Publishers (Proceedings of Symposium on Advances in Immuno Assays for Veterinary and Food Analysis, University of Surrey, July 1986)

Leiner, M. J. P. and Wolfbeis, O. S. (1991) Fibre-optic pH sensors. Chapter 8, 359–384 in Wolfbeis (1991a), vol. I

Liedberg, B., Nylander, C. and Lundström, I. (1983) Surface plasmon resonance for gas detection and biosensing. *Sensors and Actuators*, **4**, 299–304

Lübbers, D. W. and Opitz, N. (1983) Optical fluorescence sensors for continuous measurement of chemical concentrations in biological systems. *Sensors and Actuators*, **4**, 641–644

Luo, S. and Walt, D. R. (1989) Fibre-optic sensors based on reagent delivery with controlled-release polymers. *Analytical Chemistry*, **61**, 174–177

McCallum, J. J. (1989) Piezoelectric devices for mass and chemical measurements: an update. *Analyst*, **14**, 1173–1185

Mierzwinski, A. and Witkiewicz, Z. (1989) The application of piezoelectric detectors for investigations of environmental pollution. *Environmental Pollution*, **57**, 181–198

Mitchell, P. (1990) Biosensors for drug monitoring. Laboratory Equipment Digest, June 1990, 31–32 (Based on a paper written by A. Evans and S. Charles of Amersham Ltd, UK)

Moreno-Bondi, M. C., Wolfbeis, O. S., Leiner, M. J. P. and Schaffar, B. P. H. (1990) Oxygen optrode for use in a fibre-optic glucose biosensor, *Analytical Chemistry*, **62**, 2377–2380

Muramatsu, H., Tamiya, E., Suzuki, M. and Karube, I. (1988) Viscosity monitoring with a piezoelectric quartz crystal and its application to determination of endotoxin by gelation of Limulus Amebocyte Lysate. *Analytica Chimica Acta*, **215**, 91–98

Muramatsu, H., Tamiya, E. and Karube, I. (1989a) Detection of odorants using lipid coated piezoelectric crystal resonators. *Analytica Chimica Acta*, **225**, 399–408

Muramatsu, H., Tamiya, E. and Karube, I. (1989b) Determination of microbes and immunoglobulins using a piezoelectric biosensor. *Journal of Membrane Science*, **41**, 281–290

Nieuwenhuizen, M. S. and Venema, A. (1989) Surface acoustic wave chemical sensors. *Sensors Materials*, **5**, 261–300

Nomura, T. and Iijima, M. (1981) Electrolytic determination of nanomolar concentrations of silver in solution with a piezoelectric quartz crystal. *Analytica Chimica Acta*, **131**, 97–102

Nomura, T. and Mimatsu, T. (1982) Electrolytic determination of traces of iodide in solution with a piezoelectric quartz crystal, *Analytica Chimica Acta,* **143,** 237–241

Nylander, C., Liedberg, B. and Lind, T. (1982) Gas detection by means of surface plasmon resonance. *Sensors and Actuators,* **3,** 79–88

Parriaux, O. (1991) Guided wave and electromagnetism and opto-chemical sensors. Chapter 4, 111–192 in Wolfbeis (1991a)

Pettigrew, R. M. (1984) Assay technique, International Patent Application, International Publication No. W084/02578

Place, J. F., Sutherland, R. M. and Dahne, C. (1985) Optoelectronic immunosensors: a review of optical immunoassay at continuous surfaces. *Biosensors,* **1,** 321–353

Raether, H. (1977) Surface plasma oscillations and their applications, 145–261. In *Physics of Thin Films,* vol. 9, ed. G. Haas and R. E. Thun, Academic Press, New York

Roberts, G. G., Petty, M. C., Baker, S., Fowler, M. T. and Thomas, N. J. (1985) Electronic devices incorporating stable Langmuir-Blodgett films. *Thin Solid Films,* **132,** 113–123

Roederer, J. E. and Bastiaans, G. I. (1983) Microgravimetric immunoassay with piezoelectric crystals. *Analytical Chemistry,* **55,** 2333–2336

Seitz, (1984) Chemical sensors based on fibre-optics. *Analytical Chemistry,* **56,** 16A–34A

Shanks, I. A. (1986) Optical biosensor for immunoassays: the fluorescence capillary-fill device. Presentation to the Royal Society London, see Badley *et al.* 1987 for Proceedings

Tomita, Y., Ho, M. H. and Guilbault, G. G. (1979) Detection of explosives with a coated piezoelectric quartz crystal, *Analytical Chemistry,* **51,** 1475–1478

Vo-Dinh, T., Griffin, G. D. and Stepaniak, M. J. (1991) *Fiberoptic immunosensors.* Chapter 17, 217–258 in Wolfbeis (1991a), vol. II

Walt, D. R., Barnard, S. M. and Luo, S. (1991) Optical immunosensors using controlled release polymers. Symposium of the American Chemical Society, Atlanta, April 1991, Division PMSE (Polymeric Materials Science and Engineering), Session 'Biosensors: novel applications of polymeric materials'

Wang, J., Christensen, D., Bryndaa, E., Andrade, J., Ives, J. and Lin, J. (1989) Sensitivity analysis of evanescent fibre-optics sensors. Proceedings of the Society of the Photooptic Institute of Engineering, SPIE 1067, 44–52

Wohltjen, H. (1984) Chemical sensors and micro-instrumentation, *Analytical Chemistry,* **56,** 87a–103A

Wohltjen, H. and Dessy, R. (1979a, b, c) *Surface acoustic wave probe for chemical analysis.* I. Introduction and instrument description; II. Gas chromatograph detector; III. Thermo-chemical polymer analyser, *Analytical Chemistry,* **51,** 1458–1464, 1465–1470, 1470–1475

Wolfbeis, O. S. (ed.) (1991a) *Fiber Optic Chemical Sensors and Biosensors.* Vols. I and II, CRC Press, Boston, London
Wolfbeis, O. S. (1991b) Sensing schemes, Chapter 3, 61–110. In Wolfbeis (1991), vol. I
Wolfbeis, O. S. and Offenbacher, H. (1986) Fluorescence sensor for monitoring ionic strength and physiological pH values. *Sensors and Actuators,* **9,** 85–91
Wolfbeis, O. S. and Posch, H. E. (1988) Optical sensors: a fibre-optic ethanol biosensor, *Fresenius Zeitschrift für Analytische Chemie,* **332,** 255–257
Zhu, D. G., Petty, M. C. and Harris, M. (1990) An optical sensor for nitrogen dioxide based on a copper phthalocyanine Langmuir-Blodgett film. *Sensors and Actuators,* **B2,** 265–269

5 Applying Sensors

5.1 Introduction

To develop the wide range of sensors that would be desirable for the food industry, in particular, would be prohibitively expensive. Not only different analytes but different ranges and chemical and physical environments would have to be catered for. The recognition of the cost that would be associated with the development of in-line sensors optimized for a range of such diverse applications has led to an interest in techniques that make the best of the sensors that are available or that can be developed economically.

Both conventional and novel pattern analysis techniques are increasingly being suggested for the evaluation of sensor outputs. Instruments attempting to mimic the response principles of animal noses based on gas sensors with broad overlapping specificities are being developed on this basis (Section 2.4). Pattern analysis techniques, including the application of neural networks (Section 5.2), are also being suggested for the evaluation of the outputs of other sensors and instruments installed in a process line without necessarily having a well-defined relationship to the process stream characteristics of interest. While these techniques offer a considerable potential, caution is needed to avoid a breakdown of the calibration which can be less than robust.

An alternative approach uses robotic sampling and sample preparation in conjunction with sensors that would otherwise be confined to laboratory applications (Section 5.3). This has been implemented particularly in Japan. Where an automatically sampled and prepared food liquid is homogeneous, flow analysis techniques can also be used to advantage.

Instead of a chemical sensor, be it in-line or at-line, it is often possible to install a non-invasive measurement system such as a near infrared analyser, or an instrument based on the interaction of microwaves or ultrasound with the process stream. This approach can be particularly favourable, providing hygienic non-contact application and simple maintenance (Section 5.4).

Once a system of in-line and at-line sensors and instruments has been installed, together with additional signal processing where appropriate, some of the resulting output signals may be fed directly to a programmable logic controller (PLC) controlling a particular part of the process. Others may need to be fed to a process control system that may have to use a complex procedure to derive the action to be taken and then to give the appropriate signals to the actuators. It is rarely necessary for a measured variable to equate to a single target value; more often a band of values around the target value is acceptable for a particular variable. The target value itself and the width of the band of acceptable values around it may depend on the values of the other measured variables. Such relationships can be taken into account by using expert systems and fuzzy logic systems. Expert systems use reasoning based on the experience of human experts entered into an expert system 'shell'. Fuzzy logic models accommodate imprecise concepts such as 'few', 'likely', 'warm' or 'possible' (see for example Eerikainen et al. 1988). See also the entry on Fuzzy Logic in the Glossary (Appendix A on page 671).

5.2 Pattern Recognition by Multivariate Analysis or Neural Networks

The application of pattern recognition techniques to high-speed gas chromatography has been described by Kowalski and Bender (1972). The application to the recognition of gases detected by MeOx sensor arrays was then carried out by Ikegami et al. (1983). The theoretical basis for the assay with sensors having partially overlapping specificities has been examined by Zaromb and Stetter (1984). Subsequently, pattern recognition was used with acoustic devices carrying chemically sensitive films (Carey et al. 1986; Ballantine et al. 1986). Numerous reports on the combination of gas sensor arrays with pattern analysis techniques have since appeared (Section 2.4).

The algorithms used in such studies have relied largely on those of conventional multi-variate analysis. Both linear and non-linear partial least square (PLS) models are used to predict one or several variables (type and/or concentration of one or more gases) from the independent variables (that is the sensor signals) (see for example Wold et al. 1983; Sundgren et al. 1990; Abe et al. 1987).

Pattern analysis by a neural network was used instead by Ema et al. (1989) in conjunction with an array of chemically sensitized acoustic devices (coated with gas chromatographic coatings of different molecular polarities). The system was trained to recognize eleven alcoholic beverages and succeeded with a probability of 73 per cent when the training beverages were presented.

Neural networks and associated algorithms are described by Jansson (1991), Pollard (1990) and Rumelhart et al. (1986a; 1986b). The artificial neural network attempts to mimic the structure of a biological nervous system. It can learn and adapt. In the event of breakdown on the part of an input channel, the neural network will attempt to adapt and continue to operate. Conventional programming of a mathematical model is unnecessary. In the neural network computer, the processing elements are interconnected to form a network with several layers, typically an input layer to encode raw signals, a hidden layer and an output layer. Each processing element converts a set of inputs to an output with the help of transfer functions.

Algorithms have been developed for the adaptation of the combined sensor and pattern analysis system to compensate for the drift of the characteristics of the sensor elements with time, and to improve separation among the output patterns obtained for different test gases (Nakamoto et al. 1991). The resulting capability of the system is termed 'plasticity' in analogy to the self-organizing property of living systems.

Another application for pattern recognition techniques is in combination with several different types of in-line sensors and non-contact instruments that are not specific for the property of interest. For a given food processing line, the range of instruments that can be applied in-line is often quite limited. It could include, for example, probes for temperature, pressure and pH together with a near infrared analyser for the determination of water, fat, starch and protein content. For the in-line assessment of the flavour, aroma or texture of a food, none of these instruments would provide an output with a well-defined relationship to the property of interest.

In the approach discussed here, extensive output data (from the available in-line instruments) and quality control results are collected over a period of weeks or months and then subjected to analysis with software packages for pattern recognition. This can then result in a signal processing mode 'learnt' by the package that will provide an output indicative of a characteristic of the finished product such as taste or smell, although none of the contributing sensors (which can be mounted at different points of the process line) necessarily has a well-defined physicochemical relationship to that characteristic.

Although this approach can provide information that would otherwise not be available, it can be costly to implement and can be subject to frequent extensive recalibration procedures when a process condition changes. For example, the supply of ingredients from another source, or a small change in recipe, could necessitate recalibration.

To illustrate the potential pitfalls of replacing specific measurements and explicit signal analysis methods with a non-specific data acquisition and processing system, an example can be given from the application of neural networks to the evaluation of photographic images of forest areas. The neural network was set up for the recognition of tanks partially obscured by trees. The researchers used half of their set of photographs to train the neural network in distinguishing images with and without tanks. They tested the system initially with the training set of images, then with the remainder of the images. Although not having come across these before, the neural network performed satisfactorily. They subsequently acquired another set of images and, in testing the system with the new set, found that it failed abysmally. Further investigation showed that, in the forest area scanned in the first set of images, tanks had been present during a period of cloudy weather; on subsequent days, when the tanks had left, sunnier weather had prevailed. Inadvertently, the neural network had been trained to recognize cloud cover rather than tanks.

Similarly, in the application of neural networks to the evaluation of outputs from sensors and instruments with no well-defined physicochemical relationship to the property that is to be determined, any calibration will have to be very extensive, encompassing all possible circumstances that are likely to be encountered. In the event of unforeseen small changes in the process or ingredients, the calibration could break down. It is therefore important to include specific sensors and instrumental techniques in the measurement strategy. Pattern analysis techniques, including the application of neural networks, are a valuable tool in data processing, but they are not a replacement for the development of specific sensing techniques.

5.3 Robotics and Flow Injection Analysis

Many food process streams are too hot, too viscous or too rich in biofouling matter for a chemical sensor and particularly a biosensor to operate for a satisfactory lifespan. Often the product is too inhomogeneous, the analyte concentration is too high, the pH or buffer strength is unfavourable for the functioning of the sensor, or interfering substances are present. Rather than take and then prepare and analyse the samples manually, a robotic system can be installed to sample automatically from the process line and then automatically to homogenize, dilute and buffer the sample and feed it to the sensor which is situated further down the waste line. Frequent sampling is possible in this way and rapid results can be obtained without the need to transfer samples to a separate at-line instrument or to a laboratory. The higher sampling frequency means that the resulting set of values is more representative for the process stream, and the rapid results

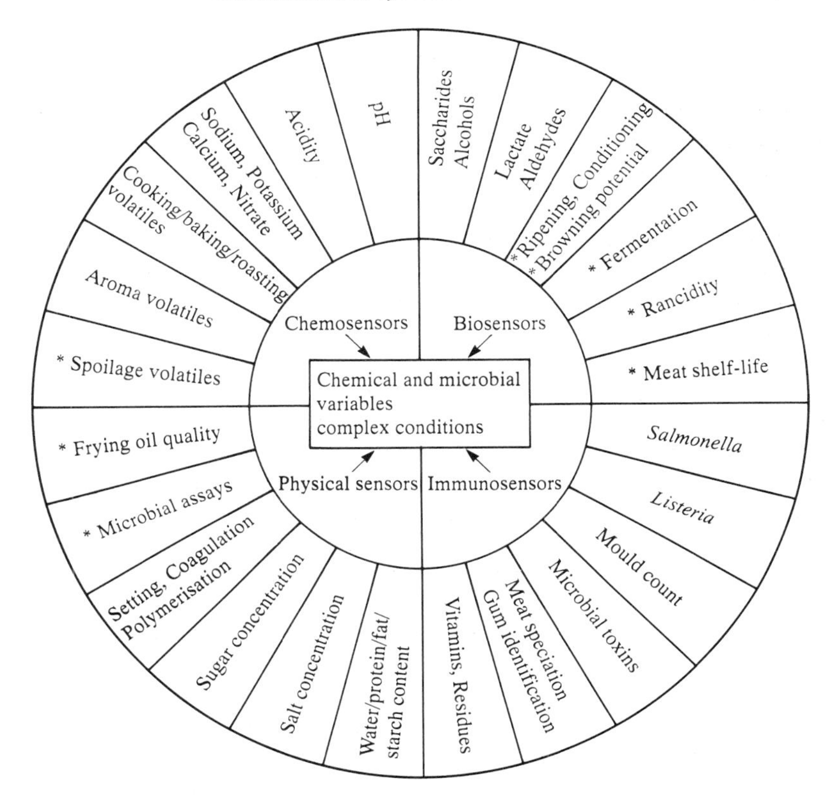

Figure 17.15 *Chemical sensors (biosensors, immunosensors and chemosensors) and physical sensors (dielectric, mechanical and ultrasonic) for the determination of chemical and microbial variables in food and for the assessment of conditions such as spoilage or ripening. For a definition of chemical sensor types see Section 1.1.*

mean that the delay for feedback and corrective action is much less than with manual sampling.

For example, in the monitoring of lactic acid fermentation, the robotic sampling system can feed to several preparation and analysis lines (see Figure 1.2). After sampling, crushing is carried out first. The flow can then be divided:

1 to a conventional pH electrode. (The pH electrode could now be replaced by an ISFET device mounted directly in the line; see Section 2.3.7)

2 (a) to a dilution/buffering system followed by an enzyme electrode for L-lactate, or (b) to a flow injection analysis system.
3 to an instrument for luminescence assay of ATP after further automatic preparation (to assess lactic acid bacteria population).

The final product, yoghurt, filled into individual pots, can be assessed for its rheological properties (setting) by a non-destructive mechanical resonance method. This can be carried out in-line, although line speeds may dictate a bypass line on which part of the product volume travels.

5.4 Choice of Instrumentation Type

Where the replacement of a current QC laboratory assay of spot-check samples by a continuous in-line measurement is intended, it is often useful to look beyond the possibility of developing an in-line instrument that measures the same variable as the QC method and is installed at the current QC sampling point in the process. For example, the concentration of a particular biochemical compound downstream may be dependent on the pH and temperature upstream. The introduction of upstream in-line pH monitoring could then be considered and compared with the cost and effectiveness of developing a biosensor for downstream measurement (directly in-line or as part of a robotic sampling system).

Alternatively, the concentration of certain compounds of interest may be correlated with the dielectric or mechanical properties of the sample. Under such circumstances, it may well be preferable to develop a mechanical or dielectric sensor that will withstand in-line conditions such as CIP procedures or hot samples. This is likely to be less expensive than the development of a chemical sensor having such specifications or being supported by an associated robotic sampling system.

An example for this approach is the development of a novel frying oil quality monitor based on *in situ* viscometry (see Chapter 16). This was adapted from a vibrating tube design developed earlier by the GEC Marconi Research Centre for applications such as boiler oil monitoring. Trials of the preliminary prototype in three frying oil types have shown a good correlation of the sensor output to the reference method for the assessment of frying oil degradation (polymeric and oxidized matter content by GLC), and a pre-commercial prototype has been developed by GEC Sensors Ltd and evaluated at the LFRA. This has now been replaced by a preproduction model.

The advantages offered by this approach compared with the development of a chemical sensor reside not only in the useful relationship of viscosity with frying-induced oil degradation, but also in the robust simple hygienic design of the probe (resistant to hot oil and cleaning procedures such as CIP) and in the convenient frequency output signal, which is less subject to corruption by electromagnetic noise and more easily and inexpensively interfaced with digital data processing systems. Mechanical resonance probes can be constructed for a range of variables including density, viscosity and flow rate (Langdon 1985; 1987). It may be possible to find other applications of the marker approach based on these sensors.

In other situations, the concentration of the target variable may be high enough for an indirect, non-invasive method more amenable to in-line implementation than a chemical sensor. Biosensors and other chemical sensors require contact with the sample (unless, for example, a volatile compound is to be assayed and can be readily isolated by a membrane). This contact can lead to problems such as biofouling or poisoning of the sensor, and the sensor design must preclude the possibility of sample contamination. Although these problems can be reduced by using the robotic approach (see Section 5.3), it will often be more efficient to install one of the instrument types providing non-contact measurements of compositional variables.

These are based on interaction with electromagnetic waves in the near infrared or microwave region or with ultrasonic waves, for example (see Chapters 5, 7–9). Near infrared reflection analysis (NIRA) has already gained widespread acceptance in the food industry for the in-line monitoring of water, fat, protein and starch content (and some other compositional measurements). To a lesser (but increasing) extent, microwave and ultrasound measurements for the determination of the moisture content or solute concentration of foods (but also of non-compositional variables such as density or liquid/solid ratio) have been implemented in the food industry. In general, such systems will be less sensitive and less specific than chemical sensors, but they can offer versatility and often operate in a non-contact configuration.

An example is the measurement of total sugar concentration in hot solutions, where an ultrasound velocity measurement can be considered provided that the composition and concentration of other solutes remains constant. Factors such as changes in the saccharide type would also have to be examined for accurate measurements. Biosensors, on the other hand, would be considered where a specific saccharide measurement was wanted and where operation at cold or tepid temperatures could be arranged.

The development of an in-line chemical sensing system to suit a particular food industry application can be difficult and expensive. It pays therefore to investigate the alternatives first (Figures 17.15).

References for Section 5

Abe, H., Yoshimura, T., Kanayama, S., Takahashi, Y., Miyashita, Y. and Sasaki, S. (1987) Automated odour-sensing system based on plural semiconductor gas sensors and computerized pattern recognition techniques. *Analytica Chimica Acta,* **194,** 1–9

Ballantine, D. S., Rose, S. L., Grante, J. W. and Wohltjen, H. (1986) Correlation of surface acoustic wave device coating responses with solubility properties and chemical sensor arrays by pattern recognition. *Analytical Chemistry,* **58,** 149–153

Carey, W. P., Beebe, K. R., Kowalski, B. R., Illman, D. L. and Hirschfeld, T. (1986) Selection of adsorbates for chemical sensor arrays by pattern recognition. *Analytical Chemistry,* **58,** 149–153; see also **58,** 3077–3084

Eerikainen, T., Linko, S. and Linko, P. (1988) The potential of fuzzy logic in optimization and control: fuzzy reasoning in extrusion cooker control. In *Automatic Control and Optimization of Food Processes,* eds M. Renard and J. J. Bimbenet, Elsevier Applied Science, London and New York, 183–200

Ema, K., Yokoyama, M., Nakamoto, T. and Moriizumi, T. (1989) Odour-sensing system using a quartz-resonator sensor array and neural-network pattern recognition. *Sensors and Actuators,* **18,** 291–296

Guilbault, G. G. and Mascini, M. (eds) (1987) Analytical Uses of Immobilized Biological Compounds for Detection, Medical and Industrial Uses. *Proc. NATO ARW Workshop,* Reidel, Dordrecht, Holland

Hall, E. A. H. (1990) *Biosensors.* Open University Press, Cambridge

Ikegami, A., Arima, H., Iwanaga, S. and Kaneyasu, M. (1983) Thick film sensors and their integration. *Proceedings of the 4th European Hybrid Microelectronic Conference,* Copenhagen, 211

Jansson, P. A. (1991) Neural networks: an overview. *Analytical Chemistry,* **63,** 357A–362A

Jones, B. E. (ed.) (1987) *Current Advances in Sensors.* Adam Hilger

Journal of Biotechnology (1990) Special issue on biosensors, vol. 15, 187–282

Karube, I. and Iwasaki, Y. (1991) Biosensors and chemical sensors. In *The Handbook of Measurement Science,* vol. 3, John Wiley & Sons Ltd

Karube, I. and Suzuki, S. (1982) Biosensors for Food Process Control. 3–12 in *Utilisation des Enzymes en Technologie Alimentaire* (Use of Enzymes in Food Technology) ed. P. Dupuy, Proc. Symp. Int. Versailles, May 1982

Karube, I. (1987) Novel biosensor systems for clinical and food analysis using micro devices. 155–164, GBF Monographs, Biosensors, International Workshop, VCH Publishers

Karube, I. and Tamiya, E. (1989) Microbiosensors for Clinical and Process Analysis, 197–306. In *Bioproducts and Bioprocesses,* eds Fiechter, Okada and Tanner, Springer-Verlag, Heidelberg

Kress-Rogers, E. (1985) Technology Transfer, Part II. The New Generation of Sensors, LFRA Scientific and Technical Surveys, No. 150

Kress-Rogers, E. and Turner, A. P. F. (1988) Immunosensors based on acoustic, optical and bioelectrochemical devices and techniques, 227–253 in *Advances in immunoassays for veterinary and food analysis*, ed. M. N. Clifford, R. Jackman, B. A. Morris and J. A. Morris, Elsevier, Barking, UK

Kowalski, B. R. and Bender, C. F. (1972) Pattern recognition: a powerful approach to interpreting chemical data. *Journal of the American Chemical Society*, **9**, 5632

Langdon, R. M. (1985) Resonator sensors: a review. *Journal of Physics E: Scientific Instruments*, **18**, 103–115

Langdon, R. M. (1987) Resonator sensors: a review. In *Current Advances in Sensors*, ed. B. E. Jones. The Adam Hilger Series on Sensors, 19–32

Moseley, P. T. and Tofield, B. C. (1987) *Solid State Gas Sensors*, Adam Hilger, Bristol

Nakamoto, T., Fukuda, T. and Moriizumi, T. (1991) Gas identification system using plural sensors with characteristics of plasticity. *Sensors and Actuators*, **B3**, 1–6

Pollard, A. (1990) What are neural networks? *Sensor Review*, July, 115–116

Roberts, G. G. (1990) Potential applications of Langmuir–Blodgett films as chemical sensors, Thin Solid Films 160, 413–429

Rumelhart, D. E., Hinton, G. E. and the PDP Research Group (1986a) *Parallel Distributed Processing*, vol. 1. MIT Press, Boston, 318–361

Rumelhart, D. E., Hinton, G. E. and Williams, R. J. (1986b) Learning representations by backpropagating errors. *Nature*, **323**, 533

Sim, K. W. (1990) Development of a sensor for ethanol. *Biosensors and Bioelectronics*, **5**, 311–325

Scheller, F. W., Hintsche, R., Pfeiffer, D., Schubert, F., Riedel, K. and Kindervater, R. (1991) Biosensors: Fundamentals, Applications and Trends. *Sensors and Actuators*, **B4**, 197–206

Schild, D. (1990) *Chemosensory Information Processing*, **H55**, Springer, Berlin

Schmid, R. D. and Scheller, F. (1989) Biosensors, Applications in Medicine, Environmental Protection and Process Control. GBF Monographs, vol. 13, VCH Verlagsgesellschaft, Weinheim; VCH Publishers, New York

Sundgren, H., Lundstroem, I., Winquist, F., Lukkari, I., Carlsson, R. and Wold, S. (1990) Evaluation of a multiple gas mixture with a simple MOSFET gas sensor array and pattern recognition. *Sensors and Actuators*, **B2**, 115–123

Wold, S., Albano, C., Dunn, W., Esbensen, K., Hellberg, S., Johansson, E. and Sjoestroem, M. (1983) Pattern recognition: finding and using regularities in multivariate data. In *Food Research and Data Analysis*, ed. H. Martens and H. Russworm, Applied Science Publishers, London 147–188

Zaromb, S. and Stetter, J. R. (1984) Theoretical basis for identification and measurement of air contaminants using an array of sensors having partly overlapping sensitivities. *Sensors and Actuators*, **6**, 225–243

General Reading on Chemical Sensors including Biosensors

Akhtar, M., Lowe, C. R. and Higgins, I. J. (eds) (1987) *Biosensors.* Proceedings of a Royal Society Meeting, London

Blum, L. J. and Coulet, P. R. (eds) (1991) *Biosensor Principles and Applications.* Marcel Dekker

Cass, A. E. G. (ed.) (1990) *Biosensors: a Practical Approach.* IRL Press

Dario, P. (ed.) (1988) *Sensors and Sensory Systems for Advanced Robots.* NATO ASI Series F43, Springer, Berlin

Edmonds, T. E. (ed.) (1988) *Chemical Sensors.* Blackie, Glasgow and London; Chapman and Hall, New York

Guilbault, G. G. and Mascini, M. (eds) (1987) *Analytical Uses of Immobilized Biological Compounds for Detection, Medical and Industrial Uses.* Proceedings of the NATO ARW Workshop, Reidel, Dordrecht

Hall, E. A. H. (1990) *Biosensors.* Open University Press, Cambridge

Jones, B. E. (ed.) (1987) *Current Advances in Sensors.* Adam Hilger, Bristol

Journal of Biotechnology (1990) Special issue on biosensors, **15,** 187–282

Karube, I. (1987) Novel biosensor systems for clinical and food analysis using micro devices. In *Biosensors,* International Workshop, GBF Monographs, VCH, 155–164

Karube, I. and Iwasaki, Y. (in press) Biosensors and chemical sensors. In *The Handbook of Measurement Science,* vol. 3, ed. P. H. Sydenham, Wiley

Karube, I. and Suzuki, S. (1982) Biosensors for food process control. In *Utilisation des Enzymes en Technologie Alimentaire* (Use of Enzymes in Food Technology), ed. P. Dupuy, Proceedings Symposium, Versailles, May, 3–12

Karube, I. and Tamiya, E. (1989) Microbiosensors for clinical and process analysis. In *Bioproducts and Bioprocesses,* eds A. Fiechter, H. Okada and R. D. Tanner, Springer, Heidelberg, 297–306

Kress-Rogers, E. (1985) Technology transfer. II: The new generation of sensors. Leatherhead Food Research Association scientific and technical survey, No 150, p 1–108

Kress-Rogers, E. and Turner, A. P. F. (1988) Immunosensors based on acoustic, optical and bioelectrochemical devices and techniques. In *Advances in Immunoassays for Veterinary and Food Analysis,* eds M. N. Clifford, R. Jackman, B. A. Morris and J. A. Morris, Elsevier, Barking, UK, 227–253

Moseley, P. T. and Tofield, B. C. (1987) *Solid State Gas Sensors.* Adam Hilger, Bristol

Roberts, G. G. (1990) Potential applications of Langmuir–Blodgett films as chemical sensors. *Thin Solid Films,* **160,** 413–429

Rumelhart, D. E., McClelland, J. L. and the PDP Research Group (1988) parallel Distributed Processing, Explorations in the Microstructure of Cognition, vol. 1: Foundations, MIT Press, Cambridge, Mass., USA

Scheller, F. W., Hintsche, R., Pfeiffer, D., Schubert, F., Riedel, K. and Kindervater, R. (1991) Biosensors: fundamentals, applications and trends. *Sensors and Actuators,* **B4,** 197–206

Scheller, F. W. and Schubert, F. (1992) *Biosensors,* Vol II of Techniques and Instrumentation in Analytical Chemistry, Elsevier Science Publishers, Amsterdam

Schild, D. (1990) *Chemosensory Information Processing.* Springer, Berlin

Schmid, R. D. and Scheller, F. (1989) *Biosensors: Applications in Medicine, Environmental Protection and Process Control.* GBF Monographs, vol. 13, VCH, Weinhein; VCH, New York

Schuetzle, D. and Hammerle, R. (eds) (1986) *Fundamentals and Applications of Chemical Sensors.* ACS Symposium Series 309, Proceedings Congress 1984, Washington

Seiyama, T. (ed.) (1988) *Chemical Sensor Technology,* vol. 1. Kodansha, Tokyo; Elsevier, Amsterdam

Turner, A. P. F. (ed.) (1991) *Advances in Biosensors,* vol. I. JAI Press

Turner, A. P. F., Karube, I. and Wilson, G. S. (eds) (1989) *Biosensors: Fundamentals and Applications.* Oxford Science (An earlier edition appeared in 1987)

Vadgama, P. (1990) Biosensors: adaptation for practical use. *Sensors and Actuators,* **B1,** 1–7

Wise, D. L. (ed.) (1989) *Applied Biosensors.* Butterworths

Wise, D. L. (ed.) (1990) *Bioinstrumentation: Research, Developments and Applications,* Butterworth–Heinemann

Wolfbeis, O. S. (ed.) (1991) *Fiber Optic Chemical Sensors and Biosensors,* vol. I. CRC Press, Boston and London

Suppliers of ISFET pH Probes

Microsens SA, Neuchatel, Switzerland
Rue Jacquet-Droz 7, CH-2007 Neuchatel

Sensoptic BV, Stedum, Netherlands
(now contact Sentron)

Sentron BV, Roden, Netherlands
Aan de Vaart 3, PO Box 125, 9300 AC Roden

UNIFET Inc., San Diego, California
11021 Via Frontera, Suite 200, San Diego, CA 92127
(In UK via ChemLab Scientific Products Ltd, Hornchurch, Essex)

Appendix A Glossary: Terms in Instrumentation and Sensors Technology

Contents

Transducer	673
Sensor	675
Actuator	675
Solid-state Transducers	676
Integrated Sensors or Smart Sensors	676
Semiconductor Devices	677
Chemical Sensors, Biosensors, Immunosensors, Physical Sensors	682
Instrument	684
On-line, In-line, At-line, Off-line	685
Bulk Measurement	685
Quality Assurance, Quality Control, Process Control	686
Fuzzy Logic	686
Terms in Food Science, Food Technology and Food Process Engineering	690
References	690

Note

Italics indicate terms defined or explained elsewhere in the Glossary, or tables and figures attached to the Glossary.

Transducer

Definition

Device which converts signals from one signal domain to another. Transducers can perform a single conversion step or a sequence of these.

Notes

A diaphragm-mounted piezoelectric crystal with oscillatory electrical excitation, for example, converts from the electrical to the acoustic signal domain. The assembly can also pick up acoustic waves and convert them into electrical signals. This transducer can thus operate in the **transmitter or receiver mode**. Alternatively, it can rapidly switch between both modes to pick up a reflected signal resulting from an acoustic pulse transmitted earlier.

The strain-gauge (on diaphragm) pressure transducer, on the other hand, is designed for the measurement of pressure changes only, not for their creation. Similarly, the thermocouple or thermopile probes are designed for the measurement of temperature, whereas different configurations would be chosen for a Peltier cooling element. It is useful to distinguish between transducers converting between the same pair of signal domains, but in the opposite direction and for a different purpose. Transducers designed to indicate a change are called *sensors*, those intended to effect a change are labeled *actuators*. In optical or microwave instruments, the terms detectors and sources are often used instead.

In recent years, advances have been made mainly in the area of *solid-state transducers*, either as a replacement for conventional transducers or larger

Table A1 *Some examples of solid-state transducers*

	Examples	
*Signal domain 1**	*Sensors*	*Actuators*
Optical	Photodiode CCD: charge-coupled device	LED: light-emitting diode LCD: liquid-crystal display
Thermal	Thermistor Thermocouple Thermopile	Thick-film thermal print head
Magnetic	Magnetoresistor Hall-effect position sensor	Read and write head
Mechanical	Thin-film strain gauge Micromachined capacitive silicon pressure gauge	Micromechanical switch
Chemical	ISFET: Ion-selective field effect transistor MeOx gas sensor: metal oxide semiconductor Capacitive polymer humidity sensor	Microelectronic coulometric titrator

* Signal domain 1: Domain of sensor input or actuator output.
Signal domain 2 (sensor output/actuator input) is electrical in these examples.

Note:
For a fuller description of transduction mechanisms and techniques, see Middelhoek and Hoogerwerf (1988).

assemblies or for a new application. Some examples of solid-state transducers are given in *Table A1*. The term sensor is used particularly for solid-state devices. Related, but less closely defined terms are gauge, meter, probe or transmitter.

An ideal transducer would convert all the energy of the input signal into energy in the output signal domain without dissipating any energy in a form outside the intended output signal domain. In practice, full conversion is not achieved. Very few transducers operate in the **passive mode**, that is without requiring any power sources other than the input signal.

Sensor

Definition

Device used to detect, locate or quantify energy or matter; giving a signal for the detection or measurement of a physical or chemical property to which the device responds.

Notes

Initially, the term was applied primarily to *transducers* in military warning and detection systems, particularly those measuring electromagnetic radiation (radar, infrared, visible), acoustic waves (seismic waves, sound, ultrasound), magnetic fields or certain gases (irritant and/or toxic).

Now, a much wider range of sensors is available, with a sensitivity towards variables in the optical, thermal, magnetic, mechanical or chemical signal domains. Output signals can be in the optical as well as the electrical signal domain to make use of signal transmission by fibre-optic guides as well as by electrical cables or radiowave transmission.

Sensors are now built into cars and lorries, domestic cookers and washing machines; they are applied in the medical field by doctors and patients; they are used to monitor industrial processes; they are still very important in the defence area, in industrial and domestic safety assurance and in environmental monitoring. Some examples are given in *Table A1* on *solid-state transducers, integrated sensors* are described below.

Actuator

Definition

Device which effects changes in physical or chemical properties in response to a signal.

Notes

Actuators can, for example, be designed to operate a mechanical switch or valve in response to an electrical signal, or to emit light or act as a coulometric titrator (electrolytic generation of protons). (See also the notes on solid-state *transducers*.)

As in the case of sensors, hygienic considerations are important in the design of actuators in contact with food. Actuators used as part of the actual food processing system (rather than as part of instrumentation aiding in process control) are described by McFarlane (1983, Section 1.5 therein).

Solid-state Transducers

In the field of electronic circuit elements, solid-state devices have already replaced many conventional devices. The fragile glass envelope valve has given way to the compact and robust transistor which has then been combined with other microelectronic circuit elements into integrated circuits (ICs) of increasing complexity. A similar trend, albeit with a time lag, is being observed in the field of *transducers*. Many of the solid-state *sensors* are based on the semiconducting and piezoelectric properties of silicon and on the properties of planar *semiconductor* devices based on this material. As in the case of the circuit elements, the advantages of the solid-state devices are robustness, compactness, low power dissipation, fast response, high device-to-device reproducibility and the possibility of integration with other sensing and electronic circuit elements (to form *integrated sensors*). As an alternative to silicon planar technology, thin and thick film technologies are used for low production volumes.

Solid-state *actuators* have advanced also, initially mainly for low power applications. More recently, power transistors have replaced glass envelope valves for many high-power applications. For the construction of compact instruments, micromechanical switches and microvalving have been developed.

Some examples of solid-state transducers are given in *Table A1* and more can be found, for example, in the chapter by Ko (1988) and references therein.

Integrated Sensors or Smart Sensors

Solid-state *sensors* can be integrated (on the same substrate) with other sensors of the same type (into **sensor arrays**), or with other sensor types for the purpose of simultaneous measurement of several variables or compensation for interfering variables. In addition, they can be integrated with interface electronics. This comprises electronic circuitry for signal conditioning and processing, so as to provide a pre-amplified, linearized output signal that can also be compensated for cross-sensitivity to temperature changes, for example. An additional function is the conversion into a form suitable for input into digital systems (current/frequency conversion, analog/digital conversion). Such an assembly of a sensor or sensor array with ancillary sensors and interface electronics on the same substrate represents an **integrated sensor** (also known as **smart sensor**). (See also *Semiconductor Devices, Solid-state Transducers.*)

In the area of volatile sensing, particularly, the combination of sensor arrays with signal processing by neural networks, is an area of active research and development (see Chapter 17, Section 5). Sensors can also be integrated with *actuators*, for example in the microelectronic device for the measurement of acidity where a coulometric micro-actuator electrode is combined with a microelectronic pH sensor (see Chapter 17, Section 2.3.9). Full sensor integration, and indeed the manufacture of a solid-state sensor in planar silicon technology is not always economical. The cost of setting up a production line for this technology is high, and the cost of the individual devices will be high if they are produced in small numbers to a particular specification. Very inexpensive devices can be produced, however, if manufactured in numbers of 100,000's per year. Integrated solid-state pressure sensors are already established in the monitoring of canning operations in the food industry, as are CCD-imaging arrays (see *Figure A1*) in automatic optical inspection systems. Solid-state microelectronic pH-sensors are now industrially used (see Chapter 17, Section 2.3.7).

An alternative to planar silicon technology are the thin film and thick film technologies. This applies both to circuits and to sensors. Hybrid circuits combine thick/thin film circuitry with planar silicon circuit elements. Similarly, semi-integrated sensors can be constructed where monolithic integrated systems are not economic (*Figure A2(b)*). Such hybrid systems still allow the sensor and some signal conditioning circuitry to be housed in a common transducer can (*Figure A2(c)*).

A more recent development area is integrated optics where optical components such as microgratings and thin-film waveguides are combined on a single substrate and can be integrated with optical sensors.

Semiconductor Devices

Many of the *sensors* developed during the last three decades are based on semiconductor devices or semiconducting materials. This is due both to the physical properties of these materials and due to the advanced technologies developed for processing them.

The electrical properties of semiconductors can be tailored over a wide range. They can have conductivities as low as insulators or as high as metals. The electrical properties can be influenced not only by the choice of

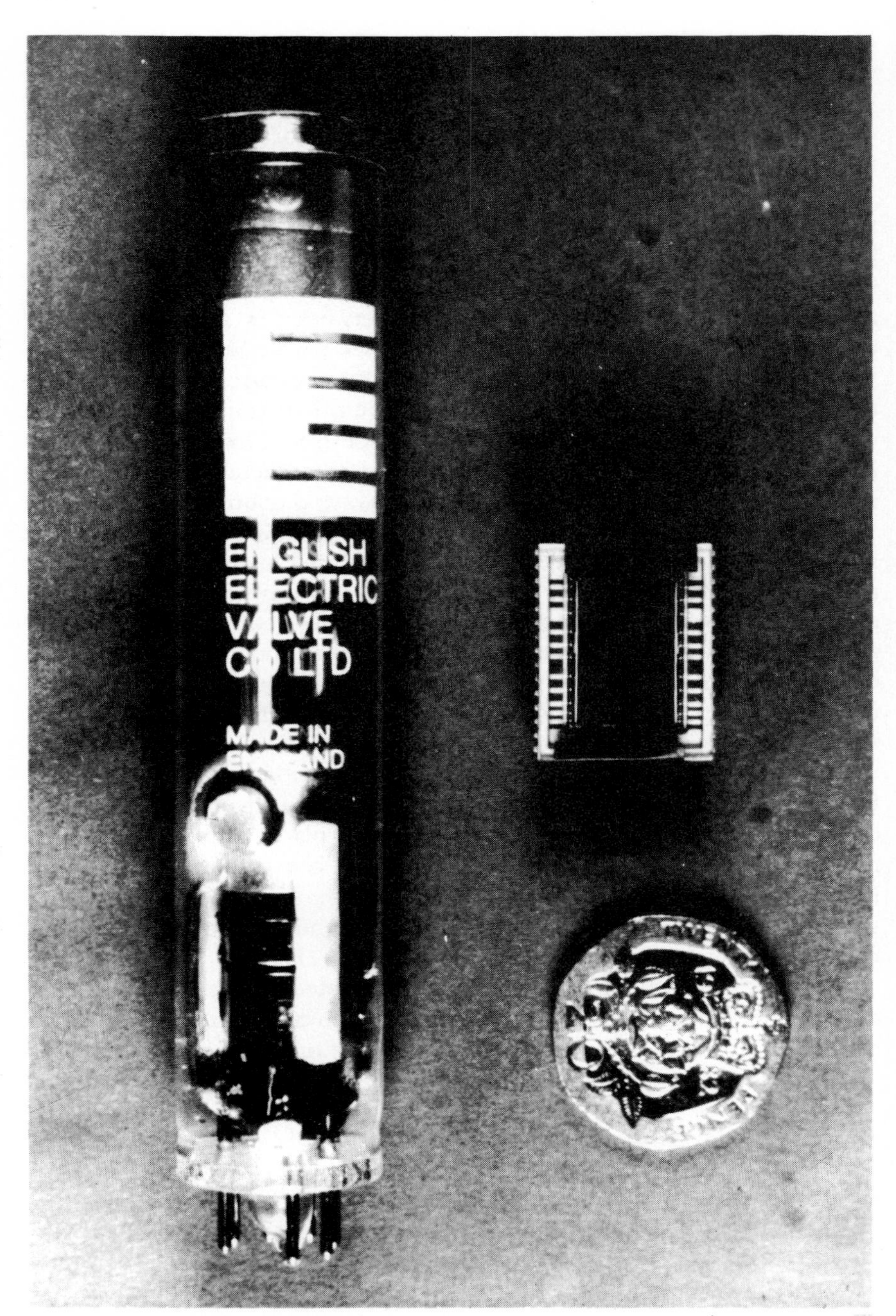

Figure A1 *CCD imaging array (top right) and conventional camera tube valve (left). (Devices from Hirst Research Centre, GEC. Photo reproduced from Kress-Rogers, 1985.)*

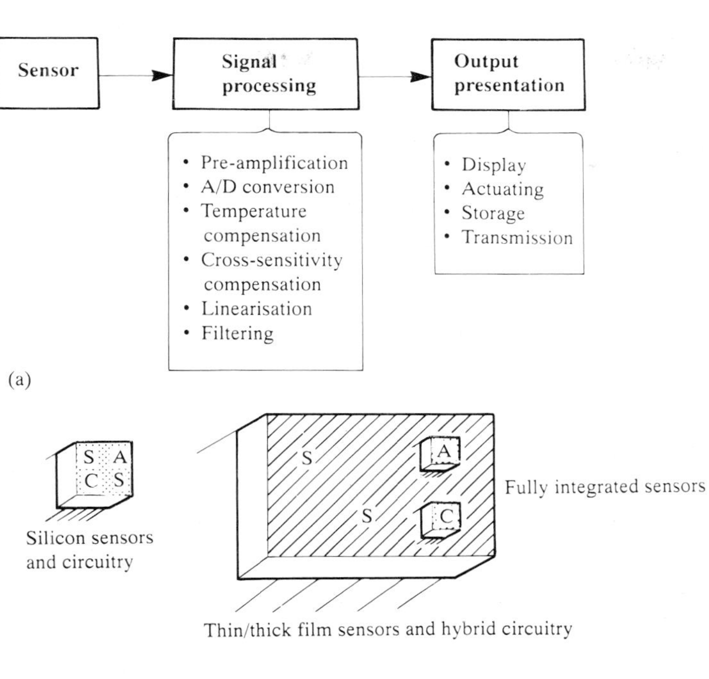

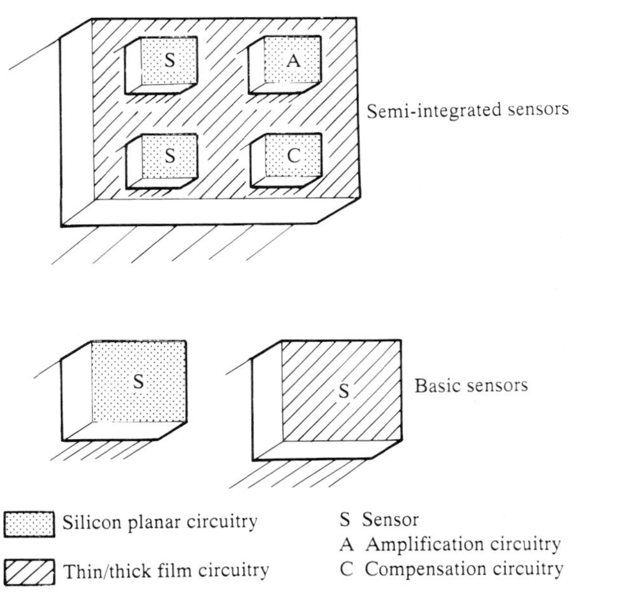

Figure A2 *(See page 680 for explanation)*

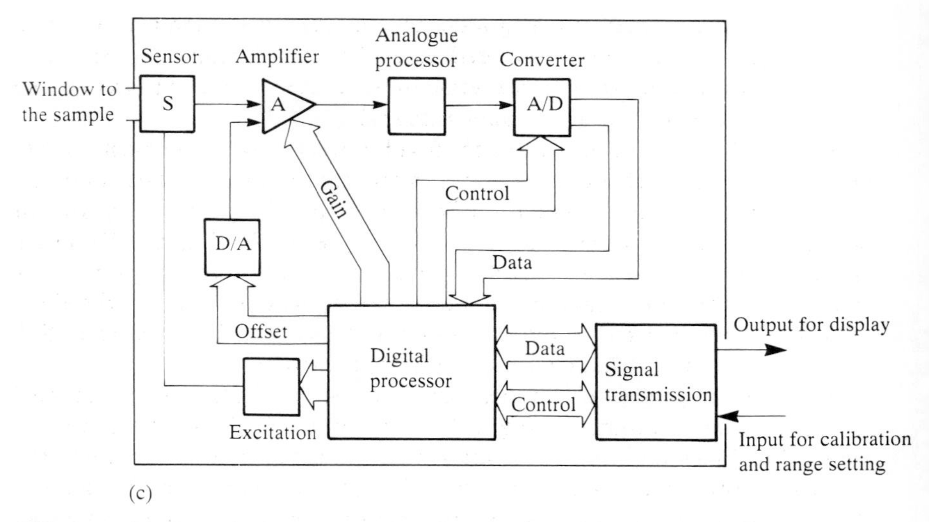

Figure A2 *Interfacing the sensor. (a) Signal processing functions; (b) Levels of integration, (after Kress-Rogers, 1985); (c) Optimized sensor system, (after Brignell and Dorey, 1983). Integrated signal processing (a) provides reliable, low noise transmission, direct communication with microprocessor or display unit and interference compensation without lag time. By integrating signal processing circuitry and sensor on one chip (b) or by enclosing a compact optimized sensor system within one housing (c), the signal conditioning and processing components are invisible to the user who is faced with an apparently ideal sensor with a linear, low-noise, temperature-independent output.*

the semiconductor material and of the crystal size and orientation, but also by doping with impurities. Depending on the choice of impurities for doping, the semiconductor will be n-type or p-type, that is with a majority of negative or positive charge carriers. These carriers are the mobile electrons and holes, the latter being crystal lattice points with missing electrons. The characteristics of the doped semiconductor material can be further influenced by creating interfaces with other materials and by the application of a bias across these interfaces.

Significant science and technology advances have already been achieved in the development of integrated electronic circuits (silicon planar technology for the construction of microelectronic devices and circuits). Homojunctions between differently doped regions based on the same host material, heterojunctions between two different semiconductor materials, and interfaces with metals and insulators can all be constructed on a microscale. A multitude of electronic device structures can be based on semiconductors as a result.

Semiconducting materials and semiconductor devices are sensitive to optical, thermal, electrical, magnetic, mechanical and chemical variables depending on the chosen material and device structure. Silicon is a piezoelectric material as well as a semiconductor and either or both of these properties have been used in many sensor designs.

Semiconductor devices such as photovoltaic silicon or germanium diodes are used as infrared detectors in remote thermometry; lead sulphide photoconductive devices are used both in infrared thermometry and in near-infrared composition analysis. (See Chapter 6, Section 2.2, Table 6.1 and Chapter 5, Section 3.1.). Microwave instrumentation (Chapter 7), for example for the determination of water content, benefits from the development of microwave semiconductor devices (Sze, 1981, Table 3 on p. 515 therein). (Microwave bands are given in Appendix B.)

In addition to the techniques used in the fabrication of integrated circuits, silicon micromachining techniques are available for the manufacture of well-defined micro-structures. These allow, for example, the fabrication of silicon-enclosed cavities with integral diaphragm. Pressure is then indicated either by a silicon strain gauge structure on the diaphragm or by a capacitor structure formed by metal films on the diaphragm and on the opposite side of the cavity enclosure. Such pressure sensors are not only more compact but also more tolerant of rapid temperature and pressure changes than transducers incorporating steel diaphragms bonded to other materials.

Ion-selective membranes (either conventional or preferably in planar technology) can be incorporated into semiconductor devices as can catalytic metals or metal alloys. A range of solid-state sensors responding to pH or to specific ions and a number of highly sensitive and selective gas sensors are based on this approach. Certain semiconductor materials are also highly sensitive to direct contact with gases and volatiles. These materials include metal oxide semiconductors such as tin oxide and conducting polymers such as substituted polypyrroles. While these two groups have a broad specificity and are often combined into sensor arrays coupled to pattern recognition systems, a high specificity and sensitivity is found with the metal phthalocyanines. (See Chapter 17, sections 2.3–2.5 and Tables 17.1 and 17.5).

Humidity-sensitive polymers can also be incorporated in semiconductor devices. The resulting humidity sensors can be applied in the monitoring of the RH (relative humidity) of air, for example to control drying processes. They can also be the basis of instruments for the measurement of water activity (Chapter 12).

Semiconductor sensors and circuit elements can be integrated on the same substrate to form *integrated sensors* ('smart sensors'). Planar technology is, however, economical for high production volumes only. Alternatives for lower production volumes are the thin film and thick film technologies. The latter generally have a lesser degree of reproducibility but lower costs for setting up a production line.

Chemical Sensors, Biosensors, Immunosensors, Physical Sensors

Many different definitions for these terms are in use. Some workers classify by the type of base device, some by the type of interaction with the chemical or physical property of the sample, some by the type of variable that is ultimately determined even if by deduction from another variable that is actually measured. A very wide definition is sometimes used for the term biosensor. This includes any measuring device, independent of its base principles or measured variable (including, for example, probes for electrical conductivity as well as miniature ion-selective electrodes or even tracers), that can be applied to animal or plant tissues.

Chemical sensors are understood by some as *sensors* measuring a chemical variable by means of an interaction of the analyte with an inorganic reagent or a sorption layer on the sensor. Others include sensors based on the interaction with enzymes or antibodies.

The definitions adopted here are based on practical considerations in the choice of a sensor for a particular application. The sensitivity and specificity of the device, the cross-sensitivities, the tolerance to environmental conditions and the operational and storage lifetimes of a sensor are to a large extent given by the type of interaction with the sample, although the type of base device will play a role. The definitions used here are:

Biosensors: Sensors that rely on the interaction of a biocatalyst, usually an isolated and purified enzyme (but can also be part of a whole microbial or tissue cell), with the analyte. Example: Sensors for the measurement of glucose based on the interaction of glucose oxidase with the analyte.

Immunosensor: Sensors that rely on the interaction of an antibody or other specific affinity binding agent with the target. **Examples:** Sensors for specific bacteria or meat species (is it pork or beef?) based on the interaction of an antibody with a specificity for the target.

Chemosensors (as a tentative label for other chemical sensors): Sensors that rely on a non-biological chemical or sorption interaction with the analyte. **Examples:** Sensors based on the interaction of catalytic metals or of redox-sensitive metal oxides with the sample.

Chemical Sensors: This group includes biosensors, immunosensors and chemosensors. The base devices can be electrical, microelectronic, optical, acoustic or mechanical devices for all these sensors. (See Chapter 17.) To qualify as a sensor, the design principle of a device must allow the provision of a continuous output signal. Dedicated configurations based on the design principle can either display all the available information to the user or restrict the user-visible display to appear under certain conditions only (for example when a set limit is exceeded or when the signal is equilibrated).

Indicator strips that give an irreversible colour change, however, are not sensors, irrespective of whether they incorporate non-biological reagents or sorption agents, enzymes or antibodies. Immunosensors are capable of giving a dynamic signal monitoring the antibody-antigen binding process although the sensing element may well need replacing or re-generating after a 'positive' sample has been assayed, or a contaminant has been detected in a sampled stream.

Physical Sensors: Sensors that rely on a physical interaction with the sample. Examples are sensors relying on the influence of sample properties (such as viscosity or density) on mechanical resonance characteristics. These sensors can be applied to the determination of correlated chemical variables under well-defined conditions. A mechanical resonance sensor sensitive to viscosity, for example, can be used to the measurement of viscosity in its own right or to the monitoring of polymerization processes and associated food conditions such as the degradation of frying oil in use (Chapter 16).

More generally, all the sensor types described above can be applied simply to the determination of the primary measured variable (which is given by the type of interaction with the sample) or, using the marker concept (Chapter 16), to the determination of another physical, chemical or microbial variable or to the assessment of complex conditions such as food freshness. A biosensor sensitive to glucose, for example, can be used to measure glucose concentrations or, for particular well-defined food groups and using suitable configurations, it can also be used to determine total reducing sugars content or microbial surface load. Examples for the sensor groups defined above are given in Figure 17.15.

Measurement of the mechanical resonance properties (indicating viscosity and density) or the ultrasonic (indicating density and solute properties, see Chapter 8) or near-infrared characteristics (indicating composition, see Chapter 5) of a culture broth allows the monitoring of microbial growth. Due to the modest sensitivity and specificity of such methods for microbial concentration, they can be applied directly to fermenter broths but not to food samples.

For foods, on the other hand, a preparation procedure is needed to supply a culture broth which, after a characteristic incubation time, can be monitored (Chapters 14 and 15) with instruments for microbial assays by electrical impedance (indicating charged solute and microbial cell properties). Such instruments are *not* associated with the description 'sensors for microbial load' whereas the instruments for the direct real-time monitoring of fermenter cultures are.

Instrument

Definition

Electrical or electronic equipment designed to carry out a specific function or set of functions.

Notes

The instrument can include, for example, one or more *sensors* and *actuators* (or conventional sources, detectors and filter wheels), components for the conversion from mains or battery power to the required form of the input signal(s), guides for electrical, microwave, optical or acoustic input and output signal(s) and interface electronics for the conditioning and processing of the output signal(s). The latter can include A/D conversion so that signals can be received from or transmitted to systems requiring digital inputs. This will facilitate interfacing, for example to process control systems, and allow low noise signal transmission. Some of the interface functions can already be present in an *integrated sensor*.

Additionally or as an alternative to an output suited to a control system, a visual display or audible indication of the output signal can be part of the instrument. The display can be quantitative or be in the form of a traffic light type LED array display indicating that the signal is falling into one of several ranges (for example, white: too low, green: good, yellow: near set limit, red: above set limit).

Sensors can be housed in the instrument box or remote from the instrument and connected by suitably robust, low-noise signal guides (electrical or fibre-optic cables, waveguides) or incorporate an RF signal transmitter for wireless communication with the instrument. Alternatively, the sensor can be connected to a small data logger that is interrogated by the instrument at a later time (for example, after sterilization or after storage and transport).

Many *on-line* instruments are of complex design and flexible enough for multi-purpose applications. After skilled installation, calibration and setting-up, the routine running and maintenance duties can then be carried out by staff having been trained more briefly. For other on-line measurements and for many *at-line* measurements, simpler instruments dedicated to a limited range of applications are used. For checks during distribution and storage, user-friendly, robust and compact instruments designed for a particular application are usually more suitable than complex, high-precision instruments. In the Quality Control and Product Development laboratories, instruments of varying complexity are used, including those requiring operation by highly skilled staff with longer training periods.

In contrast to appliances and to many *off-line* and some at-line instruments, *in-line* instruments measure non-destructively at low or zero power levels applied to the process stream. Instruments in bypass streams can be allowed to impart slight (that is, inconsequential after mixing with the main stream) changes to the bypassed part of the process stream (but must not bear the risk of contaminating the stream), those in bleed lines can be allowed to change the bled stream portion significantly during conditioning and measuring before it is discharged.

The instrument can be combined with separation or marking equipment to form an **automatic sorting or inspection system**. Sorting can be by electromagnetic induction characteristics, by optical properties (see Chapter 4) or by X-ray absorption, for example. Where spatial resolution is needed, techniques for image acquisition and automatic image analysis are applied. Simpler equipment can be based on mechanical separation by gravity or size, without the need for an instrument.

On-line, In-line, At-line, Off-line

On-line measurements provide data on a continuous production process that can be used to adjust process variables using feedback or feed-forward control. (See McFarlane, 1983, Section 1.6 on systems for automatic control.) This requires that the data on a process stream segment are available in 'real-time' or with only a short delay after the segment has passed the measuring point.

On-line measurement can be **in-line** (the measurement point is in the main process line, or the main process line is accessed through a window or container wall transparent to the radiation in question (such as microwave or infra-red). Alternatively, the measuring point can be in a bypass or bleed line. Robotic sampling, conditioning and measuring systems connected to the line can also provide a continuous or intermittent input for process control. For batch processes, instruments for *in-situ* measurements can provide real-time data.

Samples taken from the line can be analysed **at-line** with *instruments* located in the production area. If this is not possible, they can be taken to a QC laboratory for an **off-line** measurement on-site, or they can be sent to a remote central laboratory for **off-site** analysis where highly complex, expensive instrumentation is required. (See Chapter 1, Figures 1.1–1.3.)

Bulk Measurement

Depending on the measurement principle and *instrument* design, the sampled part of the tested material differs. The measurement can be of the surface layer only, alternatively it can determine an average over the sample volume or process stream segment (a bulk measurement), or it can provide a spatially resolved measurement (quantifying the variation of the measured property within the text volume).

The consideration of the penetration depth and spatial resolution is important in the choice of instrumentation and in the interpretation of readings in the measurement of food properties on-line where the material is often highly inhomogeneous. Even when the finished and equilibrated food product is homogeneous, a property (such as moisture content or temperature) measured on-line during processing can exhibit variations with depth into the material and with time. This is particularly the case during or soon after heat processing operations.

Quality Assurance, Quality Control, Process Control

Quality control allows the effectiveness of process control and good manufacturing practice to be checked and the process variables to be adjusted at intervals that can be hourly, daily, weekly or monthly, and with a delay of minutes, hours, days or even weeks after taking a sample from the line.

Quality assurance is based partly on monitoring the process with *on-line* and *at-line* measurements so that deviations from the target specification can be recognized and corrected. Thereby the characteristics of the food product remain within a preset range. By carrying out a hazard critical control point (HACCP) analysis, and then measuring the identified critical variables on-line and at-line (where possible) or *off-line* (where necessary), it is possible to reduce the risk of food contamination, spoilage or liability to spoilage during manufacture and distribution.

On-line measurement not only helps to assure the quality of the manufactured food product, but also assists process management so that safety, the effective use of resources (labour, materials and energy) and environmental protection can be effected. (See Chapter 1, Figure 1.2.)

Fuzzy Logic

Fuzzy logic is expected to be a widely used principle in the design of measurement and control systems by the end of this decade. It is based on fuzzy set theory, a mathematical system that allows calculation with quantities that are not precisely defined. The fuzzy logic program is compatible with the engineer's approach to the control of a complex technical process. For such a process, calculations using a closed and precise mathematical model are often too slow to provide results in time for control action. Instead, the engineer will combine a number of imprecise observations to decide on control actions.

This mode of control applies not only to industrial processes but also to the driving of a car, the cooking of a meal or the treatment of an illness. The driver parking a car does not use precise coordinates, velocity and acceleration figures, but turns the steering wheel 'a little more to the left', releases the gas pedal 'a little' when he is 'too far' or 'too near' the kerb, getting 'close' to the garage door, or sees a cat 'about to' run onto his intended path. The cook will stop the roasting process when the colour, texture and smell are 'right'. The doctor will prescribe a standard course of antibiotics when the patient has an 'elevated' temperature and tonsils that are a much brighter red and 'enlarged' compared with their usual state. He would gain little by measuring the oral temperature to an accuracy of several digits, determining the colour coordinates of the inflamed tonsils, weighing the patient and then calculating the antibiotics dose.

Fuzzy logic techniques accommodate the approach of the human expert and allow control systems for complex systems to be programmed with less effort (and thus less cost) than by a precise quantitative description of all variables and their relationships. Fuzzy set theory has already found practical applications in the control of auto-steady video cameras, refrigerators, microwave ovens, air conditioning systems, lifts (elevators) and of a driverless city subway systems, so far mainly by Japanese manufacturers. For the microwave oven, for example, fuzzy control is based on measurements of food weight, shape, height, thickness, temperature and of emanating volatiles.

Fuzzy-assisted sensors have just become available commercially. For the measurement of relative humidity, for example, the frequency signal from a capacitive humidity sensor together with a temperature reading are evaluated by a fuzzy logic system to provide a rapid and well-resolved (to 0.01% RH) output of the temperature-compensated humidity reading. The complete system is accommodated in a handheld instrument. (To achieve the same response time and resolution specifications with a conventional programme, would require a powerful computer and hence result in a more expensive and less compact instrument.) A colour sensor with embedded fuzzy logic has also become available. This is programmed by presenting acceptably coloured products to the sensor and setting a tolerance level for acceptable deviation. Other smart fuzzy-assisted sensors are being developed for the measurement of proximity and position. Manufacturers include Fisher Scientific and Eaton Corp.

It is expected that fuzzy logic will gain widespread applications in control systems for industrial processes within this decade. Fuzzy-logic controllers exhibit less under- and overshooting of set-points than traditional controllers. Fuzzy PID (proportional, integral or differential) loop controllers have already been industrially used for over two years to control temperatures, pressures, flow rates and air injection rates in beverage pasteurization, extrusion cooking and marshmallow production. Fuzzy-based PLC (programmable logic controller) modules have also become

Table A2 *Processes in the food industry*

- Selection of ingredients
- Cleaning of raw materials
- Water purification
- Air conditioning
- Cleaning, disinfection and rinsing of surfaces in the plant
- Waste disposal
- Sorting and grading (by size, shape, weight, density or appearance)
- Screening for contaminants or defects
- Conditioning and ripening of produce
- Refining, distilling
- Size-reduction and size-screening
- Filtration and membrane separation
- Centrifugation
- Solid-liquid extraction and expression
- Conveying, pumping and dispensing of solids, liquids and gases
- Mixing and emulsification, homogenization
- Fermenting
- Thickening, gelling
- Crystallization, tempering
- Particle agglomeration
- Coating
- Heat processing (by contact, convection, radiant heat, microwave or radiofrequency source or by direct ohmic heating)
  Pasteurization, sterilization (canned or bulk)
  Cooking (ambient or excess pressure), baking, roasting, frying
  Continuous heat processing in cooker extruder or direct ohmic heater
  Sugar inversion
  Evaporating
- Dehydration (by heating, spray drying, freeze drying, freeze concentration)
- Chilling
- Freezing and thawing
- Irradiation treatment
- Filling, closing, sealing
  Special packing methods (under modified atmosphere or vacuum)
- Product screening
- Checkweighing, packaging and labelling
- Palletizing
- Storage
- Transport

available. Fuzzy-assisted expert systems require less microprocessor hardware and less time for data exchange than their binary logic counterparts. Applications include the management of fermentation, drying or smoking processes in the food industry, the assessment of taste panel results and the deduction of rheological characteristics from mixer torque measurements. Suppliers include Omron Electronics Inc., Yokogawa Corp. and Bailey Controls Co.

A fuzzy-logic controller measures the relevant quantities and allocates values to them that express the degree to which they match 'linguistic variables'. Such a variable defines in terms of everyday language, when, for example, a temperature is to be considered as 'very high'. According to their degree of matching with the linguistic variables, the measured quantities are given values between 0 and 1. Mathematical procedures applied to these values (designed to take into account the relationships between variables) result in control instructions which can be converted into linguistic form ('turn valve A down a little and turn switch B up a little') or into input signals for actuators.

The theory of fuzzy sets was first presented by Professor Lotfi Zadeh in 1965 at Berkeley. Whereas traditional binary sets have elements that either belong or don't (either values 1 or 0), fuzzy sets allow for a degree of membership expressed by values between 0 and 1 given to the elements. It was discovered by control engineers in the mid-1980s as a technique that fitted in with their approach. The first practical application was in the plant of a Danish cement manufacturer.

Fuzzy-logic techniques are not a replacement for mathematical models, but complement them. Mathematical models will continue to be used in process optimization and, due to advances in computer technology, allow the simulation of complex systems, for example, involving chemical reactions in multi-phase flow systems with irregular geometries. For simpler relationships, mathematical models also have applications in process control. In assessing the reliability of a mathematical model, it is relevant whether the model is fully based on well-defined physico-chemical relationships, or whether a number of fitted parameters had to be used to match the model to a real system. Sometimes, the term mathematical model is misleadingly used to describe curve fitting procedures without a physico-chemical basis.

Fuzzy logic differs from neural network analysis. Fuzzy logic is based on the evaluation of inputs with a known pre-programmed, if 'fuzzy', relationship between input variables and output signals. Neural network analysis is a pattern recognition technique that relies on the examination of large numbers of representative samples of known characteristics (the learning set) to find ('learn') a set of relationships connecting the input and output signals.

On fuzzy logic applications, see Billerbeck and Bönsch (1992a, b), Oishi et al (1991), Schulz (1992) and Sperber (1991). On neural networks, see

Thai and Shewfelt (1991), and Chapter 17, Section 17.5.2 and references therein.

Terms in Food Science, Food Technology and Food Process Engineering

Definitions and information on these topics can be found in the encyclopediae by Hui (1992) and by Hall, Farrall and Rippen (1986). For terms used in this book, please consult the index.

A list of processes carried out in the food industry is given in *Table 2.* These processes are discussed in books by Spiess and Schubert (1990), Brennan, Butters, Connell and Lilley (1990), Watson and Harper (1988), Lewis (1987), McFarlane (1983) and Stewart and Amerine (1982) for a description of processes in food manufacture. Books on special processes or food products are also available, for example on extrusion cooking (O'Connor, 1987) or fermented foods (Campbell-Platt, 1987).

The sectors of the food industry are listed in Table 1.1 in Chapter 1, for an overview of measurements in the control of processes and in quality control, see Tables 1.2 and 1.3 in Chapter 1.

References

Billerbeck, J. D. and Bönsch, R. (1992a) Fuzzy-Produkte erobern zuerst die Prozessleittechnik. *VDI Nachrichten*, **46** (16) 1

Billerbeck, J. D. and Bönsch, R. (1992b) Fuzzy-Technik ist mehr als nur ein Marketing-Gag. *VDI Nachrichten*, **46** (16) 6

Brennan, J. G., Butters, J. R., Cowell, N. D. and Lilley, A. E. V. (1990) Food Engineering Operations (3rd ed). Elsevier Applied Science, London and New York

Brignell, J. E. and Dorey, A. P. (1983) Sensors for microprocessor-based applications. *Journal of Physics E: Scientific Instruments*, **16**, 952

Campbell-Platt, G. (1987) *Fermented Foods of the World: A Dictionary and Guide*, Butterworth-Heinemann, Oxford

Hall, C. W., Farrall, A. W. and Rippen, A. L. (1986) *Encyclopedia of Food Engineering* 2nd ed. AVI Publishing Company, Westport, Connecticut

Hui Y. H. (1992) *Encyclopedia of Food Science and Technology*, Vols 1–4, John Wiley & Sons, New York

McFarlane, I. (1983) *Automatic Control of Food Manufacturing Processes.* Applied Science Publishers, London and New York

Ko, W. H. (1988) Solid-state transducers. In *Sensors and Sensory Systems*, ed. P. Dario, Springer-Verlag, Berlin, Heidelberg, 219–242

Kress-Rogers, E. (1985) Technology Transfer. II: The new generation of sensors. Leatherhead Food Research Association scientific and technical survey 150

Lewis, M. J. (1987) *Physical Properties of Foods and Food Processing Systems*. Ellis Horwood, Chichester, and VCH, Weinheim

Middelhoek, S. and Hoogerwerf, A. C. (1988) Basic transduction mechanisms and techniques. In *Sensors and Sensory Systems*, ed. P Dario, Springer-Verlag, Berlin, Heidelberg, 189–199

O'Connor, C. (1987) *Extrusion Technology for the Food Industry*, Elsevier Applied Science, London and New York

Oishi, K., Tominaga, M., Kawato, A., Abe, Y., Imayasu, S., Nanba, A. (1991) Application of fuzzy control theory to the sake brewing process. *Journal of Fermentation Bioengineering*, 72 (2) 115–121

Schulz, W. (1992) Fuzzy-Konferenz läutet amerikanische Aufholjagd ein. VDI Nachrichten, **46** (16) 7

Sperber, R. M. (1991) Fuzzy logic, Food Processing (Chicago) **52** (10) 72–76

Spiess, W. E. L. and Schubert, H. (1990) *Engineering and Food*. Vol. 1: Physical Properties and Process Control, Vol. 2: Preservation and Related Techniques, Vol. 3: Advanced Processes, Elsevier Applied Science Publishers 1990, Proc 5th Int. Congress, Cologne 1989

Stewart, G. F. and Amerine, M. A. (1982) *Introduction to Food Science and Technology*. Academic Press, New York

Sze, S. M. (1981) *Physics of Semiconductor Devices*, 2nd edition. John Wiley and Sons, New York

Thai, C. N., Shewfelt, R. L. (1991) Modelling sensory colour quality of tomato and peach: neural networks and statistical regression. *Transactions of the ASAE* **34** (3), 950–955

Watson, E. L. and Harper, J. C. (1988) *Elements of Food Engineering*. AVI, Van Nostrand Reinhold Company, New York

Appendix B Ancillary Tables

Contents

B1	International System of Units	695
B2	Constants and Standard Values	697
B3	Temperature, Typical Values and Unit Conversions	698
B4	Pressure, Typical Values and Unit Conversions	699
B5	Energy and Mass	701
B6	Force, Power, Magnetic Flux and Time	702
B7	Length and Volume, Sizes	703
B8	Viscosity	705
B9	pH-Values	707
B10	Concentration	707
B11	Colour	708
B12	Near- and Mid-Infrared Range	709
B13	Microwave Bands	710
B14	Logarithms, Neper and Decibel	711
B15	Miscellaneous	712
B16	References	713

B1 International System of Units (SI)

Base Units

metre (m)	length
kilogram (kg)	mass
second (s)	time
ampere (A)	electric current
kelvin (K)	thermodynamic temperature
mole (mol)	amount of substance
candela (cd)	luminous intensity

The mole is the amount of substance of a system which contains as many elementary entities as there are atoms in 12 g of ^{12}C. The entities are specified particles or groups of particles, in the context of this book usually atoms, molecules or ions.

Supplementary Units

radian (rad)	plane angle
steradian (sr)	solid angle

The plane angle $\varphi = 1$ rad when $a = r$, where a is the arc cut off on the circumference of a circle of radius r (figure 1a).
The full plane angle (covering a full circle) is $\varphi = 2\pi$ rad $= 360°$ or simply $\varphi = 2\pi$.

The solid angle $\theta = 1$ sr when $A = r^2$, where A is the area cut out on the surface of a sphere of radius r (Figure B1b).
The full solid angle (covering a complete sphere) is $\theta = 4\pi$ sr.

Common Prefixes

T (tera/10^{12})	d (deci/10^{-1})
G (giga/10^{9})	c (centi/10^{-2})
M (mega/10^{6})	m (milli/10^{-3})
k (kilo/10^{3})	μ (micro/10^{-6})
h (hecto/10^{2})	n (nano/10^{-9})
da (deca/10)	p (pico/10^{-12})

Selected Derived Units

Hz	$= s^{-1}$	hertz (frequency)
N	$= m\,kg\,s^{-2}$	newton (force)
Pa	$= N\,m^{-2}$	pascal (pressure, stress)
J	$= N\,m$	joule (energy)
W	$= J\,s^{-1}$	watt (power)
C	$= A\,s$	coulomb (electric charge)
V	$= W\,A^{-1}$	volt (electric potential difference)
Ω	$= V\,A^{-1}$	ohm (electric resistance)
S	$= \Omega^{-1}$	siemens (electric conductance)
F	$= C\,V^{-1}$	farad (electric capacitance)
Wb	$= V\,s$	weber (magnetic flux)
T	$= Wb\,m^{-2}$	tesla (magnetic flux density)

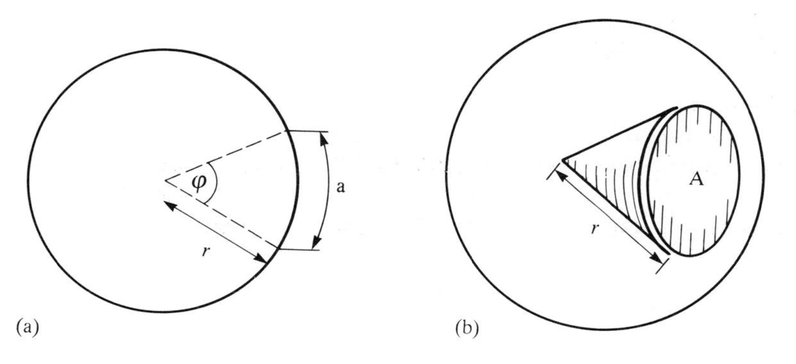

Figure B1.

H	$= \Omega\,s$	henry (inductance)
Bq	$= s^{-1}$	becquerel (activity of radionuclide)
Gy	$= J\,kg^{-1}$	gray (absorbed dose)
Sv	$= J\,kg^{-1}$	sievert (dose equivalent)
lx	$= cd\,sr\,m^{-2}$	lux (illuminance)

B2 Constants and Standard Values

Constants

Speed of light	c	$= 2.997\,924\,580 * 10^{8}\ ms^{-1}$
Planck constant	h	$= 6.626\,076 * 10^{-34}\ J\,s$
$\hbar = h/2\pi$	$\hbar$	$= 1.054\,573 * 10^{-34}\ J\,s$
Elementary charge	e	$= 1.602\,177 * 10^{-19}\ A\,s$
Avogadro constant	N_A	$= 6.022\,137 * 10^{23}\ mol^{-1}$
Gravitation constant	G	$= 6.672\,59 * 10^{-11}\ N\,m^2\,kg^{-2}$
Boltzmann constant	k	$= 1.380\,658 * 10^{-23}\ J\,K^{-1}$
Molar gas constant	R	$= 8.314\,510\ J\,K^{-1}\,mol^{-1}$
Von Klitzing constant	R_K	$= 2.581\,281 * 10^{4}\ \Omega$
Josephson constant	K_J	$= 4.835\,979 * 10^{5}\ GHz/V$

The values given for the two last-mentioned constants are as accepted in 1990 for use in the calibration of instruments.

Values of the Gas Constant for Common Gases

Gas	*Mol. weight*	*R* [$kJ\,kg^{-1}\,K^{-1}$]
air	28.97	0.287
CO_2	44.01	0.189
N_2	28.02	0.297
O_2	32.00	0.260
H_2O	18.02	0.462

Standard Values

Standard gravity acceleration:	$9.80665\ m\,s^{-2}$	$= 1\ kgf/1\ kg$ $= 9.807\ N\,kg^{-1}$
Standard atmospheric pressure:	101.325 kPa	$= 1\ atm$

B3 Temperature

SI unit
  kelvin (K)

Other units
  degree Celsius (°C)
  degree Fahrenheit (°F)
  degree Rankine (°R)

$$T\ [K] = T\ [°C] + 273.15$$
$$T\ [K] = (T\ [°F] + 459.67)/1.8$$
$$T\ [°C] = (T\ [°F] - 32)/1.8$$
$$T\ [K] = T\ [°R]/1.8$$

(An increase by 1 K used to be described as an increase by 1°C.)

Temperature values for processing instructions are often quoted to an apparent accuracy of a kelvin or even to 0.1 K. This is usually due to the conversion into °C of a rounded number originally defined in °F. A common target temperature for food freezing is −17.8°C = 0°F, for example. For the frying of pommes frites, a temperature of 188°C is sometimes recommended, corresponding to 370°F. The reference temperature in sterilization processes (see Chapter 1, Section 1.1.2) is defined as 121.1°C, corresponding to 250°F. Some typical food processing temperatures are given in the following table.

Temperatures in Food Processing and their Conversion

T [K]	T [°C]		T [°F]		Process (or calibration point)
223.15	−50			−58.0	
		−45.6	−50		Freeze-drying, freeze-concentrat.
233.15	−40			−40.0	
253.15	−20			−4.0	Target temperature for freezing
		−17.8	0		Target temperature for freezing
273.15	0			32.0	Ice point of pure water
	2			35.6	Chilled food
	4			39.2	Chilled food
283.15	10	10.0	50	50.0	Conditioning of vegetables
293.15	20			68.0	Room temperature
303.15	30			86.0	Dough fermentation, seeding (chocolate tempering)
		37.8	100		
313.15	40			104.0	Yoghurt fermentation

T [K]	T [°C]		T [°F]		Process (or calibration point)
323.15	50			122.0	
333.15	60			140.0	Milk evaporation (3rd stage) (under vacuum)
		65.6	150		
343.15	70			158.0	Milk evaporation (1st stage) (under vacuum)
353.15	80			176.0	HTST pasteurization of cream
363.15	90			194.0	
		93.3	200		
373.15	100			212.0	Steam point of pure water
		121.1	250		Reference temp. for steriliz.*
403.15	130			266.0	Low frying temperature
413.15	140			284.0	UHT processing of milk
		148.9	300		High boiling temp. for caramel
443.15	170			338.0	Medium frying temperature
473.15	200			392.0	High temp. in extrusion cooking
		204.4	400		Spray dryer air
493.15	220			428.0	Very high frying temp. (under excess pressure)
503.15	230			446.0	High baking temperature
523.15	250			482.0	Low coffee roasting temp.
		260.0	500		High oil deodorization temp. (under vacuum)
573.15	300			572.0	High oil hydrogenation temp. (under excess pressure)
		315.6	600		
593.15	320			608.0	High coffee roasting temp.
923.15	650			1202.0	Afterburners for coffee roasting exhaust gas treatment

HTST : high-temperature short-time
* : See Chapter 1, Section 1.1.2 on the F_0 value

B4 Pressure

SI unit
pascal (Pa)

Other units
atmosphere, standard (atm)
atmosphere, technical (atm/t)
($1\ \mathrm{kgf\ cm^{-2}} = 9.807\ \mathrm{N\ cm^{-2}}$)

bar (bar)
pound (force) per square inch (psi)
or (lbf/in^2)
psi absolute (psia)
psi gauge (psig)
torr (torr)
millimeter of mercury, 0°C (mmHg)

p [kPa] = p [atm] * 101.325
p [kPa] = p [bar] * 100.000
p [kPa] = p [psi] * 6.895
p [kPa] = p [mmHg] * 0.133 3
p [torr] = p [mmHg] * 1
p [kPa] = p [atm/t] * 98.067

See tables below for conversion examples.

Absolute Pressures in Food Processing and their Conversion

p [kPa]	p [bar]	p [atm]	p [torr]	p [psi]	*Conditions*
0.7	0.007	0.0069	5	0.10	milk evaporat., stage 3
4.	0.04	0.039	30	0.58	milk evaporat., stage 1
70.	0.7	0.691	525	10.15	retort, vacuum
101.325	1.013	1.00	760	14.696	ambient pressure
250.	2.5	2.47		36.3	retort, excess pressure
500.	5.0	4.93		72.5	retort, high excess pressure

p [MPa]	p [bar]	p [atm]	p [psi]	*Conditions*
1.5	15.0	14.8	218.0	oil hydrogenation
10.0	100.0	98.7	1450.0	highest pressures
13.0	130.0	128.3	1888.0	in extrusion
14.0	140.0	138.2	2030.0	cooking

Absolute and Gauge Pressures, Saturated Steam Temperatures

T (steam) [°C]	p (abs) [kPa]	p (abs) [atm]	p (abs) [psia]	p (gauge) [kPa]	p (gauge) [psig]
	0	0 (perfect vacuum)	0	101.3 vacuum (perfect vacuum)	14.70 vacuum
90	70.1	0.69	10.2	31.2 vacuum	4.53 vacuum
100	101.3	1.00	14.7	0	0
110	143.3	1.42	20.8	42.0 excess	6.09 excess
120	198.5	1.96	28.8	97.2 excess	14.10 excess
130	270.1	2.67	39.2	168.8 excess	24.48 excess
140	361.3	3.57	52.4	260.0 excess	37.71 excess
150	475.8	4.70	69.0	374.5 excess	54.31 excess

B5 Energy and Mass

Energy

SI unit
joule (J)

Other units
British thermal unit, international table (Btu)
calorie, international table (cal)
electronvolt (eV)
kilowatt-hours (kWh)

E [J] = E [cal] * 4.186 8
E [J] = E [Btu] * 1 055.1
E [J] = E [eV] * 1.6022 * 10^{-19}
E [kJ] = E [kWh] * 3 600.

The thermal energy kT at a temperature of T = 273.15 K = 0°C is E_{273} = 0.023538 eV.

Example for metabolizable energy:
100 ml milk containing 3.3 g protein, 4.7 g carbohydrate, 4.0 g total fat has an energy value of 68 kcal = 285 kJ

Mass

SI unit

kilogram (kg)

Other units

pound, avoirdupois (lbm)
ton, metric (ton_m)
ton, short, US (ton_s)
ton, long, UK (ton_l)
ounce, avoirdupois (oz_s)
ounce, troy (oz_l)
grain (gr)

m [kg] = m [lbm] * 0.4536
m [kg] = m [ton_s] * 907.18
m [kg] = m [ton_l] * 1016.05
m [kg] = m [ton_m] * 1000.
m [g] = m [oz_s] * 28.349
m [g] = m [oz_l] * 31.103
m [g] = m [gr] * 0.0648

B6 Force, Power, Magnetic Flux and Time

Force

SI unit

Newton (N)

Other units

kilogram-force (kgf)
kilopond (kp)
pound force, avoirdupois (lbm)
dyne (dyn)

F [N] = F [kgf] * 9.807
F [kp] = F [kgf]
F [N] = F [lbf] *4.448
F [N] = F [dyn] *10^{-5}

Power

SI unit

Watt (W)

Other units
horsepower (hp)

P [kW] = P [hp] * 0.745 7

Magnetic flux density

SI unit
Tesla (T)

Other unit
Gauss (Gs)

B [mT] = B [Gs] * 0.1

Time

SI unit
second (s)

Other units
hour (h)
day (d)
year (yr)

t [s] = t [h] * 3.600 * 10^3
t [s] = t [d] * 8.640 * 10^4
t [s] = t [yr] * 3.153 6 * 10^7

B7 Length and Volume, Sizes

Length

SI unit
meter (m)

Other units
Angstrom (Å)
inch (in)
foot (ft)
yard (yd)

L [m] = L [Å] * 10*$^{-10}$
L [nm] = L [Å] * 0.1
L [cm] = L [in] * 2.54

L [cm] = L [ft] * 30.48
L [m] = L [yd] * 0.9144

The description micron was used earlier for the unit micrometer (μm).

Volume

SI unit

m^3

Other units

litre (l)
gallon, US (gal_s)
gallon, UK (gal_l)
pint, US (pt_s)
pint, UK (pt_l)
fluid ounce, US (fl oz_l)
fluid ounce, UK (fl oz_s)

V [m^3] = V [dm^3] * 10^{-3}
V [dm^3] = V [cm^3] * 10^{-3}
V [dm^3] = V [l] * 1
V [dm^3] = V [gal_s] * 3.785
V [dm^3] = V [gal_l] * 4.546
V [dm^3] = V [pt_s] * 0.473
V [dm^3] = V [pt_l] * 0.568
V [cm^3] = V [fl oz_s] * 28.413
V [cm^3] = V [fl oz_l] * 29.574

Sizes

Object	*Order of magnitude*	
Atom	10^{-10} m	(0.1 nm)
Bacterial virus	10^{-7} m	(0.1 μm)
Bacterium	10^{-6} m	(1 μm)
Emulsion droplet	10^{-6} m	(1 μm)
Sugar particle in chocolate	10^{-5} m	(10 μm)
Air bubble in chocolate mousse	10^{-3} m	(1 mm)
Raisin or carrot piece	10^{-2} m	(1 cm)
Orange	10^{-1} m	(10 cm)
Milk processing tank	10^{1} m	(10 m)
Production area	10^{2} m	(100 m)

B8 Viscosity

Dynamic viscosity

SI units

pascal second (Pa s)
poiseuille (Pl)

Other unit

poise (P)
η [mPa s] = η [cP]
η [mPa s] = η [mPl]

Note: The poiseuille (Pl) is not widely used as yet, but is useful when working with viscosity data. (The pre-SI unit poise was also named after Poiseuille.)

Kinematic viscosity:

SI unit

$m^2 s^{-1}$

Other unit

stoke (St)
v [$m^2 s^{-1}$] = v [cSt] * 10^{-6}
$v = \eta / \rho$

where ρ is the density.
See table on page 706 for viscosity data.

Viscosities at standard pressure

Material	*Temperature* °*C*	*Viscosity* *mPl*	*Density* *kg dm*$^{-3}$
air	20	18.2 * 10^{-3}	1.20 * 10^{-3}
water	0	1.79 * 1	0.999 * 1
	20	1.00	0.998
	40	0.66	0.992
	60	0.47	0.983
	80	0.36	0.972
	100	0.28	0.958
ethanol	20	1.20	0.790

Material	*Temperature* °C	*Viscosity* mPl	*Density* kg dm^{-3}
glycerol	20	1490.0	1.261
milk, homogenized	20	2.0	1.032
	40	1.5	
	60	0.78	
	70	0.7	1.012
	80	0.6	
milk, raw	0	3.44	
	10	2.64	
	20	1.99	
	30	1.49	
	40	1.23	
cream (10% fat)	40	1.48	
	60	1.07	
	80	0.83	
(20% fat)	60	1.71	
(30% fat)	60	2.89	
(40% fat)	60	5.10	
sucrose sol. (20%w/w)	20	2.0	1.081
	80	0.6	1.055
sucrose sol. (40%w/w)	20	6.2	1.176
	80	1.3	
sucrose sol. (60%w/w)	20	57.0	1.287
	80	5.4	1.250
sucrose sol. (70%w/w)	20	460.0	1.347
	80	17.0	
soyabean/sunflower oil	20	~60.0	0.907
	100	~ 6.0	
cottons./groundnut oil	20	~73.0	0.904
	100	~ 7.3	
olive oil	20	~75.0	0.899
	100	~ 7.5	

For detailed data on frying oils, see Chapter 16 (Tables 16.5 and 16.6 and Figure 16.4). For each oil type, the values from different literature sources for the viscosity (taking into account the temperature dependence) are scattered with a spread of ± several per cent around the average. The averaged values given here are for fresh oils; for used oils, see Chapter 16. (For more data on oil densities, see Chapter 16, Table 16.8).

For sources of further data on the rheological properties of foods, see the bibliography by McKenna (1990). The measurement of rheological properties is described in Chapters 10 and 11.

B9 pH-Values

Lemons	2.3–2.6
Vinegar	2.4–2.8
Wine	2.8–3.2
Apples	3.0–3.3
Oranges	3.2–3.8
Peaches	3.4–3.6
Yoghurt	4.0–4.5
Beer	4.1–4.3
Potatoes	5.4–5.8
Meats	5.5–6.5 *
Cauliflower	5.6–5.7
Hard cheeses	5.6–6.2
Sardines	6.2–6.4
Poultry	6.4–6.6
Milk	6.5–6.7

* During the post-mortem period, the pH-value decreases due to the formation of lactic acid. The final pH-value depends on the initial concentrations of glucose and glycogen. These, in turn depend on the meat type and quality.

Note: Corrosion stability of construction materials; see, for example, p. 379 of Loncin and Merson (1979)

B10 Concentration

Generally, the concentration of a solute is given in % w/w (weight solute/weight solution). The concentration can be for a particular compound or for total soluble solids (TSS). On other occasions, the concentration of total solids (TS) including both solutes and non-soluble matter. In addition, the unit Brix is used in assessing saccharide solution strength, and molar concentrations are used frequently in chemical and biochemical analysis.

Brix

In the processing of sugar syrups and of fruit juices, the sugar content is frequently expressed in °Brix. This is read directly from a refractometer specially calibrated for this purpose. For pure aqueous sucrose solutions having a temperature specified on the instrument, the °Brix reading corresponds to % w/w of sucrose concentration. For solutions of other saccharides or at different temperatures, correction tables are used to relate ° Brix to concentration.

Molar

Solution strength in mol solute contained in 1 l solution.
1 mol/l (often described as 1 M)

Examples
Aqueous glucose solutions

1 mol/l = 180 g/l = 180 g/1000 ml = 18 % w/v
= 17 % w/w (at 20°C)
1 mmol/l = 0.18 mg/ml = 0.018 % w/w
1 % w/w = 55 mmol/l

Glucose in meat

1 mg glucose per g meat corresponds to approximately 1 mg glucose per 0.75 ml meat juice. The latter corresponds to a concentration of approximately 1.33 mg/ml or 7.4 mmol/l in the juice.

For data on glucose in meat, see Chapter 16, Section 16.3.2 and Figures 16.15 and 16.20.

Sugars in foods

The distinction between sugar concentration in the juice (as determined by enzyme electrodes) and the sugar concentration per weight of foods such as meat, fruit or vegetable (as usually determined by HPLC or Dionex assay) is needed to allow for the presence of non-soluble solids. When preparing calibration solutions for enzyme electrodes, note that the equilibration between the α and β forms of D-glucose can take several hours.

B11 Colour

Approximate assignation of colours

Description	*Wavelength nm*
purple	390–440
blue	440–480
turquoise	480–500
green	500–520
lime	520–575
yellow	575–585
orange	585–600
red	600–700

(Chromaticity diagrams and eye response functions: see Chapter 2, Figures 2.1–2.4. On colour measurement and sorting by colour, refer to Chapters 2, 3 and 4.)

Figure 2.1c of Chapter 2 shows the overlap of the response functions of the 'green' and 'red' eye cone pigments, with only a slight overlap for the 'blue' response function. Red-green blindness is not uncommon, particularly among men. Blue-yellow blindness, on the other hand is very rare. Engineers aiming to relate colour measurement to colour assessment by human observers, sometimes report that the discrimination between fine colour nuances diminishes with age.

For data relating to the optical properties of foods, see Chapters 2 and 3. Sources for further data are given by Kent (1989).

B12 Near- and Mid-Infrared Range

Wavelength μm	*Wavenumber* cm^{-1}	*Frequency* *THz*	*Energy* *eV*
1	10 000	299.79	1.239 2
2	5 000	149.90	0.619 6
3	3 333	99.93	0.413 1
4	2 500	74.95	0.309 8
5	2 000	59.96	0.247 8
10	1 000	29.98	0.123 9
15	667	19.99	0.082 6
20	500	14.99	0.062 0

Where

$k = 2\pi/\lambda$ is the wave vector

$\lambda = c/f$ is the wavelength

$wn = 1/\lambda$ is the wavenumber

Water absorption bands in the near infrared range used for moisture measurement:

at 1.45, 1.94 and 2.95 μm.

Most commonly used band:

at 1.94 μm

(See Chapter 5, Section 2). There is some overlap (compensated for where appropriate) with the absorption band of water vapour which is located approximately 0.1 μm below that for liquid water. (Chapter 5, Section 5.3.3, Figure 5.9).

Remote thermometry by a measurement in the near and mid-infrared range relies on black-body radiation. The maximum of the spectral energy density per wavelength interval is defined by Wien's displacement law or

Planck's formula and occurs at a wavelength of λ_{max} given by the equation $hc/(\lambda_{max}kT) = 4.965$.

Temperature	*Wavelength*
−50°C = 223.15 K	12.99 μm
20°C = 293.15 K	9.9 μm
200°C = 473.15 K	6.12 μm
650°C = 923.15 K	3.14 μm

In IR thermometry, the chosen operating wavelength depends also on the presence of a 'window' of high atmospheric transmissivity, on the availability of a suitable detector and, where appropriate, of a transparent window material to access a closed vessel (Chapter 6, Sections 6.2.2 and 6.4.3).

B13 Microwave Bands

Table B13a *Old Band Designation*

Band	Frequency range [GHz]	Wavelength [cm]	Waveguide [cm*cm]	Waveguide [in*in]
L	1.00–2.60	11.5–30.0	–	
S	2.60–3.95	7.6–11.5	7.6 * 3.8	3 * $1\frac{1}{2}$
G	3.95–5.85	5.1–7.6	5.1 * 2.5	2 * 1
C	4.90–7.05	4.3–6.1	4.4 * 2.2	$1\frac{3}{4}$ * $\frac{7}{8}$
J	5.30–8.20	3.7–5.7	3.8 * 1.9	$1\frac{1}{2}$ * $\frac{3}{4}$
H	7.05–10.00	3.0–4.3	3.2 * 1.3	$1\frac{1}{4}$ * $\frac{1}{2}$
X	8.20–12.40	2.4–3.7	2.5 * 1.3	1 * $\frac{1}{2}$
M	10.00–15.00	2.0–3.0	2.1 * 1.2	
P	12.40–18.00	1.7–2.4	1.8 * 1.0	
N	15.00–22.00	1.4–2.0	1.5 * 0.85	
K	18.00–26.50	1.1–1.8	1.2 * 0.65	
R	26.50–40.00	0.8–1.1	0.9 * 0.56	
Millim.	30.0–300.00	0.1–1.0		

Table B13b *New Band Designation (since 1970)*

A	0.1–0.25
B	0.25–0.5
C	0.5–1.0
D	1.0–2.0

Table B13b *continued*

E	2.0–3.0
F	3.0–4.0
G	4.0–6.0
H	6.0–8.0
I	8.0–10.0
J	10.0–20.0
K	20.0–40.0
L	40.0–60.0
M	60.0–100.0
Millim.	30.0–300.0

Notes
Millim. denotes the millimeter range with wavelengths from 1 to 10 mm.
At higher frequencies (300 GHz–10 THz) is the submillimeter (or far-infrared) range.
The range from 0.1 GHz (100 MHz) to 1.0 GHz includes the lower (frequency) part of the UHF band and the upper VHF band and is often included in the radiofrequency range. (See also Figure 1 of Chapter 8.)
Wavelengths in the microwave range extend from approximately 1 mm (300 GHz) to approx. 1 m (300 MHz).
Sources
Sze (1981), page 514; Hewlett-Packard (1961); Liao (1980); Kaye and Laby (1986).
For data on the dielectric properties of foods, see Chapter 7, Figures 7.2–7.15 and the tabulated data by Kent (1987).

B14 Logarithms, Neper and Decibel

Logarithms

In processes, where the growth rate dx/dt of a quantity is proportional to this quantity x, the natural logarithm of the quantity x is proportional to time t and the quantity x is proportional to the natural exponential function of time t.

From $dx/dt = a\,x$ it follows that
$dx/x = a\,dt$ and on integration of both sides that
$ln\,(x_2/x_1) = a\,(t_2 - t_1)$

By using the exponential function $\exp(X) = e^x$ of both sides, and setting $t_1 = 0$, one arrives at

$$(x_2/x_1) = \exp\,(a\,t_2)$$

Experimental data that follow such a relationship are often presented in the form of decadian logarithms and decadian exponentials. For comparison with the theoretically derived relationship, they can be converted into the natural logarithms and exponentials, respectively.

General logarithm $\log_b a$
(when $\log_b a = x$ then $a = b^x$)

Natural logarithms $\log_e = ln$
Basis e = 2.718
(when $ln\ a = x$ then $a = e^x = \exp(x)$)

Decadian logarithms $\log_{10} = \lg$
Basis 10.00
(when $\lg a = x$ then $a = 10^x$)

Conversion
$ln\ a = (lg\ a) * (ln\ 10) = 2.302\,6 * lg\ a$
$lg\ a = (ln\ a) / (ln\ 10) = 0.434\,3 * ln\ a$

Note: In contexts where the decadian logarithm only is used, this is often referred to by the notation log without explicit base indication. The abbreviation *lg* for the decadian logarithm is generally preferable as it avoids ambiguity, but is not commonly used in the presentation of microbial data. In linking the theoretical description of the relationship for microbial growth with time (in terms of the natural logarithm of the microbial cell numbers) to experimental data (generally presented in the form of the decadian logarithm), however, the distinction between the logarithmic forms as *ln* and *lg* is necessary (as in Chapter 14, Sections 14.5.1 and 14.5.2).

Neper and Decibel

The intensity ratio A for amplification or attenuation of signals can be expressed in either Decibel or Neper, where:

$A\ [dB] = 10.0\ lg\ I_2/I_1$
$A\ [Np] = 0.5\ ln\ I_2/I_1$

(Such ratios are used here in Chapters 7 and 8).

Conversion
$A\ [dB] = A\ [Np] * 20 / \ln 10 = A\ [Np] * 20 * \lg e =$
$= A\ [Np] * 8.686$

B15 Miscellaneous

Frequencies and Wavelengths for Measurements Based on Electromagnetic Waves
See Appendix to Chapter 1
Electromagnetic Spectrum
See Chapter 8, Figure 8.1

Frequencies and Wavelengths for Measurements Based on Ultrasonic Waves
See Appendix to Chapter 1
Sound velocity and acoustic impedance
See Chapter 8, Table 8.1 and Figures 8.2–8.5, also Chapter 9, Table 9.2.
Water activity
See Chapter 12, Figure 12.1 and Tables 12.1–12.11
Impedance of growth media
See Chapter 14, Table 14.2

References

Batty, J. C. and Folkman, S. L. (1983) *Food Engineering Fundamentals.* John Wiley and Sons, New York

Brennan, J. G., Butters, J. R., Cowell, N. D. and Lilley, A. E. V. (1990) *Food Engineering Operations* (3rd ed). Elsevier Applied Science, London

Hewlett-Packard (1961) Hewlett-Packard Electronic Test Instruments. Hewlett-Packard, Palo Alto, Calif

Kaye, G. W. C. and Laby, T. H. (1992) *Tables of Physical and Chemical Constants.* Longman Scientific & Technical

Kent, M. (1987) *Electrical and Dielectric Properties of Food Materials, Bibliography and Tabulated Data.* Science and Technology Publishers, (now Food Science Publishers), UK

Kent, M. (1989) *Colour and Optical Properties of Foods.* Food Science Publishers, UK

Knuth E. L. (1966) *Introduction to Statistical Thermodynamics.* McGraw-Hill, New York

Liao, S. Y. (1980) *Microwave Devices and Circuits.* Prentice-Hall, Englewood Cliffs, New Jersey

Loncin, M. and Merson, R. L. (1979) *Food Engineering, Principles and Selected Applications.* Academic Press, New York

McFarlane, I. (1983) *Automatic Control of Food Manufacturing Processes.* Applied Science Publishers

McKenna, B. M. (1990) *Solid and Liquid Properties of Foods, Bibliography.* Food Science Publishers, UK

Singh, P. R. and Heldman, D. R. (1984) *Introduction to Food Engineering.* Academic Press, New York

Sze, S. M. (1981) *Physics of Semiconductor Devices*, 2nd ed. John Wiley, New York

Watson E. L. and Harper J. C. (1988) Elements of Food Engineering, 2nd ed. AVI, Van Nostrand Reinhold Company, New York

Addendum: Trial with Prototype 2 of the *In situ* Monitor for Frying Oil (see Chapter 16, Section 2)

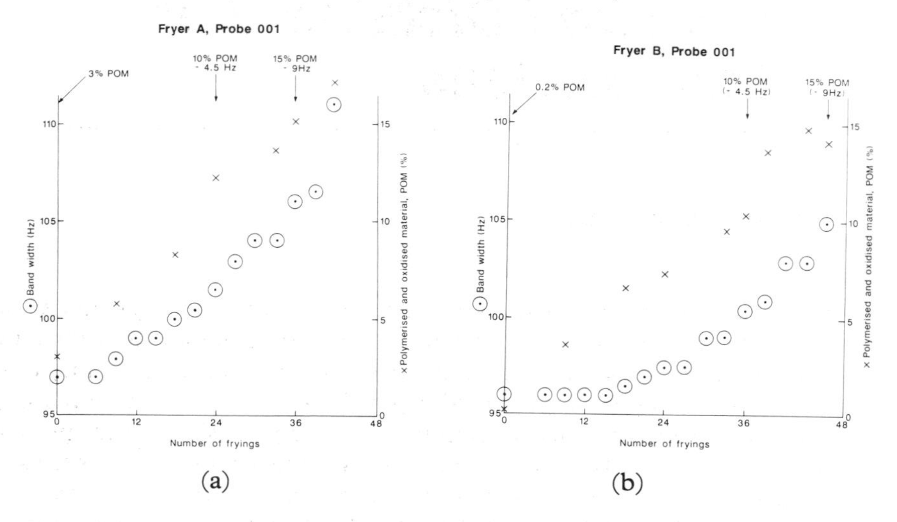

Figure 1 *Trial with prototype 2 in two samples of sunflower oil*

For both samples, the sensor output (bandwidth) increases by 4.5 Hz when the reference assay value has reached 10% POM and by 9 Hz for 15% POM, respectively. One of the samples (b) starts out with a lower POM value at the beginning of the frying trial and this shows an initial lag phase in the viscosity increase, compatible with the induction period that is thought to precede the heat-induced deterioration of frying oils. (Reproduced by permission of GEC Sensors Ltd.)

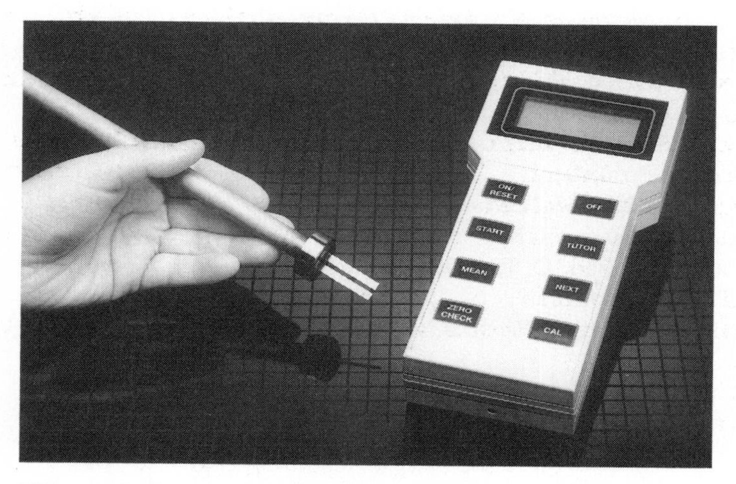

Figure 2 *Prototype 2 has now been replaced by the pre-production model shown in the photo (courtesy of GEC Sensors Ltd)*

Index and Dictionary of Abbreviations

Notes for Index Users

The index is designed to help locate information (description, comments, data or references to other publications) on measurement issues relating to a food product or ingredient, a measured variable (on-line, at-line, off-line), an interfering variable, a property to be assessed, a food processing operation to be monitored; a measurement or data processing technique, an instrument or sensor type or configuration, an instrument make or component, a window or membrane material, a relevant scientific principle or relevant data. Broader topics can be located with the help of the Contents List, pages v–xvii.

In general, measurement applications to specific foods are listed under the food name rather than under the measurement technique. Where such applications are discussed in detail, however, the specific food is also listed as sub-entry to the measurement technique entry.

When a measurement is listed for a particular food type, this is not an indication of its general applicability to this food type, nor does it necessarily imply that this is the best method for the listed food. Rather, it indicates that the application of the named method for the listed food is described or discussed on the indexed page, or that related data or references to other publications are given there.

A number of the entries are intended for users with a food science background who wish to look up instrumentation terms or vice versa. The text passages referred to under those entries give a formal definition in some instances, but more often just give an opportunity to see the term in context. The index requirements of a wide readership have been taken into account and the entries reflect the diverging interests of differing groups of readers.

A key to abbreviations of technical terms is provided in the Index. Abbreviations without key relate to companies, institutes or other organizations.

Erika Kress-Rogers

Abbeon cup hygrometer, 396
Abrasive foods, 15
Absorption/adsorption/sorption, 644
Absorption bands (NIR) of water, 126, 139
Absorption coefficient:
  see Colour measurement,
  see Microwave measurement
Absorption spectroscopy:
  see Near infrared absorption spectroscopy
AC (alternating current), 464–7
Accept/suspect/reject categories (microbial), 489, 506
Acid digestion systems for nitrogen, 445–52
Acid value (AV), 532, 536, 539
Acidity:
  acidity sensors, 586, 589, 606, 609–11
    dipstick device, 606, 611
    sample channel device, 610
  comparison with pH, 609
  significance, 24
Aclar windows, 141–2
Acoustic base transducers, 652–4
Acoustic emission monitoring (AEM), 245–6
Acoustic immunosensors and sorption sensors, 644–7, 652–4, 660
Acoustic impedance, *see* Impedance
Acoustic measurements:
  see Ultrasonic measurement
  see Ultrasound propagation
  for texture assessment, 358–9, 370, 371
Acoustic surface properties, 644
Acoustic surface waves, 644
Actuator:
  definition, 675
  notes, 11, 675
Actuators in process control systems, 659
Actuators integrated in sensors, 584, 586, 609–11
  see also Acidity sensor
Adaptation, in vision, 43–4, 64
Adaptive data processing systems, 660
Adenosine, 624, 636
Adipose tissues, *see* fatty tissues
ADP (adenosine diphosphate), 624
Adsorption/absorption/sorption, 644
Advanced Moisture Technology instrument, 231
AEM (acoustic emission monitoring), 245–6
Affinity binding, 586, 623, 644–5, 654
Affinity sensors, *see* Immunosensors
AFRC (Agriculture and Food Research Council), xxx
Ageing, ultrasound-assisted, 246
Agglutination assay, 461–3, 652
Agricultural residues, 624
AIDMEE (amperometric indirect dual membrane enzyme electrode), 585–6, 627, 633–6
  see Biosensors
Air, entrained, *see* Bubbles
Air, ultrasound data, for, 262, 293
Air content, influence on non-contact measurements, 265, 305
Air flow, control in fermentation, 389
Air purges and air curtains, 111–2, 141–2
Air slugs, 277
Alcohol assay:
  distillation, 435
  enzyme electrode analyser (bench-top), 433–5
  GLC for water samples, 427
Alcohol content monitoring:
  microwave measurement, 196, 217
  NIR measurement, 142
  significance of measurement, 24–5
  ultrasonic measurement, 251–2
Alcohols, sensors, 585, 611–3, 615, 624, 627, 636
  see also Ethanol, Methanol
Alcohols:
  dielectric data for, 196
  formation in heated frying oil, 528
  ultrasound data, for, 261, 274–5
Alcohols, polyhydric, assay, 432
Alcohols, sugar-, assay, 432
Aldehydes, 585, 526, 612, 615, 636:
  formation in heated frying oil, 528
  formation in oxidative rancidity, 525–7
  interference with Karl Fischer procedure, 438
Algorithms, 660
Alkali thermal ionization detector (ATD), 426–7
Alkaloids assay, 423

Allergens:
see Histamine
Almonds, sorting, 104
Al_2O_3 (aluminium oxide), 585, 597
Aluminium, dissolved, 651
Aluminium oxide (Al_2O_3), 585, 597
AMEE (amperometric mediated enzyme electrode), 585–6, 626–33
see Biosensors
Amersham, 646
Amines, 585, 524, 526, 559, 612, 615
in meat and fish, 524
Amines in solution, ultrasound properties, 254
Amino acids, 585, 624, 645
Amino acids in solution, ultrasound properties, 254
Ammonia, 585, 600, 606, 612, 615, 617
Ammonia, dissolved, 651
Ammonia probe as base device, 623, 636
Ammonium probe as base device, 623
Amperometric biosensors, 623–36
Amperometric detector, pulsed HPLC in, 428, 431–2
Amperometric gas probes, 623, 626, 633–6
Anaesthetics, 644, 647
Angle, plane and solid, 696
Animal feeds:
humectants in, 388
moisture monitoring, 148
water activity determination for, 400
Anions assay, 428–9
Anisidine value, 532
Anthocyanin content, 91
Antibiotics, 624
Antibiotics susceptibility of bacteria, 636
Antibodies, 585, 623, 645
Antimicrobial substances:
microbial assay (impedance) for foods with, 510
Antimony oxide (SbO_3), 585, 595
Antioxidants, 624
AnV (anisidine value), 532
AOAC (Assoc. Official Analyt. Chemists) approval, 436
AOTF (acousto-optical tunable filter), 130
AP-Elektronik instrument, 402–3
Appearance, 44–5, 64, 79
attributes, 44
gloss, 44, 72, 79
haze, 72
opacity, 44, 72
translucency, 44, 72, 79
transparency, 72, 78–9
turbidity, 72
food structure, influence, 44–5
sorting by colour, shape and size, 116
see also Colour description:
see also Colour measurement:
see also Colour vision
Apples, colour, 75, 83, 88–9
Apple sauce, viscometry for, 343
Ares-Serono, 646
Arnott, M.L., xxvii, 499–519
Aroma and odour assessment, 601, 612–7, 647, 654, 660
Aromatic hydrocarbons assay, 427
Arrays of detectors:
CCD devices, 110
photodiodes, 110
Arrays and pattern recognition, 601, 611–2, 614–7, 660–2
Arrays of sensors, 599, 601, 609, 629, 637, 639, 649, 676:
illustrations, 563, 572, 678
micro-thermocouples, 639
Ascorbate, as analyte, 624
Ascorbate, as interference, 627, 631, 633
Ash assay:
combustion systems, 436, 442–3
notes, 435, 436, 438
Ash content, phantom measurement, 152
Ash content measurement:
significance, 24
Asian products, water activity of, 389
Asparagine, 624, 636
Aspartame assay, 428
Aspartame sensors, 624, 636
Aspartate, 636
At-line instrumentation
definition, 684–5
notes, 3–7, 10
requirements, 7
ATD (alkali thermal ionization detector), 426–7

Atmosphere, infrared absorption bands, 137, 171
Atmospheric conditions:
influence on microbial deterioration, 376
Atomic absorbance spectrometry:
HPLC with, 421
metals assay, 420
Atomic emission spectrometry, 421
ATP (adenosine triphosphate):
bioluminescent assay for bacterial, 461–2
decomposition in fish, 524
microbial assay, 560
sensors, 524, 526, 624, 637, 651, 664
ATR (attenuated total reflectance), 648
Authenticity, 29
beef mince, 646
Automatic cleaning, on-line instruments, 112, 141–2
Automatic sorting, *see* Sorting by colour
AV (acid value), 532, 536, 539

Back fat thickness, determination of, 266
Back scatter configuration, 140
Bacon:
colour, 77
moisture assay, 441–2
nitrite/nitrate assay, 429
packaging materials, 157
water activity of, 386
see also Meat products
Bacteria:
Clostridium botulinum, 491
Escherichia coli/coliforms, 488, 491, 510, 512
Enterobacteriaceae, 483, 491, 510, 512
Enterococci, 491
histamine producers, 491
Lactic acid, 512
Listeria, 491, 511, 512
Pseudomonas, 512
Salmonellas, 486, 489, 491, 498, 511, 512
Shewanella putrefaciens, 487
Staphylococcus aureus, 491, 511, 512
Streptococci, 512
Bacteria contamination, removal, 104
Bacterial ATP (adenosine triphosphate):
bioluminescent assay, 461–2
Bacterial growth, 469–71, 483–6
carbon dioxide evolution, 488
gas evolution, 477
generation time, 485
growth curve, meat, 567
hydrogen evolution, 483
lag phase, 469
see also Micro-organisms, growth of
see also Microbial toxins, production of
Bacteriophage, 491
Bactometer, 463, 464, 480–3, 515–6
Baird & Tatlock instrument, 439–40
Baked products:
moisture assay, 437
Baker Compressimeter, 352
Bakery fats, quality, 534
Bakery products:
mould growth in, 388
NIR measurement for, 138, 140, 147
water activity in, 386–9
see also Biscuits, Cakes
Baking end-point, volatiles, 612
Baking oven, temperature measurement on emerged product, 182
Ball indenters, for texture assessment, 352
Banana extracts, sugar content, 632
Bandwidth, resonant probes, 546–7
Bandpass filter, 108, 110
Barley, moisture monitoring, 148
Base devices/base transducers, *see* Biosensors, Chemical sensors
Batter-coated products, 528
Bausch & Lamb, 86
Beans, green, colour measurement, 82, 87
Beans, white, sorting, 104, 117
Beans, *see also* Coffee beans
Beam-break detectors:
ultrasonic counters, 301
infrared level controls, 301
Beckman Instrument, 401
Beef:
colour, 83, 88–9
DFD (dark firm dry), 608
fat/lean separation by colour, 88
freshness, *see* Meat freshness

mince authenticity, 646
nitrogen/protein assay, 446, 453
Beef, dried, water activity of, 386
Beef, partially frozen, ice proportion in, 266
Beef tallow, water activity of, 387
Beer, *see* Brewing
Beer-Lambert transmission law, 127, 143
Benchtop enzyme electrode instrument, 636
Benson, I., B., xxvii, 121–66
Berthold, Lab. Prof., instrument, 231
Beverages:
acidity in-line, 611
alcohol monitoring in, *see* Alcohol content
aspartame in soft drinks, 428
colour, 85, 91
ethanol/methanol discrimination, 637
identification of alcoholic beverages, 616, 660
level measurement for, 308
microbial (impedance) assay, 509
see also Alcohol
see also Carbonated beverages
Beverages, flavoured (chocolate, malt)
microbial (impedance) assay, 504–6
Beynon, G., xxvii, 167–88
Bichromatic sorting, 102–4, 117
Binding of water to food matrix:
effect on dielectric properties, 196
Binding, affinity or sorption:
physical changes, 644
direct monitoring, 645
Binding proteins, endogenous, 645
Biocatalysis, isolated enzyme or whole cells: *see* Biosensors
Biocatalysts, 622, 627
BioFET, *see* ENFET
Biofouling, *see* Fouling
Bioluminescent assay, 461–2
Bioluminescence sensors, 651
Biological tissues:
microwave measurements in, 230
ultrasound data for, 258–9
Biomass, 464, 477, 652
bacterial ATP, 461–2
Biomass monitor, 483
Biomembranes, 645
Biopolymers, *see* Gels
Biosensors, 582–8, 622–39, 663, 682
adaptation to food, 629–35
preparation kit, 629–32
robotics approach, 17, 662–5
arrays, 563, 572, 629, 637
illustrations, 563, 572
artificial pancreas, 626
ascorbate, influence, 631, 633
base devices:
amperometric gas probes, 623, 626, 633–6
CHEMFET gas sensors, 636
Clark-type gas probe, 623, 626, 635–6
heat-sensitive devices, 625, 638–9
ion-selective (potentiometric) probe, 623
ISFET/REFET, 623, 625, 637–8
ISFET/microactuator, 625, 637
mediator-carrying carbon electrode, 625, 626–33
micro-peroxide probe, 636
optical fibres, 625
organic conducting electrodes, 626
oxygen/peroxide probe comparison, 635
planar micro-Clark device, 623, 635–6
thermistor, 638
thermopile, 639
biocatalysts, 622, 627
bioluminescence sensors, 651
biothermal devices, *see* TELISA
buffer strength independence, 625, 637
building blocks, 582–6
calibration:
electrode batch-specific, 627
in situ, 632
categories and types:
AIDMEE (amperometric indirect dual membrane enzyme electrode), 585–6, 627, 633–6
amperometric, 623–36
AMEE (amperometric mediated enzyme electrode), 585–6, 626–33
BioFET, *see* ENFET
biothermal, 639
direct, 626
ENFET (enzyme-sensitized FET) with REFET (reference FET), 585–6, 623, 625, 637–8

Biosensors – *continued*
categories and types – *continued*
ENTHERM (enzyme thermistor or enzyme thermopile), 585–6, 627, 638–9
first/second/third generation, 626
indirect, 623
mediated, 625–33
microelectronic, 623, 625
optical, 625
organic conducting electrode biosensors, 626
potentiometric, 623
see also ENFET
cell membrane (acetic acid bacteria), 637
commercial benchtop food analyser, 636
commercial blood glucose analyser:
description, 627–8
trial with food samples, 629–32
comparison amperometric/potentiometric, 623, 625
configuration examples:
credit-card type, 627, 629
disposable strip electrodes, 627–8
illustrations, 628, 629
in situ monitor, 632
knife type, 628–9
needle type, 635–6
pen type, 627–8
continuous operation characteristics, 632–3, 635–6
definitions, 583, 622–3, 682–3
detection types:
calorimetric, 638–9
electrochemical, 626–38
fluorescence, 625
luminescence, 625
diffusion control, 626, 633–5
disposable enzyme electrodes, 627–32
electroactive species, 627
ascorbic acid, 627
electrode strips, 627–31
electron mediators, *see* mediators
enzymes, 585–6, 622
oxidases, 627
whole cells, 585–6, 622, 637
enzyme substrates and inhibitors, 623
equilibration of current, 627, 632
extended analytical range, 632, 633–5
FIA (flow injection analysis), 637, 639
fibre-optic, *see* FOP
food analytes, 526, 624, 663
adenosine, 624, 636
asparagine, 624, 636
aspartame, 624, 636
aspartate, 636
ATP (adenosine triphosphate), 524, 526, 624, 637, 651
cadaverine, 526, 624
creatinine, 624, 636
ethanol, 526, 624, 627, 635, 651
ethanol without methanol response, 637
fructose, 624, 639
galactose, 624, 627, 639
glucose, 524, 526, 561–3, 623, 624, 626–32, 635, 636, 637, 639, 651
glucose depth profile, 559–72
glutamate, 624, 636
hexanal, 527, 624, 651
histamine, 524, 526, 624
hypoxanthine, 524, 526, 624, 635, 637
inosine, 524, 526, 624, 635, 637
lactate, 624, 632, 651
maltose, 624, 639
methanol, 624, 627
NADH (nicotinamide adenine dinucleotide), 651
nitrate, 585, 624
penicillin, 526, 624, 636
pentanal, 527, 624
putrescine, 526, 624
pyruvate, 651
starch, 624, 636
sucrose, 624, 627, 639
trimethylamine, 524, 526, 624
tyramine, 526, 624
see also food analyte groups
food analyte groups:
agricultural residues, 624
alcohols, 585, 612, 624, 627, 636
see also ethanol, methanol
aldehydes, 585, 526, 612, 636
amines, 585, 524, 526, 559, 612
amino acids, 585, 624
antibiotics, 624

antioxidants, 624
carbohydrates, 624
emulsifiers, 624
enzymes, 624
glycosides, 624
heavy metals, 624
herbicides, 585
imino acids, 624
lipids, 624
nucleosides, 624
nucleotides, 624
organic acids, 585, 624
pesticides, 624
phenols, 624
preservatives, 624
purines, 524, 526, 624
saccharides, 585, 624, 627, 636, 639
seafood toxins, 624
sweeteners, 624
veterinary residues, 624
vitamins, 624
see also food analytes
food application examples, 526, 663
banana extracts, sugar content, 632
browning potential, 526
fermentation, 526
fish freshness, 524, 526, 635, 637
fruit ripeness, 629–32
meat freshness, 524, 526, 558–75, 627–9
molasses monitoring, 632
oxidative rancidity, 525–7
past bacterial activity, 524, 526
sugar inversion, 526
vegetable conditioning, 629–32
yeast contamination, 526
see also Meat freshness probe
food ingredient properties:
chirality, 625
enzyme activity, 638
microbial activity, 625, 636
mutagenicity, 585, 625
shelf-life, 625
toxicity, 638
FOP (fibre-optic probes), 598–9, 625, 649–52
colourimetric, 644, 650–2
fluorescent, 644, 650–2
interference compensation, 650
luminescent, 644, 651
stability, 650
fruit juice glucose assay, 631
general characteristics, 626
general reading, 668–9
see also references, 639–43, (654–8)
integrated sensors, 637
kinetics, 626
marker concept, 524, 526, 628
market considerations, 629, 633, 637
mediators, 625, 627, 632–3, 635
ferrocene compounds, 627, 632–3
hexacyanoferrate, 632
improved immobilization, 633
organic dyes, 632
phenazonium, 632
polypyrrole, 632–3
quinone derivatives, 632
TTF (tetrathiafulvalene), 632
membrane technology, 633–6
analytical range extension, 633–5
fouling protection, 635
interfering species protection, 633
oxygen dependence reduction, 634–6
microbial assay, 517
multi-saccharide assay, 636
multi-saccharide sensing, 627, 639
multi-variable measurement, 637
on-line probe, 632
oxygen pressure, influence, 627, 634–5
patents, 632
pH dependence, 630–1
redox couples, 625
reference devices, 626
reference (Ag/AgCl wire) electrodes, 627
response time, 633
robotic sampling and conditioning, 662–6
secondary analytes
ammonia, 623, 636
ammonium, 623
hydrogen, 636
oxygen, 623, 626, 635
oxygen/ peroxide probe comparison, 635
peroxide, 623, 626, 633–6
pH, 623, 625, 637
technology transfer, 629

Biosensors – *continued*
waste water properties:
BOD (biological oxygen demand), 585, 625
COD (chemical oxygen demand), 625
whole cells, (622)
see also Immunosensors, Chemosensors, Chemical sensors
see also Physical sensors for chemical and microbial variables
Biothermal devices, 638–9
Biscuits:
moisture calibration samples, 160
on-line moisture measurement, 138, 140, 147, 148
sorting, 100
texture assessment for, 353, 370
ultrasound measurements, on, 259, 266, 267
Bite tenderometer, 353
Black-body radiation, 178–9, 709–10
Blackcurrant juice, level measurement, 298
Blackcurrant syrup, colour, 85, 91
Bladderworm, water activity tolerance 382–4
Blanket detection in sedimentation tanks, 313–5
Bleed-line installation, 6
Blending operations control, *see* Level measurement
Blood glucose analyser, 627–32
Bloom Gelometer, 352
Blue response of photodetectors, 110
Blueberry puree, colour, 84, 90
Blueberry beverages, colour, 85, 91
BOD (biological oxygen demand), 585, 625
Boehringer enzymic test, 430
Boilers, oil-fired, monitor for, 532
adaptation, *see* Frying oil monitor
Boiling, monitoring by AEM, 245
Bone, ultrasonic properties, 248, 258
Botulism, conditions of toxin production, 383
Bound water, *see* Water, state of
Boundary, *see* Interface
Bovine bladderworm, *see* bladderworm
Brabender viscometer, 340
Bread:
problems with moisture monitoring, 147
texture assessment for, 352
water activity in, 388
Breaded products, 528
Breakfast cereals:
microwave monitoring, 191
NIR moisture monitoring, 148
sorting, 100
Brewing:
microbial assay (impedance), 490, 512
ultrasonic measurements in, 251–2, 305, 309, 312, 314, 315
Brightness, 49
Brimelow, C.J.B., 63–96
Brix, 707–8
Bromide assay, 429
Browning potential, 526
Brookfield viscometer, 340
Browning, assessment, 54–6, 88–9
Bubbles, dispersed:
influence on ultrasound propagation, 256–7, 264, 305
volume fraction and size assessment, 264, 277
Büchi Kjeldahl system, 446, 449–52
Bugmeter, 483
Building blocks of chemical sensors, 582–6
Buffer strength independence, 625, 637
Bulk density, *see* Density, bulk
Bulk properties:
determination by microwave measurements, 192
Bulk density, *see* Density, bulk
Bulk measurement:
definition, 685
Bypass installation, 6
Butter:
colour measurement, 76
composition, microwave measurement, 191, 203
Butter oil, rancidity, 525

Cadaverine, 526, 624
Cadbury Schweppes, xxvii
Cadmium, dissolved, 651
Caffeine assay, 423
Caffeine content, 154–5

Cakes:
microwave measurements, 191
water activity in, 388–9
Calcium content measurement:
significance, 24
Calcium sensors, 585, 594, 603, 606
Calibration:
electrode batch-specific, 627
intermittent in situ, 605, 632
on-line moisture measurement, 157–64
self-calibrating instruments, 605, 606
Calomel (Hg/Hg_2Cl_2) reference electrode, 589, 590, 596–7
Calorimetric biosensors, 638–9
Camembert packaging materials, 156
Camera thermometer, 173–4
Canned foods:
colour, 90
water activity in, 385
Canned meat, moisture assay, 440
Canning, 8–9
Canonical variates analysis, 366–9
CAP (controlled atmosphere packaging), 527
gas composition, 612
CAP-meat and -fish freshness, 526–8
Capacitive contact, 597, 599
Capacitance methods in microbiology, 457–517
Capacitive hygrometers, 397
Capillary electrophoresis, 423–4, 454
Capillary sampling device, LCD-display cell, 649
Capsicum:
colour of whole, 83, 89
colour of mashed, 84, 90
Caramels, water content monitoring, 142, 147
Carbohydrate assay, 430
Carbohydrate content:
on-line, NIR, 149–56
rapid off-line, 123
significance, 24
Carbohydrates, sensors, 624
Carbon dioxide:
infrared absorption, 171
see also Gases
Carbon dioxide, sensors, 600, 612
Carbon dioxide, sensors for dissolved, 651
Carbon dioxide exclusion, 603
Carbon monoxide, 613, 615
Carbonated beverages:
fruit beverage, colour, 85, 91
level measurement for, 308
microbial (ATP) assay, 461
Carbonyl value, 532
Carcasses, *see* Meat
Carotene content, 88
Carrimed rheometer, 357
Cartridge instrument, 606
Casein, colour, 80–1
Casein content, by NIR, 128, 152
Casson equation, 330
Casing materials, moisture monitoring, 149
Catalysis in food processing, ultrasound-assisted, 246
Catalysis, metal or metal alloy catalyst: *see* Chemosensors
Catalytic metals, 600, 601, 615
palladium (Pd), 585
Catering, freshness measurements, 524, 528, 533–4, 559, 714
Cauliflower, colour, 83, 89, 90
Caustic solutions, 15:
concentration by ultrasonic measurement, 251, 317
Caution time (microbial QC), 505–7
Caution zone (microbial QC), 508
Cavitation, 246–7
CCD (charge-coupled device) array, 674, 678
in sorting, 110
imaging of TIRF arrays, 649
Cell membrane (acetic acid bacteria), 637
Cellulose, *see* Casings
CEM meat analyser, 436, 437–9, 443–4
Cereals:
compositional measurement by NIR, 122, 133
fat assay, 439
microbial (impedance) assay, 509
Cfu (colony forming units), 460, 500
see also Plate count
Chain-like structure, conducting polymers, 614
Charge-charge interactions, 645

Cheeses and cheese curds:
  amines, 526
  biomass, 477
  colour, 76
  culture activity, 517
  microbial assay (impedance), 490, 512
  microwave moisture measurement, 191
  moisture and ash assay, 443
  packaging materials, 156
  rheology, 326
  texture assessment for, 370
  ultrasound measurements on, 259, 267
  water activity in, 386–7
CHEMFET (company name, not the device), 608
CHEMFET (chemically sensitive field effect transistor):
  base device for biosensors, as, 636
  gas sensors, 527, 585–6, 599–603, 606, 611–7
  schematic diagram, 602
  see also Chemosensors
Chemical assays, interpretation, 17–9
  see also QC laboratory measurement
Chemical QC laboratory, 418–55
  dedicated instruments, 424–53
    acid digestion systems for nitrogen, 445–52
    combustion systems for ash, 436, 442–3
    combustion systems for nitrogen, 452–3
    drying systems for moisture, 436–7, 440–4
    enzyme electrode analyser for sugars and alcohols, 433–5
    GLC system for pesticides, 424–7
    HPLC system for ions and sugars, 427–32
    IC system for sulphite, 432–3
    proximate analysis systems, 435–53
    solvent extraction systems for fat, 438–9, 443–5
    titration systems for moisture, 437–8, 439–40
  methodology, 437
  versatile instruments, 419–24
    capillary electrophoresis, 423–4
    gas and liquid chromatography, 420–2
    supercritical fluid chromatography, 422–3
Chemical reactions in processing:
  pH control, *see* Chemosensors
Chemical reaction (ion exchange, complexation or redox):
  analyte with sensor, *see* Chemosensors
Chemical sensors, 581–665, 668–9, 682–3
  base transducer types:
    acoustic, 652–4
    electrical, 589–5, 596, 611–8, 626–36
    microelectronic, 589, 594, 595–8, 599–618, 636–8
    optical, 598–9, 647–52
    thermal, 638–9
  brief overview, 31
  building blocks, 582–6
    base transducer, 584
    functional membranes, 584, 587–8, 633–6, 650
      see also Functional membranes
    immobilization, 584
      see also Immobilization techniques
    integral actuators, 584
    integral signal processing, 584, 660–1, 679–80, 687
    reference element, 584
    sensing layer, 584–6
  chemically sensitive layers, 583, 585–6
  comparison with other compositional measurement techniques:
    linking to polymerization, 664
    microwave measurement, 665
    near infrared measurement, 665
    ultrasound measurement, 665
  definitions of sensor types, 583, 586, 682
    see also overview graph, 663
  fouling, 584, 587–90, 634–5, 665
  functional membranes, 584, 587, 633–6, 650
  general reading, 668–9
    see also references, 618–22, 639–43, 654–8, 665–7

hygienic design, 588
interaction of sensitive layer with analyte, 583, 585–6
affinity binding, *see* Immunosensors
biocatalysis, isolated enzyme: *see* Biosensors
biocatalysis, whole microbial or tissue cells: *see* Biosensors
catalysis, metal or metal alloy catalyst: *see* Chemosensors
chemical (ion exchange, complexation or redox) reaction: *see* Chemosensors
selective sorption, *see* Immunosensors and sorption sensors
interpretation of output signals, 659–662
neural networks, 660–2
pattern analysis techniques, 616, 659–662
see also Data Processing
market considerations, 587, 629, 633, 637, 659, 681
non-contact measurement for volatiles, 665
operating conditions, 588, 659
FIA (flow injection analysis) techniques, 600, 604, 637: 639, 664
robotic sampling and conditioning, 662–6
overview, graphical, 663
poisoning, 665
sensor principles and characteristics
see Biosensors, Chemosensors
see Immunosensors and sorption sensors
see Physical sensors for chemical and microbial variables
specifications for food applications, 588
Chemically selective sorbent coatings, 644
Chemically sensitive acoustic devices, 644–7, 652–4
see also Immunosensors
Chemically sensitive dyes, *see* FOP
Chemically sensitive layers, 583, 585–6
Chemically sensitive optical devices, 644–52
see also Immunosensors
Chemically sensitive semiconductor devices, 589–618
see Biosensors, Chemosensors
Chemosensors, 581–618, 644–54, 663, 682
acidity sensors, 586, 589, 606, 609–11
dipstick device, 606, 611
sample channel device, 610
acoustic devices, 611, 652–4
see also Immunosensors and sorption sensors
actuator-sensor combination, 586, 609–11
aroma and odour assessment, 601, 612–7, 660–2
building blocks, 582–6
catalytic metals, 600, 601, 615
palladium (Pd), 585
CHEMFET, (chemically sensitive FET), 527, 585–6, 599–603, 606, 611–7
illustration, 602
see also gas sensors
coulometric microtitrator, 609–11
cross-sensitivity compensation, 599
crown ethers, 585, 597, 603
definition, 586, 682
FET (field effect transistor) device, 597–9, 601–2
FIA (flow injection analysis), 600, 604
fibre-optic pH and ion sensors, *see* FOP
field effect, 597
FOP (fibre-optic probe) chemosensors, 598–9, 649–52
colourimetric, 644, 650–2
continuous reagent supply, 652
fluorescent, 644, 650–2
glass or plastic fibres, 650
interference compensation, 650
ionic strength and pH simultaneously, 650–1
range and stability, 650, 652
stability, 650
see also ion sensors, pH sensors, toxin
GasFET, 600

Chemosensors – *continued*
gas sensors:
alcohols in headspace, 585, 611, 613, 615
aldehydes, 585, 615
amines, 615
ammonia, 585, 600, 606, 612, 615, 617, 651
aroma, 612, 616–7, 660–2
arrays and pattern recognition, 601, 611–2, 614–7, 660–2
baking end-point, 612
beverage identification, 616
broad specificity:
see MeOx gas sensors, polymer gas sensors, Sorption sensors
CAP-meat and fish freshness, 526–8
carbon dioxide, 600, 612, 651
carbon monoxide, 613, 615
chlorinated hydrocarbons, 614–5
chlorine, 614–5
combustible gases, 613, 615
cooking end-point, 612
device principles, 598–603, 611–6
dissolved gases, 599, 651
see FOP
ethanol, 617:
see also alcohols
ethyl acetate, 615
ethylene, 526, 585, 600, 612, 615, 617
fingerprinting, 616
fire-detection, 606
high selectivity:
see CHEMFET, *see* organic semiconductor gas sensors
high sensitivity:
see CHEMFET, *see* organic semiconductor gas sensors
humidity, 612, 613, 615
hydrocarbon volatiles, 613, 615
hydrogen, 526–8, 585, 600, 612, 615, 617
hydrogen sulphide, 612, 615
methane, 585, 600, 613, 615
nitrogen, 612
nitrogen dioxide, 614–5
operating temperature, 614
overview table, 615
oxidizing and reducing gases, 585, 601
oxygen, 612, 614, 615, 651
polar compounds, 614–5
pyridine, 615
receptor discrimination, 611–2
recovery time, 614
response time, 614
ripeness, 526
roasting end-point, 612, 616
spoilage, 612
sulphur dioxide, 612, 615
taint, 617
see also odour assessment
see also CHEMFET, organic semiconductor gas sensors
see also MeOx gas sensors, polymer gas sensors, Sorption sensors
see also Near infrared, Microwave and Ultrasound measurement
general reading, 668–9
see also references, 618–22, 665–7
glass electrode, 589–93
alternatives, *see* pH sensors
IC (integrated circuit), 599, 600
impedance conversion, 597–9, 601
integrated sensors, *see* smart sensors
ionophores, 597, 603
ion-selective membranes, 594, 603–4
deposition techniques, 603
ion-sensitive insulating films, 597, 599, 601, 603–4
ISE (ion-selective electrode), 585–6
ion sensors:
aluminium, dissolved, 651
cadmium, dissolved, 651
calcium, 585, 594, 603, 606
fibre-optic, 598–9
fluoride, 585, 603
gradual solid junction, 594–5
magnesium, dissolved, 651
membrane configurations, 596–7
microelectronic, 603–5
microtip, 599
nitrate/nitrite, 594, 606
potassium, 585, 594, 606
zinc, dissolved, 651
response time, 603
sodium, 585, 594, 603, 606
see also ISFET, FOP
ISFET (ion-selective FET), 589, 594, 595–8, 599–611
channel, 597, 598, 601–2

cartridge instrument, 606
differential operation, 605, 608
gate, 601–2
industrial, 606–9
isothermal point, 603, 608
pseudo-reference electrode, 602, 605
schematic diagram, 602
Severinghouse design, 600
stab probe, 607–9
see also pH sensors, ion sensors
ISFET-microtitrator, 609–11
ISFET/REFET systems, 585–6, 589, 604–5, 607–9
commercial instruments, 607–9
LB (Langmuir–Blodgett) films, 585–6, 604–5, 615–7
membrane configurations, 596–7
capacitive contact, 597, 599
galvanic contact, 596
gradual solid junction, 596
ohmic contact, 596–7
MeOx (metal oxide) semiconductor materials, 595–7, 600–1, 611–7
MeOx gas sensors, 585–6, 600–1, 611–7, 660
FET devices incorporating, 614
tin oxide, 610, 613
underlying mechanisms, 613
zinc oxide, 601, 613
see gas sensors
metal-MeOx pH and acidity sensors, 585–6, 595–7, 611
metal phthalocyanines, 614–6
microtitrator, *see* acidity
MIS (metal-insulator-semiconductor) devices, 597
MOSFET (metal-oxide-semiconductor FET), 600
multi-ion sensing:
fibre-optic, 599
microelectronic, 599, 604, 606
nose, animal, 611
nose, electronic, 612, 616
odour assessment, 601, 612–7, 660–2
organic semiconductor gas sensors, 614–5
lead phthalocyanine (PbPc), 614
oxides, electronically conducting, 595
oxides, insulating, 585, 597, 603
pH significance, 589, 609, 664
pH sensors, 589–608, 650–2
conventional (glass membrane electrode), 589–91, 592–3
conventional, armed pistol type, 609
conventional, characteristics, 590
development history, 591–4, 606–9
electronic/ionic conduction regions, connecting, 591, 594–6 599
FOP (fibre-optic probe) pH sensors, 598–9, 650–2
hydration time, 590, 604
impedance, 590, 604
industrial ISFET/REFET probe, 607–9
ion sensitivity, 608
ISFET/REFET, characteristics, 604
meat industry applications, 606, 608–9
membrane configurations, 596–7
metal/ metal oxide electrodes, 585–6, 595
microelectronic, 585, 589, 594, 595–8, 599–611
microelectronic with integral reference device: *see* ISFET/REFET
microenvironment, 589, 599, 604
molecular sieves, 617
REFET, microelectronic reference device, 605
response time, 589, 599, 603, 604
robustness, 589
self-calibrating ISFET probe, 606, 610
solid contact to conventional membrane, 591, 594–5
solid-state, 589–611
stab probes, 607, 609
stable potential, 595, 605
summary, 617
polymer (conducting) gas sensors, 611–7
polyanilines, 614–5
polypyrroles, 614–7
REFET (reference FET), 597, 605
semiconductor devices, chemically sensitive, 589–618

Chemosensors – *continued*
semiconductor gas sensors
see MeOx gas sensors, *see* organic semiconductor gas sensors, *see* polymer gas sensors
semiconductor devices, gas sensitive, *see* CHEMFET
signal to noise, 599, 604
smart sensors, 600, 604, 614, 616
solid-state sensors for pH, acidity and ions, 589–611, 618–9
solid-state sensors for gases and volatiles, 599–603, 611–8
stab probes, 607, 609
sorption coatings, *see* Immunosensors and sorption sensors
temperature-independent operation, 603–4, 608–9
test signal generation, 610
titration, real-time, 610
toxin, paralytic shellfish poison, 651
volatiles, *see* gas sensors
see also Biosensors, Chemical sensors, Immunosensors:
see also Physical sensors for chemical variables
Cherries, sorting, 100, 117
Chewing:
behaviour, 363
biting force assessment, 364
muscles activity monitoring, *see* Electromyography
Chilled foods:
microbial assay (impedance), 509, 512
water activity, 390–2
Chilled foods:
see also Meat freshness probe
Chirality, 625
Chloride assay, 428–9
Chlorinated hydrocarbons, 614–5
Chlorine, 614–5
Chocolate:
infrared emissivity of, 182
slabs, problems with monitoring, 147
ultrasound measurements in, 258, 276
see also Cocoa
Chocolate crumb:
level measurement, 299
moisture and fat content monitoring, 148, 153–4
sugar content monitoring, 154
Chocolate-based products, monitoring of water & fat content, 153
Chocolate manufacture, monitoring of water & fat content in, 153
Chocolate, molten:
enrobing material, as, 326, 330
flow metering, for, 277
rheology of 324, 326
rheological measurement for, 338, 343
texture assessment for, 370
Christel Texture Meter, 352
Chroma, *see* Colour description systems
Chromatographic coatings in sensors, 660
Chromatography, *see* GLC, HPIC, HPLC, HRGC, IC, SFC
see also Columns
Chrompack, 421
CIE (Commission Internationale de l'Eclairage):
see Colour description systems
CIELAB space, *see* Colour description systems
CIP, *see* Cleaning-in-place
CISE (cell sensitized ion-selective electrode), 585–6
Citric acid crystals, moisture measurement, 147
Citrus juice, quality, 65
see also Orange juice
Clamp-on transducers, 304, 305–6, 311, 318
Clark-type gas probe:
basis for biosensors, 623, 626, 635–6
Clark device, planar micro device, 623, 635–6
Cleaning, automatic of on-line instruments, 112, 141–2
see also Self-cleaning transducers
Cleaning, ultrasound-assisted, 246
Cleaning-in-place, 15
monitoring of cycles by ultrasound, 251–2
monitoring of spray impact by AEM, 245
see also Caustic solutions

Cleaning-in-place tolerance to, 192, 298, 306
  mechanically resonant probes, 548
  microwave instrumentation compatibility, 192
  viscometry, 342
Coagulation, *see* Agglutination
Coal, microwave moisture measurement, for, 221
Coated wire electrodes, 595
Coating thickness (confectionery), 326, 330
Coating weight (packaging), 156
Coatings, rheology of confectionery, 326, 330
Coatings, chemically selective sorbent, 644
Cobalt phthalocyanine, 614
Cocoa, microbial assay (impedance), 510
Cocoa butter, content monitoring, 148
Cocoa powders, moisture monitoring, 148
COD (chemical oxygen demand), 625
Coextrudates, 157
Coffee, assays for:
  caffeine assay, 423
  moisture assay, 437
Coffee beans:
  moisture monitoring, 148
  sorting, 100, 102–4, 117
  spectral curves, 102–3
Coffee, extract:
  concentration monitoring, 317
Coffee, ground:
  caffeine monitoring, 154–5
  colour, 80–1
  moisture monitoring, 148
Coffee, instant:
  microwave moisture monitoring, 209
  moisture assay, 18–9
  moisture calibration, 160
  NIR moisture monitoring, 141, 148
Coffee roasting, odour assessment, 612–3
Collagen, *see* Casings
Colloidal particles, settling, 277
Colour, relationship to quality, 98
Colours, wavelengths, 709
Colour as indicator:
  carotene content, 88
  deterioration, 44
  frying oil quality, 529–32, 552–3
  heat treatment, 90
  jelly ageing, 89
  kernel maturity, 82
  pigment content, 64–5, 80, 88–9
  ripening, 44, 89, 289
  shelf-life, 91
  storage effects on yoghurt, 91
Colour as quality attribute, 64
  assessment environments, 64–5
Colour blindness, 709
Colour description systems, 45–51, 66–8
  chroma, 48–51
  chromaticity coordinates, 46, 48, 51
  chromaticity diagram, CIE, 48, 68
  chromaticness coordinates, 51
  CIE, 40, 65
  CIE 2° standard observer, 42, 46, 65–6
  CIE 10° standard observer, 42, 46
  CIELAB space, 49–51
  CIELUV space, 49
  colour differences, 50
    maximum acceptable, 68
    small differences, scales for, 49, 68
  colour matching functions, *see* CIE standard observer
  hue angle, 48–51, 67, 70
  Hunter space, 49, 51
  illuminants, 46–8
    colour rendering index, 47, 55
    colour temperature, 46
  Judd-Hunter solid, 67, 69
  lamps, *see* illuminants
  lightness coordinate, 48–51, 67, 70
  lightness stimulus, 46
  luminous reflectance and transmittance, 45, 51
  Munsell colours, 48, 49, 50
  objective terms, 49
  opponent coordinates, 48–51
  overview, 51
  primaries, 46, 65
  saturation, 49, 67, 70
  standard observer, *see* CIE
  subjective terms, 49
  uniform colour space, 49–51, 66
  see also Colour vision
Colour image, sorting by, 104, 116
Colour measurement, 39–93
  absorption coefficient, 45, 58

Colour measurement – *continued*
aperture, *see* geometry
colour standards, 69
diagnostic, 71
hitching-post, 70, 90–1
instrument, 70
master, 70
perfect white diffuser, 69, 80
primary, 69
secondary, 70–1
maintenance of, 71
specific, 71
traceability, 71
working, 70
colour tiles, *see* colour standards
enhancement of colour differences, 73, 80
examples:
liquid foods, 91–2
meat and produce, 88–9
orange juice concentration, 56–9
particulate and lumpy solid foods, 87–8
pastes and slurries, 90–1
pigment oxidation in fresh meat, 54–6, 88–9
powdered, granulated and flaked foods, 80–7
food structure, relevance, 44–5
geometry, 45, 52–3, 56, 72, 78–9
glossy objects, 45, 52, 72, 79
hand-held instruments
hazy samples, 56–8, 72
illuminants, 46–8, 50–3, 56, 78
influence, meat display, 54–6
influence, orange juice assessment, 56–9
prediction, for other illuminants, 79
in-line, 52
influence:
browning, 73
film-formation, 73
particle size, 75, 80
settling, 73
temperature and humidity, 73, 91
see also thermochromicity
integrating sphere, 52
Kubelka-Munk coefficients, 45, 56, 90
large area solid foods (meat and produce), for, 54–6, 76, 83: 88–9
liquid foods, for, 56–9, 77–8, 85, 91–2
Lovibond tintometer, 50
methodology, 68–80
instrument standardization, 69–71
instrument variables setting, 78–9
interpretation, 92–3
sample selection, preparation and presentation, 72–8, 92
signal processing, 79–80
metamerism, 78
oily materials, 76
opaque objects, 45, 52, 72, 76, 79
optical classification, 78–9
particulate and lumpy solid foods, for, 76, 82, 87–8
pastes and slurries, for, 77, 84, 90–1
photoelectric instruments, 50–3, 66–7
pigment content, rapid assessment of, 80–1, 88–91, 92–3
powdered, granulated and flaked foods, for, 75–6, 80–1, 87
principles, 39–59, 65–8
reasons, for, 64
scatter coefficient, 45, 58
significance, 24
small colour differences for, 52
small samples, 75
spectrophotometers, 50–3
commercial instruments, 86
reflectance instruments, 52, 88–9
specular component, 45, 52–3, 79
sphere-type instruments, 77
thermochromicity, consideration of, 71, 77
translucent objects, 45, 53, 56, 72, 76–7, 79, 90–1
transparent objects, 72, 77, 78, 89, 91
trichromatic colorimeters, 50–3, 66
commercial instruments, 86
early instrument, 50
reflectance instrumentation, 66–7
transmittance measurement, 77, 89, 91
tristimulus colorimeters, *see* trichromatic colorimeters
viewing angle, *see* geometry
visual assessment, 50, 78
relationship to measurement,

examples, 54–9
wavelength interval, 53
see also Appearance
see also Colour description systems
see also Colour vision
Colour, relationship to pH for meat, 609
Colour perception, 98
see also Colour vision
Colour scales, *see* Colour description systems
Colour sorting, *see* Sorting by colour
Colour space, *see* Colour description systems
Colour television cameras, 104, 116
Colour temperature, 44, 46, 56
Colour, translucent materials (CTM) probe, 609
Colour vision, 40–4, 49
adaptation, 43–4, 64
ambient light, influence of, 43–5
brightness, 49
chroma, 49
colour constancy, 43
colourfulness, 43, 49
contrast, 43
dullness, 55
fading, 54
fidelity, 56
flattery, 55–6
hue, 49
human eye:
cone sensitivity curves, 42
photoptic cones, 42, 65
scotoptic rods, 42, 65
structure, 40–2
lightness, 43, 49
opponent mechanism, 43
light/dark, 43
blue/yellow, 43
red/green, 43
red enhancement, 55
saturation, 49
subjective colour terms, 49
trichromatic detection in the eye, 42–3, 65–6
see also Appearance
Colourmaster, 86
Columns for HPLC, 428
Combustible gases, 613, 615
Combustion systems for ash, 436, 442–3
Combustion systems for nitrogen, 452–3
Compact instrument designs: *see* Configuration
Complex formation, affinity binding, 645
Complexation, analyte-sensor, *see* Chemosensors, Crown ethers
Compliance, related measurements, 24–5
Compositional measurements:
chemical sensors versus physical sensors, 664–5
linking to polymerization, 664
on-line monitoring
comparison of methods, 264–6
microwave, by, *see* Microwave measurements
near infrared, by, 149–56
on-line/off-line comparison, 122–3, 128
ultrasound, by *see* Ultrasonic measurements
rapid at-line by NIR, 123
techniques:
see Biosensors
see Chemical sensors
see Chemosensors
see Microwave measurement
see Near infrared measurement
see Ultrasound measurement
see also Concentration, Water content, Moisture
Compressibility, 248
Computer vision systems, 114–6
Computing and microprocessor applications, 40, 52, 66, 79, 101–2, 113–6, 133, 164, 173, 207, 209, 291, 294, 296, 299, 301, 360, 366, 389, 397, 491–2
see also Data processing
Computrac analyser, 441–2
Concentration by microwave measurement:
dissolved sucrose, for, 196, 199
Concentration by ultrasonic measurement:
concentration, solutes of, 244, 250–6, 317
concentration, suspended solids, of, 256–7, 277, 312–5
Concentration units, 707

Concentric cylinder viscometer, 337
Condensation:
  interference in microwave measurement, 192
Conditioner control, 123
Conditioning of vegetables, 629–32
Conductance methods in microbiology, 457–517
Conductance of liquids, measurement, *see* Electrical impedance of liquids
Conductimetric detector:
  HPLC in, 428
Conducting polymer gas sensors, 611–7
Conductivity, electrical, *see* DC conductivity
Conductivity hygrometers, 399–403
Conductivity measurement, 33
Cone and Plate viscometers, 338–9
Confectionery:
  microbial assay (impedance), 490, 509, 512
  microwave moisture measurement, 191
  moisture assay, 438
  problems with moisture measurement, 147
  sorting, 100
Configuration examples for sensors:
  cartridge multi-ion sensor instrument, 606
  credit-card type, 627, 629
  dipstick, 27, 309, 606, 611, 714
  disposable strip electrodes, 627–8
  in-line monitor:
    biosensors, 632
    see also Near infrared measurement
    see also Mechanically resonant probes
    see also Microwave measurement
    see also Sorting by colour
    see also Ultrasonic measurement
  knife type, 572, 629
  needle type, 635–6
  pen type, 628
  pistol type, 609
  stab probe, 607–9
Consistent product:
  achieving, 10, 64
  relevance, 2
  system for, 10
  see also Repeatability
Constants, 697
Construction and packaging materials:
  infrared emissivity data, for, 182
  ultrasound data for, 258–9, 275
Contamination, detection and removal:
  see Sorting by colour
Contrast, of grey shades, 43
Controlled-release polymer, continuous reagent supply, 652
Controllers, fuzzy-logic, 687
Conveyor belts, instrumentation in, *see* NIR, MW, US, ColSort
Cook value, 9
Cooker extruder:
  see Extrusion cooker
Cooking end-point, volatiles, 612, 647, 654
Coolant supply monitoring, 308
CoPc (copper phthalocyanine), 614
Copper as catalyst of fat oxidation, 548
Copper phthalocyanine, 614
Coriolis flow meter, 27
Corn starch, modified, concentration monitoring, 317
COST90 (COST: European Coop. in Scientific and Technological Research; (9: Food, 0: First project/physical properties), xxviii, xxix
Cottage cheese, moisture measurement, 147
Cottonseed oil, 537, 539–40, 544
Cotyledons, *see* Hulls, 88
Coulometric microtitrator, 609–11
Counting of items passing, 301
Cranberries, sorting, 104
Cranfield Biotechnology Centre, xxv, 559, 561, 563–70, 575–6: 627–33
Cream:
  microbial assay (impedance), 490
  ultrasonic properties, of, 278–80, 310
  viscosity, 706
Cream layers, detection, 316
Creamers, non-dairy:
  fat and moisture monitoring, 153–4
  moisture monitoring, 148
  sugar content monitoring, 154
Creaming, *see* Droplets

Creatinine, 624, 636
Crispness, 266, 359
Critical angle, 647
Cross-checking of calibration used on-line, 5, 18–9
Cross-correlation flow meters, 278
Cross-sensitivity compensation, 679–80
Cross-sensitivity (ions) compensation, 599
Crown ethers, 585, 597, 603
Crunchiness, 359
Cryogenic liquid, level measurement, 308
Cryoscopic method of water activity meas., 403
Crystal microbalance, 644
Crystal seeds, volume fraction of dispersed:
  see Solid/liquid ratio
Crystal size, relationship to rheology, 326
Crystalline materials, NIR monitoring for, 147
Crystallization:
  ultrasound-assisted, 246
  ultrasonic monitoring of, 269
CTM (colour of translucent materials) probe, 609
Cured products:
  phosphate assay, 428
  see also Meat products
Curing:
  as preservation process, 378, 381–6, 389
  control of, 402
  prediction, raw ham, 609
  ultrasound-assisted, 246
CuPc (copper phthalocyanine), 614
Customer specifications, 24–5
CV (carbonyl value), 532
Cyanidin–3-galactoside content, 89
Cyclic compounds, 529
Cyclic voltammetry, 431
Cyclone separator control, 143

Dairy products, *see* Milk, Milk products, Cream, Cheese, Yoghurt
Damaged food particles, detection and removal:
  see Sorting by colour
Damp materials, sample scanning NIRA for, 135
Dampness as indicator of ripening, 389
Data processing approaches, 20
  adaptive, 660
  algorithms, 660
  canonical variates analysis, 366–9
  expert systems, 11, 659–60, 687
  fuzzy logic, 659–60, 686–9
  learning systems, 660
  mathematical models, 20, 689
  multi-variate analysis, 366–9, 660–3
  neural networks, 20, 246, 616, 660–2, 689
    potential pitfalls, 661–2
  pattern analysis techniques, 616, 659–662
  plasticity, 661
  PLS (partial least square models), 660
Db, *see* Decibel
DC (direct current) conductivity:
  contribution to dielectric loss, 201
  non-contact measurement of, 202, 231
Decaffeination control, 155
Decibel, 712
Defective level (microbial QC), 506
DEFT (direct epifluorescent filter technique), 461–3
Degassing, ultrasound-assisted, 246
Dehydrated vegetables, sorting, 100
Denbow, N., xxviii, 289–320
Density, bulk:
  interference from changes of in microwave measurements, 200–1
  microwave measurement of, 192, 201, 204–11
  significance, 24
  ultrasonic measurement of, 250, 270
  vibrating rod probe, 27
  vibrating tube densitometer, 27
Density and moisture measurement, by microwave, 204, 205–11
Density determination by Gamma-ray absorption, 206
Density (specific gravity):
  frying oils, 546, 548
Denture tenderometer, 353

Deposition techniques, ion-sensitive membranes, 603
MBE (molecular beam epitaxy), 615
see also LB films
Derivative spectral analysis, 134
Detection time (microbial assay), 485–7, 491–2, 500, 506
Detectors, on-line instrumentation, 674
see Infrared detectors
see Photodetectors
Detectors used with chemical laboratory instruments
HPLC systems, 421
conductimetric, 428
pulsed amperometric, 428
refractive index, 431
UV-VIS, 428
GLC systems:
alkali thermal ionization detector (ATD), 426–7
electron capture detector (ECD), 426–7
Detergent powders, moisture measurement, 137
Development of novel instrumentation:
contributions to development, 558, 575
examples, 528–76, 714
history, pH sensors, 591–4, 606–9
market consideration, 557, 574, 587, 629, 633, 637, 659, 681
phases of development, 556–7, 573–4
Dew-point hygrometer, 395, 402–3
Dextran label, 651
DFD (dark firm dry) beef, 608
Diabetes, 627
Dielectric constant (permittivity), 194
frying oil quality, 531–2
high frequency value, 194
Dielectric properties:
see also Microwave measurement
alcohols, data for, 196
binding of water, effect on relaxation properties, 196, 218
complex permittivity, 194, 210
critical frequency, 195
data for polar liquids, 196
Debye dispersion, 195
density, bulk, influence of, 192, 201
dispersion, 194–8
dispersion of water in food, 196–200
ice, of, 198
loss factor, 194, 210
loss tangent, 194
permittivity, 194, 210
refractive index, 194
relaxation time, 194, 196, 200
rotational hindering of water dipoles, effect on, 201
state, solid-liquid, influence on diel. relaxation time, 196
influence of viscosity on dielectric relaxation time, 196
water, pure of, 194
water in foods, of, 194–200
water in foods containing protein, 198, 200
Dielectric material:
polar, 193, 196
polarization, of, 193
water, as, 195
Dielectric spectroscopy, 196
Dielectric surface properties, 644
Diffration grating, 647
Diffusion control in biosensors, 626, 633–5
Diffusion rate, glucose in meat, 572
Dimethylamine content, 524
Dionex instruments, 424–32, 631
Dipole layer, electrode surface, 465
Dipstick (level measurement), ultrasonic replacement, 309
Dipstick probes:
chemical sensor, acidity, 606, 611
mechanical resonance, 27
ultrasonic, 309
see also Microwave measurement, Stripline
Direct biosensors, 626
Disaccharides concentration:
relationship to speed of sound, 251–4
see also Sugar solutions
Discoloured food particles, detection and removal:
see Sorting by colour
Discrimination of volatiles, molecular size, by, 644, 645
molecular shape, by, 645

polarity, by, 644, 660
Disposable cartridge instrument, 606
Disposable enzyme electrodes, 627–32
Disposable tip FOP, 650
Dissolved gases, 651
Dissolved metal ions (aluminium, cadmium, magnesium, zinc), 651
Distillery wastes, water content monitoring, 147, 148
DNA (deoxyribo-nucleic acid) probes, 585, 624
Doppler flow meter, 277, 304–7
portable, 305
time-gated, 278
Dough, rheology of, 327
Doughnuts, 528
Dried products, NIR measurements for, 138, 143, 147, 161–3
Droplet size, relationship to rheology, 326, 327
Droplets, dispersed, influence on ultrasound prop., 256–7, 264, 277
Droplet sizing, 278–280
significance, 24
Drugs, 645, *see* also Veterinary residues
Drying:
as preservation process, 378, 389
ultrasound-assisted, 246–7
Drying processes, monitoring of:
AEM, by, 245
NIR, by, 123
see also Moisture measurement
see also Water activity
Drying systems for moisture assay, 436–7, 440–4
Drying technology optimization
role of water activity profile, 393
Dual monochromatic sorting, 103–4, 117
Dublin, University College, xxxi
DU COLOR, 86
Dullness, 55
Durum semolina, *see* Semolina
Dust explosion, 15–6
Dust extraction systems, 112
Dust tolerance, 16 , 128, 132, 141, 192, 301
Dyes, chemically sensitive, *see* FOP
ECD (electron capture detector), 426–7
Edible oils:
see Oils, Frying oils
Effervescence monitoring by AEM, 245
Effluent, *see* Waste water
Egg albumen, *see* egg white
Egg shell thickness, 267
Egg white thickness, determination by ultrasound, 254–5, 260, 267
Egg yolk, colour, 77, 85, 91–2
Elastic modulus, 247
Elasticity as indicator of ripening, 389
ELBOS (electrolyte/LB-film/oxide/ semiconductor) device, 605
Electrical base transducers, 589–5, 596, 611–8, 626–36
Electrical conductance/ capacitance, microbial assay, 457–517
Electrical impedance methods for microbial assay, 457–517
see Impedance microbiology
Electrical impedance of liquids:
instrument design for measuring, 478–83
principles of measurement, 464–73
Electrical impedance of fluid-electrode system, 464–9
capacitance, electrode , 465
dipole layer, electrode, 465
oxide layer, electrode, 465
phase angle, 466
reactance, electrode, 466
resistance, fluid, 465
Electrical signals from chewing muscles, *see* Electromyography
Electroactive species, 627
ascorbic acid, 627
Electrochemical biosensors, 626–38
Electrode strips, 627–31
Electrodes:
enzyme, *see* Enzyme electrode analyser, Biosensors
gold, 431
Electrode materials in impedance measurement, 467–9, 516
oxide formation, 467
platinum, 467–9
polarization effects, 467, 469
stainless steel, 467–9
Electrolytic cells for water activity meas., 399

Electromagnetic induction, 33
Electromagnetic noise, immunity, 599
Electromagnetic measurement applications:
  comparison with ultrasonic measurements, 240–2
Electromagnetic spectrum, 29–30, 32–3, 123–4, 241
Electromagnetic waves, 125, 665
  non-contact measurement with, 16
  overview table, 29–30, 32–3, 241
Electromyography (EMG), 359–71
  choice of subjects, 363
  comparison with other methods of texture assessment, 371
  data interpretation, 364–8
  multivariate analysis, 366–9
  parameters, 362
  reproducibility, 363–4
Electron capture detector (ECD), 426–7
Electron mediators, *see* mediators
Electronic gating, 140
Electronic/ionic conduction regions, connecting, 591, 594–6, 599
Electronic nose, 612, 616
Electronic sorting, *see* Sorting by colour
Electro-osmosis, 423
Electrostatic charges, 15–6
Electrostimulation control, 608
EMG, *see* Electromyography
Emissivity, *see* Infrared thermometry
Emulsifiers, 624
Emulsifier sprays, influence on fat distribution, 156
Emulsions:
  considerations in viscometry for, 334
  distinction from single-phase liquids, 313, 316
  monitoring of:
    see Ultrasound scattering
    see also Droplets
    see also Interface
  specific food emulsions, *see* entries for named foods
  ultrasound scattering, in, 278–280
  ultrasonic determ. of droplet size and volume fraction, 278–80
  ultrasonic treatment of, 248
Endotoxin, 652
Energy saving, 389
Energy units, conversion, 701
ENFET (enzyme-sensitized field effect transistor), 585–6, 623, 625, 637–8
  see Biosensors
ENFOP (enzyme-sensitized fibre-optic probe), 585–6
ENISE (enzyme-sensitized ion-selective probe), 585–6
Enrobing thickness, 326, 330
Enterobacteriaceae:
  hydrogen production, 483
  microbial assay, *see* Impedance microbiology
Enterotoxins, production conditions, 383
ENTHERM (enzyme-sensitized thermistor or thermopile):
  585–6, 627, 638–9
Enzymes, 585–6, 622, 624
  oxidases, 627
  whole cells, 585–6, 622, 637
Enzyme activity, 638
Enzyme electrode analyser for sugars and alcohols, 433–5
Enzyme-related deterioration:
  water activity, role of, 377–8
    Enzyme substrates and inhibitors as analytes, 623, 624
Enzymic test, 430
Epoxy lacquer thickness, 156
Equilibration of biosensor current, 627, 632
Equilibrium relative humidity, 21–2
  see also Water activity
Ergotamine assay, 423
ESR (Electron spin resonance), 26
Esters, ultrasonic properties, 273, 274
Ethanol, 526, 617, 624, 627, 635
  see also Alcohols
Ethanol without methanol response, 637
Ethyl acetate, 615
Ethylene, 526, 585, 600, 612, 615, 617
Ethylene vinyl alcohol film thickness, 157
Evanescent wave devices, 647–9
Eves, A., xxviii, 349–74
EVOH, *see* Ethylene vinyl alcohol
ExacTech, 627–32
Exotherm reactions, monitoring, 638–9
Expert systems, 11, 659–60

fuzzy-assisted, 687
Explosives, 652
Extended analytical range of biosensors, 632, 633–5
Extraction, ultrasound-assisted, 246
Extruded products, NIR measurements for, 147
Extrusion cooker, 15, 327
control by NIR measurement, 147
moisture determination in, 205, 224
performance monitoring by AEM, 245
thermometry in, 186
Extrusion tests, for texture assessment, 352

F_0 value, 8–9, 376
FAC (fatty acid composition), 551
Fading, 54
Falling cylinder viscometer, 342
Fast response, sensors with, 676
Fat assay:
hydrolytic methods for complexed fats, 439
notes, 435, 436
solvent extraction systems, 438–9, 443–5
problems with complexed or unsaturated fats, 439
Fat content, measurement:
microwave on-line, 196
NIR on-line, 140, 149–56
rapid off-line, 123, 128
relationship to water activity, 389
significance, 24
see also Emulsions, Meat
Fat distribution, influence of emulsifier sprays, 156
Fat, *see* also Fouling
Fat-rich foods:
shelf-life in frozen storage, 531
Fats:
infrared emissivity data for, 182
solid/liquid ratio determination, 257
texture assessment for, 352
ultrasound propagation in, 257, 260
Fatty acid composition, 551
Fatty acids, free, *see* FFA
Fatty acid methyl ester assay, 427
Fatty samples:
colour measurement, 76
microbial assay, 509
moisture assay, 437
Fatty tissues:
colour, 76–7
ultrasound data, for, 259
water activity of, 387
Federal Centre for Meat Research, xxxi
Feedforward control, 19
Feed systems, in colour sorters, 98–100, 111
Feeds:
see Animal feeds
see also Cereals
Fermentation, 379, 526
control of, 389
see also Alcohols, Biomass, Brewing, Yeast, Yoghurt
Ferrocene compounds, 627, 632–3
Fermenters:
see Biomass
FET (field effect transistor) device, 597–9, 601–2
FFA (free fatty acids):
content, 529–32, 536, 537–9, 551–3
formation, 528
FIA (flow injection analysis) techniques, 600, 604, 637, 639, 664
Fibre-optic cables:
access to hostile environments, 186–8
glass, 188
silica, 188
zirconium fluoride, 188
Fibre-optic immunosensor devices:
SPR or TIR, 644, 651
TIRF, 649
Fibre-optic probes:
covalent bonding, 598
pH and multi-ion sensing, 598–9
polymer films, grown onto, 598
thermometers in microwave ovens
Fibre-optic probes (FOP), chemically sensitive:
see FOP chemical sensors
see Fibre-optic immunosensor devices
Fibre-optic probes (FOP), extrinsic, 650
Fibre-optic probes, pH and ions, 598–9, 650–2
see also FOP chemical sensors

Fibre-optic sensors, intrinsic, 649–50
Fibre-optic thermometer, fluorescent, 599, 650
Fibrousness, 352
Fibrous materials, NIR monitoring of, 147
FID (flame ionization detection), 422
Fidelity, 56
Field effect, 597
Figaro sensor, 601, 613
Fill level indication, *see* Level measurement
Film thickness measurement, 156–7
Filters, *see* Optical filters
Fingerprinting, 616
FIRA Jelly Tester, 352
FIRA-NIRD Extruder, 352
Fire-detection, 606
Firefly luciferase, 651
Firmness, 352
First/second/third generation biosensors, 626
Fish:
  amines, 524
  ATP, 524
  CAP-packed, freshness, 526–8
  freshness, 486, 524, 526, 635, 637
  glucose, 524
  histamine in dark-fleshed, 524
  hypoxanthine, 524
  infrared emissivity data, for, 182
  inosine, 524
  microbial assay (impedance), 487, 489, 490, 493, 509, 512
  packaging materials, 157
  purines, 524
  shelf-life prediction, 515
  ultrasound data, for, 259
Fish cakes, 528
Fish products:
  batter-coated or breaded, 528
Fishmeal:
  microwave moisture monitoring, 200, 207, 209
  NIR moisture monitoring, 148
  water content determination, 200, 207
Fisons, 421
Flaked foods, *see* Powdered foods
Flaky materials, NIR monitoring of, 147
Flame ionization detection (FID), 422
Flattery, 55–6
Flavour, role of microbes, 379
Flavour compounds assay, 427
  see also Aroma
Flavour extraction control, 316
Flavour preparations:
  moisture assay, 438
Flavours and off-flavours:
  fried foods, 528–9, 552
  see also Acidity
Flaw detector, 243
Flour:
  moisture assay, 437, 442
  moisture monitoring, 148
  nitrogen/protein assay, 453
Flow curve, 328, 329
Flow profile, influence on flow metering, 278
Flow rate profiling, by time-gated Doppler flow meter, 278
Flow rate of air, control in fermentation, 389
Flow rate of liquids, ultrasonic meas., 242–3, 277–8, 302–6, 318
  clamp-on transducers, in, 304
  Cross-correlation flow meter, 278
  Doppler meter, 277, 304–6
  solids loading, influence on, 303, 304, 306
  Transit-time (time-of-flight) flow meter, 278, 303–4, 318
Flow rate of solids by AEM, 245
Flow rate by pressure drop, 332
Flow velocity profile, influence of rheology, 326
Fluid bed dryer control, 143
Fluorescence techniques, in sorting, 104
Fluorescein, 651
Fluorescence-based biosensors, 625, 651
Fluorescence measurement, 649, 650
Fluorescent rare earth compounds, 650
Fluoride sensors, 585, 603
FMBRA (Flour Milling & Baking Res. Assoc.) Biscuit Texture Meter, 353
FMC Pea Tenderometer, 352
Foam:
  distinction from liquid, 308–9
  influence on ultrasound propagation, 277

Foam persistence, 531–2
Fondants, moisture monitoring, 147
Food poisoning, 379:
  see also Micro-organisms, growth of
Food processing, instrument compatibility, 13
Food quality, *see* Quality
Food safety, *see* Safety
Food spoilage, *see* Spoilage
Food stability, *see* Stability
Food structure, *see* Structure
FOP (fibre-optic probe) chemical sensors, 598–9, 649–52
  bioluminescence sensors:
    ATP (adenosine triphosphate), 651
    glutamate, 651
    NADH (nicotinamide adenine dinucleotide), 651
  biosensors, 625
    ethanol, 651
    glucose, 651
    lactate, 651
    pyruvate, 651
  chemosensors:
    dissolved gases (ammonia, carbon dioxide, oxygen), 651
    dissolved metal ions (alum., cadmium, magnesium, zinc), 651
    paralytic shellfish poison, 651
  detection modes:
    colourimetric, 644, 650–2
    fluorescent, 644, 650–2
    interference compensation, 650
    luminescent, 644, 651
    stability, 650
  immunosensors:
    ConA (concanavalin A) binding agent
    continuous antibody supply, 651–2
    glucose analysis with ConA, 651
    see also FOP (fibre-optic probes), chemically sensitive, pH sensors, 598–9, 650–2
    continuous reagent supply, 652
    glass or plastic fibres, 650
    ionic strength and pH simultaneously, 650–1
    range and stability, 650, 652
Force-distance relationship, 354
Force units, conversion, 702
Foreign body hazard, 590
  elimination in in-line pH-measurement, 589
Foreign matter, *see* Sorting by colour
Foss-Let fat analyser, 436, 439, 445
Fouling of sensors, 584, 587–90, 634–5, 665
  approaches to reduction, 589
    see Non-contact measurements, Functional membranes
Fourier transform infrared analysis, 125–6
Fourier transform analysis, in NIR measurement, 134
Frequency ranges:
  near- and mid-infrared, 709
  NIR spectroscopy, 123–4
  overview, 29–30, 33
Fragrance extraction control, 316
Frankfurter, water activity of, 385
Freezing:
  frozen/liquid water ratio, 266
  preservation process, as, 378
  proportion of ice monitoring, 266
  see also Cryogenic liquids
Freezing point, as indicator of water activity, 392
French fries (US), Potato chips (UK), 528
Free fatty acids, *see* FFA
Freeze-thaw information, 572
Frequency ranges:
  electromagnetic waves, 240–2
  sound waves, 240–2
Freshness assessment, 523
  eggs of, 254–5
  significance of measurement, 24–5
  see Microbial spoilage, Rancidity, Frying oil degradation
  see Meat freshness probe, Frying oil monitor
  see Marker concept
Fritters, 528
Frozen desserts, texture assessment for, 370
Frozen foods:
  colour, 90
  deterioration of high fat, 531
  microbial assay (impedance), 509
  progress of rancidity, 531
  sorting, 100, 117
  water activity of, 390–2

Fructose, 624, 639
Fructose assay, 430
Fructose concentration:
  relationship to speed of sound, 251–5
  see also sugar solutions
Fruit:
  colour, 76–7, 83, 88–9
  freezing, measurements in controlling, 308–9
  ripeness assessment, 44, 89, 266, 277, 526, 629–32
  skin texture, 277
  sorting, 100, 117
  texture assessment for, 352
  water activity in, 385, 388
  see also entries for specific fruits
Fruit jelly, *see* Jelly, Gels
Fruit juice, *see* Juice
Fruit serum, colour, 72
Frying, deep fat:
  catering, 528
    batter-coated or breaded fish or meat products
    doughnuts
    fritters
    potato chips (UK)/ French fries (US)
  freezer storage of fried food, rancidity, 531
  industrial, 528
    fish cakes
    potato crisps (UK)/ chips (US)
    potato chips (US)/ French fries (US)
  temperatures, 529
  see Frying oil monitor
  see also Browning potential
Frying oils:
  density, 546
  free fatty acids, 165
  polyunsaturated, 528
  palm olein, 529, 551
  PHVO (partially hydrogenated vegetable oil), 529, 551
  sunflower oil, 529, 551
  viscosity, *see* Frying oil viscosity
Frying oil degradation, heat-induced, 523–6, 528
  assays, sophisticated:
    see Frying oil quality indicators
  batch/ continuous frying comparison, 531
  density change, 548
  cyclic compounds, 529
  copper as catalyst, 548
  flavours and off-flavours, 528–9, 552
  free fatty acids formation, 528
  hydroperoxide formation and decomposition, 528
  in situ probe, *see* Frying oil monitor
  oxidation, 528–9
  polymerization, 528–9
  silicone additive, 529, 531, 551
  simple tests, *see* Frying oil quality indicators
  viscosity increase, 529
    see also Frying oil viscosity
  volatile formation, 528
Frying oil monitor, in situ, 525, 526, 528–58, 575–6, 714
  advantages, 548–9
  bandwidth output, 546–7
  calibration, 548–9, 555, 714
  cleaning, 548
  compensation for oil temperature, 549
  concept, 531–2
  contributions to development, 558
  development phases, 556–7
  feasibility study, 532–46, 556
    data used, 537–42
    reference limit values used, 536
    summary graph, 543
    variation with laboratory and oil type, 544–5
    see also Frying oil quality indicators
    see also Frying oil viscosity
  need for, 528–31
  pre-production model, photo, 714
  probe adapted from, 533, 535
  probe characteristics, prototypes 0 to 1, 545–9
  prototypes, photo and diagram, 547
    see also pre-production model, 714
  prototype 1, protocol, 549–51
  prototype 2, 714
  ruggedness, 546
  specifications, 545
    formulation, 532–45, 556
  trial with prototype 1, conclusions, 553
    oils and fried food used, 551–2

protocol for prototype 1, 549–51
reference method (POM) and other oil assessments used, 552
results, 530, 552–5
trial with prototype 2, 714
tuning fork-type operation, 546
update, 714
viscosity and density dependence, 546, 548
Frying oil quality indicators:
AnV (anisidine value), 532
AV (acid value), 532, 536, 539
colour indication, 529–32, 552–3
CV (carbonyl value), 532
dielectric constant (permittivity), 531–2
FAC (fatty acid composition), 551
FFA (free fatty acids content), 529–32, 536, 537–9, 551–3
foam persistence, 531–2
hydroperoxide content, 531
IV (iodine value), 532, 538, 540, 551
limit values, 533, 536
NAF (non-urea-adducting fatty acid content), 532, 534, 536: 537, 539–40
NAFE (NAF esters), 539, 541
NMMCC (new chemical compounds content), 532
odour, 531–2
OFA (oxidised fatty acids), 539, 541
PC (polar components content), 532–4, 536, 538–9
PFA (polymeric fatty acids content), 532
PIOFA (petroleum-ether-insoluble oxidized fatty acids content), 532, 536, 539–40
Pol (polymeric triglycerides content), 532, 536, 537, 539–40
POM (polymerized and oxidized fatty material.), 529–32: 552–5, 714
PV (peroxide value), 531–2
SP (smoke point), 529, 531–2, 539
taste, 531–2, 552
viscosity, *see* Frying oil viscosity
Frying oil viscosity, 529, 532–4, 537–45
bakery fats, 534
catering fats, 533–4
cottonseed oil, 537, 539–40, 544
dependence on temperature, degradation and oil type, 532–45
summary graph, fresh and discard-point oils, 543
variation with laboratory and oil type, 544–5
groundnut oil, 537–40, 544
hydrogenated fats, 537–8
hydrogenated lard, 534
lard, 537–9
maize oil, 538–40, 544
olive oil, 537–40, 544
palm oil, 538, 545
palm oil fraction, 539
palm olein, 538, 545
peanut oil, *see* groundnut oil
rapeseed oil, (probably LEAR: low erucic acid rapes.) 539, 545
safflower oil, 539, 544
severely degraded oils, 533
soya-bean oil, 537–40, 544
sunflower oil, 537–8, 540, 544
tabulated data, 537–42
temperature dependence, 533, 543, 550
trilinolein, 539
see also Frying oil monitor
FTIR, *see* Fourier transform infrared analysis
Functional membranes, 584, 587–8, 633–6, 650
analytical range extension, 633–5
fouling protection, 635
interfering species protection, 633
oxygen dependence reduction, 634–6
Fuzzy-assisted expert systems, 687
Fuzzy-assisted sensors, 687
Fuzzy control, 687
Fuzzy logic, 686–90
Fuzzy logic systems, 659–60
Fuzzy set theory, 687, 689

Galactose concentration:
relationship to speed of sound, 254
see also Sugar solutions
Galactose sensors, 624, 627, 639
Galvanic contact, 596
Gamma-rays, 33
absorption for density determination, 206

Gardner colorimeter, 86
Gas and liquid chromatography, *see* GLC systems
Gas barrier properties, 157
Gases:
  oxidizing and reducing gases, sensors for, 585, 601
  partial pressures during food storage:
    influence on microbial deterioration, 376, 382
  polyatomic:
    ultrasound attenuation in, 271
    speed of sound in, 250
  speed of sound, in, 250
  ultrasound data, for, 262, 293
  see also Ultrasound propagation
  see also Volatiles
GasFET, 600
Gas sensors:
  broad specificity:
    see MeOx gas sensors, Polymer gas sensors, Sorption sensors
  dissolved gases, 651
  high selectivity:
    see CHEMFET, *see* Organic semiconductor gas sensors
  high sensitivity:
    see CHEMFET, *see* Organic semiconductor gas sensors
  overview table (not including sorption sensors), 615
  see Biosensors, Chemosensors
  see Immunosensors and sorption sensors
  see Odour assessment
  see also Near infrared, Microwave and Ultrasound measurement
Gate-controlled diode, 600
GC–MS (Gas chromatography-mass spectrometry), 454
GC–MS study, rancidity, 525–7
GE spectrophotometer, 86
GEC Marconi Research Centre, xxv, 532, 535, 545–8, 558, 576, 714
GEC Sensors Ltd, 576, 714
Gelatin gels, texture assessment for, 354, 356, 367
Gels:
  texture assessment for, 352, 370
  ultrasonic properties of, 248, 267, 269, 273
  see also Jelly
  see also Setting
General Foods Texturometer, 354
Germanium photodiode, 170–2
Germanium window, 186–8
Gibson, D., M., xxviii, 457–98
Glass electrode, *see* Chemosensors
Glass windows, 141–2
  laminated, 142
GLC (gas–liquid chromatography) systems, 419–22
  detectors used with, 426–7
  pesticides assay, 424–7
  saccharides assay, 432
GLC measurement of POM (polymerized and oxidized material), 529–32, 552–5, 714
Glossary, 673–90
Glossy foods:
  colour measurements for, 45, 52, 72, 79
  NIR measurement for, 140–1, 147
Glucose assay, 430, 433
Glucose concentration:
  relationship to speed of sound, 251–5
  see also Sugar solutions
Glucose sensors, 524, 526, 561–3, 623, 624, 626–32, 635, 636:
  637, 639, 651
Glucose depth profile, 559–72
Glucose decomposition in meat and fish, 524
Glutamate, 624, 636
Glycerine, *see* Glycerol
Glycerol, adjustment of water activity with, 381, 388
Glycogen, 562, 571
GMP (good manufacturing practice), 8
Grain:
  moisture monitoring by microwave meas., 191, 201, 209, 218, 227
  moisture monitoring by NIR, 148
  level measurement for, 299, 301
  sorting, 100, 117
Granulated foods, *see* Powdered foods
Grape beverage, colour of carbonated, 85, 91
Grating, diffration, 647
Gravimetric method for water activity meas., 395
Grinding, monitoring by AEM, 245

Groesbeck, C.A., xxviii–xxix, 63–96
Groundnut, *see* Peanut
Groundnut oil, 537–40, 544
Gum identification, 663
Gums, texture assessment for, 365, 367

HACCP (hazard analysis critical control point), 8, 14, 376, 686
Hagburg factor, phantom measurement, 152
Hagen–Poiseuille law, 332
Halothane, 647
Ham:
  cooked, yield prediction, 609
  moisture and ash assay, 443
  raw, curing prediction, 609
  role of water activity, 382, 384–6
  see also Meat products
Hamburger packaging materials, 156
Hamworthy Engineering instrument, 231
Hardness, 353, 360
Harwell ultrasonic systems, 302, 311
Hazard Analysis Critical Control Point, *see* HACCP
Hazardous environment:
  see Hostile conditions, Dust explosions
Hazelnuts, sorting, 117
Headspace measurement, 7, 292
  see also Alcohols, Aldehydes, Amines
Heat processing, 8–9
Heat-sensitive devices, 625, 638–9
Heat treatment, indication of severity, 90
Herbicides, 585
Herbs, dried:
  colour, 81, 87
  moisture monitoring, 148
Herbs, microbial assay (impedance), 509, 510
Herring, histamine formation, 524
Herschel–Buckley equation, 330
Heterogeneous materials:
  see Microwave measurement
Hewlett-Packard, 421
Hexacyanoferrate, 632
Hexanal, 527, 624
High-pass filter, 108–9
High performance liquid chromatography, *see* HPLC systems
Hilger Biochem, 86
Histamine, 524, 526, 624
  indicator of past bacterial activity, 524
  scombroid poisoning, 524
Hitching-post standard, 70, 90–1
Homogenization, *see* Emulsions
Hormones, 638, 645
Horn antennae, *see* Microwave measurement
Horseshoe crab, assay based on, 461–3, 652
  see also Limulus Amebocyte Lysate
Hostile environments, 14–6
  example, *see* Extrusion cooker
Hostile environments, tolerance, 16
  fibre-optic probes, 651
  microwave instrumentation, 192
  NIR instrumentation, 141–2
  ultrasonic instrumentation, 16, 306, 311
  thermometry, 186–8
  pH measurement, 594, 595
  viscosity measurement, 545
  see also Electromagnetic noise, Dust explosions, Dust tolerance
Hot wire viscometer, 342
Howard mould count replacement, 646
H_2Pc (hydrogen phthalocyanine), 614
HPIC (high performance ion chromatography), 427
HPLC (high performance liquid chromatography) systems, 419–22
  detectors used with, 421, 428, 431–2
  ions and sugars assay, 427–32
  sulphite assay, 432–3
HPLC-MS (HPLC-mass spectrometry), 421
HRGC (high resolution gas chromatography), 421
Hue angle, 48–51, 67, 70
Human eye, *see* Colour vision
Humectants, 388
  see also Glycerol
Humidity, 612, 613, 615, 645
  fermentation, control in, 389
  microbial deterioration, influence on, 376

Humidity – *continued*
sensors, 28, 33, 681
see also Water activity measurement
Hulls, colour, 82, 88
Hunter colorimeter, 86
HunterLab 86, 87, 89, 90
Hunter space, 49, 51
Hurdle effect, 382
Hybrid circuits, 677, 679
Hydration monitoring:
by AEM, 245
by ultrasound velocity, 254
Hydrocarbon volatiles, 613, 615
Hydrogen bonding, 645
Hydrogen phthalocyanine, 614
Hydrogen probe as base device, 636
Hydrogen production:
Enterobacteriaceae, 483
Hydrogen sensors, 526–8, 585, 600, 612, 615, 617
Hydrogen sulphide, 612, 615
Hydrogenated fats, 537–8
Hydrogenation, ultrasound-assisted, 246
Hydrolysis:
relationship to water activity, 378
Hydrolytic methods:
assay of complexed fats, 439
Hydronix instrument, 231
Hydroperoxides in frying oil, 528, 531
Hygiene requirements, 14–6
Hygienic design:
chemical sensors, 588
NIR instruments, 141
viscometry, in, 342
see also Microwave measurement
see also Infrared thermometry
see also Ultrasound measurement
Hygrometric method for water activity meas., 396–403
Hygroscopic materials, 396–9
Hypoxanthine in fish, biosensors for, 524, 526, 624, 635, 637

IC (ion chromatography) system:
anions, 428–9
sulphite, 432–3
Ice:
dielectric properties of, 198
proportion in partially frozen foods, 266
ultrasound data, for, 258
Icecream, rheology of, 326
ICS spectrometer, 55
ICMSF two-class specification (microbial), 506
Illuminants, 46–8, 50–6, 78–9
Illumination:
colour measurement, for, 46–8, 50–3, 56, 78
colour sorting, for, 101, 105–7
colour rendering index, 47, 55
colour temperature, 46
spectral curves of fluorescent lamps, 47, 106
spectral curves of incandescent lamps, 107
ultraviolet for fluorescence techniques, 104
Image, sorting by colour, 104, 116
IMFET (immunosensitive field effect transistor), 638
Immiscible liquids interface detection, 316
Immobilization techniques:
colorimetric reagents, 598
enzyme electrodes, for, 633, 636
FOP, for, 650
ion-sensitive membranes, 603
MBE (molecular beam epitaxy), 615
see also LB films
improved for mediators, 633
Immunoassays:
see Immunosensors, Indicator strips
Immunological agglutination assay, 461–3
LAL (limulus amoebocyte lysate), 461–3, 652
Immunosensors and sorption sensors, 582–6, 623, 644–54, 663, 682
acoustic devices, 644–7, 652–4, 660
affinity binding, 586, 623, 644–5, 654
antibodies, 585, 645
complex formation mechanisms, 645
endogenous binding proteins, 645
lectins, 585, 645
reversible binding, 645
agglutination assay, 652

array (TIRFs) with CCD imaging, 649
ATR (attenuated total reflectance), 648
binding, direct monitoring, 645
biomembranes, 645
biothermal devices, *see* TELISA
building blocks, 582–6
calorimetric devices, *see* TELISA
capillary tube devices, 649
commercial immunosensor instruments, 646
continuous immunosensor operation:
  continuous supply of antibody, 645
  real-time monitoring of binding process, 645, 647
definition, 586, 623, 645, 682
determinands (analytes or identifications), affinity sensing:
  agricultural residues, 624
  amino acids, 645
  drugs, 645, *see* also veterinary residues
  gases, 645
  glucose, 651
  gum identification, 663
  hormones, 638, 645
  Howard mould count replacement, 646
  meat speciation, 585, 646
  microbe identification, 585, 625, 645, 646
  (bacteria, moulds, yeasts, virus)
  mycotoxins, 624, 646
  pathogens, 646
  (Listeria, Salmonella)
  pesticides, 585, 624
  polysaccharides, 645
  proteins, 638
  saccharides, 645
  small ions, 645
  sourcing (origin of meat or produce), 585
  odours, 652
  toxins, 646
  veterinary residues, 624
  vitamins, 585, 624, 646
determinations, agglutination sensing:
  endotoxin, 652
determinands (analytes or identifications), sorption sensing:
  aroma, 647, 654, 660
  cooking progress, 647, 654
  iodide traces, 652
  mercury traces, 652
  odours, 654
  organophosphorous compounds, 644
  silver traces, 652
  solvents, organic, 644
  volatiles, 644, 646–7, 652, 654
DNA probes, 585, 624
electrochemical immunosensors, 645–6
evanescent wave devices, 647–9
FOP (fibre-optic probe) immunosensors:
  ConA (concanavalin A) binding agent
  continuous antibody supply, 651–2
glucose analysis with ConA, 651
  see also FOP (fibre-optic probes) chemical sensors
fibre-optic devices:
  SPR or TIR, 644, 651
  TIRF, 649
fluorescence measurement:
  enhanced signal, 649
  time-constant mode, 650
general reading, 668–9
  see also references, 654–8
humidity, 645
incubation time, elimination or reduction, 645–6
interferometric technique (phase measurement), 649
kinetic mode, 645, 647
LB (Langmuir–Blodgett films), 645, 647, 654
  odorant recepting, 654
lipid coatings, odorant sensitive, 652
multi-variable (TIRF/CCD), 649
odorant sensitive films, 652, 654, 660
optical devices, 644–52
physical changes on binding, 644
pattern recognition, 654, 660–2
piezocrystal microbalance, 644, 652
potentiometric devices:
  IMFET, 638

Immunosensors – *continued*
reference device, 652
refractive index, *see* SPR, TIR
enhancement labels, 647
profiling, 649
remote immunosensing, 649
separation step, elimination, 646
spatially resolved, 649
sorption, selective, 611, 623, 644, 654, 660
chromatographic coatings, 660
selectivity mechanism, 644
SAW (surface acoustic wave devices), 585, 644, 652–4
SPR (surface plasmon resonance) devices, 585, 644, 647, 648
TELISA, 638, 645
TIR (total internal reflection) devices, 585, 644, 647–9
TIRF (total internal reflection fluorescence), 649
ultrarapid immunoassay, 645
see also Biosensors, Chemosensors, Chemical sensors
see also Physical sensors for chemical and microbial variables
Impedance:
acoustic, 258–62, 267–70, 311
definition, 267
see also Ultrasonic measurement
see also Ultrasound propagation
matching in microwave measurement, 204, 212, 215
matching in ultrasonic measurement, 265, 268–9
Impedance (electrical), pH sensors, 590, 604
Impedance (electrical), liquids:
see Electrical impedance of liquids
accept/suspect/reject categories, 506
Impedance conversion, 597–9, 601
Impedance microbiology, oscillometric techniques, 5
Impedance microbiology, 457–517, 560
accept/reject decisions, 489
accuracy, 485
anaerobic organisms, 510
applications overview, 490
bacteria:
Clostridium botulinum, 491
E. coli/coliforms, 488, 491, 510, 512
Enterobacteriaceae, 483, 491, 510, 512
Enterococci, 491
histamine producers, 491
Lactic acid, 512
Listeria, 491, 511, 512
Pseudomonas, 512
Salmonellas, 486, 489, 491, 498, 511, 512
Shewanella putref., 487
Staph. aureus, 491, 511, 512
Streptococci, 512
bacterial growth, 469–71, 483–6
carbon dioxide evolution, 488
gas evolution, 477
hydrogen evolution, 483
lag phase, 469
bacteriophage, 491
baseline, 501
biomass, 464, 477
biosensors, use, 517
calibration, 485, 503–7, 511
caution time, 505–7
caution zone, 508
cell design, 467–9, 492, 516
commercial instruments, 480–3, 515–6
conductance data evaluation, 469–71, 483–91
conductance methods, principles, 464–73
conductance/capacitance methods comparison, 461, 474–8, 503
dead cells, influence, 475
early in the growth cycle, 474
later in the growth cycle, 474–7
correlation with plate count, 488–91, 500, 503, 509–10
cut-off time, 505–7
defective level, 506
detection curve, 501
detection time, 485–7, 491–2, 500, 506
electrode material, 467–9, 516
oxide formation, 467
platinum, 467–9
polarization effects, 467, 469
stainless steel, 467–9
electronics, advances, 491–2
fatty samples, 509
food poisoning organisms, 491, 511
food type, influence, 488, 510

frequency, influence, 464, 469
generation time, 485
growth media:
bacteriophage addition, 511–2, 514
development, 501–3, 511
influence, 468, 472, 486–8, 492:
selective, 511–4
histogram presentation, 507
impedance as indicator of microbial load, 464
impedance of fluid-electrode system, 464–9
capacitance, electrode , 465
dipole layer, electrode, 465
oxide layer, electrode, 465
phase angle, 466
reactance, electrode, 466
resistance, fluid, 465
incubation temperature, 502
indicator organisms (hygiene), 510
instrument design, 478–83
circuits, 478–9
metabolic activity, 461, 464, 500
multiplexed cell switching, 479
nomenclature for chapter 14, 458–9
official method for salmonellas, 498
particulate samples, 509
pre-enrichment, 514
pre-incubation, 508, 509, 515
protocols, development, 501–3, 516
reject level, 506
resolution, 471–3
resuscitation period, 514
sample size, 460, 489
selective procedures, 510–4, 517
shelf-life estimate, 488, 490, 493, 514–5
stabilization period, 510, 514
sterile products, procedure for, 508
temperature control, 472–3
threshold, 492, 510
two-class (ICMSF) specification, 506
yeasts and moulds, 464, 477, 491, 512
Impedance techniques for microbial assay, 457–517
see Impedance microbiology
Incidence angle for total absorption, SPR device, 647
Incidence angle for total reflection,
TIR device, 647
Incubation time, elimination or reduction, 645–6
Indicator approach, *see* Marker concept
Indicator strips, 623, 645
Indicator organisms (hygiene), 510
Indicators:
colour, indicating:
carotene content, 88
deterioration, 44
heat treatment, 90
jelly ageing, 89
kernel maturity, 82
pigment content, 64–5, 80, 88–9
ripening, 44, 89, 289
shelf-life, 91
storage effects on yoghurt, 91
water activity, indicating:
ripening of fermented sausages, 389
Indirect biosensors, 623
Industrial measurements:
consistent product objective, 64
repeatability, 181
resolution, 52
target value band, 659
see also Consistent product
Infections, food-induced, 379
Infragauge, multi-filter, 149
Infralab, 135
Infrared:
brief overview, 241
see also Near infrared, Mid infrared
Infrared beam-break level controls, 301
Infrared detectors, 170–2
germanium photodiode, 170–1
lead selenide photoconductor, 176
lead sulphide photoconductor, 129, 133, 170, 172, 174, 184, 188
mercury cadmium telluride photoconductor, 176
pyroelectric device, 170, 172, 173, 184
response times, 171, 172, 173
silicon photodiode, 133, 170–1
thermopile, 170, 172, 184
Infrared drying, moisture assay, 437
see, in contrast, Near infrared measurement (non-drying)

Infrared Engineering, xxvii, 129, 135, 150–1, 231
Infrared measurements:
  see Infrared thermometry
  see Near infrared absorption spectroscopy
  see Near infrared instruments
Infrared response of detectors in Sorting by colour, 110
Infrared sources:
  acousto-optical tunable filter (AOTF), 130
  IREDs (infrared emitting diodes), 130
  laser diodes, 130
  quartz halogen lamp, 128–9
Infrared thermometry, 167–188
  accuracy, 183–4
  aiming system, 169
  air purge attachment, 175, 177
  auxiliary sensor, 179
  baked products, emerging from oven, 182
  background radiance, *see* background temperature
  background temperature, 178–81
  black body radiator, 178–9
  camera, 173–4
  cold-wall conditions, 182–4
  collection system, 168–9
  configurations, 173–5
  data processing, 172–3
  detectors, 170–2
    germanium photodiode, 170–1
    lead selenide photoconductor, 176
    lead sulphide photoconductor, 170, 172, 174, 184, 188
    mercury cadmium telluride photoconductor, 176
    pyroelectric device, 170, 172, 173, 184
    response times, 171, 172, 173
    silicon photodiode, 170–1
    thermopile, 170, 172, 184
  difficult targets, 182
  emissivity, 172, 178–9, 181–6
    dependence on surface composition and structure, 182
    dependence on wavelength, 182
    enhancement, 185
    examples for cold-wall and isothermal conditions, 182–4
    values for foods and construction materials, 182
  emissivity-free measurement, 186
  fibre-optic cables, 186–8
  field of view, 169, 174
    partially obstructed, 185
  filter, 172
    polarizing, 186
  focus, 169
  extrusion cookers, in, 186
  hostile conditions, 186–8
  imagers, 175–8
  inaccessible targets, 186–8
  isothermal conditions, 184
  laser reflectometer, 186
  line scanners, 175–8
  low cost instruments, 174
  microprocessors used, 173
  microwave oven, in, 177, 182
  miniature, 177
  penetration depth, 169
  Planck equation, 178
  portable instruments, 173, 174
  principles, 178–84
  ratio thermometer, 185
  target spot, 169
  temperature range, 170, 172
  two-colour thermometer, 185
  wavelengths used, 168, 170–2, 176
  windows, 186–8
    polyethylene, 188
    synthetic sapphire, 186–7
Infrared transmission windows of the atmosphere, 170–1
In-line and in-situ measurements:
  colour, 52
  composition and other variables:
    see Infrared thermometry
    see Microwave measurement
    see Near-infrared measurement
    see Ultrasonic measurement
  viscosity, 340–2
  see also Frying oil monitor
  see also Meat freshness probe
In-line instrumentation:
  definition, 684–5
  illustrations, 6, 10
  notes, 3–9
  requirements, 13, 14–6
  see also On-line measurement
In-situ instrumentation:
  definition, 685

illustration, 7
notes, 11
Inosine in fish, biosensors for, 524, 526, 624, 635, 637
Inspection systems, optical, 98–100, 104–10
Instron, 355
Instrument:
definition, 684
notes, 684–5
Instrument development:
contributions to development, 558, 575
examples, 528–76, 714
phases of development, 556–7, 573–4
role in quality assurance, 3, 1–34
Instrumentation terminology, 673–90
Integral actuators, 584
Integral signal processing, 584, 660–1, 679–80, 687
Integrated circuit, 599, 600
Integrated optics, 677
Integrated sensors, 637, 676–7, 679
see also Smart sensors
Integrating sphere, 52, 126–7
Interaction of sensitive layer with analyte, 583, 585–6
Intercept, 158, 161
Interdisciplinary collaboration, 21–2, 558, 575
Interface:
see Impedance matching
see Surface potential
Interface:
influence on non-contact measurements, 267, 271
location by ultrasound, 312–3, 315–6 (liquid/second liquid boundary)
see also Impedance matching
Interfacing the sensor, 679–80
Interferometric technique (phase measurement), 649
Intermediate moisture foods:
see Moisture, intermediate
Intermittent calibration, 605
International system of units, 695–7
conversion, 698–705
Interpretation of readings, 5, 17–22
Interpretation of output signals, 659–662
see also Data Processing
Inversion, sugar, 526
Iodide traces, 652
Iodine value, 532, 538, 540, 551
Ion chromatography, *see* IC
Ion-controlled diode, 600
Ion exchange, analyte-sensor:
see Chemosensors
Ion-fibre suppressor column, 428
Ionizing radiation, 32–3
Ionophores, 597, 603
Ion-selective membranes, 594, 603–4
deposition techniques, 603
Ion-sensitive insulating films, 597, 599, 601, 603–4
Ion sensors, 585, 595–607, 649–52
see Chemosensors
Ion-selective probe, *see* ISE
Ions assay:
HPLC system, 427–32
Ions, dissolved, influence on dielectric properties, 191, 201
IRED (infrared emitting diode), 130
Iridium oxide (IrO2), 585, 595, 611
Irrigation control, by AEM, 245
ISE (ion-selective electrode), 585–6, 603
ISE-based biosensors, 623
ISFET (ion-selective field effect transistor), 585, 589, 594, 595–8, 599–611
industrial pH stab probe, 606–9
meat industry applications, 606, 608–9
schematic diagram, 602
see Chemosensors
ISFET/REFET systems, 585–6, 589, 604–5, 607–9
ISFET/REFET-based biosensors, 623, 625, 637–8
ISFET/ microactuator-based biosensors, 625, 637
Isopiestic method of water activity determination, 395
Isothermal point, ISFET, 603, 608
IV (iodine value), 532, 538, 540, 551

Jams, flow rate measurement, 304
Jason, A.C., xxix, 457–98
Jelly:
colour, 72, 89
ageing, 89
see also Gels

Jelly sweets:
  moisture assay, 440
Joule, conversion, 701
Judd-Hunter solid, 67, 69
Juice:
  colour, 77
  flow rate measurement, 305
  glucose assay, 629–32
  level measurement, 298
  microbial (ATP) assay, 461
  microbial assay (impedance), 490, 512
  packaging materials, 156
  stability monitoring, 277
  ultrasound data for, 260

Karl Fischer procedure, 437–8
Karl Fischer systems, 437–40
Kaymount instrument, 400
Kay-Ray Moisture Meter, 231
KDG Mobrey, xxviii
Kelvin, conversion, 698
Kent, M., xxix, 189–235
Ketones, formation in heated frying oil, 528
Kilcast, D., xxix, 349–74
Kilogram, conversion, 702
Kinetic mode immunoassay, 645, 647
Kjeldahl nitrogen assay procedure, 445
Kjel-Foss nitrogen analyser, 436, 446–8
Kjeltec analyser, 446, 448–9
Knife-type probe, 572
  see also Meat freshness probe
Kramer Shear Cell, 352, 359
Kress-Rogers, E., xvix–xxv, xxx, 1–36, 237–288, 523–714
Kubelka-Munk scattering analysis:
  NIR measurement, 134, 158
  colour measurement, 45, 56, 90
Kulmbach, Federal Centre for Meat Research, xxxi

Labelling, acceptable assays, 436
Lactate sensors, 624, 632
Lactose assay, 430, 435
Lactose content, by NIR, 128
Lactose, alpha-, 136
Lactose, influence on moisture assay, 18–9, 437, 438
LAL (limulus amoebocyte lysate), immunological agglutination assay, 461–3, 652
LPS (lipopolysaccharide), 463
Lamb, *see* Meat freshness
Lamb waves, 249, 312
Laminated packaging materials, film thickness, 156
Lamp colour temperature, compensation for in NIR measurement, 130
Lamps, *see* Illuminants
Land Infrared, xxvii
Land Infrared instruments, 173–7, 187
Langford, Bristol, xxx
Langmuir-Blodgett films, *see* LB films
Lard:
  heat-induced deterioration, 537–9
  oxidative rancidity, 525
Laser reflectometer, 186
LB (Langmuir–Blodgett) films, 585–6, 604–5, 615–7, 645, 647, 654
  ELBOS device, 605
  non-amphiphilic, 604
  odorant recepting, 654
LCD-display cell, as capillary tube, 649
Lead phthalocyanine (PbPc), 614–5
Lead selenide photoconductor, 176
Lead sulphide photoconductor, 170–88
Lead zirconate titanate (PZT), 291, 297
Leak detection by AEM, 245
LEAR (low erucic acid rapeseed), 539, 541
Learning systems, 20, 113, 660
Leco analyser, 427, 442–3, 452–3
Lectins, 585, 645
LED (light emitting diode), 108
Legal requirements:
  pH value, 384–5
  related measurements, 24–5
  water activity, 384–5
Length units, conversion, 703–4
Lethality, 9
Level alarms:
  high level, 298, 300, 306, 308
  low level, 306, 308
  pump control point, 308
Level measurement:
  ultrasonic, 242–3, 291–301, 306–12
    echo-ranging, by, 291–301

Lamb waves, using, 249, 312
transmission, by, 306–12
Level switches, 306–11, 318
comparison with echo-ranging, 306
LFRA (Leatherhead Food Research Association):
xxv-xxxi, 525–576, 628–31, 664
Lie detector, *see* Polygraph
Ligands, 645
Lightness, 46–51, 67, 70
Lime addition monitoring, 305, 315
Limit values, frying oil quality indicators, 533, 536
Limulus Amebocyte Lysate, 461–3
piezocrystal assay for endotoxin, 652
Line scanners, thermal, 175–8
Linguistic variables, 689
Lipid coatings, odorant sensitive, 652
Lipids, 624
Liquid foods:
colour, 56–9, 77–8, 85, 91–2
NIR measurement for, 142, 147–9
Liquid identification, 251–2, 315–6
Liquid level, *see* Level
Liquified gases, *see* Cryogenic liquids
Listeria, 491, 512, 646
Lithium ions, in glass membranes, 594
Lithium niobate SAW device, 652
Load cells, for texture assessment, 356
Logarithms, 711–2
Londreco, xxvii
Longissimus dorsi, 564
Loss factor, *see* Dielectric properties
Loss tangent, *see* Dielectric properties
Lovibond colour, frying oil assessment, 529–32, 552
Lovibond tintometer, 50
Low, J.M., xxx, 97–119
Low-pass filter, 108–9
Luciferase, 651
Lufft hygrometer, 396
Luminescence-based biosensors, 625, 651, 664
see also ATP
Lumpy solid foods:
colour measurement, for, 76, 82, 87–8
moisture assay, 437

Macbeth spectrophotometer, 86, 91, 92
MacDougall, D.B., xxx, 39–62
MAFF (Ministry of Agriculture, Fisheries and Food), xxv 575
Magnesium, dissolved, 651
Magness-Taylor Tester, 352
Magnetic flux units, conversion, 703
Magnetic resonance, *see* NMR
Mahalonobis distances analysis, 134
Maize flakes, grits and meals, moisture monitoring, 141, 147
Maize kernels, sorting from beans, 104
Maize, fat assay, 439
Maize oil, 538–40, 544
Malthus analyser, 464, 473, 480–3, 488, 515–6
Maltose, 624, 639
Mannitol assay, 432
Mannose concentration:
relationship to speed of sound, 254
see also Sugar solutions
Manometric method for water activity meas., 394–5
MAP (modified atmosphere packaging), 527
gas composition, 612
MAP-meat and -fish freshness, 526–8
Mapping techniques, in sorting, 114–5
Marconi, *see* GEC
Margarine, microwave measurements, 191
Margules equation, 337
Marker approach, 523–77, 664–5
detailed examples:
frying oil monitor, 528–58, 575–6, 714
meat freshness probe, 558–76
overview, 523–8
browning potential, 526
CAP-meat and -fish freshness, 526–8
chilled meat freshness, 524, 526
fish freshness, 524, 526
frying oil, heat-induced deterioration, 525–6
microbial assessment, 560
oxidative rancidity, 525–7
past bacterial activity, 524, 526
ripening, 526
yeast contamination, 526
Market considerations, chemical sensors development, 587, 629, 633, 637, 659, 681

Market sectors, food and drink (UK), 3, 4
Mass flow meter, Coriolis force, 27
Mass spectrometry (MS), 421, 423
Mass units, conversion, 702
Mathematical models, 20, 689
Maturity, *see* Ripening
Maughan, W.S., xxx, 97–119
Mayonnaise, moisture assay, 440
MBE (molecular beam epitaxy), 615
McKenna, B.M., xxxi, 323–47
Measured properties and their significance, 12–4, 24–5
Measurement types, 22–6
  overview table, 29–34
Meat:
  colour, 54–6, 76–7, 83, 88–9
  display, 54–6
  fat/lean ratio, determination, 266
  infrared emissivity data, for, 182
  microbial assay (impedance), 490, 509, 512
  moisture and fat assay, 427, 436, 437
  packaging materials, 157
  partially frozen, frozen/liquid water ratio, in, 266
  ultrasound data, for, 259
  water activity, levels in, 385–7, 390–2
  see also Beef
Meat content assay, 445–53
  dependence on species, 446
Meat freshness probe (knife-type), 524, 526, 558–76
  biosensor principle, 561, 626–33
  comparison with other microbial assessment types, 560
  concept, 558–61
  contributions, 575
  development phases, 573–4
  feasibility study, 561–3
  freeze-thaw information, 572
  future development, 571–3
    estimate of characteristics, 568
    illustration, 572
  glucose depth profile, 559–70
    lamb model system, 559, 561
    pork, beef and lamb with native flora, 562
  knife-configuration, 571–2
  pH probe, combination with, 572
  pH depth profile, 563
  probe adapted from, 559–60, 571, 627–31
  prototype probe characteristics, 563–4
  specifications formulation, 561–3, 572
  trials:
    biosensor/enzyme kit comparison, 564–6, 568
    glucose near the surface, 566
    glucose depth profile, 568–70
    main trial, 564–70
    microbial growth, 567
    preliminary trial, 564
    probes, illustration, 563, 565
Meat products:
  batter-coated or breaded, 528
  cured, phosphate assay, 428
  meat content assay, 445–53
  microbial (ATP) assay of cooked, 461
  microwave compositional measurements, 191, 211–215
  moisture and fat assay, 436, 440
  packaging materials, 157
    see also Casing materials
  pH stab probes, 608–9
  raw ham curing prediction, 609
  ultrasound treatment of, 246
  water activity, control in, 388–90
  water activity, levels in, 385–93
  water activity, determination in, 390–3, 394–404
  water activity, role of, 384–90
  water activity profile in dried, 393
  yield prediction for cooked ham, 609
  see also Bacon, Ham, Sausages
Meat quality:
  amines, 524
  CAP-packed, freshness, 526–8
  colour, relationship to pH, 609
  colour, translucent materials (CTM) probe, 609
  DFD (dark firm dry) beef, 608
  diffusion rates, 572
  electrostimulation control, 608
  freeze-thaw information, 572
  freshness, 524, 526, 558–75, 627–9
  glucose concentration, 524, 561–3
  glucose depletion, 559–70
  glycogen, 562, 571
  pH measurement, 606, 608–9

post mortem period, pH measurement, 609
PSE (pale soft exudative) pork, 609
quality assessment in the slaughterhouse, 609
trial with ISFET/REFET pH probe, 608–9
Meat speciation, 585, 646
Meat, texture assessment for, 352
Mechanically resonant probes, 27, 31
boiler oil monitor, 532, 535
example of advantages, 548–9
frying oil monitor, 546–7, 714:
marker approach, 664–5
vibrating rods (tuning fork type) viscometers, 546:
vibrating tube (vibrating pipe section) densitometer, 27
vibrating table, yoghurt setting, 664
viscosity and density dependence, 546
see also Frying oil monitor
Mechanical stress:
measurements of properties under, 31
Mechanical surface properties, 644
Mediated biosensors, 625–33
Mediators, 625, 627, 632–3, 635
Mediator-carrying carbon electrode, 625, 626–33
MediSense, *see* ExacTech
Melting:
see Phase changes
see Solid/liquid ratio
see Ice
Membrane configurations, pH and ion-selective, 596–7
see Chemosensors
Membrane technology, sensor optimization:
see Functional membranes
MeOx (metal oxide) semiconductor materials, 595–7, 600–1, 611–7
see Chemosensors
MeOx gas sensors, 585–6, 600–1, 611–7, 660
see Chemosensors
Mercury cadmium telluride photoconductor, 176
Mercury compounds, *see* Calomel
Mercury traces, 652
Metabolic activity of microbes, 461, 464, 500
Metals assay, atomic absorbance spectrometry, 420
Metal ions, dissolved, 651
Metal-MeOx pH and acidity sensors, 585–6, 595–7, 611
see also Chemosensors
Metal oxide, *see* MeOx
Metal phthalocyanines, 614–6
Metamerism, 78
Methane, 585, 600, 613, 615
Methanol, 624, 627
see also Alcohols
Methanol, *see* Alcohols
Methodology in the chemical QC laboratory, 437
Microbe concentration, significance of measurement
microbial load, 24–5
biomass, 24–5
Microbial assay:
conventional assay, 460
electrical impedance techniques, 457–517
see Impedance microbiology
rapid methods, 460–3
Microorganisms of public health significance, 491, 511
Microbalance, crystal, 644
Microbe identification, 625
Microbe identification, 585, 625, 645, 646 (bacteria, moulds, yeasts, virus)
Microbial activity, 625, 636
Microbial deterioration
see Micro-organisms, growth of
Microbial growth curve, meat, 567
Microbial safety and stability:
see Safety of food
Microbial spoilage assessment, 523–8
comparison of conventional, rapid and ultrarapid methods, 560
impedance microbiology, 457–517, 560
pre-spoilage stage, 524
see also Meat freshness probe
Microbial toxins:
histamine, 524, 526
indicators of past bacterial activity, 524, 526
production of, 382–3
Micro-Clark device, planar, 623, 635–6

Microelectronic base transducers, 589, 594, 595–8, 599–618, 636–8
Microelectronic biosensors, 623, 625
Microelectronic devices, 680
Microelectronic gas sensors, *see* CHEMFET, ELBOS, Chemosensors
Microelectronic ion sensors, 603–5
Microelectronic pH sensors, 585, 589, 594, 595–8, 599–611
Microenvironment, 589, 599, 604
Micromachining, 681
Micro-organisms, growth of:
  desirable effects, 379
  factors determining, 376
  minimum water activity, 379–82
  mould growth in bakery products, 388
  undesirable effects, 379
  water activity, role of, 377–90
  see also Water activity, pH sensors, pH value
Micro-peroxide probe, 636
Microprocessor applications, *see* Computing and microprocessor applications
Microsens, 608
Microtip ion sensors, 599
Microtitrator, 609–11
Microvalves, 605, 606
Microwaves:
  attenuation, 202
  band designation, old, 710
  band designation, new, 710–11
  brief overview, 29–30, 241
  frequency range, 190, 710
  impedance matching, 204, 212, 215
  phase shift, 203
  propagation characteristics, *see* Dielectric properties
  reflections, 203, 226
  Q-factor, 203
  resonant cavity, 203
  resonant frequency, 203
  S and X bands, 190, 710
  scattering and interference, 190
  scattering parameter, 212
  standing waves, 203, 204, 226
  wavelengths, 190, 710
  window materials, 214
Microwave drying moisture assay, 437, 443–4
  see, in contrast, Microwave measurement, non-drying
Microwave heating/ thawing, 8, 27, 198, 201
Microwave measurement, 189–232
  see also Dielectric properties
  advantages and disadvantages, 191–3, 198, 665
  antennae, *see* sensors
  applicators, *see* sensors
  attenuation and phase shift, 207–11, 202–3, 221–3
  avoidance of perturbing effects, 191
    see also multi-variable measurement
    see also product and process variables, relationship to
  bulk properties determination with, 192
  choice of frequency, 191
  circuits, 219–25
  cleaning-in-place compatibility, 192
  coal, for, 221, (for foods, *see entries for named foods*)
  commercial instruments, 219, 230, 231
  comparison with ultrasonic measurement, 240–3
  cost, 192
  DC conductivity, non-contact measurement of, 202, 231
  density and moisture measurement, 204, 205–11
  dip-probe, 227
  echo ranging, 243
  extrusion cooker, for, 205, 224
  foods, specific, for, *see entries for named foods*
  frequency, choice of, 211
  future development, 230–1
  hazardous environment, for, 192
  heterogeneous materials, for, 192, 198, 207
  homodyne and heterodyne circuits, 221–3
  horn antennae, 192, 224–7
  hygienic design, 192
  measurement cell, 192, 225–6
  moisture determination, 190–231
  moisture and density determination, 204, 205–11
  multiplex measurement heads, 192

multi-variable measurement, 202, 204, 205–17, 230–1
non-linear dependence on hydration, 196, 207
particulate materials, for, 205–11
phase shift and attenuation, 202–3, 207–11, 221–3
power levels, 192
precision:
single microwave variable methods, for, 192
dual microwave variable methods, for, 192
enhancement by iterative method, for, 192, 209
process conditions, tolerance to:
condensation, 192
dust, 192
vapours, 192
product and process variables, relationship to:
air content, *see* density, bulk
alcohol content, 196, 217
density, bulk, 192, 201, 204–12
fat content, 196, 211–5
particle size and shape 207, 217–9
protein content, 211–5
salt content, dissolved, 191, 201–2, 217
structure, 217–9
sucrose content, dissolved, 196, 199
temperature, 196, 198–200
water content, 190–231
proteinaceous food properties, 198, 200
ratio method, 209–11
reflectance techniques, 223–5
remote instrument, 192
resonant systems, 215, 223
sensors:
coaxial line, open-ended, 204, 217, 229–30
definition, 224
horn antennae, 214, 219, 224–7
stripline 227–9
transmission lines, open-ended, 204, 217, 229–30
simultaneous determination of several variables:
see multi-variable measurement
solutions, water content, for, 215–7
ternary mixtures, for, 211–7
tissues, biological, for, 230
transmission, 212, 214
tobacco, for, 207, 221 (for foods: *see entries for named foods*)
two-variable measurement, *see* multi-variable measurement
Microwave oven, non-contact thermometry in, 177, 182
Microwave ovens, thermometry, 599, 650
Mid Infrared spectroscopy:
comparison NIR/MIR spectroscopy, 125, 139
FTIR, 125–6
qualitative analysis, 125
Milk:
colour measurement for, 53, 77, 85, 91–2
flow rate measurement, for, 305
lactose assay, 435
level measurement, for, 298
microbial assay (impedance), 490, 509, 512
microbial (DEFT) assay for raw, 461
packaging materials, 156
rheology of, 329
ultrasonic properties, of, 260, 278–80, 310
viscosity, 706
Milk concentrates, rheology, relevance of, 327
Milk, heat-treated:
microbial (LAL) assay, 463
shelf-life prediction, 515
Milk powder:
compositional analysis by NIR, 128
fat and water content by NIR, 149–50, 153, 156
microbial assay (impedance), 490, 509, 512
microbial (DEFT) assay for raw, 461
moisture calibration, 160
moisture assay, 18–9, 437
moisture and ash assay, 443
moisture content by microwave meas., 209–11
moisture content by NIR, 126–8, 141, 143–5, 148
protein monitoring, 152

Milk powder – *continued*
reflectance spectra (NIR), 126
seasonal variation of composition, 152
Milk products:
colour measurement for, 85, 91–2
colour of translucent materials probe, 609
pH measurement, 608
water activity in, 385
water content by microwave meas. during processing, 227
Milling, monitoring by AEM, 245
Milltronics instrument, 299
Ministry of Agriculture, Fisheries and Food, xxv, 575
Minolta instrument, 52, 76, 86, 87, 89
Minute changes, determination
acoustic properties, *see* SAW device
chemical composition of gases and liquids:
see CHEMFET, Organic semiconductors:
see Biosensors, Immunosensors and sorption sensors
mass of thin film, *see* Piezocrystal balance
optical refractive index, *see* SPR, TIR
MIR, *see* Mid Infrared
MIS (metal-insulator-semiconductor) devices, 597
Mixing, monitoring by AEM, 245
Mobrey HiSens, 310
Moisture assay:
drying systems, 436–7, 440–4
infrared drying, 437
microwave drying, 437, 443–4
oven drying, 18–9, 437, 440–3
vacuum oven drying, 437
Karl Fischer procedure, 437–8
Karl Fischer systems, *see* titration systems
notes, 435, 436
titration systems, 18–9, 437–8, 439–40
interference from aldehydes, 438
sensitivity, 439
turbotitrator, 439–40
see also Lactose, influence
Moisture conditioning for calibration, 159–60
Moisture content, intermediate, 382, 388
Moisture retainers, 388
Moisture measurement on-line:
comparison of techniques, 21, 138
microwave, by, 190–231
extrusion cookers, in, 205
high moisture materials, for, 227–8
low moisture materials, for, 202–4
low and medium moisture content materials, for, 204–5
slurries, for, 204
particulate materials, for, 205–11
see also Microwave measurement
NIR (near infrared), 123–149
low moisture, 140, 147
high moisture, 140, 142, 147
see also Near infrared instrumentation (on-line)
significance, 21, 24
ultrasonic water content determination:
solutions, 244, 250–6, 317
slurries, 256–7, 277, 312–5
see also Ultrasonic measurement
see also Water content
Molar, 708
Molasses:
glucose monitoring, 632
level measurement for, 298
Molecular sieves, 617
Molecular size, discrimination by, 644, 645
Molecular shape, discrimination by, 645
Molecular vibration, 123–6
Molecular weight of compounds, choice of assay system, 422
MOMCOLOR, 86
Monochromatic sorting, 102–4, 117
Monosaccharides concentration:
relationship to speed of sound, 251–5
see also Sugar solutions
MOSFET (metal-oxide-semiconductor FET), 600
multi-ion sensing:
fibre-optic, 599
microelectronic, 599, 604, 606
Mould count, 646

Mould growth:
  see Micro-organisms, growth of
  see Microbial toxins, production of
Moulds and yeasts, assay, 464, 477, 491, 512
Mouthfeel, 325
MOX devices, 613
  see MeOx (metal oxide) devices
MR, *see* NMR (nuclear magnetic resonance)
MS (Mass spectrometry) systems, 421, 423
Multi-ion sensing, 589, 599, 604, 606
  see also Chemosensors
Multiple regression analysis, applied to NIR measurement, 151, 153
Multiplex measurement heads, microwave measurement, 192
Multiplexed cell switching, 479
Multiplexing circuit, internal, 110
Multi-saccharide assay, 636
Multi-saccharide sensing, 627, 639
  see also Biosensors
Multivariable measurement
  FET-based chemical sensors, 637
  TIRF/CCD, 649
  see also FOP, Infrared thermometry, Microwave measurements
  see also Near infrared instruments, Ultrasonic measurements
Multivariate analysis, 366–9, 660–3
Muscle foods, *see* Fish, Meat
Muscles, chewing, *see* Texture assessment
Mustard seeds, sorting, 100
Mutagenicity, 585, 625
Mycotoxins, sensors, 624, 646
Mycotoxins, production conditions, 383
Myoglobin, 54, 88

NAF (non-urea-adducting fatty acid content), 532, 534, 536, 537, 539–40
NAFE (NAF esters), 539, 541
Nagy Mess-systeme, 403
National Dairy Research Centre, Ireland, xxxi
NCC (new chemical compounds) content, 532
Near- and mid-Infrared range:
  brief overview, 29
  wavelength, wavenumber and frequency, 709
Near infrared absorption spectroscopy:
  absorbing molecular groups, 126, 139, 149, 157
  Beer-Lambert transmission law, 127, 143
  carbohydrate content, 128, 149, 154
  comparison NIR/MIR spectroscopy, 125, 139
  compositional analysis:
    on-line, 149–56
    on-line/off-line comparison, 122–3, 128
  conductivity independence, 136
  control of moisture, 123–49
  diffuse reflectance, 126
  fat content, 128, 149–54
  frequency range, 123–4
  integrating sphere, 126–7
  molecular vibration:
    fundamental, 123–4
    modes, 125
    overtones, 124, 126
    water molecule, 125
  organic coatings, 156
  principles, 123–8
  principle absorption bands of water, 126, 139
  protein content, 128, 149–54
  reflectance measurements, 122, 127
  reflectance spectra, milk powder, 126
  scanning spectrophotometer, 126, 149
  scattering effects, 127
  starches, 149
  state of water, influence, 136–7
  sugar content, 128, 149, 154
  temperature of product, independence of, 136
  transmission law, 127, 143
  transmittance spectrum of atmospheric water, 137
  transmittance spectrum of liquid water, 137, 139
  water content, 121–49
  wavelength range, 123–4
Near infrared instruments (on-line):
  accuracy, 148
  alignment, 128

Near infrared instruments – *continued*
applications overview, 148
atmospheric humidity, 137
automatic cleaning, 141–2
back scatter configuration, 140
baked products, 138, 140, 147
bound water, influence, 136
caffeine content, 154–5
calibration for composition measurement, 152–3, 157–8
pre-calibration, 153, 158
proving sample set, 153
training sample set, 152
calibration for moisture measurement, 144, 157–64
carbohydrates, 149–56
coating weight, 156
colour, compensation for, 142, 153
comparison to microwave measurement, 138
composition:
compensation in moisture measurement, 144
determination of, 149–56, 164
continuous monitoring, 136
cross-checking procedures, 161–4
crystalline materials, 147
density of product, effect, 136
detectors:
improved lead sulphide, 133
lead sulphide, 129
primary, 129
secondary, 129
distance, sensor-product, 132, 140, 161
dried products, 138, 143, 147, 161–3
dust, compensation for, 132
dust tolerance, 128, 141
electronic gating, 140
extruded products, 147
fat content and type, compensation for, 144
fat content measurement, 140, 149–56
feedback control, for, 132
film thickness, 156–7
filters, 128–9
glossy foods, 140–1, 147
hygienic design, 141
incident energy, compensation for, 131–2
industrial characteristics:
see operating conditions
infragauge, multi-filter, 149
installation:
composition monitoring, for, 156
moisture monitoring, for, 140–2
Kubelka-Munk scattering analysis, 134, 158
lamp colour temperature, compensation for, 130
liquid foods, 142, 147–9
moisture, 121–49
applications overview, 148
compensation for in composition measurement, 150
measurement of, 128, 140, 142–9
moisture range, 147
multiple regression analysis, 151, 153
multi-wavelength measurement, 133, 143–6, 150–6
nicotine content, 133
non-linearity, 144
non-reflective materials, 147
operating conditions, 128, 141, 151
packaging materials, screening, 156–7
particle size and shape, compensation for, 142–6, 151, 153
particulate (coarse) materials, 138, 140, 147, 160
pass height, 140, 151
penetration, 137, 140
phantom components, 152
powdered and granulated materials, 138, 140, 147, 160
product identification, 164
protein content measurement, 140, 149–56
reference wavelength, 132
response time, 162–3
product scanning, 141
robustness:
of calibration, 152–3
of sensing heads, 141, 151
salt independence, 136
sensing head temperature, compensation for, 130
shielding, ambient light from, 136
solid foods, 140
solid state instrument components, 130

source:
acousto-optical tunable filter (AOTF)
IREDs (infrared emitting diodes)
laser diodes, 130
quartz halogen lamp, 128–9
screw conveyor, 141
starch content measurement, 149–56
sugar content measurement, 149–56
transmission configuration, 142
two-wavelength moisture measurement, 131–2, 142–3
water content, *see* moisture
wavelength selection, 139–40
window materials, 141–2
window influence on calibration, 161
see also Near infrared absorption spectroscopy
see also entries for specific foods
Near infrared instruments (off-line):
derivative spectral analysis, 134
Fourier transform analysis, 134
full spectrum analysers, 133
Mahalonobis distances analysis, 134
operator dependence, 134
avoidance, 135
partial least square analysis, 134
principal component analysis, 134
reflectance mode, 133
sample preparation and presentation, 134–5
sample scanning for reduced preparation, 135
scanning monochromator, 133
silicon detectors, 133
tilting filter system, 133
transmission mode, 133
typical laboratory analyser, 134
see also Near infrared absorption spectroscopy
Near infrared measurement, 665
Near infrared reflection analysis (NIRA), *see* Near infrared absorption spectroscopy
Near infrared spectrum:
use in sorting, 101, 104, 117
use in composition and film thickness measurement, 121–65
Neper, 712
Nestec, Nestlé/Westreco, xxvii–xxix
Neukum, 609
Neural networks, 20, 246, 689, 616, 660–2
potential pitfalls, 661–2
New chemical compounds content, 532
Newton, conversion, 702
Newtonian liquid:
definition, 327
viscometer for, 333
Niacin, microbial assay (impedance), 490
Nickel phthalocyanine, 614
Nicotine content by NIR, 133
NiPc (copper phthalocyanine), 614
NIR, *see* Near infrared
NIRA (near infrared reflection analysis), *see* Near infrared absorption spectroscopy
Nitrate sensors, 585, 624
Nitrate/ nitrite sensors, 594, 606
Nitrogen, 612
Nitrogen dioxide, 614–5, 647
Nitrate assay, 429
Nitrite assay, 429
Nitrogen assay (for protein and meat content):
acid digestion systems, 445–52
combustion systems for assay of, 452–3
Kjeldahl procedure, 445
NMR (Nuclear magnetic resonance), 26
NMRI (NMR Imaging), 26
Non-contact measurement, 16, 17
alternatives to chemical sensors, 588, 665
CAP-food freshness, 527–8
illustration, 6–7
rancidity, 527
volatiles, 665
see Infrared thermometry
see Microwave measurements
see Near infrared instruments (on-line)
see Ultrasonic measurements
see also Colour measurement, Sorting by colour
see also Gas sensors
Non-destructive measurement:
illustration, 6–7
kernel maturity, for, 88
see also Non-contact measurements
Non-destructive testing, ultrasonic, 243–4, 291

Non-invasive measurements:
  see Non-contact measurements
Non-Newtonian liquid:
  description, 329
  flow metering, 9
  viscometers for, 336–40
Non-sacrificial, *see* Non-destructive
Non-urea-adducting fatty acid content,
    see NAF
Nordion analyser, 424–7
Nose, animal, 611
Nose, electronic, 612, 616
Novasina instrument, 401
Novel chemical sensors, 21
Novel sensors for the food industry,
    521–669
  contributions to development, 558,
    575
  examples (marker approach),
    528–76, 714
  phases of development, 556–7,
    573–4
Np (Neper), 712
Nucleosides, 624
Nucleotides, 624
Nuts:
  microwave measurements, 191
  sorting, 100, 102, 117
Nut shells, removal, 104, 117
Nylon (polyamide) film thickness, 157

Oats, moisture monitoring, 148
Odorant sensitive films, 652, 654, 660
Odour assessment, 601, 612–7, 652,
    654, 660–2
  see also Chemosensors
OFA (oxidised fatty acids), 539, 541
Off-line instrumentation:
  definition, 684–5
  notes, 3–8, 10
Off-site instrumentation:
  definition, 685
Ohmic heating, 8
Ohmic contact, 596–7
Oils (edible):
  colour, 72
  density, 546, 706
  flow rate measurement, 305
  level measurement, 297, 298
  moisture assay, 437, 438, 440–1
  rancidity, 525–7
  rheology of, 329, 331
  ultrasound propagation in, 257,
    260
  viscosity, 537–45, 706
    see also Frying oil viscosity
  viscometry for, 333, 343
  see also Frying oils
Oil, palm acid, moisture assay, 440
Oilseeds, moisture assay, 442
Oily foods:
  colour measurement for, 76
  microbial assay, 509
  moisture assay, 437
Oligosaccharides assay, 432
Olive oil, 537–40, 544
On-chip integration, 600
On-line biosensor probe, 632
On-line measurements:
  colour, 52
  composition and other variables:
  definition, 684–5
  notes, 3–9, 18–9
  operating conditions, 11–6, 128, 141,
    151
  viscosity, 340–2
  see also Infrared thermometry
  see also Measurement types
  see also Microwave measurement
  see also Near-infrared measurement
  see also Ultrasonic measurement
On-line instrumentation:
  benefits, 5
  cross-checking of calibration, 5,
    18–9
  illustrations, 6, 10
  interdisciplinary cooperation,
    relevance to, 21–2
  interlinking with off-line analysis, 5,
    18–9, 22
  notes, 3–9
  requirements, 13, 14–6
  see also On-line measurements
Opaque objects, 45, 52, 72, 76, 79
Open-ended coaxial RF probes, 15,
    204, 217, 229–30
Operating conditions for sensors, on-
    line, 11–6, 128, 141, 151
Operator dependence, 134
  avoidance, 135
Opiates assay for drugs, 423
Opponent coordinates, 48–51
Opponent mechanism, 43

Optical base transducers, 598–9, 647–52
Optical biosensors, 625, 644–52
Optical classification of foods, 78–9
Optical fibres:
 see Fibre-optic cables
 see Fibre-optic sensors, FOP (fibre-optic probes)
Optical devices, 644–52
Optical filters, 105, 108–10, 128–9
Optical inspection systems, 98–100, 104–10
Optical sorting, *see* Sorting by colour
Optical surface properties, 644
Optically active materials, 130
Optodes, 651
Opto-electronic dew point hygrometer, 403
Optrodes, 651
Orange juice:
 colour, 56–9, 65, 72, 74, 85, 91
 flow rate measurement, 305
 microbial assay (impedance), 512
 separation monitoring, 277
Orange skin, smoothness, 277
Organic acids assay, 424
Organic acids, sensors, 585, 624
Organic conducting electrode biosensors, 626
Organic dyes, 632
Organic semiconductor gas sensors, 614–5
Organochlorine pesticides assay, 427
Organophosphorous compounds, 644
Organophosphorous pesticides assay, 426–7
Osmotic pressure, 378
Ostwald-Viscometer, 335
Oven (baking)
Oven drying, moisture assay, 18–9, 437, 440–3
 control, 138, 140
 see also Baking oven, *see* Microwave oven
Overrun, relationship to rheology, 326
Oxidases, 627
Oxidative rancidity, *see* Rancidity
Oxidation of heated oil, 528–9
Oxidation processes:
 pigment in fresh meat, 54–6, 88–9
 relationship to water activity, 378
 ultrasound-assisted, 246
Oxides, electronically conducting, 595
Oxides, insulating, 585, 597, 603
Oxidized fatty acids, 539, 541
Oxygen, 612, 614, 615
 see also Gases
Oxygen, dissolved, 651
Oxygen pressure, influence on biosensors, 627, 634–5
Oxygen probe as base device, 623, 626, 635

Packaging materials:
 coextrudates, 157
 laminates, 156
 organic coating weight, 156
 plastic film thickness, 157
 screening, 156–7
 see also Construction and packaging materials
Palladium (Pd), 585
Palm oil, 538, 545
Palm oil fraction, 539
Palm olein, 529, 538, 545, 551
Panels, 23
 Paprika, *see* Capsicum
Paralytic shellfish poison, 651
Parasites, water activity tolerance, 382–4
Parenteral feeds, pyrogen assay, 463
Parsley, colour, 80–1
Partial least square analysis, 134
Particles, dispersed:
 assessment by ultrasonic measurement, 257, 277, 312–5
 influence on ultrasound propagation, 256–7, 273–8
 significance of size distribution, 24
Particles, colloidal, monitoring of, 277
Particulate materials:
 colour measurement, for, 76, 82, 87–8
 microbial assay, 509
Particulate (coarse) materials, NIR measurement for, 138, 140, 147, 160
Pascal, conversion, 699
Pascal second, conversion, 705
Pass height, variation tolerance:
 NIR measurement, 140, 151
 microwave measurement, 205–11

Passive mode measurement, 675
acoustic emission monitoring, by, 245–6
Pasta:
colour, 76, 82, 88
microbial assay (impedance), 509
microwave measurements, 191
problems with moisture monitoring, 147
Pastes:
colour measurement, for, 77, 84, 90–1
infrared emissivity data, for, 182
level measurement for, 299
Pastilles, texture assessment for, 363, 365, 366
Patents, 632
Pathogens:
Listeria, 646
Salmonella, 646
see Safety of food
Pattern analysis techniques, 616, 659–662
Pattern recognition, 654, 660–2
piezocrystal microbalance, 644, 652
PbPc (lead phthalocyanine), 614–5
PC (polar components content), 532–4, 536, 538–9
PCB (polychlorinated biphenyls) assay, 427
Pd (palladium), 585
PE, *see* polyethylene
Peaches, sorting, 100
Peanut oil, *see* groundnut oil
Peanut processing, ultrasonic measurements in, 315
Peanuts:
bromide assay, 429
colour of hulls and kernel maturity, 82
sorting, 102, 104
Pears, colour, 83, 88
Peas:
level measurement for, 299
sorting, 117
texture assessment for, 352
Peck, black, detection of, 102
Peltier cooled element, 395, 401–3
Penetration depth:
infrared thermometry for, 169:
NIR measurement, 137, 140
non-contact composition measurement, for, 266
Penicillin, 526, 624, 636
Pentanal, 527, 624
Peppers, bell, *see* Capsicum
Perkin-Elmer, 421
Permittivity, *see* Dielectric properties
Peroxide probe in enzyme analyser, 434
Peroxide micro-probe, 636
Peroxide probe as base device, 623, 626, 633–6
Peroxide value, 531–2
Pesticides assay, GLC system, 424–7
Pesticides sensors, 585, 624
Pet foods:
humectants in, 388
water activity measurement, 400
Petroleum-ether-insoluble oxidized fatty acids content, *see* PIOFA
PFA (polymeric fatty acids content), 532
pH dependence of biosensors, 630–1
pH sensors, 589–608, 650–2
biosensors based on, 623, 625, 637
characteristics, 590, 604
conventional (glass membrane electrode), 589–91, 592–3
armed, pistol type, 609
characteristics, 590
development history, 591–4, 606–9
FOP (fibre-optic probe) pH sensors, 598–9, 650–2
ISFET, 585, 589, 594, 595–8, 599–611
industrial probe, 607–9
meat industry applications, 606, 608–9
metal/ metal oxide electrodes, 585–6, 595
microenvironment, 589, 599, 604
REFET reference device, microelectronic, 605
stab probes, 607, 609
pH value:
comparison with acidity, 609
examples, 707
indicator of fermentation process, 389
legal requirements for, 384–5
meat quality control, 608–9
microbial deterioration, influence on, 376, 382

significance, 24, 589, 609, 664
Phantom measurements, 152
Pharmacia, 646
Phase changes, monitoring of, 269
see also Solid/liquid ratio, Ice
Phenazonium, 632
Philips, 592
Phosphates assay, 428
Photoconductors, in thermometry, 170–2
Photodetectors:
in HPLC, 421
in sorting, 110
in thermometry, 170–1
Photoelectric instruments for colourimetry, 50–3, 66–7
Photomultiplier, in sorting, 110
Photopolymerized membrane, 603
Phthalocyanines, 614–6
Physical changes on binding, 644
Physical sensors for chemical and microbial variables, 663
definitions, 682–3
see Agglutination assay
see Frying oil monitor
see Impedance microbiology
see Near infrared measurement
see Marker approach
see Mechanically resonant probes
see Microwave measurement
see Ultrasound measurement
PHVO (partially hydrogenated vegetable oil), 529, 551
PID (proportional integral differential), 687
Piezocrystal microbalance, 644, 652
affinity-, enzyme- and sorbent coatings, 652
coating mass sensitivity, 652
liquid density and viscosity response, 652
Piezocrystal transducers, 291, 294–5, 535, 546–9, 673, 676, 681
biomass monitoring, ultrasonic, 652
see also Mechanically resonant probes
Piezoelectric materials, *see* Quartz, Lithium niobate, PZT
Piezoelectric oscillator, 652
Pigment content, rapid assessment:
apples, 89
beverages, 91
capsicum, 89–90
meat, 54–6, 88–9
pasta, 88
semolina, 80–1
Pigment oxidation in fresh meat, 54–6, 88–9
PIOFA (petroleum-ether-insoluble oxidized fatty acids content), 532, 536, 539–40
Pipelines, 325–6
Pixels number, 110
Planar micro-Clark device, 623, 635–6
Planar silicon technology, 600, 676–7, 681
Planck's formula, 178, 710
Plant design, relevance of rheology, 325–6
Plasmons, 647
Plate count, 460, 560
correlation with impedance methods, 488–91, 500, 503, 509–10
Platinum electrode, *see* Electrodes
Platinum oxide, PtO_2, 585, 595
PLC (Programmable logic controller), 11, 659, 687
Pleva moisture meter, 231
PLS (partial least square models), 660
pattern analysis techniques, 616, 659–662
Plasticity in data processing, 661
Pockets, high pH, 604
Poise, conversion, 705
Poiseuille, conversion, 705
Poisoning:
chemical sensors, 665
humidity sensors, 393, 399–400
Pol (polymeric triglycerides content), 532, 536, 537, 539–40
Polar components content, 532–4, 536, 538–9
Polar compounds, 614–5
Polar liquids, Dielectric data, 196
Polarity of compounds, choice of assay system, 422, 423
Polarity, discrimination by, 644, 660
Polyamide, *see* Nylon
Polyanilines, 614–5
Polycarbonate membrane, 634
Polychlorinated biphenyls assay, 427
Polydimethyl siloxane additive, 551
see also silicone additive

Polyester windows, 141
Polyester/ glass laminate windows, 142
Polyethylene film thickness, 156
Polyethylene, as infrared window material, 188
Polygraph, in electromyography, 360
PolyHEMA membrane, 603
Polyhydric alcohols assay, 432
Polymers, *see* Gels
Polymer film thickness, 121, 156–7
Polymer films, grown onto fibre-optic probes, 598
Polymer (conducting) gas sensors, 611–7
Polymer membranes, 603
Polymer sensing tip (FOP), 650
Polymeric fatty acids content, 532
Polymeric triglycerides content, 532, 536, 537, 539–40
Polymerization:
  of hot oil, 528–9
  linking to chemical changes, 664
  see also Agglutination
Polymerized and oxidized fatty material (POM), 529–32, 552–5, 714
Polyphosphates assay, 428
Polypropylene film thickness, 121
Polypyrroles, 614–7, 632–3
Polysaccharides concentration:
  relationship to speed of sound, 251–5
  biosensors, 645
  see also Sugar solutions
Polysaccharides, bacterial, 463
Polyunsaturated oils, 528
Polyurethane membrane, 603, 636
Polyvinyl chloride film thickness, 156
POM (polymerized and oxidized fatty material), 529–32, 552–5, 714
Pork:
  PSE, 609
  freshness, *see* Meat freshness
Pork fat, water activity of, 387
Porphyrins, 616
Post mortem period, pH measurement, 609
Potassium content measurement, significance, 24
Potassium sensors, 585, 594, 606
Potato crisps (UK), Potato chips (US), 528
Potato chips (UK), French fries (US), 528, 552
Potato flakes, rancidity, 525–7
Potato chips/crisps:
  colour, 81, 87
  microwave measurements, 191
  NIR moisture monitoring, 148, 162
  texture assessment for, 359
Potato mash, colour, 84, 90
Potato pieces:
  colour measurement, 82, 83, 87, 89
  stacking level measurement for, 299
Potato, mashed, rheology of, 324
Potatoes, sorting, 100
Potatoes, sugar content, 165
Potentiometric biosensors, 623
  see also ENFET
Potentiometric immunosensor (IMFET), 638
Potentiometric pH and ion sensors, *see* ISFET, ISE
Powdered and granulated materials:
  colour measurement, for, 75–6, 80–1, 87
  moisture assay, 438
  moisture and density by microwave meas., 205–11, 217–9
  NIR measurement for, 138, 140, 147, 160
  specific food powders, grains, etc.: *see* entries for named foods
Power law equation, in rheology, 329
Pre-amplifier, 601
Pre-amplification, intrinsic, 636–7
  see also FET, ISFET, ENFET, CHEMFET
Precipitation monitoring by AEM, 245
Precursor variables, 5, 19, 588, 664
Preservation processes, 378–9
  see also Curing, Sugaring
Preservatives, 24–5, 382, 624
  fungal growth inhibitor, 389
Pressure drop, as indicator of flow rate, 332
Pressure sensors, 653, 681
Pressure units, conversion, 699–700
Pressures in food processing, 15, 700–1
Preventative maintenance, application of AEM, 245
Primaries, 46, 65
Principal component analysis, 134

Process control, 686
rheology, role of, 326–7
Process engineering and control, references, 28, 690
Processes in the food industry, 688
associated temperatures, 698–9
associated pressures, 700–1
Process stream characteristics, relationship to rheology, 326
Produce, *see* Fruit, Vegetables
Product identification, 164
Propylene glycol, as humectant, 388
Protective lacquer coating weight, 156
Proteins, sensors, 638
Protein content:
assay, 445–53
notes, 435
see also Nitrogen assay
on-line by NIR measurement, 140, 149–56
rapid off-line, 123, 128
significance, 24
see also Fouling
Protimeter, 402
Proving sample set, 153
Proximate assay systems, 435–53
Proximates measurement, significance, 24
PSE (pale soft exudative) pork, 609
Pseudoplastic materials, 329
Pseudo-reference electrode, 602, 605
Psychrometric method for water activity meas., 395–6
PTFE (polytetrafluoroethylene, Teflon):
membrane, 636
microwave window, 214
PtO_2 (platinum oxide), 585, 595
Pulsed amperometric detector, pulsed HPLC in, 428, 431–2
Pulses, sorting, 100
Purines, 524, 526, 624
in fish, 524
Putrescine, 526, 624
PV (peroxide value), 531–2
PVC (polyvinylchloride) membranes, 603
PVC film thickness, 156
Pyridine, 615
Pyroelectric devices, in thermometry, 170–2
Pyrogen assay, 463
PZT, *see* Lead zirconate titanate

QA, *see* Quality Assurance
QC, *see* Quality Control
QC laboratory measurement:
brief overview, 32
chemical laboratory, 418–54
interlinking with on-line instrumentation, 5, 18–9, 22
interpretation of readings, 17–9
official methods, 18
pH measurement, 589–594, 607–9
sample and method dependence, 17–9
sampling and transfer, 5, 18
see also Chemical QC laboratory
QC laboratory, microbiological, *see* Impedance microbiology
QDA (quantitative descriptive analysis), 351
Quality of food, 2
relationship to colour, 64–5, 98
relevant properties, 12, 22–5
rheological properties, role of, 326
texture, effect on, 349
water activity, effect on, 377–8, 389
see also Frying oil degradation, heat-induced
Quality assurance:
instrumentation for, 1–34, 686
measurements in, 12–3
role, 2–3
system, 10
Quality control, 10, 499–517, 686
measurements in, 12–3
chemical laboratory, 418–54
see Chemical QC laboratory
Quartz, 28, 142, 223, 291
Quartz microbalance, 652
Quartz SAW device, 652
Quasi-one-dimensional conductivity, 614
Quinone derivatives, 632

RABIT analyser, 481–2, 488, 515–6
Radar, 242
Radiation thermometer, *see* Infrared thermometry
Radiofrequency excitation, *see* SAW, *see* NMR, *see* Open-ended coaxial probes

Radiofrequency heating, 8, 27
Radiowaves:
  brief overview, 30, 241
Raffinose concentration:
  relationship to speed of sound, 251–5
  see also Sugar solutions
Rancidity (oxidative), progress in frozen food, 531
Rancidity assessment, 523–7
  GC-MS study, 525–7
Rapeseed oil, (probably LEAR: low erucic acid rape), 539, 545
Rapid methods, comparison, 560
  ultrarapid methods, *see* Marker concept, Meat freshness probe
Rapid composition analysis, 133–5
Rapid microbial methods:
  impedance techniques, 457–517
    see Impedance microbiology
  overview, 460–3
    bacterial ATP (adenosine triphosphate):
      bioluminescent assay, 461–2
    DEFT (direct epifluorescent filter technique), 461–3
    electrical methods, 460–2, 517
    LAL (limulus amoebocyte lysate), immunological agglutination assay, 461–3
Rare earth compounds, fluorescent, 650
Raspberries, freezing control, 309
Ratatouille, texture assessment for, 369
Rayleigh waves, 249
Reading, IFR, xxx
Reading, Unversity of, xxx
Reagent supply, continuous by controlled-releasė polymer, 652
Receiver mode, 673
Receptor-based chemical sensors, 645
Receptor discrimination, 611–2
Recovery time, gas sensors, 614
Red enhancement, 55
Redox couples, 625
Redox potential, influence on microbial deterioration, 376, 382
Redox reaction, analyte-sensor:
  biological, *see* Biosensors
  non-biological, *see* Chemosensors, MeOx sensors
Redox sensitivity of metal/ metal oxide pH electrodes, 595
Redox sensitivity of metal/ metal oxide gas sensors, 613
Reference assays, 164
Reference element, 584
Reference (Ag/AgCl wire) electrodes, 627
REFET, microelectronic potentiometric reference device, 597, 605
Reflectance measurements:
  near infrared, 122, 127, 133
  see also Colour measurement
  see also Infrared thermometry
  see also Microwave measurement
  see also Ultrasound measurement
Reflectance spectra (NIR), milk powder, 126
Reflection spectrophotometer, 101
Refractive index, *see* Dielectric properties
Refractive index changes on binding, 644
  detection, *see* SPR, TIR
  enhancement labels, 647
  profiling, 649
Refractive index detectors:
  HPLC in, 421, 431
  sugar concentration for, *see* Brix
Refractometric method for water activity meas., 395
Refrigerant level control, 311
Reject level (microbial QC), 506
Relaxation:
  dielectric, 194, 196, 200, 218
  microwave measurement, relevance to, *see* dielectric
  ultrasound attenuation, contribution to, 271–3, 274–7
Relaxation time:
  acoustic measurements, relevant for, 271, 274
  dielectric measurements, relevant for, 194, 196, 200
Remote instrument:
  immunosensor, 649
  IR thermometer, *see* Infrared thermometry
  microwave, 192
  ultrasound, 249
Remote measurements by fibre-optics:

immunosensing (TIRF/FOP), 649
IR thermometry, 187–8
Repeatability, *see* Industrial measurements
Reproducibility, sensors with high, 676
Requirements for on-line instrumentation, 13, 14–6
Residence time (distribution), relationship to rheology, 326
Residual moisture monitoring, 149
Residues, 424–7, 526, 585, 624, 636
Resonance frequency, piezocrystal, 644
Response time:
biosensors, 633
chemosensors, 614
on-line instruments, 162–3
pH and ion sensors, 603
Retrofitting, *see* Clamp-on
RF (radio frequency) probes, *see* Open-ended coaxial probes
Rheological measurements, 323–72
conditions to be selected, 324, 357
emulsions, for 334
foods, for specific, *see entries for named foods*
relevance, 324–7
solutions, for, 333
suspensions, for, 334, 339, 343
see also Texture assessment
Rheological properties, 323–57
Rheology:
enrobing (or coating) materials, 326
food properties, relationship to, 326
crystal size, 326
droplet size, 326, 327
mouthfeel, 325
overrun, 326
sensory attributes, 325
plant design, relevance for, 325
process stream characteristics, relationship to, 326
flow velocity profile, 326
residence time (distribution), 326
significance, 24
see also entries for specific foods
Rheometer, 357
Riboflavin content, 92
Rice, microwave moisture measurement for, 227
Rice grains, sorting, 100, 102, 117
Rigidity, 248
Ripeness of fruit and vegetables, 266, 277, 526, 629–32
Ripening, indicated by colour, 44, 82, 89
Ripening sausages:
control of, 389
indicators of, 389
Roasting end-point, volatiles, 612, 616
Robot graders and sorters, 116
Robotic sampling and conditioning, 16, 17, 662–6
example: dairy fermentation, 663–4
Robust calibration, 152–3
Robustness of sensing heads:
density, *see* Mechanically resonant probes
gases and volatiles, *see* Chemosensors
microwave measurement, 192
near infrared measurement, 123, 141–2, 151
photodetectors, 110
pH and ions, 589, 604, 607–8
viscosity, *see* Mechanically resonant probes
see also Fibre optics
Robust transducers, 676, 681
Rödel, W., xxxi, 375–415
Roentgen rays, 33
Rotronic Hygroskop, 400
Rotten beans or nuts, removal, 102, 104, 117
RSSL (Reading Scientific Services Ltd), xxv
Ruggedness, *see* Robust sensing heads

Saccharides assay:
enzyme electrode analyser (bench-top), 433–5
enzymic test, 430
GLC by, 432
HPLC system, 427–32
titrimetric, 432
see also Biosensors
Saccharides, sensors, 585, 624, 627, 636, 639, 645
Saccharide solutions:
saccharide content and composition, significance of measurement, 24–5

Saccharide solutions – *continued*
non-invasive concentration measurement, *see* Sugar solutions
Safety of food, 376
microorganisms of public health significance, 486, 491, 511–2
relevant properties, 12, 24–5
water activity, role of, 378–85
see also pH sensors, pH value, Water activity
see also Robustness of sensing heads
Safflower oil, 539, 544
Salad cream:
droplet sizing, 279
moisture assay, 440
ultrasonic properties, of, 260, 278–80
Salad dressings, rheology of, 324
Salami:
colour, 82, 87
water activity, 386
Salmonellas, 486, 491, 512, 646
Salt content, relationship to water activity, 381
see also Sodium, Potassium
Salt slurries, for calibration in water activity measurement, 393–4
Salt solutions, concentration, by ultrasonic measurement, 251–6, 317
Salt/filter paper method, for water activity measurement, 396
Salting as preservation process, *see* Curing
Salts, dissolved, effect on dielectric properties, 201
Sample preparation and presentation, 134–5
for moisture calibration, 159–60
Sample volume metering, automatic, 649
Sampling and conditioning, robotic, 662–6
Sampling regime, 161–4
Sapphire windows, 141, 186–7
Saturation, colour of, 49
Sauce, pH measurement, 608
see also Acidity
Sausage, pH stab probes, 609
Sausage ripening, control of, 389
Sausages:
moisture and ash assay, 443
nitrogen/ protein assay, 453
role of water activity, 382–93
see also Meat products
SAW (surface acoustic wave device), 249, 585, 644, 652–4
chemical sensors, 652–4
pressure sensors, 653
temperature sensors, 653
VHF circuit component, 653
see also Immunosensors and sorption sensors
SbO_3 (antimony oxide), 585, 595
Scaling, 158–9
Scanning monochromator, 102, 133
Scanning spectrophotometer (NIR), 126, 149
Scattering:
colour measurement, in, 45, 58
microwave measurement, in, 190, 212
NIR measurement, in, 127
ultrasound attenuation, contribution to, 273–80
Schottky diode, 600, 615
Scombroid poisoning, 524
Scraped vessels, 16
Screw conveyor, instrumentation in, 141
Seafood toxins, 624, 651
Seasonal variations, 14, 152
Second phase inclusions, *see* particles, droplets, bubbles, crystal seeds
Sedimentation tank, blanket detection, 313–5
Seeds, dielectric properties of, 201
Seeds, sorting, 100
Selective sorption, *see* Immunosensors and sorption sensors
Self-calibrating instruments, 605, 606
Self-cleaning transducers, 301
Self-organizing instruments, 20, 113, 660
Semiconductor devices, 677–81
chemically sensitive, 589–618
see also Biosensors, Chemosensors
gas sensitive, *see* CHEMFET
photosensitive, *see* Infrared thermometry, Detectors
Semiconductor gas sensors:
see MeOx gas sensors, *see* Organic semiconductor gas sensors:

see Polymer (conducting) gas sensors
Semi-solid, 324
Semi-liquid, 324
Semolina:
colour, 80–1
moisture measurement for, 147
Sensitivity, 158
see also Sorting by colour
Sensitivity, *see* also Minute changes, determination
Sensoptic (now Sentron), 607–9
Sensors:
arrays, *see* Arrays of sensors, Arrays of detectors:
see Arrays and pattern recognition
configurations and installation, 6–7
see also Configuration examples
definition, 675
fuzzy-assisted, 687
notes, 675
role in quality assurance, 3, 1–34
see also Non-contact measurement, Novel sensors:
see also Chemical sensors, Physical sensors
Sensor/ reference pairs, 584, 605
Sensor technology, terminology, 673–90
Sensory quality of peas, as phantom measurement, 152
Sensory attributes, 64, 91, 349
indicators of ripening, 389
rheology, relationship to, 325
texture, as, *see* Texture
Sentron, *see* Sensoptic
Separation step, elimination, 646
Separation, ultrasound-assisted, 246
Separation, monitoring by ultrasound, 277, 316
Separation systems, in colour sorters, 99–100, 111–2
Set point, 158
Setting, influence on ultrasonic properties of gels, 267
Setting of yoghurt in pot, 664
Settling, monitoring of, 277
Severinghouse gas sensor, 600
SFC (supercritical fluid chromatography) systems, 422–3, 454
SFE (supercritical fluid extraction), 423, 454
Shear thinning, shear thickening, 329, 343
Shelf-life:
limiting factors, 23, 378
relationship to water activity, 378
see also Stability
Shelf-life assessment:
colour as indicator, 91
impedance methods, 488, 490, 493, 514–5
marker concept, 524, 625
see also Meat freshness probe
Shell fragments, removal, 104, 117
Shimadzu, 86
SI units (The International System of Units), 695–7
conversion, 698–705
Sigma assay, 561, 564–5
Signal conditioning, 600
Signal domains, 674
Signal to noise, 599, 604
Signal processing:
integrated with sensor, 584, 660–1, 679–80, 687
see also Data processing
Significance of measured variables, 24–5
Silicon detector (photodiode):
colour measurement, 52
remote thermometry, 170–2
NIR measurement, 133
Silicon dioxide (SiO_2), 585, 597
Silicon sensors, 674, 676–81
see Silicon detectors
see also Semiconductor devices, chemically sensitive
Silicon technology, planar, 600, 676–7, 681
Silicone additive to frying oil, 529, 531, 551
Silicone-rubber membrane, 603
Silver traces, 652
Silver/ silver chloride reference electrode, 589–90, 595, 597
SiO_2 (silicon dioxide), 585, 597
Size distribution, *see* Particles, Droplets, Bubbles, Crystal seeds
Skin texture, 277
Slack, P.T., xxxi, 417–55
Slurries:
flow rate measurement for, 304, 306

Slurries – *continued*
infrared emissivity of, 182
moisture/ solids determination for, 204, 277, 312–5
Smart sensors, 600, 604, 614, 616, 676–7, 681
Smoke point, 529, 531–2, 539
Snack foods, relevance of texture, 350
Snacks, extrusion-cooked, microwave measurements, 191
Snack food mix, during extrusion:
rheology of, 327
moisture determination in, 205, 224
Snack products:
fat and water content monitoring, 153
moisture monitoring, 148
see also Frying oils
SnO2 (tin oxide), 585, 601, 613
Soap, microwave moisture measurement, for, 227
Sodium, sensors, 585, 594, 603, 606
Sodium chloride concentration, *see* Salt
Sodium content measurement, significance, 24
Sodium hydroxide solutions, *see* Caustic solutions
Soft drinks, *see* Beverages
Solenoid valve, 111–2
Solid contact to conventional membrane, 591, 594–5
Solid foods of large area, colour measurement, for, 54–6, 76, 83, 88–9
Solid/liquid ratio:
solid/liquid fat by NMR, 26, 33
solid/liquid fat by ultrasound, 257
significance, 24
ice/water by ultrasound, 266
Solids level, *see* Level
Solid-state actuators, 674, 676
Solid-state instrument components, NIR, 130
Solid-state sensors, 674, 676–81
see also Infrared detectors, Semiconductor devices
Solid-state sensors for pH, acidity and ions, 589–611, 618–9
Solid-state sensors for gases and volatiles, 599–603, 611–8
Solid-state transducers, 674, 676–81
Solubility:
assessment for triglycerides, 257
discrimination by sorption devices, 644
Solutions:
rheology of dilute, 329
specific food solutions, *see* entries for named foods
viscometer for dilute, 333
water content determination by microwave measurement, 215–7
Solute concentration, significance of measurement, 24–5
Solvents, organic, 644
Solvents, residual moisture monitoring, 149
Solvent extraction systems for fat, 438–9, 443–5
Sonar, 242
Sonochemistry, 247
Sonoprocessing, 27, 246–7
Sorbent coatings, chemically selective, 644
Sorbitol assay, 432
Sorption isotherms, 395
Sorption, selective, 611, 623, 644, 654, 660
chromatographic coatings, 660
selectivity mechanism, 644
Sorption/absorption/adsorption, 644
Sorption sensors, 644–54
see Immunosensors and sorption sensors
Sortex, xxx
Sorting by colour, 98–118
assessment of foods to be sorted, 100–4
reflection spectrophotometer, 101
illumination, 101
scanning monochromator, 102
computer vision systems, 114–6
fluorescence techniques, 104
machines, *see* Sorting machine
significance, 24
sorted foods:
beans, 100, 102–4, 117
biscuits, 100
cereals, 100
fruit, 100, 117
grains, 100, 102, 117
nuts, 100, 102, 117
seeds, 100

vegetables, 100
sorted unwanted materials:
bacteria, 104
damaged food particles, 100, 117
discoloured food particles, 100, 102–3, 117
foreign matter (e.g. sticks and stones), 100, 104, 117
rotten beans or nuts, 102, 104, 117
shells, 104, 117
wrong food particles, 104
television cameras, special applications, 104, 116
wavelenghts used:
near infrared, 101, 104, 117
ultraviolet, 104
visible, 101–4, 117
Sorting machine, 98–100
aperture, 108
background matching, 107
broad belt machine, 100–1, 111
chute machine, 100–1, 111
colour tracking, 113
detectors, 110
dust extraction, 112
efficiency, 117
feed system, 98–100, 111
acceleration, 111
alignment, 111
channels, 111, 117
delay, 111–2
metering, 111
presentation speed, 111
filters, 105, 108–10
illumination, 105–7
spectral curves of fluorescent tubes, 106
spectral curves of incandescent lamps, 107
UV for fluorescence techniques, 104
inspection system, 98–100, 104–10
learning facility, 113
mapping techniques, 114–5
running, 116–8
sensitivity, 118
separation system, 99–100
setting up, 113, 118
signal processing system, 99–100
sorting criteria, 101, 113
sorting modes:
bichromatic, 102–4, 117
dual monochromatic, 103–4, 117
monochromatic, 102, 117
trichromatic, 104
specular reflection, 105
throughput, 100, 111, 117
optimization, 118
user interface, 113–4
viewing conditions, 101, 105
wavelengths used, 101
window maintenance, 112
see also Sorting by colour
Sound waves, overview, 241
Soups, flow rate, 304
Source/ drain contacts, FET devices, 601–2
Sources, 674
see Illuminants
see Infrared sources
Sourcing (origin of meat or produce), 585
Soya-bean oil, 537–40, 544
Soya flakes and meals, moisture monitoring for, 141, 148
Soya products, protein monitoring, 152
Soya protein in beef mince, 646
SP (smoke point), 529, 531–2, 539
Span adjustment, 158, 162
Spatially resolved, 649
Speciation of meat, 585, 646
Specific gravity, *see* Density (specific gravity)
Specifications for chemical sensors, 588
Spectral curves, for coffee beans, 102–3
Spectral curves, illuminants, 106–7
Spectral reflectivity, as sorting criterion, 101
Spectrophotometers, 50–3, 86–9
chemical QC laboratory, in, 419–20
metals assay with, 420
Specular component, 45, 52–3, 79
Speed of sound, 247–67
see also Ultrasound propagation
Spent wash, water content monitoring, 147
Spices:
microbial assay (impedance), 509, 510
Spillage, prevention, *see* Level alarms

Spinach puree, colour, 84, 90
Spoilage, 612
  see Microbial deterioration
  see Oxidation, Hydrolysis
  see Freshness, Meat freshness
SPR (surface plasmon resonance) devices, 585, 644, 647, 648
  see Immunosensors and sorption sensors
Spray dryer, 327
Spray dryer control, 143, 161
Spreads, fat based:
  rheology of 324, 326
Squash puree, colour, 84, 90
Stab probe, 607, 609
Stability of food, 3, 8, 10, 376, 384
  relevant properties, 12, 23, 24–5
  water activity, effect on, 378–89
  see also Micro-organisms, growth of
  see also Microbial stability
Stability of juices, 277
Standard observer, 42, 46, 65–6
Standard values, 697
Starch, 624, 636
  microwave measurements, 191
  NIR moisture monitoring, 148
  see also Fouling
Starch content measurement, 149–56
Starch gels, viscometers for, 340
Starchy foods, moisture assay, 439
State, solid–liquid, relationship to dielectric relaxation time, 196
State of water, influence on NIR measurement, 136–7
Steam tolerance, 15–6, 141
Steel electrodes, *see* Electrodes
Steinecker Hygromess, 398
Sterility, commercial, 9
Sterilizable probes, *see* Hostile environments
Sticks, detection and removal, *see* Sorting by colour
Sticky pastes, measurements for, 299
  see also Non-contact Measurement
Sticky materials, moisture monitoring for, 141
Stinkers, removal, 104
Stirred tanks, measurements in, 299, 302
Stoke, conversion, 705
Stones, detection and removal, *see* Sorting by colour
Strain gauge force sensor, 364
Strawberry wine, colour, 85, 91
Stress-strain relationships, 327–31
Stretchability, 248
Stripline sensors, 227–9
Structure of food, 11
  relevance for colour measurement, 44–5
Sucrose assay, 430
Sucrose, sensors, 624, 627, 639
  see also Saccharides, Sugar
Sucrose solutions:
  density and viscosity, 706
  viscometry for, 343
Sucrose solutions, non-invasive concentration measurement:
  by microwave, 196, 199
  by ultrasound, 251–5
  see also Sugar solutions
Sugar:
  acoustic impedance of, 258
  moisture assay, 438
  moisture monitoring, 147, 148
Sugars, total reducing, significance of measurement, 24–5
Sugar alcohols assay, 432
Sugar beet products, microwave measurements, 191, 227
Sugar beet processing, lime addition monitoring, 315
Sugar beet slurry, flow rate, 304
Sugar cane processing, blanket detection in, 314–5
Sugar confectionery:
  relevance of texture, 350
  colour, 72
Sugar content monitoring, 128, 149–56
  see also Sugar solutions
Sugar inversion, 526
Sugar refining, lime milk flow rate control, 305
Sugar solutions:
  concentration by microwave measurement, 196, 199, 215–7, 221
  concentration by ultrasonic measurement, 251–5, 260, 317
    dependence on saccharide type, 251–5
  concentration, comparison of methods, 665
  level measurement for, 297

Sugaring as preservation process, 378, 388–9
Sulphate assay, 429
Sulphite assay, IC system, 432–3
Sulphur dioxide, 612, 615
Sunflower oil, 529, 537–8, 540, 544, 551
Supercritical fluid chromatography, 422–3
Surface acoustic wave (SAW) device, 249, 585, 644, 652–4
Surface charge density, 597
Surface potential, 597, 599
Surface properties:
  acoustic properties, 644
  dielectric properties, 644
  density, elasticity, 654
  electrical conductivity, 654
  mechanical properties, 644
  optical, 644
Suspensions:
  considerations in viscometry for, 334, 339, 343
  solids concentration, *see* Concentration
Syrups:
  colour, 77
  moisture assay, 438
  spent wash, water content monitoring, 147
  water content monitoring, 142, 147
  see also Blackcurrant syrup
Sweet potato mash, colour, 84, 90
Sweet corn, cob quality, 277
Sweeteners, 624, 636
Sweets:
  moisture assay, 438, 440
  see also Sugar confectionery

Tables, ancillary, 693–712
Taguchi sensor, 601, 613
Taint, 617
Tank:
  ultrasonic transducer for, 295, 298, 300, 307–18
  viscometer in, 341
  see also NIR, microwave, ultrasound instrumentation
Tantalum oxide (Ta_2O_5), 585, 597
Ta_2O_5 (tantalum oxide), 585, 597
Tapeworm, *see* bladderworm
Tar content, phantom measurement, 152
Target value, 158, 161
Target values in process control, 158, 161, 659
Taste as quality indicator, 531–2, 552
Tea leaf or powder, moisture measurement for, 144, 146, 148
Technology transfer, 9:
  adaptations illustrated, 535, 547, 607, 628–629, 714
  potential, 527, 629
Teflon, *see* PTFE
Television cameras, 104, 116
TELISA, 638, 645
Temperature:
  dielectric relaxation time, influence of, 196
  fermentation, control in, 389
  microbial deterioration, influence of, 376, 382
  microwave measurement, as perturbing factor in, 198–200
  NIR measurement, independence, 136
  rheological measurements, control of, 343
  rheological properties, dependence on, 331
  tolerance of high, *see* Hostile environments
  ultrasound velocity measurement, relationship to, 250–6, 260–2
  upstream control, 664
  water activity, influence on, 390–2
Temperature dependence, frying oil viscosity, 533, 543, 550
Temperature-independent operation, 603–4, 608–9
Temperature measurement, 33–4, 167–88
  contour map, 176
  hostile environments, for, 186–8
  interpretation, 178
  non-contact measurement, *see* Infrared thermometry
  position trace, 176
  quartz devices, 28
  remote measurement, 167–88
  significance, 24
  ultrasound velocity measurement, as indicator, 255–6

Temperature units, conversion, 698
Temperatures in food processing, 15, 698–9
  see also Time-temperature profile
Tempering, *see* Crystallization
Tenderizing, by ultrasound, 246
Test signal generation, 610
Texture:
  change in the mouth, 358, 370
  perception, 350, 358
  physiological aspects, 358
  space, 368
  see also Texture assessment
Texture assessment, 349–72
  acoustic measurements, 266, 358–9, 370, 371
  assessment classes, 371
  ball indenters, 352
  comparison of methods, 371
  cutting devices, 352
  electromyography, 359–71, *see* also Electromyography
  empirical methods, 352–2, 369, 371
  extrusion tests, 352
  flow devices, 353
  foods, specific, for, *see entries for named foods*
  fundamental methods, 357, 371
  future developments, 369–72
  general purpose testing machines, 353, 370
  imitative methods, 353–6, 371
  liquid and semi-liquid foods, for, 353, 357
  mixing devices, 353
  panels, 351, 370, 371
  penetration or puncture devices, 352
  profiling, 351, 370
  significance, 24
  sensory methods, *see* panels
  shearing devices, 352
  sound emission monitoring, *see* acoustic measurements
  time-intensity testing, 370
  viscous and semi-solid foods, 370
Thermal base transducers, , 638–9
Thermal devices, *see* Calorimetric devices
Thermal imaging, 175–8
Thermal lability of compounds, choice of assay system, 422, 423
Thermistor, 638
Thermochromicity, 71, 77
Thermocouple array, 639
Thermometers, *see* Infrared thermometry
Thermometry:
  fluorescent fibre-optic, 599, 650
  SAW sensors, 653
Thermometric method of water activity meas., 403–4
Thermopile:
  in biosensor, 639
  in thermometry, 170–2
Thermospray devices, 421
Thick film technology, 679, 681
Thick film devices, 601, 613, 615
Thickness gauge, ultrasonic, 243
Thickness of egg shells by ultrasonic measurement, 267
Thin film technology, 679, 681
Thin film devices, 614, 615, 616
Thin films, mass monitor (microbalance) for, 644
Thin metal film, surface plasmons, 647
Thorn EMI, xxv, 576, 606–7
Tilting filter system, 133
Time-dependent behaviour in rheology, 330, 343
Time-of-flight flow meter, *see* Transit time flow meter
Time-temperature integrals and profiles, 8, 23, 24
Tin oxide (SnO_2), 585, 601, 613
Tintometer, *see* Lovibond
TIR (total internal reflection) devices, 585, 644, 647–9
  see Immunosensors and sorption sensors
TIRF (total internal reflection fluorescence), 649
Tissues, biological, *see* Biological tissues
Titration, real-time, 610
Titration systems for moisture assay, 18–9, 437–8, 439–40
Titrimetric saccharides assay, 432
Tobacco:
  identification, 616
  moisture monitoring by microwave meas., 207, 221, 227
  moisture monitoring by NIR, 148
  nicotine content on-line, 133

sample scanning NIRA for damp, 134
Toffees, texture assessment for, 363
Tomato ketchup, rheology of, 325
Tomato paste, 65, 71, 84–5, 91
Tomato skin, cracks in, 277
Tomatoes:
colour of whole, 83, 89
sorting, 100
Tooling-up costs, 600
Toppings, *see* Blueberry puree
Torque, as indicator of viscosity, 340
Torry Research Station, xxv-xxix
Total absorption of incident light, 647
Total reflection of incident light, 647
Tough foods, pH sensors for, 589, 607, 609
Toughness, 352
Toxicity, 638
Toxins, 646
endotoxin, 652
paralytic shellfish poison, 651
see also Histamine
Toxin production by microbes, minimum water activity, 382–3
Traceability, colour standards, 71
Training sample set, 152
Transit-time flow meter, 278, 303–4
Translucent foods, 45, 53, 56, 72, 76–7, 79, 90–1
Translucent materials colour (CTM) probe, 609
Transmission mode, NIR, 133
Transmission law, Beer-Lambert, 127, 143
Transmission configuration, NIR, 142
Transmittance measurement:
see Colour measurement
see Microwave measurement
see Near-infrared measurement
see Ultrasound measurement
Transparent foods, 72, 77, 78, 89, 91
Trichinae, water activity tolerance 382–4
Trichromatic (Tristimulus) colorimeters, *see* Colour measurement
Trichromatic sorting, 104
Trichromatic (Tristimulus) vision, *see* Colour vision
Triglycerides, solubility assessment, 257
Trimethylamine, 524, 526, 624
Trilinolein, 539
Tripolyphosphates, 137
Trouble-shooting, *see* Clamp-on transducers
TRS, *see* Sugars, total reducing
TS (total solids) content, 707
TSS (total soluble solids) content, 707
Transducer:
definition, 673
notes, 673–5
Transmitter mode, 673
Truffle pig, electronic, 612
TTF (tetrathiafulvalene), 632
Tuna, histamine formation, 524
Tuning fork-type probe, 535, 546–7, 714
Turbotitrator, 439–40
TVC (total viable count), 460
Two-class (ICMSF) specification (microbial), 506
Tyramine, 526, 624

Ullage space, level measurements through, 293
Ultrarapid immunoassay, 645
Ultrasonic appliances (high power), 27, 246–7
Ultrasonic measurement, 237–83, 289–319, 665
acoustic emission monitoring, 245–6
acoustic noise management, 291, 306
applications, full overview, 239–46, 281–3
blanket detection, 313
brief overview, 30
carbon dioxide in headspace, problems with, 273, 298
comparison with electromagnetic measurements, 240–2, 243
concentration, solutes of, 244, 250–6, 317
concentration, suspended solids, of, 256–7, 277, 312–5
contacts, 318
counting with beam-break detectors, 301
couplant, 268
cross-correlation flow meter, 278
crystallization, of, 269
density of, 250, 270

Ultrasonic measurement – *continued*
- Doppler flow meter, 277, 304–6
- dust tolerance, 301
- echo detection, 291, 301
- echo ranging, 291–301
- fermenting liquids, problems with, 273
- flaw detector, 243
- flow rate, of, 242–3, 302–6, 318
- frequency, choice of, 242–3, 270, 276, 290–1, 299
- frequency ranges, 240–2
- guiding tube, use of, 302
- headspace, in, 292
  - foam, influence, 297, 298, 312
  - vapours, influence, 250, 294
- impedance matching, 265, 268–9
- impedance measurement for attenuative samples, 265, 269–70
- impedance mismatch level meter, 311
- impedance mismatch probe, 269–70
- interface detection systems, 312–3, 315–6
- Lamb wave probe, 312
- level, of, 242–3, 249, 291–301, 306–12, 318
- level alarms:
  - high level, 298, 300, 306, 308
  - low level, 306, 308
  - pump control point, 308
- liquid identification, 251–2, 315–6
- microprocessors, advances, through, 291, 294, 296, 299, 301
- multi-variable, 280
- non-destructive testing, 243–4, 291
- passive, 245–6
- phase changes, monitoring of, 269
- pressure limits, 293, 298
- pulse-echo, 269
- pulse-echo level measurement, transducer for, 294 pulse-echo systems design, 295–7, 299, 302
- pulse-echo reflectometer, 270
- pulse-echo ranging, 291–301
- reference pin reflector, 294–5
- remote, 249
- resolution in distance determination, 242
- size distribution for second phase inclusions, 256–7, 264
- slurries, for, 277, 312–5, 304, 306
- soft-tipped probes, 267
- solid/liquid ratio, of, 257
- solute concentration, of, 250–5, 317
- solute changes, compensation for, 315
- solvent concentration, of, 250–4, 317
- specific foods, for, *see entries under food types*
- stirred tanks, for, 299, 302
- suspended solids concentration, *see* concentration
- temperature dependence, 250–6, 292, 317
  - compensation for, 292, 299–300, 315
  - PZT crystal, response of, 297
- temperature determination, 255–6
- temperature tolerance, 302
- thickness gauge, 243
- time-of-flight, *see* transit time
- transducer materials and designs, 290–1, 295, 308
  - clamp-on, 304–6, 311, 318
  - cleaning tolerance, 298, 306
  - gap-type, 307–10, 313
  - pressure limits, 293, 298, 302
  - retrofitting, *see* clamp-on
  - ringing-after, 295
  - self-cleaning, 301
- transit time, 239
- transit time flow meter, 278, 303–4, 318
- transit time level meter, 297
- ultrasound variables measured, 239–40, 247–80, 291, 302
- *see* also Ultrasound propagation
- velocity measurements, considerations in, 268–9
- volume fraction of second phase inclusions, 256–7, 264, 312–5
- wavelength ranges, 240–2, 290
- *see also* Ultrasound propagation

Ultrasound propagation, 237–83
- absorption coefficient, values for, 274–5
- attenuation, 270–80
  - carbon dioxide, high for, 250
  - classical prediction, 271, 272
  - relaxation, by, 271–3, 274–6
  - scattering, by 273–80
  - sources of, 270–1

bubbles, dispersed, influence on, 256–7, 264
characteristics:
bone, of, 248
carbon dioxide, 250
gels, of, 248
compressional waves, *see* longitudinal waves
crystal seeds, dispersed, influence on, 257
dispersive velocity, 250, 271
droplets, dispersed, influence on, 256–7, 264
elastic modulus, effect on, 247–8
gels, in, 248, 267, 269, 273
impedance, acoustic, 267–70
definition, 267
gases, values for, 262
implications for velocity measurement, 268–9
liquids and semi-solids, values for, 260–1
relationship to velocity, 269–70
solid materials and tissues, values for, 258–9
Lamb waves, 249
longitudinal waves, 239, 247–8, 250
nature of, 239, 247–9
particles, dispersed, influence on, 256–7
Rayleigh waves, 249
relaxation frequency, 272, 276
relaxation time, 271, 274
resonant scattering, 264
scattering, 264, 273–80
shear waves, *see* transverse waves
speed of sound, 247–67
gases, in, 250, 292
gases, values for, 262, 293
liquids, in, 250–66
liquids and semi-solids, values for, 260–1
longitudinal waves, for, 247–8
solids, in, 266–7
solid materials and tissues, values for, 258–9
temperature dependence of, 260–2
transverse waves, for, 248
vapours, in, 250
Surface Acoustic Wave (SAW) Device, 249
surface waves, 249
transmission, 268–9
transverse waves, 248
velocity, *see* speed
wave types, 247–9
see also Ultrasonic measurement
Ultrasound treatment, in meat processing, 246
Ultraviolet detectors:
HPLC in, 421
see also Sorting by colour, Authenticity
Ultraviolet response of photodetectors, 110
Ultraviolet spectrum, use in sorting, 104
Ultraviolet/visible/NIR, application examples, 29
UNIFET, 592, 594, 608
Units, international, 695–7
conversion, 698–705
Urea sensing, 636, 637
UV-VIS (Ultraviolet-visible) detector, HPLC in, 428

Vacuum oven drying, moisture assay, 437
Valinomycin, 585
Van der Waals interactions, 645
Vapours:
speed of sound, in, 250, 262, 293
tolerance by microwave instrumentation, 192
Variables:
measured, 17–8
precursor, 5, 19
significance, 12–4, 24–5
target, 17–8, 22–3
Vegetables:
colour, 76–7, 88–9
conditioning, 629–32
glucose assay, 629–32
microbial assay (impedance), 490, 509
ripeness assessment for by ultrasound, 266
skin texture, 277
sorting, 100, 117
water activity in, 385, 388
see also Browning potential
Veterinary residues, 624
Vibrating pipe section densitometer, 27

Vibrating probes, 27
see Mechanically resonant probes
Vinegar, colour, 78
Viscoelastic properties, 326, 358
Viscometers:
capillary, 331–6
empirical, 340
falling cylinder, 342
hot wire, 342
ring test of, 343
rotary, 336–40
selection of, 342–3
shear-rate selection with, 336, 338, 343
vibrational, 342
Viscosity:
definition, 327
influence on dielectric relaxation time, 196
measurement, *see* Rheological measurements
oils, *see* Frying oil viscosity
units, conversion, 705
values, examples, 705–6
Viscosity probes, mechanically resonant, 27
Visible light detectors:
HPLC in, 421
see also Sorting by Colour, Colour measurement
titrimetric, 432
Visible spectrum:
colours, 709
use in sorting, 101–17
Vision, *see* Colour vision
Vision systems, computer, 114–6
Visual assessment, comparison with colorimetry, 50–9, 78
Vitamins, 585, 624, 646
Vitek Systems, 507:
see also Bactometer
Volatile organic substances:
effect on hygrometers, 393, 399 *see also* Vapours
Volatiles:
frying oil flavours and off-flavours, 528–9, 531–2, 552
meat and fish spoilage, 524–8, 559
oxidative rancidity, 525–7
significance of measurement, 24–5
see Gas sensors
Volatiles measurement, 644, 646–7, 652, 654
near infrared spectroscopy, 612
see also Chemosensors, Gas sensors
see also Immunosensors and sorption sensors
see also Odour assessment, Aroma
Volatile-rich foods, moisture assay, 437
Voltammetry, cyclic, 431
Volume fraction dispersed in liquid:
bubbles, assessment of, 264
crystal seeds (fat), 257
droplets, 277–80
ice, 266
particles, 277, 312–5
Volume units, conversion, 704

Warner-Bratzler Shear, 352
Waste water:
BOD (biological oxygen demand), 585, 625
COD (chemical oxygen demand), 625
microbial assay (impedance), 512
sugar contamination, 526
Water:
bound, dielectric relaxation properties, 196, 218
content, *see* Water content monitoring
crystallization, of, 136
influence on moisture assay, 437, 438
density and viscosity, 705
dielectric properties of, pure, 194
dielectric properties of, in foods, 194–200
dielectric properties of, in foods containing protein, 198, 200
principle NIR absorption bands, 126, 139
rheology of, 329, 331
rotational hindering, effect on dielectric properties, 201
state of water, 11–2
influence on NIR measurement, 136–7
transmittance spectrum of atmospheric water, 137
transmittance spectrum of liquid water, 137, 139
ultrasound data, for, 261, 274–5

vapour, *see* also Humidity
Water activity, 376–405
adjustment, 381
chilled foods of, 390–2
control, 385–90
definition, 377
enzyme-related deterioration, influenced by, 377–8
fermentation, role in, 389
Fett-Vos method, 395
frozen foods of, 390–2
legal requirements, for, 384–5
levels in foods, 385–92
measurement, *see* Water activity measurement
meat and meat products, levels in, 385–93
meat products, role in, 382–90
microbial deterioration, influenced by, 376–90
minimum levels for specific microbial growth, 379–82
minimum levels for toxin production, 382–3
parasites, tolerance of, 382–4
quality of food, effect on, 377–378, 389
reference materials, 395
refractometric method, 395
ripening, control by, 389
safety of food, effect on, 378–85
shelf-life limiting factors, relationship to, 378
significance, 377–90
stability of food, effect on, 378–89
temperature, influence on, 390–2
Water activity measurement, 390–404
calibration, 390, 393–4, 397
capacitive hygrometers, 397
conductivity hygrometers, 399–403
considerations in, 390–4
cryoscopic method, 403
curing, during, 402
dew-point hygrometer, 395, 402–3
dried products, for, 393
electrolytic cells, 399
equilibration, 392–3
electric hygrometers, 397–403
freezing point method, 392
gravimetric method, 395
instrumentation for, 394–404
isopiestic method, 395
heterogeneity of samples, for, 393
hygrometric methods, 396–403
hygroscopic thread, 396
manometric method, 394–5
measuring period, 392–3, 404
opto-electronic dew point hygrometer, 403
poisoning of sensor, 393, 399–400
proximity equilibration cell, 395
psychometric method, 395–6
reaction of sensor with sample ingredients, *see* poisoning
salt slurries for calibration, 393–4
salt/filter paper method, 396
significance, 24, 377–85
see also HACCP, Stability
state of, 11–2
temperature:
influence on water activity, 390–2
sensor and electronics response to, 392, 397, 399, 401
thermometric method, 403–4
thread hygrometers, 396–7
volatile organic substances, effect of, 393, 399–402, 404
Water content measurement, significance, 21, 24
Water content monitoring:
see Microwave measurements
see Near-Infrared measurements
see Ultrasonic measurements
see also Moisture
comparison of methods, 264–6
fish meal, in by microwave meas., 200
foods, in by microwave meas., 190–231:
see also entries for named foods
Water supply pipes, flow rate measurement, 303–4
Water treatment, floc blanket alarm, 314
Water vapour, infrared absorption, 171
Wavelength ranges:
colours, 708
electromagnetic waves, 240–2
microwave bands, 710–11
near- and mid-infrared, 123–4, 709
overview, 29–30, 33, 241
infrared thermometry, 168, 170–2, 176
sound waves, 240–2

Wavelengths, used in sorting, 101–17
  near infrared, 101, 104, 117
  ultraviolet, 104
  visible, 101–4, 117
Wavenumber, 709
Wave types, acoustic, *see* Ultrasound propagation
Weight loss, as indicator of ripening, 289
Weissenberg rheogoniometer, 357
Wescan analyser, 432–3
Wheat:
  fat assay, 439
  microwave moisture measurements for, 209, 218, 226
  moisture monitoring by NIR, 148
  protein monitoring, 152
  see also Grain
Whey powder, water of crystallization, 136
Whey powder, moisture measurement, 136
Wien's displacement law, 709
Window cleaning, automatic, 141–2
Whisky blending, level control, 308
White diffuser, perfect, 69, 80
Whole cell membranes, 585–6, 622, 637
  acetic acid bacteria, 637
Wilmer instrument, 231
Window maintenance, colour sorters, 112
Windows:
  colour sorters, for, 112
  configurations, 6–7
  infrared thermometry, for, 186–8
  microwave measurements for, 214
  near infrared spectroscopy, for, 141–2
  non-contact measurements, for, 266
Window materials:
  acoustic, metal, 266
  dielectric, PTFE (polytetrafluoroethylene, Teflon) 214
  optical (visible, NIR, MIR):
    Aclar (PTFE polymer), 141–2
    glass, 112, 141–2
    glass fibre, 188
    glass, laminated, 142
    polyester, 141–2
    polyethylene film, 188
    sapphire, 141, 186
    silica fibre, 188
    zirconium fluoride fibre, 188
Wine:
  adulterated, 217
  colour, 72, 78
  microbial assay (impedance), 490, 512
Wood, detection and removal, *see* Sorting by colour

Xanthophyll content, 89, 90
Xenon arc lamps, 52
X-rays, 33
Xylitol assay, 432
Xylose concentration, relationship to speed of sound, 254
  see also Sugar solutions

Yam mash, colour, 84, 90
Yeast, dried, moisture monitoring, 148
Yeast growth, *see* Micro-organisms, growth of
Yeast recirculation, monitoring, 315
Yeast tanks, level measurement, 298
Yeast-loaded liquids, flow rate measurement, 305
Yeasts and moulds:
  microbial assay, 464, 477, 491, 512
Yeast contamination, 526
Yield prediction, cooked ham, 609
Yield stress, significance of, 325, 326, 330
Yoghurt:
  biomass, 477
  colour, 84, 91
  fermentation monitoring, 663–4
  microbial assay (impedance), 490, 509, 512
  packaging materials, 156
  relevance of rheology, 326
Yoghurt in pot, setting, 664
YSI (Yellow Springs Instruments) analyser, 433–5, 631, 636

Zero adjustment, 158
Zinc, dissolved, 651
Zinc oxide (ZnO_2), 585, 601, 613
Zinc phthalocyanine, 614
Zirconium fluoride fibre-optic cable, 188
ZnO_2 (zinc oxide), 585, 601, 613
ZnPc (zinc phthalocyanine), 614

HPLC : 427-32

Spectroscopy : 126

NMR ; 26

Instron 355